DEPARTMENT LIBRARY

Molecular Electronics

ST ANDREWS UNIVERSITY
LIBRARY
OFFICIALLY WITHDRAWN
FROM STOCK
BY COLLECTIONS

MULTIPLE COPY

AF869801

Wiley Series in Materials for Electronic and Optoelectronic Applications

Series Editors

Dr Peter Capper, SELEX Sensors and Airborne Systems Infrared Ltd, Southampton, UK
Professor Safa Kasap, University of Saskatchewan, Canada
Professor Arthur Willoughby, University of Southampton, Southampton, UK

Published Titles

Bulk Crystal Growth of Electronic, Optical and Optoelectronic Materials, Edited by P. Capper
Properties of Group-IV, III–V and II–VI Semiconductors, S. Adachi
Charge Transport in Disordered Solids with Applications in Electronics, Edited by S. Baranovski
Optical Properties of Condensed Matter and Applications, Edited by J. Singh
Thin Film Solar Cells: Fabrication, Characterization and Applications, Edited by J. Poortmans and V. Arkhipov
Dielectric Films for Advanced Microelectronics, Edited by M. R. Baklanov, M. Green and K. Maex
Liquid Phase Epitaxy of Electronic, Optical and Optoelectronic Materials, Edited by P. Capper and M. Mauk

Forthcoming Titles

Luminescent Materials and Applications, Edited by A. Kitai
Photovoltaic Materials: From Crystalline Silicon to Third-Generation Approaches, Edited by G. J. Conibeer
CVD Diamond for Electronic Devices and Sensors, Edited by R. S. Sussmann
Silicon Photonics: Fundamentals and Devices, J. Deen and P. K. Basu
Zinc Oxide Materials for Electronic and Optoelectronic Device Applications, Edited by C. Litton, D. C. Reynolds and T. C. Collins

Molecular Electronics
From Principles to Practice

Michael C. Petty
School of Engineering and Centre for Molecular and Nanoscale Electronics, Durham University, UK

John Wiley & Sons, Ltd

BIBLIOTHECA UNIVERSITATIS ANDREANAE

Copyright © 2007 John Wiley & Sons Ltd. The Atrium, Southern Gate, Chichester, West Sussex PO19 8SQ, England
Telephone (+44) 1243 779777

Email (for orders and customer service enquiries): cs-books@wiley.co.uk
Visit our Home Page on www.wileyeurope.com or www.wiley.com

All Rights Reserved. No part of this publication may be reproduced, stored in a retrieval system or transmitted in any form or by any means, electronic, mechanical, photocopying, recording, scanning or otherwise, except under the terms of the Copyright, Designs and Patents Act 1988 or under the terms of a licence issued by the Copyright Licensing Agency Ltd, 90 Tottenham Court Road, London W1T 4LP, UK, without the permission in writing of the Publisher. Requests to the Publisher should be addressed to the Permissions Department, John Wiley & Sons Ltd, The Atrium, Southern Gate, Chichester, West Sussex PO19 8SQ, England, or emailed to permreq@wiley.co.uk, or faxed to (+44) 1243 770620.

Designations used by companies to distinguish their products are often claimed as trademarks. All brand names and product names used in this book are trade names, service marks, trademarks or registered trademarks of their respective owners. The Publisher is not associated with any product or vendor mentioned in this book.

This publication is designed to provide accurate and authoritative information in regard to the subject matter covered. It is sold on the understanding that the Publisher is not engaged in rendering professional services. If professional advice or other expert assistance is required, the services of a competent professional should be sought.

This Publisher and the Author make no representations or warranties with respect to the accuracy or completeness of the contents of this work and specifically disclaim all warranties, including without limitation any implied warranties of fitness for a particular purpose. The advice and strategies contained herein may not be suitable for every situation. In view of ongoing research, equipment modifications, changes in governmental regulations, and the constant flow of information relating to the use of experimental reagents, equipment, and devices, the reader is urged to review and evaluate the information provided in the package insert or instructions for each chemical, piece of equipment, reagent, or device for, among other things, any changes in the instructions or indication of usage and for added warnings and precautions. The fact that an organization or Website is referred to in this work as a citation and/or a potential source of further information does not mean that the author or the publisher endorses the information the organization or Website may provide or recommendations it may make. Further, readers should be aware that Internet Websites listed in this work may have changed or disappeared between when this work was written and when it is read. No warranty may be created or extended by any promotional statements for this work. Neither the Publisher nor the author shall be liable for any damages arising herefrom.

Other Wiley Editorial Offices

John Wiley & Sons Inc., 111 River Street, Hoboken, NJ 07030, USA

Jossey-Bass, 989 Market Street, San Francisco, CA 94103-1741, USA

Wiley-VCH Verlag GmbH, Boschstr. 12, D-69469 Weinheim, Germany

John Wiley & Sons Australia Ltd, 42 McDougall Street, Milton, Queensland 4064, Australia

John Wiley & Sons (Asia) Pte Ltd, 2 Clementi Loop #02-01, Jin Xing Distripark, Singapore 129809

John Wiley & Sons Ltd, 6045 Freemont Blvd, Mississauga, Ontario, L5R 4J3, Canada

Wiley also publishes its books in a variety of electronic formats. Some content that appears in print may not be available in electronic books.

Anniversary Logo Design: Richard J. Pacifico

British Library Cataloguing - in - Publication Data

Petty, Michael C.
Molecular electronics : from principles to practice / Michael C. Petty.
p. cm. – (Wiley series in materials for electronic and optoelectronic)
Includes bibliographical references and index.
ISBN 978-0-470-01307-6 (cloth) – ISBN 978-0-470-01308-3 (pbk.)
1. Molecular electronics. I. Title.
TK7874.P457 2007
621.381–dc22

2007041701

A catalogue record for this book is available from the British Library
ISBN 978-0-470-01307-6 (H/B)
978-0-470-01308-3 (P/B)

Typeset in 10/12 pt Times by Thomson Digital, Noida, India
Printed and bound in Great Britain by Antony Rowe Ltd, Chippenhan, Wiltshire, England
This book is printed on acid-free paper responsibly manufactured from sustainable forestry in which at least two trees are planted for each one used for paper production.

Physics
DEPARTMENT LIBRARY
1078936-4

Dedicated to Gareth Gwyn Roberts
(1940–2007)

Contents

Series Preface xv

Preface xvii

Acknowledgements xix

Symbols and Abbreviations xxi

1 Scope of Molecular Electronics 1
- 1.1 Introduction 1
- 1.2 Molecular Materials for Electronics 2
- 1.3 Molecular-Scale Electronics 5
  - 1.3.1 Evolution of Microelectronics 5
  - 1.3.2 Moore's Laws 7
  - 1.3.3 Beyond Moore 8
- 1.4 The Biological World 12
- 1.5 Future Opportunities 13
- 1.6 Conclusions 15
  - Bibliography 15
  - References 16

2 Materials' Foundations 19
- 2.1 Introduction 19
- 2.2 Electronic Structure 19
  - 2.2.1 Atomic Structure 19
  - 2.2.2 Electrons in Atoms 20
  - 2.2.3 Filling of Orbitals 24
  - 2.2.4 The Periodic Table 24
- 2.3 Chemical Bonding 26
  - 2.3.1 Bonding Principles 26
  - 2.3.2 Ionic Bond 27
  - 2.3.3 Covalent Bond 29
  - 2.3.4 Metallic Bonding 32
  - 2.3.5 Van der Waals Bonding 33
  - 2.3.6 Hydrogen Bonding 34

2.4 Bonding in Organic Compounds 35
2.4.1 Hybridized Orbitals 35
2.4.2 Isomers 37
2.4.3 Double and Triple Bonds 42
2.5 Crystalline and Noncrystalline Materials 45
2.5.1 States of Matter 45
2.5.2 Phase Changes and Thermodynamic Equilibrium 47
2.5.3 The Crystal Lattice 48
2.5.4 Crystal Systems 49
2.5.5 Miller Indices 50
2.5.6 Distance Between Crystal Planes 52
2.5.7 Defects 52
2.5.8 Amorphous Solids 56
2.6 Polymers 57
2.6.1 Molecular Weight 57
2.6.2 Polymer Structure 59
2.6.3 Polymer Crystallinity 60
2.7 Soft Matter: Emulsions, Foams and Gels 63
2.8 Diffusion 63
Bibliography 64
Reference 64

3 Electrical Conductivity 65

3.1 Introduction 65
3.2 Classical Theory 65
3.2.1 Electrical Conductivity 66
3.2.2 Ohm's Law 68
3.2.3 Charge Carrier Mobility 69
3.2.4 Fermi Energy 70
3.3 Energy Bands in Solids 72
3.3.1 Quantum Mechanical Foundations 73
3.3.2 Kronig–Penney Model 80
3.3.3 Conductors, Semiconductors and Insulators 85
3.3.4 Electrons and Holes 86
3.3.5 Intrinsic and Extrinsic Conduction 88
3.3.6 Quantum Wells 92
3.3.7 Disordered Semiconductors 93
3.3.8 Conductivity in Low-dimensional Solids 94
3.4 Organic Compounds 96
3.4.1 Band Structure 96
3.4.2 Doping 107
3.4.3 Solitons, Polarons and Bipolarons 109
3.4.4 Superconductivity 112
3.5 Low-Frequency Conductivity 112
3.5.1 Electronic Versus Ionic Conductivity 113
3.5.2 Quantum Mechanical Tunnelling 114

3.5.3 Variable Range Hopping 116
3.5.4 Space-Charge Injection 118
3.5.5 Schottky and Poole–Frenkel Effects 120
3.6 Conductivity at High Frequencies 121
3.6.1 Complex Permittivity 121
3.6.2 Impedance Spectroscopy 125
Bibliography 127
References 127

4 Optical Phenomena **129**

4.1 Introduction 129
4.2 Electromagnetic Radiation 129
4.3 Refractive Index 130
4.3.1 Permittivity Tensor 132
4.3.2 Linear and Nonlinear Optics 133
4.4 Interaction of EM Radiation with Organic Molecules 135
4.4.1 Absorption Processes 135
4.4.2 Aggregate Formation 140
4.4.3 Excitons 141
4.4.4 Effect of Electric Fields on Absorption 143
4.4.5 Emission Processes 144
4.4.6 Energy Transfer 147
4.5 Transmission and Reflection from Interfaces 149
4.5.1 Laws of Reflection and Refraction 149
4.5.2 Fresnel Equations 150
4.5.3 Ellipsometry 152
4.5.4 Thin Films 152
4.6 Waveguiding 154
4.7 Surface Plasmons 156
4.7.1 The Evanescent Field 156
4.7.2 Surface Plasmon Resonance 157
4.8 Photonic Crystals 162
4.8.1 Subwavelength Optics 165
Bibliography 166
References 166

5 Electroactive Organic Compounds **169**

5.1 Introduction 169
5.2 Selected Topics in Chemistry 169
5.2.1 Moles and Molecules 169
5.2.2 Acids and Bases 170
5.2.3 Ions 171
5.2.4 Solvents 173
5.2.5 Functional Groups 173
5.2.6 Aromatic Compounds 178
5.3 Conductive Polymers 180

5.4 Charge-Transfer Complexes 185
5.5 Buckyballs and Nanotubes 188
5.5.1 Fullerenes 188
5.5.2 Carbon Nanotubes 191
5.6 Piezoelectricity, Pyroelectricity and Ferroelectricity 194
5.6.1 Basic Principles 194
5.6.2 Organic Piezoelectric, Pyroelectric and Ferroelectric Compounds 197
5.7 Magnetic Materials 201
5.7.1 Basic Principles 201
5.7.2 Organic Magnets 208
Bibliography 210
References 211

6 Tools for Molecular Electronics **213**

6.1 Introduction 213
6.2 Direct Imaging 214
6.2.1 Optical Microscopy 214
6.2.2 Electron Microscopy 216
6.3 X-ray Reflection 218
6.3.1 Electron Density Profile 221
6.3.2 Keissig Fringes 222
6.3.3 In-Plane Measurements 223
6.4 Neutron Reflection 223
6.5 Electron Diffraction 223
6.6 Infrared Spectroscopy 224
6.6.1 Raman Scattering 231
6.7 Surface Analytical Techniques 232
6.8 Scanning Probe Microscopies 233
6.9 Film Thickness Measurements 236
Bibliography 238
References 238

7 Thin Film Processing and Device Fabrication **241**

7.1 Introduction 241
7.2 Established Deposition Methods 242
7.2.1 Spin-Coating 242
7.2.2 Physical Vapour Deposition 243
7.2.3 Chemical Vapour Deposition 251
7.2.4 Electrochemical Methods 252
7.2.5 Inkjet Printing 253
7.2.6 Sol–Gel Processing 255
7.2.7 Other Techniques 258
7.3 Molecular Architectures 258
7.3.1 Langmuir–Blodgett Technique 258
7.3.2 Chemical Self-Assembly 268
7.3.3 Electrostatic Layer-by-Layer Deposition 270

7.4 Nanofabrication 275
7.4.1 Photolithography 275
7.4.2 Nanometre Pattern Definition 276
7.4.3 Soft Lithography Techniques 278
7.4.4 Scanning Probe Manipulation 278
7.4.5 Dip-Pen Nanolithography 280
7.4.6 Other Methods 282
Bibliography 283
References 283

8 Liquid Crystals and Devices 287

8.1 Introduction 287
8.2 Liquid Crystal Phases 287
8.2.1 Thermotropic Liquid Crystals 287
8.2.2 Lyotropic Liquid Crystals 293
8.3 Liquid Crystal Polymers 295
8.4 Display Devices 297
8.4.1 Bifrefingence 297
8.4.2 Freedericksz Transition 299
8.4.3 Twisted Nematic Display 300
8.4.4 Passive and Active Addressing 302
8.4.5 Full-Colour Displays 303
8.4.6 Super-Twisted Nematic Display 304
8.5 Ferroelectric Liquid Crystals 305
8.6 Polymer-Dispersed Liquid Crystals 306
8.7 Liquid Crystal Lenses 308
8.8 Other Application Areas 309
Bibliography 311
References 311

9 Plastic Electronics 313

9.1 Introduction 313
9.2 Organic Diodes 313
9.2.1 Schottky Diode 313
9.2.2 Ohmic Contacts 318
9.3 Metal–Insulator–Semiconductor Structures 318
9.3.1 Idealized MIS Devices 319
9.3.2 Organic MIS Structures 320
9.4 Field Effect Transistors 321
9.5 Integrated Organic Circuits 327
9.5.1 Radiofrequency Identification Tags 328
9.6 Organic Light-Emitting Displays 330
9.6.1 Device Efficiency 334
9.6.2 Methods of Efficiency Improvement 336
9.6.3 Full-Colour Displays 343
9.6.4 Electronic Paper 345

9.7 Photovoltaic Cells 345
9.7.1 Organic Semiconductor Solar Cell 346
9.7.2 Dye-Sensitized Solar Cell 348
9.7.3 Luminescent Concentrator 350
9.8 Other Application Areas 351
9.8.1 Conductive Coatings 351
9.8.2 Batteries and Fuel Cells 351
9.8.3 Xerography 354
Bibliography 356
References 356

10 Chemical Sensors and Actuators 359

10.1 Introduction 359
10.2 Sensing Systems 360
10.3 Definitions 361
10.4 Chemical Sensors 363
10.4.1 Calorimetric Gas Sensors 364
10.4.2 Electrochemical Cells 365
10.4.3 Resistive Gas Sensors 369
10.4.4 Dielectric Sensors 376
10.4.5 Acoustic Devices 380
10.4.6 Optical Sensors 382
10.5 Biological Olfaction 388
10.6 Electronic Noses 390
10.7 Physical Sensors and Actuators 391
10.7.1 Touch Sensors 391
10.7.2 Polymer Actuators 392
10.7.3 Lab-on-a-Chip 395
10.8 Smart Textiles and Clothing 399
Bibliography 399
References 400

11 Molecular-Scale Electronics 403

11.1 Introduction 403
11.2 Nanosystems 403
11.2.1 Scaling Laws 403
11.2.2 Interatomic Forces 404
11.3 Engineering Materials at the Molecular Level 405
11.3.1 Polar Materials 406
11.3.2 Nonlinear Optical Materials 408
11.3.3 Photonic Crystals 410
11.4 Molecular Device Architectures 411
11.5 Molecular Rectification 415
11.6 Electronic Switching and Memory Devices 417
11.6.1 Resistive Bistable Devices 418
11.6.2 Flash Memories 421

11.6.3 Spintronics 424
11.6.4 Three-Dimensional Architectures 426
11.7 Single-Electron Devices 427
11.8 Optical and Chemical Switches 429
11.8.1 Fluorescence Switching 430
11.8.2 Photochromic Systems 431
11.8.3 Chemical Control 435
11.9 Nanomagnetic Systems 436
11.10 Nanotube Electronics 437
11.11 Molecular Actuation 440
11.11.1 Dynamically Controllable Surfaces 440
11.11.2 Rotaxanes 442
11.11.3 Optical Tweezers 443
11.12 Logic Circuits 443
11.13 Computing Architectures 447
11.14 Quantum Computing 449
Bibliography 450
References 451

12 Bioelectronics 455

12.1 Introduction 455
12.2 Biological Building Blocks 455
12.2.1 Amino Acids and Peptides 455
12.2.2 Proteins 458
12.2.3 Enzymes 459
12.2.4 Carbohydrates 461
12.2.5 Lipids 462
12.3 Nucleotides 465
12.3.1 Bases 465
12.3.2 DNA 466
12.3.3 RNA 467
12.3.4 ATP, ADP 467
12.4 Cells 468
12.5 Genetic Coding 469
12.5.1 Replication, Transcription and Translation 470
12.6 The Biological Membrane 474
12.6.1 Transport Across the Membrane 475
12.7 Neurons 480
12.8 Biosensors 482
12.8.1 Biocatalytic Sensors 483
12.8.2 Bioaffinity Sensors 484
12.9 DNA Electronics 487
12.10 Photobiology 487
12.10.1 Bacteriorhodopsin 487
12.10.2 Photosynthesis 493

12.11 Molecular Motors 498
12.11.1 Nature's Motors 498
12.11.2 Artificial Motors 501
Bibliography 502
References 503

Index **505**

Series Preface

WILEY SERIES IN MATERIALS FOR ELECTRONIC AND OPTOELECTRONIC APPLICATIONS

This book series is devoted to the rapidly developing class of materials used for electronic and optoelectronic applications. It is designed to provide much-needed information on the fundamental scientific principles of these materials, together with how these are employed in technological applications. The books are aimed at postgraduate students, researchers and technologists, engaged in research, development and the study of materials in electronics and photonics, and industrial scientists developing new materials, devices and circuits for the electronic, optoelectronic and communications industries.

The development of new electronic and optoelectronic materials depends not only on materials engineering at a practical level, but also on a clear understanding of the properties of materials, and the fundamental science behind these properties. It is the properties of a material that eventually determine its usefulness in an application. The series therefore also includes such titles as electrical conduction in solids, optical properties, thermal properties and so on, all with applications and examples of materials in electronics and optoelectronics. The characterization of materials is also covered within the series in as much as it is impossible to develop new materials without the proper characterization of their structure and properties. Structure–property relationships have always been fundamentally and intrinsically important to materials science and engineering.

Materials science is well known for being one of the most interdisciplinary sciences. It is the interdisciplinary aspect of materials science that has led to many exciting discoveries, new materials and new applications. It is not unusual to find scientists with a chemical engineering background working on materials projects with applications in electronics. In selecting titles for the series, we have tried to maintain the interdisciplinary aspect of the field, and hence its excitement to researchers in this field.

Peter Capper

Safa Kasap

Arthur Willoughby

Preface

Molecular electronics is an exciting and challenging sphere of activity, enabling the fabrication of a wide range of electronic devices based around organic materials and low-cost technologies. Taken to its limit, molecular electronics also offers unlimited computational power. However, the subject is not without its critics, who are quick to point out that organic materials have restricted stability and will find difficulty competing with their inorganic counterparts. The International Semiconductor Roadmap also predicts that silicon microelectronics still has some way to go before it runs out of steam.

Supporter or sceptic – much depends on your definition of molecular electronics. Are we dealing with materials science or molecular manipulation? The purpose of this book is to provide some clarity here. My intention is first to give some insight into the physics and chemistry of organic materials, and to explore the means that are now available to manipulate these materials and to measure their properties. An overview is then provided of what already has been achieved, in terms of technological applications, and what may be accomplished in the foreseeable future. I have written the book from the viewpoint of a final-year science or engineering undergraduate; I hope it is accessible to readers from a wide range of backgrounds.

Examples of 'applications' have been taken from the portfolio of molecular electronics research that was available as this book was being written (2006) – not because I am convinced that they will all definitely become future technologies (many certainly will not) but because these form a vision of molecular electronics and indicate important directions in research. Some of the current activity in molecular electronics (or plastic electronics, as many refer to it) is focused on replacing the silicon and other inorganic semiconductors, which are currently used in devices, with organic materials. The motivation is the reduced cost and large-area capability of organic thin film technologies. Over the last 30 years, tremendous progress has been achieved. However, much more of the work demonstrated in the laboratory needs to be translated into reliable and reproducible products, which can be manufactured cheaply.

The path towards molecular-scale electronics will probably be found in other directions. In particular, I believe that valuable lessons can be taken from the natural world. I have therefore included a final chapter in this book on the topic of bioelectronics.

However, can anyone really predict the future? Growing up in the UK in the 1960s, Thursday evenings were special. On BBC television, there was 'Top of the Pops,' and before this, 'Tomorrow's World'. The host was Raymond Baxter, a genial ex-RAF man and qualified Spitfire pilot. He spoke eloquently and with great authority on the technology to come. He broadcast from Concorde in its early days and he introduced us to the pocket calculator, microwave oven and the barcode (but curiously did not foresee the Internet). In several programmes, predictions were made on the development of household appliances. By the end of the 20th century, our houses would be run by robots, Raymond informed us confidently. These metal and plastic creatures would wander around doing all the household chores such as cleaning, vacuuming, gardening and cooking.

We know this did not happen, but not for the problems with developing the technology. After all, Neil Armstrong stepped on to the surface of the Moon in 1969 and unmanned spacecraft have landed on Venus and Mars. The explanation lies in a mixture of politics and cost: apart from what is technically feasible, political and economic matters can play an important role in the progress of technology. Environmental issues will certainly have an increasing influence on the developments in the 21st century. So, be wary of predictions of nano-robots wandering around your bloodstream (we already have these anyway – in the form of bacteria). I suspect that the future will reveal things that are far more intriguing.

Mike Petty
Durham University, UK

Acknowledgements

First and foremost, I owe an enormous debt of gratitude to my friend and colleague Gareth Roberts, who sadly died as this book was nearing completion. Some 30 years ago, Gareth steered me in the direction of organic materials research. Without his influence and enthusiasm, this book would never have been possible. I remember vividly that day, late in the hot summer of 1976, in Collingwood College, Durham, when Gareth was 'bending my ear', extolling the virtues of both Langmuir–Blodgett films and the Tottenham Hotspur football team (he was, at least, 50% correct!). From his briefcase, he produced a bundle of papers on molecular films and metal–insulator–semiconductor devices, talked about strange chemicals and spoke excitedly about the experiments of Hans Kuhn in Germany. This was molecular electronics – and my adventure had begun.

Safa Kasap (founder member of the Imperial College II–VI Club) talked me into this project and I am indebted to Durham University for providing me with research leave to get the whole thing started. I am ever thankful to my academic colleagues at Durham who have provided invaluable input to this book (although they might not realize this): Jas Pal Badyal, David Bloor, Martin Bryce, Stewart Clark, Karl Coleman, Graham Cross, Ken Durose, Jim Feast, Gordon Love, Andy Monkman, Chris Pearson, David Wood and Jack Yarwood. Some of the results presented are from students and postdocs in my own research group: Jin Ahn, Paul Barker, John Batey, Duncan Cadd, Riccardo Casalini, Phil Christie, John Cresswell, Ajaib Dhindsa, Carole Jones, Dan Kolb, Igor Lednev, Mohammed Mabrook, Marco Palumbo and Shashi Paul.

Further afield, numerous colleagues have provided me with original data, for which I am extremely grateful: Campbell Scott, David Morris, Dimitris Tsoukalas, Fraser Stoddart, Geoff Ashwell, Graham Leggert, Harold Zandvliet and Yuri Lvov Jeremy Burroughes, Martin Taylor, Phaedon Avouris, Richard Tredgold, Tim Richardson, Wolfgang Knoll. I must also acknowledge the many other workers worldwide whose data are reproduced in the book.

Drawing all the diagrams has been a joint effort. I did the simple bits (lines and rectangles), Julie Morgan-Dodds filled in the gaps, and Skimble the cat revealed the mysteries of the keyboard shortcuts in CorelDraw while I endeavoured to edit everything into the same format.

Last and by no means least, a big thank you is due to my wife, Anne (who always asks the best questions) for her never ending support and infinite patience.

Symbols and Abbreviations

amu	atomic mass units
au	arbitrary units
A	acceptor
AC	alternating current
ANN	artificial neural network
B, $\mathbf{B}$	magnetic field [T]
c	molecular concentration [mol m^{-3}]
C	capacitance [F]
CB	conduction band
d_{hkl}	interplanar spacing [m]
D	donor
D	diffusion coefficient [m^2 s^{-1}]
D, $\mathbf{D}$	electric displacement [C m^{-2}]
DC	direct current
E	energy [J]
E, $\mathbf{E}$	electric field [V m^{-1}]
E_c	conduction band edge [J or eV]
E_v	valence band edge [J or eV]
E_F	Fermi energy [J or eV]
EL	electroluminescence
EM	electromagnetic
EIL	electron injection layer
ETL	electron transport layer
EML	emissive layer
FET	field effect transistor
G	conductance [S or Ω^{-1}]
GPS	global positioning system
GSM	global system for mobile communications
H, $\mathbf{H}$	magnetizing field [A m^{-1}]
HIL	hole injection layer
HTL	hole transport layer
HOMO	highest occupied molecular orbital
I	electric current [A]
ISFET	ion-sensitive field effect transistor
j	$\sqrt{-1}$
J	electric current density [A m^{-2}]
k	wavevector [m^{-1}]
LUMO	lowest unoccupied molecular orbital

M	molecular weight
M, $\mathbf{M}$	magnetization [A m^{-1}]
MOSFET	metal–oxide–semiconductor field effect transistor
MIS	metal–insulator–semiconductor
MPP	maximum power point (for a PV device)
MWNT	multiwall nanotube
n	refractive index
n'	real part of refractive index
n''	imaginary part of refractive index
n	number per unit volume or concentration [m^{-3}]
NDR	negative differential resistance
NLO	nonlinear optics
OLED	organic light-emitting device
p, $\mathbf{p}$	electric dipole moment or transition dipole moment [C m]
P, $\mathbf{P}$	polarization [C m^{-2}]
PDA	personal digital assistant
q	charge [C]
R	resistance [Ω]
RF	radiofrequency
RAM	random access memory
ROM	read only memory
RMS	root-mean-squared
Re	Reynolds number
S/N	signal-to-noise ratio [dB]
SAW	surface acoustic wave
SET	single electron transistor
SPR	surface plasmon resonance
SWNT	single wall nanotube
t	time [s]
T	temperature [K or °C]
T_g	glass transition temperature [K or °C]
T_m	melting point [K or °C]
TE	transverse electric
TM	transverse magnetic
v	phase velocity of light in material [m s^{-1}]
v_d	drift velocity [m s^{-1}] (charge carrier velocity resulting from an applied electric field)
v_t	thermal velocity [m s^{-1}] (charge carrier velocity resulting from temperature)
V	voltage [V]
V	potential energy [J or ev]
VB	valence band
XGA	eXtended Graphics Array
α	absorption coefficient [m^{-1}]
β	phase change [°]
γ	surface tension [N m^{-1}]
ϵ_r, $\hat{\epsilon}_r$	relative permittivity

η	viscosity [P]
η	efficiency
θ	angle [°]
θ	fraction of surface sites occupied (Langmuir isotherm)
θ_B	Brewster angle [°]
θ_c	critical angle [°]
λ	wavelength [m]
μ	charge carrier mobility [$m^2\,V^{-1}\,s^{-1}$]
μ_r, $\hat{\mu}_r$	relative permeability
ν	frequency [Hz]
$\Delta\nu$	frequency shift [Hz]
Π	surface pressure [$N\,m^{-1}$]
ρ	density [$kg\,m^{-3}$]
ρ	resistivity [$\Omega\,m$]
σ	electrical conductivity [$S\,m^{-1}$] or [$\Omega^{-1}\,m^{-1}$]
τ	lifetime [s]
Φ	work function [J or eV]
χ_e, $\hat{\chi}_e$	electric susceptibility
χ_m, $\hat{\chi}_m$	magnetic susceptibility
X	electron affinity [J or eV]
ω	angular frequency or velocity [$rad\,s^{-1}$]

1 Scope of Molecular Electronics

What's in a name?

1.1 INTRODUCTION

Molecular electronics is a relatively new and emerging technology, relating to the exploitation of organic and biological materials in electronic and opto-electronic devices. The subject has taken on a variety of definitions in different parts of the world, and at different times, but as it has matured over the last 30 years, molecular electronics can be divided broadly into two themes. The first, *molecular materials for electronics*, has its origins in materials science and concerns the development of electronic and opto-electronic devices that utilize the unique macroscopic properties of organic compounds. The most successful commercial product to date is the liquid crystal display, LCD. However, following many years of research, organic light-emitting devices based on dyes and polymers, plastic electronics circuitry and chemical and biochemical sensors are beginning to make their technological marks. The Nobel Prize in Chemistry for 2000 was awarded to three scientists working in this area: Alan Heeger, Alan MacDiarmid and Hideki Shirakawa, who have made significant contributions to the development of electrically conductive polymers.

More challenging is *molecular-scale electronics*. Here, the focus is on the behaviour of individual organic molecules or groups of molecules and the precise three-dimensional positional control of individual atoms and molecules. Topics as diverse as molecular switching, DNA electronics and molecular manufacturing have all been described in the literature. Much of the research activity is directed towards computational architectures that may, one day, rival silicon microelectronics. However, even the most optimistic researchers recognize that this is going to be some time away!

The two approaches to molecular electronics outlined above, to some extent, exemplify the 'top-down' and 'bottom-up' themes of nanotechnology. The former refers to making nanoscale structures, for example by machining, whereas bottom-up, or molecular, nanotechnology applies to building organic and inorganic architectures atom-by-atom, or molecule-by-molecule. Top-down methods for manufacturing involve the construction of parts through methods such as cutting, milling and moulding, or etching methods linked to photolithographic pattering. The progress of the microelectronics industry is an excellent example of the top-down approach. On the other hand, bottom-up manufacturing would provide components made of single molecules, held together by covalent forces that are far stronger than the forces that hold together macro-scale components.

The physicist Richard Feynman was one of the first to predict a future for molecular-scale electronics. In a lecture in December 1959, at the annual meeting of the American Physical Society, entitled 'There's Plenty of Room at the Bottom', he described how the laws of physics do not limit our ability to manipulate single atoms and molecules. Instead, it was our

Molecular Electronics: From Principles to Practice M. C. Petty
© 2007 John Wiley & Sons, Ltd

lack of the appropriate methods for doing so. Feynman correctly predicted that the time would come in which atomically precise manipulation of matter would be possible. Several advances have been made over the last 20 years to suggest that the prophesy was correct. The most notable of these has been the introduction of scanning probe microscopy.

1.2 MOLECULAR MATERIALS FOR ELECTRONICS

Liquid crystals represent a remarkable molecular electronics success story. However, the transformation of these organic compounds into the established display technology of today took many decades. In the latter half of the 19th nineteenth century, researchers discovered several materials whose optical properties behaved in a strange way near their melting points. In 1922, a liquid crystal classification scheme was presented by George Friedel, but it took until the 1960s for the potential of liquid crystals in display devices to be recognized. From this point, research into liquid crystals and their applications burgeoned. It is encouraging for workers in molecular electronics that the relatively unstable (thermally and chemically) liquid crystal compounds came to form the foundation of such a substantial worldwide industry.

Much of the 'molecular materials' theme to molecular electronics derives from the intriguing electrical and opto-electrical behaviour of organic materials. Two distinct groups of organic compounds have been studied – low molecular weight crystalline compounds (molecular crystals) and polymers. In the former category, the photoconductivity of anthracene was discovered in 1906 [1]. However, systematic study of the electrical behaviour of organic molecular solids did not begin until the 1950s [2–4]. The phthalocyanine compounds were one of the first classes of organic molecular crystals to be investigated [2, 3]. These large, flat, ring-shaped structures are relatively stable organic materials and demonstrate that the words 'organic' and 'thermally unstable' need not always go hand-in-hand.

The first synthetic polymers were produced in the late 19th century. These were eventually developed into useful products in the 1940s and 1950s (exploiting their toughness, strength to weight ratio, low cost and ease of fabrication). At this time, polymeric materials were all good insulators, and the idea that a plastic material might conduct electricity was not generally considered. Polyacetylene is the simplest conductive polymer. Its chemical structure (Figure 1.1), consists of a hydrocarbon chain with the carbon atoms connected together by a system of alternating single and double bonds. This particular chemical bonding arrangement confers polyacetylene with its conductive properties.

Reports on acetylene polymers date back to the 19th century. However, the polymer was first prepared as a linear, high molecular weight polymer of high crystallinity and regular structure in 1958, by Giulio Natta (Nobel Prize for Chemistry in 1963) and co-workers [5].

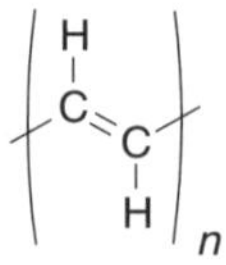

Figure 1.1 Chemical structure of polyacetylene.

The material was discovered to be a semiconductor, with an electrical conductivity between 7×10^{-11} and 7×10^{-3} S m^{-1}, depending on how the polymer was produced. Some years layer, in 1967, a student of Hideki Shirakawa, working at the Tokyo Institute of Technology, produced, by accident, a polyacetylene film with a conductivity value similar to that of the best Natta material. Furthermore, the conductivity could be increased by a factor of about 10^9 by exposure to halogens. However, in the form in which the polymer was discovered, the material was practically useless; it was insoluble in any solvent and so could not be processed into any kind of useable structure, such as a thin film or a wire. This problem was solved by Jim Feast, a chemist at Durham University. This involved making a soluble precursor. Rather than directly synthesizing polyacetylene, a polymer with easily removable groups attached to the side is first synthesized. These groups have the effect of making the polymer soluble, so the material can be formed into a film, a fibre or a wire. Only then, after processing, are the groups removed to leave the conductive polymer in the desired form.

In the 1980s, conductive polymers based on polyheterocyclic compounds (e.g. polyaniline, polypyrrole, polythiophene) were produced that were soluble in organic solvents and, consequently, could be processed into the form needed for many applications. Figure 1.2 depicts the very wide range of conductivity values that can now be found in conductive polymers.

Polymeric semiconductors can have significant advantages over their inorganic counterparts. For example, thin layers of polymers can easily be made by low-cost methods such as

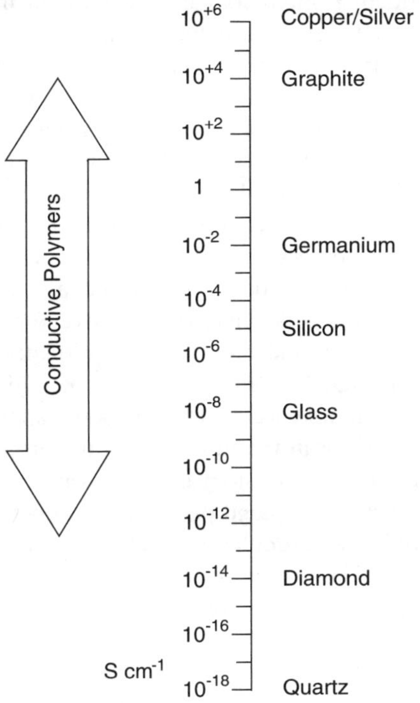

Figure 1.2 Range of conductivities for conductive polymers compared with various inorganic materials.

spin-coating. High-temperature deposition from vapour reactants is generally needed for inorganic semiconductors. Since polymers are lightweight and can be manufactured into many different shapes, obvious uses are as components in portable batteries or as electrostatic or electromagnetic shielding, conducting adhesives, printed circuit boards and replacements for conventional electrolytes in electrolytic capacitors. Other electroactive properties of these organic compounds can also be exploited in electronic devices. Examples include components in photocopying machines, organic light-emitting displays, gas, vapour and biosensors and plastic transistors. The electrical properties of the last devices cannot be directly compared with those based on single crystal silicon and gallium arsenide. The mobilities (carrier velocity per unit electric field) of the charge carriers in organic field effect transistors are low and similar to those found in amorphous silicon. Nevertheless, the simple fabrication techniques for polymers have attracted several companies to work on polymer transistor applications, such as data storage and thin film device arrays to address liquid crystal displays. The long-term stability of organic materials has often been considered as a limitation to their exploitation in electronic devices. To some extent, this difficulty has now been resolved and some conductive polymers are now displaying adequate storage and operating lifetimes, e.g. estimated operating lifetimes of over 10^6 hours for some organic light-emitting devices.

Certain organic compounds are superconducting, but usually at very low temperatures. A challenge has been to try to increase the transition temperature. Although there have been some theoretical predictions on the type of organic molecules that would be superconductive at room temperature, this 'Holy Grail' remains elusive. At the opposite extreme of the conductivity scale, nonconductive polymers can also play an important role in molecular materials for electronics. Pyroelectric, piezoelectric and ferromagnetic materials may find use in infrared detection, intruder alarms and nonlinear optics (e.g. second-harmonic generation). Similarly, some effort is being focused on organic magnetic materials. Due to the high density of iron, such molecular devices are generally inferior to conventional magnets on a 'magnet per unit weight or unit volume' basis. However, for more specialist applications, such as data storage, molecular systems may become important.

The development of effective devices for the identification and quantification of chemical and biochemical substances for process control and environmental monitoring is a growing need. Many sensors do not possess the specifications to conform to existing or forthcoming legislation; some systems are too bulky and/or expensive for use in the field. Inorganic materials such as the oxides of tin and zinc have traditionally been favoured as sensing elements. However, one disadvantage of sensors based on metallic oxides is that they usually have to be operated at elevated temperatures, limiting some applications. As an alternative, there has been considerable interest in trying to exploit the properties of organic materials. Many such substances, in particular phthalocyanine derivatives, are known to exhibit high sensitivity to gases. A significant advantage of organic compounds is that their sensitivity and selectivity can be tailored to a particular application by modifications to their chemical structure. Moreover, thin film technologies, such as self-assembly or layer-by-layer electrostatic deposition, enable ultra-thin layers of organic materials to be engineered at the molecular level.

This transition of materials from the micro-scale to the nano-scale offers several advantages. First, an enormous increase in surface area may be achieved; bulk properties become governed by surface properties. A material can be made light absorbing by coating its constituent particles with a dye. The particle size reduction also induces both mechanical advantages and quantum effects. For instance, the hindered propagation of lattice defects leads to strong and hard metals, and enhanced diffusional creep leads to 'super-plastic' ceramics during processing at elevated temperatures. Quantum confinement allows for control over material 'constants', as demonstrated by the blue shift of the optical spectrum of nanoparticles. Colours can be controlled and manipulated. For example, titanium dioxide in nano-material form absorbs much more UV light than its bulk counterpart.

1.3 MOLECULAR-SCALE ELECTRONICS

The name 'molecular electronics' dates back to the 1950s in the USA, although its meaning was somewhat different to that accepted today. Although all branches of the US Military were interested in miniaturization, the form that interest took varied among the Services. The Air Force supported an idea known as 'molecular electronics'. The concept was radically different to any other attempt at miniaturization – to build a circuit in the solid without reproducing individual component function. The whole was to do more than the sum of the parts. Molecular electronics demanded a technological leap beyond even integrated circuits, which were really just discrete components formed in the same piece of silicon (with the two important problems of isolation and interconnection overcome). The idea was ahead of its time and certainly in advance of the technology available, and no significant progress was made.

Serious research on molecular electronics started in the USA in the 1970s with pioneering work by Ari Aviram at IBM, who proposed a structure for a molecular rectifier [6]. Impetus was provided by the enthusiasm of Forrest Carter working in the US Naval Research Laboratory [7] and the field was epitomized by the elegant experiments on monolayer films undertaken by Hans Kuhn in Göttingen [8].

However, emerging technologies must compete with existing ways to doing things. The microelectronics industry, based largely on the inorganic semiconductor silicon, has been developing since the 1950s and is continuing to make dramatic progress. To put molecular-scale electronics into context, a review of these developments is provided in the next section.

1.3.1 Evolution of Microelectronics

The workhorse of today's electronic computer is the metal–oxide–semiconductor transistor, or MOSFET; the basic structure of a silicon MOSFET is shown in Figure 1.3. This is a three-terminal device, comprising the source, drain and gate. A conductive channel in the silicon semiconductor is formed beneath the insulating gate oxide and between the source and drain contacts. Application of an electric field to the gate influences the conductivity of the channel; this is called the field effect. Because there are charge carriers with both positive and negative charges, there are two kinds of MOS transistor – n-channel and p-channel. The technology is therefore referred to as complementary MOS, or CMOS. Table 1.1 shows some of the significant dates in the evolution of electronics during the 20th century [9, 10]. Although a patent for a MOSFET was filed in 1930, most of the developments in electronics around this time exploited the valve (or vacuum tube) as a signal processing device. By the late 1940s, after 40 years of development, valve technology was mature.

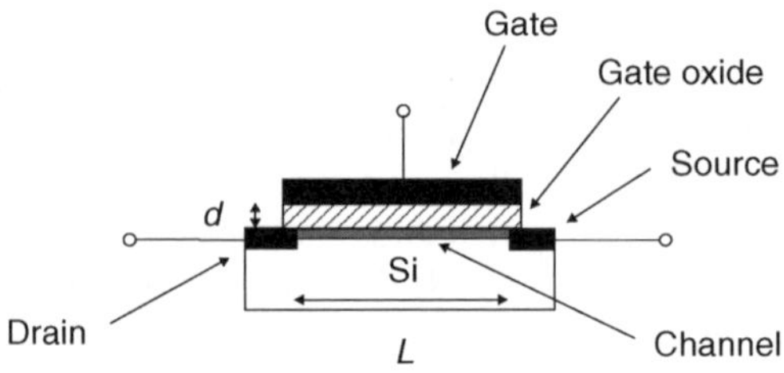

Figure 1.3 Schematic diagram of a metal–oxide–semiconductor transistor or MOSFET. The length of the conductive channel is L and the thickness of the insulating gate oxide is given by d.

The first large-scale digital electronic computer that could be reprogrammed, ENIAC (Electronic Numerical Integrator And Computer), was built in 1946, in the Moore School of Electronic Engineering, University of Pennsylvania (although earlier computers had been built with some of these properties). ENIAC could add 5000 numbers per second. It could calculate the trajectory of an artillery shell in only 30 s (in contrast, a human would need about 40 h). ENIAC contained about 17 000 valves, weighed 27 000 kg, occupied about 500 m^3 and consumed 174 kW. Its programme was wired into the processor and had to be mechanically altered.

The junction transistor was invented at Bell Laboratories, USA, in 1947 by John Bardeen, Walter Brattain and William Shockley. The three inventors shared the 1956 Nobel Prize for Physics. Even though transistors as discrete devices had significant advantages over vacuum tubes, and progress on transistors was steady during the 1950s, the directors of many large electronics companies believed that vacuum tubes held an unassailable competitive position. Arguably, the most significant breakthrough for technology came late in the 1960s, when the integrated circuit, IC, was introduced. The patent for the first IC was filed in February 1959 by Jack Kilby of Texas Instruments. His IC was two circuits constructed in a piece of germanium. Kilby, the acknowledged inventor of the microchip, died in 2005, during the

Table 1.1 Dates of key inventions in microelectronics [9, 10].

Date	Milestone
1930	MOSFET concept patent (Lilienfield, University of Liepzig, Germany)
1946	Stored-program computer (ENIAC, University of Pennsylvania)
1947	Bipolar transistor (Bardeen, Brattain, Shockley, Ball Laboratories, USA)
1952	IC concept (Dummer, Royal Radar Establishment, UK)
1959	Planar process (Hoerni, Fairchild, USA)
1959	IC patent (Kilby, Texas Instruments, USA)
1960	MOSFET (Kahng and Atalla, Bell, USA)
1962	MOS IC (Hofstein and Heiman, RCA, USA)
1968	CMOS (Westing house, GT&E, RCA, Sylvania, USA)
1969	Internet (ARPAnet, USA)
1971	Microprocessor (Hoff, Intel, USA)
1972	1024 bit DRAM (Intel, USA)
1980	256 k DRAM (NEC-Toshiba, NTT-Musashino, Japan)
1981	MS-DOS (Gates, Microsoft, USA)

writing of this book. He was awarded the Nobel Prize for Physics in 2000. The microprocessor became a reality in the mid-1970s with the introduction of the large-scale integrated circuit (LSI) and later the very large-scale integrated circuit (VLSI), with many thousands of interconnected transistors etched into a single silicon substrate.

1.3.2 Moore's Laws

A common barrier to sustaining a high rate of growth in any industry is the existence of a fundamental limitation to the basic technologies on which the industry depends. A good example of this is the commercial aerospace industry, where fundamental limitations in the strength of materials and the cost of energy have resulted in a current generation of aircraft no faster or more comfortable than those of 40 years ago. Improvements have instead been evolutionary in nature and focused largely around operating cost reduction. By contrast, no fundamental limitation in the technologies enabling the electronics industry has yet arisen as a significant barrier to its continued growth.

The first microprocessor chip (the 4004), manufactured in 1971 by Intel using 10 μm process technology, had a clock speed of 108 kHz and contained 2300 transistors. There are 169×10^6 transistors on the Pentium® 4 chip (February 2005), fabricated using 90 nm technology, and operating at a clock speed of 3.73 GHz. The 2006 version of the Titanium® 2 Processor contains over 10^9 transistors [11]. Furthermore, Intel is on track to manufacture chips on 300 mm Si wafers using 45 nm process technology by 2007.

Gordon Moore of Intel was the first to quantify the steady improvement in gate density when he noticed that the number of transistors that could be built on a chip increased exponentially with time. His observation, made in 1965, was that the number of transistors per unit area on integrated circuits, or functionality per chip, had doubled every year since the integrated circuit was invented. Moore predicted that this trend would continue for the foreseeable future. This is Moore's Law, or Moore's First Law. Although not strictly a 'law', this rule has been a consistent trend and key indication of successful leading-edge semiconductor products. In recent years, the pace slowed a little: the scaling of the microprocessor unit, MPU, is based more on a 2.5 year cycle (i.e. the functionality per chip doubles every 2.5 years) while the scaling of a dynamic random access memory, DRAM, is a 3 year cycle. Most experts expect Moore's Law to hold for at least until 2020.

The semiconductor industries have produced an *International Technology Roadmap for Semiconductors* on the future of CMOS technology [12]. Figure 1.4 shows the anticipated growth in the density of the transistors in both the MPU and the DRAM of a CMOS chip. The prediction is for a 13 nm minimum feature size (gate length, L in Figure 1.3) for the MPU and 4×10^{10} transistors per cm^2 for the DRAM (multi-level cell) by 2013. Longer term predictions give these values as 6 nm and $2 \times 10^{11}\ cm^{-2}$, respectively. All these figures are regularly updated. The transistors in Intel's new 65 nm process technology have gates measuring 35 nm. About 100 of these gates could fit inside the diameter of a human red blood cell.

Intel has already published experimental results on a transistor with 10 nm gate length [11]. Researchers from Intel and QinetiQ have also jointly developed prototype transistors with indium antimonide (InSb is compound semiconductor from based on elements from Groups 3 and 5 of the periodic table – a so-called III–V compound) which show promise for future high-speed and very low-power logic applications. These transistors could be used in

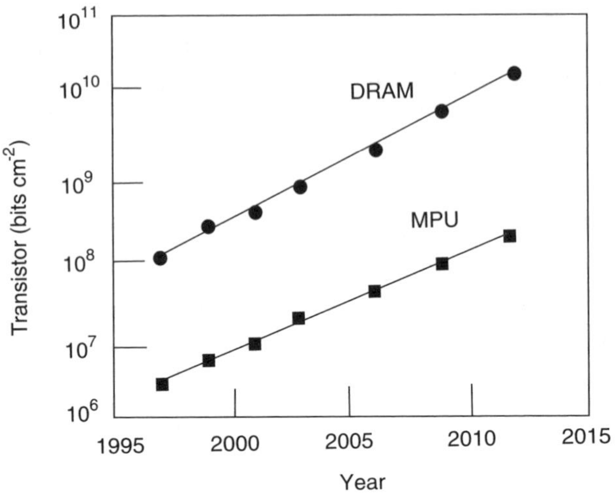

Figure 1.4 Predicted scaling for transistors in a dynamic random access memory (DRAM) and microprocessor unit (MPU). Taken from the *International Technology Roadmap for Semiconductors* [12].

Intel's logic products in the second half of the next decade and could be a factor in the continuation of Moore's Law well beyond 2015.

There are many recognized factors that could bring the Moore's Law scaling to an end. Some of the technical issues are explored in the next section. However, one significant factor is the economics of chip production. The cost of building fabrication facilities to manufacture chips has been increasing exponentially, by a factor of two for every chip generation. This is sometimes known as Moore's Second Law. The cost of manufacturing chips is increasing significantly faster than the market is expanding. Intel's Fab 22, a chip-fabrication facility which opened in Chandler, AZ, in October 2001, cost \$2 billion to construct and equip. The cost of building a Fab is projected to rise to \$15–30 billion by 2010 and could be as much as \$200 billion by 2015 [13]. This significant increase in cost is due to the extremely sophisticated tools that will be needed to form the increasingly small features of the devices.

1.3.3 Beyond Moore

As the dimensions of MOSFET devices decrease, the effect of the operating characteristics can be calculated [14]. The scaling can be done either to keep the electric field constant or to maintain the same operating voltage. For a scaling factor $K(K > 1)$, Table 1.2 shows the effect on the operating parameters of the MOSFET for a constant field. The benefits of shrinking the device in terms of packing density, speed and power dissipation are evident. The scaling of depletion widths is achieved indirectly by scaling up the doping concentrations (which keeps the resistance unchanged). For ideal scaling, power supply voltages should also be reduced to keep the internal electric fields reasonably constant from one device generation to the next. In practice, however, power supply voltages are not scaled with the device dimensions, partly because of other system-related constraints.

One parameter that does not move in a favourable direction as the device dimensions are reduced is the current density, which is increased with the scale factor. Metal conductors

Table 1.2 Scaling rules for MOSFETs according to a scale factor K. The voltages are scaled to keep the electric fields constant [14].

Parameter	Scaling factor
Device dimensions (channel length, gate width, oxide thickness)	$1/K$
Current, voltage	$1/K$
Current density	K
Impurity concentration	K
Gate capacitance	$1/K$
Time constant	$1/K$
Switching energy	$1/K^3$
Power per device	$1/K^2$

have an upper current density limit imposed by the process of electromigration. This is a diffusive process in which the atoms of a solid move under the influence of electrical forces. This effect limits the maximum current that can be carried by a conductor without its rapid destruction. For example, the current density for aluminium conductors in integrated circuits must be kept below $10^{10}\,\mathrm{A\,m^{-2}}$.

Limits that are closely related to the basic physical laws are called fundamental limits. A basic concept of quantum mechanics is that a physical measurement performed in a time Δt (and computing may be considered as a 'measurement') must involve an energy ΔE:

$$\Delta E \geq \frac{h}{\Delta t} \tag{1.1}$$

where h is Planck's constant ($= 6.63 \times 10^{-34}\,\mathrm{J\,s}$). The energy is dissipated as heat. The power $P(=$ energy per unit time) dissipated during the measurement, or switching, process is

$$P = \frac{\Delta E}{\Delta t} \geq \frac{h}{(\Delta t)^2} \tag{1.2}$$

This can be considered as the lower band of power dissipation per unit operation. Equation (1.1) predicts a minimum energy dissipated in a nanosecond switching device (i.e. operating at 1 GHz) is about $10^{-25}\,\mathrm{J}$, which is currently orders of magnitude below the actual switching energy in an MOS device (the switching energy of silicon MOSFET devices that will be produced in 2016 is estimated as $4 \times 10^{-18}\,\mathrm{J}$ [12]).

The switching energy must generally be greater than the thermal energy, otherwise the device will switch on and off randomly. This requires

$$\Delta E > k_B T \tag{1.3}$$

where k_B is Boltzmann's constant ($1.38 \times 10^{-23}\,\mathrm{J\,K^{-1}}$) and T is the absolute temperature. At room temperature, $k_B T = 4 \times 10^{-21}\,\mathrm{J}$, once again many orders of magnitude below the actual switching energy in MOSFETs.

Equation (1.3) is not regarded as a fundamental limit to computing operations, as many methods are available to reduce noise in electronic systems [15]. An assertion, put forward by John von Neumann, in a 1949 lecture, was that a computer must dissipate an energy per reversible operation (or per bit in a binary computer) given by

$$\Delta E = k_B T \log_e 2 \tag{1.4}$$

which is about 3×10^{-21} J at room temperature. This can be considered as a fundamental limit.

A further issue is the time taken for a charge carrier to acquire energy from the applied electric field. The maximum rate of energy transfer (power) to an electron with an electrical charge $e(= 1.6 \times 10^{-19}\ \text{C})$ moving at a speed v in an electric field $\boldsymbol{E}$ is $ev\boldsymbol{E}$ [i.e. = (force × distance)/time]. The electron energy must be greater than the thermal energy $k_B T$. The time to achieve this is therefore given by

$$\text{time} = \frac{\text{energy}}{\text{power}} = \frac{k_B T}{ev\boldsymbol{E}} \tag{1.5}$$

The maximum electron velocity in silicon is about $10^5\ \text{m s}^{-1}$ for an applied field of $5 \times 10^7\ \text{V m}^{-1}$. This gives a limiting response time of approximately 5×10^{-15} s (5 fs) at room temperature. Devices will not respond so quickly as this the electrons will need to acquire energies $\gg k_B T$ ($10k_B T$ or $100k_B T$). However, this principle imposes a restriction on the operating frequency of silicon MOSFETs.

There are also a number of technological problems issues that will need to be overcome for the predictions for the CMOS-based roadmap to be realized. Not least are the materials limitations of the silicon/silicon dioxide system. For example, charge leakage becomes a problem when the insulating silicon dioxide layers are thinned to a few nm. Intel's 65 nm transistors have a gate oxide thickness of 1.2 nm. Quantum mechanical tunnelling through the oxide can become a serious problem if the tunnel currents become comparable to the other circuit currents. Alternative insulating materials, such as hafnium oxide, are currently under investigation by the large electronics companies.

Heat dissipation is a further important factor. A Pentium® chip, with around 10^8 transistors operating at the nanosecond rate, can emit up to 100 W of heat. Although the simple scaling shown in Table 1.2 predicts that the power per device decreases with the square of the scaling factor, the power density (power per unit area) will remain constant. This power density p is given by

$$p = \Delta E \nu n X \tag{1.6}$$

where n is the device density, ν is the frequency of operation and X is the probability that the device switches in a clock cycle, typically $X \sim 0.1$. The maximum tolerable power density is about $100\ \text{W cm}^{-2}$. This means that the switching energy, operating frequency and device density are limited by

$$\Delta E \nu n \approx 1000\ \text{W cm}^{-2} \tag{1.7}$$

At high frequency, a high device density is desirable, and therefore a low energy per bit is desirable. These three limits are depicted in the switching energy–time diagram in Figure 1.5.

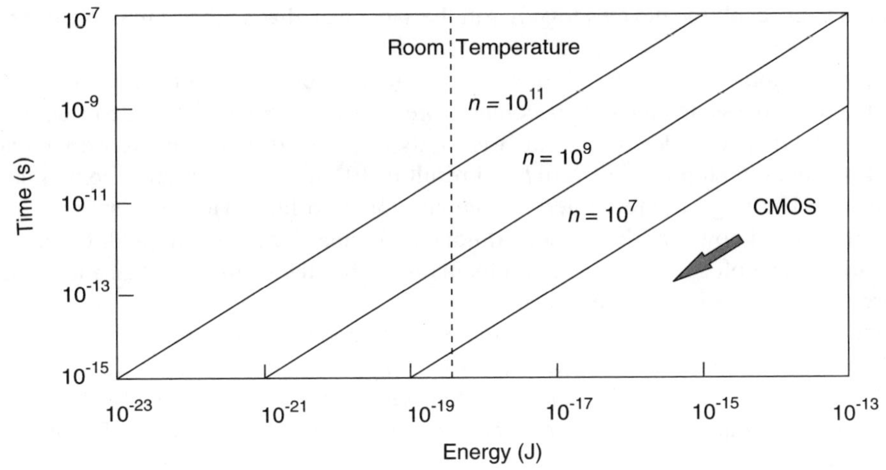

Figure 1.5 Switching time versus switching energy for MOSFETs based on (Equation (1.7)). The three different curves correspond to different device densities n (in units of cm^{-2}). The vertical line corresponds to $100k_{B}T(= 4 \times 10^{-19}$ J), the optimum value for the energy to write a bit of information at room temperature.

Beyond 2016 and the demise of the top-down lithographic technologies which are currently employed, bottom-up nanomaterials fabrication for nanoelectronics has the potential to move electronic materials, computers and devices into a new era of sustainable growth.

The promise of molecular-scale electronics is compelling. The device density offered by silicon technology and outlined above is remarkable and offers the means to store significant amounts of data (Table 1.3) [16]. However, molecular-scale electronics has the potential for further increases in device density. For example, using 1–3 nm organic molecules as the processing elements, 10^{13}–10^{14} 'devices' could be fitted into 1 cm^2. Discrete organic processing devices (diodes and transistors) and simple circuits based on these already exist. What is now needed is to make them work faster (the maximum operational speed of such plastic circuits is around a few MHz) and more reliably.

Molecular-scale technology will, of course, also need to address all the problems of the silicon microelectronics industry – plus many more! New approaches are needed. It will not be sufficient to reproduce the functionality of discrete silicon devices in molecules and then

Table 1.3 Information content of various sources [16]. (1 byte = 8 bits)

Application	Typical information content (bytes)
Colour photograph	10^5
Average book	10^6
Desktop computer	10^8
Genetic code	10^{10}
Human brain	10^{13}
Library of Congress	10^{15}

attempt to connect all the devices together in the same way that a silicon integrated circuit is built.

As computational devices shrink in size, the tolerance of their architectures to defects in the individual processing elements becomes more important. Even if the rate of defects in a chemically fabricated molecular circuit were only one per 10^9 components, which exceeds the best practices in chip fabs, it would still result in 10^6 defects in a system containing 10^{15} components. The largest defect-tolerant computer built so far is Hewlett Packard's experimental machine known as Teremac. Although Teremac was constructed by means of conventional technology, many of its problems resemble the challenges that face scientists who are exploring molecular electronics.

Teremac was built from a large number of components that had significant defect probabilities. To keep the construction costs reasonable, the machine was built using components that were defective and inexpensive. Furthermore, the techniques used to connect all the components together were error prone. Teremac is a reconfigurable multi-architecture computer with 10^6 gates that operate at 1 MHz or a total of 10^{12} bit operations per second (hence 'tera'). It is based on field-programmable gate arrays (FPGAs). These are essentially lookup tables connected by a huge number of wires and switches that are arranged to form crossbars, which permit the connection of any input to any output. In principle, FPGAs substitute memory for logic whenever possible. Perhaps the most significant fact about Teremac is that it was comatose at birth. Three-quarters of the FPGAs contained defects that would be fatal to an isolated chip. Teremac contained 220 000 wiring and gate defects – a total of 3% of all its resources. For the first 24 h of its existence, Teremac was connected to a workstation to undergo a series of tests in order to find out where the defective resources were. These locations were then written to a configuration table as being 'in use' to ensure that the defective components would not be assessed by a running programme.

1.4 THE BIOLOGICAL WORLD

Scientists and technologists working in the field of molecular electronics frequently draw attention to the analogy between the devices they are working on and those found in Nature. There are very many examples. For instance, the brain is nature's computer. Although the brain works in a different way to silicon microprocessors (exploiting ions rather than electrons, and utilizing parallel processing, instead of the serial approach in silicon systems), it is nonetheless constructed from a large number of individual processing elements. Nature's 'gates' are the neurons. There are about 10^{11} neurons in the human brain, and each is connected to 10^3–10^4 others. This gives a crude 'bit count' of 10^{11}–10^{15}. An equivalent artificial 'brain' might therefore be built from 10^5 8 Gbit chips.

Living systems are able to exist because of the vast amount of highly ordered molecular machinery from which they are built. The information required to build a living cell or organism is stored in the DNA and is then transferred the proteins by the processes called transcription and translation. These are all executed by various biomolecular components, mostly protein and nucleic acids. Such molecular-scale machinery is highly sophisticated and has evolved over millions of years. For example, bacteria may be considered as 'micro-robots,' moving throughout the human body and taking part in highly complex biochemical processes.

Electronics at the molecular scale is concerned with atomic precision and molecular manipulation and, in principle, has much to learn from Nature. This rapidly developing field is known as biomimetics. However, it might be argued that attempts to reproduce Nature's complex molecular architectures in the laboratory or factory using mechanical and chemical processes are too ambitious and doomed to failure (after all, humans' most successful flying machines are constructed from metal and not feathers!). A better approach for molecular-scale electronics is to learn from Nature rather than to copy it. For example, fundamental studies into the operation of Nature's molecular motors may, one day, lead to a new technology for energy production.

The other field in which molecular-scale manipulation of matter is receiving considerable attention is medicine. Since all living organisms are composed of molecules, molecular biology has become the primary focus of biotechnology. Countless diseases have been cured by our ability to synthesize small molecules – drugs – that interact with the protein molecules which make up the molecular machinery that keeps us alive. Our understanding of how proteins interact with DNA, phospholipids and other biological molecules is fundamental to progress.

1.5 FUTURE OPPORTUNITIES

The growth of the electronics industry has been driven by the development of disruptive technologies in parallel with evolutionary improvements in enabling device technologies. The development of personal computers, PCs, was a disruptive applications technology. Its progress, however, was greatly accelerated by rapid evolutionary improvements in technology for manufacturing semiconductor devices (increasing significantly the number of transistors on a chip). Other 'killer' applications have included cell phones, personal electronics (e.g. games), digital imaging and video, along with plasma and LCD flat panel displays. Digital video (or versatile) discs, DVDs, have enjoyed the most rapid and unexpected acceptance of any new home entertainment technology. The introduction of advanced optical discs with the required storage capacity, along with increasingly widespread adoption of high-definition television, HDTV, will help drive a high rate of growth in this sector of consumer electronics for the immediate future. Indeed, management of the current menagerie of discrete personal electronic devices may well be the most important driver for early implementation of some of the concepts embodied in 'ambient intelligence'.

The concept of ambient intelligence, AmI, was first outlined by Marc Weiser [17]. The idea was that small computer chips would be embedded in everyday objects all around us ('ubiquitous computing') and would respond to our presence, needs and wishes by way of wireless connectivity. It was imagined that these devices would undertake important tasks without active manipulation and so unobtrusively that we would notice only their effects. Weiser referred to this as 'calm technology', since it would allow us to focus on our work and on our social and recreational activities without requiring that we 'interact' directly with these devices as we currently do today with PCs, TVs, VCRs (video cassette recorders) and other home and workplace electronic items and appliances. Ambient intelligence has the potential to fulfil an important societal need – the requirement to simplify human interactions with the plethora of electronic devices that surround us today and to take advantage of the unfulfilled capabilities of electronic devices to make work and home life easier and more productive. If this technology is widely accepted and adapted, it will certainly establish rapid and sustainable growth in the electronics industry.

Since the original concept paper by Weiser, the vision of calm technology has expanded to include context awareness (i.e. analysing context, adapting to people who live in it, learning from their behaviour and eventually recognizing and demonstrating emotion). Such AmI systems would be aware of your physical and emotional state and react accordingly; for example, suggesting the need for, or requesting, medical attention, or simply adjusting ambient light, sound and colour to suit your mood.

The *International Technology Roadmap for Semiconductors* offers a similar viewpoint on the way ahead [12]. Figure 1.6 shows how non-CMOS devices can be integrated with the Roadmap. The essential functions of a silicon integrated circuit – or system-on-chip (SoC) – are those of data storage and digital signal processing. However, many functional requirements, such as wireless communication, sensing and actuating and biological functions do not scale with Moore's Law. In these cases, non-CMOS solutions can be used. In the future, the integration of CMOS and non-CMOS technologies within a single package – or system-in-package (SiP) – will become increasingly important.

A good example of the diversity of electronic systems is the 'lab-on-a-chip' technology, under development by many commercial organisations [18]. There is considerable interest in reducing the size of sensing systems and integrating them with some functionality. This has

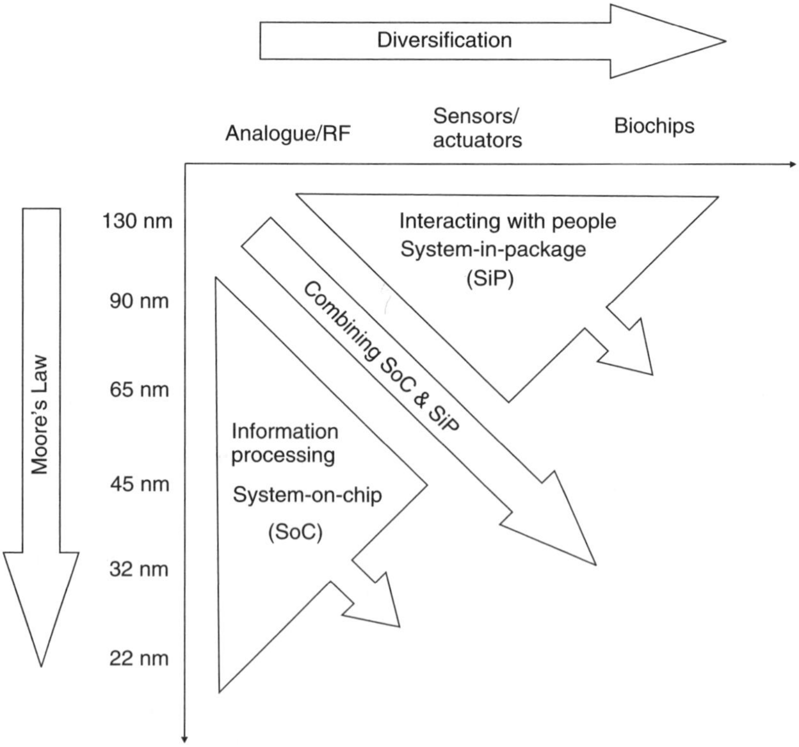

Figure 1.6 Moore's Law and diversification. Taken from the *International Technology Roadmap for Semiconductors* [12].

been made possible by the emergence of novel sensors. A completed system would, in essence, reduce the functionality of a complete chemistry laboratory to a single chip. For example, Sandia's handheld μChemlab system is about the size of a palmtop computer but is capable of using both gas and liquid chromatographic techniques to separate the various constituents of complex chemical mixtures [19]. Samples are drawn into the device by a micropump and then passed through a series of channels where the mixture is separated into its basic components. A microcomputer measures and identifies the components and displays results on the device screen. Advanced microsensor technology is used to perform chemical analysis after separation of the mixture. The system incorporates miniaturized surface acoustic wave detectors, laser-induced fluorescence analysis and electrochemical detection components. Future lab-on-a-chip technologies may allow the manipulation and measurement of individual molecules. This would provide many new opportunities for studying biochemical reactions and life processes [20].

1.6 CONCLUSIONS

Organic compounds possess a wide range of fascinating physical and chemical properties that make them attractive candidates for exploitation in electronic and opto-electronic devices. It is not anticipated, however, that these materials will displace silicon in the foreseeable future as the dominant material for fast signal processing. It is much more likely that organic materials will find use in other niche areas of electronics, where silicon and other inorganic semiconductors cannot compete. Examples already exist, such as liquid crystal displays and certain chemical sensors. Organic light-emitting structures and plastic circuitry for identification tags and smart cards are likely to make a major impact in the market place over the next 10 years.

Over the first decades of the 21st century, classical CMOS technology will come up against a number of technological barriers. The bottom-up approach to molecular electronics provides an alternative and attractive way forward and, as such, it is currently an area of exciting interdisciplinary activity. However, the challenges in fabricating molecular switches and connecting them together are formidable. Living systems use a different approach. These assemble themselves naturally from molecules and are extremely energetically efficient when compared with man-made computational devices. More radical approaches to materials fabrication and device design, exploiting self-organization, may be needed to realize fully the potential offered by molecular-scale electronics.

BIBLIOGRAPHY

Amos M. *Genesis Machines.* London: Altantic Books; 2006.

Bar-Cohen Y (Editor). *Biomimetics: Biologically Inspired Technologies.* Boca Raton, FL: Taylor and Francis; 2006.

Braun E, MacDonald S. *Revolution in Miniature.* Cambridge: Cambridge University Press; 1977.

Cuniberti G, Fagas G, Richer K (Editors). *Introducing Molecular Electronics.* Berlin: Springer; 2005.

Drexler KE. *Engines of Creation.* London: Fourth Estate; 1990.

Feast WJ, Tsibouklis J, Pouwer KL, Groenendaal, Meijer EW. Synthesis, processing and material properties of conjugated polymers. *Polymer* 1996; **37**: 5017–5047.

Ferraro JR, Williams JM. *Introduction to Synthetic Electrical Conductors.* Orlando, FL: Academic Press; 1987.
Feynman RP. There's plenty of room at the bottom. Lecture of the American Physical Society at California Institute of Technology, December 1959. Available from URL http://www.zyvex.com/nanotech/feynman.html.
Flood AH, Stoddart JF, Steuerman DW, Heath JR. Whence molecular electronics. *Science* 2004; **306**: 2055–2056.
Gardner JW. *Microsensors – Principles and Applications.* Chichester: John Wiley & Sons, Ltd; 1994.
Gardner JV, Varadan VK, Awadelkarim OO. *Microsensors, MEMS and Smart Devices.* Chichester: John Wiley & Sons, Ltd; 2001.
Gershenfeld N. *The Physics of Information Technology.* Cambridge: Cambridge University Press; 2000.
Greer J, Korkin A, Labanowski J (Editors). *Nano and Giga Challenges in Microelectronics.* Amsterdam: Elsevier; 2003.
Jones RAL. *Soft Machines.* Oxford: Oxford University Press; 2004.
Leff HS, Rex AF (Editors). *Maxwell's Demon: Entropy, Information, Computing.* Bristol: Adam Hilger; 1990.
Mansoori GA. *Principles of Nanotechnology.* Singapore: World Scientific; 2005.
Petty C, Bryce MR, Bloor D (Editors). *An Introduction to Molecular Electronics.* London: Edward Arnold; 1995.
Reitman EA. *Molecular Engineering of Nanosystems.* New York: Springer; 2001.
Wallace GG, Spinks GM, Kane-Maguire LAP, Teasdale PR. *Conductive Electroactive Polymers: Intelligent Materials Systems.* Boca Raton, FL: CRC Press; 2002.
Whitesides GM. The once and future nanomachines. *Sci. Am.* 2001; **285**: 78–83.
Whitesides GM, Love JC. The art of building small. *Sci. Am.* 2001; **285**: 38–47.
Wright JD. *Molecular Crystals.* Cambridge: Cambridge University Press; 1995.

REFERENCES

[1] Pocchetino A. Sul comportamento foto-elettrico dell'antracene. *Atti Acad. Naz. Lincei Rend.* 1906; **15**: 355–368.
[2] Eley DD. Phthalocyanines as semiconductors. *Nature* 1948; **162**: 819.
[3] Varanyan AT. Poluprovodnikovye svoistva organicheskikh krasitelei. 1. Ftalotsianiny. *Zh. Fiz. Khim.* 1948; **22**: 769–789.
[4] Akamatu H, Inokuchi H. On the electrical conductivity of violanthrone, isoviolanthrone, and pyranthrone. *J. Chem. Phys.* 1950; **18**: 810–811.
[5] Natta G, Mazzanti G, Corradini P. Polimerizzazione stereospecifica dell'acetilene. *Atti Acad. Naz. Lincei, Cl. Sci. Fis. Mat. Tend.* 1958; **25**: 3–12.
[6] Aviram A, Ratner MA. Molecular rectifiers. *Chem. Phys. Lett.* 1974; **29**: 277–283.
[7] Carter FL. Molecular-level fabrication techniques and molecular electronic devices. *J. Vac. Sci. Technol. B* 1983; **1**: 959–968.
[8] Kuhn H. Organized monolayers – building blocks in constructing supramolecular devices. In FT Hong (editor). *Molecular Electronics: Biosensors and Computers Symposium.* New York: Plenum Press; 1989, pp. 3–24.
[9] Dummer GWA. *Electronic Inventions and Discoveries*, 4th ed. Bristol: Institute of Physics; 1997.
[10] Braun E, MacDonald S. *Revolution in Miniature.* Cambridge: Cambridge University Press; 1977.
[11] Intel. Available from URL http://www.intel.com/.
[12] *International Technology Roadmap for Semiconductors.* Semiconductor Industries Association. Available from URL http://public.itrs.net/.
[13] Dettmer R. Grand designs. *IEE Rev.* 2003; **49**: 30–33.

[14] Kasap SO. *Principles of Electrical Engineering Materials and Devices.* Boston: McGraw-Hill; 1997.

[15] Landauer R. Computation: a fundamental physical view. In HS Leff and AF Rex (Editors). *Maxwell's Demon: Entropy, Information, Computing.* Bristol: Adam Hilger; 1990, pp. 260–267.

[16] Tour JM. *Molecular Electronics.* Singapore: World Scientific; 2003.

[17] Weiser M. The computer for the 21st century. *Sci. Am.* 1991; **265**: 66–75.

[18] Janasek D, Franzke J, Manz A. Scaling and the design of miniaturized chemical-analysis systems. *Nature* 2006; **442**: 374–380.

[19] Research and technology development performed by Sandia's California laboratory for the Department of Energy. Available from URL http://www.ca.sandia.gov/casite/index.html.

[20] Craighead C. Future lab-on-a-chip technologies for interrogating individual molecules. *Nature* 2006; **442**: 387–393.

2 Materials' Foundations

In nature's infinite book of secrecy

2.1 INTRODUCTION

All physical properties of materials ultimately depend on how the constituent atoms and molecules are held together – the chemical bonds. Therefore, this chapter begins by looking at the various types of bond that can form between atoms. These ideas are then developed with a particular emphasis on the forces that hold organic materials together. The morphological properties of crystalline and noncrystalline materials are also described and the common defects that pervade materials (all too often ignored) are introduced.

2.2 ELECTRONIC STRUCTURE

2.2.1 Atomic Structure

Each atom in a material consists of a very small nucleus composed of *protons* and *neutrons*, which is encircled by moving *electrons*. Both electrons and protons are electrically charged. The magnitude of the charge, e, is 1.60×10^{-19} C, negative in sign for the electrons and positive for the protons. Neutrons are electrically neutral. Although there is *Coulombic repulsion* between the protons, all the protons and neutrons are held together in the nucleus by the *strong force*, which is a powerful, fundamental natural force between particles. The force has a very short range of influence, typically less than 10^{-15} m. Masses for the subatomic particles are infinitesimally small: protons and neutrons have approximately the same mass, 1.67×10^{-27} kg, significantly larger than that of an electron, 9.11×10^{-31} kg.

Each chemical element is characterized by the number of protons in the nucleus or the *atomic number*, Z. For an electrically neutral or complete atom, the atomic number also equals the number of electrons. This atomic number ranges from 1 for hydrogen to 92 for uranium, the highest of the naturally occurring elements.

The atomic mass of a specific atom can be expressed as the sum of the masses of protons and neutrons within the nucleus. Although the number of protons is the same for all atoms of a given element, the number of neutrons may be variable. Thus atoms of some elements have two or more different atomic masses, called *isotopes*. The *atomic weight* of an element corresponds to the weighted average of the atomic masses of the atom's naturally occurring isotopes. The *atomic mass unit*, amu, may be used for computations of atomic weight. A scale has been established whereby 1 amu is defined as 1/12 of the atomic mass of the most common isotope of carbon, carbon-12 (^{12}C). Within this scheme, the masses of protons and

Molecular Electronics: From Principles to Practice M. C. Petty
© 2007 John Wiley & Sons, Ltd

neutrons are slightly greater than unity. The *atomic weight* of an element or the *molecular weight* of a compound (also called *molar weight* or *molar mass*) may be specified on the basis of amu per atom (molecule) or mass per *mole* of material. In one mole of substance there are 6.02×10^{23} (*Avogadro's number*) atoms or molecules.

2.2.2 Electrons in Atoms

During the latter part of the 19th century, it was realized that many phenomena involving electrons in solids could not be explained in terms of classical mechanics. What followed was the establishment of a set of principles and laws that govern systems of atomic and subatomic entities which came to be known as *quantum mechanics*. One early outcome of quantum mechanics was the simplified *Bohr atomic model*, in which the electrons are assumed to revolve around the atomic nucleus in discrete orbitals. An important assumption in the Bohr model is that only certain orbits with fixed radii are stable around the nucleus. For example, the closest orbit of the electron in the hydrogen atom can only have a radius of 0.053 nm. A Bohr atom is depicted in Figure 2.1.

The Bohr model was eventually found to have some significant limitations. Since the electron is constantly moving around an orbit with a given radius, over a long period of time (perhaps 1 ps on the atomic time scale), the electron would appear as a spherical negative-charge cloud around the nucleus and not as a single dot representing a finite particle. The electron can therefore be viewed as a charge contained within a spherical shell at a given radius. This *wave-mechanical model* of an atom describes the position of an electron in terms of a probability distribution or electron cloud.

The electronic orbital is therefore a region of space in an atom or molecule where an electron with a given energy may be found. Due to the requirement of stable orbits, the electrons occupy well-defined spherical regions, distributed in various *shells* and *subshells*.

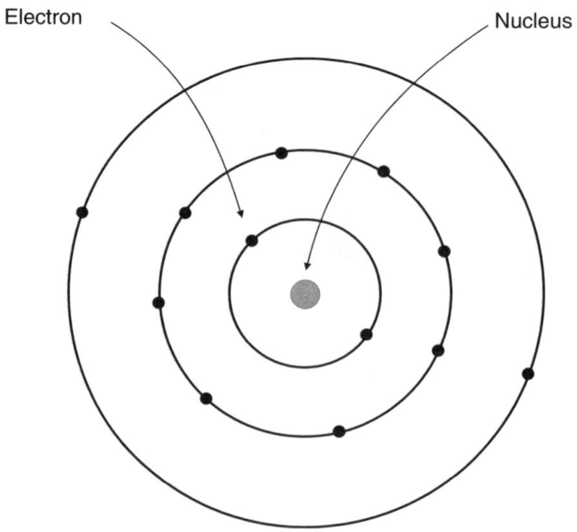

Figure 2.1 Schematic representation of the Bohr atom.

Table 2.1 Summary of quantum numbers of electrons in atoms.

Name	Symbol	Permitted values	Property
Principal	n	Positive integers (1, 2, 3, ...)	Orbital energy (size)
Angular momentum	l	Integers from 0 to $n - 1$	Orbital shape (the l values 0, 1, 2 and 3 correspond to s, p, d and f orbitals, respectively)
Magnetic	m_l	Integers from $-l$ to 0 to $+l$	Orbital orientation
Spin	m_s	$+\frac{1}{2}$ or $-\frac{1}{2}$	Direction of electron spin

In this model of the atom, the shells and subshells are spatial regions around the nucleus where the electrons are most likely to be found. Using wave mechanics, every electron in an atom is characterized by four parameters called *quantum numbers*. The size, shape and spatial orientation of an electron's probability density are specified by these numbers.

The first three quantum numbers are the principal, n, orbital (angular momentum), l, and magnetic, m_l, numbers (Table 2.1). The principal quantum number is related to the distance of the electron from the nucleus and gives the total energy of an electron in an atom, the energy increasing with n. This number may take on integral values beginning with unity and specifies the main shell where the electron is located. The shells are denoted by the capital letters K, L, M, N, O and so on, which correspond to $n = 1, 2, 3, 4, 5 \ldots$, as indicated in Table 2.2. The orbital quantum number specifies the magnitude of the orbital momentum of the electron; the l values 0, 1, 2, 3 indicate the electron subshell and are labelled lowercase s, p, d, f states. This second quantum number is related to the shape of the electron subshell. Furthermore, the number of these subshells is restricted by the magnitude of n; the allowable subshells for several n values are shown in Table 2.2. The number of energy states for each subshell is determined by the third quantum number, m_l. For an s subshell, there is a single energy state, whereas for p, d and f subshells three, five and seven states exist, respectively.

Table 2.2 Maximum number of available electrons in the shells and subshells of an atom.

Principal quantum number n	Shell designation	Subshells	Number of states	Number of electrons: Per subshell	Number of electrons: Per shell
1	K	s	1	2	2
2	L	s	1	2	8
		p	3	6	
3	M	s	1	2	18
		p	3	6	
		d	5	10	
4	N	s	1	2	32
		p	3	6	
		d	5	10	
		f	7	14	

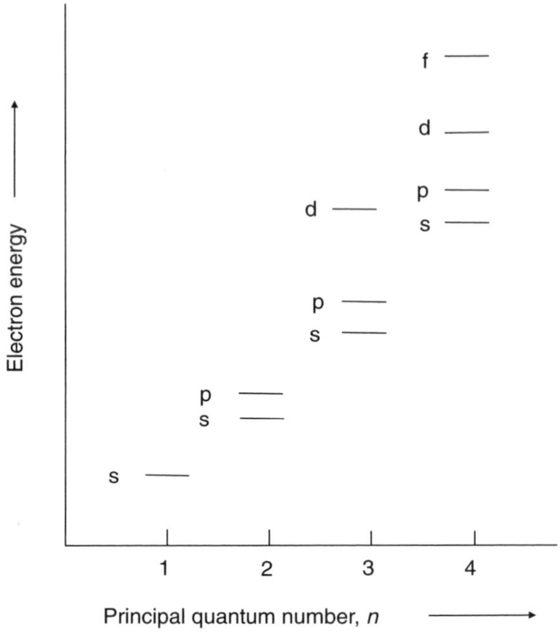

Figure 2.2 Relative electron energies for shells and subshells.

In the absence of an external magnetic field, the states within each subshell are identical. However, when a magnetic field is applied, these subshell states split, each state assuming a slightly different energy. For this reason, m_l is known as the magnetic quantum number (in fact, m_l determines the orientation of the orbital magnetic moment relative to the magnetic field). Electrons also have a fourth quantum number, m_s, associated with the characteristics of spin, which may have a value of either $+½$ or $-½$ (Table 2.1).

A complete energy level diagram for the various shells and subshells using the wave-mechanical model is shown in Figure 2.2. Several features of the diagram are worth noting. First, the smaller the principal quantum number, the lower is the energy level; for example, the energy of the 1s state is less than that of the 2s state, which in turn is less than the 3s state. Second, within each shell, the energy of a subshell level increases with the value of the l quantum number. For example, the energy of a 3d state is greater than that of a 3p state, which is larger than that of a 3s state. Finally, there may be overlap in energy of a state in one shell with states in an adjacent shell, which is especially true of d and f states; for example, the energy of a 3d state is greater than that of a 4s state.

In quantum mechanics, an atomic orbital is defined as a one-electron *wavefunction*, ψ (x, y, z, t), which is function of both position (x, y and z using Cartesian coordinates) and time (t). Wavefunctions are discussed further in Chapter 3, Section 3.3.1. For each point in space, there is associated a number whose square is proportional to the probability of finding an electron at that point, i.e. $|\psi(x, y, z, t)|^2$ is the probability of finding the electron per unit volume at x, y, z and at time t.

Only $|\psi|^2$ has a physical meaning as the function ψ can be a complex mathematical quantity, with real and imaginary parts. The probability distribution has all the properties associated with waves. It has a numerical magnitude (its amplitude) which can be either

Table 2.3 Relationship between type of orbital and the number of nodes in the corresponding electron wavefunction.

Orbital	Nodes
1s	No nodes
2s	One spherical node
$2p_x$	One node, the *yz* plane
$2p_y$	One node, the *xz* plane
$2p_z$	One node, the *xy* plane

positive or negative (corresponding to a wave crest or a wave trough) and nodes. A node is a region where a crest and a trough meet. In three-dimensional waves characteristic of electron motion, the nodes are two-dimensional surfaces at which $\psi = 0$. Consequently, atomic orbitals may be characterized by their corresponding nodes, related to the quantum numbers. For example, the nodes associated with the 1s, 2s and 2p orbitals (there are three of these) are given in Table 2.3. The 1s orbital has no nodes. The wavefunction is a spherically symmetrical function whose numerical value decreases exponentially from the nucleus.

The orbitals in Table 2.3 are depicted schematically in Figure 2.3. The plus and minus signs in the figure have no relationship to electric charge. They are simply the arithmetic signs associated with the wavefunction, e.g. a positive sign denotes a wave crest and a negative sign a wave trough. The signs associated with the wavefunctions determine how two or more wavefunctions will combine when they interact (Section 2.4). The dashed lines in Figure 2.3 represent nodal surfaces. These nodes are a sphere for the 2s orbital and a plane for the 2p orbital.

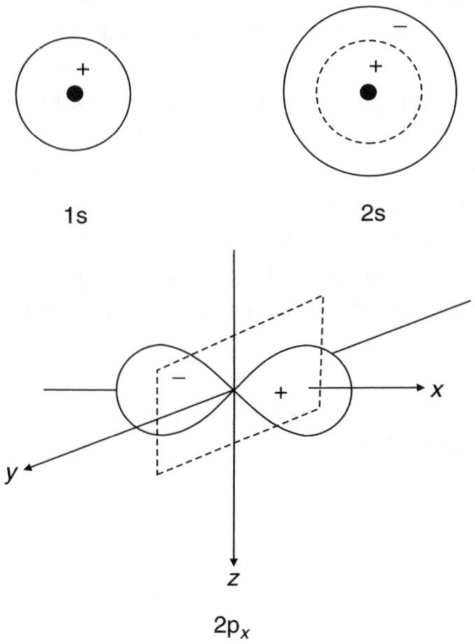

Figure 2.3 Schematic representations of s and p orbitals.

In summary, the various *n* quantum numbers define electron shells, within which the electrons are most likely to be located. The shells may be further divided into subshells, according to their *l* values. The first shell contains just one s orbital, denoted the 1s. The second shell has one spherical orbital, the 2s, and three p orbitals, the 2p orbitals, which are arranged at right-angles to each other. The third shell has one s orbital, the 3s, three p orbitals, the 3p orbitals, and a group of five d orbitals, the 3d orbitals. The pattern is that each successive shell comprises all the orbital types of the previous shell (but of greater size) plus a new group of orbitals that the previous shell does not possess. The energy of an electron depends both on the shell and, in general, on the nature of the orbital.

2.2.3 Filling of Orbitals

To determine the way in which the electron states are filled with electrons, the *Pauli Exclusion Principle* is used. This states that no two electrons in an atom can have identical quantum numbers, i.e. each electron in an atom has a unique set of quantum numbers. If this principle did not hold, all of the electrons in an atom would be located in the lowest energy state (the K shell). The Pauli Exclusion Principle holds for not just electrons but for any *fermions* (half-integer spin particles such as electrons, protons, neutrons, muons and many more). It does not apply to particles of integer spin (*bosons*). In filling up the various orbitals, *Hund's rule* is also important. This states that when several orbitals of equal energy are available, electrons enter singly with parallel spins. In other words, one can add one electron to each orbital of equal level (such as the five d orbitals) and these will all have the same spin. Electrons are only paired when there are no available orbitals of the same energy. When all the electrons occupy the lowest possible energies according to the foregoing restrictions, an atom is said to be in its *ground state*. However, electron transitions to higher energy states are possible. The electron configuration or structure of an atom represents the manner in which the electron states are occupied. In the conventional notation, the number of electrons in each subshell is indicated by a superscript after the shell-subshell designation. For example, the electron configurations for hydrogen, carbon and silicon are $1s^1$; $1s^2 2s^2 2p^2$; and $1s^2 2s^2 2p^6 3s^2 3p^2$.

The outermost filled shell is occupied by the *valence electrons*. These participate in the bonding between atoms and are responsible for many of the physical and chemical properties of the resulting solids. In addition, some atoms have what are termed 'stable electron configurations', i.e. the states within the outermost or valence electron shells are completely filled. Such elements are neon, argon, krypton and helium. These are the inert gases and are very unreactive chemically.

2.2.4 The Periodic Table

All the elements have been classified according to their electron configurations in the *periodic table* as shown in Figure 2.4. The similarity of elements in each group (vertical column) of the periodic table is the result of similarity in the outer electron or valence electron configuration. All the alkali metals – Group 1A – (lithium, sodium, potassium, rubidium, caesium and francium) have 'inert-gas-plus-one-configurations'. The alkaline earth metals make up Group 2A; these elements (beryllium, magnesium, calcium, strontium,

	1A	2A	3A	4A	5A	6A	7A	8			1B	2B	3B	4B	5B	6B	7B	0
1	1 H																	2 He
2	3 Li	4 Be											5 B	6 C	7 N	8 O	9 F	10 Ne
3	11 Na	12 Mg											13 Al	14 Si	15 P	16 S	17 Cl	18 Ar
4	19 K	20 Ca	21 Sc	22 Ti	23 V	24 Cr	25 Mn	26 Fe	27 Co	28 Ni	29 Cu	30 Zn	31 Ga	32 Ge	33 As	34 Se	35 Br	36 Kr
5	37 Rb	38 Sr	39 Y	40 Zr	41 Nb	42 Mo	43 Tc	44 Ru	45 Rh	46 Pd	47 Ag	48 Cd	49 In	50 Sn	51 Sb	52 Te	53 I	54 Xe
6	55 Cs	56 Ba	L	72 Hf	73 Ta	74 W	75 Re	76 Os	77 Ir	78 Pt	79 Au	80 Hg	81 Tl	82 Pb	83 Bi	84 Po	85 At	86 Rn
7	87 Fr	88 Ra	A															
		L	57 La	58 Ce	59 Pr	60 Nd	61 Pm	62 Sm	63 Eu	64 Gd	65 Tb	66 Dy	67 Ho	68 Er	69 Tm	70 Yb	71 Lu	
		A	89 Ac	90 Th	91 Pa	92 U	93 Np	94 Pu	95 Am	96 Cm	97 Bk	98 Cf	99 Es	100 Fm	101 Md	102 No	103 Lr	

Metals	
Metalloids	
Non-metals	
Transition metals	

Figure 2.4 The periodic table of the elements. The symbol for each element is shown together with its atomic number.

barium and radium) have 'inert-gas-plus-two' configurations, and the halogens that make up Group 7B (fluorine, chlorine, bromine, iodine and astatine) possess 'inert-gas-minus-one' structures. The horizontal rows in the periodic table are the periods. The elements in the three long periods, Groups 3A to 2B, are termed the transition metals, which have partially filled d electron levels and, in some cases, one or two electrons in the next higher energy shell. Groups 3B, 4B and 5B (boron, silicon, germanium, arsenic, etc.) are metalloids, displaying characteristics that are intermediate between those of metals and nonmetals by virtue of their outer electron structures. Most of the elements in the periodic table really come under the metal classification. These are sometimes called *electropositive* elements, indicating that they are capable of giving up their few outermost electrons to become positively charged ions. However, the elements situated on the right-hand side of the periodic table are *electronegative*, i.e. they readily accept electrons to form negatively charged ions, or sometimes they share electrons with other atoms. Figure 2.5 shows how the *electronegativity* of the elements varies across the periodic table. As a general rule, electronegativity increases on moving from left to right and from bottom to top. Atoms are more likely to accept electrons if their outer shells are almost full and if they are less 'shielded' from (i.e. closer to) the nucleus. On the Pauling scale of electronegativity (Linus Pauling, 1932), the most electronegative chemical element (fluorine – see Figure 2.4) is given an electronegativity value of 3.98 (textbooks often state this value to be 4.0); the least electronegative element (francium) has a value of 0.7, and the remaining elements have values in between. On this scale, hydrogen is arbitrarily assigned a value of 2.1 or 2.2. The *valence*, *valency* or *valency number*, is a measure of the number of chemical bonds formed by the atoms of a given element.

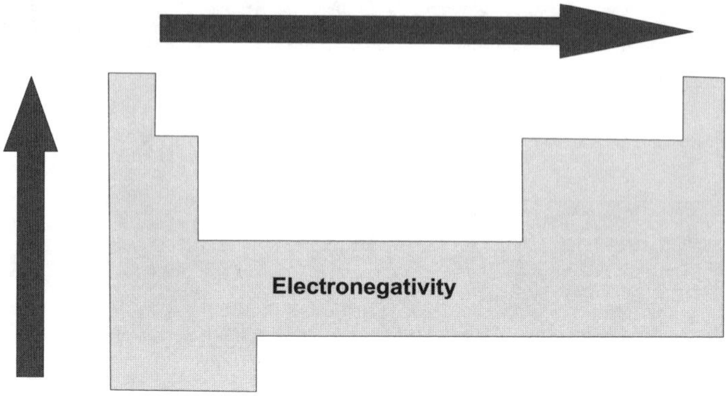

Figure 2.5 Variation of electronegativity for elements across the periodic table.

A slight complication occurs with the M and N shells because the 3d and the 4s subshells have similar energies, Figure 2.2 (a result of electrostatic screening of the outermost electron by the inner ones). Argon ($Z = 18$) has all the 1s, 2s, 3s and 3p subshells filled, but in potassium ($Z = 19$) the additional electron goes into a 4s energy level rather than a 4d level because the 4s level has a slightly lower energy.

Something similar happens for elements with atomic number $Z = 57-71$, which have two electrons in the 6s subshell but only partially filled 4f and 5d subshells. These are the rare earth elements, the lanthanides (L); they all have similar properties. Yet another such series, called the actinide series (A), starts with $Z = 89$.

2.3 CHEMICAL BONDING

In this section, the main types of chemical bonding that are responsible for the cohesive forces between atoms and molecules are described. This list is not exhaustive. The particular types of bond that form in organic compounds are described in more detail in Section 2.4.

2.3.1 Bonding Principles

All of the mechanisms that cause bonding between atoms derive from electrical attraction and repulsion. The general principle is illustrated in Figure 2.6, which shows what happens to the energy of the system as two isolated atoms approach each other. As the atoms become closer, they exert attractive and repulsive forces on each other as a result of mutual electrostatic interactions. Initially, the attractive force dominates over the repulsive force but, as the atoms become close, the electron shells of the individual atoms overlap and the repulsive force dominates. Figure 2.6 shows that the force is zero at an interatomic separation of $r = r_0$. The potential energy U is related to the force F by

$$F = \frac{dU}{dr} \tag{2.1}$$

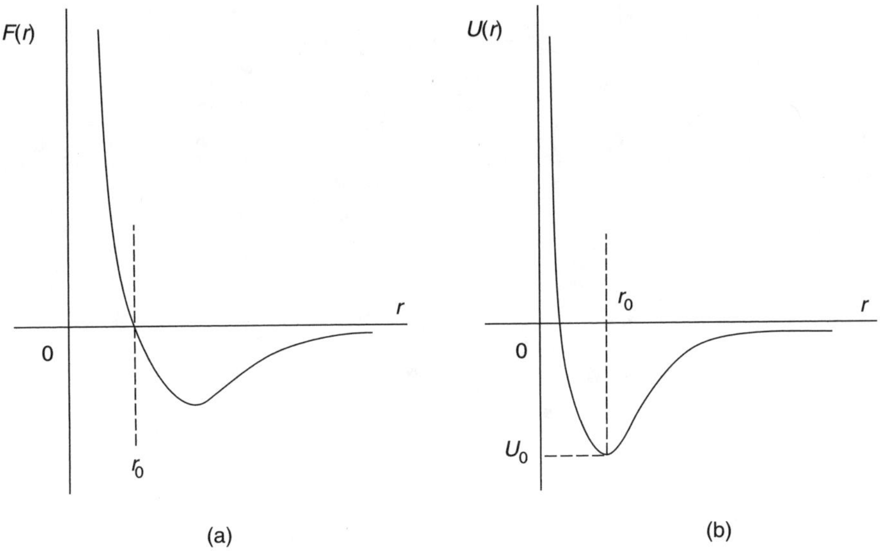

Figure 2.6 (a) Interaction force $F(r)$ between two atoms as a function of their interatomic separation r. (b) Potential energy $U(r)$ as a function of interatomic separation r. The equilibrium separation is r_0.

When the force between the atoms is zero ($r = r_0$), the potential energy will be a minimum. This corresponding minimum energy U_0 defines the bond energy of the molecule, representing the energy needed to separate the atoms. The differing strengths and differing types of bond are determined by the particular electronic structures of the atoms involved. The weak *van der Waals* bond provides a universal weak attraction between closely spaced atoms. Generally, the effect is dominated by *ionic*, *covalent* or *metallic* bonding.

The existence of a stable bonding arrangement (whether between a pair of otherwise isolated atoms or throughout a large, three-dimensional crystalline array) implies that the spatial configuration of positive ion cores and outer electrons has less total energy than any other configuration (including infinite separation of the repulsive atoms). The energy deficit of the configuration compared with isolated atoms is called the *cohesive energy* and ranges from 0.1 eV per atom for solids that are held together by van der Waals forces to 7 eV per atom in some covalent and ionic compounds, and in some metals. (An electronvolt, eV, is a unit of energy equal to the energy acquired by an electron falling through a potential difference of 1 V, approximately 1.60×10^{-19} J.)

In covalent bonding the angular placement of bonds is very important, whereas in some other types of bonding it is important to have the largest possible *coordination number* (number of nearest neighbours). For some solids, two or more completely different structures would result in nearly the same energy. A change in temperature or pressure can then provoke a change, a *phase change*, from one form of the solid to another.

2.3.2 Ionic Bond

The ionic bond is perhaps the simplest chemical bond to understand. This results from the mutual attraction of positive and negative charges. The atoms of certain elements lose or gain

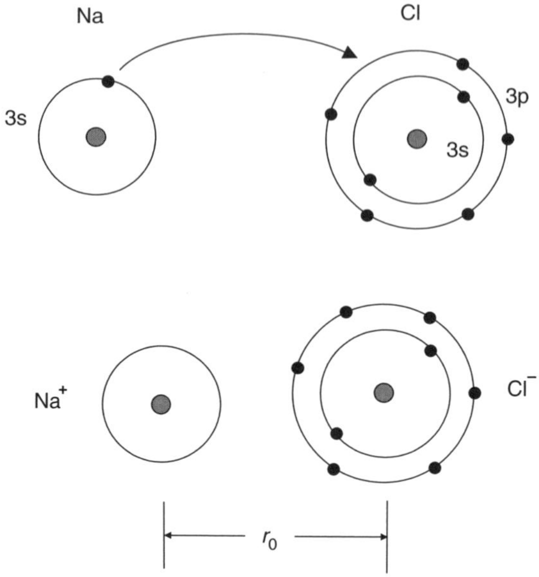

Figure 2.7 Formation of an ionic bond between Na and Cl atoms in NaCl. The equilibrium separation is r_0.

electrons very easily because, in doing so, they acquire a completely filled outer electron shell, which provides the atom with great stability. For example, sodium (Na), which has only one electron in its outer occupied shell (3s shell), will readily lose this electron to become a singly charged positive ion (Na^+). Similarly, calcium (Ca) will readily lose the two electrons in its outer occupied 4s shell to become the doubly charged calcium ion (Ca^{2+}).

Atoms such as those of chlorine, which have an almost completely filled outer shell, readily accept additional electrons to complete their outer shell and become negative ions. The formation of a molecule of sodium chloride can be depicted as shown in Figure 2.7. In fact, the negative charge associated with the chlorine ion possesses an attraction for all positive charges in its vicinity; this will also be the case for the positive sodium ion. Consequently, Na^+ ions will surround themselves with Cl^- ions and vice versa. The result, shown in Figure 2.8, is a regular array of sodium and chlorine ions in a crystalline lattice (two interpenetrating cubic networks of sodium and chorine ions are formed, Section 2.5.4).

As shown in Figure 2.5, the elements are arranged in the periodic table in order of their ability to loose their electrons or *ionization energy*. Generally, ionic bonds are formed between elements with a large difference in their ionization energies, e.g. between elements of Groups 1A (metal) and 7B (non-metal) and between elements of Groups 2A (metal) and 6B (nonmetal), for example LiF and MgO. These ionic crystals show many similar physical properties. They are strong, brittle materials with high melting points compared with metals. Most become soluble in polar liquids such as water. Since all the electrons in ionic crystals are within rigidly positioned ions, there are no free electrons to move in response to an applied electric field and to contribute to the electrical conductivity (in contrast to metals). Therefore, ionic solids are typically electrical insulators. Moreover, compared with metals, ionic solids have poor thermal conductivity since the constituent ions cannot readily pass vibrational kinetic energy to their neighbours.

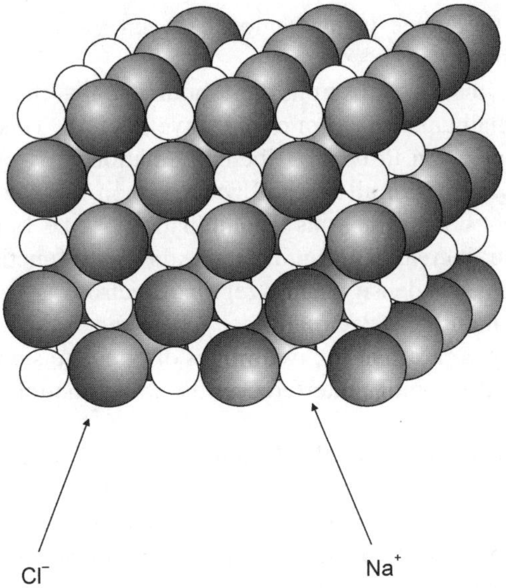

Figure 2.8 Solid NaCl crystal with Na^+ and Cl^- ions arranged close to one another.

2.3.3 Covalent Bond

Another way in which the outer electron shell of atoms can be effectively filled to achieve a stable configuration is by sharing of electrons, which results in the covalent bond. A simple example is that of the hydrogen molecule (Figure 2.9). When the 1s subshells overlap, the electrons are shared by both atoms and each atom now has a complete subshell. As shown in the figure, the two electrons must now orbit both atoms; they therefore cross the overlap region more frequently, indeed twice as often. Thus, electron sharing results in a greater

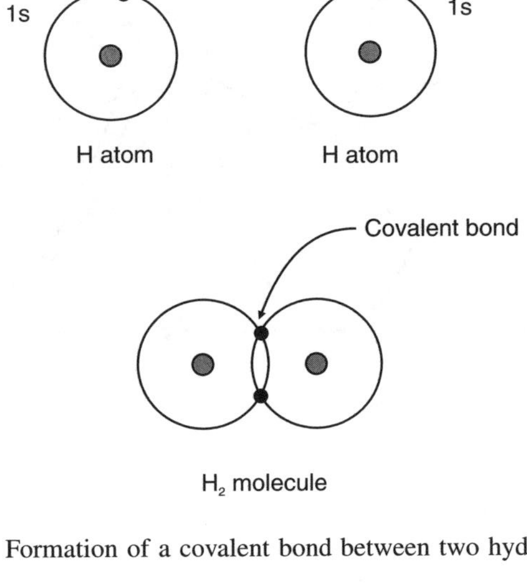

Figure 2.9 Formation of a covalent bond between two hydrogen atoms.

concentration of negative charge in the region between the two nuclei, which keeps the nuclei bonded to each other.

In the ground state, the two electrons will both occupy the same state and can be described by the same mathematical form of the wavefunction. This, of course, is allowable in terms of the Pauli Exclusion Principle. The combination of two waves having the same sign is reinforcing. This is true for all types of wave motion, from sound waves to the waves in an ocean. The covalent bond represents a combination of two 1s atomic orbitals to give a new wavefunction. The increased magnitude of the wavefunction between the atoms corresponds to higher electron density in this region. Electrons are attracted electrostatically to both nuclei and the increased electron density between the nuclei counterbalances the internuclear repulsion. In contrast, when two waves of opposite sign interact, they interfere, or cancel each other. At the point of interference, the wavefunction has the value zero, i.e. a node has been created.

A mathematical method known as the *Linear Combination of Atomic Orbitals* may be used to find the form of the electron wavefunctions in the hydrogen molecule. If the 1s atomic orbitals of two nuclei A and B are denoted ψ_A and ψ_B then, as the nuclei are brought together to form the molecule, two linear combinations of ψ_A and ψ_B are found to give satisfactory molecular orbitals. These are

$$\begin{aligned} \psi_S &= N_S(\psi_A + \psi_B) \\ \psi_{AS} &= N_{AS}(\psi_A - \psi_B) \end{aligned} \tag{2.2}$$

where N_S and N_{AS} are normalizing constants. The two wavefunctions, ψ_A and ψ_B, are symmetrical and antisymmetric, respectively, about the mid-point of a line between the two nuclei A and B. As noted earlier, the square of the wavefunctions gives the probability of finding an electron or the electron density. This is shown in Figure 2.10. It is evident that the electron density corresponding to the antisymmetric wavefunction is low in the region between the nuclei. The energy of this orbital increases continuously with decreasing separation of the nuclei and the absence of a minimum leads to a repulsive energy and what

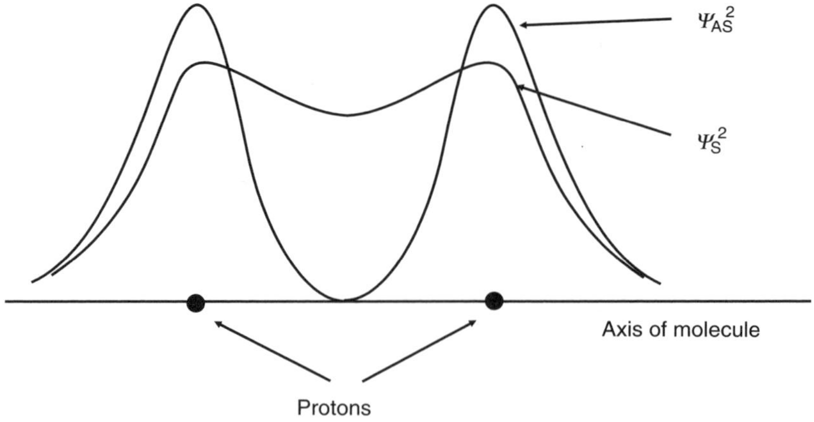

Figure 2.10 A comparison of the electron densities of the ψ_S and ψ_{AS} wavefunctions in the hydrogen molecule. The densities are symmetrical about the axis of the molecule.

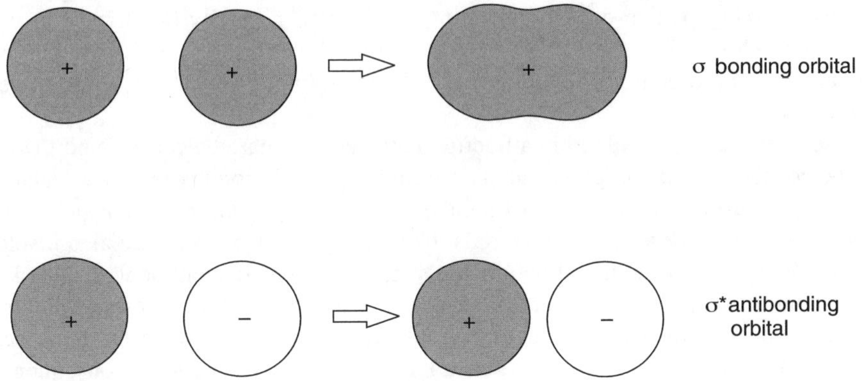

Figure 2.11 The formation of the σ bonding orbital and the σ^* antibonding orbital for the hydrogen molecule. The $+$ and $-$ refer to the signs of each orbital.

is called an *antibonding* orbital. The electron in the ground state of the positively charged *hydrogen molecule ion* is therefore located in the *bonding* ψ_S orbital. These orbitals are cylindrically symmetric with respect to the axis of the molecule and are referred to a *sigma bond*, or *σ-bond*.

A further illustration of the bonding and antibonding situations in the hydrogen molecule is given in Figure 2.11. In the ground state of the hydrogen molecule, the two electrons will both occupy the symmetrical wavefunction. This will have a lower energy than the asymmetric orbital and, in accordance with the Pauli Exclusion Principle, the electrons will have opposite spins. Each wavefunction corresponds to two quantum states so that the total number of states for the two atoms has remained unaltered.

The bonding combination corresponds to a decrease in energy (greater stability); the antibonding combination corresponds to an increase in energy (lower stability). Two atomic orbitals give rise to two molecular orbitals. The two paired electrons of opposite spin available for the bond can be put into the bonding molecular orbital. The energy relationships are summarized in Figure 2.12. Note how the energies of the two starting orbitals separate or spread apart when they interact. The amount of the separation depends on the degree to which the orbitals overlap. A slight overlap gives two molecular orbitals that differ little in energy; a large overlap results in strong separation. For axially symmetric orbitals, such as p orbitals, the greatest overlap occurs when the bonds are allowed to interact along the nuclear axis.

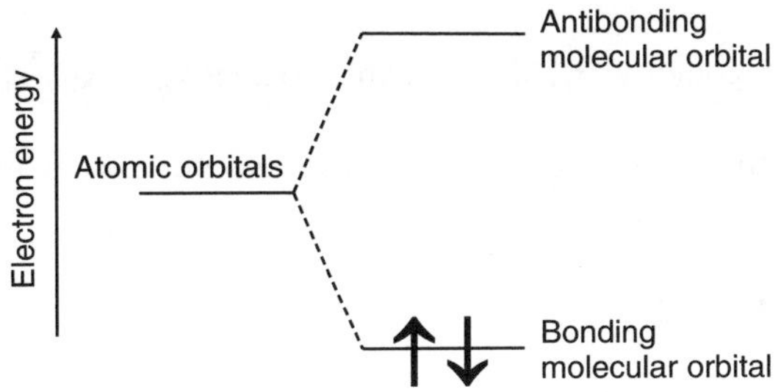

Figure 2.12 Energy relationships of atomic orbitals before (left) and after bonding (right).

The atoms taking part in covalent bonding need not necessarily be alike. For example, hydrogen and chlorine may be combined via covalent bonding to form hydrogen chloride. Most bonding *within* organic compounds is covalent bonding; this is discussed in some detail in Section 2.4.

Because of the strong Coulombic attraction between the shared electrons and the positive nuclei, the covalent bond energy is usually the highest for all bond types. This leads to high melting temperatures and to very hard solids; for instance, diamond, a covalently bonded solid, is one of the hardest known materials. Covalently bonded solids are also insoluble in nearly all solvents. The directional nature and strength of the covalent bond also make these materials nonductile (or nonmalleable). Under a strong force they exhibit little fracture. Furthermore, since all the valence electrons are locked in the bonds between the atoms, they are not free to drift in the crystal when an electric field is applied. Consequently, the electrical conductivity of such materials is very poor.

It is possible to have interatomic bonds of intermediate character between ionic and covalent. Complete ionic bonding requires the presence of an extremely electropositive component (which can be ionized easily to form a cation) and of an extremely electronegative component (for which the electron affinity to form an anion is as large as possible). These requirements are well satisfied in the alkali metal halides, in which there is a strong encouragement for electron transfer. In compounds with less extreme electropositive and electronegative character, however, there is less than 100% charge transfer from cation to anion. For example, the noble metals have larger ionization energies than alkali metals, and silver halides are less ionic in nature than the corresponding alkali metals halides.

There is a continuous progression from purely ionic character to purely covalent character as the electronegativity difference in the resulting compounds becomes smaller. When there is a partial tendency towards electron sharing, the optimum bonding can be considered as arising from a resonance between ionic and covalent charge configurations. The resulting time-averaged wavefunction for a bonding electron is then

$$\psi = \psi_{\text{cov}} + \lambda\psi_{\text{ion}} \tag{2.3}$$

where ψ_{cov} and ψ_{ion} are normalized wavefunctions for completely covalent and ionic forms and λ is a parameter which determines the degree of ionicity:

$$\%\ \text{ionicity} = \frac{100\lambda^2}{1 + \lambda^2} \tag{2.4}$$

It is possible to estimate the percentage ionic character of a bond between elements A and B (A being the most electronegative) from the following expression:

$$\%\ \text{ionicity} = 100\{1 - \exp[-(0.25)(X_{\text{A}} - X_{\text{B}})^2]\} \tag{2.5}$$

where X_{A} and X_{B} are the electronegativities of the respective elements (Section 2.2.4).

2.3.4 Metallic Bonding

In contrast to ionic and covalent bonds, the properties of the metallic bond cannot be inferred from the nature of bonding in isolated atoms or molecules and a model for bonding in metals

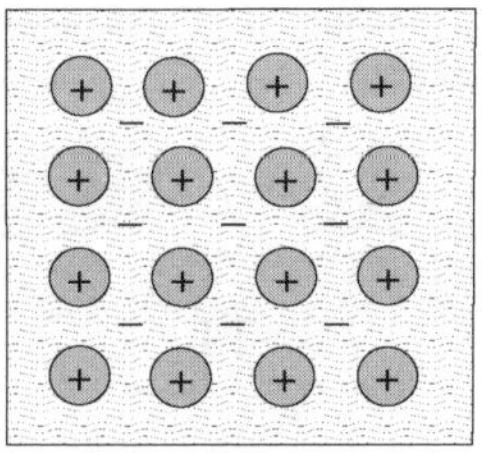

Figure 2.13 Schematic diagram of metallic bonding showing the positive ion cores of the metal atoms surrounded by negative electrons.

is not easy to construct. In some respects, it may be regarded as intermediate in character between that of ionic and covalent bonds.

The atoms of metallic elements readily lose their valence electrons (thereby acquiring the stable closed-shell electron configuration of the inert gases) to form positive ions. These valence electrons become delocalized and may be regarded as being shared by all of the ions in the crystal. The free valence electrons thereby form an electron 'gas', which may be regarded as permeating the entire crystal, as shown in Figure 2.13. Metallic bonding is essentially an electrostatic attraction between the array of positive ion cores and the electron gas; it is non-directional. Consequently, the metal ions try to approach as close as possible, which leads to close-packed crystal structures. The free electrons of this model explain the high electrical and thermal properties of metals, together with their high ductility.

2.3.5 Van der Waals Bonding

Since the atoms of inert elements have full shells and cannot accept any electrons or share any electrons, it might be thought that no bonding is possible between them. However, a solid form of argon exists at low temperatures. The forces that hold the argon atoms together are due to van der Waals bonding. This is electrostatic in nature and results from an attraction between the electron distribution of one atom and the nucleus of another.

When averaged over time, the centre of mass of electrons in the closed shell of an element such as argon coincides with the location of the positive nucleus. However, at any instant, the centre of mass is displaced from the nucleus as a result of the various motions of the individual electrons around the nucleus. This creates an instantaneous separation of the centres of positive and negative charge in the atom – an *electric dipole*. The macroscopic manifestation of an atomic dipole on the surface of a material is called the *polarization*. This is defined as the dipole moment (Chapter 4, Section 4.3) per unit volume (or charge per unit surface area).

When two such atoms approach each other, the rapidly fluctuating electronic charge distribution on one atom can induce a dipole on another atom (Figure 2.14). This type of attraction between two atoms is due to induced synchronization of the electron motions around the nucleus. This is referred to as dipole–dipole attraction. The magnitude of the attractive force drops very sharply with atomic separation, scaling as $(\text{distance})^{-7}$. The energy of van der Waals bonds is at least an order of magnitude lower than those of ionic, covalent and metallic bonding, which explains why the inert gases only solidify at very low temperatures.

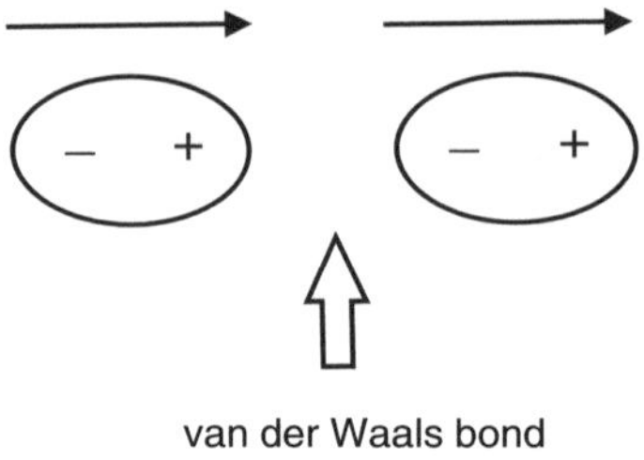

Figure 2.14 Van der Waals bonding between two dipoles.

2.3.6 Hydrogen Bonding

It is also possible to have an attraction and bonding between molecules that already possess a permanent electric dipole, such as H_2O. This is depicted in Figure 2.15. Each hydrogen atom in water shares its electrons with the oxygen atom, which has four unshared electrons. As a result, the negative charge distribution is shifted more towards the oxygen atom and the positive hydrogen proton is relatively exposed. This means that the H_2O molecule has a permanent charge separation, resulting in an electric dipole, shown in Figure 2.15(c).

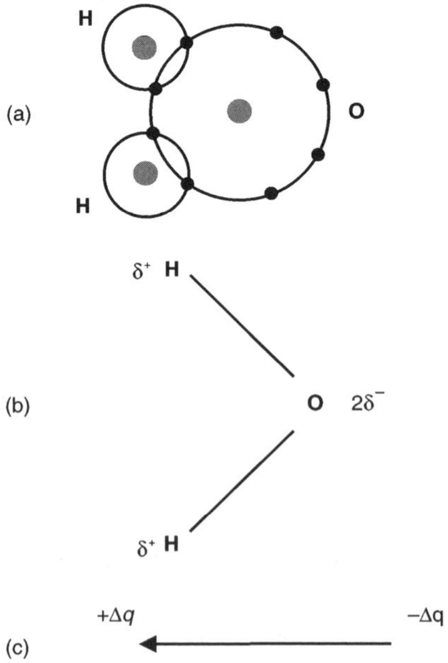

Figure 2.15 Hydrogen bonding in water. (a) Relative positions of hydrogen and oxygen atoms in an H_2O molecule. (b) Charges on hydrogen and oxygen atoms. (c) Resulting electric dipole in a molecule of water.

Table 2.4 Comparison between different types of chemical bonding.

Bond type	Examples	Bond energy (eV per atom)	Melting temperature (°C)
Ionic	NaCl	3.3	801
	MgO	5.2	2852
Covalent	Si	4.0	1410
	C (diamond)	7.4	3550
Metallic	Cu	3.1	1083
	Al	3.4	660
Van der Waals	Ar	0.08	−189
	Cl_2	0.32	−101
Hydrogen	H_2O	0.52	0
	NH_3	0.36	−78

When two water molecules are close together, the unshared electrons on the oxygen will be attracted to the exposed proton of the hydrogen in the neighbouring molecule, leading to what is known as *hydrogen bonding*. The true attraction is actually between two permanent dipoles. Hydrogen bonding is very important in organic materials. For example, it is responsible for holding the carbon chains in polymers together. Although the carbon–carbon bond in a chain is due to covalent bonding, the interaction between the chains arises from hydrogen bonds. For example in poly(vinyl chloride) (PVC), the Cl atom has five unshared electrons and therefore has a slightly more negative charge distribution than the hydrogen atom, in which the proton is relatively exposed. Consequently, a hydrogen bond develops between the Cl and the H of the neighbouring chain. These bonds are weak and can easily be stretched or broken. Polymers, therefore, have lower elastic moduli and melting temperatures than metals or ceramics.

Table 2.4 contrasts the energies involved in the five mains types of chemical bonding discussed in this section.

2.4 BONDING IN ORGANIC COMPOUNDS

2.4.1 Hybridized Orbitals

Carbon has an atomic number of six and is found in Group 4B of the periodic table. Its electron configuration is $1s^2$, $2s^2$, $2p^2$, i.e. the inner s shell is filled and the four electrons available for bonding are distributed two in s orbitals and two in p orbitals. As the s orbital is spherically symmetrical, it can form a bond in any direction. In contrast, the p orbitals are directed along mutually orthogonal axes and will tend to form bonds in these directions (Figure 2.3).

When two or more of the valence electrons of carbon are involved in bonding with other atoms, the bonding can be explained by the construction of *hybrid orbitals* by mathematically combining the wavefunctions of the 2s and 2p orbitals. In the simplest case, the carbon 2s orbital hybridizes with a single p orbital. Two *sp hybrids* result by taking the sum and difference of the two orbitals, shown in Figure 2.16, and two p orbitals remain. The sp orbitals are constructed from equal amounts of s and p orbitals; they are linear and 180° apart.

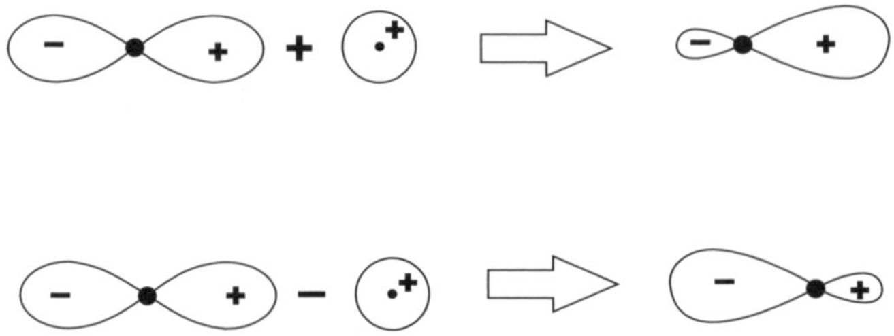

Figure 2.16 Mathematical combination of s and p orbitals to yield two sp hybridized orbitals.

Other combinations of orbitals lead to different hybrids. For example, in the case of three groups bonded to a central carbon atom, three equivalent *sp*^{2} *hybrids* may be constructed from the 2s orbital and two p orbitals (e.g. a p_x and a p_y). Each orbital is 33.3% s and 66.7% p. The three hybrids, shown in Figure 2.17, lie in the *xy* plane (the same plane defined by the two p orbitals) directed 120° from each other, and the remaining p orbital is perpendicular to the sp^2 plane. Four *sp*^{3} *hybrids* may be derived from an s orbital and three p orbitals. These are directed to the corners of a tetrahedron with an angle between the bonds of 109.5°; each orbital is 25% s and 75% p. Figure 2.18 shows the tetrahedral structure of methane, CH_4. Each bond is derived by the interaction of a C(sp^3) hybrid orbital with a hydrogen 1s and may be described as a C(sp^3)–H(1s) bond.

Long-chain organic compounds, such as many of those referred to this book (e.g. in Chapter 7, Section 7.3) involve C–H bonds that are all approximately C(sp^3)–H(s) with a bond length of 0.110 nm and a strength of 4.26 eV (98×10^3 kcal kmol^{-1}). All the C–C

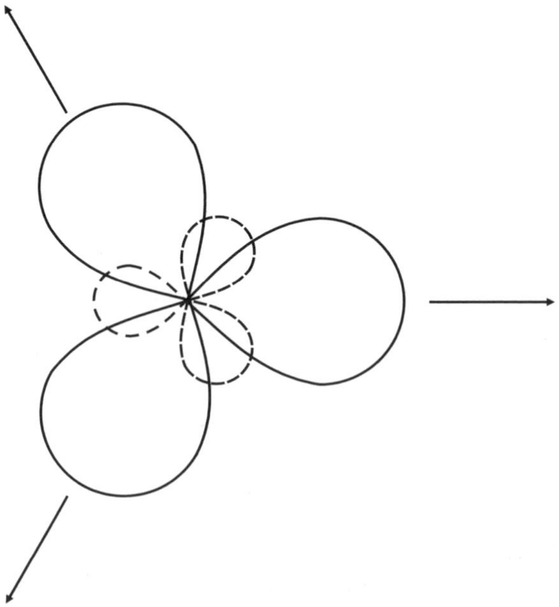

Figure 2.17 Three sp^2 hybridized orbitals.

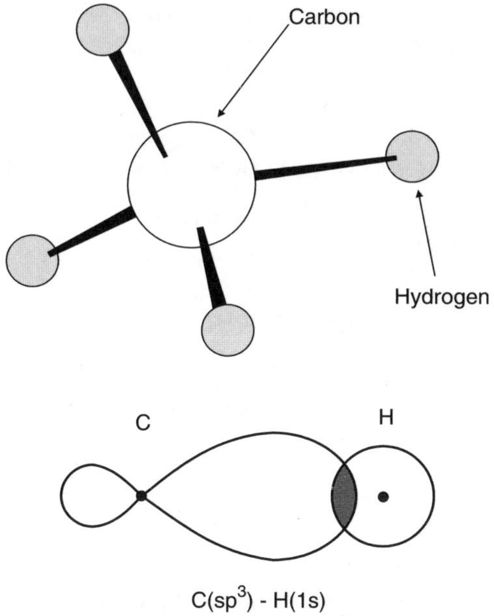

Figure 2.18 (a) Tetrahedral arrangement of carbon and hydrogen atoms in a methane CH_4 molecule. (b) One of the four C(sp^3)–H(1s) bonds in methane.

bonds are approximately C(sp^3)–C(sp^3) with a length of 0.154 nm. The tetrahedral geometry of the sp^3 orbitals leads to an alkyl chain with a zig-zag conformation.

It is important to note that a chemical bond can be formed from a mixture of the above hybrid orbitals. In the case of a molecule of ammonia, NH_3, the H–N–H bond angle of 107.1° does not correspond to any simple hybrid. Thus, sp, sp^2 and sp^3 hybrids must be considered as limiting cases only. In addition to its three N–H bonds, ammonia also has a *nonbonding* pair (or lone pair) of electrons on the nitrogen. These electrons are in an orbital which is has slightly more s character than in a simple sp^3 orbital. Consequently, the three hybrid orbitals that overlap with the three hydrogen atoms contain slightly less s character than in an sp^3 orbital (in fact 23%s and 77%p).

2.4.2 Isomers

Isomeric compounds are those which have identical chemical formulae (e.g. C_4H_{10}) but differ in the nature or sequence of bonding of their atoms or in the arrangement of atoms in space. Isomers may be subdivided into *constitutional* (or structural) isomers and *stereo-isomers*. The former group differ in the order in which the atoms are connected, so they contain different functional groups and/or bonding patterns. For example, ethyl methyl ether and isopropyl alcohol can both be described by the chemical formula C_3H_8O. There are three structural isomers of the alkane pentane, all with the chemical formula C_5H_{12}; these are depicted in Figure 2.19. The isomers – pentane, isopentane and neopentane – differ in the spatial arrangement of the atoms. As the number of carbons in these alkanes increases, so

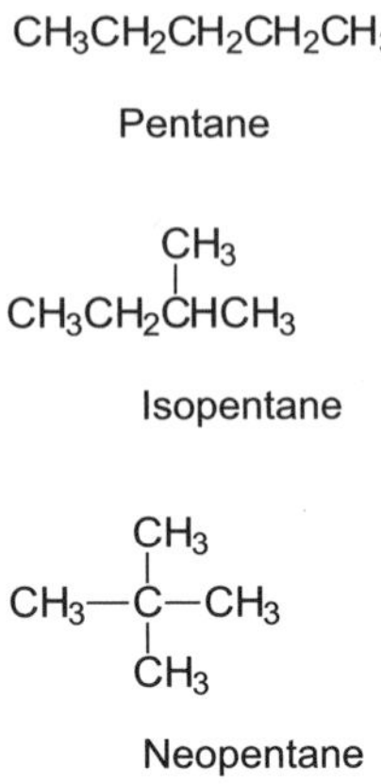

Figure 2.19 Isomers of pentane. The compounds have the same chemical formula C_5H_{12} but different bonding arrangements between the atoms.

does the number of possible isomers; for example, there are nine possible heptanes (C_7H_{16}) and 75 decanes ($C_{10}H_{22}$). In general, structural isomers have different physical and chemical properties. Branched-chain hydrocarbons are more compact than straight-chain isomers and tend to possess lower boiling points and higher densities.

Stereoisomers are compounds that possess the same sequence of covalent bonds but differ in the arrangements of their atoms in space. For this group, a distinction may be made between isomers that rapidly interconvert at ordinary temperatures, known as *conformational isomers* or *conformations*, and those that interconvert only at elevated temperatures or not at all are called *configurational isomers*.

Conformational Isomers

Figure 2.20 shows the most stable structure of pentane. In this representation, a dashed bond projects away from the reader, a heavy wedge bond projects towards the reader and a normal bond lies in the plane of the page. Rotation about one or more of the single bonds in this molecule leads to different conformations. These isomers cannot usually be isolated as they interconvert too rapidly. The conformation in Figure 2.20 is called *staggered* and has all the carbon–hydrogen bonds as far away from each other as possible. The least stable conformation is called *eclipsed* and has the carbon–hydrogen bonds as close as possible.

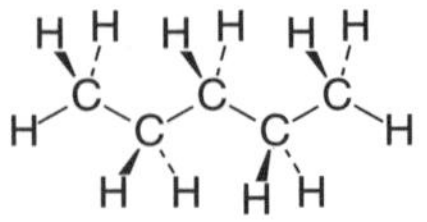

Figure 2.20 Covalent bonding in pentane. In the representation shown, a dashed bond points away from the viewer, a heavy wedge bond projects towards the viewer and a normal bond lies in the plane of the page.

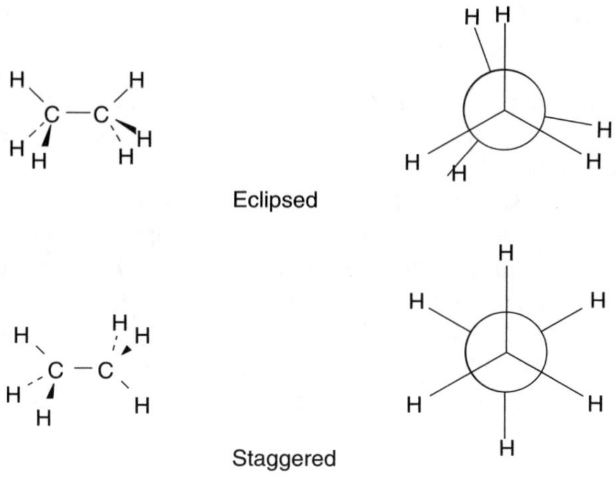

Figure 2.21 Eclipsed and staggered conformations of ethane. The Neuman projections are shown on the right.

Figure 2.21 gives a way of representing both the eclipsed and staggered conformations of ethane (C_2H_6). These diagrams are called *Newman projections*. The C–C bond is viewed end-on; the nearest carbon is represented by a point and the three groups attached to the carbon radiate as three lines from this point. The furthest carbon is represented by a circle with its bonds radiating from the edge of a circle.

For longer chain alkanes, the conformational situation becomes complex. There are different staggered and eclipsed conformations, not all of equal energy. The lowest energy conformation is the one in which the two large methyl groups are as far apart as possible – 180°; this is the *anti conformation*. A higher energy *gauche conformation* exists when the methyl groups are 60° apart. Newman projections are shown for both these arrangements in Figure 2.22. Note that there are no eclipsing interactions for the *gauche* conformation; the increased energy simply arises because the large methyl groups are close together.

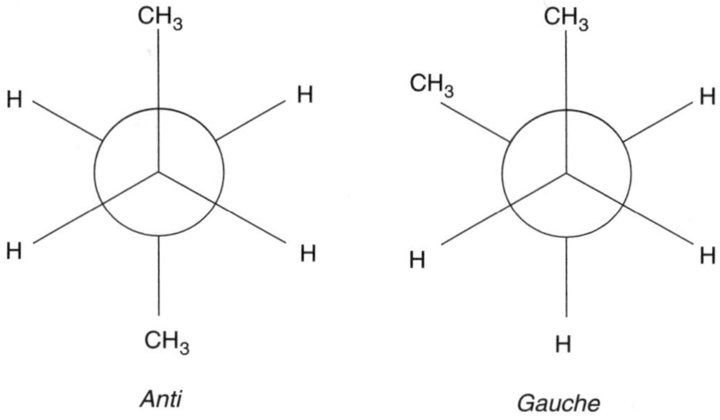

Figure 2.22 *Anti* and *gauche* conformations of alkanes.

Configurational Isomers

Interconversion between configurational isomers requires the breaking of bonds rather than rotation about bonds and requires high temperatures or special catalysts. Further examples of configurational isomers – *trans* and *cis isomers* – are described in the following section dealing with double bonds.

Some configurational isomers differ in their three-dimensional relationship of the substituents about one or more atoms; these are called optical isomers. *Enantiomers* are optical isomers that are nonsuperimposable mirror images. These exhibit the property of *chirality*. Figure 2.23 shows the example of the two amino acids (Chapter 12, Section 12.2.1) glycine and alanine. Alanine, which has four different chemical groups bonded to the central carbon atom, is a chiral molecule – it cannot be superimposed on its mirror image. There are two enantiomers of alanine. Molecules which contain one such asymmetric carbon atom are always chiral. In contrast, glycine, with only three different groups attached to the carbon atom, is achiral.

The easiest way to understand the above is to think of carbon as a central atom in a pyramid with other atoms at each apex. When the four atoms are all different (*W*, *X*, *Y* and *Z*, say), it is impossible for the original structure to be superimposed on its mirror image [Figure 2.24(a)]. However, when two of the groups attached to the central carbon are the same, group *W* in Figure 2.24(b), then the mirror image can be superimposed on the original structure by selecting the other *W* group to be the top of the pyramid.

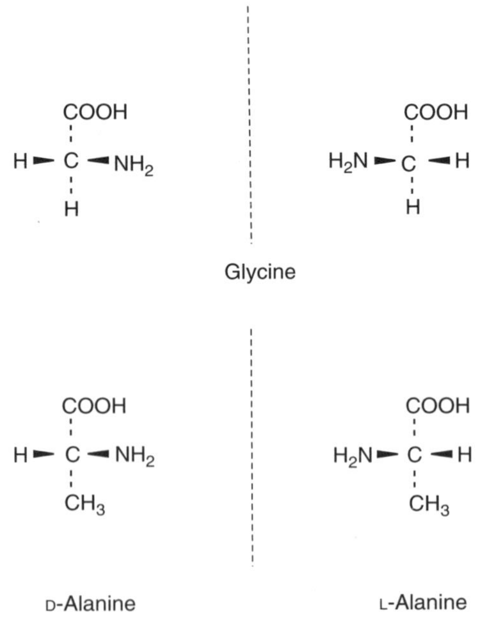

Figure 2.23 Stereoisomerism. Top: the glycine molecule is superimposable on its mirror image; glycine is achiral. Bottom: the molecule of alanine cannot be superimposed on its mirror image; alanine is chiral. The two stereoisomers (enantiomers) of alanine are D-alanine and L-alanine.

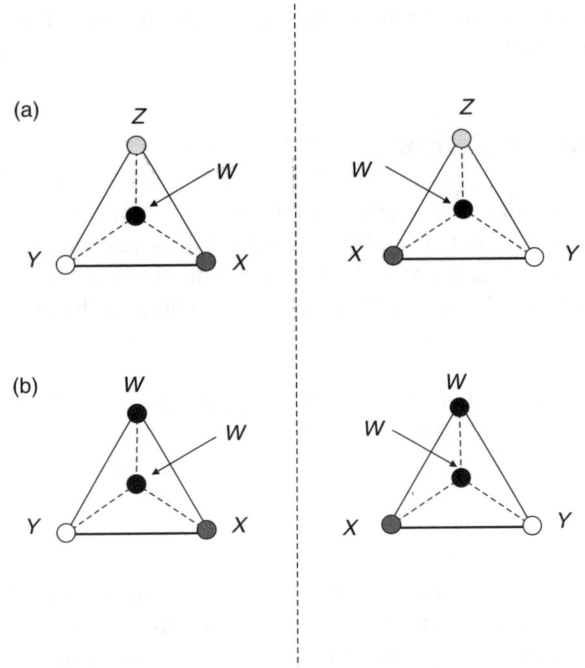

Figure 2.24 Schematic representation of an asymmetric carbon atom. In the figures the carbon atom is not shown but assumed to be at the centre of the pyramids. The vertical dashed line represents a mirror plane. In (a) four different groups are attached to the carbon. The carbon is asymmetric as this particular molecular arrangement cannot be superimposed on its mirror image (shown on the right). Different stereoisomers are therefore possible. In (b) the carbon with two identical groups is not asymmetric as either clockwise sequence of the groups in the pyramid's base can be obtained by choosing which *W* group to be at the top. No isomerism is possible.

Most of the physical properties (boiling points, densities, refractive indices, etc.) of the two of stereoisomers of alanine are identical. However, they differ in one important respect, the way in which they interact with polarized light (Chapter 4, Section 4.2). One enantiomer will rotate the plane of polarized light to the left (anti-clockwise) and is called the *lavorotatory* isomer, whereas the other will rotate the plane to the same extent to the right, and is called the *dextrorotarory* isomer. By convention, the two sterioisomers of alanine, shown in Figure 2.23, are designated L and D (small capital letters). It is important to note that these L and D symbols refer to the absolute configuration of the steioisomer rather than to the direction of rotation of the plane polarized light, i.e. some L isomers can be lavorotatory and some dextrorotatory.

Many medicines are produced as a mixture of enantiomers, only one of which is pharmacologically active, as it can prove costly to separate the isomers. One of the enantiomers of salbutamol, used in the treatment of asthma, is 68 times more effective than the other. One of the most dramatic effects of chirality was seen with the disastrous introduction of thalidomide in the 1950s and 1960s. This drug was intended as a sleeping aid for pregnant women and to combat morning sickness and other symptoms. It was later found to be responsible for causing birth defects; this was related to one of the activity of one of the enantiomers of thalidomide.

2.4.3 Double and Triple Bonds

Despite the propensity of carbon atoms to form four covalent bonds, in the ethylene molecule, C_2H_4, each of the two carbons is attached to just three atoms. The apparent deficiency in bonds to carbon is avoided by the bonding of two carbon atoms by two bonds, or what is more properly called a *double bond*. The bonding for hydrocarbons with carbon–carbon double bonds (*alkenes*) and *triple bonds* (*alkynes*) involve sp^2 and sp hybrids, respectively.

In ethylene ($CH_2{=}CH_2$), two sp^2 hybrids on each carbon bond with the hydrogens. A third sp^2 hybrid on each carbon forms a $C(sp^2)$–$C(sp^2)$ single bond, leaving a p orbital 'left over' on each carbon. This orbital lies perpendicular to the plane of the six atoms. The two p orbitals are parallel to each other and have regions of overlap above and below the molecular plane. This type of bond in which there are two sideways bonding regions above and below a nodal plane is called a *pi* or *π-bond*. In contrast, the bond formed by the head-on overlap of the two carbon sp^2 orbitals is the σ-bond, already described in Section 2.3.3. The π-bonds are the result of overlap of dumbell-like p orbitals lying adjacent to each other, shown in Figure 2.25. One consequence of this structure is that the orbitals of the π-bond act like struts to stop the atoms at each end from rotating; the molecule stays more or less flat.

The C=C double bond distance in ethylene is 0.133 nm, less than the value of 0.154 nm given above for a C–C single bond, and the C–H bond is 0.108 nm long. Strengths of the C=C and C–H bonds in ethylene are 6.61 eV (152×10^3 kcal kmol^{-1}) and 4.48 eV (103×10^3 kcal kmol^{-1}), respectively.

As one might expect, double bonds are stronger than single bonds: to break the carbon–carbon link in ethylene requires more energy than is needed to do the same in ethane, C_2H_6, in which two CH_3 groups are linked by a single bond between carbons. The double bond is, however, considerably less than twice as strong; breaking open the π component of the double bond is easier than breaking a single bond. For this reason, ethylene reacts with other compounds more readily than does ethane. Carbon compounds that contain π-bonds are said

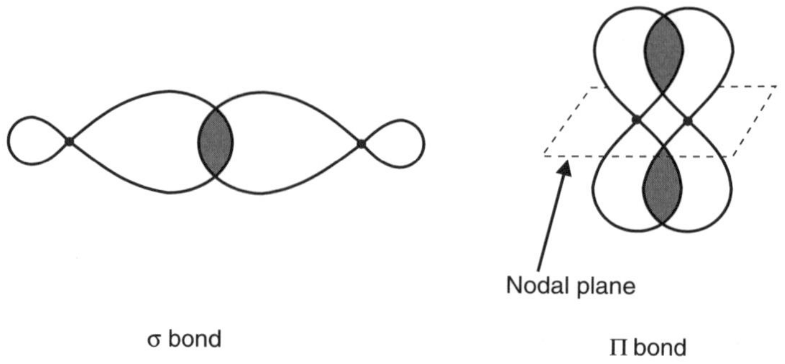

Figure 2.25 σ- and π-bonds.

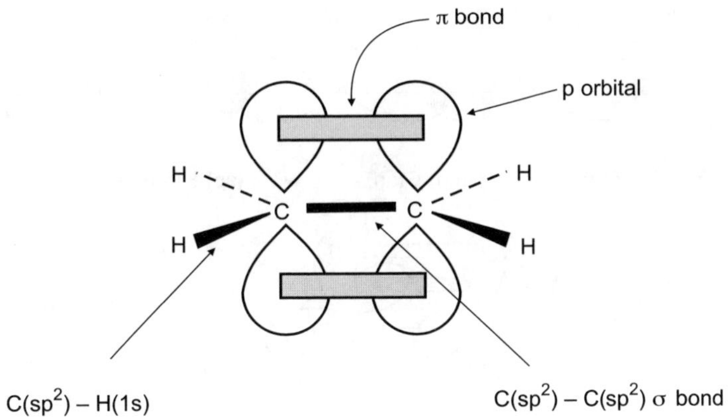

Figure 2.26 σ- and π-bonding in ethylene, $CH_2{=}CH_2$.

to be *unsaturated*, meaning that the carbon atoms, while having formed the requisite number of bonds, are not fully saturated in terms of their number of potential neighbours. *Saturated* carbon molecules, on the other hand, contain only single bonds.

The noncylindrically symmetric electron density about the C=C bond axis in ethylene, shown in Figure 2.26, and related compounds leads to a barrier to rotation about this axis. Therefore, two isomers exist. These particular isomers are not easily interconverted and are configurational isomers, as discussed in Section 2.4.2. These are known by the prefixes *cis* (from the Latin for 'on this side' – both bonds are on the same side of the alkene plane) and *trans* ('across' – the bonds are on opposite sides of the alkene). Figure 2.27 illustrates the *cis* and *trans* forms for a long-chain compound containing a C=C double bond.

In acetylene (HC≡CH), which contains a C≡C triple bond, the C–H bonds are shorter than in acetylene (0.106 nm) and are C(sp)–H(s) in character. The bond strength is 5.43 eV (125 kcal kmol^{-1}). The molecule has a linear structure with a C≡C bond of strength 8.7 eV (200×10^3 kcal kmol^{-1}) and length 0.120 nm, the shortest carbon–carbon bond distance known. A σ-bond is formed from the head-on overlap of the carbon sp hybrid orbitals. The two orthogonal p orbitals give rise to cylindrically symmetrical π-bonds, as shown in Figure 2.28.

Table 2.5 compares the carbon–carbon and carbon–hydrogen bond characteristics for ethane, ethylene and acetylene. In unsaturated alkyl chains, both –C= and –C≡ bonds are

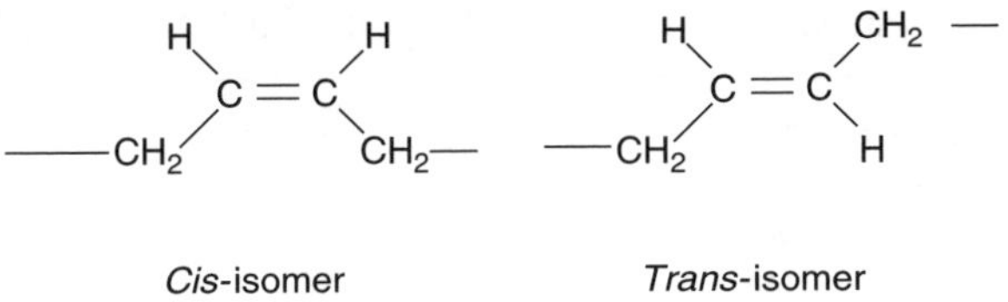

Figure 2.27 The *cis* and *trans* isomers for a long-chain compound containing a C=C double bond.

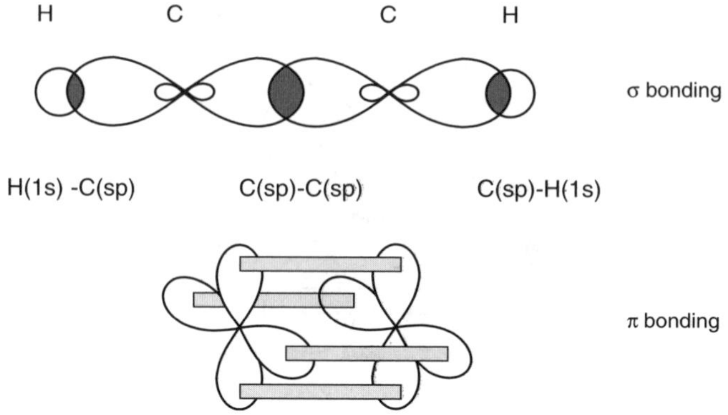

Figure 2.28 Bonding in acetylene, CH≡CH. Top, σ-bonding; bottom, π-bonding. The electrons in the two orthogonal p orbitals form a cylindrically symmetric torus of doughnut-like electron distribution.

associated with small, but finite, dipole moments (Section 2.3.5). This is simply because of the different amounts of s character of the carbon orbitals making up the bond. Symmetrically substituted alkenes and alkynes, of course, possess no overall moment.

Benzene, C_6H_6, is a particularly interesting molecule, with an important relevance to molecular electronics. X-ray crystallography reveals that this compound has a regular hexagonal structure and that the carbon–carbon bond distance is 0.140 nm, intermediate between those for a single bond (0.154 nm) and a double bond (0.133 nm) (Table 2.5). The carbon atoms form a hexagon with one hydrogen atom attached to each carbon, as depicted in Figure 2.29. The bonding between the carbon atoms takes the form of $C(sp^2)$–$C(sp^2)$ σ-bonds and a π-bond formed from the p orbitals extending around the hexagon. Benzene can be considered as two equivalent structures, shown in Figure 2.30(a), in resonance. The resulting *resonance hybrid* is conveniently shown with a circle (or dashed hexagonal line) inside a hexagon to indicate the partial double bond character of the bonds.

Table 2.5 Comparison of carbon–carbon and carbon–hydrogen bonds in ethane, ethylene and acetylene.

Molecule	Bond	Bond strength (eV)	Bond length (nm)
Ethane,	$C(sp^3)$–$C(sp^3)$	3.83	0.154
CH_3CH_3	$C(sp^3)$–H(1s)	4.26	0.110
Ethylene,	$C(sp^2){=}C(sp^2)$	6.61	0.133
$H_2C{=}CH_2$	$C(sp^2)$–H(1s)	4.48	0.108
Acetylene,	C(sp)≡C(sp)	8.70	0.120
HC≡CH	C(sp)–H(1s)	5.43	0.106

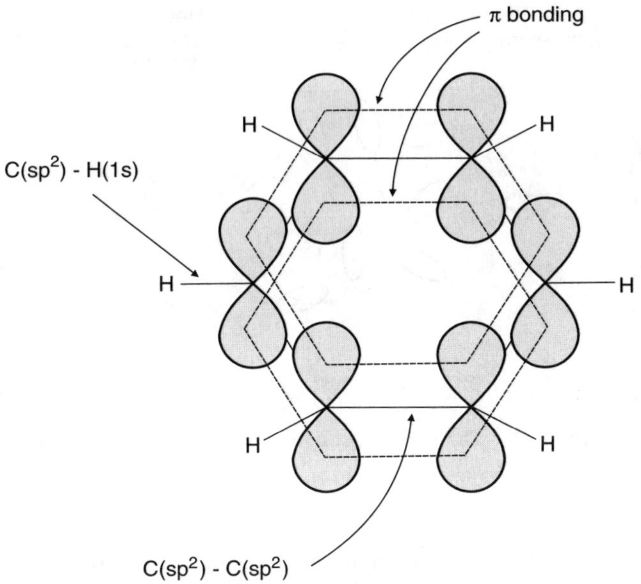

Figure 2.29 Orbital structure of benzene.

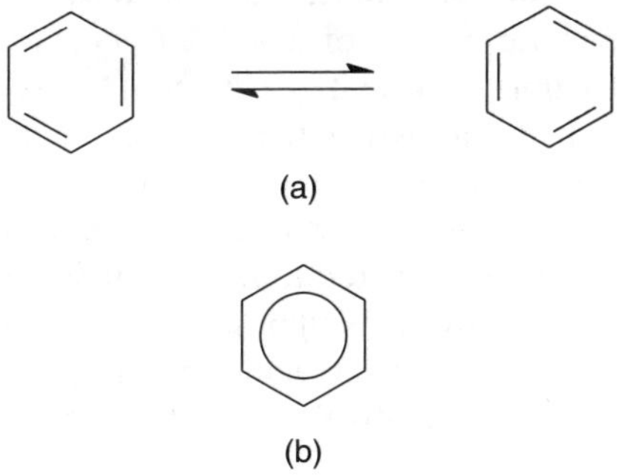

Figure 2.30 (a) Two equivalent resonance structures for benzene. (b) Method of indicating partial double bond character of the carbon–carbon bonds in benzene.

2.5 CRYSTALLINE AND NONCRYSTALLINE MATERIALS

2.5.1 States of Matter

The three most common states, or phases, of matter – gases, liquids and solids – are very familiar. Phases that are not so well known are plasmas and liquid crystals (described in Chapter 8). All these states are generally distinguished by the degree of translational and orientational order of the constituent molecules. On this basis, some phases may be further subdivided. For example, solids, consisting of a rigid arrangement of molecules, can be crystalline or amorphous. In an amorphous solid (a good example is a glass), the

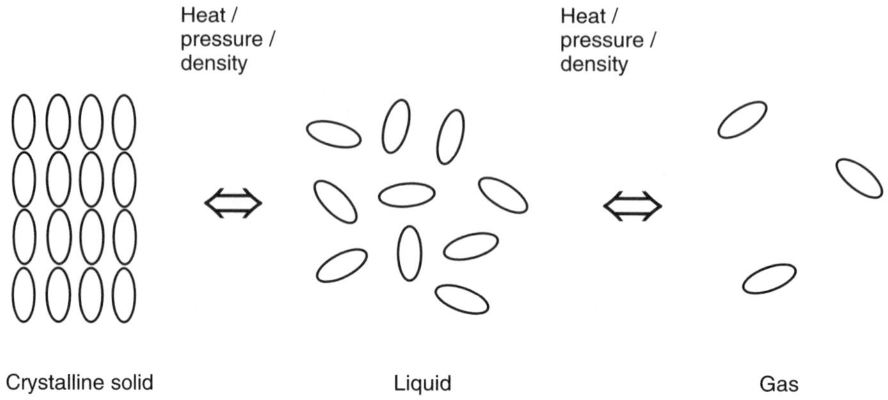

Figure 2.31 The three most common bulk phases of matter.

molecules are fixed in place, but with no pattern in their arrangement. As shown in Figure 2.31, the crystalline solid state is characterized by long-range translational order of the constituent molecules (the molecules are constrained to occupy specific positions in space) and long-range orientational order (the molecules orient themselves with respect to each another). The molecules are, of course, under a constant state of thermal agitation, with a mean translational kinetic energy of $3k_BT/2$ ($\frac{1}{2}k_BT$ for each component of their velocity, where k_B = Boltzmann's constant – Chapter 3, Section 3.2.1). However, this energy is considerably less than that associated with the chemical bonds in the material and the motion does not disrupt the highly ordered molecular arrangement.

In the gaseous state, the intermolecular forces are not strong enough to hold the molecules together. These are therefore free to diffuse about randomly, spreading evenly throughout any container they occupy, no matter how large this is. The average interatomic distance is determined by the number of molecules and the size of the container. A gas is easily compressed, as it takes comparatively little force to move the molecules closer together.

On the microscopic level, the liquid state is generally thought of as a phase that is somewhere between that of a solid and that of a gas. The molecules in a liquid neither occupy a specific average position nor remain oriented in a particular way. They are free to move around and, as in the gaseous state, this motion is random. The physical properties of both liquids and gases are *isotropic*, i.e. they do not depend on direction (directional-dependent properties are called *anisotropic*). Therefore, there is similarity between liquids and gases and, under certain conditions, it is impossible to distinguish between these two states. When placed in a container, the liquid will fill it to the level of a free surface. Liquids flow and change their shape in response to weak outside forces. The forces holding a liquid together are much less than those in a solid. Liquids are highly incompressible, a characteristic that is exploited in hydraulic systems.

Almost all elements and chemical compounds possess a solid, a liquid and a gaseous phase. The gaseous phase is favoured by high temperatures and low pressures. A transition from one phase to another can be provoked by a change in temperature, pressure, density or volume.

2.5.2 Phase Changes and Thermodynamic Equilibrium

When an arbitrary *thermodynamic system* (e.g. a fixed amount of matter) is left to itself, its properties (pressure, volume, temperature, etc.) will generally change. However, after a sufficiently long period, the properties will be invariant with time. This final state will depend on the nature of the system. For example, a simple mechanical arrangement, such as a ball rolling down a hill, consistently comes to rest in the state of lowest energy (with the ball at the bottom of the hill). Any thermodynamic system, left for long enough, will reach a state of *thermodynamic equilibrium*. Here, thermodynamic equilibrium is achieved when the *free energy* is minimized. For systems at constant volume, this is given by the *Helmholtz function*, F:

$$F = U - TS \tag{2.6}$$

and for constant pressure changes, the free energy is given by the *Gibbs function*, G:

$$G = H - TS = U + PV - TS \tag{2.7}$$

In both of the above equations, U is the *internal energy*, the sum of all the potential and kinetic energies of the system, H is the *enthalpy* or *latent heat* of transformation (e.g. heat associated with a process such as solidification or vaporization) and S is the *entropy*. Classically, entropy is the heat into, or out of, the system divided by temperature:

$$S = \frac{H}{T} \tag{2.8}$$

Entropy is also a measure of the disorder of a thermodynamic system. The specific molar entropy change (S/n, where n is the number of moles) for vaporization is normally much greater than that for solidification, confirming the increased disorder produced in the former phase transformation.

Equations (2.6) and (2.7) reveal that free energy can be minimized either by reducing the internal energy or by increasing the entropy term. At low temperature, the internal energy of the molecules makes the greater contribution to the free energy, so the solid phase is the favoured state. At higher temperature, the entropy of the system becomes the predominant influence. As a result, the fluid phases are stable at elevated temperatures, although they constitute higher internal energy configurations than the solid phase.

Many states of matter that are commonly encountered are not in a true state of thermodynamic equilibrium. Equilibrium may be approached very slowly (e.g. an amorphous glass will eventually crystallize over a period of many hundreds of years) or the system may be in a *metastable* state (e.g. a supercooled liquid or vapour). However, as such systems will have directly measurable properties (P, H, S, etc.) that are stable during an experiment, it is appropriate to assume these properties can be related in the same way as for a true equilibrium state.

Thermodynamically, phase changes of matter can be divided into *first-order* and *higher-order transitions*. In the former case, the specific molar Gibbs function $g(= G/n)$ is continuous but there is a discontinuity in the first derivative of g across the transition. An example of such a change is that of melting or evaporation: in these cases the discontinuity in $\partial g/\partial T$ is simply equal to the entropy of transformation (H/T). For second-order thermodynamic transitions, both the specific Gibbs function and its first derivative are continuous, but the second derivative changes discontinuously. In such transitions, the enthalpy of transformation is zero and the molar specific volume does not change. The transition of a liquid to a vapour at the critical point, ferromagnetic to paramagnetic transitions (Chapter 5, Section 5.7) and the change of a superconductor from the superconducting state to the normal state in zero magnetic field are examples of second-order phase transitions.

2.5.3 The Crystal Lattice

An ideal crystal contains atoms arranged in a repetitive three-dimensional pattern. If each repeat unit of this pattern, which may be an atom or group of atoms, is taken as a point then a three-dimensional *point lattice* is created. A *space lattice*, such as that shown in Figure 2.32, is obtained when lines are drawn connecting the points of the point lattice. The space lattice is composed of box-like units, the dimensions of which are fixed by the distances between the points in the three non-coplanar directions x, y and z. These are known as unit cells and the crystal has a periodicity (based on the contents of these cells) represented by the translation of the original unit of pattern along the three directions x, y and z. These directions are called the *crystallographic axes*. Any directions may, in principle, be chosen as the crystallographic axes, i.e. not necessarily orthogonal axes. It is most convenient to select a set of axes which bears a close resemblance to the symmetry of the crystal. In Figure 2.32, the angle between the y and z axes is designated α, between the z and x axes β

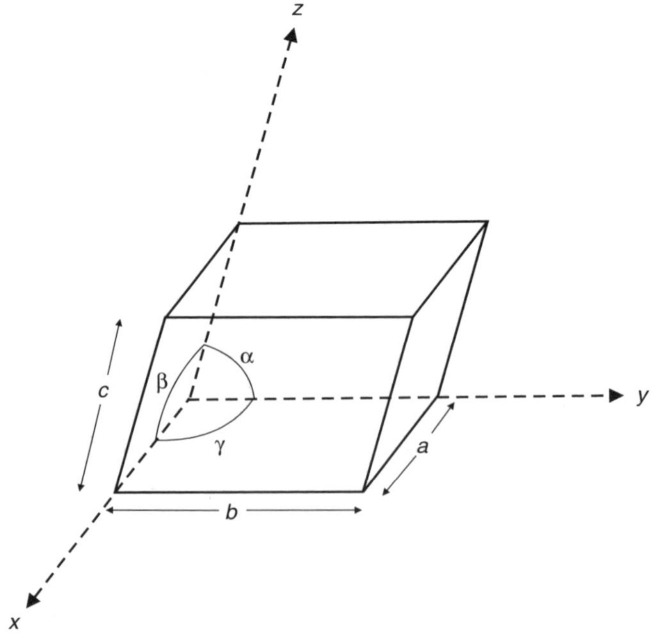

Figure 2.32 A unit cell with x, y and z coordinate axes, showing cell parameters a, b and c and interaxial angles α, β and γ.

and between the x and y axes γ. The measured edge lengths of the unit cell along the x, y and z axes are commonly given the symbols a, b and c.

2.5.4 Crystal Systems

As noted above, the unit cells of which a space lattice is composed do not necessarily have their three axes at right-angles. The lengths of the sides can also vary from the case where they are all equal to the case where no two of them are the same. Crystals can belong to seven possible *crystal systems*, characterized by the geometry of the unit cell. These are listed in Figure 2.33: *cubic*, *tetragonal*, *orthorhombic*, *hexagonal*, *rhombohedral*, *monoclinic* and *triclinic*. In the simplest lattices based on these crystal systems (primitive unit cells), the lattice points are positioned at the corners of the cell. However, the monoclinic, orthorhombic, tetragonal and cubic systems can also have cells that possess additional lattice points. These can occur at the centre of faces or in the middle of the body diagonal, leading to

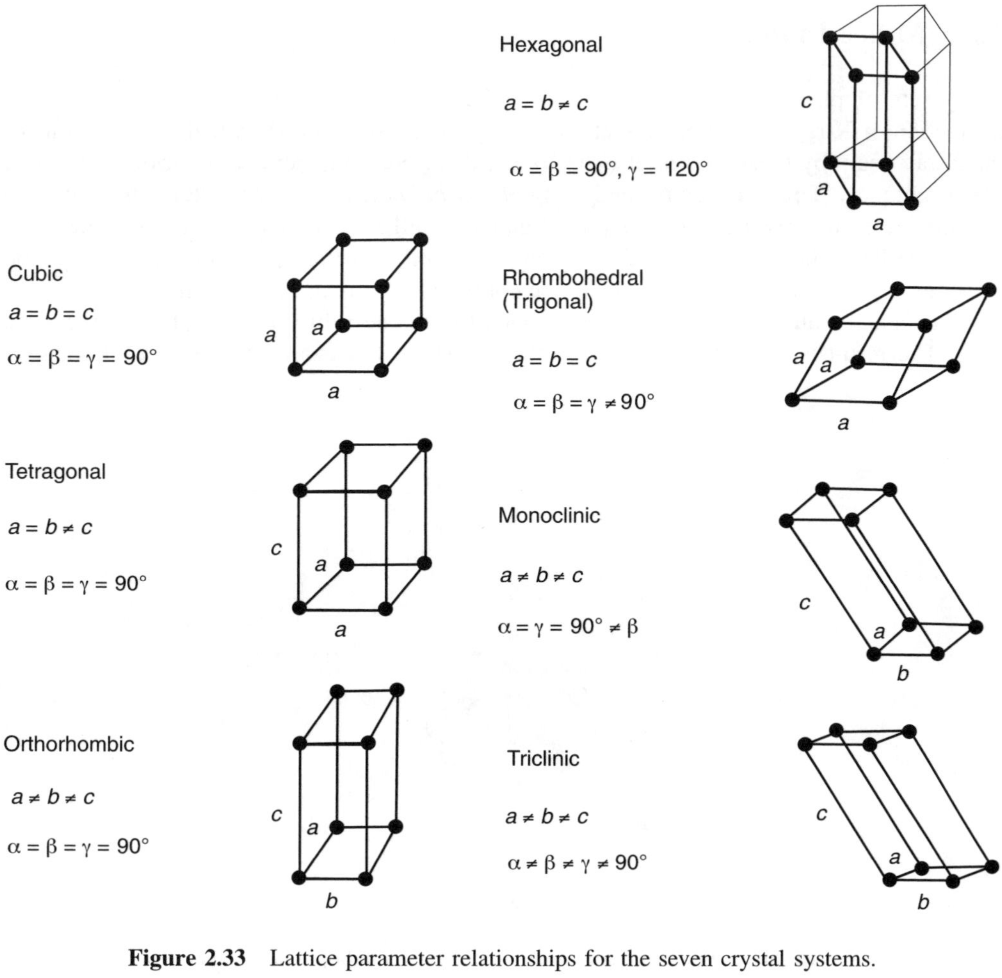

Figure 2.33 Lattice parameter relationships for the seven crystal systems.

based-centred (*C*), *face-centred* (*F*) and *body-centred* (*I*) unit cells. There are seven of these lattices and with the seven primitive (*P*) lattices they constitute 14 distinct *Bravais lattices*. Whereas a primitive cell possesses one lattice point per unit cell (each of the eight corners being shared by eight cells), body-centred and base-centred cells possess two points and face-centred cells four points.

The crystal system described above is the point group of the lattice (the set of rotation and reflection symmetries which leave a lattice point fixed) not including the positions of the atoms in the unit cell. In contrast, the *crystal structure* is one of the lattices with a unit cell that contains atoms at specific coordinates, at every lattice point. Because it includes the unit cell, the symmetry of the crystal can be more complicated than the symmetry of the lattice. The *crystallographic point group* or *crystal class* is the set of non-translational symmetries that leave a point in the crystal fixed. There are 32 possible crystal classes (i.e. 32 possible combinations of symmetry elements in three dimensions). Finally, the *space group* of the crystal structure is composed of the translational symmetries in addition to the symmetries of the point group. There are 230 distinct space groups.

2.5.5 Miller Indices

Many sets of planes can be drawn through the lattice points of a crystal structure and diffraction of X-rays (Chapter 6, Section 6.3) by the crystal can be treated as reflections of the X-ray beam by these planes. It is desirable, therefore, to be able to describe each set of planes uniquely. This is accomplished using *Miller indices*, which were originally derived to describe crystal faces, but can be applied equally well to any plane or set of planes in a crystal. Miller indices are allocated to a plane in a crystal by first specifying its intercepts on the three crystal axes, in terms of the lattice constants. The reciprocals of these numbers are then taken and reduced to three integers, usually the smallest three integers, having the same ratio. The result is then enclosed in parentheses: (*hkl*). For example, Figure 2.34 shows a

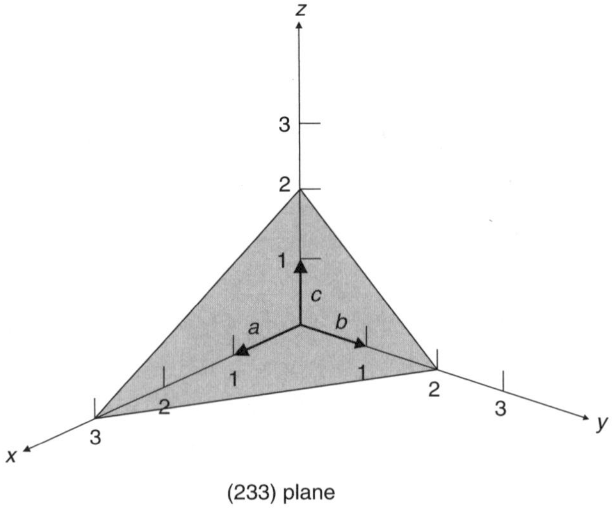

Figure 2.34 Representation of a (233) crystallographic plane.

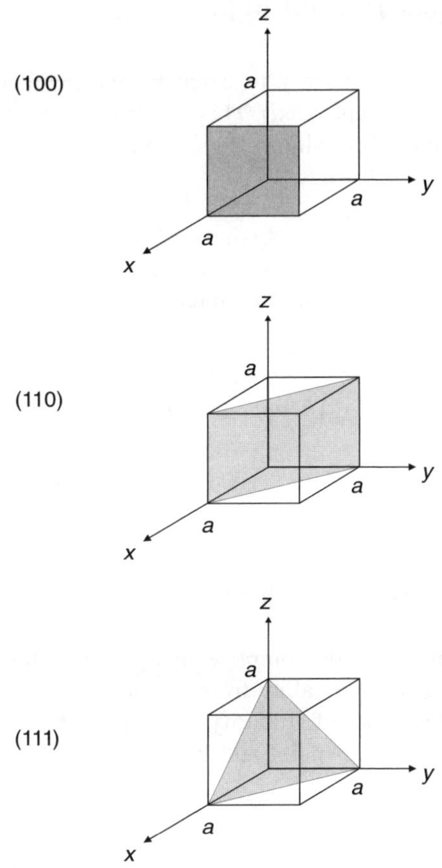

Figure 2.35 Miller indices of some important planes in a cubic crystal (lattice constant = a).

crystal plane intercepting the x, y, z axes at $3a$, $2b$, $2c$. The reciprocals of these numbers are ⅓, ½, ½. The smallest three integers having the same ratio are 2, 3, 3. Therefore, the Miller indices of the plane are (233). If one or more of the intercepts are at infinity (i.e. for planes parallel to a crystallographic axis), then the corresponding index is zero. Figure 2.35 shows some important planes in a cubic crystal. Other conventions that are used include the following:

(*hk*l): for a plane that intercepts the x-axis on the negative side of the origin.

{*hkl*}: for planes of equivalent symmetry, e.g. {100} for (100), (010), (100), (010) and (001) in a cubic crystal.

[*hkl*]: for a direction in a crystal, e.g. [100] is the x axis in a cubic crystal.

<*hkl*>: for a full set of equivalent directions.

2.5.6 Distance Between Crystal Planes

In crystallography, it is often necessary to calculate the perpendicular distances between successive planes in a series of planes (*hkl*). This is called the d_{hkl} spacing. For a cubic crystal system, this is easily obtained from simple geometry:

$$d_{hkl} = \frac{a_0}{\sqrt{(h^2 + k^2 + l^2)}} \tag{2.9}$$

The equivalent relationship for the orthorhombic systems is

$$d_{hkl} = \frac{1}{\sqrt{(h/a_0)^2 + (k/b_0)^2 + (l/c_0)^2}} \tag{2.10}$$

2.5.7 Defects

No crystal is perfect. Thermodynamic equilibrium requires that certain defects are always present in crystals. The surface of a crystal also sets an upper limit on its perfection. Macroscopic samples of most solids comprise many crystallites, randomly oriented, with grain boundaries separating one crystallite from the next. Such grain boundaries can act as locations for many mobile forms of impurity. Within each crystalline region, there will certainly be finite concentrations of point and line defects.

Point Defects

The point defects in a solid can be grouped into two principal categories. These originate either from foreign atoms (impurities) or from native atoms. Certain atomic sites may not be occupied. This results in point defects known as vacancies, illustrated for a cubic lattice in Figure 2.36. Figure 2.36(a) shows a perfect lattice. Removal of an atom from one of the

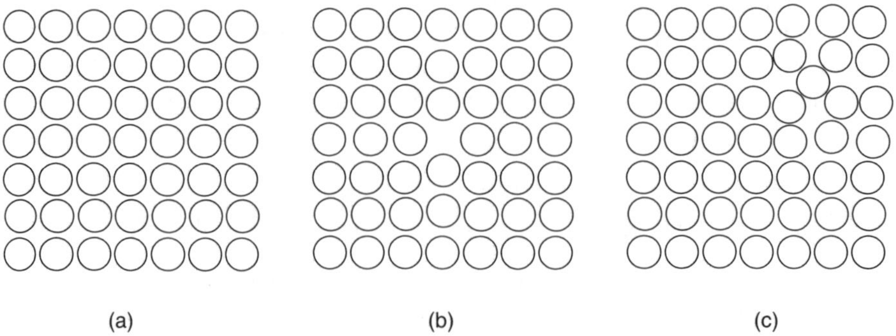

Figure 2.36 (a) Perfect cubic lattice. (b) Vacancy or Schottky defect. (c) Interstitial or Frenkel defect. The defects cause a deformation of the regular lattice.

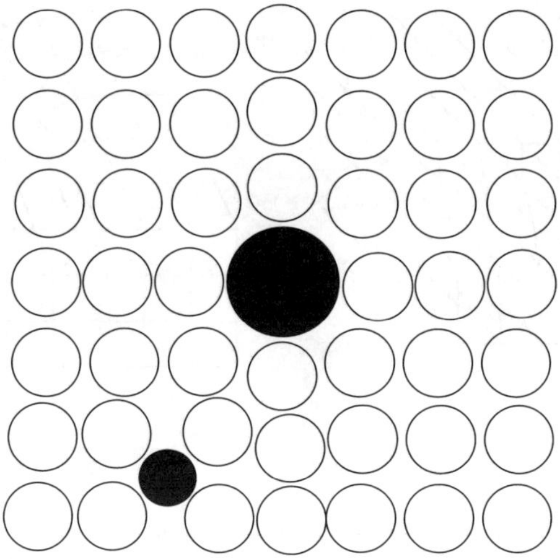

Figure 2.37 Substitutional and interstitial impurity atoms (shown in black) in a cubic lattice.

lattice sites produces a vacancy, Figure 2.36(b). If the native atom is relocated into a nonregular atomic site, this produces an interstitial defect, Figure 2.36(c). These two types of defect formed by native atoms are also referred to as *Schottky* (vacancy) and *Frenkel* (interstitial) defects. In both cases, the defects produce a deformation of the regular lattice.

All crystalline solids contain vacancies and it is not possible to create such a material that is completely free of these defects. The necessity for the existence of vacancies is explained using the principles of thermodynamics. In essence, the presence of vacancies increases the entropy (i.e. randomness) of the crystal. Impurity atoms in otherwise pure crystals can be also considered as point defects, and they play an important role in the electronic and mechanical properties of materials. Such atoms may either occupy a normal atomic site in the host lattice, a substitutional impurity, or a nonregular atomic site, an interstitial impurity. These are shown in Figure 2.37.

All point defects produce a local distortion in the otherwise perfect lattice, the degree of distortion depending on the crystal structure, parent atom size, impurity atom size and crystal bonding. These local distortions act as additional scattering centres to the flow of charge carriers through the crystal, thereby increasing its electrical resistance. In the case of ionic crystals, the removal of an ion produces a local charge as well as distortion of the lattice. To conserve overall charge neutrality, the vacancies occur either in pairs of oppositely charged ions or in association with interstitials of the same sign (e.g. in Na^+Cl^-, an Na^+ion may move into an interstitial site).

Line Defects

Line defects in a crystalline solid are called *dislocations*. There are two basic types of dislocation, *edge* and *screw* dislocations, depicted in Figure 2.38. The edge dislocation, Figure 2.38(a), can be visualized as an extra plane of atoms inserted part way into the crystal lattice. The edge of this extra plane is the actual dislocation. There is severe distortion in the region around the dislocation and the lattice planes are bent. The presence of the edge

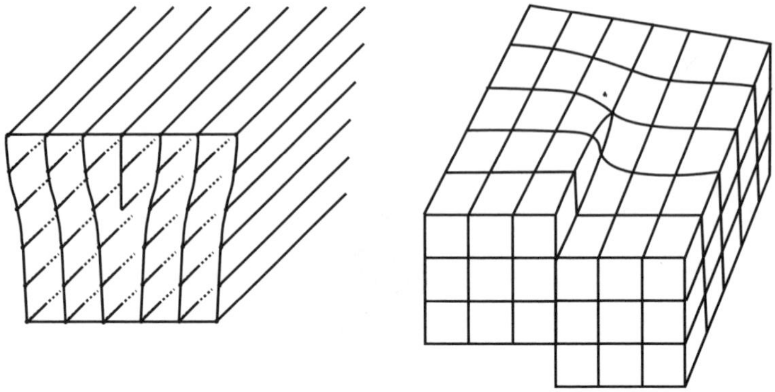

(a) (b)

Figure 2.38 Two idealized forms of dislocation: (a) edge dislocation; (b) screw dislocation. In general, a dislocation follows a curved path which has varying components of these two extremes.

dislocation greatly facilitates slip in the crystal when a shear force is applied, the slip occurring normal to the line marking the end of the extra plane of atoms. In a screw dislocation, Figure 2.38(b), part of the lattice is displaced with respect to the other point, and the displacement is parallel to the direction of the dislocation. A general dislocation in a real crystal is likely to have both edge and screw character.

The dislocation density is an indication of crystal perfection. For single crystals of highly perfect solids such as silicon, it is possible to create crystals in which the dislocation density is less than $100\,cm^{-2}$. A more typical value for other crystals is around $10^4\,cm^{-2}$, whereas dislocation densities for metallic crystals are usually at least $10^7\,cm^{-2}$. Unlike vacancies, thermal equilibrium does not require the presence of dislocations in a single crystal. The application of mechanical stress encourages dislocations to sweep through a crystal, generating point defects until the movement is pinned either by impurities or by the intersection with the path of other dislocations (a process known as work hardening).

Plane Defects

The predominant plane defect is the *grain boundary*, illustrated in Figure 2.39. Most crystalline solids do not consist of a large single crystals (an exception is perhaps the semiconductor silicon, the crystal growth of which has been perfected over the latter half of the 20th century) but of many randomly oriented crystallites. The junction of these crystallites or grains results in grain boundaries, which represent mismatches in the rows and planes in the adjoining crystallites. Each grain itself is an individual crystal and probably contains the point and line defects described above. Various degrees of crystallographic misalignment between adjacent grains are possible. When this orientation mismatch is slight, on the order of a few degrees, then the term low-angle grain boundary is used. These boundaries can be described in terms of dislocation arrays.

The atoms are bonded less regularly along a grain boundary and there is an interfacial or grain boundary energy (similar to surface energy) as a consequence of this boundary energy.

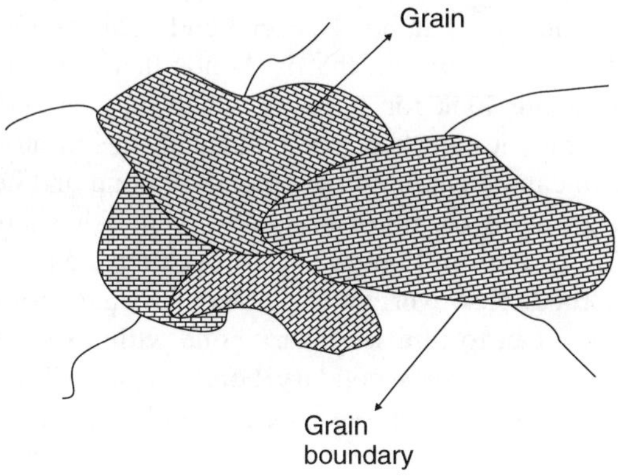

Figure 2.39 Individual grains in a polycrystalline material. The grains are separated by grain boundaries.

Grain boundaries are more chemically reactive than the grains themselves and impurity atoms often preferentially segregate in these regions.

A further plane defect is the *stacking fault*, which occurs when mistakes are made in the sequence of stacking the crystal planes. The plane separating two incorrectly juxtaposed layers is the stacking fault. These defects occur most readily in crystals in which the layer sequence is ABCABC..., for example the face-centred cubic structure.

Surfaces

Crystal surfaces represent a special type of plane defect. Much of the understanding of solids is based on the fact that these are perfectly periodic in three dimensions; for example, the electronic and vibrational behaviour can be described in great detail using methods that rely on this periodicity. The introduction of a surface breaks this periodicity in one direction and can lead to structural and also to subsequent electronic changes. As technology moves further in nanoscale dimensions, surfaces are becoming increasingly important in the determination of the behaviour of related devices. In the limiting cases, devices built using nanotechnology will be dominated by surface effects.

When the crystal lattice is abruptly terminated by a surface, the atoms cannot fulfil their bonding requirements, as illustrated for a silicon surface in Figure 2.40. Each silicon atom in

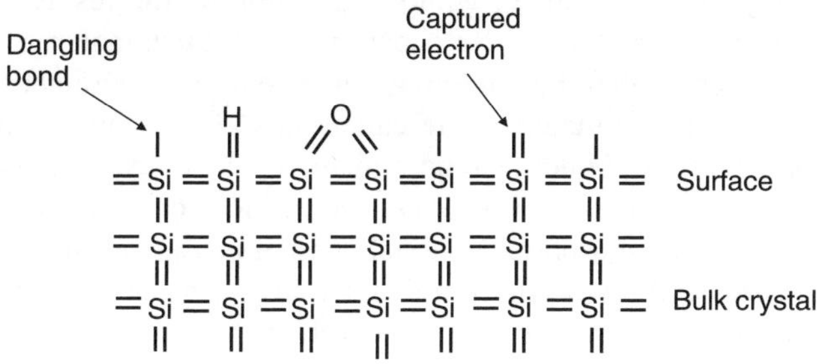

Figure 2.40 Schematic representation of possible defects occurring at the surface of silicon.

the bulk of the crystal has four covalent bonds, each bond with two electrons. The Si atoms at the surface are left with *dangling bonds*, that is, bonds that are half full, having only one electron. These dangling bonds look for atoms to which they can bond.

Atoms from the environment can therefore bond with the Si atoms on the surface. For example, a hydrogen atom can be captured by a dangling bond and hence become absorbed, or *chemisorbed*. A foreign atom or molecule is absorbed if it chemically bonds with the atoms on the surface. The H atom in Figure 2.40 forms a covalent bond with a silicon atom and hence becomes absorbed. However, an H_2 molecule cannot form a covalent bond but, due to hydrogen bonding, it can form a secondary bond with a surface Si atom and become *adsorbed*. Molecules that only form secondary bonds with surface atoms are said to be adsorbed. Similarly, a water molecule in the air can readily become adsorbed at the surface of the crystal.

The difference between chemisorption and physisorption (adsorption) therefore lies in the form of the electronic bond between the adsorbate and the substrate. If an adsorbed molecule or atom suffers significant electron modifications relative to its state in the gas phase to form a chemical bond with the surface (covalent or ionic), it is said to be chemisorbed. If one, the other hand, it is held on the surface only by van der Waals forces, relying on the polarizability of the otherwise undisturbed molecule, it is said to be physisorbed. Clearly, physisorption produces weak bonds while chemisorption often produces strong bonds. It is usual to regard the upper limit of the bond strength in physisorption as around 0.6 eV per atom or molecule, or $60\,\text{kJ mol}^{-1}$ ($1\,\text{eV molecule}^{-1} = 96.5\,\text{kJ mol}^{-1}$). Thermal considerations lead to the conclusion that such weekly bonded species would be desorbed from a surface at a temperature much in excess of 200 K. Adsorbates stable on a surface above this sort of temperature are therefore likely to be chemisorbed [1]. Also, if there are any free electrons, these can be captured by dangling bonds. The dangling bonds represent surface traps for carriers and can affect the performance of electronic devices (Chapter 9, Section 9.3).

When left unprocessed, the surface of a crystal will have absorbed and adsorbed atoms and molecules from the environment. This is one of the reason why the surface of a silicon wafer in microelectronics technology is first etched and then oxidized to form SiO_2, a passivating layer on the crystal surface. Many substances have a natural oxide layer on their surfaces.

2.5.8 Amorphous Solids

When the periodic and repeated arrangement of atoms in a solid is perfect and extends throughout the entirety of the specimen without interruption, the result is a single crystal. Noncrystalline or *amorphous* materials also occur in nature. Such materials are often formed by rapidly cooling a liquid so that the random arrangement of the atoms or molecules in the liquid phase is frozen-in. These materials are called *glasses*. The amorphous structure may have short-range order, as the bonding requirement of the individual molecules must be satisfied. The structure is a *continuous random network*, CRN, of atoms. As a consequence of the lack of long-range order, amorphous materials do not possess such crystalline imperfections as grain boundaries and dislocations, a distinct advantage in some applications. Some materials can exist in both the crystalline and amorphous forms. Figure 2.41 contrasts these two forms of silicon dioxide, SiO_2. The amorphous phase is vitreous silica and is formed by rapidly cooling molten SiO_2.

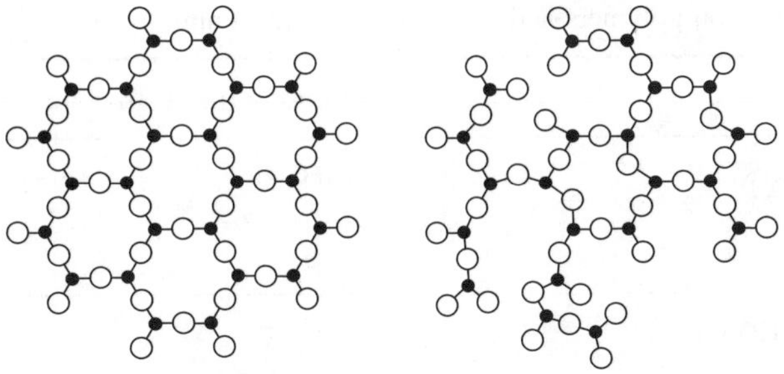

Figure 2.41 Silicon dioxide. (a) Crystalline form of SiO_2. (b) Amorphous SiO_2.

2.6 POLYMERS

Polymeric materials can have massive molecular structures in comparison with the carbon-based compounds described earlier in this chapter. Most polymers are in the form of long and flexible chains composed of structural entities called *mer* units (derived from the Greek word *meros*, meaning part) which are successively repeated along thee chain. The term *monomer* refers to the basic molecule from which the polymer is formed. Table 2.6 gives a list of some common monomer units and the corresponding polymers. When all the repeating units along a chain are of the same type, the resulting polymer is called a *homopolymer*. There is no restriction in polymer synthesis that prevents the formation of compounds other than homopolymers. Chains composed of two or more different mer units are termed *copolymers*.

The process of polymerization may proceed in a number of ways. For example, if ethylene gas is subjected catalytically to appropriate conditions, it will transform to polyethylene. The process, illustrated in Figure 2.42, begins when an active mer is formed by the reaction between an initiator or catalyst species ($R^{\bullet}$) and the ethylene mer unit. The polymer chain then forms by the sequential addition of polyethylene monomer units to this active initiator–mer centre. The active site, or unpaired electron (denoted by $^{\bullet}$) is transferred to each successive end monomer and it is linked to the chain. The final result is the polyethylene molecule. This is an example of *addition polymerization* (or chain reaction polymerization). The growth of the chain can be extremely rapid; the time required to grow a molecule consisting of, say, 1000 mer units is on the order of 10^{-2}–10^{-3} s. A further important method of polymerization is *condensation polymerization*. Here, the polymer is formed by stepwise intermolecular chemical reactions that normally involve more than one monomer species. There is usually a small molecular weight by-product such as water, which is eliminated.

2.6.1 Molecular Weight

High molecular weights are to be found in polymers with very long chains. During the polymerization process in which these large macromolecules are synthesized from smaller

Table 2.6 Some common polymers and their repeating (mer) units.

Polymer	Repeating (mer) structure
Polyethylene (PE)	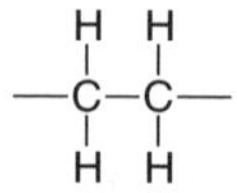
Poly(vinyl chloride) (PVC)	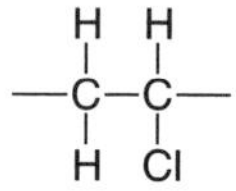
Polytetrafluoroethylene (PTFE)	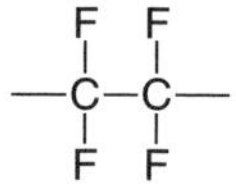
Polypropylene	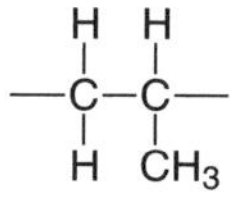
Polystyrene	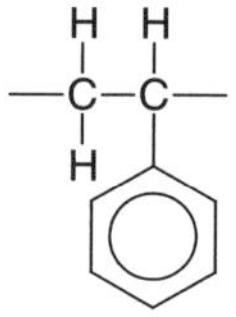
Polycarbonate	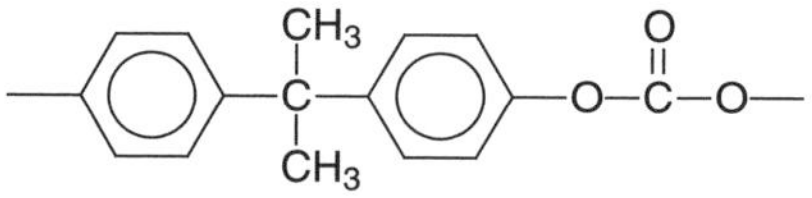

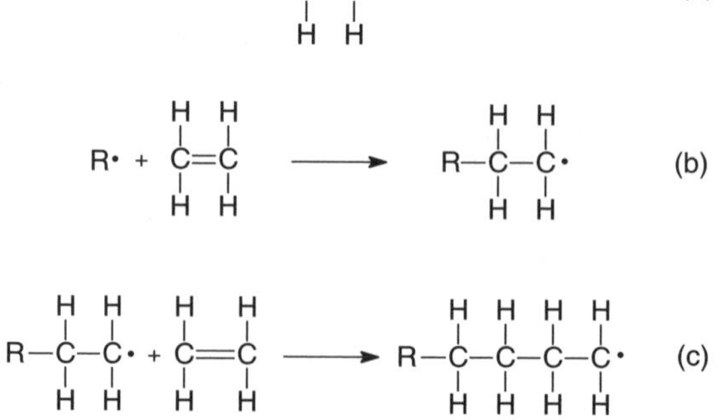

Figure 2.42 Polymerization of ethylene, C_2H_4, by the sequential addition of polyethylene monomer units. (a) Ethylene molecule. (b) Reaction between an initiator or catalyst species ($R^\bullet$) and the ethylene mer unit. (c) Further mer units are added to increase the length of the polymer chain.

molecules, not all the polymer chains will grow to the same length. This results in a distribution of chain lengths or molecular weights. There are several ways in defining the average molecular weight. The number-average molecular weight $\bar{M}_n$ is expressed as

$$\bar{M}_n = \sum x_i M_i \tag{2.11}$$

where M_i represents the mean (middle) molecular weight of size range i and x_i is the fraction of the total number of chains within the corresponding size range. A weight-average molecular weight M_w, based on the weight fraction of molecules within the various size ranges, may also be defined.

An alternative way of expressing average chain size of a polymer is as the *degree of polymerization n*, which represents the average number of mer units in a chain. Both number-average and weight-average degrees of polymerization are possible. In the former case

$$n_n = \frac{\bar{M}_n}{\bar{m}} \tag{2.12}$$

where $\bar{m}$ is the mer molecular weight. Equation (2.12) can also be used for copolymers, but the value of $\bar{m}$ will be calculated as a summation over the different mer units.

2.6.2 Polymer Structure

Polymers can be subdivided into three, or possibly four, structural groups. Examples of linear, branched and cross linked polymers are shown in Figure 2.43. The molecules in linear polymers consist of long chains of monomers joined by bonds that are rigid to a certain

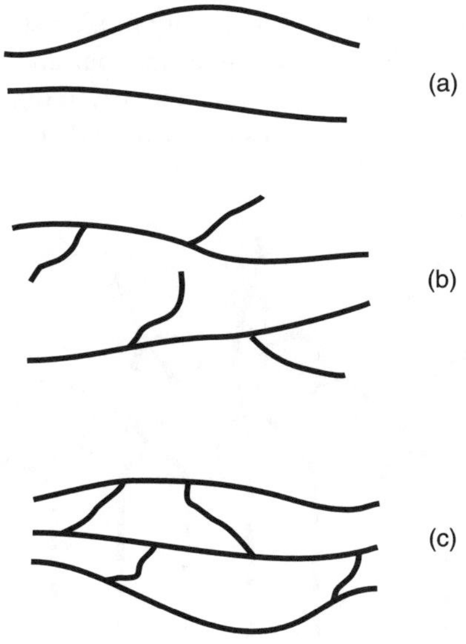

Figure 2.43 Schematic representations of polymer chains: (a) linear; (b) branched; and (c) cross-linked.

degree – the monomers cannot rotate freely with respect to each other. Typical examples are polyethylene, poly(vinyl alcohol), and poly(vinyl chloride) (PVC).

Branched polymers have side chains that are attached to the chain molecule itself. Branching can be caused by impurities or by the presence of monomers that have several reactive groups. Chain polymers composed of monomers with side groups that are part of the monomers, such as polystyrene or polypropylene (Table 2.6), are not considered branched polymers. When branching is prevalent, it can have a marked effect on the polymer properties. For example, the polymerization of ethylene under high-pressure conditions gives a product that has many side chains. This material is softer and much less crystalline (Section 2.6.3) than the non-branched form, which is synthesized from a catalytic low-pressure process. The two forms may be distinguished by a difference in density, the more crystalline material being the more dense.

In cross-linked polymers, two or more chains are joined together by side chains. With a small degree of cross-linking, a loose network is obtained that is essentially two dimensional. High degrees of cross-linking result in a tight three-dimensional structure. Cross-linking is usually caused by chemical reactions. An example of a two-dimensional cross-linked structure is vulcanized rubber, in which cross-links are formed by sulfur atoms. Thermosetting plastics are examples of highly cross-linked polymers; their structure is so rigid that when heated they decompose or burn rather than melt.

2.6.3 Polymer Crystallinity

The crystalline state may exist in polymeric materials. If the chains of polymeric molecules were all of the same length and completely stretched out, or if the chains were folded back and forth in a symmetrical manner, as depicted in Figure 2.44, then these molecules would form completely crystalline structures. However, most polymeric materials comprise regions that are both crystalline and amorphous. The density of a crystalline polymer will be greater than that of an amorphous polymer of the same material and molecular weight, since the chains are more closely packed together for the crystalline structure. The degree of crystallinity may therefore be determined from accurate density measurements.

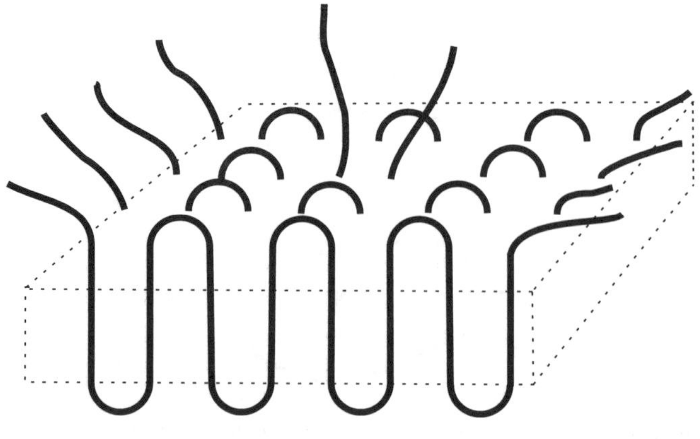

Figure 2.44 Chain-folded lamella example of polymer crystallinity.

The molecular chemistry, in addition to chain configuration, also influences the ability of a polymer to crystallize. Crystallization is not favoured in polymers that are composed of chemically complex mer structures. On the other hand, crystallization is not easily prevented in chemically simple polymers such as polyethylene, even for very rapid cooling rates. When a polymer is stretched or drawn, the internal shearing action tends to align the long molecules preferentially in the stretch direction. Not surprisingly, such one-dimensional ordering tends to induce crystallization.

An interesting example of a polymer exhibiting various forms of crystallinity is poly(vinylidene difluoride) (PVDF). This polymer, shown in Figure 2.45(a), consists of the mer unit CH_2CF_2. PVDF has at least four different crystalline phases, two of which, the α-phase and the β-phase, are shown in Figure 2.45(b) and (c), respectively. These are both based on an orthorhombic unit cell (Figure 2.33). The figures show the projection of the carbon (small open circles) and fluorine (large filled circles) on to the *ab* planes of the unit cells. The α-phase is the most common structure and the other forms can be obtained from this parent phase by applications of mechanical stress, heat and electric field. The polymer chain in the α-phase results in the dipole moments associated with the carbon–fluorine bonds arranged in opposite directions so that there is no net polarization within the crystal.

When the α-phase of PVDF is mechanically deformed by stretching or rolling at temperatures below 100 °C, the β-phase of PVDF is formed. The unit cell of this structure has a net dipole moment (normal to the chain direction). However, because of the random

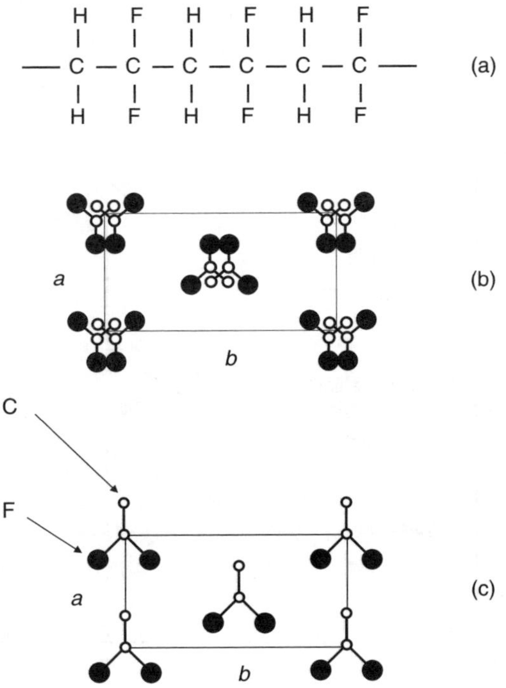

Figure 2.45 Structural forms of poly (vinylidene fluoride), PVDF. (a) Section of $(CH_2CF_2)_n$ chain; (b) α-phase; (c) β-phase. The projections of C atoms (small open circles) and F atoms (large full circles) on to the *ab* planes are shown. The H atoms have been omitted.

orientation of the crystallites, there is no net polarization in the material. The application of a strong electric field, a process called *poling*, is needed to confer the PVDF with an overall dipole moment. The β-phase of PVDF can be exploited in piezoelectric and pyroelectric applications and is discussed further in Chapter 5, Section 5.6.

Because of their chain-like nature and semicrystalline structure, the concept of defects in polymeric materials is somewhat different to those described in Section 2.5.7 for crystalline materials. However, important defects do occur in polymers, particularly in their crystalline regions. As will be discussed in Chapter 3, structural defects along the chains of conductive polymers can be very important in molecular electronics applications.

When a polymer is heated towards its melting point, several distinct features may be observed. First, and in contrast to most other materials, the melting takes place over a range of temperatures. At low temperature, the amorphous regions of a polymer are in a glassy state in which the molecules are effectively frozen in place (apart from thermal vibrations). As the material is heated, it will eventually reach its *glass transition temperature*, T_g. At this point, large segments of the molecular chains can start to move. A rubbery-like phase is formed. It is important to note that the glass transition temperature is not the same as the melting point of the polymer, T_m. The value of T_g for a particular polymer is related to its amorphous regions and will depend on its molecular weight, on its thermal history and on the rate of heating or cooling. The approximate glass transition temperature is 150 °C for polycarbonate, whereas its melting point is 265 °C. The glass transition (Figure 2.46), is a second-order thermodynamic transition (Section 2.5.2), in contrast to melting. In a

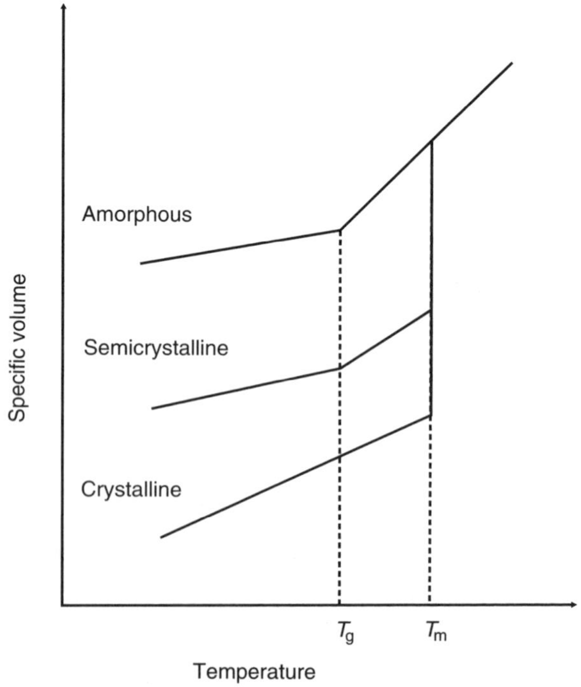

Figure 2.46 Specific volume versus temperature for amorphous, semicrystalline and crystalline materials. T_g = glass transition temperature; T_m = melting temperature.

second-order transition, there is no associated heat of transformation as in the case of melting or vaporization. The volume of the polymer will change to accommodate the increased motion of the chains, but this does not happen in a discontinuous manner.

2.7 SOFT MATTER: EMULSIONS, FOAMS AND GELS

It is likely that some of the applications of molecular electronics will make use of 'soft' materials such as rubbers and plastics. Other states of matter that might be exploited are *emulsions*, *foams* and *gels*. The liquid crystalline state is discussed separately in Chapter 8. Many of the interesting states of soft matter are based on *colloids*. Nature, of course, makes extensive use of materials in these soft forms.

A colloid is a suspension in which the dispersed phase is so small (1–1000 nm) that gravitational forces are negligible and interactions are dominated by short-range forces, such as van der Waals attraction (Section 2.3.5) and surface charges. The molecules in suspension are in continual collision with solvent molecules and the walls of the container, and display a random motion through the liquid known as *Brownian motion* (discovered by the botanist Robert Brown in 1827). A *sol* is a colloidal suspension of solid particles in a liquid, whereas an *aerosol* describes a colloidal suspension of particles in a gas (the suspension may be called a *fog* if the particles are liquid and a *smoke* if they are solid). An emulsion is a suspension of liquid droplets in another liquid; the two common types of emulsions are oil-in-water and water-in-oil (the term 'oil' in this context is a general word denoting the water-insoluble fluid). If two pure, immiscible liquids, such as benzene and water, are vigorously shaken together, they will form a dispersion. However, on stopping the agitation, phase separation will quickly occur. A surfactant component is always needed to obtain a reasonably stable emulsion. In contrast, microemulsions are generally thermodynamically stable. The dispersed microdroplets are kinetically stabilized because they are embedded inside surfactant micelles (Chapter 12, Section 12.2.5).

A foam can be considered as a type of emulsion in which the inner phase is a gas. As with emulsions, a surfactant compound is needed to provide stability. A gel is a porous three-dimensionally interconnected solid network that expands in a stable fashion throughout a liquid medium. The sol–gel technology that is used to form nanostructured materials is described in Chapter 7, Section 7.2.6.

2.8 DIFFUSION

Many reactions that are important in the processing of materials rely on the transfer of mass within the solid state or from a liquid, gas or another solid phase. This is accomplished by *diffusion*. Whenever there is a concentration gradient of particles, there is a net diffusional motion of particles in the direction of decreasing concentration. The origin of diffusion lies in the random motion of particles. This process is a type of transport phenomenon and can also apply to the spontaneous spreading of other physical properties, such as heat, momentum or light. The different forms of diffusion can be modelled quantitatively using the diffusion equation. In all cases, the net flux of the transported quantity (atoms, energy or electrons) is equal to a physical property (diffusivity, thermal conductivity, electrical conductivity) multiplied by a gradient (concentration, thermal or electric field gradient).

For transport in a single (x) direction, the flux of particles, J (i.e. the number of particles diffusing through and perpendicular to a unit cross-sectional are per unit time) is given by

$$J = -D\frac{dC}{dx} \tag{2.13}$$

where C is the particle concentration. The constant of proportionality D is called the diffusion coefficient, which has units of $m^2\,s^{-1}$. The negative sign in Equation (2.13) indicates that the direction of diffusion is down the concentration gradient. This expression is called *Fick's first law.*

BIBLIOGRAPHY

Blythe T, Bloor D. *Electrical Properties of Polymers.* Cambridge: Cambridge University Press; 2005.

Callister WD. *Materials Science and Engineering: an Introduction*, 6th ed. New York: John Wiley & Sons, Inc.; 2003.

Hamley IW. *Introduction to Soft Matter.* Chichester: John Wiley & Sons, Ltd.; 2000.

Jones RAL. *Soft Machines.* Oxford: Oxford University Press; 2004.

Kasap SO. *Principles of Electrical Engineering Materials and Devices.* Boston: McGraw-Hill; 1997.

Pauling L. *The Nature of the Chemical Bond*, 3rd ed. Ithaca, NY: Cornell University Press; 1960.

Petty MC, Bryce MR, Bloor D (Editors). *An Introduction to Molecular Electronics.* London: Edward Arnold; 1995.

Schwoerer M, Wolf HC. *Organic Molecular Solids.* Berlin: Wiley-VCH; 2006.

Sears FW, Salinger GL. *Thermodynamics, Kinetic Theory and Statistical Thermodynamics.* Reading, MA: Addison-Wesley; 1975.

Streitwieser A, Heathcock CH. *Introduction to Organic Chemistry.* New York: Macmillan; 1976.

REFERENCE

[1] Woodruff DP, Delchar TA. *Modern Techniques of Surface Science*, 2nd ed. Cambridge: Cambridge University Press; 1994.

3 Electrical Conductivity

Two truths are told

3.1 INTRODUCTION

Metallic and semiconductive properties are not restricted to inorganic materials. For example, Figure 1.2, in Chapter 1, shows that the room temperature conductivity values for organic polymers can extend over the entire spectrum of electrical conductivity, from insulating behaviour, as shown in diamond, to highly conducting, as for copper. The explanation can be found in the nature of the chemical bonds that hold solids together. In this chapter, the physical principles underlying some of the important electrical conductivity processes in crystalline inorganic solids are first developed. These ideas are then extended to encompass organic solids, which are generally less ordered.

3.2 CLASSICAL THEORY

The idea of electrons in a metal being free to move about as if they constituted an 'electron gas' was first put forward by Drude in 1900, soon after the discovery of the electron. The theory was later extended by Lorentz (1905) who applied classical *kinetic theory of gases* to the free electron gas in a metal. According to this hypothesis, these electrons would possess a range of speeds (or energies) distributed about a mean velocity, $\bar{v}$, as depicted in Figure 3.1. The resulting curve is known as the *Maxwell–Boltzmann distribution* and has the mathematical form

$$n(v) = Av^2 T^{-3/2} \exp(-Bv^2/T)\mathrm{d}v \tag{3.1}$$

where $n(v)$ is the number of molecules having a velocity in the range v to $v + dv$, T is the absolute temperature and A and B are constants. If the temperature is increased, the maximum and mean velocities shift to higher values and the Maxwell–Boltzmann curve flattens out. If the temperature is reduced to zero, $\bar{v}$ becomes zero.

In the Drude theory, the behaviour of free electrons in a metal is, in many respects, analogous to that of gaseous molecules. There is one significant difference between the two cases. In the free electron theory, collisions between electrons are neglected and the electrons only undergo collisions with the metal ions. Drude also assumed that the distance travelled by an electron between collisions, the mean free path (see the following section), is governed by the lattice spacing in the crystal and is independent of the electron's speed.

Molecular Electronics: From Principles to Practice M. C. Petty
© 2007 John Wiley & Sons, Ltd

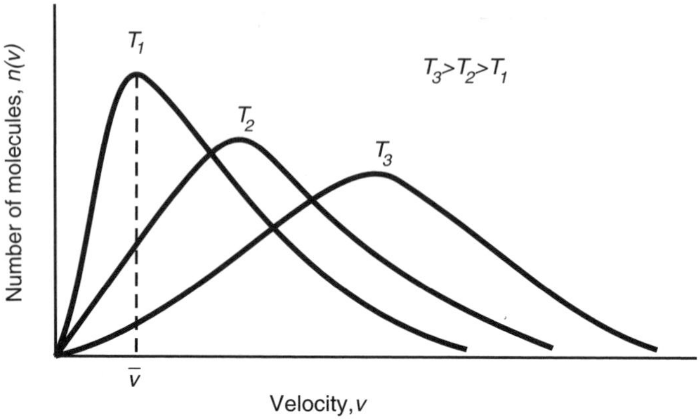

Figure 3.1 Maxwell–Boltzmann distribution of the electron velocities in a metal at different temperatures. Mean velocity $= \bar{v}$.

3.2.1 Electrical Conductivity

The *electric field strength*, E, is the force on a unit positive electric charge placed in that field. It is related to the gradient of the *electric potential*. For an electric field applied in one dimension, say the x-direction:

$$E_x = \frac{dV}{dx} \tag{3.2}$$

where V is the potential in volts. The units of E are therefore $V\,m^{-1}$.

The *electrical conductivity* (measured in units of siemens per metre, $S\,m^{-1}$, or reciprocal ohms per metre, $\Omega^{-1}\,m^{-1}$) of a solid is the rate at which charge is transported across unit area of the solid as a result of a unit applied electric field. If the current per unit area, or *current density*, is J_x (units $A\,m^{-2}$), the conductivity, σ, is given by

$$\sigma = \frac{J_x}{E_x} \tag{3.3}$$

The electric field and current density have a direction associated with them; these are therefore both vector quantities. In this book, when the direction is important such quantities are denoted using a bold typeface. Hence **J** is the current density vector, whereas J simply represents the magnitude of the current density.

The basic Drude model is that of a random motion of the electrons before any electric field is applied, i.e. as many electrons are moving in any one direction as in the opposite direction, so that there is not net flow in any direction and the current is zero. An estimate of the average *thermal velocity* of an electron, $\bar{v}_t$, can be estimated from the Maxwell–Boltzmann distribution law:

$$\frac{1}{2} m \bar{v}_t^2 = \frac{3}{2} k_B T \tag{3.4}$$

where m is the electron mass and k_B is a constant, known as *Boltzmann's constant* ($1.38 \times 10^{23}\,\mathrm{J\,K^{-1}}$).

The direction of electron motion may be altered when it collides with a metal ion. The average distance travelled by an electron before it experiences such a collision is called the *mean free path*, λ. For a metal at room temperature, the average electron thermal velocity is about $10^6\ \mathrm{m\,s^{-1}}$ and the mean free path is approximately 10 nm. In some special cases, the electrons are not subject to scattering events. This situation is referred to as *ballistic* transport and occurs, for example, when the mean free path is long or when electrons are confined in ultra-small regions in semiconductor structures. Ballistic transport is determined by the electronic structure of semiconductors, and allows ultra-fast devices to be fabricated.

When an electric field is applied across the sample, the electrons will be accelerated during their free periods between collisions. This acceleration is generally taken to be in a direction opposite to that of the field since the charge on the electron is negative. Simultaneously, collisions of the electrons with the ions will tend to restore the condition in which all the electron velocities are random. The situation is illustrated in Figure 3.2. In equilibrium, the situation is equivalent to the valence electrons in the sample all possessing a common *drift velocity* due to the applied electric field. This electron drift constitutes the electric current. It is important to realize that this drift velocity is superimposed on the random electron velocities due to thermal motion and its value will be many orders of magnitude less than that of the thermal electron velocities.

A simple relationship for the conductivity of a material may be evaluated using Newton's laws. With an electric field applied, the force, F, acting on an individual electron of charge, e

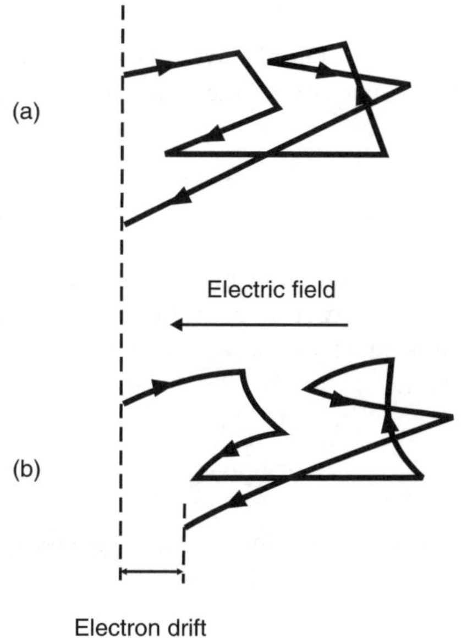

Figure 3.2 Model for electron drift. (a) In the absence of an applied electric field a conduction electron moves about randomly in a metal being frequently and randomly scattered by thermal vibrations of the atoms. There is no net electron drift in any direction. (b) With an applied field there is a net drift of the electrons along the direction of the field. After many scattering events the electron has been displaced a small distance from its original position towards the positive electrode.

($e = 1.60 \times 10^{-19}$ C), will be proportional to its charge and also to the applied field, i.e. $F = eE$. Therefore, the acceleration, a, experienced by the electron is (Newton's second law)

$$am = eE \tag{3.5}$$

If the average time of flight between collisions is denoted by τ, an appropriate time constant (or relaxation time), and assuming that all the extra energy gained from the applied field is lost upon collision, the average velocity in the electric field, $\bar{v}_d$, is given by $a\tau$, i.e.

$$\bar{v}_d = \frac{e}{m}E\tau \tag{3.6}$$

This represents the drift velocity of the electron. If there are a total of n electrons per unit volume taking part in the conduction, the total current density will be the product of the total charge density and the drift velocity, i.e.

$$J = ne\bar{v}_d \tag{3.7}$$

3.2.2 Ohm's Law

Substituting the value of $\bar{v}_d$ from Equation (3.6) into Equation (3.7) gives

$$J = \frac{ne^2\tau}{m}E \tag{3.8}$$

which can be recognized as a form of Ohm's law – the current density is proportional to the electric field. The electrical conductivity of the material can then be calculated by noting Equation (3.3):

$$\sigma = \frac{ne^2\tau}{m} \tag{3.9}$$

The electrical behaviour of a material may also be characterized in terms of its *resistance*, R, measured in ohms (Ω), or its *resistivity*, ρ, measured in Ω m. The latter quantity provides a useful parameter that is independent of the dimensions of the sample and is simply the reciprocal of the conductivity, i.e.

$$\rho = \frac{1}{\sigma} = \frac{m}{ne^2\tau} \tag{3.10}$$

For a wire of cross-sectional area, A, length, l, with a resistance, R, the resistivity is given by

$$\rho = \frac{RA}{l} \tag{3.11}$$

Noting that

$$E = \frac{V}{l} \text{ and } J = \frac{I}{A}$$

it is evident that Equation (3.9) can also be written as

$$V = IR \tag{3.12}$$

which is the most familiar form of Ohm's law. In some situations (in particular, when dealing with electrical conductivity at high frequencies, Section 3.6) a parameter called the *conductance*, G, is used; this is simply the reciprocal of the resistance (i.e. $G = 1/R$) with units Ω^{-1} or S.

For many applications in molecular electronics, the conductive materials are in the form of thin films and it is the *sheet resistance* (or surface resistance) R_s that is important. This is because the sheet resistance is easier to measure than the resistivity. The sheet resistance of a layer with a thickness t is given by

$$R_s = \frac{\rho}{t} \tag{3.13}$$

Strictly, the units of sheet resistance are ohms. However, it is normal to quote the sheet resistance in units of ohms per square (although this can give rise to some confusion). This nomenclature is particularly useful when the resistance of a rectangular piece of material, with length l (between the electrodes) and width w is needed. The resistance is simply the product of the sheet resistance and the number of squares, or

$$R = R_s \frac{l}{w} \tag{3.14}$$

3.2.3 Charge Carrier Mobility

Equation (3.9) can also be written as

$$\sigma = ne\mu \tag{3.15}$$

where

$$\mu = \frac{e\tau}{m} \tag{3.16}$$

The parameter μ is called the *carrier mobility* of the electrons and is equal to the average drift velocity per unit applied electric field [Equation (3.6)], i.e.

$$\mu = \frac{\bar{v}_d}{E} \tag{3.17}$$

As noted above, the velocity of the electrons acquired as a result of their finite temperature (thermal velocity) is much larger than the carrier drift velocity.

The concept of the mobility of a charge carrier is an important one in semiconductor device physics. It is a measure of how quickly the carriers (electrons) will respond to an applied electric field and provides an indication of the upper frequency limit of the material if it is used in a device such as a transistor (Chapter 9, Section 9.4). The mobility of charge carriers in organic molecular and polymeric compounds is generally lower than in inorganic semiconductors such as silicon and gallium arsenide (see Table 3.1, Section 3.4.1). A high mobility in a material does not necessarily imply a high electrical conductivity. Equation (3.15) shows that σ also depends on the concentration of the charge carriers (electrons in our example).

Equation (3.17) suggests that the carrier drift velocity will continue to increase as the applied electric field increases. At very high electric fields, the carrier energy can become larger than the normal thermal energy. The carriers are referred to as *hot* and this situation leads to a reduction in the carrier mobility and a saturation of the drift velocity. In silicon, the maximum drift velocity for both electrons and holes is approximately $10^5\ \mathrm{cm\,s^{-1}}$, corresponding to an applied field of $5 \times 10^7\ \mathrm{V\,m^{-1}}$.

If an electron concentration gradient exists in a sample, then the carrier motion will also be affected by diffusion (Chapter 2, Section 2.8). For the one-dimensional case, the total current density is therefore given by an equation of the form

$$J = ne\mu \mathsf{E} + eD\frac{\mathrm{d}n}{\mathrm{d}x} \tag{3.18}$$

where D is the electron diffusion coefficient. The first term represents the drift process and the second term reflects diffusion.

3.2.4 Fermi Energy

The classical theory for electrical conductivity can provide reasonable 'order of magnitude' values for the conductivity of some metals. However, the theory does fail in predicting the correct temperature dependence of conductivity.

The Drude model treats electrons as classical indistinguishable particles having velocities or energies governed by the Maxwell–Boltzmann distribution law. No restriction is placed on the number of electrons that can possess any particular value of energy. The quantum mechanical approach, first put forward by Sommerfeld in 1928, postulated that the valence electrons in a metal obey *Fermi–Dirac quantum statistics*. In accordance with the theory, the electrons obey the Pauli Exclusion Principle (Chapter 2, Section 2.2.3) and, in contrast to classical theory, they must be regarded as indistinguishable particles. The Fermi–Dirac distribution function $f(E)$ governing the occupation of the electron levels (or states) over energy may be shown to be

$$f(E) = \frac{1}{\exp[(E - E_{\mathrm{F}})/k_{\mathrm{B}}T] + 1} \tag{3.19}$$

where $f(E)$ is the probability of a level at energy E being filled with an electron and E_{F}, which has the dimensions of energy, is termed the *Fermi level* or *Fermi energy.*

When $E = E_{\mathrm{F}}$, then from Equation (3.19), $f(E) = 0.5$, i.e. *the Fermi level is the energy at which there is a 50:50 chance of finding an electron.*

The Sommerfeld model predicts a Fermi energy in metals at low temperatures of several electronvolts per electron. Figure 3.3 shows how the occupation of electron states $f(E)$ varies

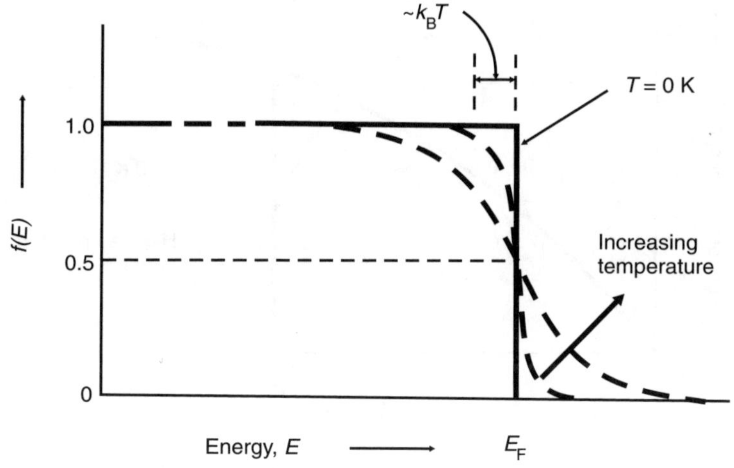

Figure 3.3 The Fermi–Dirac function $f(E)$ at $T = 0\,\text{K}$ and at $T > 0\,\text{K}$.

with energy. At $T = 0\,\text{K}$, all energy levels below E_F are completely filled $[f(E) = 1]$ whereas the levels are completely empty $[f(E) = 0]$ above E_F. Thus, at absolute zero, the Fermi energy represents the demarcation between filled and empty states. At higher temperatures, the Fermi–Dirac distribution function becomes less step-like and some of the levels below E_F become depopulated and some above become populated. However, the effect fairly small except at high temperatures. The function $f(E)$ drops from unity to zero as the energy increases by a few $k_B T$. Since the value of $k_B T$ is only about 0.026 eV at 300 K, the transition range is very narrow.

For values of energy that are more than a few $k_B T$ from the Fermi level [i.e. $(E - E_F)/k_B T \gg 1$], Equation (3.19) reduces to

$$f(E) = \exp\left(\frac{E_F}{k_B T}\right)\exp\left(-\frac{E}{k_B T}\right) \tag{3.20}$$

which is the same form as the Maxwell–Boltzmann law given by Equation (3.1).

The Sommerfeld model also predicts how the number of electron states will vary with energy. This is given by a function called the *density of states*, $S(E)$, where

$$S(E) = CE^{\frac{1}{2}} \tag{3.21}$$

and C is a constant.

The number of occupied states within an incremental energy range can be found by multiplying the density of states function by the probability of their occupancy, i.e.

$$N(E)\mathrm{d}E = S(E)f(E)\mathrm{d}E \tag{3.22}$$

where $N(E)\mathrm{d}E$ is the number of occupied states in the energy range E to $E + \mathrm{d}E$. Figure 3.4 reveals how the theoretical distribution of electrons varies with energy for the cases $T = 0\,\text{K}$ and $T > 0\,\text{K}$

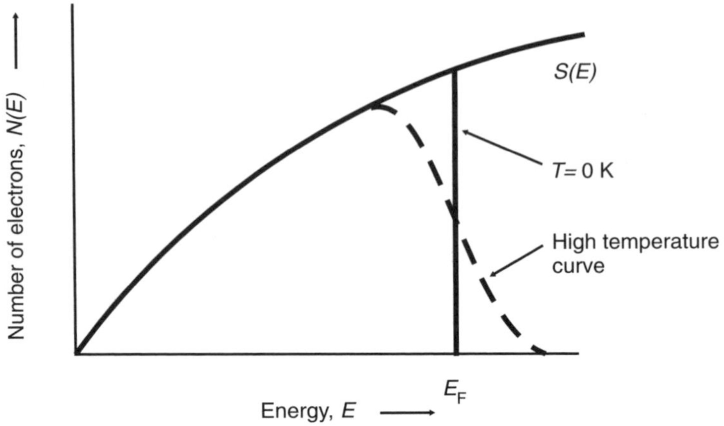

Figure 3.4 Theoretical distribution of the electrons as a function of energy at temperature 0 K and at a higher temperature. E_F is the Fermi energy or Fermi level. $S(E)$ is the density of states function.

3.3 ENERGY BANDS IN SOLIDS

One of the main problems with the Drude and Lorentz approaches is the postulate that the mean free path of the electrons is assumed to be of the same order as the lattice spacing. In reality, mean free paths in metals are much larger (~50 nm in copper at 300 K). The free electron theories of Drude, Lorentz and Sommerfeld also fail to predict why some materials are metals, others are good insulators and some are of intermediate conductivity (semiconductors). The band theory of solids, described in the following sections, addresses these issues.

3.3.1 Quantum Mechanical Foundations

Before band theory can be developed, some insight into some of the more important physical foundations is needed.

Electromagnetic Waves

Besides what is known as light, *electromagnetic radiation* includes radiation of longer (infrared, microwave, terahertz) and shorter (ultraviolet, X-ray) wavelengths. As the name implies, electromagnetic (EM) radiation contains both electric E and magnetic field B components. The relationship between these fields is best illustrated by considering *plane-polarized* radiation in which the electric field is confined to a single plane. Figure 3.5 depicts such radiation of wavelength λ, frequency ν and travelling with *phase velocity* v (in a vacuum, the phase velocity $= c$, the velocity of light, $= \lambda\nu = 3.00 \times 10^8\ \mathrm{m\,s^{-1}}$) along the x-axis. The electric component of the radiation is in the form of an oscillating electric field while the magnetic component is an oscillating magnetic field. The vectors representing these fields are orthogonal to each other and, in free space, are also orthogonal to the direction of propagation

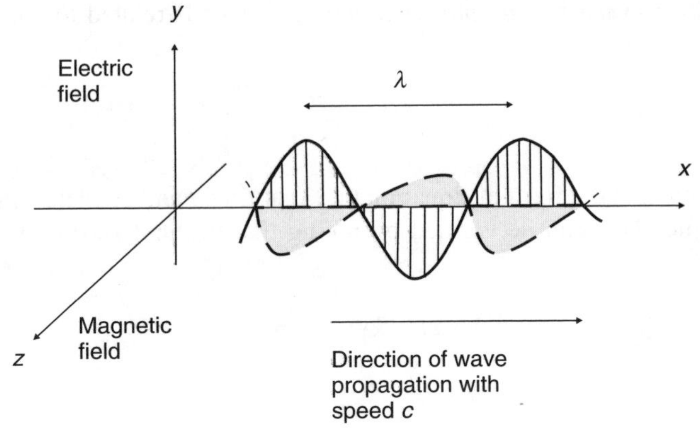

Figure 3.5 Schematic representation of an electromagnetic wave travelling in free space with velocity c, the speed of light. Wavelength $= \lambda$. The wave consists of oscillating electric and magnetic field components at right-angles to each other and orthogonal to the direction of propagation x.

of the wave. The amplitude of each field component can be described mathematically by a simple sinusoid. For example, the variation of the electric field E may be written as

$$E = E_0 \cos(\omega t - kx) \tag{3.23}$$

where E_0 represents the maximum amplitude of the wave. Equation (3.23) indicates that the amplitude of the electric field depends both on time t and position x, as shown in Figure 3.6. The constant ω is called the *angular frequency* of the wave ($= 2\pi\nu$, where ν is in hertz) and k

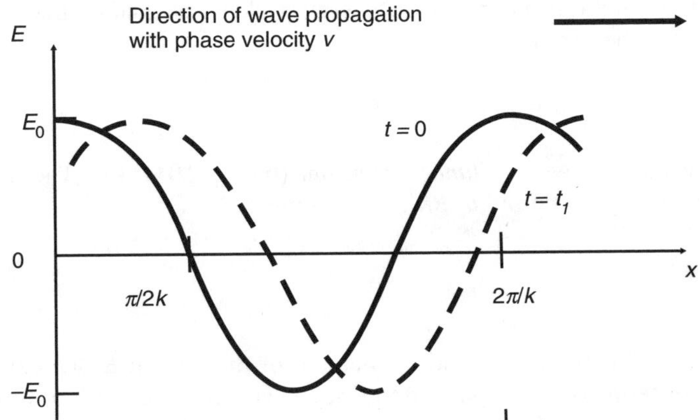

Figure 3.6 Variation of the electric field E of an electromagnetic wave, given by Equation (3.23), with position x for time $t = 0$ and a later time $t = t_1$.

is known as the *wavevector* or *propagation constant*. This is related to the wavelength λ of the wave by

$$k = \frac{2\pi}{\lambda} \tag{3.24}$$

Figure 3.6 compares the form of Equation (3.23) at time t and at a later time $t = t_1$, shown by the broken line. This can be verified by noting that the first zero of Equation (3.23) is given by

$$(\omega t - kx) = -\pi/2$$

or

$$x = (\pi/2k) + (\omega t/k) \tag{3.25}$$

Hence the wave representing the electric field is moving in the positive x-direction with a constant phase velocity v given by

$$v = \frac{\omega}{k} \tag{3.26}$$

Photons as Particles

The wave nature of light can account for many of its characteristics, such as interference and diffraction. Other features can be explained only if light can also be treated as particles. The most important phenomenon is the photoelectric effect by which electrons are emitted from a clean metal surface in vacuum when irradiated by light of an appropriate wavelength. An increase in light intensity is accompanied by an increase in electron emission but not in electron energy. The successful interpretation of the photoelectric effect, given by Einstein in 1905, proposed that light also consists of 'energy packets' or quanta. Each of these has an energy E of magnitude given by

$$E = h\nu \tag{3.27}$$

where h is a constant known as *Planck's constant* (6.63×10^{-34} J s). The light quanta also possess a momentum p given by *de Broglie's relationship*:

$$p = \frac{h}{\lambda} \tag{3.28}$$

The expressions for the energy and momentum of the photon $E = h\nu$ and $p = h/\lambda$ can also be written in terms of the angular frequency ω and wavevector k as follows:

$$E = \hbar\omega \quad \text{and} \quad p = \hbar k \tag{3.29}$$

where $\hbar = h/2\pi$.

The wave nature of particles such as photons cannot be represented adequately by an equation of the form of Equation (3.23). Instead, the particle is represented by a group or

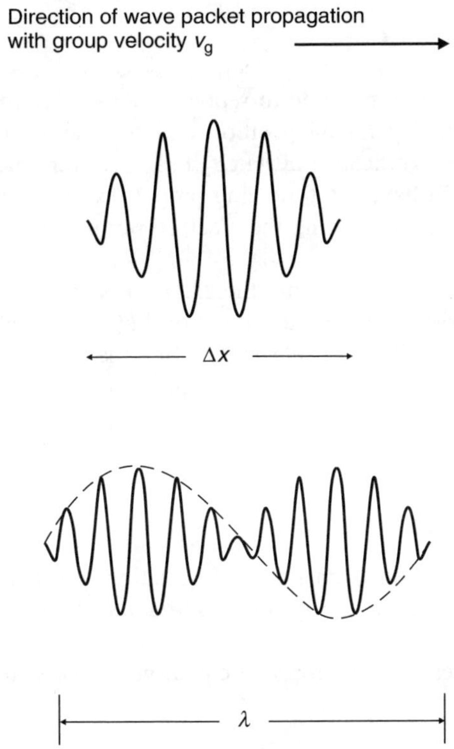

Figure 3.7 Illustration of a wave packet or group of width Δx. The wave packet travels with a group velocity v_g (top). The packet can result from the interference of two waves of similar wavelengths (bottom). The wavelength of the wave packet is λ.

packet of a large number of waves, spread over a distance Δx; each of the waves has a slightly different frequency and velocity from the other members of the group. The waves within the group interfere with one another to give a localized entity that moves through space with a wavelength λ and a *group velocity* v_g, as shown in Figure 3.7. This group velocity represents the speed at which the wave packet, or particle, propagates and is given by

$$v_g = \frac{d\omega}{dk} \tag{3.30}$$

Group velocity is therefore not generally the same as the phase velocity v of a wave given by Equation (3.26). The group velocity is the speed at which the energy of a wave packet propagates. This velocity cannot exceed that of light in free space. In contrast, the phase velocity is the speed at which the crests (or troughs) of a wave moves through space. In certain circumstances it is possible for $v > c$. Only in the case of a linear relationship between the angular frequency ω and the wavevector k (as, for example, the case for EM waves propagating in vacuum) will $v = v_g$.

Electron Wavefunction

Following de Broglie's suggestion that electrons possessed a wave nature, Schrödinger (1926) argued that it should be possible to represent these electron waves mathematically. Such a wavefunction will depend on position and time and can be represented in one dimension by $\psi(x,t)$. The significance of $\psi(x,t)$ has been introduced in Chapter 2, Section 2.2.2. To recall, $\psi(x,t)$ will have the following properties:

$|\psi(x,t)|^2$ is the probability of finding the electron per unit length at x at time t

and

$|\psi(x,t)|^2 \mathrm{d}x$ is the probability of finding the electron between x and $x + \mathrm{d}x$ at time t.

Since the electron carries a charge, the function $|\psi(x,t)|^2 \mathrm{d}x$ also represents the charge distribution associated with the electron. In the case of three dimensions, $|\psi(x,y,z,t)|^2 \mathrm{d}x\mathrm{d}y\mathrm{d}z$ is the probability of finding the electron in a small elemental volume $\mathrm{d}x\mathrm{d}y\mathrm{d}z$ at x, y, z at time t.

Schrödinger Wave Equation

The Schrödinger wave equation is an important relationship involving the electron wavefunction. In essence, it is a statement of the *conservation of energy*, i.e. that the total of the kinetic energy, E, and potential energy, V, is a constant. It is useful to draw on an analogy with the classical case. Here, the appropriate equation can be written as

$$\frac{1}{2m}p^2 + V = E \tag{3.31}$$

i.e. kinetic energy + potential energy = total energy.

In the quantum mechanical formulation for one dimension and for a potential energy that is only dependent on space, i.e. $V = V(x)$, the relevant equation becomes

$$\frac{\mathrm{d}^2\psi}{\mathrm{d}x^2} + \left(\frac{2m}{\hbar^2}\right)\psi[E - V(x)] = 0 \tag{3.32}$$

Essentially, the classical quantities of Equation (3.31) have been replaced by quantum mechanical *operators*, which operate upon the wavefunction (representing a particular state of the system) to extract observable value quantities such as position, momentum and energy. For example, the momentum p in Equation (3.31) is replaced by the momentum operator $(\hbar/j)(\partial/\partial x)$. where $j = \sqrt{-1}$. Equation (3.32) is called the *time-independent Schrödinger wave equation*. Many of the problems in solid-state physics are concerned with solving this equation for various forms of the potential energy. Once the wavefunction has been determined, then the probability distribution and energy of the electron can be determined from ψ^2. The solutions for ψ are known as *eigenfunctions* (characteristic functions) and the corresponding energies are called *eigenenergies*.

The solutions to the Schrödinger wave equation for the case of (a) free electrons and (b) electrons confined to a one-dimensional 'box' are illustrated below.

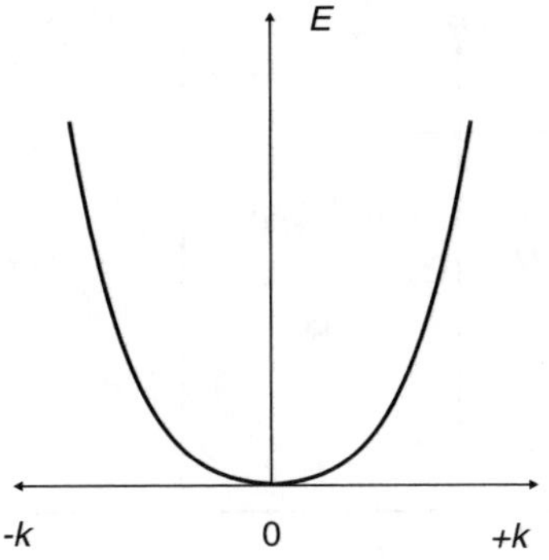

Figure 3.8 Electron energy E versus wavevector k for a free electron.

(a) Free electrons

In this simple case, the value of the potential energy is zero and Equation (3.32) becomes

$$\frac{d^2\psi}{dx^2} + k^2\psi = 0 \tag{3.33}$$

where $k^2 = (2m/\hbar^2)E$. Solutions to this differential equation yield

$$\psi(x) = A\exp(jkx) \quad \text{or} \quad \psi(x) = B(-jkx) \tag{3.34}$$

where A and B are constants. Each of the above equations represents a travelling wave. The first solution represents a wave travelling in the $+x$ direction and the second a wave travelling in the $-x$ direction. The total energy E of the electron is simply its kinetic energy. Therefore,

$$E = \frac{\hbar^2 k^2}{2m} \tag{3.35}$$

As noted in Equation (3.29), the wavevector k is proportional of momentum p. Hence, the energy versus momentum relationship for a free electron is essentially a parabolic one, as shown in Figure 3.8.

(b) Electrons confined to a one-dimensional box

A further important solution to the Schrödinger wave equation considers electrons confined by an infinite potential energy barrier V_x to a one-dimensional 'box' of length L (Figure 3.9). The solution yields eigenenergies E_n:

$$E_n = \frac{\hbar^2\pi^2}{2mL^2}n^2 \tag{3.36}$$

where n is a quantum number (Chapter 2, Section 2.2.2). For each value of n, there will be an electron wavefunction (the eigenfunction) and a corresponding eigenenergy. This result is

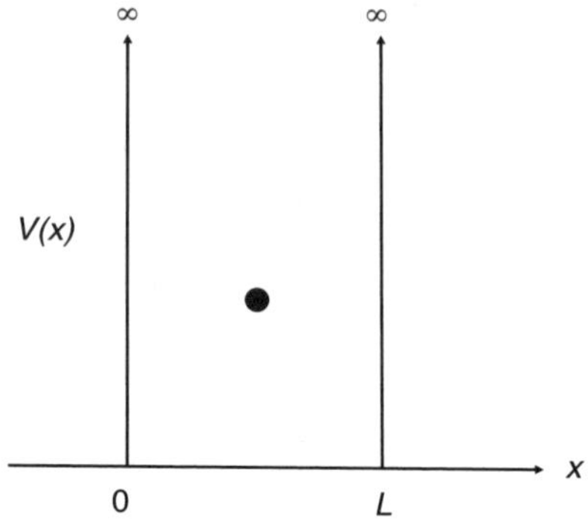

Figure 3.9 Electron confined by infinite potential energy barriers to a one-dimensional 'box' of dimension L.

summarized in Figure 3.10. The wavefunction and energy of the electron both depend on the quantum number n. The picture is that of a large number of possible standing waves, analogous to standing waves on a vibrating string, each particular wavefunction having its own specific energy. The energy of the electron increases as n^2, so the minimum energy of the electron corresponds to $n = 1$. This is called the *ground state*. Note that the energy of the electron in this potential well cannot be zero, even though the potential energy is zero. The electron always has kinetic energy, even when it is in its ground state.

This result above demonstrates clearly that the energetic spacing of the eigenenergies depends on the dimensions of the box: the smaller the box, the larger the spacing. If the length of the box is of atomic dimensions, say 0.1 nm, the energy states are widely spaced (in energy) and a large amount of energy (hundreds of electronvolts) will be required to promote

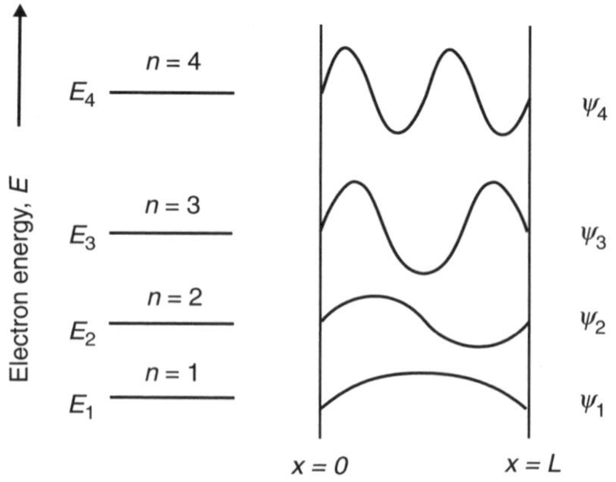

Figure 3.10 Solutions to electron in a one-dimensional 'box' problem for confinement by infinite potential energy barriers. The allowed energy levels for the electron are shown on the left and the corresponding wavefunctions on the right.

an electron to a higher energy state. Electrons in a crystal or polymer (dimensions millimetres to centimetres) will have much more closely space eigenenergies.

Heisenberg's Uncertainty Principle

The wavefunction of a free electron corresponds to a travelling wave with a single wavelength λ. The travelling wave extends over all space, for instance along all the x direction, with the same amplitude, so the probability distribution function is uniform throughout the whole of space. The 'uncertainty' Δx in the position of the electron is therefore infinite. However, the uncertainty in the momentum Δp_x of the electron is zero, because λ is well defined, which means that we know p_x exactly from the de Broglie relationship (Equation 3.28).

For an electron trapped in a one-dimensional potential energy well, as described in the previous section, the wavefunction extends from $x = 0$ to $x = L$, so the uncertainty in the position of the electron is L. We know that the electron is within the well, but cannot pinpoint with certainty where exactly it is. The momentum is either $p_x = \hbar k$ in the $+x$ direction or $-\hbar k$ in the $-x$ direction. The uncertainty Δp_x in the momentum is therefore $2\hbar k$, i.e. $\Delta p_x = 2\hbar k$. These results are summarized in the *Heisenberg uncertainty principle*, which states the product of the position and momentum uncertainties must be equal to or greater than Planck's constant h:

$$\Delta x \Delta p_x \geq h \tag{3.37}$$

Therefore, the position and momentum of a particle along a given coordinate cannot be known simultaneously. There is a corresponding relationship between the uncertainty ΔE in the energy E (or angular frequency ω) of the particle and the time duration Δt during which it possesses the energy (or during which the energy is measured):

$$\Delta E \Delta t \geq h \tag{3.38}$$

The uncertainty relationships in Equations (3.37) and (3.38) are sometimes written in terms of $\hbar$ (i.e. $h/2\pi$). There is also a numerical factor which comes about when a Gaussian spread for all possible position and momentum values is considered. Details can be found in advanced quantum mechanics books.

Quantum Mechanical Tunnelling

An interesting result is found if the problem depicted in Figure 3.9, i.e. an electron confined by infinite potential energy barriers, is replaced by one in which the potential energy is finite. Solutions for wavefunctions with $n = 1$ and $n = 2$ are shown superimposed on the one-dimensional potential well in Figure 3.11. In both cases, ψ penetrates the barrier, leading to a finite probability of finding the electron on the outside the box. This mechanism by which the electron moves through the barrier is known as *quantum mechanical tunnelling*. The probability of finding the electron outside the box diminishes very rapidly with distance so that tunnelling is only important over small, nanometre, dimensions. A

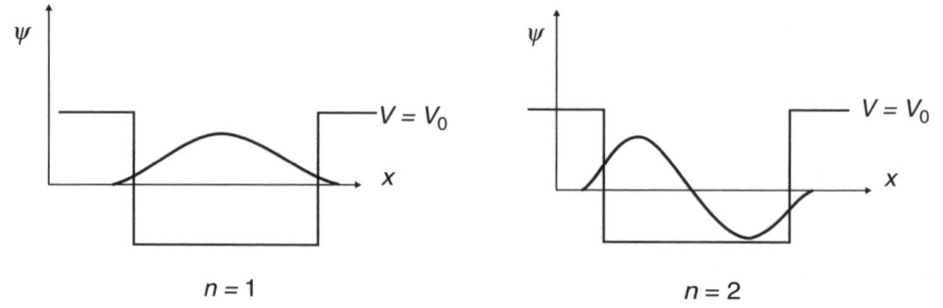

Figure 3.11 Wavefunctions for $n = 1$ and $n = 2$ are superimposed on a one-dimensional well of finite height. Penetration of the wavefunctions outside the well represent quantum mechanical tunnelling.

number of electron devices can exploit the tunnelling phenomenon (see, for example, Chapter 11, Section 11.5).

3.3.2 Kronig–Penney Model

A useful model for the propagation of electron waves in a crystal was first formulated by Kronig and Penney in 1930. The periodic potential associated with the positive ion cores is represented by a series of rectangular barriers of width b, height V_0 separated by distance a, as shown in Figure 3.12. The mathematical form of the repeating energy barrier is

$$V(x) = V_0 \quad -b < x < 0$$
$$V(x) = 0 \quad 0 < x < a$$

The Kronig–Penney model solves the Schrödinger wave equation in the regions of the barriers and also between the barriers. A solution is then found that is appropriate for both of these regions which will satisfy the boundary conditions at the barriers, i.e. at $x = 0$ and at $x = a$ in Figure 3.12. The detailed mathematics can be found in a number of texts (e.g. [1]).

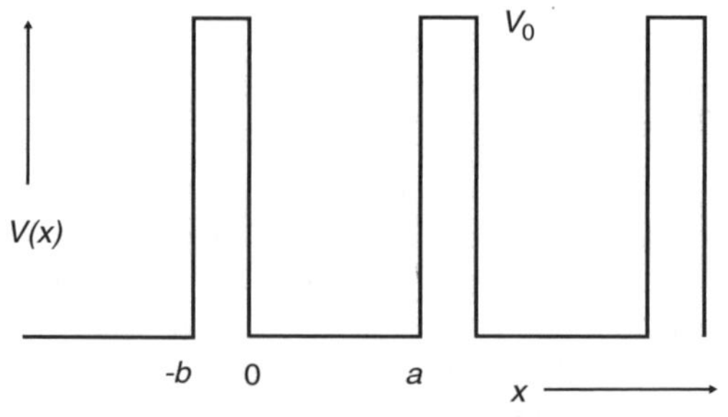

Figure 3.12 The Kronig–Penney model. Potential distribution in a one-dimensional lattice. The energy barriers have a height V_0, a width b and spacing a.

The result is the following equation which expresses conditions for solutions to the Schrödinger wave equation to exist:

$$\cos ka = \cos \alpha a + \gamma a\left(\frac{\sin \alpha a}{\alpha a}\right) \tag{3.39}$$

where k is the wavevector, γ is a variable related to the 'strength' of the barrier and given by

$$\gamma = \frac{mV_0 b}{\hbar^2} \tag{3.40}$$

and α is a further variable related to energy:

$$\alpha = \frac{(2mE)^{\frac{1}{2}}}{\hbar} \tag{3.41}$$

Equation (3.39) therefore gives the values of α that permit solutions of the electron wave to exist. The significance of this relationship can be best understood by reference to Figure 3.13. Here, the right-hand side of Equation (3.39) is plotted as a function of αa for a particular value of $\gamma (= \pi/a)$.

Since values of cos(ka) – the left-hand side of Equation (3.39) – can only lie between $+1$ and -1, only certain ranges of αa will satisfy Equation (3.39). These ranges are indicated in Figure 3.13. Moreover, since αa is proportional to energy, the Kronig–Penney model predicts that the motion of electrons in a periodic lattice is characterized by bands of allowed energies separated by forbidden regions (i.e. energy ranges for which no solutions to the Schrödinger wave equation exist).

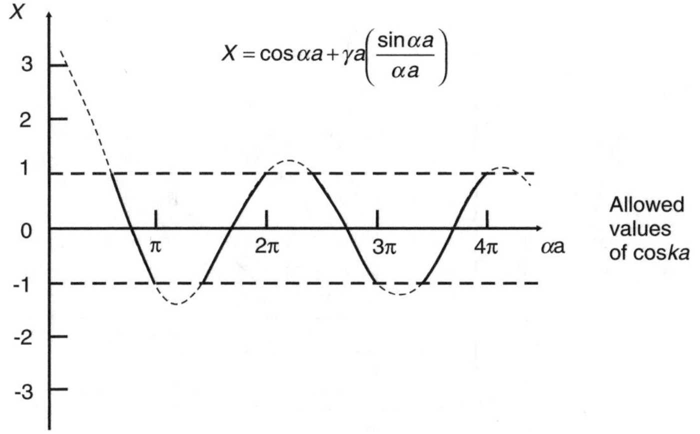

Figure 3.13 Solutions to the Kronig–Penney model. Allowed values of the parameter αa are restricted to those which yield values of the function X lying between $+1$ and -1. In the figure, the parameter $\gamma a = \pi$.

The width of the energy bands is governed by the parameter γ. When γ is very small, i.e. for a very 'weak' barrier, Equation (3.39) gives $\alpha = k$, so that

$$E = \frac{\hbar^2 k^2}{2m} \tag{3.42}$$

which is the energy versus wavevector relationship derived from the free electron theory outlined in the previous section [Equation (3.35)].

In contrast, when γ is large, solutions to the Schrödinger wave equation become restricted to those for which $\sin(\alpha a) = 0$, i.e.

$$\alpha a = \pm n\pi \tag{3.43}$$

so that in accordance with Equation (3.41)

$$E = \frac{\hbar^2 \pi^2}{2ma^2} n^2 \tag{3.44}$$

Therefore, for a tightly bound electron, the Kronig–Penney model predicts a series of discrete levels, as determined for an electron in a one-dimensional box [Equation (3.36)], with L replaced by a. Under such conditions, the electron may be considered to be confined to a single cell of the potential barrier model depicted in Figure 3.12.

For intermediate values of the barrier strength parameter γ, the electrons having energies within the appropriate bands can propagate freely throughout the crystal. This occurs even though the electron energy is less than the barrier height (via quantum mechanical tunnelling).

The variation of electron energy E with wavevector k is shown in Figure 3.14. This diagram again shows the formation of allowed bands of energies separated by forbidden energy ranges or *energy gaps* or *band gaps*. The figure should be contrasted to the continuous

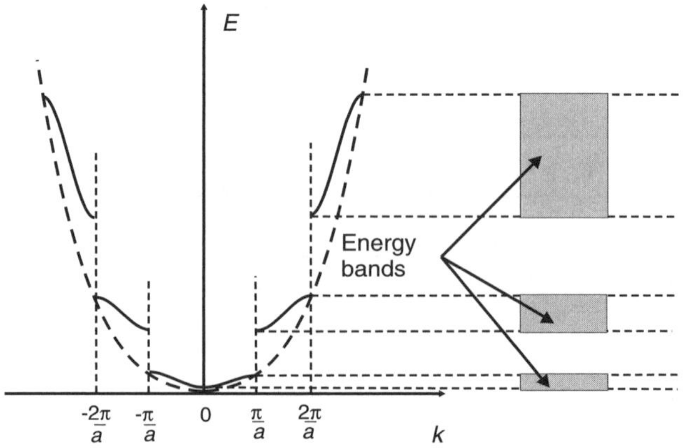

Figure 3.14 The variation of the electron energy E with the wavevector k for the Kronig–Penney model showing the formation of bands of allowed energies separated by forbidden energy regions.

parabolic E versus k curve for a free electron, in Figure 3.8. The discontinuities in Figure 3.14 occur when

$$k = \pm \frac{n\pi}{a} \tag{3.45}$$

where $n = 1, 2, 3 \ldots$. These k values define the boundaries of what are termed *Brillouin zones*. For example, the first zone contains k values lying between $-\pi/a$ and $+\pi/a$ and the second zone contains values between $-2\pi/a$ and $+2\pi/a$. At the boundary between the zones, $k = \pm n\pi/a$ and since $k = 2\pi/\lambda$ [Equation (3.24)]

$$n\lambda = 2a \tag{3.46}$$

which, as will be discussed in Chapter 6, Section 6.3, represents Bragg reflection for normal incidence of the electron waves on a set of planes with separation a. Thus, electrons with energies at the boundaries of the Brillouin zones undergo multiple reflections and the travelling wave is replaced by equal components of waves travelling in the positive and negative directions. The wavefunctions representing solutions to the Schrödinger wave equation become standing waves.

Within an energy band the electron energy is a periodic function of k. This means that if k is replaced by $k + 2\pi n/a$, where n is an integer, the left-hand side of Equation (3.39) remains the same. For convenience, E–k curves are often depicted in what is known as the *reduced wavevector representation*, in which all the energy bands are shown in the first Brillouin zone (Figure 3.15). In three-dimensional solids, the E–k curves depend on the direction of the electron wavevector with respect to the crystallographic axis, and the shape of the Brillouin zones will reflect the symmetry of the crystal structure.

In Chapter 2, Section 2.3.3, we saw how the electron energy levels in two isolated hydrogen atoms split (the bonding and antibonding molecular orbitals) when the atoms joined in a covalent bond to form a hydrogen molecule. An appropriate engineering analogy is that given by Kasap [2] for the resonant frequency in an *RLC* (resistor–inductor–capacitor) circuit. In isolation, the circuit will possess a characteristic resonant frequency ω_0. However, when two

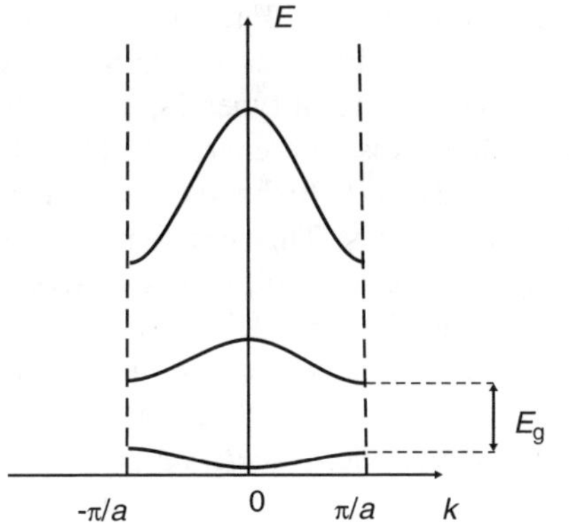

Figure 3.15 Reduced zone representation of the E versus k curves. E_g = band gap.

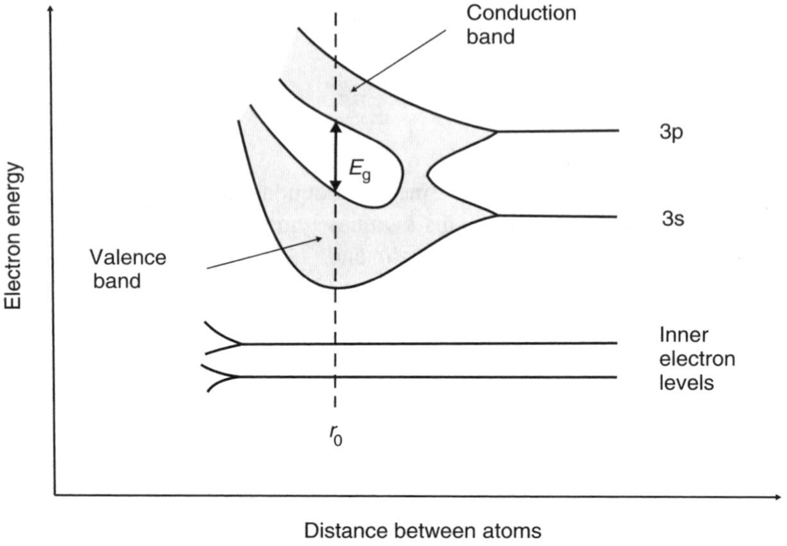

Figure 3.16 Energy levels in silicon as a function of interatomic spacing. The inner electron levels are completely filled with electrons. At the actual atomic spacing of the crystal r_0 the electrons in the 3s subshell and the electrons in the 3p subshell undergo sp^3 hybridization and all are accommodated in the lower valence band, while the upper conduction band is empty. The energy separation is the band gap E_g.

identical circuits are brought together, the circuits become coupled via mutual inductance. The effect is the formation of two resonant frequencies ω_1 and ω_2, below and above ω_0.

When many isolated atoms are brought together to form a complete crystal, multiple splitting of the electron energy levels associated with the isolated atoms occurs. Figure 3.16 illustrates the process of formation of a crystal of silicon, which shows the electron energy levels as a function of interatomic separation. The equilibrium separation of silicon atoms in the crystal is shown as r_0. As the isolated atoms become closer together, their electron orbitals overlap and discrete energy levels associated with the isolated atoms split to form bands of energies. Isolated silicon atoms possess the electronic structure $1s^2 2s^2 2p^6 3s^2 3p^2$ in their ground state (Chapter 2, Section 2.2.3). Each atom has available two 1s states, two 2s states, six 2p states, two 3s states, six 3p states and higher states. If we consider N atoms, there will be $2N$, $2N$, $6N$, $2N$ and $6N$ states of types 1s, 2s, 2p, 3s and 3p, respectively. As the interatomic spacing decreases, these discrete energy levels split into bands beginning with the outer $(n = 3)$ shell. As the 3s and 3p bands grow, they merge into a single band composed of a mixture of energy levels. This new band of 3s and 3p levels contains $8N$ available electron states. The chemical bonding in a silicon crystal consists of four sp^3 hybridized covalent bonds associated with each atom (Chapter 2, Section 2.4.1) As the distance between the atoms approaches the interatomic spacing of silicon, the band splits into two bands separated by an energy gap E_g. The upper band, called the *conduction band*, contains $4N$ states, as does the lower energy *valence* band. In general, the total number of levels in a band must be the total number of levels in an individual atom multiplied by the total number of atoms in the solid. Thus an s band will have $2N$ levels, a p band $6N$ levels, a d band $10N$ levels and so on (Chapter 2, Table 2.2).

The $4N$ electrons in the original isolated silicon atoms ($2N$ in 3s states and $2N$ in 3p states) must occupy states in the valence or conduction bands. At 0 K, the electrons will occupy the lowest energy states available to them. For the silicon crystal, there are exactly $4N$ states in the valence band. Hence at 0 K, every state in the valence band will be filled, whereas the conduction band will be completely empty.

An estimate of the average energy level separation within a band can be obtained from the fact that there are approximately 10^{26} atoms in a macroscopic (kg) quantity of a crystal. Energy bands are typically 1 eV wide, giving an energy separation of 10^{-26} eV. To a first approximation, therefore, the variation in electron energy across a band can be treated as being continuous.

3.3.3 Conductors, Semiconductors and Insulators

Although the motion of an electron through a crystal lattice as described by the Kronig–Penney model is an oversimplification, it does explain qualitatively the electrical behaviour of most materials. Every solid has its own characteristic energy band structure. After the allowed and forbidden energy regions have been determined, the occupation of the available energy levels is decided by the Fermi–Dirac distribution function [Equation (3.19)]. At absolute zero, all the states lying below the Fermi level are filled and all those above are empty.

Either the energy bands overlap or they do not and the Fermi level may lie within one of the bands or in the forbidden energy region. This leads to essentially three cases, which are illustrated in Figure 3.17.

For conductivity to be possible, there must be available, in the same band, energy states that are occupied and others that are empty. This is clearly the case in Figure 3.17(a) and (b);

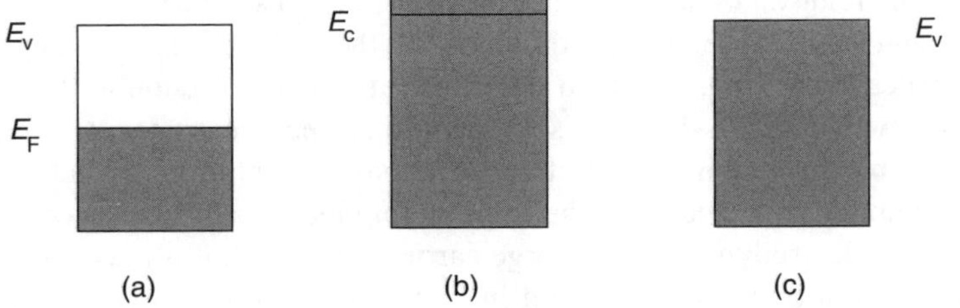

Figure 3.17 Possible energy band structures of crystalline solids. (a) Energy gap between the conduction and valence bands. The valence band is partly filled. (b) Overlapping valence and conduction bands. (c) Energy gap between the valence and conduction bands. The valence band is completely filled. Parts (a) and (b) represent the band structures of metals and (c) is the band structure for an insulator. E_v = top of valence band; E_c = bottom of conduction band; E_F = Fermi energy.

these represent the energy band structures for metals. Metals generally have bands that are only half-filled. For example, in a piece of sodium comprising N atoms, there will be $2N$ available states in the uppermost occupied energy band. As sodium has one valence electron (Group 1A in the periodic table; Chapter 2, Section 2.2.4) there are only N electrons to be accommodated in the band, which is therefore half full.

In Figure 3.17(c), the filled valence band (the highest energy of which is represented E_v in the diagram) is separated from the next highest band. Conductivity is not possible and the situation corresponds to that of an insulator. If a valence band electron were able to move as a consequence of an applied field, it would take energy from the field and change its energy state, i.e. move to a higher energy in the valence band. This is not possible since no higher energy levels are available for the electron to move into without violating the Pauli Exclusion Principle.

It is possible for the forbidden energy gap in a crystal to be fairly narrow. At a sufficiently high temperature, some of the valence electrons can gain enough thermal energy to transfer to states lying in the lowermost empty band, the conduction band. When an external electric field is applied, these electrons can gain additional energy from the field, and conduction is possible. The currents produced in such crystals are necessarily small, so they are distinguished from metallic conductors and true insulators by being called semiconductors. The energy gaps in semiconductors are mostly in the range 0.1–3 eV at room temperature (for Si, $E_g = 1.1\,\text{eV}$; for GaAs, $E_g = 1.4\,\text{eV}$).

Note that the conductivity of a semiconductor will increase as the temperature increases since more valence electrons can transfer to empty states lying in the conduction band. This is exactly opposite to the temperature dependence of conductivity in metals, in which the resistivity increases with increasing temperature. The conductivity of a semiconductor will also be affected by radiation, i.e. it will be photoconductive. Incident radiation with energy equal to or greater than the band gap will be absorbed, promoting electrons from the valence band to the conduction band and thereby increasing the conductivity.

3.3.4 Electrons and Holes

The thermal or optical excitation of an electron from the valence band to the conduction band, as depicted in Figure 3.18, leaves behind a valence band state with a missing electron. This unoccupied electron state has an apparent positive charge, because the crystal region was neutral prior to the removal of an electron. The valence band state with the missing electron is called a *hole* and can 'move' in the direction of the field by exchanging places with a neighbouring valence electron. Hence, it contributes to the conduction in the semiconductor.

From the previous discussion, it is evident that the electrons in a crystal are not completely free, but instead interact with the periodic potential of the lattice. As a result, their motion cannot be expected to be the same as for electrons in free space. In applying the usual equations of electrodynamics to charge carriers in a solid, it is reasonable to expect the electron mass will not be the same as that in free space. The relationship between electron energy, E, and wavevector, k, for a free electron has already been given by Equation (3.35) and may be modified by replacing the free electron mass m with an effective mass m^*, i.e.

$$E = \frac{\hbar^2 k^2}{2m^*} \tag{3.47}$$

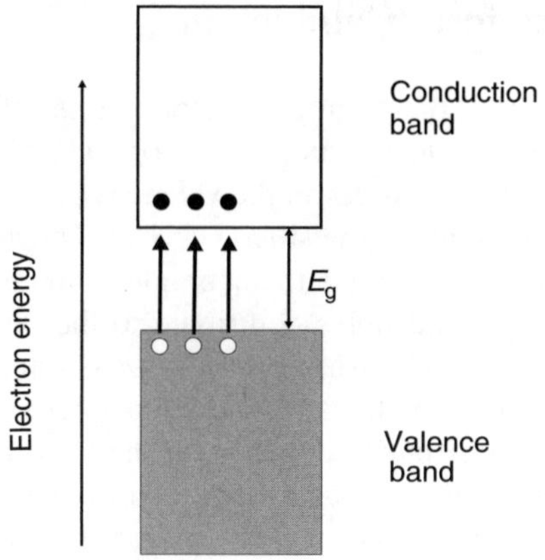

Figure 3.18 Model for intrinsic conduction. Thermal excitation of electrons across the band gap E_g produces an increase in electron population of the conduction band and an increase in the hole population of the valence band.

The effective mass will be related to the second derivative of the *E*–*k* curve, since

$$\frac{d^2E}{dk^2} = \frac{\hbar^2}{m^*} \tag{3.48}$$

or

$$m^* = \frac{\hbar^2}{d^2E/dk^2} \tag{3.49}$$

Therefore, the electron's mass is determined by the curvature of the energy bands. This idea can now be applied to the energy bands shown in Figures 3.14 and 3.15. Near the bottom of a band d^2E/dk^2 is a positive quantity, and so is the effective mass. However, near the top of a band, the curvature is such that the effective mass of the electrons is now negative.

A negative effective mass implies that if electrons near the top of the valence band have their energies increased by the application of an electric field, then this results in a reversal of their momenta (i.e. the momentum transfer from the crystal lattice to the electron is opposite and larger than the momentum transfer between the applied field and the electron).

For nearly filled bands, such as the valence band, it is much easier to deal with the behaviour of vacant sites rather than that of electrons. To achieve this, the electron effective mass, which is negative near the top of a band, is replaced by a hole effective mass, which is positive. If there is an electron missing from a given state then the state is deemed to be occupied by a positive charge. Holes lower in the valence band possess a greater energy and it follows that representation of hole energy in a band must be opposite to that of electron energy. Valence band electrons with negative charge and negative mass move in an electric field move in the same direction as holes with a positive charge and positive mass. We can therefore account fully for charge transport in the valence band by considering hole motion.

3.3.5 Intrinsic and Extrinsic Conduction

As noted above, the conductivity of a semiconductor can be affected by thermal or optical processes. Heating a pure semiconductor such as Si from 0 K will produce equal numbers of electrons in the conduction band and holes in the valence band. This process is equivalent to breaking some of the covalent bonds in the silicon crystal. The vacant states or holes that are produced in the valence band correspond to incomplete bonds. On the application of an electric field, both the electrons and holes contribute to the resulting electric current. The conductivity is referred to as *intrinsic*. An intrinsic semiconductor is therefore a material in which the electronic properties (such as the DC conductivity) are not determined by impurities. The necessary level of purity depends on the band gap and on the temperature. The semiconductor germanium, which has a band gap of 0.67 eV at room temperature, is intrinsic provided that the impurities are present at less than about 1 part in 10^{10}. On the other hand, at the same temperature, intrinsic silicon with the larger band gap of 1.1 eV would require a purity level of better than 1 part in 10^{13}.

At room temperature, the value of the thermal energy k_BT is approximately 0.025 eV and the electrons in the crystal receive energies distributed about this mean value. Since the band gaps of semiconductors lie in the range 0.1–3 eV, the proportion of electrons excited into the conduction band will be very small. This will be particularly so for semiconductors having the larger energy gaps.

The intrinsic carrier concentration n_i will depend on both the band gap and temperature. It is given by

$$n_i = AT^{\frac{3}{2}} \exp\left(\frac{-E_g}{2k_BT}\right) \quad (3.50)$$

where A is a constant. This variation of n_i with temperature will be dominated by the exponential term in the above equation. The carrier concentration can be substituted into Equation (3.15) to provide an expression for the temperature dependence of the conductivity. As both electrons and holes are present in the semiconductor, the more general form of this relationship is

$$\sigma = ne\mu_e + pe\mu_h \quad (3.51)$$

where n, p and μ_e, μ_h are the carrier concentrations and mobilities of the electrons and holes, respectively. Of course, for intrinsic material $n = p = n_i$ and so the intrinsic conductivity σ_i can be written as

$$\sigma_i = n_i(\mu_e + \mu_h) \quad (3.52)$$

If the temperature dependences of the carrier mobilities are slowly varying functions of temperature, then the conductivity will exhibit an almost exponential increase with temperature.

n-Type Doping

Whereas the conductivity of an intrinsic semiconductor is a function only of temperature, that of an *extrinsic* semiconductor is determined by the impurity content. The process of

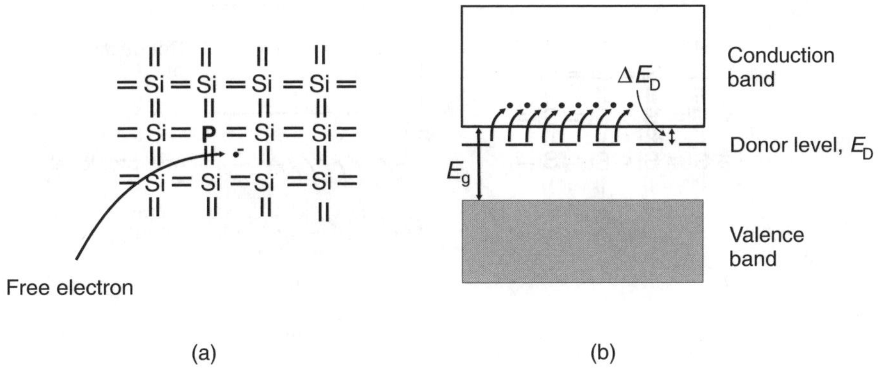

Figure 3.19 Donor impurity, such as phosphorus, in silicon. (a) A phosphorus atom substitutes for a silicon atom in the lattice, leaving a free electron. (b) Energy band diagram showing the formation of a donor level with activation energy ΔE_D associated with the phosphorus impurity, close to the conduction band of the silicon.

adding small amounts of impurity to control the electrical conductivity of semiconductors is called *doping*. In an extrinsic semiconductor, a specific impurity element is added in controlled amounts; its concentration determines the conductivity of the sample and the type of carriers present. When these are electrons, i.e. negative carriers, the sample is termed *n-type*; when these are holes, i.e. positive carriers, the sample is *p-type*.

A semiconductor such as silicon has tetrahedral covalent bonding in which each atom shares its four valence electrons with each of its four nearest neighbours. If a small amount of a Group 5 element, such as phosphorus, is incorporated into a silicon crystal, the phosphorus impurity atom is substituted for a silicon atom on the lattice so that it is tetrahedrally bound to its nearest neighbours. In the free atom state, phosphorus has two 3s electrons and three 3p electrons so that in the substitutional alloy the atom has one surplus electron over and above that necessary to form the covalent bonds. At very low temperatures, close to 0 K, this electron is loosely bound to its parent atom due to the Coulombic attraction of the nucleus. However, if the temperature of the semiconductor is increased, this electron is released from the attraction of the parent impurity atom by thermal energy and is then free to move throughout the crystal. Each impurity atom gives rise to one electron in the conduction band and such atoms are called *donors*. With a suitable number of donor atoms in a sample, the electron concentration may be many orders of magnitude greater than that of an intrinsic sample.

The band representation of an n-type semiconductor is shown in Figure 3.19. At room temperature, each donor atom is ionized and provides an electron in the conduction band. In addition to phosphorus, the other Group 5 elements, arsenic, antimony and bismuth, also behave as donor impurities in silicon. The ionization energies, ΔE_D, of these donors are all around 0.05 eV.

p-Type Doping

The isolated atoms of the Group 3 elements – boron, aluminium, gallium and indium – have outermost electronic shells of two electrons in s states and one electron in a p

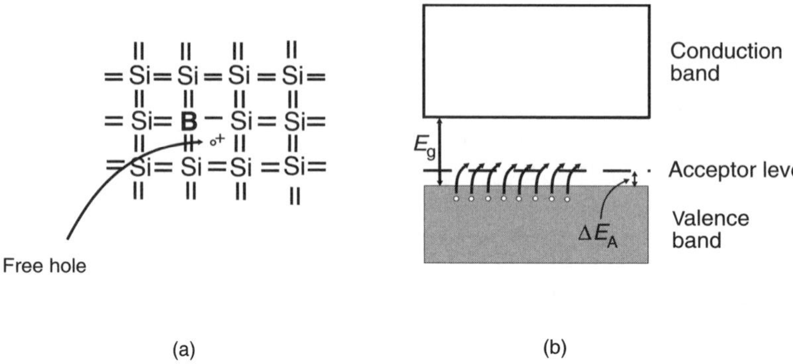

Figure 3.20 Acceptor impurity, such as boron, in silicon. (a) A boron atom substitutes for a silicon atom in the lattice, leaving a free hole. (b) Energy band diagram showing the formation of an acceptor level with an activation energy ΔE_A associated with the boron impurity, close to the valence band of the silicon.

state. Consequently, when one of these elements is added to silicon there is a deficiency of one electron for tetrahedral bonding of the impurity atom into the crystal lattice. At very low temperatures, this missing bond is localized in the Coulombic field of the impurity atom. At ordinary temperatures, however, the impurity atom, which is called an *acceptor*, receives an electron from another atom in the crystal and the missing bonding electron, or hole, becomes mobile. This band representation is shown in Figure 3.20. The acceptor atom becomes ionized, with the simultaneous creation of a mobile hole in the valence band. The ionization energies of the acceptor atoms ΔE_A are 0.05–0.16 eV in silicon.

The temperature dependence of the carrier concretion of an extrinsic semiconductor n_e depends on the relationship between the magnitude of the ionization energies for donors (or acceptors) and the absolute temperature. For example, for an n-type material at sufficiently high temperature, all the donor atoms will be ionized and

$$n_e = N_D \tag{3.53}$$

where N_D is the concentration of donor atoms. At lower temperatures, n_e is given by

$$n_e = BT^{\frac{3}{4}} \exp\left(\frac{-\Delta E_D}{2k_B T}\right) \tag{3.54}$$

where B is a constant. In the case of p-type material, a similar equation holds, but with the ionization for donor atoms ΔE_D replaced by ΔE_A, the acceptor ionisation energy. Figure 3.21 reveals how the carrier concentration in an n-type semiconductor is expected to vary with temperature. In the figure, log(n) is plotted as a function of reciprocal temperature. At low temperatures (and above 0 K) there is sufficient thermal energy to ionize the donor atoms and the carrier concentration versus temperature relationship is that of Equation (3.54). This is the ionization region. At intermediate temperatures, all the donor atoms are ionized and the carrier concentration is constant [Equation (3.53)]. A further temperature increase will take the semiconductor into the intrinsic conduction region [Equation (3.50)] where there is sufficient thermal energy available to promote electrons from the valence band to the conduction band.

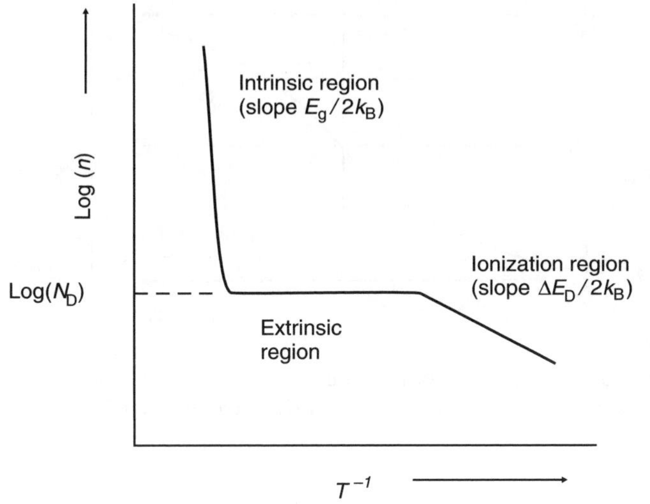

Figure 3.21 Variation of the carrier concentration, n, in an n-type semiconductor with temperature, T. Concentration of donor atoms $= N_D$.

A semiconductor may also contain simultaneously both donor- and acceptor-type impurity atoms. When the donor and acceptor levels are approximately equal, the material is referred to as being *compensated*. In other cases, the conductivity is governed by the net excess impurity of one type. For silicon, when an impurity element from a group other than Group 3 or 5 is present, it may give rise to one or more separate levels lying within the forbidden energy gap and many such levels may lie deep within the gap.

The above principles may be applied to the doping of other inorganic semiconductors, including compound semiconductors such as GaAs. In such compounds, a departure from a strict *stoichiometric* composition (i.e. Ga : As = 1 : 1) leads to either an n-type or p-type sample depending on which element is present in the greater concentration. The III–V semiconductor gallium arsenide can also be doped n-type by adding tellurium from Group 6 as a donor or n-type by adding zinc from Group 2 as an acceptor.

Traps and Recombination Centres

Many impurities give rise to one or more localized energy levels within band gap of the semiconductor. These levels can exchange charge with the conduction and valence bands. For example, such states can temporarily remove electrons or holes from the conduction or valence bands, respectively, acting as carrier *traps*. Alternatively, the localized level can attract first an electron (hole) and subsequently a hole (electron). This is the process of *recombination* and the localized level is referred to as a *recombination centre*. Trapping involves the temporary removal of a carrier (which is subsequently excited back into the valence or conduction band) whereas, in the case of recombination, the carriers are permanently removed from the bands. Traps can be viewed as defects in the regular crystalline structure of the semiconductor; these can be the main limiting factor on the performance of many semiconductor devices.

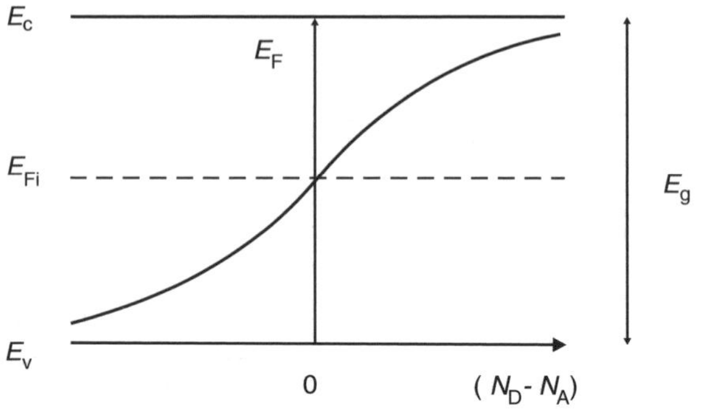

Figure 3.22 Variation in the position of the Fermi level E_F with net donor or acceptor concentration. For $N_D = N_A$ the position of the Fermi level lies close to the middle of the band gap, at energy E_{Fi}.

Fermi Level Position

There is an interdependence of the position of the Fermi level and the carrier concentration for both intrinsic and extrinsic semiconductors. At 0 K, the Fermi level represents the energy below which all electron states are occupied and above which all electron states are filled (Section 3.2.4). In a metal, E_F coincides with the top occupied level in a partially filled band. When the temperature rises, some electrons will move from energy band levels below E_F to levels above E_F, thereby conforming to the distribution shown in Figure 3.3. The Fermi energy no longer separates filled and unfilled states, but is still a useful reference level. For intrinsic semiconductors (and insulators) at 0 K, the Fermi level can still be positioned somewhere between the valence band (all states filled) and the conduction band (all states empty). It can be shown that the Fermi level is located very nearly midway between the valence and conduction bands. As the temperature rises, E_F remains positioned approximately at the mid-point in the energy gap, at energy E_{Fi}.

In the case of extrinsic material, the Fermi level will move towards the conduction band for n-type material and towards the valence band for p-type, as indicated in Figure 3.22. For samples containing both donor and acceptor impurities, the position of the Fermi level will be determined by the net donor or acceptor concentration $(N_D - N_A)$. As an extrinsic material is heated and it moves into the region of intrinsic conductivity, E_F will again move towards the middle of the band gap, to E_{Fi}.

3.3.6 Quantum Wells

A single crystal of an inorganic semiconductor contains a continuum of electron energy levels in the valence and conduction bands and discrete energy levels in the band gap arising from doping. Discrete energy levels for electrons and holes may also be obtained by quantum-mechanical confinement. This process is particularly important in nanotechnology. For example, Figure 3.23 shows the spatial variation in the conduction and valence bands for a multilayer structure in which a thin (~5 nm) layer of GaAs is sandwiched between two layers of AlGaAs, which has a wider band gap than the GaAs. The consequence of confining

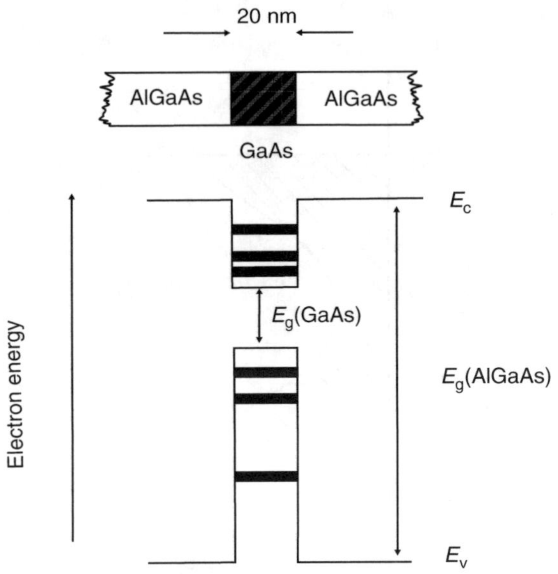

Figure 3.23 Example of a quantum well. Discontinuities in the conduction and valence band edges in the case of a thin layer of GaAs sandwiched between layers of wider band gap AlGaAs. Quantum states are formed in the valence and conduction bands of the GaAs.

electrons and holes in such a thin layer is that these particles behave according to the electron in a box model (Section 3.3.1). Instead of having the continuum of states normally available in the conduction band, the conduction band electrons in the narrow-gap material are confined to discrete quantum states as described by Equation (3.36). Similarly, the states in the valence band available for holes are restricted to discrete levels in the quantum well.

3.3.7 Disordered Semiconductors

The energy band model for semiconductors is a consequence of the ordered arrangement of the atoms on a lattice, i.e. a crystalline structure. Interesting questions then arise: what happens if the regular periodicity of the lattice is disrupted? Will the material still possess a recognizable band structure? As noted in Chapter 2, Section 2.5.8, the atoms in an amorphous solid [Figure 2.41(b)] are often arranged in a continuous random network. Amorphous pure silicon contains numerous dangling bonds similar to those found at the surface of single crystal silicon (Chapter 2, Section 2.5.7). However, provided that the short-range order present in the crystalline phase is essentially unchanged (i.e. similar bond lengths, bond angles and local coordination), the main features of the density of states function is preserved. The overall result is that energy bands in amorphous materials are generally less defined than in their crystalline counterparts. *Tails* in the density of states function can extend into the band gap, as depicted in Figure 3.24, and localized states can be found within the band gap [3]. For certain materials, the tails may extend so far into the energy gap that they partially overlap. The concept of a *mobility edges* within the band tails is introduced for amorphous materials. These are associated with the critical energies

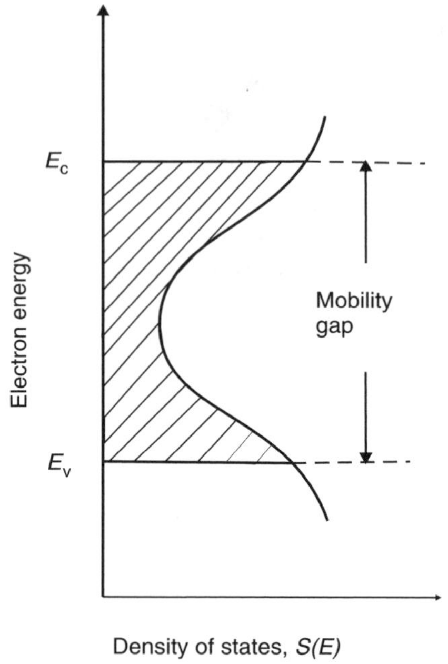

Figure 3.24 Electronic energy versus density of states for an amorphous semiconductor showing overlapping conduction and valence band states. The shaded area represents localized states. A mobility gap separates the extended conduction and valence band states from the localized states.

separating localized from extended states. The difference between the energies of the mobility edges is called the *mobility gap*, and this parameter is often used instead of the energy gap.

3.3.8 Conductivity in Low-dimensional Solids

As the size associated with materials and devices moves into the nanoscale, dimensional effects become increasingly important. There are essentially two influences that need to be considered, determined by external (e.g. geometric shaping) and internal (increase in anisotropy – Chapter 2, Section 2.5.1) factors. An example of the first would be to take a metallic wire and to draw it out until it becomes sufficiently thin to be considered one-dimensional. How thin does the wire have to be? Important parameters are the mean free path of an electron (Section 3.2.1) and the *Fermi wavelength.* The latter is the wavelength of the electrons at the Fermi energy. If the electrons in a thin metal wire can be considered by the electron in a box model (Section 3.3.1), and if the size of the box is just the Fermi wavelength, only the first eigenstate is occupied. If the energy difference to the next level is much larger than the thermal energy ($\gg k_B T$), there are only completely occupied and completely empty levels and the system is an insulator.

A thin wire represents a small box for electronic motion perpendicular to the axis, but it is a very large box for motion along the wire. Hence in two dimensions (radially) it is an insulator, and in one-dimension (axially) it is a metal. If there are only a few electrons, the Fermi energy is small and the Fermi wavelength is large. This is the case for semiconductors at low doping concentrations. Wires of such semiconductors are already one-dimensional if their diameter is around 10 nm.

If the dimensions of a conductor are large, then its resistance will scale with its length and cross-sectional area according to Equation (3.11). For very small dimensions, this should lead to a zero resistance (infinite conductance). However, experiment reveals that the resistance approaches a limiting value when the length of the conductor becomes much shorter than the mean free path. The 'excess' resistance can be interpreted in terms of localization of the electron states. The effect occurs for resistance values that are greater than about $\hbar/e^2$. This combination of fundamental constants is the *von Klitzing constant* and has a value of 4.11 kΩ.

One-dimensional (and even zero-dimensional) systems may also be achieved by selecting the anisotropy of the material. For example, certain materials (conductive polymer chains and carbon nanotubes) may be regarded as natural (quasi) one-dimensional. Examples of two-dimensional solids include graphite and the phthalocyanine compounds, which are described in Chapter 5. Semiconductor *quantum dots* may be considered as zero-dimensional objects. These structures are small discs of material which are small compared with the Fermi wavelength, so that the electrons are restricted in all three dimensions. The quantum dots of the organic world are the fullerenes, regular clusters of carbon atoms (Chapter 5). A C_{60} cluster (60 carbon atoms) has a diameter of about 1 nm.

Further peculiarities of one-dimensional systems are band-edge singularities in the density of states function. The density of states function $S(E)$ [Equation (3.21)] depends on the crystal structure and, near the band edge, it reflects the dimensionality of the system. The parabolic nature of $S(E)$ in the case of a three-dimensional crystal has been shown in Figure 3.4. Figure 3.25 contrasts the $S(E)$ function for one-, two- and three-dimensional systems. For one dimension, a singularity to infinity is predicted. If the Fermi level is within such a region, high-temperature superconductivity is favoured (Section 3.4.4).

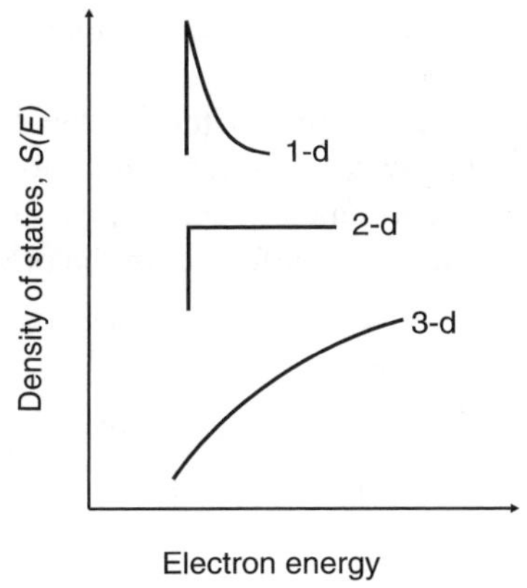

Figure 3.25 Density of states functions at the band edge in one-, two- and three-dimensional electronic systems.

3.4 ORGANIC COMPOUNDS

3.4.1 Band Structure

A periodic variation in the potential experienced by an electron moving in a solid results in discontinuities in the energy versus momentum relationship (*E–k* curve), and to a series of energy bands separated by energy gaps. Although these ideas have been developed in order to explain the electrical behaviour of inorganic materials, such as silicon or gallium arsenide, there is no reason why the physical basis is not equally applicable to other substances in the solid state. Organic solids possess varying degrees of crystallinity. Many, such as molecular crystals and charge-transfer complexes, can be grown in the form of single crystals while an elongated polymer chain is perhaps the best natural manifestation of the Kronig–Penney model. The following sections will examine the electronic structure of a number of important groups of organic solids.

Molecular Crystals

Molecular crystals can possess a high degree of both short- and long-range order. However, they differ from crystals of metals and inorganic materials because they are made up of discrete molecules. The compound pentacene (Figure 3.26), comprising a linear arrangement of five interconnected benzene rings, is a good example of a molecular crystal. The molecule itself may be regarded as a miniature lattice, with a precisely spaced series of atoms. The atoms are in close proximity, giving rise to a good overlap of their atomic orbitals. The intramolecular interactions between the atoms lead to a splitting of the carbon $2p_z$ orbitals (Chapter 2, Section 2.4.3, Figure 2.29) and to a localization of the π electrons over the molecule. As a result, bonding orbitals take up these electrons, whereas the antibonding orbitals remain empty. On a downward positive electron energy scale there is a *highest occupied molecular orbital*, known by the acronym *HOMO* and a *lowest unoccupied molecular orbital*, *LUMO*, with an energy separation of about 2 eV. In the simplest approximation of an energy-level description, internal electron–electron interactions are neglected, with the consequence that the LUMO describes both the lowest energetic position of an electron excited out of one of the occupied lower levels of a neutral molecule and also the lowest possible energetic position of an additional electron brought in from outside the molecule.

The relatively strong intramolecular forces within molecules such as pentacene must be contrasted to the weaker van der Waals intermolecular forces that hold the molecules together in the solid crystalline state. As a consequence, the qualitative description provided by the molecular orbital model is largely unaffected. In the organic crystalline solid, there is

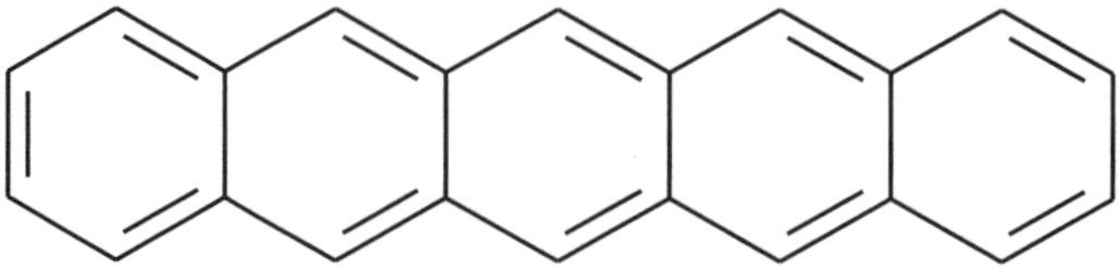

Figure 3.26 Chemical structure of pentacene.

a moderate splitting of the molecular energy levels by these intermolecular interactions into narrow ‘bands’. However, there is an important shift in these levels as neutral molecules become embedded in the solid-state environment. The resulting energy bands are very different from those found in inorganic semiconductors such as silicon. Their small width and high associated effective mass lead to very small mobilities for the charge carriers. The electronic energy levels in an isolated pentacene molecule (gas phase) and in the solid crystalline phase are contrasted in Figure 3.27 [4]. The energy required to remove an electron from the HOMO level in the crystal is I, the ionization energy (Chapter 2, Section 2.3.2), and the empty LUMO level can be similarly characterized by a binding energy or *electron affinity*, X, which, like I, is measured with respect to the vacuum level. Clearly, the HOMO–LUMO separation or energy gap ΔE is given by

$$\Delta E = I - X \tag{3.55}$$

The value of ΔE varies with the molecular weight of the organic compound. For example, for the acene series from benzene to pentacene, this energy separation decreases with increase in the number of benzene rings, and hence the size of the π electron system.

It is important to note that the Fermi level of an organic semiconductor is often not located halfway between the HOMO and LUMO levels (NB, as noted in Section 3.3.5, the position of E_F in an intrinsic semiconductor is about the mid-point of the band gap). This is the result of inadvertent doping.

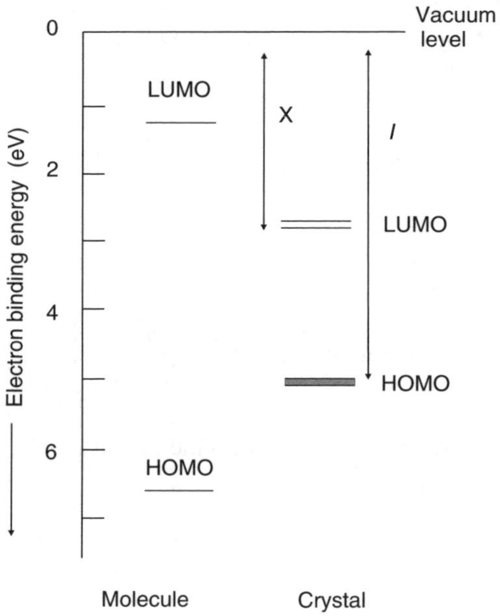

Figure 3.27 Electronic energy levels associated with pentacene. The levels on the left are for free molecules in the gaseous state and those on the right are for the crystal. I = ionization energy; χ = electron affinity [4].

Polymers

A linear arrangement of atoms, as might be found in a polymer chain, provides an excellent experimental basis for the application of the Kronig–Penney model. An important feature of the band model is that the electrons are delocalized or spread over the lattice. The strength of the interaction between the overlapping orbitals determines the extent of delocalization that is possible for a given system. For many polymeric organic materials, the molecular orbitals responsible for bonding the carbon atoms of the chain together are the sp^3 hybridized σ bonds, which do not give rise to extensive overlapping. The resulting band gap is large, as the electrons involved in the bonding are strongly localized on the carbon atoms and cannot contribute to the conduction process. This is why a simple saturated polymer such as polyethylene, $-(CH_2)_n-$, is an electrical insulator.

A significant increase in the degree of electron delocalization may be found in unsaturated polymers, i.e. those containing double and triple carbon–carbon bonds. If each carbon atom along the chain has only one other atom, e.g. hydrogen, attached to it, the spare electron in a p_z orbital of the carbon atom overlaps with those of carbon atoms on either side forming delocalised molecular orbitals of π symmetry.

Suppose that such as a polymer is composed of a linear chain of atoms, with N atoms, each separated by a distance d, the total length of the chain is $(N-1)d$, which for a large number of atoms approximates to Nd. The eigenenergies, given by Equation (3.36), are then

$$E_n = \frac{\hbar^2 \pi^2 n^2}{2m(Nd)^2} \tag{3.56}$$

Assuming that the π electrons from the N p orbitals are available, with two electrons per molecular orbital (according to Pauli), the HOMO will be that given by $n = N/2$, and the corresponding energy will be

$$E_{\mathrm{HOMO}} = \left(\frac{N}{2}\right)^2 \left[\frac{\hbar^2 \pi^2}{2m(Nd)^2}\right] \tag{3.57}$$

The LUMO has the energy

$$E_{\mathrm{LUMO}} = \left(\frac{N}{2} + 1\right)^2 \left[\frac{\hbar^2 \pi^2}{2m(Nd)^2}\right] \tag{3.58}$$

The energy required to excite an electron from the HOMO to the LUMO level is the band gap of the polymer, E_g, i.e.

$$E_g = E_{\mathrm{LUMO}} - E_{\mathrm{HOMO}} = (N+1)\left[\frac{\hbar^2 \pi^2}{2m(Nd)^2}\right] \approx \left(\frac{\hbar^2 \pi^2}{2md^2}\right)\left(\frac{1}{N}\right) \tag{3.59}$$

The band gap is therefore predicted to decrease with increasing length of the polymer chain, and will practically vanish for macroscopic dimensions. For example, if $d = 0.3$ nm and $N = 100$, then $E_g = 42$ meV.

From the above, it might be expected that a linear polymer backbone consisting of many strongly interacting coplanar p_z orbitals, each of which contributes one electron to the resultant continuous π-electron system, would behave as a one-dimensional metal with a half-filled conduction band. In chemical terms, this is a *conjugated chain* and may be represented by a system of alternating single and double bonds. It turns out that, for one-dimensional systems, such a chain can more efficiently lower its energy by introducing bond alternation (alternating short and long bonds). This limits the extent of electronic delocalization that can take place along the backbone. The effect is to open an energy gap in the electronic structure of the polymer. All conjugated polymers are large band gap semiconductors, with band gaps more than about 1.5 eV, rather than metals. The effect, known as *Peierls distortion*, is described in the following section.

Peierls Distortion

In 1955, Peierls showed that a monatomic metallic chain is unstable and will undergo a metal-to-insulator transition at low temperature. In Figure 3.28(a), such a chain, for example sodium atoms, is shown. (This is simply a thought experiment, because sodium atoms will not arrange in chains – they tend to form clusters.) An S-shaped energy versus wavevector relation, parabolic on both sides, will result. The density of states has square root

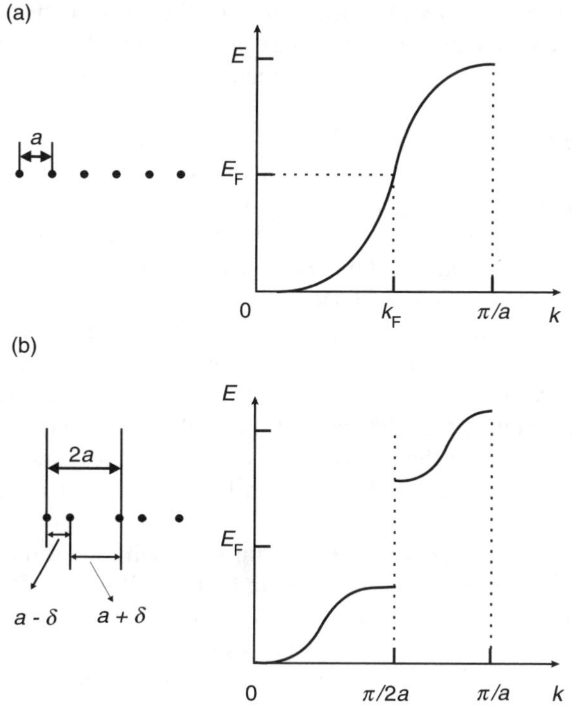

Figure 3.28 Peierls distortion in an isolated chain of equidistant monatomic sodium atoms. Energy versus wavevector diagrams, (a) without distortion and (b) with distortion. The distorted lattice has a periodicity of half of that of the undistorted lattice.

singularities at the top and at the bottom of the band (which are of no significance in this particular context). The band is half-filled because each sodium atom contributes one delocalized electron to the solid and two electrons – spin-up and spin-down – can be accommodated in each state. The Fermi energy E_F is at the band centre [the corresponding Fermi wavevector k_F is shown in Figure 3.28(a)].

Now suppose that the arrangement of the atoms is changed and every second atom is displaced by a small amount, δ, so that the atoms are no longer equidistant; short spacings, $a - \delta$, and long spacings, $a + \delta$, alternate [Figure 3.28(b)]. Again, the arrangement is periodic, but now with a repeat distance $2a$ instead of a. In the *E–k* diagram in Figure 3.28(a), the boundary of the first Brillouin zone occurs at $k = \pi/a$; in the case of Figure 3.28(b), it is at half that distance, just at the Fermi wavevector k_F. The system has thereby been transformed from a metal with no gap at the Fermi level into a semiconductor with a gap at $\pi/2a$. All states below the gap are filled at absolute zero and all above are empty.

The important question is, what is the most stable (lowest energy) state for the one-dimensional array of atoms? It turns out that the answer is the distorted case. This is because the states in the gap have been accommodated above and below the gap. States cannot disappear and there are as many states as atoms in the lattice (because the states are formed from atomic orbitals). Consequently, when summing the electronic energies from zero to E_F, it is evident that the creation of the gap had reduced the electronic energy. However, to achieve this, work has to be done (against the interatomic forces) to displace the atoms. A full analysis reveals that the electronic energy is approximately linear in δ whereas the elastic energy depends quadratically on δ. For sufficiently small displacements the gain in electronic energy predominates over the elastic term. Consequently, under Peierls' assumptions there is always a gap at absolute zero. Electron–lattice coupling will drive the one-dimensional metal into an insulator.

We shall now return to our discussion of organic polymers. The link between the preceding discussion for a linear array of sodium atoms and a chain of carbon atoms is perhaps unclear. The polymer polyethylene consists of long chains of carbon atoms with two hydrogen atoms per carbon atom. As noted above, the sp^3-hybridized σ-bonds do not give rise to extensive overlapping and the band gap is large. Removing one hydrogen atom from each carbon (i.e. to provide an unsaturated polymer chain) would leave unbonded electrons everywhere. The result is a chain of $CH^{\bullet}$ *radicals*. The superscript dot denotes a chemical group carrying an odd number of electrons. It is these radicals that have some similarity with an alkali metal atom – both have an extra electron. Hence the electrons in polyacetylene are not completely delocalized along the chain. There is an alternation of short and long bonds between the carbon atoms (where short bonds are drawn as double bonds and long bonds are single bonds), which leads to a semiconductive rather than a metallic band structure.

In summary, a completely delocalized electron system in one dimension is expected to lead to the metallic state. However, as depicted in Figure 3.29(a), the Peierls transition leads to a bond alternation, a doubling of the unit cell and a semiconductive state.

Figure 3.29(b) shows the electronic energy band structure for *trans*-polyacetylene; the band structure for the *cis* isomer (Chapter 2, Section 2.4.3) is somewhat different. The valence and conduction bands of semiconductive polymers are often referred to as the π and π^* bands, respectively. In theory, the π-electron band structure extends over a band width *W*, given by

$$W = 2zt \tag{3.60}$$

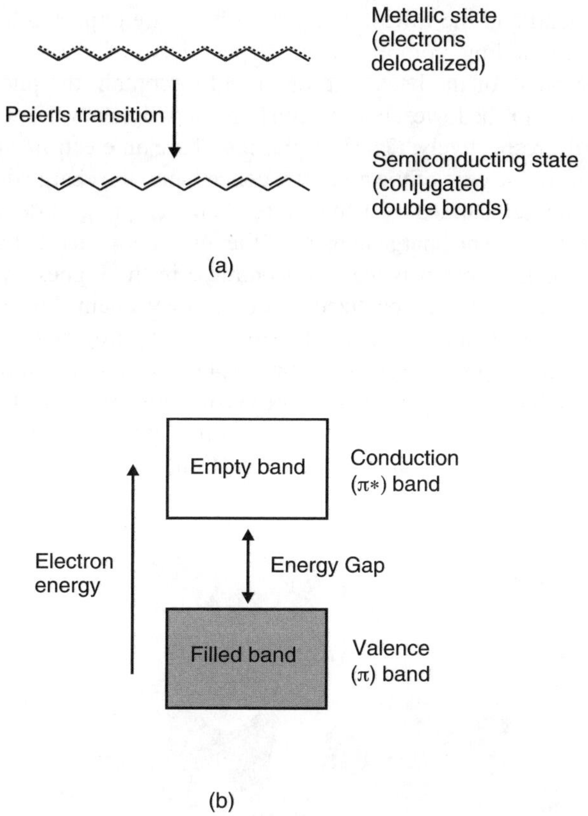

Figure 3.29 (a) The electrons associated with the CH$^{\bullet}$ radicals in *trans*-polyactetyene will delocalize over the chain. The Peierls distortion leads to single/double bond alternation. (b) Resulting energy bands.

where z is the number of nearest neighbours and t is known as the *transfer integral*. This is a measure of the π wavefunction overlap between neighbouring carbon sites along the backbone, i.e. the degree of delocalization. For linear polymers such as polyacetylene, $z = 2$ and $t \approx 2.5\,\text{eV}$, giving $W \approx 10\,\text{eV}$.

The distribution of electronic charges in an isolated chain of a conductive polymer can be illustrated with reference to some band structure calculations for poly(p-phenylenevinylene), PPV [5]. The repeat unit is shown in Figure 3.30(a) (Chapter 5, Section 5.3). There are eight carbon atoms in a PPV repeat unit but only one electron per atom takes part in the conjugated π-bonding. Since each band can hold exactly two electrons, the four π valence bands with the lowest energy are filled and the four π* conduction bands at higher energies are empty. Three of the four π bands belong to the benzene ring, and the corresponding wavefunctions are localized along the chain. The other π band derives mainly from the vinylene states, and the corresponding wavefunctions are localized on the vinyl C=C double bond.

The total electronic charge density for an isolated PPV chain is shown in Figure 3.30(b). The total charge density in the plane of the backbone is equally distributed in the benzene ring and the vinyl linkage, as expected. The pattern is consistent with the delocalization of electrons suggested in the bonding diagram [Figure 3.30(a)]. The wavefunctions of the uppermost valence band and the lowest conduction band are expected to be pure π and π^*, i.e. bonding and antibonding combinations, respectively, of carbon p orbitals directed perpendicular to the plane of the backbone (assumed to contain the phenyl ring).

The charge densities for the lowest conduction band and highest valence bands are shown in Figure 3.30(c) and (d), respectively. (These represent where an electron would be if it were in the first excited state.) In the case of the highest valence band, the charge density is spread over the molecule, therefore delocalized over the entire chain. The largest density is in the double bond of the vinylene group. The charge density of the lowest conduction band [Figure 3.30(d)] is highest on those bonds where it is least concentrated in the highest valence band [Figure 3.30(c)]. However, the charge is delocalized along the PPV chain. For both the valence and conduction bands, there is a smaller charge density on the hydrogen atoms.

The energy gap of *trans*-polyacetylene is about 1.4 eV, which is comparable to the values of 1.1 eV for single crystal silicon and 1.4 eV for GaAs. However, the electrical conductivity in inorganic crystals and organic polymers is very different, because the conductivity also depends on the mobility of the charge carriers [Equation 3.52)]. The band model of

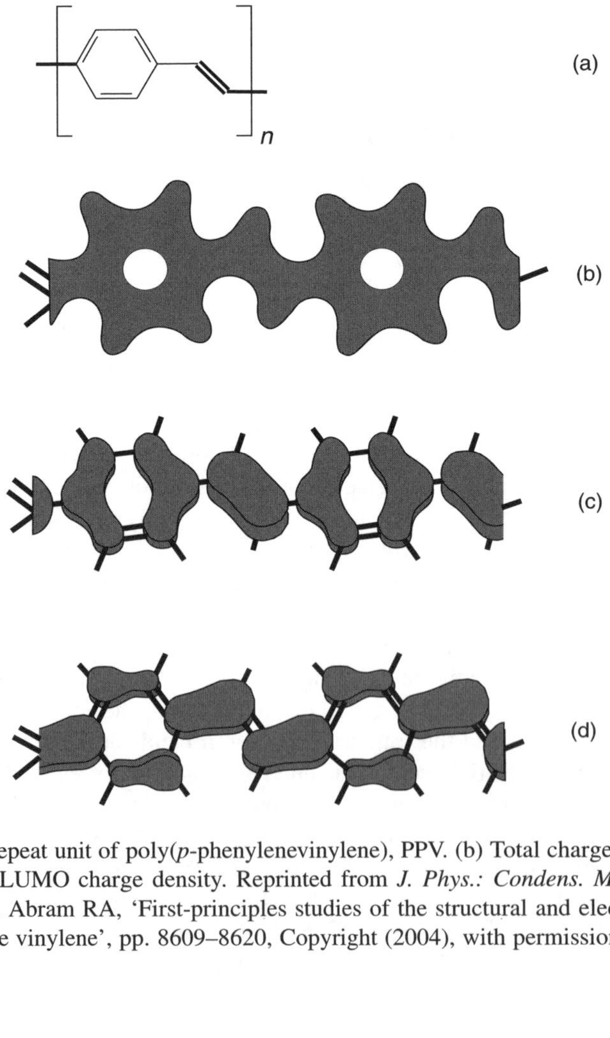

Figure 3.30 (a) Repeat unit of poly(*p*-phenylenevinylene), PPV. (b) Total charge density. (c) HOMO charge density. (d) LUMO charge density. Reprinted from *J. Phys.: Condens. Matter*, **16**, Zheng G, Clarke SJ, Brand S, Abram RA, 'First-principles studies of the structural and electronic properties of poly-*para*-phenylene vinylene', pp. 8609–8620, Copyright (2004), with permission from IOP Publishing Limited.

semiconductors predicts that the greater the degree of electron delocalization, the larger is the width of the bands (in energy terms) and the higher is the mobility of the carriers within the band. For inorganic semiconductors such as silicon or gallium arsenide, the three-dimensional crystallographic structure provides for extensive carrier delocalization throughout the solid, resulting in relatively high carrier mobilities.

Electrical conduction in polymers not only requires carrier transport along the polymer chains but also some kind or transfer, or 'hopping', between these chains, which tend to lie tangled up like a plate of spaghetti. The charge carrier mobilities in organic polymers are therefore fairly low, making it difficult to produce very high-speed electronic computational devices that are competitive with those based on silicon and gallium arsenide. However, some improvement in the carrier mobility can be achieved by both increasing the degree of order of the polymer chains and improving the purity of the material. Table 3.1 contrasts the room temperature carrier mobility values (in the commonly used units of $cm^2\ V^{-1}\ s^{-1}$) for Si, GaAs, pentacene, polyacetylene and a number of other organic compounds (which will be described further in Chapter 5) [6, 7]. Most conductive organic compounds are predominantly p-type conductors and the mobility values will refer to holes. Although the mobility values for the organic materials listed in Table 3.1 are fairly low, other features make them attractive for certain types of electronic device, as discussed in later chapters.

Charge-transfer Complexes

A further important class of electrically conductive organic materials are *charge-transfer* compounds. These materials are formed by the combination of two (or more) types of neutral molecules, one of which is an electron donor, D, i.e. has a low ionization energy and can be easily oxidized (electron removal), and the other is an electron acceptor, A, i.e. has a high electron affinity and can easily be reduced (electron addition). The transfer of an electron from the donor molecule to the acceptor molecule can be represented as follows:

$$D + A \rightarrow [D^{+\bullet}][A^{-\bullet}] \tag{3.61}$$

Table 3.1 Room temperature carrier mobilities for field effect transistors based on organic semiconductors. After Dimitrakopoulos and Mascaro [6] and Pearson *et al.* [7]. The electron mobilities in single crystal silicon and gallium arsenide are also given.

Material	Carrier mobility ($cm^2\ V^{-1}\ s^{-1}$)
Si single crystal (electrons)	1500
GaAs single crystal (electrons)	8500
Pentacene	10^{-3}–1
Polyacetylene	10^{-4}
Polythiophene	10^{-5}
Phthalocyanine	10^{-4}–10^{-2}
Thiophene oligomers	10^{-4}–10^{-1}
Organometallic dmit complex	0.2
C_{60}	0.3

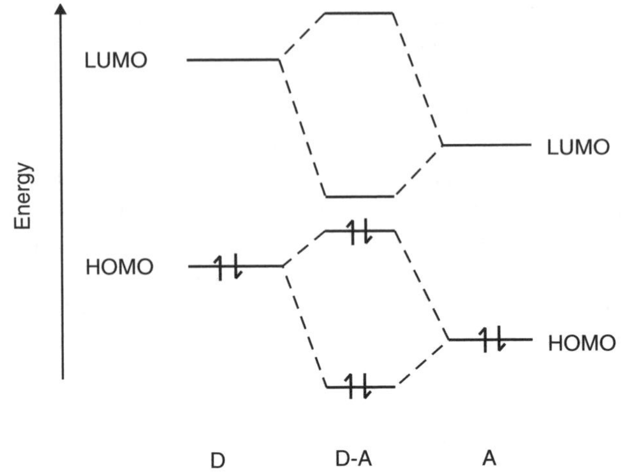

Figure 3.31 Molecular orbitals formed in a charge-transfer complex. The HOMO and LUMO levels in the isolated donor D and acceptor A molecules interact to produce new energy levels in the D–A complex.

The transfer process leaves behind an organic cation $[D^{+\bullet}]$ that has a 'free' electron, i.e. an electron that is not strongly involved in the chemical bonding. At the same time, the acceptor molecule gains an electron to become the anion radical $[A^{-\bullet}]$. Under certain electronic and structural criteria these electrons can become conduction electrons, just like those of traditional metals. In terms of the energy band diagram, Figure 3.31 shows the energy level arrangement in the separated donor and acceptor molecules compared to the D–A complex. The LUMO and HOMO levels of both the D and A molecules split. As a consequence, there is a smaller energy gap for excitation of an electron from the HOMO to the LUMO level in the complex.

A good example of a charge-transfer process is the reaction between pyrene (conductivity $10^{-12}\,\mathrm{S\,m^{-1}}$) and iodine (conductivity $10^{-7}\,\mathrm{S\,m^{-1}}$) to give a complex with a conductivity of about $1\,\mathrm{S\,m^{-1}}$, as shown in Figure 3.32. In many cases, partial transfer of charge occurs between the donor and acceptor molecules; say six electrons in every 10 donor atoms are

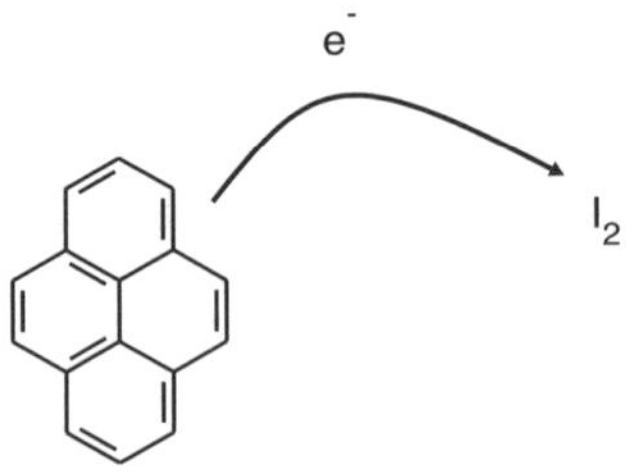

Figure 3.32 Charge-transfer reaction between pyrene and iodine to give a high-conductivity complex.

transferred. This leads to *mixed valence states* in the complex. Charge-transfer complexes generally pack closely in their crystalline phase through the formation of rigid multi-sandwich stacks, resulting in a rather brittle solid.

The stacking can be of two types: mixed stacks in which the donors and acceptors stack alternately ...*ADADADAD*... or segregated stacks in which the donors and acceptors form separate donor stacks (...*DDDDDDDD*...) and acceptor stacks (...*AAAAAAAA*...). These are illustrated in Figure 3.33. Molecular compounds with mixed stacks are not highly conductive because of electron delocalization on the acceptor species. However, for segregated stacks, the π overlap and charge-transfer interaction between adjacent molecules in the stacking directions are strong, causing the unpaired electrons to delocalize partially along these one-dimensional molecular stacks, resulting in a high conductivity in this direction. This overlap is different from the p-orbital overlap forming the π bands in conjugated polymers. Overlapping in conjugated polymers occurs sideways, in the direction of the polymer axis, and leads to very wide bands, $\sim$ 10 eV. Overlapping in charge-transfer salts is top to bottom, along the stacking axis, and leads to rather narrow

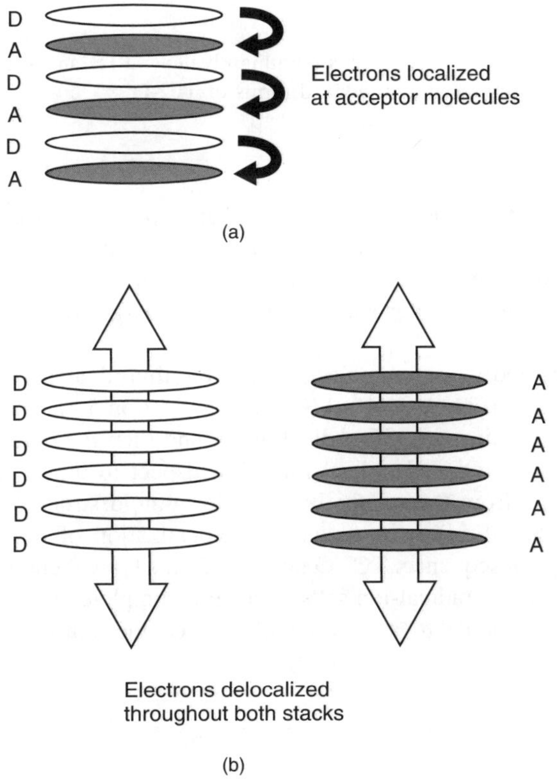

Figure 3.33 Possible stacking sequences for donor D and acceptor A molecules. (a) In mixed stacks, the electrons become localized on the acceptor molecules and are unable to move through the stack. (b) For segregated stacks, the electrons are delocalized throughout both stacks, resulting in high conductivity.

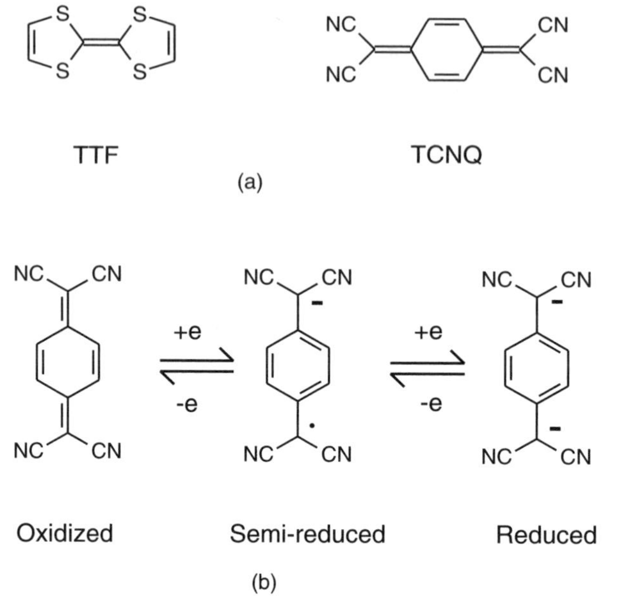

Figure 3.34 (a) Charge-transfer compounds tetrathiafulvalene, TTF, and tetracyanoquinodimethane, TCNQ. (b) Oxidized, semi-reduced and reduced forms of TCNQ.

bands, with widths of the order of 1 eV. Conjugated polymers are *intramolecular* one-dimensional conductors, charge-transfer salts are *intermolecular* conductors (there is also an intermolecular contribution in polymers, for example inter-chain overlapping with band width < 1 eV, and an inter-stack overlap in charge-transfer salts, with a band width $\ll 1$ eV).

Well-known donor and acceptor molecules are tetrathiafulvalene (TTF) and tetracyano-quinodimethane (TCNQ) [Figure 3.34(a)] (Chapter 5, Section 5.4). The latter compound is a very strong acceptor forming first the radical anion and then the dianion [Figure 3.34 (b)]. The stability of the semi-reduced radical ion with respect to the neutral molecule mainly arises from the change from the relatively unstable quinoid structure to the aromatic one (Chapter 5, Section 5.2.6), allowing extensive delocalization of the π-electrons over the carbon skeleton. As a consequence, TCNQ not only forms typical charge-transfer complexes but is also able to form true radical-ion salts, incurring complete one-electron transfer. Thus, on addition of lithium iodide to a solution of TCNQ, the simple lithium TCNQ salt is formed:

$$\mathrm{TCNQ} + \mathrm{LiI} \rightleftharpoons \mathrm{Li^+TCNQ^{-\bullet}} + \frac{1}{2}\mathrm{I_2} \tag{3.62}$$

Following removal of the free iodine precipitate, the TCNQ salt may be crystallized. The crystals show an electronic conductivity of about 10^{-5} S cm^{-1}. A 1:1 TCNQ:TTF salt exhibits a high room temperature conductivity (5×10^2 S cm^{-1}) and metallic behaviour is observed as the temperature is reduced to 54 K.

3.4.2 Doping

The principle of doping in organic semiconductors is similar to that in their inorganic counterparts. Impurities are added which either transfer an electron to the electron-conducting (LUMO or π^*) states (n-type doping) or remove an electron from the hole-conducting (HOMO or π) states to generate a free hole (p-type doping). However, the term 'doping' can be a misnomer as it tends to imply the use of minute quantities, parts per million or less, of impurities introduced into a crystal lattice, as for inorganic semiconductors. In the case of conductive polymers, typically 1–50% by weight of chemically oxidizing (electron-withdrawing) or reducing (electron-donating) agents are used to alter physically the number of π-electrons on the polymer backbone, leaving oppositely charged counter ions alongside the polymer chain. These processes are *redox* (reduction–oxidation) chemistry. For example, the halogen doping process that transforms polyacetylene to a good conductor is oxidation (or p-doping):

$$[\mathrm{CH}]_n + \frac{3ny}{2}\mathrm{I}_2 \rightarrow [(\mathrm{CH})^{y+}(\mathrm{I}_3^-)_y]_n \tag{3.63}$$

Reductive doping (n-doping) is also possible, e.g. using an alkali metal:

$$[\mathrm{CH}]_n + ny\mathrm{Na} \rightarrow [(\mathrm{CH})^{y-}y\mathrm{Na}^+]_n \tag{3.64}$$

In both cases, the doped polymer is a salt. The counter ions, I_3^- or Na^+, are fixed in position while the charges on the polymer backbone are mobile and contribute to the conductivity. The doping effect can be achieved because a π-electron can be removed (or added) without destroying the σ backbone of the polymer. In this way, the charged polymer remains intact. The resulting increase in conductivity can be many orders of magnitude.

A completely different type of doping is possible with the polymer polyaniline (PANi). This material, the chemical structure of which is shown in Figure 3.35(a), can be considered as being derived from a polymer which consists of alternating reduced [Figure 3.35(b)] and oxidized [Figure 3. 35(c)] repeat units [8,9]. The average oxidation state can be varied continuously from the completely reduced polymer to the fully oxidized material. Complete protonation of the imine nitrogen atoms in the emeraldine base form of polyaniline (for which $x = 0.5$ in Figure 3.35) by aqueous HCl, for example, leads to a structural change with one unpaired spin per repeat unit, but with no change in the number of electrons. The result is a half-filled band and a conductive state where there is a positive charge in each repeat unit (from protonation) and an associated negative charge (e.g. Cl^-). The protonation is accompanied by an increase in conductivity of about around 11 orders of magnitude.

Doping of molecular crystals can be achieved in a similar way to that described above. For example, high conductivities can be achieved when organic dyes with a weak donor character (such as phthalocyanine, Chapter 5, Section 5.4) are exposed to strongly oxidizing gases such as iodine or bromine. Alternatively, relatively large aromatic molecules, which are strong π-electron donors or π-electron acceptors, can be used. Figure 3.36 shows the conductivity of zinc phthalocyanine doped with the strong acceptor F_4-TCNQ (p-type doping) as a function of the molecular doping ratio [10]. The conductivity of can be controlled over two orders of magnitude by changing the amount of dopant; furthermore, the conductivity is many orders of magnitude higher than the background conductivity of undoped zinc phthalocyanine ($\sim 10^{-10}\ \mathrm{S\,cm^{-1}}$).

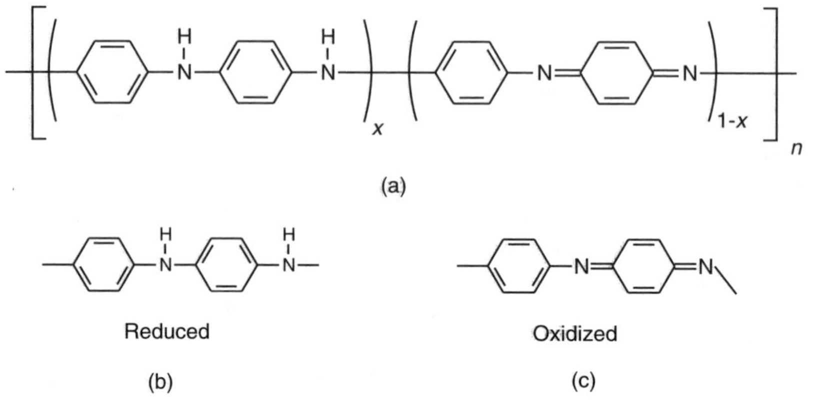

Figure 3.35 (a) Generic form of polyaniline. The polymer consists of alternating reduced (b) and oxidized (c) units. The average oxidation state $(1 - x)$ can be varied continuously from 0 (completely reduced) to 1 (fully oxidized).

Finally, it should be noted that the doping of organic semiconductive materials can be achieved in two further ways: photo-doping and charge injection. In the former case, the material is locally oxidized and reduced by photo-absorption. In the second case, electrons and holes can be injected from suitable metal contacts directly into the π^* and π bands, respectively.

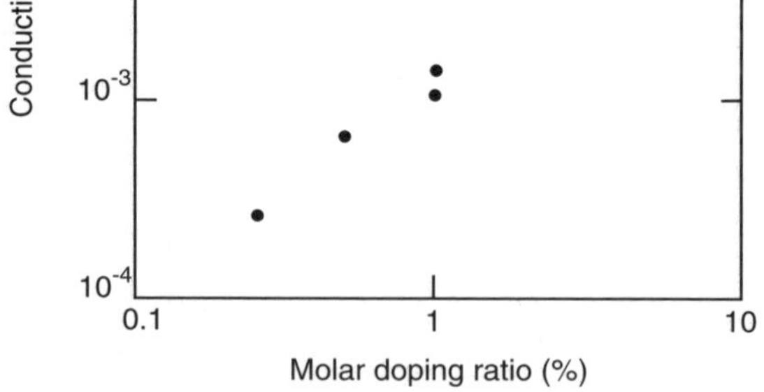

Figure 3.36 Conductivity of p-doped zinc phthalocyanine as a function of the doping concentration with the molecular dopant F_4-TCNQ. Reprinted from *Organic Electronics*, **4**, 2–3, Pfeiffer M, Leo K, Zhou X, Huang JS, Hofmann M, Werner A, Blockwitz-Nimoth J, 'Doped organic semiconductors: physics and application in light emitting diodes', pp. 89–103, Copyright (2003), with permission from Elsevier.

3.4.3 Solitons, Polarons and Bipolarons

In Chapter 2, Section 2.5.7, we considered some of the defects that can be associated with crystalline materials. There are also a number of specific defects associated with polymers that can influence their electrical and optical properties. An overview of these is given in this section.

Solitons

First, consider the bonding structure in the *trans* form of polyacetylene. The double and single bonds can be interchanged without affecting the overall electron energy. This was also seen to be the case for benzene (Chapter 2, Section 2.4.3, Figure 2.29) Therefore, there are two *degenerate* lowest energy states A and B, shown in Figure 3.37, possessing distinct bonding structures. A simple defect can occur when these two degenerate forms of the polymer are joined, as shown in Figure 3.38. In this instance, a bond mismatch occurs, i.e. a carbon site with one too few π-electrons so that it cannot form a double bond. This results in a dangling bond. This concept has already been introduced when dealing with defects occurring at the surfaces of crystalline semiconductors (Chapter 2, Section 2.5.7) and in amorphous silicon (Section 3.3.7). The defect results in one unpaired π-electron, but as the entire system is electrically neutral, it has the same number of protons as electrons.

The topological defects described above are called *solitons*, due to their non-dispersive nature (an everyday manifestation of a soliton, or solitary wave, is the bow wave of a boat). The soliton will be mobile due to the translational symmetry of the chain. Although the soliton shown in Figure 3.38 is shown to occur as an abrupt change from the A to the B form of polyacetylene, the evidence is that the defect extends over about seven carbon atoms [8]. Solitons can become positively or negatively charged. If the localized state contains one electron, the soliton is neutral and can be associated with an energy level half way between the valence and conduction bands [Figure 3.39(a)]. The unpaired electron will have a spin of ½ and can be detected by *electron spin resonance*, ESR. When the electron in the localized state is removed, for example by p-doping, the soliton is positively charged with spin = 0 [Figure 3.39(b)] and is no longer detectable by ESR. Similarly, if n-doping occurs, a negative soliton is obtained with spin = 0 [Figure 3.39(c)]. The ESR studies reveal that neutral solitons are highly mobile, whereas positive and negative solitons are believed to be localized over a number of carbon atoms.

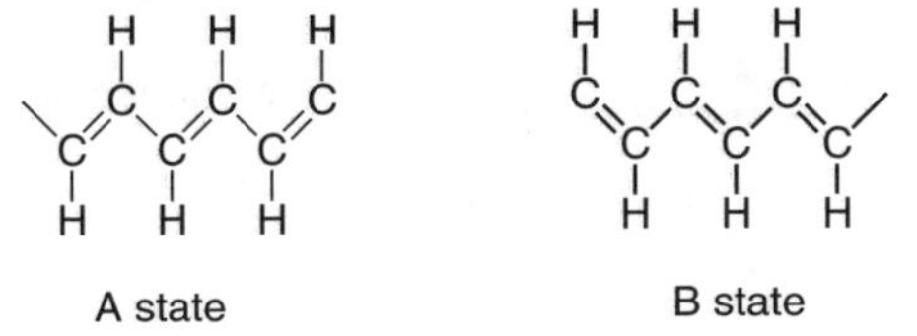

Figure 3.37 Two degenerate forms (A and B states) of *trans*-polyacetylene.

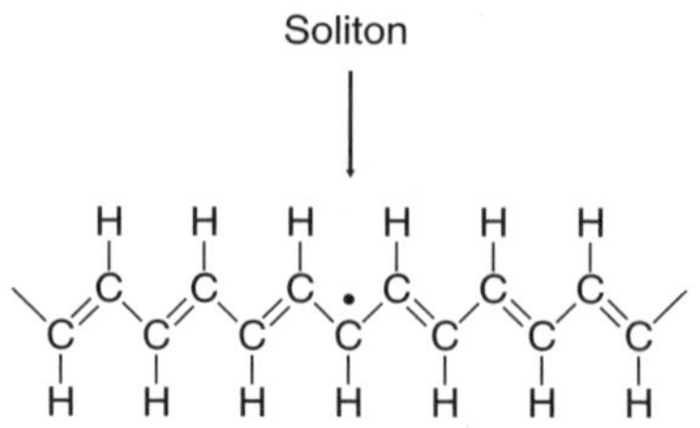

Figure 3.38 Soliton formation in a *trans*-polyacetylene chain.

Polarons and Bipolarons

As noted above, polyacetylene is a conjugated polymer with degenerate ground states. However, most conjugated polymers have *non-degenerate* ground states. In such materials, solitons are not stable and double-conjugational defects are found instead. Such a species is termed a *polaron* if it is singly charged and a *bipolaron* if it is doubly charged. In essence, a

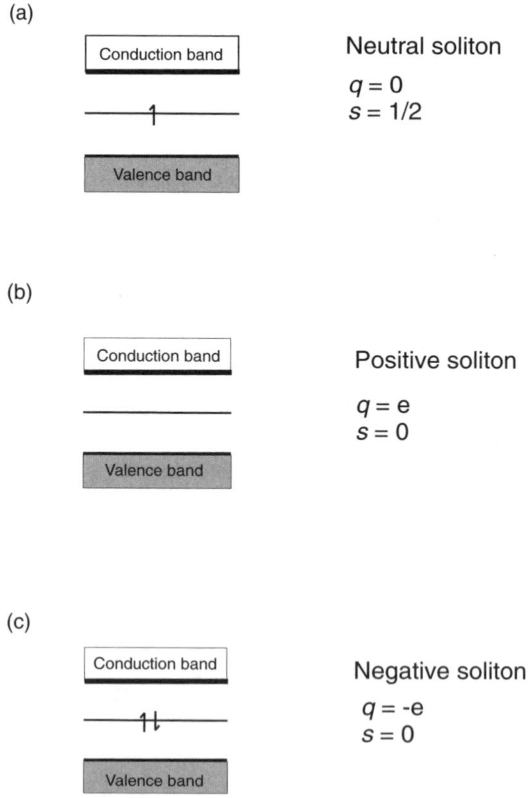

Figure 3.39 Energy levels associated with (a) a neutral soliton, (b) a positively charged soliton and (c) a negatively charged soliton. The charge q and spin s of the defects are indicated.

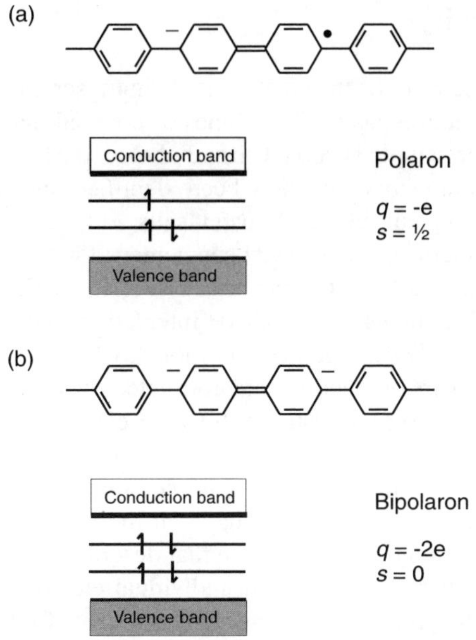

Figure 3.40 Schematic diagram of (a) a polaron and (b) a bipolaron in poly(*p*-phenylene). The charge *q* and spin *s* of the defects are indicated.

polaron can be thought of as the bound state of a charged soliton and a neutral soliton, whose mid-gap energy states hybridize to form bonding and antibonding levels [e.g. a 'combination' of the levels depicted in Figure 3.39(a) and (c)]. Consequently, a polaron is characterized by two states in the gap. Polarons are also encountered in inorganic semiconductor physics: an electron moves through a lattice and affects the constituent ions. However, the resulting distortion of the lattice in inorganic semiconductors is small compared with the polaron defect in conjugated polymers

The polaron and bipolaron are illustrated schematically for poly(*p*-phenylene) (PPP) in Figure 3.40. The positive (negative) polaron is a radical cation (anion), an entity consisting of a single electronic charge associated with a local geometric relaxation of the bond lengths. Similarly, a bipolaron is a bound state of two charged solitons of like charge (or two polarons whose neutral solitons annihilate each other) with two corresponding mid-gap levels. The formation of these defect states leads to new localized energy levels in the band gap; consequently, characteristic optical absorption signatures will be observed (Chapter 4, Section 4.4). For further information, the reader is referred to the excellent book by Roth, which contains a more detailed description of the 'menagerie' of conjugational defects found in conductive polymers [11].

Soliton and polaron states play an important role in the conductivity of conjugated polymers. For example, it is generally believed that electrical conduction in doped polyacetylene proceeds by a hopping mechanism that may involve the capture of a mobile electron from a neutral soliton in an adjacent chain. However, due to disorder, solitons and polarons lose many of their characteristic features in real polymer samples.

3.4.4 Superconductivity

Only a decade after the discovery of the electron (by Thompson in 1887) and the introduction of Drude's model of the electron gas (1900), Onnes discovered that the resistivity of mercury suddenly vanishes when the sample is cooled slightly below the boiling point of liquid helium. A large number of *superconductors* have now been identified, including elements, alloys and intermetallic compounds, organic charge-transfer salts, fullerenes and oxides. The highest superconducting transition temperatures have been achieved with the oxides (around 135 K at atmospheric pressure). The Bardeen, Cooper and Schrieffer (BCS) theory for superconductivity formulated in 1957 assumes an attractive interaction between the electrons. This is provided by the exchange of *phonons* (lattice vibrations), which bind two electrons together to form *Cooper pairs*. These pairs of electrons are correlated over fairly large distances (100 nm to 1 μm) compared with the average distance between electrons. Consequently, the Cooper pairs interpenetrate highly. The two electrons involved in a Cooper pair have opposite spin and the quasi-particle representing the pair has no net spin. Hence the Cooper pairs do not obey Fermi–Dirac statistics (Section 3.2.4). They can therefore all 'condense' to the lowest energy state and possess one single wavefunction that describes the entire assembly of Cooper pairs. Since the electrons operate as a pair, an individual electron cannot gain or lose small amounts of energy from an applied field or through collisions. Cooper pairs are able to move through the lattice without any energy exchange in collisions, which of course is the origin of electrical resistance.

A superconductor below its critical temperature expels all the magnetic field from the bulk of the sample as if it were a perfectly diamagnetic substance (Chapter 5, Section 5.7.1). This phenomenon is called the *Meissner* effect and is often used as a test for superconductivity.

Organic superconductors can be subdivided into one-, two- and three-dimensional solids and encompass the charge-transfer salts, discussed briefly above in Section 3.4.1, and the fullerenes. Perhaps the 'Holy Grail' in superconducting materials research is to identify a compound that exhibits the superconducting state at room temperature. In 1964, Little envisaged a superconducting polymer with a conjugated backbone and dye groups attached regularly to the chain (Figure 3.41) [12]. The dye groups are characteristically polarizable. Little suggested that the formation of Cooper pairs amongst the π-electrons of the chain might occur through such polarization of the side groups. Taking into account that the electronic mass is much smaller than any atomic mass, the coupling might be expected to be very large indeed in comparison with that in conventional semiconductors. As a consequence, Little suggested a critical temperature for the transition from the superconducting state to the normal sate of about 2000 K. This dramatic prediction – a room temperature organic superconductor – met with considerable criticism and has not yet been realized in practice. Organic superconductors do exist, but with transition temperatures considerably below room temperature. The highest transition temperature for a charge-transfer salt is about 13 K and for fullerenes about 33 K (Chapter 5, Section 5.5.1).

3.5 LOW-FREQUENCY CONDUCTIVITY

In the following sections, the theoretical ideas that have been presented to this point in this chapter are extended to include the effects that are observed in the laboratory when a

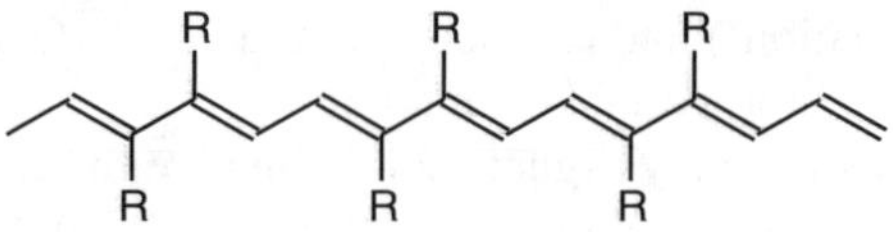

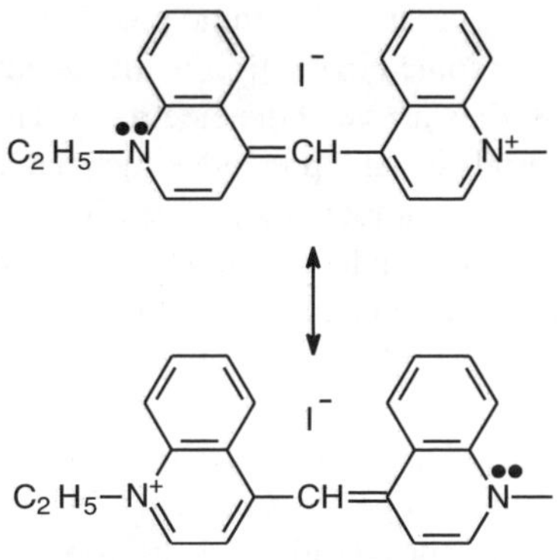

Figure 3.41 (a) Proposed structure of a room temperature polymeric superconductor by Little [12]. Side groups, designated R, are attached to a polyacetylene backbone. (b) Suggestion for R substituent. These side groups are resonating hybrids of the two extreme structures depicted.

voltage is applied to macroscopic samples and a resulting electrical current is produced. At a fundamental level, the presence of a DC voltage will lead to a drift in the free charges within the specimen. For those materials with no or very few such charges, the main effect of the applied field is to polarize the solid. The result is a separation of the centres of positive and negative charge and electric dipoles (Chapter 2, Section 2.3.5) are produced. For such cases, the application of a step DC voltage across the sample can produce an initial *displacement current* that can dominate over the small ionic and electronic contributions for a long time. In extreme cases, a steady current reading will never be obtained and a direct measurement of conductivity will not be possible. It can therefore be more appropriate to use AC methods to study the electrical processes in the sample; this is discussed in Section 3.6 on AC conductivity.

Some of the important low-frequency electrical conductivity processes observed in organic solids are outlined in the following sections. These processes can occur simultaneously in the samples and each may dominate at different values of the applied electric field and/or over different temperature ranges.

3.5.1 Electronic Versus Ionic Conductivity

Although most of the electrical processes described focus on electrons as the transporters of electric charges, nature favours ionic conduction. Ions possess both shape and charge (and thereby carry more 'information') and can be exploited very effectively in natural processes involved in the transfer and storage of information.

The preparation methods of many organic solids involve the use of aqueous solvents and other substances containing ions. Ionic conduction can also be a significant factor in polymeric materials, particularly those that contain counter ions to balance the charges on the polymer backbone. An ionic contribution to the measured conductivity is therefore to be expected. It is crucial to be able to separate these ionic processes from electronic conduction.

The most definitive evidence for ionic conduction is the detection of electrolysis products formed on discharge of the ions as they arrive at the electrodes. However, the very low level of conductivity in most organic materials usually precludes such detection. Even at a conductivity of $10^{-9}\,\Omega^{-1}\,\mathrm{m}^{-1}$ (and many organic compounds possess lower ionic contributions to their conductivity), 100 V applied across a sample $100\,\mathrm{mm}^2$ in area and 1 mm thick will produce only about $10^{-11}\,\mathrm{m}^3$ of gas at standard temperature and pressure (STP) per hour [13].

Ions travelling through a sample under the influence of an applied electric field will accumulate at defects (e.g. grain boundaries) or at one of the solid electrodes. The resulting polarization will reduce the ionic current to zero over a period. Ionic conductivity is therefore expected to give rise to time-dependent currents. Furthermore, ions do not respond as readily as electrons to high-frequency fields and the ionic contribution to conductivity should be reduced at high frequencies. For organic materials based on charge-transfer complexes, the presence of a charge-transfer band in the infrared part of the EM spectrum can be used to identify electronic conductivity.

A strong correlation between the measured conductivity of a sample and its permittivity indicates the presence of ionic conductivity [13]. This can be explained by the reduction of the Coulomb forces between ions in a high-permittivity medium. For example, the absorption of water, which has a relatively high permittivity, generally enhances the conductivity of a polymer significantly and polymer–water systems frequency conform to the equation

$$\log \sigma = -\frac{A}{\varepsilon_s} + B \tag{3.65}$$

where A and B are constants and ε_s is the low-frequency (or static) value of the real part of the relative permittivity (Section 3.6.1).

Further evidence of ionic conduction may be obtained from studies of the dependence of current on applied voltage. A simple model, based on the work of Mott and Gurney, gives [13]

$$J \propto \sinh \frac{eaE}{2k_B T} \tag{3.66}$$

where a is the distance between potential wells associated with the ionic movement.

3.5.2 Quantum Mechanical Tunnelling

If the energy of an electron is less than the interfacial potential barrier at a metal/insulator interface upon which it is incident, classical physics predicts reflection of the electron at the interface. The electron cannot penetrate the barrier and its passage from one electrode to the other is inhibited. Quantum mechanics contradicts this view. The wave nature of the electron

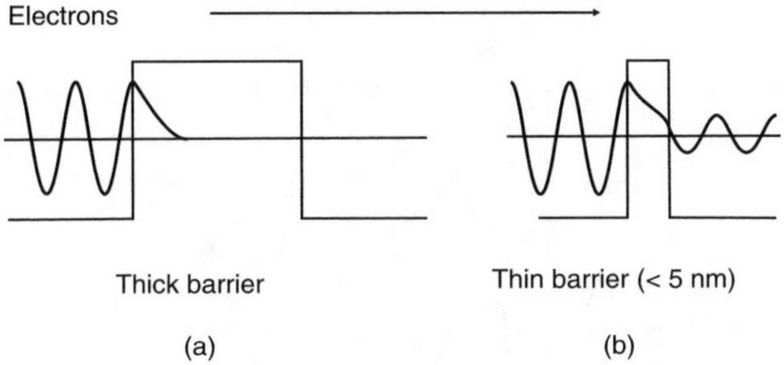

Figure 3.42 Schematic representation of the quantum mechanical tunnelling of an electron wave: (a) thick barrier; (b) thin barrier.

allows penetration of the forbidden region of the barrier. This is the origin of quantum mechanical tunnelling described earlier in Section 3.3.1. The wavefunction associated with the electron decays rapidly with depth of penetration from the electrode/insulator interface and, for barriers of macroscopic thickness, is essentially zero at the opposite interface, as shown in Figure 3.42(a) for a simple metal/insulator/metal (MIM) structure. This indicates a zero probability of finding the electron here. However, if the barrier is very thin (<5 nm), the wavefunction has a non-zero value at the opposite interface. For this case, there is a finite probability that the electron can pass from one electrode to the other by penetrating the barrier [Figure 3.42(b)].

The current versus voltage relationships for tunnelling are complex and depend on the magnitude of the applied voltage and whether the tunnel barrier is symmetric or asymmetric (i.e. whether the two electrodes are similar or different metals) [14]. For very low applied voltages (much less than the energy barrier height divided by the electronic charge), the tunnelling probability varies exponentially with the barrier thickness, and the tunnelling conductivity, σ_t, may be given by an equation of the form

$$\sigma_t = A\exp(-Bd) \tag{3.67}$$

where d is the thickness of the tunnel barrier and A and B are constants. However, at higher voltages, the conductivity data for insulating films have been shown to deviate from linearity. For example, one theoretically predicted current versus voltage dependence for a symmetrical rectangular barrier takes the form [15]

$$I = 2I_0\left[\frac{\pi CkT}{\sin(\pi CkT)}\right]\exp(-BV^2)\sinh\left(\frac{CV}{2}\right) \tag{3.68}$$

where I_0 is a constant and B and C are coefficients related to the tunnelling barrier height.

The ability to form thin organic films with precisely defined thickness, such as Langmuir–Blodgett films (Chapter 7, Section 7.3.1), should offer an excellent basis for studying quantum mechanical tunnelling. Even so, experimental data must be treated with caution because of the presence of oxide and other layers on the metallic electrodes that will have thicknesses comparable to that of the organic film that is supposedly being tested [16]. However, with careful attention to experimental detail, it is possible to observe some of the theoretical predictions for tunnelling. Figure 3.43 shows the results of experiments using

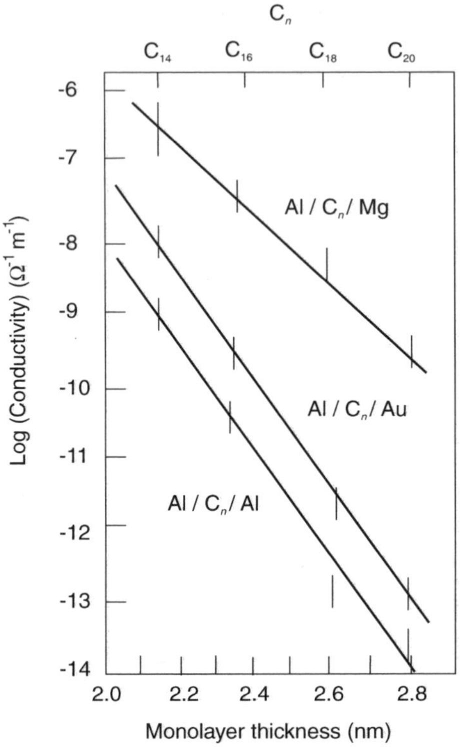

Figure 3.43 Logarithm of conductivity versus monolayer thickness for different metal/monolayer/metal structures. C_n is the number of carbon atoms in the monolayer compound. Reprinted with permission from Polymeropoulos EE, *J. Appl. Phys.*, **48**, 'Electron tunnelling through fatty-acid monolayers', pp. 2404–2407, Copyright 1977, American Institute of Physics.

monolayer films of fatty acids of different chain length sandwiched between metallic electrodes [17]. The dependence of the tunnelling conductivity (measured at an applied voltage of 10 mV) on film thickness is clearly of the form of Equation (3.67). For larger applied voltages, the current versus voltage behaviour for fatty acid monolayer films is also similar to that predicted by theory [18].

3.5.3 Variable Range Hopping

Polymers contain both crystalline and amorphous regions (Chapter 2, Section 2.6). Disorder in a doped conductive polymer can also arise from the random distribution of doping ions, or the clustering of doping ions. Consequently, the electronic band structure may be more like that of an amorphous semiconductor rather than its crystalline counterpart. States in the band tails are localized and there exists a mobility edge (Section 3.3.7) which separates the region of the extended states in the interior of the band from the region of the localized states. Whether the doped polymer behaves as a metal or an insulator is determined by the relative position of the Fermi level with respect to the mobility edge.

At high temperature, the conductivity will be determined by electrons (or holes) thermally activated to extended states above (below) the mobility edge (E_c for electrons and E_v for holes, Figure 3.24). In this case, the conductivity can be expressed in the form of Equation (3.15), i.e. as for the crystalline case. However, for disordered materials, the carrier mobility μ cannot be calculated assuming the electrons are free (with their paths interrupted only by occasional scattering events) because the mean free path is of the order of an interatomic distance. It is generally assumed that the transport is diffusive (Chapter 2, Section 2.8) and the mobility is given by the *Einstein relation*:

$$\mu = \frac{eD}{k_B T} \tag{3.69}$$

where D, the diffusion coefficient is $\nu_{el}a^2/6$, ν_{el} being an electronic frequency which, from quantum theory, is roughly $\hbar/ma^2$, where a is taken to be an interatomic spacing or as the coherence length of an electron wave. Therefore,

$$\mu \approx \frac{e\hbar}{6mk_B T} \approx 6\,\text{cm}^2\,\text{V}^{-1}\,\text{s}^{-1} \tag{3.70}$$

Carrier mobilities of around $1\,\text{cm}^2\,\text{V}^{-1}\,\text{s}^{-1}$ are measured in organic molecular materials; the values for conductive polymers are usually much less (Table 3.1).

If the localized energy levels in a disordered material are unoccupied, these provide a possible alternative conduction path for carriers, in parallel with the band (extended state) conduction discussed above. Clearly, the electron motion in localized states is much smaller than in extended states, but at low temperatures the number of electrons activated to the conduction band can be sufficiently small to allow localized state conduction to become the predominant conduction mechanism.

Electrons can move between localized states in three different ways. First, they can be thermally activated over the potential barrier separating the two states. However, the barrier height is usually of the same order as the energy separating the localized states from the extended states. If the temperature is sufficiently high to facilitate jumping, extended state conduction will be the more favoured process. The emission at lower temperatures can be aided by the electric field and will be discussed in Section 3.5.5. Second, electrons may tunnel through the potential barrier and third they can move by a combination of activation and tunnelling. Since, in the disordered solid, the localized states will not be degenerate, some measure of activation will generally be required to allow the full state and the empty state to resonate and hence allow tunnelling. This thermally assisted tunnelling is called *hopping* and is an important feature of conduction in non-crystalline materials.

There have been many theoretical treatments of the tunnelling process [3, 19]. It is generally assumed that the states are distributed randomly in energy and space and that the Fermi level lies within these localized states. The *variable range hopping* model of Mott [3] gives the following equation for the temperature dependence of conduction:

$$\sigma = \sigma_0 \exp\left[-\left(\frac{T_0}{T}\right)^{\frac{1}{4}}\right] \tag{3.71}$$

where σ_0 and T_0 are constants. The term variable range hopping comes from the fact that the average hopping distance varies with temperature and is not necessarily to the closest site. If the dimensionality of the transport is reduced, then the factor ¼ in Equation (3.71) is replaced by ⅓ in the two-dimensional case and ½ for one-dimensional hopping.

3.5.4 Space-Charge Injection

If the electrical contacts to an insulating or semiconducting sample are Ohmic, these allow electron transfer between the electrodes and the sample and the resulting current is proportional to the applied voltage [Equation (3.8)]. Under certain conditions, however, the contacts can become 'super-Ohmic' and the current is only limited by the *space-charge* between the electrodes. This conductivity regime is called *space-charge limited* [20]. In the simplest case, the current density J varies with applied voltage V and sample thickness d as follows:

$$J \propto \frac{V^2}{d^3} \tag{3.72}$$

Figure 3.44 shows the expected current versus voltage behaviour under these conditions. The lowest voltage region of the curve corresponds to the situation in which the injection of excess carriers in negligible. At these voltages the volume conductivity dominates (i.e. Ohm's law). Only when the injected carrier density exceeds the volume generated carrier density will space-charge effects be seen and the quadratic current versus voltage dependence become evident.

If the insulator or semiconductor contains traps then the injected charge will fill these. When sufficient charge has been injected, the traps will become saturated. The voltage at which this occurs corresponds to the *trap-filled limit* and is shown as V_{TFL} in Figure 3.44. Beyond V_{TFL}, the material behaves as if it was trap free and the quadratic current versus voltage relationship given in Equation (3.72) is again observed.

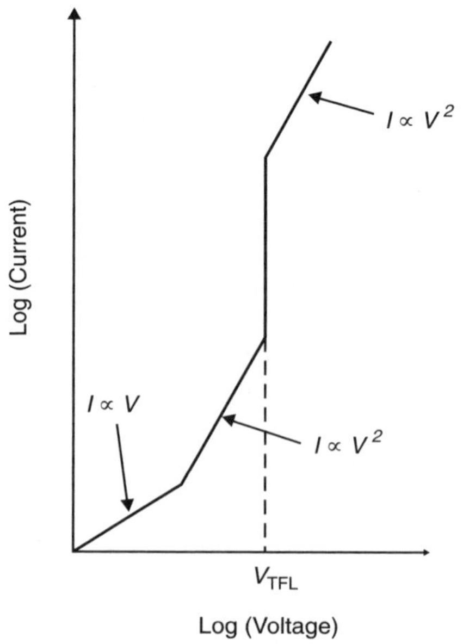

Figure 3.44 Space-charge-limited current versus voltage characteristics for a semiconductor (or insulator) containing traps. V_{TFL} is the trap-filled limit.

Equation (3.72) is a special case of the general scaling law for bulk for bulk space-charge currents in a homogeneous medium [19]:

$$J \propto d\left(\frac{V}{d^2}\right)^m \tag{3.73}$$

where m is a constant, which need not necessarily be an integer. For example, in the trap-free insulator case above, $m = 2$. In the case of 'double' injection, i.e. electrons injected at one electrode and holes at the other, $m = 3$, and for recombinative space-charge injection $m = ½$. Figure 3.45 shows current versus voltage data, in the form of a log(I) versus log(V) plot, for a sample of thin-film copper phthalocyanine, a molecular crystalline compound (Chapter 5, Section 5.4), sandwiched between gold electrodes [21]. The transitions from Ohmic conductivity, where $m = 1$, to regions for which $m > 1$ are clearly evident. The values of m in the high-voltage regions are shown in the figure. The dependence of m on the temperature is an indication that the organic solid contains an exponential distribution of trapping centres within the energy band gap.

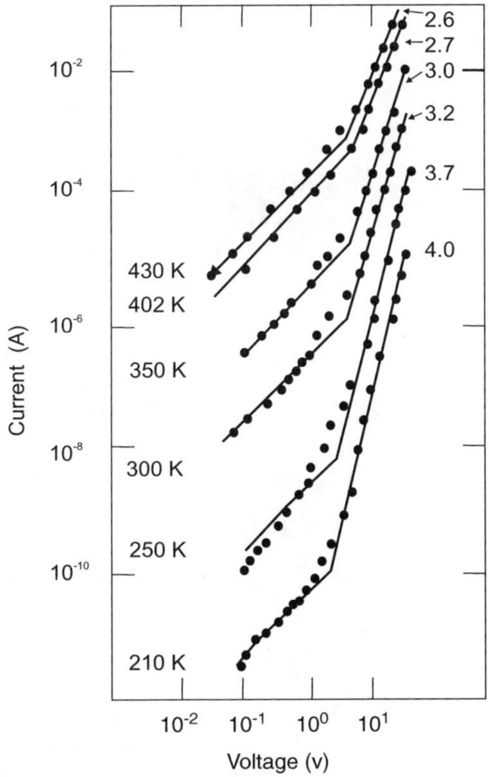

Figure 3.45 Current versus voltage characteristics of an Au/copper phthalocyanine/Au sandwich at different temperatures. The linear dependence of current on voltage at low bias is the result of Ohmic conductivity. The super-Ohmic regions at higher bias indicate space-charge limited conductivity. The value of m from Equation (3.73) is indicated for each curve. Reprinted from *Coordination Chemistry Reviews*, **156**, Gould RD, 'Structure and electrical conduction properties of phthalocyanine thin films', pp. 237–274, Copyright (1996), with permission from Elsevier.

3.5.5 Schottky and Poole–Frenkel Effects

When an insulator (or semiconductor) is placed in contact with a metal, a redistribution of charge will occur in the interface region (to minimize the free energy of the system). This will produce a distortion or bending of the energy bands. Abrupt changes in the potential energy do not occur, as these would imply infinite electric fields. The potential step changes smoothly because of the *image force* effect. This arises because of the metal's surface becoming positively charged by an escaping electron. The resulting potential barrier at the metal/insulator interface is called a *Schottky barrier* [Figure 3.46(a)]. The electrons in the metal are filled to the Fermi level E_F and they must gain enough energy to overcome the barrier Φ_B to get into the conduction band of the semiconductor. This process is called *Schottky emission* and has a current density versus electric field dependence:

$$J \propto \exp\left(\frac{E\beta^{0.5}}{k_B T}\right) \tag{3.74}$$

where β is a constant.

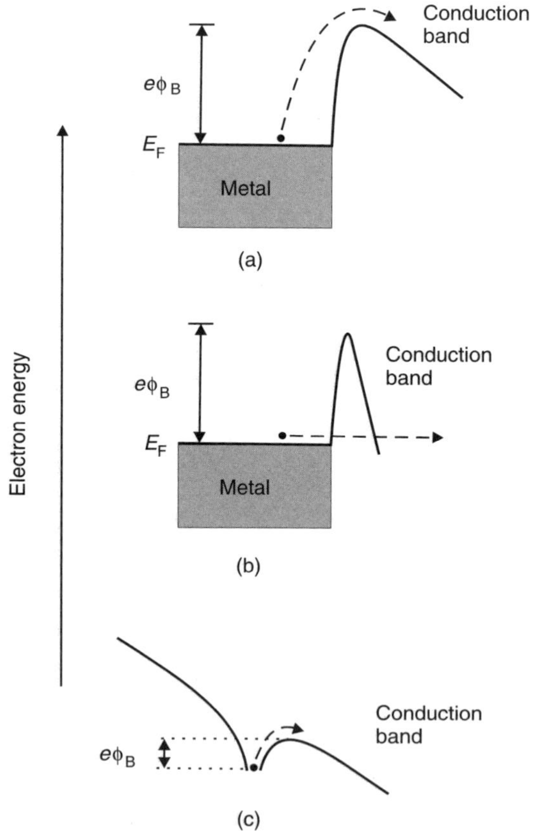

Figure 3.46 (a) Schottky emission of an electron from the Fermi level E_F in a metal into the conduction band of a semiconductor (or insulator); (b) Fowler–Nordheim tunnelling; (c) Poole–Frenkel effect: the applied electric field lowers the barrier surrounding a trapped carrier and helps it to move into the conduction band.

If the applied electric field is large enough, the barrier to electrons at the interface becomes very thin and electrons can tunnel directly from the metal to the conduction band of the insulator. This conductivity mechanism is termed *field emission* and is illustrated in Figure 3.46(b). The resulting current density versus voltage behaviour is of the form

$$J \propto E^2 \exp\left(-\frac{\gamma}{E}\right) \tag{3.75}$$

where γ is another constant. The process is also referred to as *Fowler–Nordheim tunnelling*. This process can also be thermally assisted, the electrons being excited to an energy state where they are then able to tunnel through the barrier.

A similar process to Schottky emission can take place at impurity centres in the bulk of the material. This is the *Poole–Frenkel effect* and is illustrated in Figure 3.46(c).The application of a high electric field will result in the lowering of the potential barrier around the centre, allowing a carrier to escape into the conduction band of the insulator. The current versus voltage behaviour for Poole–Frenkel conduction has the same form as Equation (3.74) for Schottky emission (although the β coefficient is different) and experimentally it can be difficult to distinguish between the two processes. However, Poole–Frenkel conduction is essentially a bulk effect and should show little dependence upon electrodes or upon the polarity of the field. Poole–Frenkel and Schottky emission processes are usually observed at high applied electric fields ($>10^7\ \mathrm{V\,m^{-1}}$).

3.6 CONDUCTIVITY AT HIGH FREQUENCIES

DC and high-frequency phenomena are essentially separate and, for the most part, independent quantities. However, at low frequencies, both can contribute to a measured current. The frequency-dependent conductivity $\sigma(\omega)$ of a sample may be expressed in the form

$$\sigma(\omega) = \sigma_{\mathrm{DC}} + \sigma_{\mathrm{AC}} \tag{3.76}$$

where σ_{DC} is the DC conductivity discussed in the previous sections and σ_{AC} is the AC component.

The AC conductivity of a sample is generally measured by applying a voltage at an appropriate frequency and then measuring both the in-phase and 90° out-of-phase components of the current. The equipment required depends on the frequency of measurement: simple bridge circuits can be used over the frequency range 1–10^7 Hz; for lower frequencies, step-response methods are used (e.g. a voltage step is applied and the subsequent current monitored over time).

3.6.1 Complex Permittivity

The AC conductivity is related to the *permittivity* of the material. The *relative permittivity* ε_{r} of a material is a complex quantity, i.e.

$$\varepsilon_{\mathrm{r}} = \varepsilon_{\mathrm{r}}' - j\varepsilon_{\mathrm{r}}'' \tag{3.77}$$

where ε_r' and ε_r'' are the real and imaginary parts, respectively, both being frequency dependent. Often, the real part of the relative permittivity, ε_r', is called the *dielectric constant*. Application of an electric field to a material will create electric dipoles (Chapter 2, Section 2.3.5). Some materials already contain dipoles by virtue of their chemical bonding; these are *polar* compounds. In this case, the application of the electric field tries to align the dipoles in the field direction. This is opposed by thermal motion. The real part of the permittivity is a measure of the ability of an electric field to polarize the medium whereas the imaginary part represents losses in the material as an AC field attempts to orient the electric dipoles within the material.

The polarization mechanism described above, i.e. the reorientation of dipoles in an electric field, is called *orientational* (or *dipolar*) *polarization*. In general, a dielectric medium will also exhibit other polarization mechanisms on application of a field. These are *electronic polarization*, the displacement of the electron clouds around the nucleus (described more fully in Chapter 4, Section 4.3), *ionic polarization*, due to the displacement of positive and negative ions, and *interfacial polarization*, resulting from the accumulation of charges at interfaces within the material. Each process will produce a change in ε_r' and a peak in ε_r'' over a particular frequency range. For example, in ionic solids this corresponds to the infrared part of the electromagnetic spectrum where the energy stored in the induced dipoles is the same as the lattice vibrational frequency. At higher frequencies, the positive and negative ions in the solid will not be able to respond sufficiently fast to the applied AC electric field and the ionic contribution to the material's permittivity is lost.

Figure 3.47 shows the frequency dependence of the real and imaginary parts of the permittivity. Although the figure shows distinct peaks in ε_r'' and transitions in ε_r', in real materials these features can be broader and overlap with one another. For single crystals, the polarization features also depend on the orientation of the electric field with respect to the crystallographic axes. At low frequencies, the interfacial (or space-charge) features are very

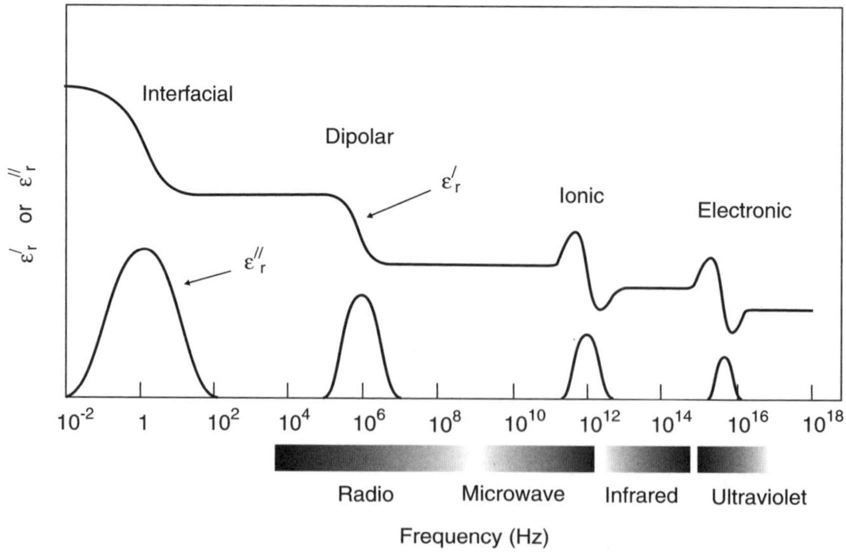

Figure 3.47 Frequency dependence of the real, ε_r', and imaginary, ε_r'', parts of the permittivity in the presence of interfacial, orientational, ionic and electronic polarization mechanisms.

broad because there can be a number of different conduction mechanisms contributing to the charge accumulation. At high frequencies – usually microwave and beyond – the polarization processes that take place are undamped and are invariably observed as resonances.

The real part of permittivity is the value that is used when calculating the capacitance of a parallel plate capacitor, e.g. for a parallel plate capacitor

$$C = \frac{\varepsilon_r \varepsilon_0 A}{d} \tag{3.78}$$

where ε_0 is a constant called the *permittivity of free space* ($= 8.85 \times 10^{-12}\,\mathrm{F\,m^{-1}}$) and A and d are the plate area and dielectric thickness, respectively, of the capacitor. Application of an AC voltage of RMS amplitude V to this capacitor will result in an AC current, I, given by

$$I = YV = \frac{V}{Z} = j\omega CV \tag{3.79}$$

where Y is called the electrical *admittance* and Z is the electrical *impedance*. The presence of j in the above equation indicates a 90° phase shift between the applied voltage and the current. Substituting Equation (3.78) into Equation (3.79) and replacing with the complex quantity of Equation (3.77) then gives

$$I = V(j\omega C + G) \tag{3.80}$$

where

$$C = \frac{A\varepsilon_0 \varepsilon_r'}{d} \quad \text{and} \quad G = \frac{\omega A \varepsilon_0 \varepsilon_r''}{d} \tag{3.81}$$

Therefore, the admittance of the dielectric medium is a parallel combination of an ideal, lossless capacitor C, with a relative permittivity ε_r' and a conductance G (or resistance $R = 1/G$) proportional to ε_r''. This is illustrated in Figure 3.48.

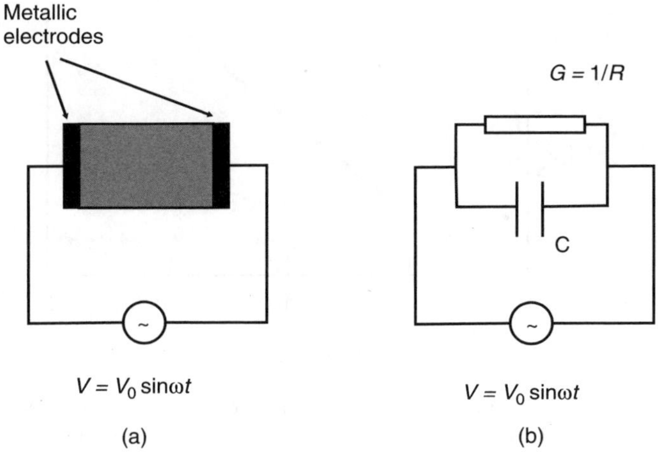

Figure 3.48 (a) A dielectric stimulated by an AC voltage. (b) Equivalent electrical circuit.

From the above, it is expected that the measured AC conductivity, σ_{AC}, will increase with frequency:

$$\sigma_{AC} = \omega \varepsilon_0 \varepsilon_r'' \tag{3.82}$$

If the sample is lossless over the frequency range, then ε_r''will be constant and the AC conductivity will increase linearly with frequency. However, many inorganic and organic solids exhibit a simple power-law relationship of the form

$$\sigma_{AC} \propto \omega^n \tag{3.83}$$

where n is less than unity (usually in the range $0.7 < n < 1$). This is often referred to as the *universal law* for the response of dielectrics [22]. Many theories have been developed to relate this observed AC electrical behaviour to the microscopic properties of the solid. Such models are often based on the hopping of charge carriers.

Figure 3.49 shows the AC conductivity data for a thin film of an anthracene derivative [23]. The low frequency behaviour $\sigma \propto \omega^{0.85}$ agrees with that expected from the discussion above. However, at frequencies above 10^4 Hz, the power dependence changes to approximately 1.8. The explanation is that at these high frequencies contact resistances associated with the metallic electrodes become important. A full analysis of the equivalent circuit of the

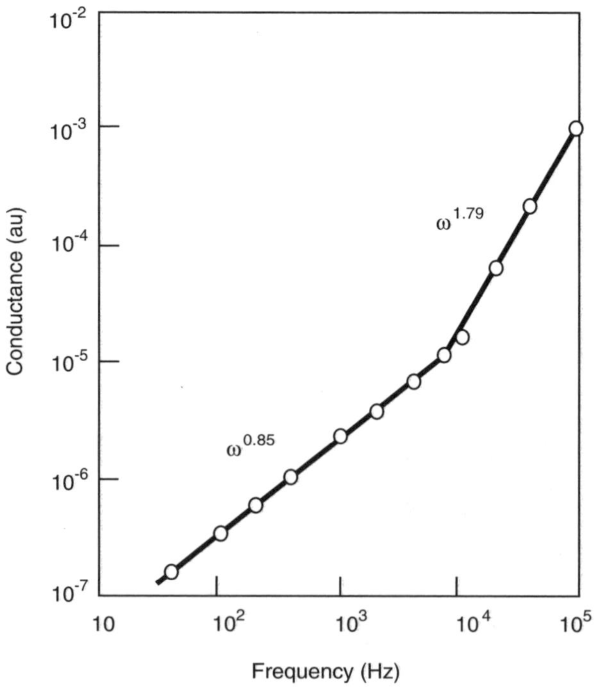

Figure 3.49 AC conductivity data showing the variation in conductance with frequency for a thin film of an anthracene derivative. Reprinted from *Thin Solid Films*, **68**, Roberts CG, McGinnity JM, Barlow WA, Vincent PS, 'Ac and dc conduction in lightly substituted anthracene Langmuir films', pp. 223–232, Copyright (1980), with permission from Elsevier.

sample including these additional resistances reveals the $\sigma \propto \omega^{2n}$ relationship. In the case $n = 1$, a square law dependence of conductance on frequency is seen at high frequencies.

3.6.2 Impedance Spectroscopy

The measurement of the electrical impedance of a sample as a function of frequency is often referred to as *impedance spectroscopy* (or *admittance spectroscopy* in the case of a measurement of Y with frequency). To investigate the equivalent electrical circuit of the material under investigation, the measured data are plotted in the form of the real part of the impedance Z' versus the imaginary part Z'', a so-called *Cole–Cole diagram.* Semicircles are revealed in the complex impedance plane. These result from the classical *Debye equations*, which relate the real and imaginary parts of the permittivity to frequency:

$$\varepsilon_r'' = \varepsilon_s + \frac{\varepsilon_s - \varepsilon_\infty}{1 + \omega^2\tau^2} \quad \text{and} \quad \varepsilon_r'' = \frac{(\varepsilon_s - \varepsilon_\infty)\omega\tau}{1 + \omega^2\tau^2} \tag{3.84}$$

where ε_s and ε_∞ represent the low- and high-frequency values of the real parts of the relative permittivity, respectively, and τ is a characteristic time constant of the system.

Figure 3.50(a) and (b) show two simple resistor/capacitor circuits and their resulting impedance spectra. The arrows indicate the direction of increasing frequency. The circuit with two resistor/capacitor combinations [Figure 3.50(b)] will possess two distinct time constants, R_1C_1 and R_2C_2, and shows distinct two semicircles in impedance plot [Figure 3.50(d)], if these

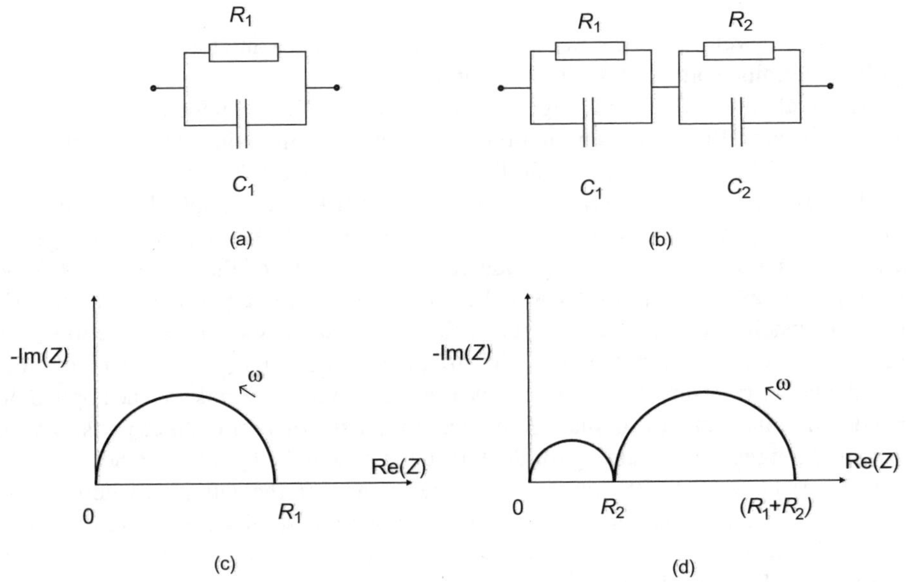

Figure 3.50 Diagrams (a) and (b) represent two simple resistor/capacitor circuits, and (c) and (d) are plots in the complex impedance plane of the imaginary part of impedance, Im(Z), versus the real part, Re(Z), for circuits (a) and (b), respectively.

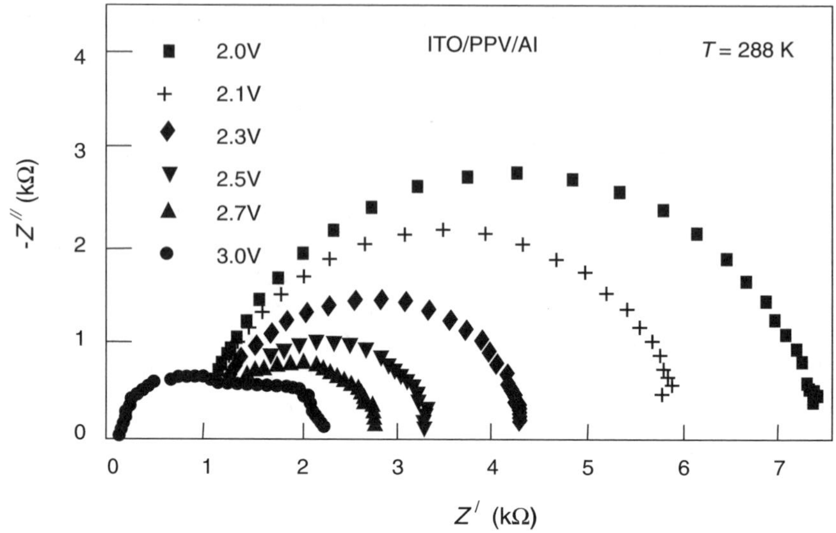

Figure 3.51 Imaginary (Z'') versus real (Z') parts of the complex impedance for an ITO/PPV/Al sandwich structure for various values of the forward bias. The frequency increases around the semicircles from right to left. The left semicircles are assigned to the polymer bulk and the right semicircles to a Schottky barrier. Reprinted with permission from Scherbel J, Nguyen PH, Paasch G, Brüting W, Schwoerer MK, *J. Appl Phys.*, **83**, pp. 5045–5055, Copyright (1998), American Institute of Physics.

time constants are well separated, i.e. $R_1C_1 \gg R_2C_2$. The high- and low-frequency limits can be used to determine some of the circuit components.

Experimental data are shown in Figure 3.51 for a layer of the conductive polymer poly(*p*-phenylenevinylene), PPV, between aluminium and indium–tin–oxide (ITO) electrodes [24]. This type of structure forms an organic light-emitting device (Chapter 9, Section 9.6) and will emit visible light when a negative voltage, or forward bias, is applied to the Al contact. The impedance spectra are shown for different values of forward bias voltages. Two semicircles are evident, suggesting that the equivalent circuit of the Al/PPV/ITO device is more complex than that of a lossy parallel plate capacitor, i.e. a parallel combination of a resistor and capacitor as depicted in Figure 3.50(a). The small semicircle close to the origin in Figure 3.51 (corresponding to the high-frequency region) is independent of voltage, whereas the larger right-hand semicircle (low-frequency region) shrinks as the applied bias is increased. The interpretation is that the device comprises of two distinct parts, a voltage-independent bulk region accounting for the left semicircle in the figure and a Schottky barrier (Section 3.5.5) at the Al/PPV interface. The resistance of the latter will depend on the applied voltage, accounting for the behaviour of the right semicircle in Figure 3.51. The equivalent circuit of the Al/PPV/ITO sandwich structure can therefore be modelled by the circuit shown in Figure 3.50(b).

Conductivity measurements over a wide frequency range provide a very useful insight into the electrical networks associated with these and other devices. However, the interpretations are not all straightforward. In some cases, impedance spectroscopy reveals arcs of semicircles rather than full semicircles. This can be attributed to a distribution of relaxation times for the material under study.

BIBLIOGRAPHY

Barford W. *Electronic and Optical Properties of Conjugated Polymers*. Oxford: Oxford University Press; 2005.

Blakemore JS. *Solid State Physics*. Philadelphia: Saunders; 1969.

Blythe T, Bloor D. *Electrical Properties of Polymers*. Cambridge: Cambridge University Press; 2005.

Cuniberti G, Fagas G, Richer K (Editors). *Introducing Molecular Electronics*. Berlin: Springer; 2005.

Farchioni R, Grosso G (Editors). *Organic Electronic Materials*. Berlin: Springer; 2001.

Ferraro JR, Williams JM. *Introduction to Synthetic Electrical Conductors*. Orlando, FL: Academic Press; 1987.

Harrop PJ. *Dielectrics*. London: Butterworth; 1972.

Hummel RE. *Electronic Properties of Materials*, 2nd ed. Berlin: Springer; 1992.

Kasap SO. *Principles of Electrical Engineering Materials and Devices*. Boston: McGraw-Hill; 1997.

Macdonald JR (Editor). *Impedance Spectroscopy: Emphasizing Solid Materials and Systems*. New York: John Wiley & Sons, Inc.; 1988.

Mott NF, Davis EA. *Electronic Properties in Non-crystalline Materials*, 2nd ed. Oxford: Clarendon Press; 1979.

Nabook A. *Organic and Inorganic Nanostructures*. Boston: Artech House; 2005.

Roberts GG, Apsley N, Munn RW. Temperature dependent electronic conduction in polymers. *Phys. Rep.* 1980; 60: 59–150.

Roth S. *One-dimensional Metals*. Weinheim: VCH; 1995.

Rudden MN, Wilson J. *Elements of Solid State Physics*. Chichester: John Wiley & Sons, Ltd; 1993.

Salaneck WR, Lundström I, Rånby B (Editors). *Conjugated Polymers and Related Materials*. Oxford: Oxford University Press; 1993.

Simmons JG. *DC Conduction in Thin Films*. London: Mills and Boon; 1971.

Stubb H, Punkka E, Paloheimo J. Electronic and optical properties of conducting polymer thin films. *Mater. Sci. Eng.* 1993; **10**: 85–140.

Wright JD. *Molecular Crystals*. Cambridge: Cambridge University Press; 1995.

REFERENCES

[1] Blakemore JS. *Solid State Physics*. Philadelphia: Saunders; 1969.

[2] Kasap SO. *Principles of Electrical Engineering Materials and Devices*. Boston: McGraw-Hill; 1997.

[3] Mott NF, Davis EA. *Electronic Properties in Non-crystalline Materials*, 2nd ed. Oxford: Clarendon Press; 1979.

[4] Karl N. Low molecular weight organic solids. Introduction. In R Farchioni, G Grosso (Editors). *Organic Electronic Materials*. Berlin: Springer; 2001, pp. 215–239.

[5] Zheng G, Clark SJ, Brand S, Abram RA. First-principles studies of the structural and electronic properties of poly-*para*-phenylene vinylene. *J. Phys.: Condens. Matter* 2004; **16**: 8609–8620.

[6] Dimitrakopoulos CD, Mascaro DJ. Organic thin-film transistors: a review of recent advances. *IBM J. Res. Dev.* 2001: **45**; 11–27.

[7] Pearson C, Moore AJ, Gibson JE, Bryce MR, Petty MC. A field effect transistor based on Langmuir–Blodgett films of an $Ni(dmit)_2$ charge transfer complex. *Thin Solid Films* 1994; **244**: 932–935.

[8] Heeger AJ. Semiconducting and metallic polymers: the fourth generation of polymeric materials. *Synth. Met.* 2002; **125**: 23–42.

[9] MacDiarmid AG. Synthetic metals: a novel role for organic polymers. *Synth. Met.* 2002; **125**: 11–22.

[10] Pfeiffer M, Leo K, Zhou X, Huang JS, Hofmann M, Werner A, Blochwitz-Nimoth J. Doped organic semiconductors: physics and application in light emitting diodes. *Org. Electron.* 2003; **4**: 89–103.

[11] Roth S. *One-dimensional Metals*. Weinheim: VCH; 1995.

[12] Little WA. Possibility of synthesizing an organic superconductor. *Phys. Rev. A* 1964; **134**: 1416–1424.

[13] Blythe T, Bloor D. *Electrical Properties of Polymers*. Cambridge: Cambridge University Press; 2005.

[14] Simmons JG. *DC Conduction in Thin Films*. London: Mills and Boon; 1971.

[15] Stratton R. Volt–current characteristics for tunnelling through insulating films. *J. Phys. Chem. Solids* 1962; **23**: 1177–1190.

[16] Petty MC. *Langmuir–Blodgett Films*. Cambridge: Cambridge University Press; 1996.

[17] Polymeropoulos EE. Electron tunnelling through fatty-acid monolayers. *J. Appl. Phys.* 1977; **48**: 2404–2407.

[18] Petty MC. Characterization and properties. In GG Roberts (Editor). *Langmuir–Blodgett Films*. New York: Plenum Press; 1992, pp. 133–221.

[19] Roberts GG, Apsley N, Munn RW. Temperature dependent electronic conduction in polymers. *Phys. Rep.* 1980; **60**: 59–150.

[20] Tredgold RH. *Space Charge Conduction in Solids*. Amsterdam: Elsevier; 1966.

[21] Gould RD. Structure and electrical conduction properties of phthalocyanine thin films. *Coord. Chem. Rev.* 1966; **156**: 237–274.

[22] Jonscher AK. *Dielectric Relaxation in Solids*. London: Chelsea Dielectric Press; 1983.

[23] Roberts GG, McGinnity TM, Barlow WA, Vincett PS. Ac and dc conduction in lightly substituted anthracene Langmuir films. *Thin Solid Films* 1980; **68**: 223–232.

[24] Scherbel J, Nguyen PH, Paasch G, Brütting W, Schwoerer MK. Temperature dependent broadband impedance spectroscopy on poly-(*p*-phenylene-vinylene) light-emitting diodes. *J. Appl. Phys.* 1998; **83**: 5045–5055.

4 Optical Phenomena

What light through yonder window breaks?

4.1 INTRODUCTION

The response of materials to high-frequency electromagnetic radiation is generally very different to their response to lower frequency (e.g. DC) electric fields. The behaviour is conveniently described in terms of a refractive index, rather than a conductivity or permittivity. In this chapter, the properties of electromagnetic radiation, introduced in Chapter 3, Section 3.3.1, will be further discussed. The linear and nonlinear interactions of EM radiation with organic materials and, in particular, the effects that occur at the various interfaces, which may be present in molecular electronic devices, are then examined. Selected specialist topics – waveguiding, surface plasmon resonance and photonic crystals – are introduced towards the end of the chapter.

4.2 ELECTROMAGNETIC RADIATION

In free space, the electric and magnetic components of an EM wave are at right-angles to each other and also to the direction of propagation. *Plane polarized* EM radiation is radiation in which the electric (and magnetic) field is confined to a plane. The plane of polarization is conventionally taken to be the plane containing the direction of the electric field. Unpolarized radiation, or radiation of an arbitrary polarization, can always be resolved into two orthogonally polarized waves. If the two electric field components possess a constant phase difference and equal amplitudes, the resultant EM wave is said to be *circularly polarized*. If the amplitudes differ, then the wave is *elliptically polarized*. (Plane and circular polarizations are special cases of elliptical polarization.) Plane polarized radiation can be produced from unpolarized radiation using specials sheets of material – usually in the form of aligned long-chain molecules – that absorb one polarized direction almost completely.

In many experiments, plane polarized EM radiation is used to study organic solids. These are often in the form of thin films deposited on planar substrates (Chapter 6, Section 6.6). Two important measuring arrangements may be distinguished: *p-polarized* or *transverse-magnetic* (TM) incident radiation, in which the electric field vector is in the plane of incidence of the EM wave, and *s-polarized* or *transverse-electric* (TE) incident radiation, where the electric vector is perpendicular to the plane of incidence. These are contrasted in Figure 4.1

Molecular Electronics: From Principles to Practice M. C. Petty
© 2007 John Wiley & Sons, Ltd

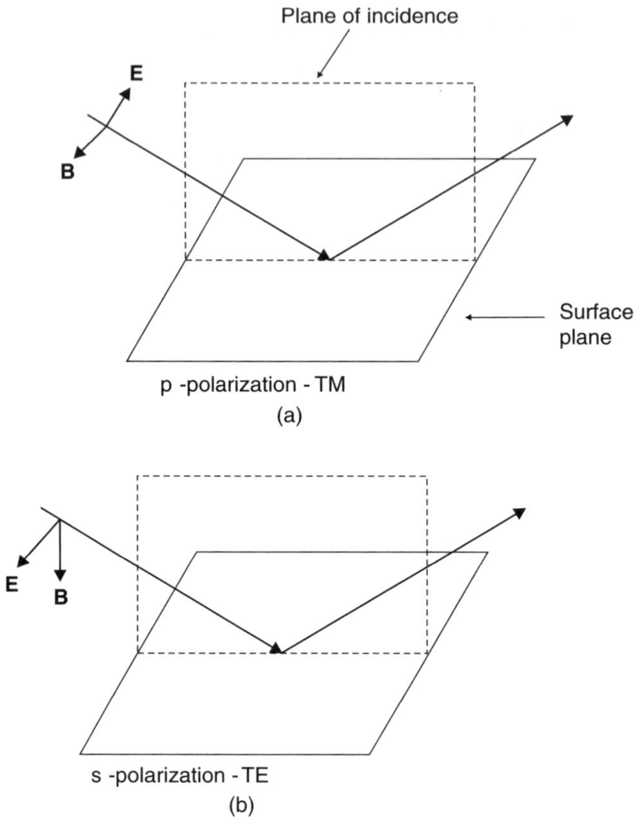

Figure 4.1 Diagrams showing s- and p-polarized EM waves incident on a surface. (a) For p-polarized (TM, transverse magnetic) waves, the electric field vector is in the plane of incidence. (b) For s-polarized (TE, transverse electric) waves, the electric field is perpendicular to the plane of incidence of the EM wave.

4.3 REFRACTIVE INDEX

The speed of light in vacuum is $3.00 \times 10^8\,\mathrm{m\,s^{-1}}$. However, when passing through a material, the light appears to slow down. The origin of this effect is the interaction of the EM wave with the atoms and molecules that make up the medium, and is described in terms of the *refractive index.*

A static electric field can induce an electric dipole (Chapter 2, Section 2.3.5) in a solid material; this is depicted in Figure 4.2, which illustrates the application of an electric field to a neutral atom. Before the electric field is applied, the centres of positive and negative charge coincide [Figure 4.2(a)]. The field then produces a very small shift [much exaggerated in Figure 4.2(b)] in the electron cloud with respect to the nucleus. This results in the formation of an induced dipole. This is the origin of the electronic polarization, described in Chapter 3, Section 3.6.1 (although the discussion concerned AC fields). The induced dipole can be

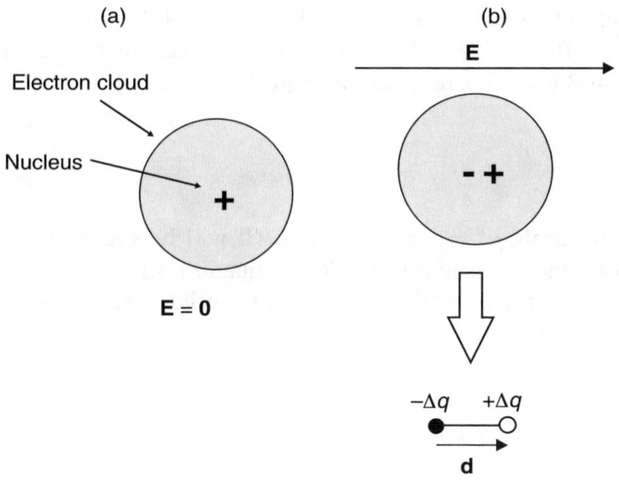

Figure 4.2 Formation of a dipole moment. (a) The centre of the negative electron cloud in a neutral atom coincides with the nucleus. (b) Application of an electric field, **E**, results in a shift in the centre of positive and negative charge and creates an electric dipole, which can be represented by an equal positive and negative charge Δq at a distance d apart.

modelled as two equal and opposite charges $(\pm\Delta q)$ at a distance apart of d. The *dipole moment*, **p**, can then be defined:

$$\mathbf{p} = \mathbf{d}\Delta q \tag{4.1}$$

where the use of the bold typeface indicates that dipole moment is a vector quantity, directed from negative to positive charge. The distance d (a scalar quantity) is therefore replaced by the vector **d** from negative to positive charge in Equation (4.1). As noted in Chapter 3 (Section 3.6.1), the application of an electric field to a polar compound (i.e. one that already contains dipoles by virtue of the chemical bonding) will cause the dipoles to attempt to orient in the field direction. This *ground-state* dipole results from the electronegativities of the constituent atoms. A simple example is the carbonyl group (C=O) which has a small positive charge on the carbon atom and a small negative charge on the oxygen (the latter is more electronegative than the former).

The oscillating electric field of an EM wave produces oscillating dipoles in any matter with which it is able to interact. The electron cloud in an atom is relatively light with respect to the nucleus and can respond to EM fields up to very high frequencies (visible to ultraviolet light). The oscillating dipoles, in turn, produce their own oscillating electric and magnetic fields which radiate EM waves – the dipoles acting as 'molecular aerials'. These combine with the incident wave to produce an electromagnetic wave that travels through the material with a velocity slower than the velocity of light in vacuum. If v is the phase velocity of light in the material and c is the velocity of light in vacuum, then

$$n = \frac{c}{v} \tag{4.2}$$

where n is the refractive index. Typical values are 1.0003 for air, 1.33 for water and about 1.5 for glass. Refractive index may also be linked to the relative permittivity ε_r (Chapter 3, Section 3.6.1) and relative *permeability* μ_r (Chapter 5, Section 5.7.1) of the medium:

$$n = \sqrt{\mu_r \varepsilon_r} \tag{4.3}$$

For nonmagnetic materials $\mu_r = 1$ and $n = \sqrt{\varepsilon_r}$. (It will be shown in the next section that both the permittivity and permeability are tensor quantities.)

Like the permittivity, the refractive index of a material is also a complex quantity:

$$n = n' - jn'' \tag{4.4}$$

where the real part n' controls effects such as reflection, refraction (by *Snell's law*) and the velocity of propagation, and the imaginary part n'' represents loss (i.e. absorption in the material). Because light of various wavelengths produces different effects on the charges present in matter, the velocity at which light propagates through a material depends on the wavelength of the incident light. This means that the refractive index also depends on wavelength (the frequency dependence of ε_r has already been noted in Chapter 3, Section 3.6.1).

4.3.1 Permittivity Tensor

The response of a material to the electric and magnetic components of an EM wave is determined by its permittivity and permeability. For most of the organic compounds discussed in this book, the relative permeability μ_r is equal to unity (magnetic organic compounds are discussed in Chapter 5, Section 5.7), so the interaction with EM waves is due solely to the dielectric properties.

The electric field, **E**, induces an *electric displacement*, **D**, in a medium such that

$$\mathbf{D} = \hat{\varepsilon}_r \varepsilon_0 \mathbf{E} \tag{4.5}$$

The electric field and resulting displacement are both vectors quantities, each with three components along three mutually orthogonal directions. This makes the permittivity is a *second-rank tensor* (formally written as $\hat{\varepsilon}_r$) and its components may be represented by a 3×3 matrix array.

For any non-optically active dielectric, it is possible to choose the x, y and z axes so that the off-diagonal elements in the 3×3 permittivity tensor array are zero. This process is called *diagonalizing* the matrix and the resulting directions are called the *principal axes*. In matrix terms, Equation (4.5) can then be written as

$$\begin{pmatrix} D_x \\ D_y \\ D_z \end{pmatrix} = \varepsilon_0 \begin{pmatrix} \varepsilon_{11} & 0 & 0 \\ 0 & \varepsilon_{22} & 0 \\ 0 & 0 & \varepsilon_{33} \end{pmatrix} \begin{pmatrix} E_x \\ E_y \\ E_z \end{pmatrix} \tag{4.6}$$

i.e.

$$D_x = \varepsilon_0 \varepsilon_{11} E_x$$
$$D_y = \varepsilon_0 \varepsilon_{22} E_y \quad (4.7)$$
$$D_z = \varepsilon_0 \varepsilon_{33} E_z$$

In an isotropic (Chapter 2, Section 2.5.1) medium, the induced polarization is independent of the electric field direction, so that $\varepsilon_{11} = \varepsilon_{22} = \varepsilon_{33}$. Therefore, all propagation directions experience the same refractive index.

For an anisotropic medium, the situation changes. In general, there are two possible values of the phase velocity for a given direction of propagation. These are associated with mutually orthogonal polarization of the light waves. The two polarizations define the *ordinary* and *extraordinary* rays and possess distinct refractive indices, and therefore different angles of refraction at an interface. Hence, when light of an arbitrary polarization propagates through an anisotropic medium, it can be considered to consist of two independent waves, which travel with different velocities. Media exhibiting such effects are said to be *birefringent*. Two levels of anisotropy may be distinguished.

1. uniaxial medium: $\varepsilon_{11} = \varepsilon_{22} \neq \varepsilon_{33}$

   For most ray directions and polarizations, both ordinary (refractive index $= \sqrt{\varepsilon_{11}} = \sqrt{\varepsilon_{22}}$) and extraordinary rays (refractive index $= \sqrt{\varepsilon_{33}}$) are generated. The exception is for light travelling in the z direction (the *optic axis*), where all polarizations are governed by the ordinary refractive index.

2. biaxial medium: $\varepsilon_{11} \neq \varepsilon_{22} \neq \varepsilon_{33}$

   Here, there are two optic axes of transmission along which the velocities of the two orthogonally polarized waves are the same. Otherwise, ordinary and extraordinary rays are produced.

Many liquid crystalline molecules (Chapter 8, Section 8.2) and organic molecules that are used for self-assembly (Chapter 7, Section 7.3) are rod-like. The films will possess a biaxial symmetry. However, $\varepsilon_{11} \approx \varepsilon_{22}$ and the situation will approximate to case 1, with two refractive indices – along the rod and perpendicular to it.

4.3.2 Linear and Nonlinear Optics

A second definition of electric displacement, **D**, to that provided by Equation (4.5) is

$$\mathbf{D} = \varepsilon_0 \mathbf{E} + \mathbf{P} \quad (4.8)$$

where **P** is the *polarization* induced in the material by the electric field. Polarization is the induced charge per unit area or the dipole moment per unit volume; the units are C m^{-2}.

In *linear optics*, the electric field is linearly related to the polarization, i.e.

$$\mathbf{P} = \varepsilon_0 \hat{\chi}_e \mathbf{E} \quad (4.9)$$

where $\hat{\chi}_e (= \hat{\varepsilon}_r - 1)$ is the second-rank *susceptibility* tensor. For *nonlinear optics* (NLO), the above equation is replaced by a series expansion in the electric field:

$$\frac{\mathbf{P}}{\varepsilon_0} = \hat{\chi}_e^{(1)}\mathbf{E} + \hat{\chi}_e^{(2)}\mathbf{E}\cdot\mathbf{E} + \hat{\chi}_e^{(3)}\mathbf{E}\cdot\mathbf{E}\cdot\mathbf{E} + \ldots \tag{4.10}$$

where $\hat{\chi}_e^{(2)}$ is called the second-order susceptibility tensor (SI units = C V^{-2}) and $\hat{\chi}_e^{(3)}$ the third-order susceptibility tensor. The second-order term is responsible for effects such as *second-harmonic generation* and the *linear electro-optic effect* (*Pockels' effect*), while the third-order term is responsible for *third-harmonic generation* and the *quadratic electro-optic effect* (*Kerr effect*). An important consequence of crystal symmetry is that in a material possessing a *centre of inversion* (a centrosymmetric material), the second-order nonlinear susceptibility $\hat{\chi}_e^{(2)}$ is zero.

Nonlinear properties are normally measured on macroscopic samples that consist of many individual molecules. On the microscopic level, a molecule placed in an electric field experiences a polarizing effect through a change in its dipole moment, **p**. Its relation to electric field may also be expressed by a power series:

$$p = \hat{\alpha}\mathbf{E} + \hat{\beta}\mathbf{E}\cdot\mathbf{E} + \hat{\gamma}\mathbf{E}\cdot\mathbf{E}\cdot\mathbf{E} + \ldots \tag{4.11}$$

where $\hat{\alpha}$ is the linear *polarizability* and $\hat{\beta}$, $\hat{\gamma}$, etc. (all tensor quantities) are the higher order polarizabilities, or *hyperpolarizabilities*.

The SI units for the β coefficients are C m^3 V^{-2}. However, workers in the nonlinear optics often quote values using the CGS (centimeter gram second) system (in common with many workers on magnetic materials – Chapter 5, Section 5.7.1), which can be obtained by multiplying the SI values by 2.7×10^{20}. Similarly, the second-order nonlinear susceptibility coefficients $\chi_e^{(2)}$ measured in the SI units of C V^{-2} can be converted to CGS units by multiplying by 2.7×10^{14}.

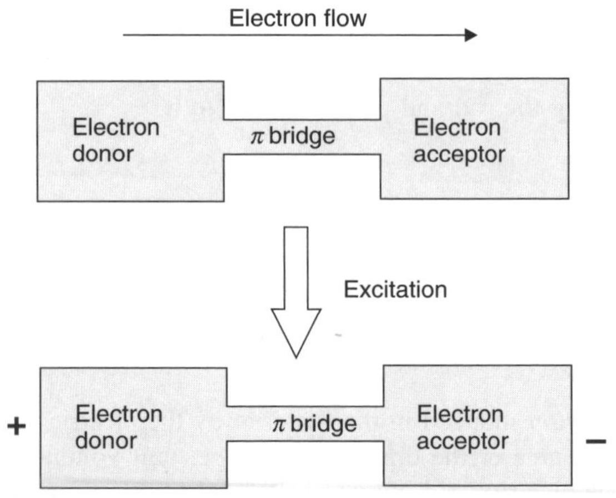

Figure 4.3 A generalized second-order NLO organic material. Electron-rich (donor) and electron-deficient (acceptor) groups are joined by a π-conjugated bridge.

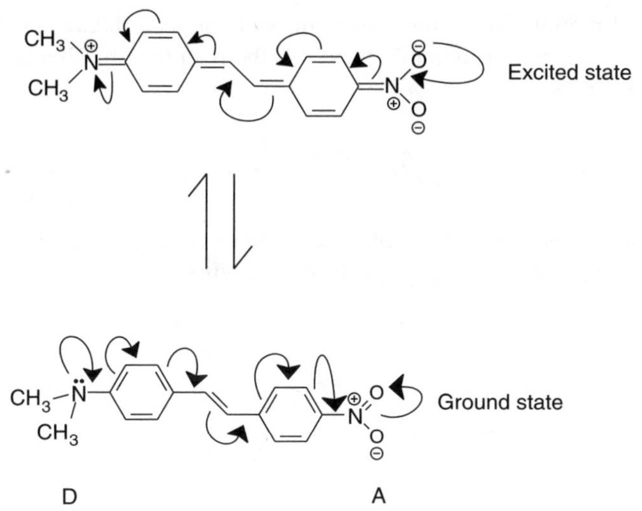

Figure 4.4 Structure of 4-(dimethylamino)-4′-nitrostilbene in its ground (bottom) and excited state (top). The arrows indicate a 'push–pull' sequence of π-electron movement.

Centrosymmetric molecules will respond to the electric field of the optical wave to give equal and opposite polarization as the phase of the wave changes through 180° and therefore will have zero β coefficients. Large β coefficients arise through the presence of an asymmetric mobile π-electron system and low lying charge-transfer states in molecules. A substantial proportion of NLO materials is represented by organic donor and acceptor moieties linked via aromatic, alkenyl or alkynyl spacer groups. A generalized model of a typical second-order NLO molecule is shown in Figure 4.3.

The donor unit, which should be electron rich, normally pushes electrons towards the electron-deficient acceptor via a π-resonance effect. Thus, the donor moiety is usually loaded with heteroatoms bearing lone pairs of electrons (preferred to alkyl donating groups, which are weaker electron contributors). Typical examples of strong electron-donating groups include alkoxy (RO–), alkylthio (RS–) and amino (R_2N–) subunits. An efficient acceptor is one which is electron deficient and can easily accommodate a negative charge. Common electron-withdrawing groups include nitro (–NO_2), cyano (–CN) and carbonyl [–C(O)–] functionalities. An example of an efficient second-order NLO material is 4-(dimethylamino)-4′-nitrostilbene, shown in its two resonance forms in Figure 4.4. The arrows indicate how electrons move along the length of the molecule.

4.4 INTERACTION OF EM RADIATION WITH ORGANIC MOLECULES

4.4.1 Absorption Processes

The intensity of light passing through an absorbing material is reduced according to *Beer's law*:

$$I = I_0 \exp(-\alpha d) \tag{4.12}$$

where I is the measured intensity after passing through the material, I_0 is the initial intensity, α is the *absorption coefficient* (units m^{-1}) and d is the path length. This is often written in a more convenient form as

$$A = -\log_{10}\left(\frac{I_0}{I}\right) \tag{4.13}$$

where A is the *absorbance* of the sample and is given by $A = (\log_{10} e)\alpha d = 0.43\alpha d$. The absorption coefficient of a solid is related to the imaginary part of the complex refractive index by

$$\alpha = \frac{4\pi n''}{\lambda} \tag{4.14}$$

A straightforward test of the reproducibility of the transfer of a dye chromophore from the water surface to a solid substrate by, say by the Langmuir–Blodgett (LB) technique (Chapter 7, Section 7.3.1), is to monitor the optical absorption (at a particular wavelength) as a function of the number of LB layers deposited. Figure 4.5 shows the result of such an experiment, using an organotransition metal complex [1]. The increase in absorbance with increasing film thickness agrees with Beer's law.

The *Beer-Lambert law* introduces the concentration of an absorbing species into the above relationship and is used in work with solutions. The absorbance of a sample A can be expressed as

$$bcd = -\log_{10}\left(\frac{I_0}{I}\right) = A \tag{4.15}$$

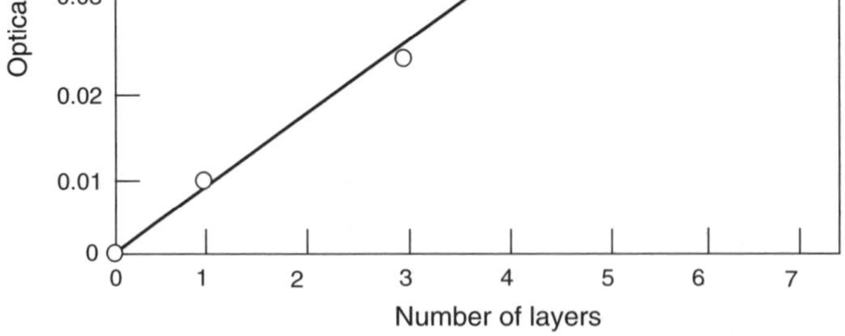

Figure 4.5 Optical absorbance at 332 nm versus number of Langmuir–Blodgett layers for an organotransition metal complex. Reprinted from *Thin Solid Films*, **160**, Richardson T, Roberts GG, Polywka MEC, Davies SG, 'Preparation and characterization of organotransition metal Langmuir–Blodgett films', pp. 231–239. Copyright (1988), with permission from Elsevier.

where b is the molar absorption coefficient ($\mathrm{l\,mol^{-1}\,m^{-1}}$) and c is the molar concentration of the solution ($\mathrm{mol\,l^{-1}}$ or M). The value of b is usually quoted at an absorption maximum. Since the absorbance is proportional to the concentration of the solution, a plot of A versus c should yield a straight line. This relationship is observed in many materials. However, at high concentrations, changes in the position or intensity of the absorption maximum, caused by the formation of molecular aggregates such as *dimers* or *trimers* (Section 4.4.2) often results in a deviation from the Beer–Lambert law. The interaction between the dye molecules and the solvent can also give deviations from this law as a result of *solvatochromism*, which is the ability of a chemical substance to change colour due to a change in solvent polarity. Negative solvatochromism corresponds to a shorter wavelength or *hypsochromic* shift, whereas positive solvatochromism corresponds to a longer wavelength, or *bathochromic*, shift with increasing solvent polarity. The sign of the solvatochromism depends on the difference in dipole moment of the molecule of the dye between its ground state and excited state.

In terms of electronic energy levels, a molecule in its ground state can absorb a photon of light if the photon energy is equal to the difference between the two energy levels in the system. When this occurs, the molecule is excited into a higher energy state so that

$$E_1 - E_0 = h\nu \tag{4.16}$$

where ν is the frequency of the radiation in Hz and E_1 and E_0 are the excited and ground states of the molecule, respectively. Thus, for molecular crystals, the minimum photon energy required for absorption will be when E_0 corresponds to the HOMO level and E_1 to the LUMO level. In the case of a conductive polymer, absorption will occur when the incident radiation has sufficient energy to promote electrons from the valence (π) band to the conduction (π^*) band. The optical absorption spectra for *trans*-polyacetylene are shown in Figure 4.6. The data are shown for both the undoped polymer and following doping with

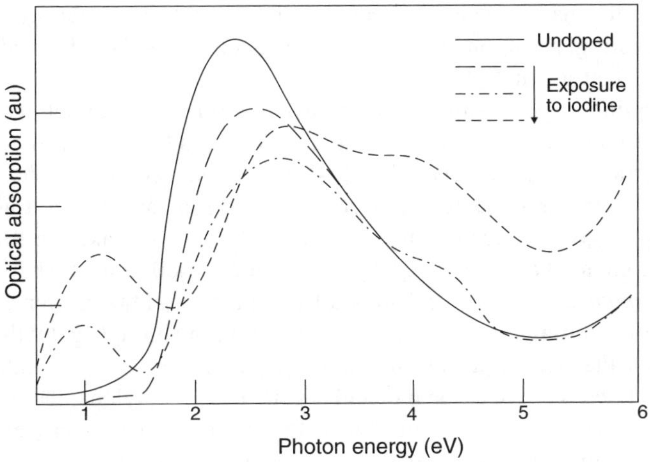

Figure 4.6 Optical absorption of *trans*-polyacetylene as a function of doping level. As the doping increases a mid-gap soliton level forms below the band edge. Reproduced from *Phys. Trans. R. Soc. London A*, **314**, Friend RH, Bott DC, Bradley DDC, Chai CK, Feast WJ, Foot PJS, Giles JRM, Norton ME, Pereira CM, Townsend PD, 'Electronic properties of conjugated polymers', pp. 37–49, Copyright (1985), with permission from The Royal Society.

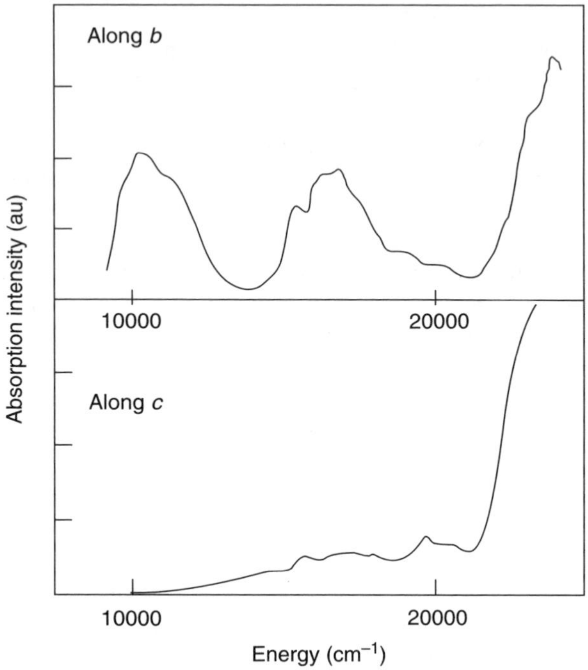

Figure 4.7 Polarized optical absorption spectra along different crystallographic axes for the charge-transfer compound perylene/tetracyanoethene. Reprinted from *J. Mol. Spectrosc.* **22**, Kuroda H, Kunii T, Hiroma S, Akamatu H, pp. 60–75, Copyright (1967), with permission from Elsevier.

iodine [2]. For the undoped material, there is a rapid increase in the optical absorption when the incident radiation has an energy exceeding 1.4 eV, the band gap of the polymer. On iodine doping, a mid-gap soliton state (Chapter 3, Section 3.4.3) is formed, accounting for the absorption peak at about 1 eV.

Optical absorption experiments can also reveal information about the charge-transfer bands in certain complexes. For example, Figure 4.7 shows the polarized absorption spectra of a perylene/tetracyanoethene crystal [3]. The low-energy transitions in the 10 000–20 000 cm^{-1} wavenumber (Chapter 6, Section 6.6) region are charge-transfer transitions and are clearly strongly polarized along the *b*-axis, the stacking axis, in the crystal.

The dipole moment that is induced by the applied oscillating EM field is called the *transition dipole moment*. It can be calculated from an integral taken over the product of the wavefunctions of the initial and final states of a spectral transition. Light will not be absorbed completely unless the oscillating electric field is parallel to the transition moment. This dipole moment is fixed relative to the molecular structure. In ethylene, the $\pi \rightarrow \pi^*$ transition, which is associated with the C=C bond, has its transition moment polarized along the bond. Complex molecules may have more than one transition moment.

For dye molecules on a solid plate (substrate), the orientation (if any) can be conveniently investigated by using polarized light. For instance, if a wavelength at which a molecule absorbs is chosen, then direct comparison of the absorption intensities of the s- and p-polarized radiation (at the same angle of incidence) enables the average orientation of the transition moments to be determined.

Most organic molecules have an even number of electrons, with all electrons paired. Within each pair, the opposing electron spins cancel and the molecule has no net electronic spin. Such an electronic structure is called a *singlet* state. When a ground-state singlet absorbs a photon of sufficient energy, it is converted to an excited singlet state in which the spin of the excited electron is not altered. The process is so fast that the excited state has the same geometry of bond distances and bond angles. This is the *Franck–Condon principle*. However, the most stable geometry of the excited state often differs from that of the ground state, so that the excited electronic state is formed in an *excited vibrational state*.

Shapes of absorption bands can be explained by reference to *Morse curves*. These are plots of potential energy as a function of the nuclear distance *r*. In Figure 4.8, Morse curves corresponding to a ground and an electronically excited state of a particular molecule are shown. The horizontal lines represent the vibrational levels of the electronic states and each of these has a vibrational quantum number associated with it, $j = 0, 1, 2, 3, \ldots, n$; these levels may have energy spacings of ~0.01 eV. The excited state is represented by a Morse

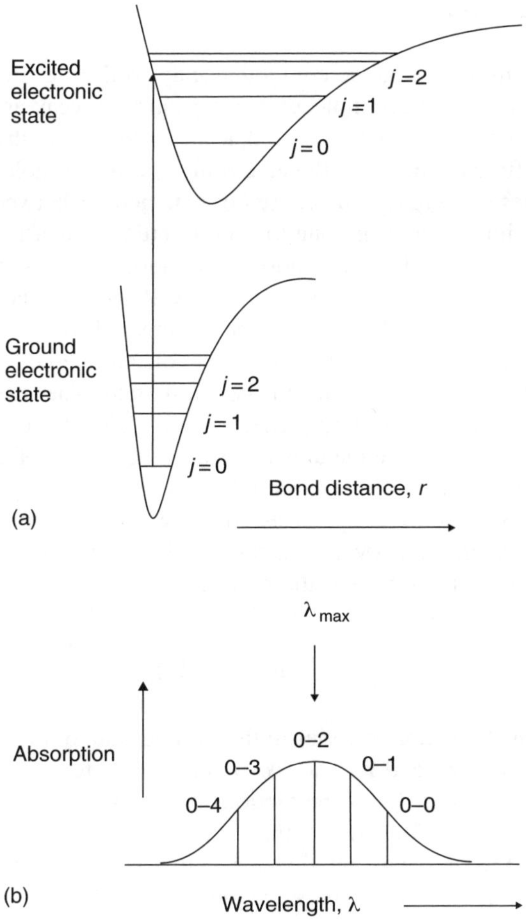

Figure 4.8 (a) Morse curves for the ground and excited states of a molecule. (b) The origin of the shape of absorption bands.

curve which is displaced vertically (higher energy) and horizontally (due to vibration increased bond length) from that of the ground state. By the Franck–Condon principle, electronic transitions are represented on the Morse curve by vertical lines.

The probability and intensity of an electronic transition will not be a maximum between the $j = 0$ levels of the ground and excited states. Because of the relative displacement of the two Morse curves, the maximum absorption will occur between the $j = 0$ level of the ground state and a higher energy level in the excited state [$j = 2$ in Figure 4.8(a)]. This corresponds to the wavelength for maximum absorption, λ_{max}. On each side of this, the absorption intensities will decrease to zero, producing the familiar bell-shaped absorption band shown in Figure 4.8(b). In the case of polyatomic molecules, the Morse curves are replaced by polydimensional surfaces and the number of allowed transitions will become very large. Consequently, the absorption bands become smooth curves. Sometimes, however, vibrational fine structure can be seen.

4.4.2 Aggregate Formation

The absorption spectrum of an organic compound may well differ in the solid and solution states. Although the forces between molecules in crystals are weak and short range (van der Waals interactions) and the overlap between adjacent molecules in the lattice is small, there is still a substantial difference between the electronic spectra of molecular crystals and free molecules. Some of these differences are caused by interactions between the electronic states of molecules in the vicinity; others are due to crystal lattice properties. Crystal spectra have absorption bands that are broader than those in solution. This is because the molecular interactions are affected by thermal vibrations as well as the relative orientation of the molecules. Bands are often shifted and split when compared to solution.

The formation of aggregates is fairly common in concentrated dye solutions and is often difficult to avoid when the molecules are in the solid state. The interactions between dye molecules can be explained by considering energetically delocalized states, termed *excitons*. In Figure 4.9, $E_{excited}$ and E_{ground} are the unperturbed energy levels of an isolated molecule in dilute solution. When these molecules are brought close together [Figure 4.9(a)], the dipoles interact and multiple excitation energy levels are observed. The splitting of energy levels, *Davydov splitting*, is determined by the difference between interacting transition dipole moments, their relative orientations and the number of interacting molecules. For a simple dimer, the energies are given by

$$\Delta E_{dimer} = \Delta E_{monomer} + D \pm \varepsilon \tag{4.17}$$

where ΔE represents the transition energies to the excited state of the dimer or monomer, ε is the exciton interaction energy and D is a *dispersion energy* term, given by $W - W'$ in the figure; this depends on van der Waals interactions between the molecules.

Equation (4.17) shows that for a dimer the excited energy state has two possible levels. These can be attributed to the phase relationship between the transition dipole moments. If the dipole moments of the two molecules are parallel, as shown in Figure 4.9(b), the net transition moment of the dimer is zero in the lower energy state E' and this transition is forbidden. Only transitions to the higher energy state E'' are allowed, leading to a hypsochromic shift, *H-band*, which is often observed in organized thin films. If the transition

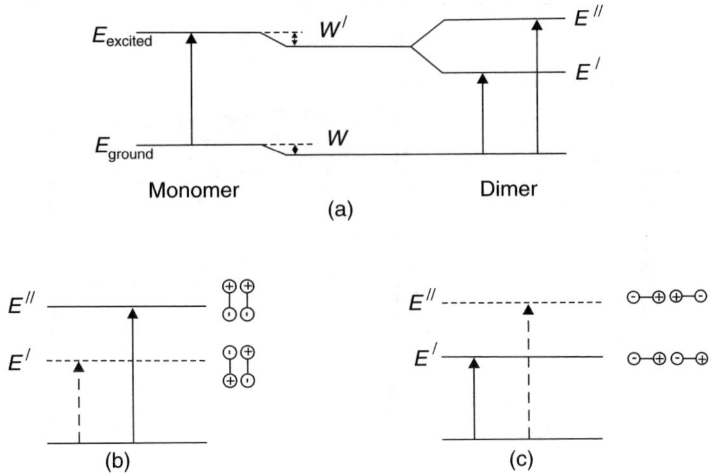

Figure 4.9 (a) Aggregation effects on the energy levels of molecules. (b) Parallel transition dipoles (hypsochromic shift). (c) In-line transition dipoles (bathochromic shift).

dipoles are in-line rather than parallel [Figure 4.9(c)], transitions to the higher energy state are forbidden and a bathochromic shift is observed in the spectrum. Higher aggregates of this type exhibit a large bathochromic shift, with an intense, narrow absorption; these are termed *J-aggregates*. Of course, parallel and in-line are the two extreme forms of molecular aggregation and, in general, the transition dipoles are oriented at some angle to each other, resulting in more complex spectral changes.

The formation of J-aggregates is illustrated in Figure 4.10(a), which shows the absorption spectrum of a monolayer of an amphiphilic cyanine dye mixed in a 1:1 molar ratio with octadecane [4]. A very narrow absorption band is due to an in-phase relationship between the oscillators corresponding to the dye chromophores. The closest packing of the chromophores is achieved with the brickwork arrangement of dye molecules in which the octadecane chains fit into the cylindrical holes left by the hydrocarbon chains of the dye [Figure 4.10(b)].

4.4.3 Excitons

Extending the above dimer model to an infinite three-dimensional lattice is a similar exercise to the extension of a covalent bond between two atoms to the delocalized band structure of a metal or semiconductor (Chapter 3, Section 3.3). Sharply defined energy levels associated with the two-body system become a band of energy levels in the infinite lattice, with a bandwidth dependent on overlap within the lattice. Thus Figure 4.9 can be applied to a crystal if each of the levels E' and E'' is replaced by a narrow band of levels. These bands are known as exciton bands. In solids with strong intermolecular interactions, an exciton can be delocalized over a number of molecules. Depending on the degree of delocalization, the excitons are identified as *Frenkel*, charge-transfer or *Wannier–Mott*. These two extreme cases are depicted in Figure 4.11. The Frenkel exciton [Figure 4.11(a)],

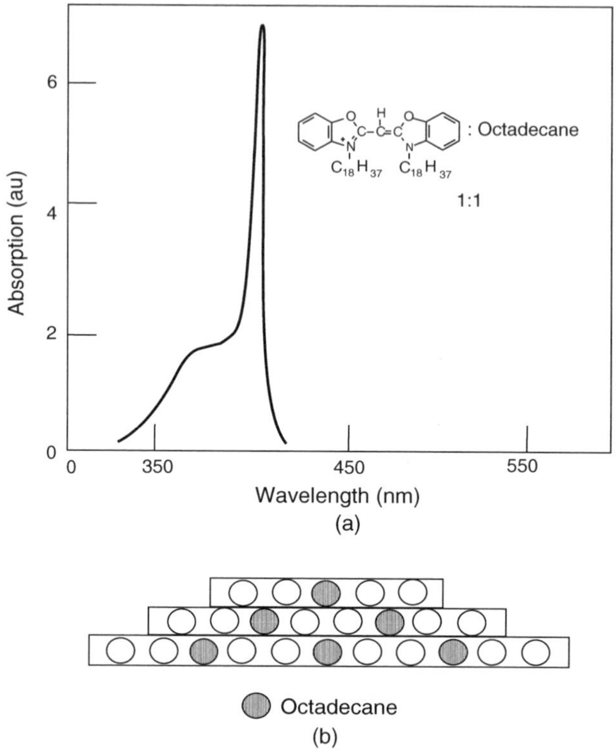

Figure 4.10 (a) Narrow absorption band resulting from *J*-aggregate formation in a mixed monolayer of an amphiphilic cyanine dye and octadecane. (b) The molecules are assumed to be packed in a brickwork arrangement shown (viewed from above). Reproduced with permission from Kuhn H, Möbius D, Bücher H, 'Spectroscopy of monolayer assemblies', in A Weissberger, B Rossiter (Editors), *Techniques of Chemistry*, Vol. 1, Part 3B, pp. 577–702, Copyright (1972), John Wiley & Sons Inc.

corresponds to a correlated electron–hole pair localized on a single molecule. Its radius is therefore comparable to the size of the molecule (<0.5 nm) or is smaller than the intermolecular distance. A Frenkel exciton can be considered as a neutral particle that can diffuse from site to site, perhaps moving hundreds of molecules away from its origin. Excitons can therefore transport energy without involving the migration of net electric charges. In contrast, Wannier–Mott excitons [Figure 4.11(b)], occur in crystalline materials in which overlap between neighbouring lattice atoms reduces the Coulombic interaction between the electron and the hole of the exciton (large dielectric constants). This results in a large radius, 4–10 nm, many times the size of the lattice constant. This type of exciton is not found in van der Waals-bonded organic solids, but is more typically found in inorganic semiconductors such as silicon or gallium arsenide. Figure 4.11 also indicates the energy levels associated with the different types of excitons. The Frenkel exciton can be considered as two distinct localized states, located above the HOMO and below the LUMO levels of the molecule, whereas the Wannier–Mott excitons are associated with the extended band structure of the crystal in which they are formed.

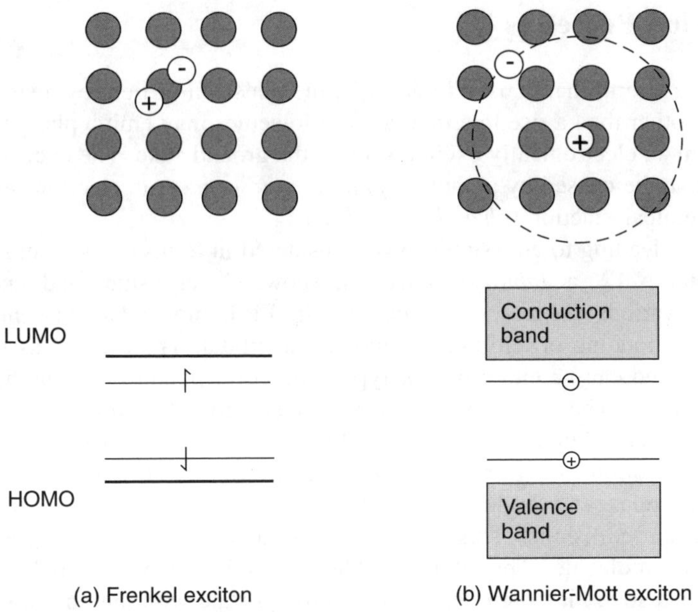

Figure 4.11 (a) Frenkel and (b) Wannier–Mott excitons with their respective electron energy diagrams.

4.4.4 Effect of Electric Fields on Absorption

Changes in the optical properties of organic molecules may be produced by applying large electric fields. *Electroabsorption* is one technique that has been used to study materials. This is a particular branch of the group of experimental methods known as modulation spectroscopy and involves monitoring small changes in a sample's optical transmission that result from the application of an external field. For molecular compounds, these changes are usually the result of the *Stark effect*. The change in potential energy ΔV that arises when a molecule is placed in an electric field may be written as

$$\Delta V = -pE\cos\theta + \frac{\alpha}{2}E^2 \tag{4.18}$$

where the dipole moment p makes an angle θ to the applied field and α is the polarizability [Equation (4.11)]. The first term on the right-hand side of Equation (4.18) represents the linear Stark effect and the second term the quadratic effect. In general, noncentrosymmetric crystals with polar space groups exhibit a first-order Stark effect. This is usually recognized in electroabsorption spectra as the second derivative of the zero-field absorption curve. Crystal structures having centrosymmetric crystal structures may still exhibit a second-order Stark effect. This process is observed in electroabsorption as a first-order derivative of the zero-field absorption curve.

4.4.5 Emission Processes

Luminescence is a term that is used to describe the emission of light by a substance caused by any process other than a rise in temperature. Molecules may emit a photon of light when they decay from an electronically excited state to the ground state. The excitation leading to this emission may be caused by a photon (*photoluminescence*), an electron (*electroluminescence*) or a chemical reaction (*chemiluminescence*).

The processes leading to emission can be considered in terms of the energy levels in the molecule. Figure 4.12, a *Jablonski diagram*, shows a very simplified energy diagram illustrating the various processes that can occur. Each molecular state in the diagram corresponds to a bonding or antibonding molecular orbital. The orbitals associated with a carbon–carbon bond can be either σ or π type, with corresponding antibonding orbitals σ^* and π^*, respectively. There are also valence-shell electrons that do not participate in the formation of molecular bonds. These nonbonding orbitals are designated n.

Following promotion to a vibrationally excited singlet state (S_1 or S_2 in the figure), the molecule usually relaxes to the lowest vibrational level of S_1, the energy being lost as heat via intermolecular collisions. This process takes about 10^{-11} s, or approximately 10^2 vibrations of the molecule. The lifetime of S_1 in its lowest vibrational state is longer, 10^{-8}–10^{-7} s. This state may then decay to the ground state with the emission of a photon.

This emission is termed *fluorescence*. Alternatively, the energy may be transferred by *intersystem crossing* to a *triplet state* (the *multiplicity* of such a state $= 2S + 1$, where S is the algebraic sum of the electron spins, is 3). In such a state the spins of the two electrons are parallel and a transition to the ground state, with the emission of a photon, involves a change of spin. This is a *spin-forbidden* transition; quantum mechanical selection rules forbid transitions in which the electron spin changes, i.e. only singlet to singlet or triplet to triplet transitions are allowed. Triplet states are fairly long lived, with lifetimes of greater than 10^{-5} s. The triplet

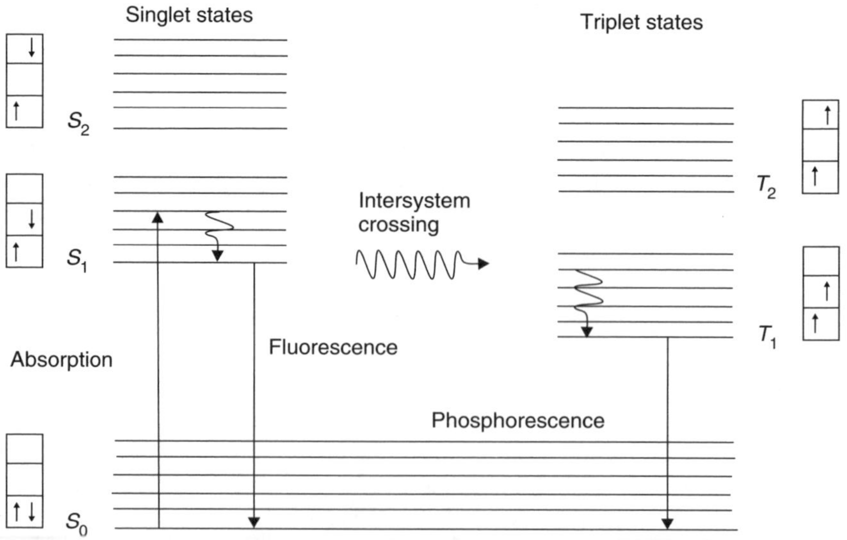

Figure 4.12 Jablonski diagram showing the radiative and nonradiative decay of an excited molecule. The orientations of the electron spins are shown in the boxes next to each state.

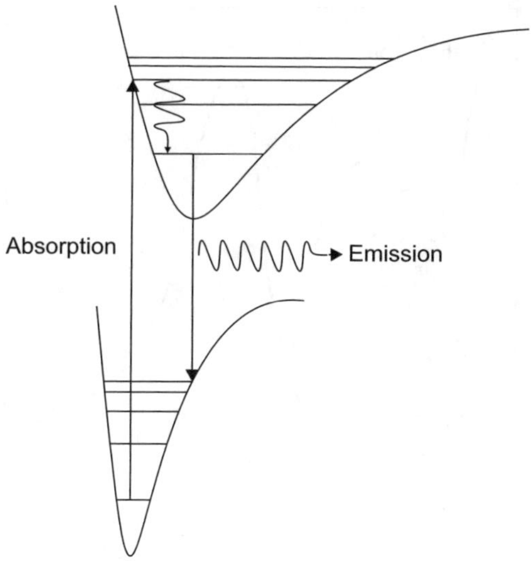

Figure 4.13 Absorption and emission processes compared using the Morse curve. The emission occurs at a lower energy (longer wavelength) than the absorption.

state is typically at a lower energy than the corresponding singlet state. The triplet state can decay to S_0, emitting a photon. This process is known as *phosphorescence* and may persist for seconds or even longer after the incident excitation has ceased. Nonradiative transitions from the excited singlet and triplet states to the ground state are also possible. In these instances the original light quantum is converted into heat.

The energy of an emitted photon is lower than that of the absorbed photon because the excited state has lost energy before the emission occurs. This is evident by reference to the Morse curves (Figure 4.13). The energetic red shift is often referred to as the *Stokes shift*. Almost all molecular organic compounds show this energy shift. For example, Figure 4.14 compares the absorption and emission spectrum in a thin film of the organic dye compound aluminium *tris*(8-hydroxyquinoline) (Alq_3) [5]. There is almost no overlap between these spectra and the Alq_3 is therefore transparent to its own emission radiation. The absorption and emission spectra of organic solids often exhibit a mirror-image relationship to one another.

Another important interaction of excited states is *excimer* formation. An excimer is an excited-state dimer and results from the interaction between an excited singlet state and an unexcited molecule. Excimer emission is usually broad, structureless and somewhat red shifted from the normal molecular fluorescence. In molecular crystals, excimer emission can be observed when the crystal structure is such that adjacent molecules form parallel-plane dimers which allow appreciable overlap of the delocalized π-electrons. An excellent example of the correlation of crystal structure to excimer emission is provided by work with perylene [6]. Perylene exists in two crystal forms, the dimeric α-form as shown in Figure 4.15(a) and the monomer β-form shown in Figure 4.15(b). The emission spectra of perylene from solution and the two crystal forms are shown in Figure 4.15(c). In solution, the fluorescence emission is highly structured and corresponds to a mirror image of the solution absorption spectrum. The β crystal form also shows a structured emission and a good mirror image relationship with the crystal absorption spectrum. The excimer emission spectrum of the α crystal form is in marked contrast; it is broad, structureless and considerably red shifted

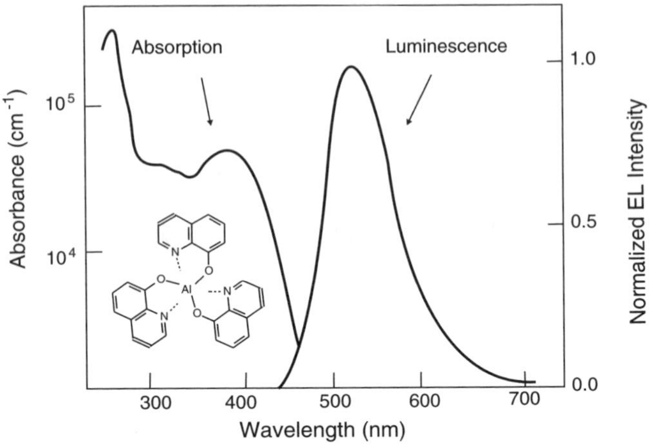

Figure 4.14 Absorption and luminescence of aluminium *tris*(8-hydroxyquinoline) with the chemical formula shown. Reproduced from *Organic Electronic Materials*, Bulović VG, Baldo MA, Forrest SR, 'Excitons and energy transfer in doped luminescent molecular organic materials', p. 397, Copyright (2001), with kind permission of Springer Science and Business Media.

from the monomer emission [note the abscissa in Figure 4.15(c) is in terms of wavenumber $\bar{\nu}$ – Chapter 6, Section 6.6].

The light emitted from an organic solid as described above will generally be *incoherent*, which means that there is no ordered phase relationship between the EM waves emitted.

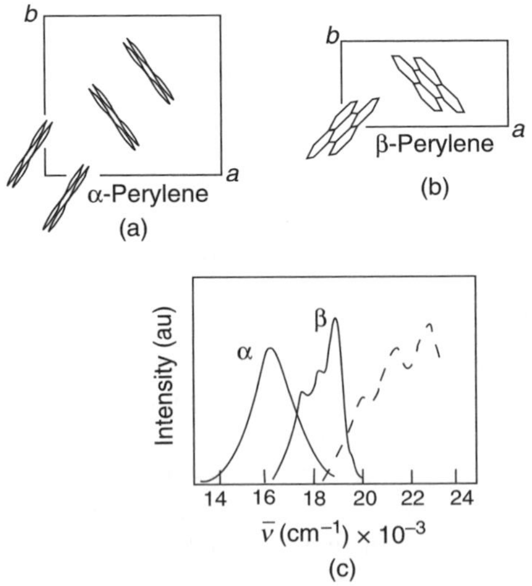

Figure 4.15 (a) Crystal structure of α-perylene. (b) Crystal structure of β-perylene. (c) Fluorescence spectra of perylene in solution (dashed line) and α-perylene and β-perylene at 77 K; the abscissa has units of wavenumber $\bar{\nu}$. Reprinted from Tanaka J, *Bull. Chem. Soc. Jpn.*, **36**, 'The electronic spectra of aromatic molecular crystals. The crystal structure and spectra of perylene', p. 1234, Copyright (1963), with permission from the Chemical Society of Japan.

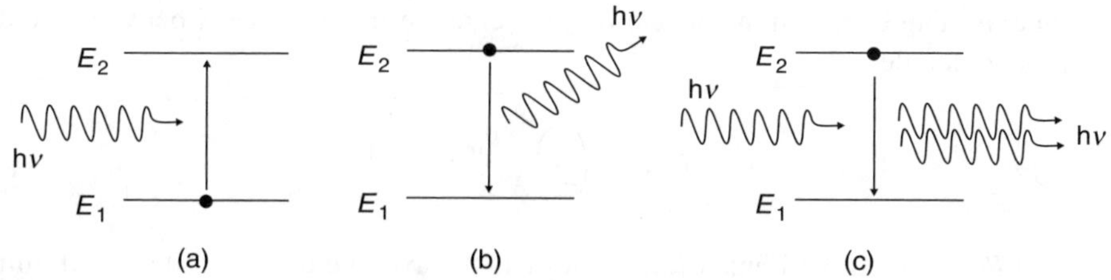

Figure 4.16 Absorption and emission processes associated with two electron energy levels E_1 and E_2. (a) Absorption of a photon with energy $h\nu$ promotes an electron from level E_1 to E_2. (b) Spontaneous emission; the photons are emitted randomly (c) Stimulated emission; the emitted photons are coherent with the stimulating photons.

The electron in the excited state E_1 spontaneously falls back into the ground state E_2, with the emission of a photon of energy $h\nu = E_2 - E_1$. If there are many such excited electrons, they will emit photons at random times and so generate incoherent light. This is *spontaneous emission*, illustrated in Figure 4.16. However, it is possible for an incoming photon to trigger the emission process. This situation is called *stimulated emission* [Figure 4.16(c)]. The emitted photon is in-phase with the incoming photon, it is going in the same direction and it has the same frequency (since it must also have the same energy $E_2 - E_1$). The result is the emission of *coherent* radiation. Stimulated emission is the basis for the operation of the *laser* (an acronym for *l*ight *a*mplification by *s*timulated *e*mission of *r*adiation). This device acts as a photon amplifier since one incoming photon results in two outgoing photons. An important requirement is that there are more electrons in energy state E_2 than in E_1, a situation referred to as *population inversion*. It is not possible to achieve this state with only two energy levels as, in the steady state, the incoming photon flux will cause as many upward transitions as downward stimulated transitions. To create a population inversion, an additional level(s) is needed and energy is *pumped* into the lasing medium by some process such as the passage of a current, the creation of an electrical discharge or illumination with EM radiation.

4.4.6 Energy Transfer

The energy of excited molecules can also be transferred to other molecules. In general terms, during the energy transfer process, an excited donor molecule D* transfers its energy to an acceptor A which, in turn is promoted to an excited state:

$$D^* + A \rightarrow D + A^* \tag{4.19}$$

For example, in some doped organic films, the exciton energy can be transferred from the host to the guest molecules, quenching the luminescence of the host while increasing that of the guest. Two important and distinct energy transfer mechanisms are recognized: *Förster transfer* and *Dexter transfer*. The former is a mechanism of excitation transfer that can occur between molecular species which are separated by relatively large distances (i.e. exceeding their van der Waals radii). It is described in terms of an interaction between the transition dipole moments. The energy transfer rate K_{ET} depends on the distance R between the donor and acceptor molecules:

$$K_{ET} = \left(\frac{1}{\tau}\right)\left(\frac{R_0}{R}\right)^6 \quad (4.20)$$

The constant R_0 is called the Förster radius and τ is the average donor excitation lifetime for recombination in the absence of energy transfer. When $R = R_0$, the probability than an exciton will recombine at the donor is equal to its transfer probability. In Förster energy transfer, the spin of both D and A is conserved. Dexter excitation transfer is the result of an electron exchange mechanism. This requires an overlap of the wavefunctions of the donor and acceptor. Dexter transport occurs over short distances, typically 0.1 nm, and is effectively restricted to neighbouring molecules. In the Dexter transfer process, only the total spin of the D^*A system is conserved and triplet–triplet energy transfer, which is forbidden in the Förster process, is now allowed. Schematic representations of both Förster and Dexter processes for the case of singlet–singlet transfer are shown in Figure 4.17.

An elegant demonstration of Förster energy transfer is provided by work with multilayer films, such as Langmuir–Blodgett film [4]. The film deposition method allows molecular assemblies to be built up in which individual layers of molecules are positioned at a very precise distance from one another (Chapter 7, Section 7.3.1). Figure 4.18 illustrates the type of experiment that can be undertaken. A donor (sensitizer) molecule within a monolayer is represented by D and an acceptor molecule by A. The D molecules absorb in the ultraviolet part of the EM spectrum and fluoresce in the blue, whereas molecules of A absorb in the blue and fluoresce in the yellow. If there is sufficient distance between A and D, the fluorescence

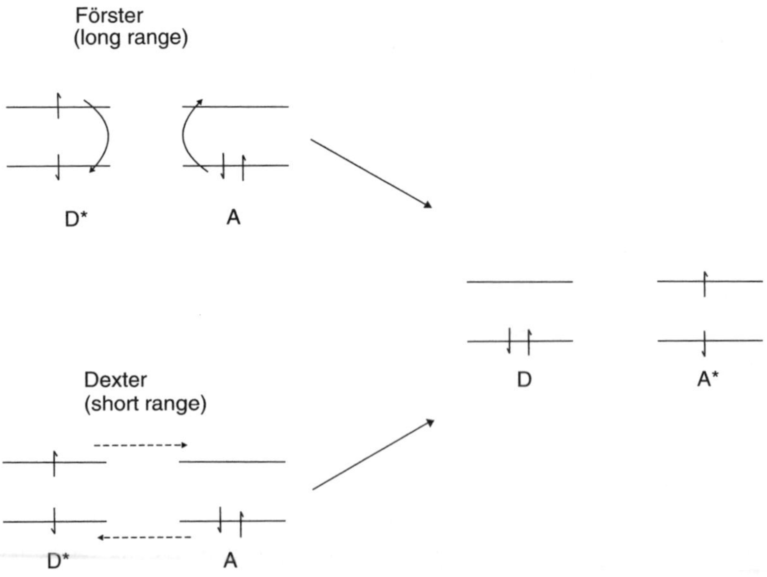

Figure 4.17 Schematic representation of Förster and Dexter energy transfer processes from an excited donor D* to acceptor A in single–singlet energy transfer.

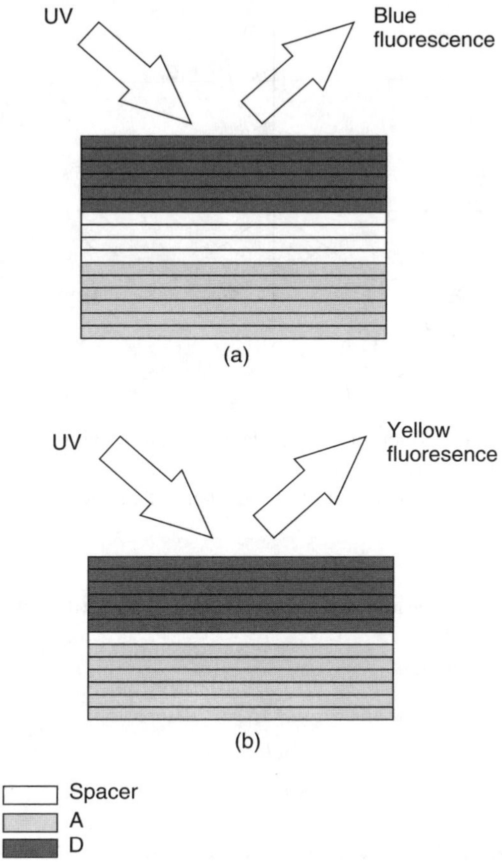

Figure 4.18 Monolayers of a sensitising donor D and an acceptor molecule A are separated by monolayers of a spacer material. In (b) the spacing between D and A is sufficiently close for energy transfer to occur between the D and A molecules.

of D only appears. However, if the distance between the monolayers of D and A is reduced sufficiently, the fluorescence of D is absorbed by A and yellow fluorescence is observed.

4.5 TRANSMISSION AND REFLECTION FROM INTERFACES

4.5.1 Laws of Reflection and Refraction

When electromagnetic waves are incident on the interface between two dielectrics, the familiar phenomena of reflection and refraction take place. Figure 4.19 shows a simple case in which a p-polarized EM wave is incident, at an angle θ_i, to the boundary between two dielectric media of refractive indices (real quantities) n_1 and n_2. Usually there will be a reflected beam at an angle θ_r and a transmitted (refracted) beam at θ_t. Application of the boundary conditions that the normal components of the magnetic fields and the tangential

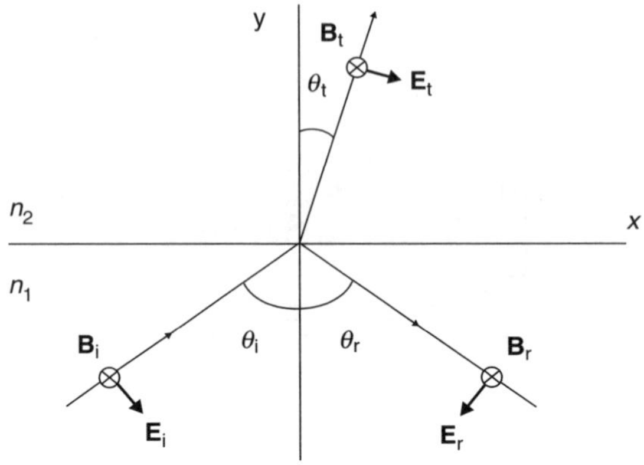

Figure 4.19 Reflection and refraction of an electromagnetic wave at a boundary between two dielectric media. The incident wave is p-polarized with its electric vector E_i in the plane of incidence. ⊗ represents a magnetic vector perpendicular to the plane of incidence.

component of the electric fields are continuous at the interface gives the following relations:

$$\theta_i = \theta_r \tag{4.21}$$

$$\frac{\sin\theta_i}{\sin\theta_t} = \frac{n_2}{n_1} \tag{4.22}$$

Equations (4.21) and (4.22) are, of course, familiar as the *law of reflection* and *Snell's law of refraction* (these two equations also hold for s-polarized incident radiation).

4.5.2 Fresnel Equations

The proportions of the incident electric field amplitude that are reflected may also be evaluated. For p-polarized (TM) radiation, the ratio E_r/E_t defines the *reflection coefficient* $r_{\|}$ and E_t/E_t is the *transmission coefficient* $t_{\|}$

$$r_{\|} = \frac{E_r}{E_i} = \frac{\tan(\theta_i - \theta_t)}{\tan(\theta_i + \theta_t)} \tag{4.23}$$

$$t_{\|} = \frac{E_t}{E_i} = \frac{2\cos\theta_i \sin\theta_t}{\sin(\theta_i + \theta_t)\cos(\theta_i - \theta_t)} \tag{4.24}$$

Equations (4.23) and (4.24) are known as *Fresnel equations* and apply at optical frequencies to transparent nonmagnetic media where the refractive indices are real quantities [i.e. in Equation (4.4), $n'' = 0$). Equivalent relationships for $r_{\perp}$ and $t_{\perp}$ may be obtained for s-polarized incident radiation. The r and t coefficients also give the phases of the beams. If positive, there is no change in phase; if negative, the phase changes by π.

There is a particular angle of incidence at which $r_{\parallel}$ becomes zero while $r_{\perp}$ remains finite. This is called the *Brewster angle*, θ_B, and is given by

$$\tan\theta_B = \frac{n_2}{n_1} \tag{4.25}$$

For incidence from air to glass ($n_1 = 1$; $n_2 = 1.5$), $n_2/n_1 = 1.5$ and θ_B is approximately 56°. At this angle, any incident light can only be reflected with the electric field vector perpendicular to the plane of incidence, giving a method of producing linearly polarized light from an unpolarized beam.

The amount of power (or intensity) reflected or transmitted from an interface is proportional to the square of the electric field amplitude. *Reflecting*, *R*, or *transmitting power coefficients* (or intensities), *T*, may be defined:

$$R = r^2 \quad \text{and} \quad T = t^2 \tag{4.26}$$

where *T* and *R* are also known as the *transmittance* and *reflectance*, respectively. At normal incidence (i.e. $\theta_i = 0$), and for the case $n'' = 0$, *R* and *T* are given for both states of polarization by

$$R = \left(\frac{n_1 - n_2}{n_1 + n_2}\right)^2 \tag{4.27}$$

$$T = 1 - R = \frac{4n_1 n_2}{(n_1 + n_2)^2} \tag{4.28}$$

Equation (4.27) predicts that *R* is approximately 4% for radiation incident from air to a glass surface.

When radiation is incident from a dense to a less dense medium, Snell's law gives a *critical angle of incidence* θ_c at which there is 100% reflection (for both polarizations). The situation is referred to as *total internal reflection*.

Figure 4.20 summarizes the main features for the reflection of p- and s-polarized EM radiation at a vacuum/dielectric interface ($n_1 = 1$; $n_2 = n$).

If the EM radiation is incident from a transparent to an absorbing medium, then the real refractive index n_2 in Equation (4.22) is replaced by $n_2' - jn_2''$, i.e.

$$\sin\theta_t = \frac{n_1 \sin\theta_i}{n_2' - jn_2''} \tag{4.29}$$

Therefore, θ_t is complex and does not represent the angle of refraction, except for the special case $\theta_i = \theta_t = 0$. Here, the Fresnel refection coefficients (identical for both components of polarization) are given by

$$r_{\parallel} = r_{\perp} = \frac{n_1 - n_2' + jn_2''}{n_1 + n_2' - jn_2''} \tag{4.30}$$

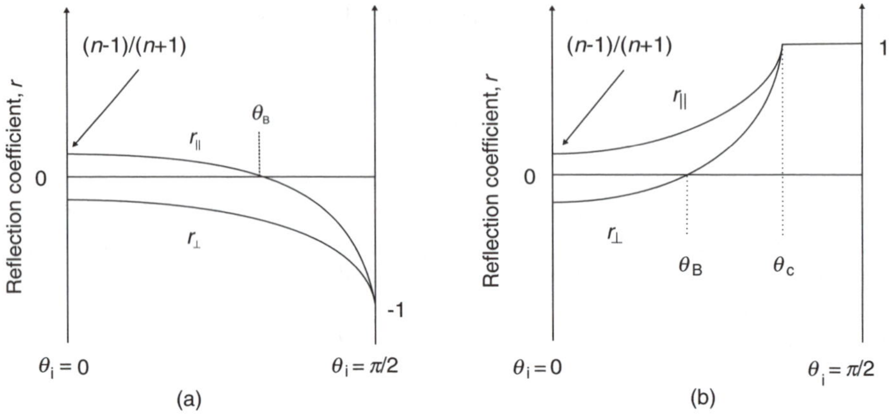

Figure 4.20 Variation of reflected wave amplitude with angle of incidence for (a) waves incident from vacuum ($n = 1$) to a dielectric with refractive index n and (b) waves incident from inside the dielectric to a vacuum: $r_{\|}$, reflection coefficient for p-polarized incident radiation; $r_{\perp}$, reflection coefficient for s-polarized incident radiation; θ_B, Brewster angle; θ_c, critical angle.

4.5.3 Ellipsometry

As we have seen above, the reflectivity from a surface differs in both amplitude and phase for s- and p-polarized incident radiation. *Ellipsometry* is a comparison of these reflectivities and it is the basis for a powerful method for measuring the thickness and optical constants of thin films. The presence of a surface layer alters the ratio of the electric field vectors vibrating in the plane of incidence and perpendicular to it, also their difference in phase Δ. In the former case, an angle ψ is defined as $\tan^{-1}$ (reflectivity amplitude ratio). The theory of ellipsometry correlates the parameters Δ and ψ with the optical thickness of the layer and the optical constants of the surface.

If there is no mixing between the polarizations, the reflection behaviour can be fully described by the two Fresnel coefficients above for p-polarized input to p-polarized output $r_{\|}$ and the equivalent for s-polarization:

$$\frac{r_{\|}}{r_{\perp}} = \tan\psi \exp j\Delta \tag{4.31}$$

It is usual to measure indirectly the ratio of these quantities by finding ψ and Δ. Using the Fresnel equations, it is relatively straightforward to calculate ψ and Δ. However, to evaluate the refractive index and thickness of a thin film from measured values, the situation is more complicated. Computer programs do this for commercial ellipsometers.

4.5.4 Thin Films

The above ideas can now be applied to the situation in which there is more than one interface, for example that of a thin film on a solid substrate. Figure 4.21 shows a schematic diagram for this. Radiation will first be reflected at the air/film boundary; the Fresnel equations for the

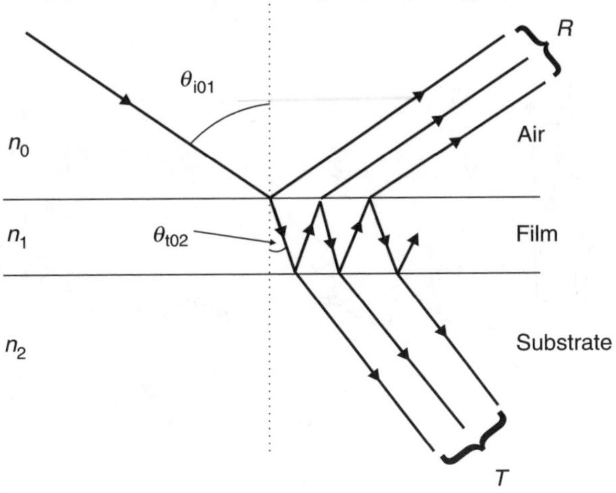

Figure 4.21 Multiple transmission and reflection from an air/thin film/substrate combination. T = transmitted intensity; R = reflected intensity.

appropriate polarization will govern the amounts of light transmitted and reflected. The transmitted radiation will undergo further reflection and transmission at the film/substrate interface. The overall air/film/substrate electric field amplitude reflection coefficient r is obtained by combining the Fresnel coefficients (for the appropriate polarization) after allowing for a change of phase β across the film (thickness d). The expression is given by

$$r = \frac{r_{01} + r_{12}\exp(-2j\beta)}{1 + r_{01}r_{12}\exp(-2j\beta)} \tag{4.32}$$

where r_{01} and r_{12} are the Fresnel coefficients for the air/film and film/substrate boundary, respectively. The phase change β across the film of thickness d is

$$\beta = 2\pi\left(\frac{d}{\lambda}\right)n_1\cos\theta_{t01} = 2\pi\left(\frac{d}{\lambda}\right)(n_1^2 - n_0^2\sin^2\theta_{i01})^{0.5} \tag{4.33}$$

The transmission coefficient t is given by

$$t = \frac{t_{01}t_{12}\exp(-j\beta)}{1 + r_{01}r_{12}\exp(-2j\beta)} \tag{4.34}$$

Equations (4.32) and (4.34) are generally valid. For non-normal incidence, each takes two possible forms, depending on the state of polarization of the incident EM radiation. If the film is absorbing, or if it is bound by absorbing media, then the values of n_0, n_1 and n_2 are replaced by the corresponding complex quantities. The resulting equations for r and t become somewhat cumbersome, although readily calculable using a computer.

The oscillatory nature of Equations (4.32) and (4.34) is a result of the constructive and destructive interference that results from the multiple reflected and transmitted EM waves. By appropriate choice of the thin film thickness ($d = \lambda/4$) and refractive index ($n_1 = n_2$), it

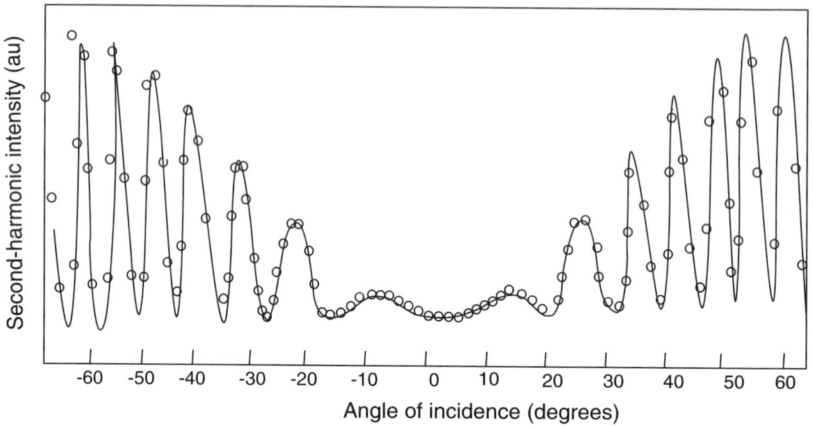

Figure 4.22 Experimental (points) and theoretical (solid curve) data for second-harmonic intensity of an oligomeric Langmuir–Blodgett film as a function of the angle of incidence of the fundamental laser beam. Reproduced from Allen S, Ryan TG, Hutchings MG, Devonald JP, Ferguson I, Swart RM, *et al.*, 'Characterization of nonlinear optical Langmuir–Blodgett oligomers', in GJ Ashwell, D Bloor (Editors), *Organic Materials for Non-linear Optics III*, pp. 50–60, Copyright (1993), with permission from the Royal Society of Chemistry.

is possible to obtain zero reflectance at normal incidence for a particular wavelength (the principle of an anti-refection coating).

Under certain conditions the equations for the transmitted and reflected intensities (powers) from a thin absorbing film/transparent substrate combination may be considerably simplified. For normal incidence and if multiple reflections and interference can be neglected (which will be the case if the film is sufficiently absorbing):

$$T = (1 - R)^2 \exp\left(\frac{-4\pi n''}{\lambda}\right) \tag{4.35}$$

Independent measurements of T and R then enable values of n' and n'' to be determined. Interference effects are found with EM radiation of wavelengths other than visible; the interference of X-rays passing through thin films provides a precise means to measure their thickness (Chapter 6, Section 6.3.2). Figure 4.22 shows a set of interference fringes that result from the second-harmonic generation in a thin organic film of a dye, which possesses the general chemical structure shown in Figure 4.3 [7]. In the figure, the intensity of the generated second-harmonic radiation is plotted as a function of the angle of incidence of the fundamental laser beam. In this case, the fringing effect is due to interference between the harmonic signals produced from the front and back surfaces of the organic film. The amplitude of the fringes can be used to determine the electro-optic coefficient of the film.

4.6 WAVEGUIDING

Electromagnetic waves can be confined (or guided) within thin films or channels of material. Such *waveguides* play an important role in optical communications. A well-known example

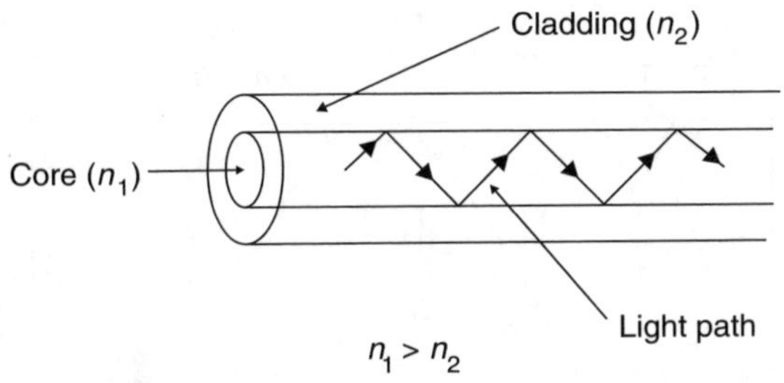

Figure 4.23 The basic structure of an optical fibre. A core region (refractive index n_1) is surrounded by a cladding region (refractive index n_2) where the refractive index of the core material is greater than that of the cladding.

is that of an optical fibre shown in Figure 4.23. The central region is called the *core* and the surrounding regions are referred to as the *cladding*. To confine an EM wave fully within the core, it must be unable to escape, i.e. it must be totally internally reflected. The conditions for total internal reflection are that the core medium must be of the greater refractive index, i.e. $n_1 > n_2$ in Figure 4.23. However, the total internal reflection condition does not specify fully the requirements for waveguiding.

Figure 4.24 shows an example of a planar waveguide formed by a thin film of refractive index n_f on top of a substrate of refractive index n_s. The topmost layer (refractive index n_0) is often air and has a much lower refractive index than the other two. Waveguiding will occur in the film providing that its thickness *d is* given by

$$d \geq \frac{\lambda(m + 0.5)}{2\sqrt{(n_f^2 - n_s^2)}} \tag{4.36}$$

where λ is the wavelength in the guide and m is an integer. Each value of m is associated with a distinct wave pattern or *mode* within the waveguide. Transverse magnetic and transverse electric radiation (Section 4.2) give rise to two types of mode, called TM_m and TE_m. The cut-off thickness for waveguiding in the asymmetric structure shown in Figure 4.24 is usually about one-third of the wavelength of the light. Figure 4.25 shows how the optical electric field intensity, E^2, varies with position in a 2 μm thick guiding layer for the TE_0 and TE_1 modes [8]. The $m = 1$ mode has two maxima in the field intensity within the guided layer

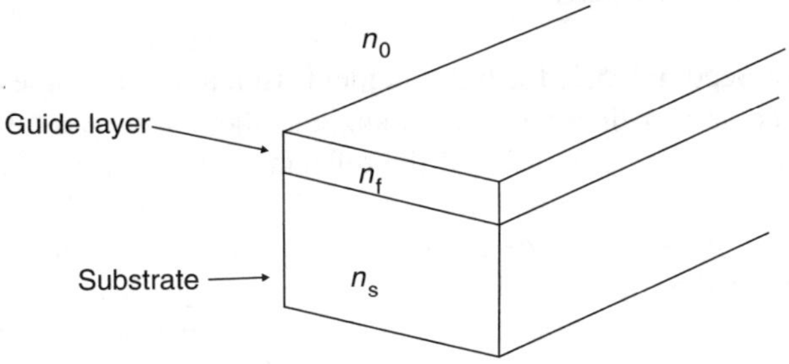

Figure 4.24 Schematic diagram of a thin-film planar waveguide.

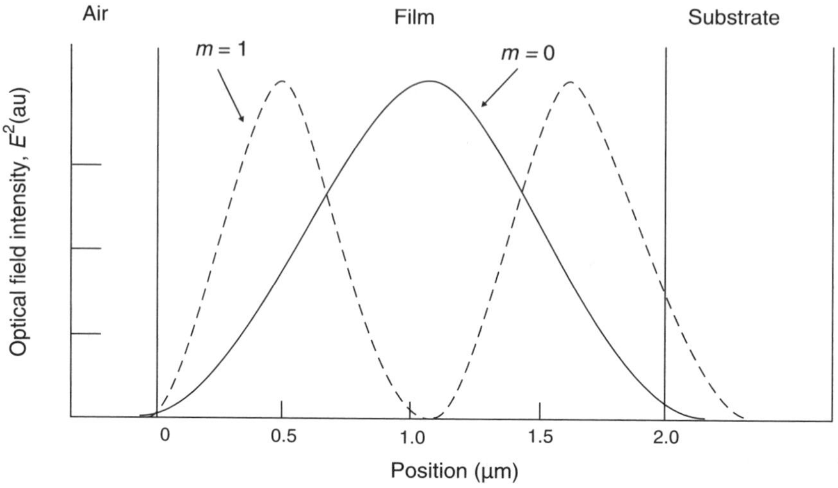

Figure 4.25 The optical electric field intensity E^2 for two modes propagating in a 2 μm thick film. The solid line is the $m = 0$ mode. The $m = 1$ mode (dashed line) has a minimum in the electric field intensity near the middle of the waveguide [8].

and a minimum near the middle of the guide. It is also important to note that the electric field penetrates both the air and substrate regions. These fields are *evanescent* in nature and will be discussed more fully in the following section. Different modes propagate along a waveguide with different velocities, even if they are generated by monochromatic radiation. This phenomenon is called *mode dispersion*.

4.7 SURFACE PLASMONS

Surface plasmons are collective oscillations of the free charges at a metal boundary which propagate along the interface; these therefore represent a special type of guided wave. The intensity maximum of these waves is a maximum in the surface and decay exponentially perpendicular to the surface. Surface plasmons can be produced by electrons and by light.

4.7.1 The Evanescent Field

As we have noted in Section 4.5.2, for light incident from a dense to a less dense medium at an angle of incidence greater than the critical angle θ_c there will be total internal reflection. This situation is sketched in Figure 4.26. If the reflectance R is recorded as a function of the angle of incidence, R becomes unity at the critical angle. At angles greater than θ_c, the reflected beam has the same amplitude as the incident beam, but a phase difference between 0 and π. The effect on the transmitted beam is rather strange. This takes on the form of a wave that travels along the boundary with an amplitude that decays perpendicular to the boundary. The wave carries no energy away from the boundary and is the evanescent wave we have already encountered in our discussions on waveguiding (Section 4.6). Figure 4.27

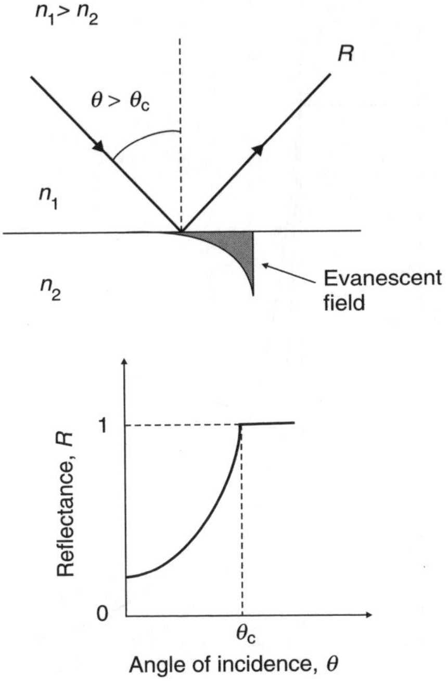

Figure 4.26 The total internal reflection from light incident from a medium 1 with refractive index n_1 to an interface with a material 2 with refractive index n_2, where $n_1 > n_2$. For angles of incidence θ greater than the critical angle θ_c the reflectance is unity. However an evanescent field penetrates into medium 2.

shows how both the amplitude and intensity (square of the amplitude) of a typical evanescent wave vary with distance from a glass/air interface [8]. It is evident that the penetration depth (into the air medium in this example) is similar to the wavelength of light. The electric field associated with evanescent waves is therefore a convenient way exciting molecules close to a surface.

4.7.2 Surface Plasmon Resonance

Under certain conditions, electromagnetic surface waves can propagate along the interface between two media. The requirement is a combination of a lossless medium with a positive permittivity with another of negative real part and positive imaginary part. Although this is an idealized situation (as all materials exhibit loss) it is close to the practical system of a dielectric and metal at frequencies below the *plasma frequency*, ω_p. This frequency corresponds to that at which the permittivity of the metal is equal to zero; for $\omega < \omega_p$, ε_r' is negative. The modes are called *surface plasmons* or *surface plasmon polaritons* (SPPs) – plasmon because the optical properties of the metal are consistent with its electrons behaving as a free electron plasma and polariton to suggest the coupling of photons with polar excitations in the metal.

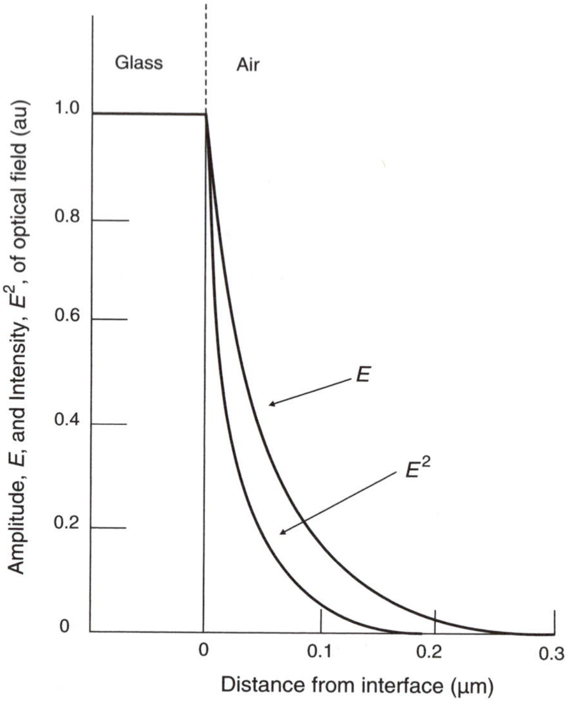

Figure 4.27 Decay of amplitude E and intensity E^2 of an evanescent wave at a glass/air interface [8].

Surface plasmons can be generated on a metal surface using an electron beam. However, photons do not possess sufficient energy to couple directly to these surface oscillations. Optical excitation of SPPs requires matching of both energy (i.e. optical frequency) and momentum. The latter quantity is related to the wavelength of the light by de Broglie's relationship in Equation (3.28) (Chapter 3, Section 3.3.1), $p = h/\lambda$. Figure 4.28 contrasts the wavevector versus frequency relationships for light in air [$\omega/k =$ constant = phase velocity of light; Chapter 3, Section 3.3.1, Equation (3.26)] and surface plasmons; such curves are called *dispersion curves*. These two curves do not intersect at any frequency. To be able to

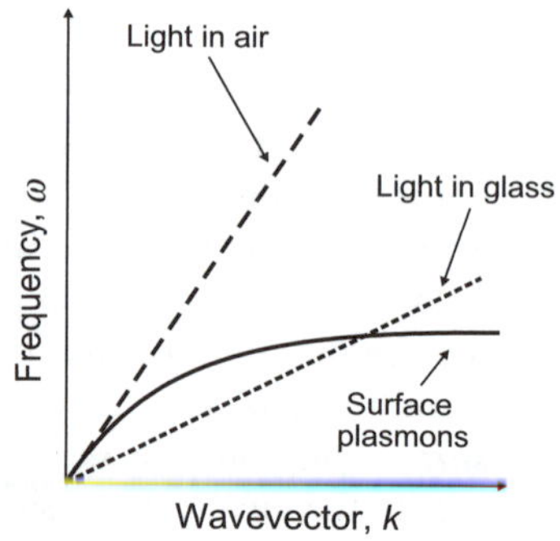

Figure 4.28 Dispersion curves for light in air and glass and for surface plasmons.

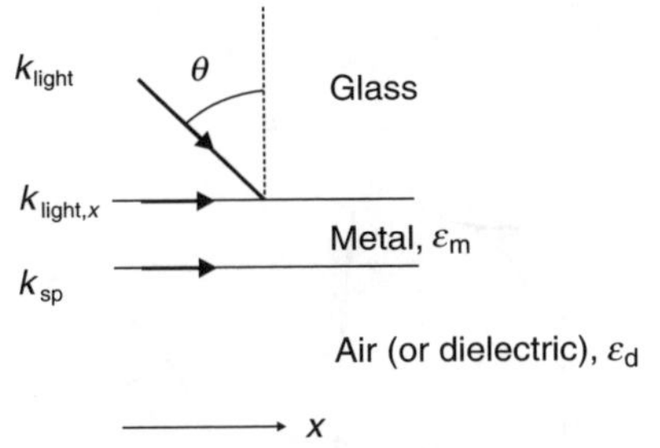

Figure 4.29 Coupling conditions between light incident from glass to a boundary with a thin metal film. k_{light} is the wavevector of the incident light; $k_{light,x}$ is the component of the wavevector in the x-direction; and k_{sp} is the surface plasmon wavevector.

couple the light to the SPPs, the momentum of the photons at a particular frequency must be increased. This is accomplished by decreasing the velocity of the light, i.e. slowing the photons. As Figure 4.28 reveals, the dispersion curve for light in glass now intersects that of the surface plasmons.

An experimental arrangement to realize the above conditions is shown schematically in Figure 4.29. This is known as the *Kretschmann* configuration and exploits evanescent electromagnetic waves. Monochromatic light (p-polarized) travels through glass and is reflected from a metal coating. If this coating is sufficiently thin, the resulting evanescent field penetrates to the opposite side of this layer to an interface between the metal and air (or a dielectric coating). The incoming EM radiation will have a wavevector k_{light}, as indicated in Figure 4.29. Changing the angle of incidence of the light θ alters the component of the wavevector parallel to the prism base $k_{light,x}$. When this component matches the real part of the surface plasmon wavevector, k_{sp}, then surface plasmons are excited. This condition requires that

$$k_{sp} = \frac{\omega}{c}\sqrt{\frac{\varepsilon_m \varepsilon_d}{\varepsilon_m + \varepsilon_d}} \tag{4.37}$$

where ε_m and ε_d are the relative permittivities of the metal and the dielectric, respectively. Equation (4.37) describes a surface plasmon wave if the real part of ε_m is negative and its absolute value is smaller than ε_d. At optical wavelengths this condition is fulfilled by several metals; among these, silver and gold are the most commonly used.

A curve resulting from the measurement of the reflectance, R, against angle of incidence, θ, is shown in Figure 4.30. As the angle of incidence is increased past the critical angle θ_c, all of the incidence light will be reflected ($R = 1$). When the component of the light's wavevector parallel to the interface is equal to that of the surface plasmons ($k_{light,x} = k_{sp}$) then all of the optical energy will couple into the SPPs and the reflected intensity will go to zero. This condition is known as *surface plasmon resonance* (SPR). The angle at which this occurs is shown as θ_{SPR} in Figure 4.30. The reflectance curve is often referred to as an SPR curve.

The reflectance versus angle relationship can be calculated using the Fresnel equations (Section 4.5.2). The reverse problem of finding optical parameters of a metal film from experimentally measured SPR curves has no analytical solution. However, it can be solved numerically by fitting the experimental SPR data to theory using a least-squares technique.

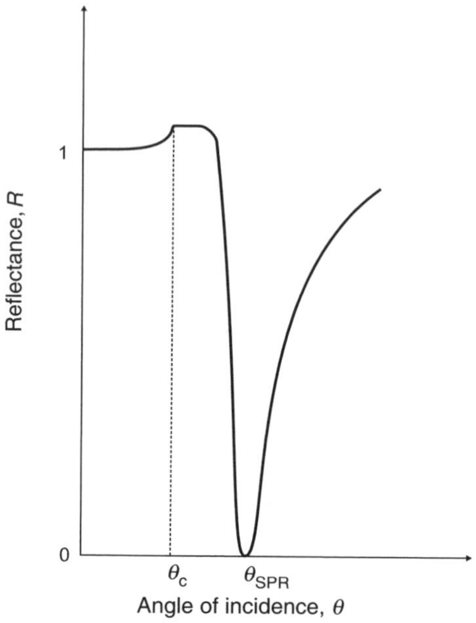

Figure 4.30 Surface plasmon resonance curve. θ_c = critical angle; θ_{SPR} = surface plasmon resonance angle.

The Kretschmann configuration method described above is only one of the techniques available to enhance the momentum of the incident optical wave. In the *Otto* arrangement, the prism is separated by an air gap from a metal plate. The coupling takes place between the evanescent field in the gap and the plasmons on the metal surface. An advantage of the Otto approach is that there is no restriction on the thickness of the metal. Diffraction gratings and optical waveguides can also be used. Figure 4.31 contrasts the coupling to SPPs using the Kretschmann and diffraction grating approaches. In the former [Figure 4.31(a)], a prism or semi-cylinder may be used to change the momentum of the incident photons. The thin metal

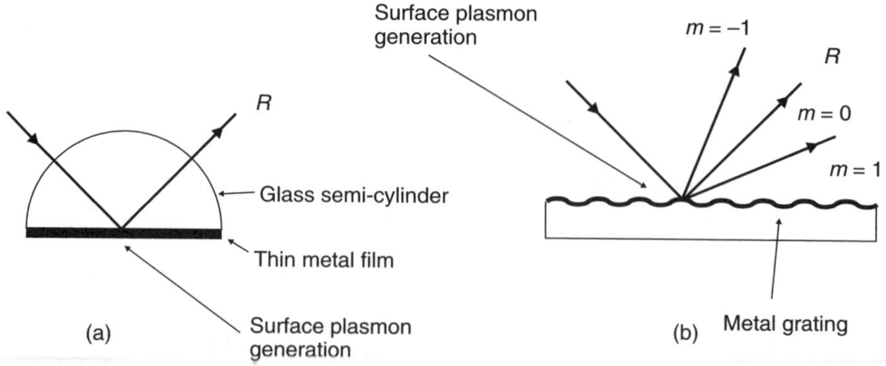

Figure 4.31 Methods for coupling light to surface plasmons. (a) Kretschmann configuration; (b) grating coupling.

layer (~50 nm) can either be deposited directly onto the base of the semi-cylinder or deposited on a glass slide which is placed in contact with the semi-cylinder using a refractive index matching fluid. The use of a diffraction grating to couple light to the SPPs [Figure 4.31(b)] enables the plasmons to be generated on the same side of the metal coating as the incident light. The periodicity of the grating effectively enhances the momentum of the incoming photons. However, the reflected beam will consist of different *orders of diffraction*, different values of m shown in the figure. This results in the appearance of more than one minimum in the SPR curve.

A thin dielectric coating on top of the metal film will shift the surface plasmon dispersion curve to higher momentum. Consequently, the SPR will move to a higher angle. This can be conveniently demonstrated using Langmuir–Blodgett films. Figure 4.32 shows the effect of the multilayers on the SPR curves. In Figure 4.32(a), the organic film is transparent to the incoming EM wave. The effect of increasing the number of layer in the LB array is to shift the SPR curves to higher angles; the minimum in R remains at zero, but the increasing film thickness produces a slight broadening of the resonance curves. The effects are somewhat different in the case of an absorbing organic film [Figure 4.32(b)]. Here, the resonance depth is decreased and the resonance width is significantly increased as the film thickness increases. Surface plasmon resonance can be used to determine the thickness and refractive index of organic layers. The sensitivity of the method is high and changes in refractive index of the overlayer of about 10^{-5} may be monitored. In this respect, the technique compares favourably with ellipsometry (Section 4.5.3). In Chapter 10 (Section 10.4.6), the application of SPR to chemical sensing is described.

The shift of the SPR conditions with film thickness can be also used as the basis for a high-contrast microscope. The reflected and scattered plasmonic light is converted by a lens to form an image of the interface on a television camera, for example. Only the areas at resonance appear dark. Variations in film thickness of as little as a few tenths of a nanometre

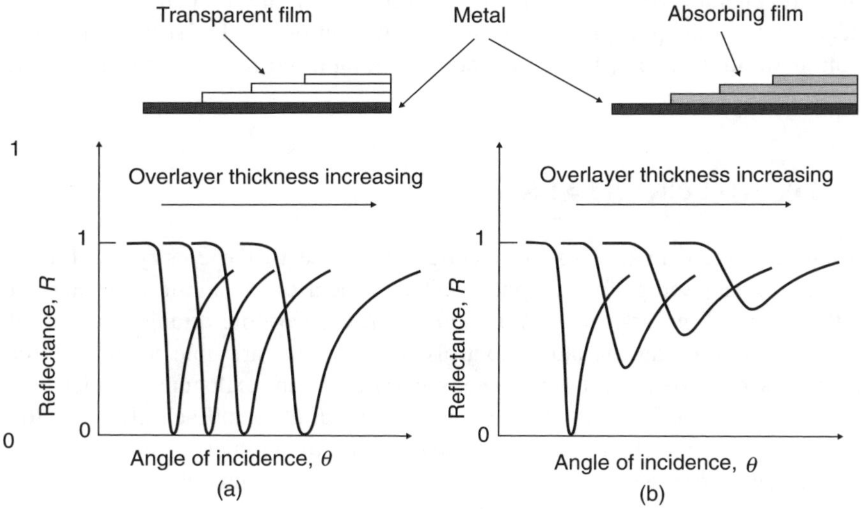

Figure 4.32 Effect of overlayers on SPR curve. The curves are shown for different thicknesses of overlayer. (a) Overlayers transparent to wavelength of incident radiation; (b) overlayers absorbing at wavelength of incident radiation.

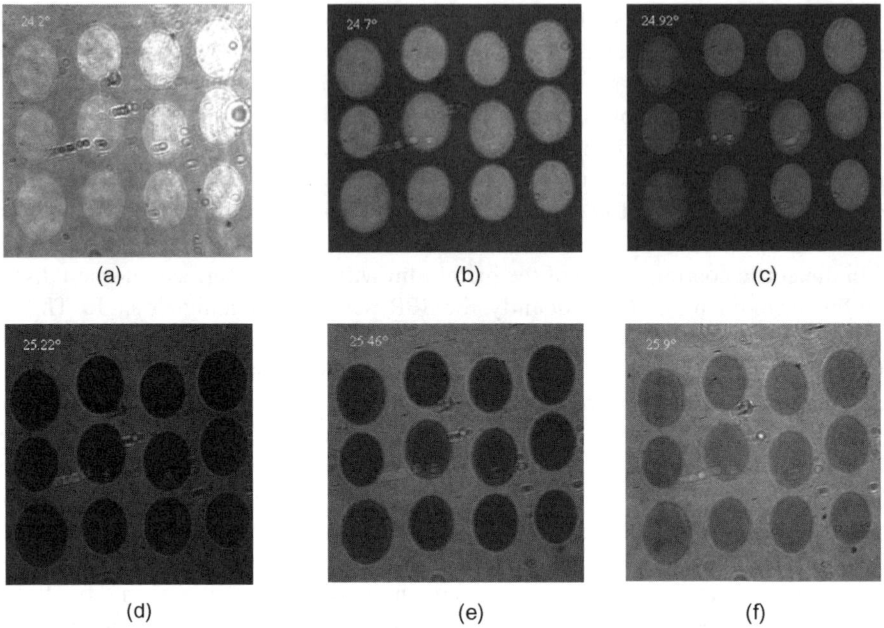

Figure 4.33 Surface plasmon microscopy images taken from a 3 × 4 matrix of identically functionalized sensor elements at various angles of incidence: (a) 24.2°; (b) 24.7°; (c) 24.92°; (d) 25.22°; (e) 25.46°; (f) 25.9°. Reproduced from *Prog. Colloid Polym. Sci.*, **109**, Zizlsperger M, Knoll W, 'Multispot parallel on-line monitoring of interfacial binding reactions by surface plasmon microscopy', p. 249, Copyright (1998), with kind permission of Springer Science and Business Media.

are enough to generate sufficient contrast for an image. Figure 4.33 shows examples of the SPR images from multispot, parallel on-line monitoring of interfacial binding reactions by surface plasmon microscopy [9]. The various photographs correspond to different incidence angles.

4.8 PHOTONIC CRYSTALS

The periodic arrangement of ions on a lattice gives rise to the energy band structure in semiconductors. The energy bands that result then control the motion of charge carriers through the crystal. Similarly, in a *photonic crystal*, the periodic arrangement of refractive index or dielectric constant variation controls how photons are able to move through the crystal. Photons react to the refractive index contrast in an analogous manner to the way electrons react when confronted with a periodic potential of ions. Each gives rise to a range of allowed energies and a band structure characterized by an energy gap or photonic band gap. A band gap forms when the electron wavelength is comparable to the inter-atomic spacing.

Photonic band gaps were first predicted in 1987 by Yablonovitch, at Bell Communications Research. A few years later, Yablonovitch and co-workers produced a photonic crystal using an array of 1 mm holes milled into a slab of material of refractive index 3.6. The resulting

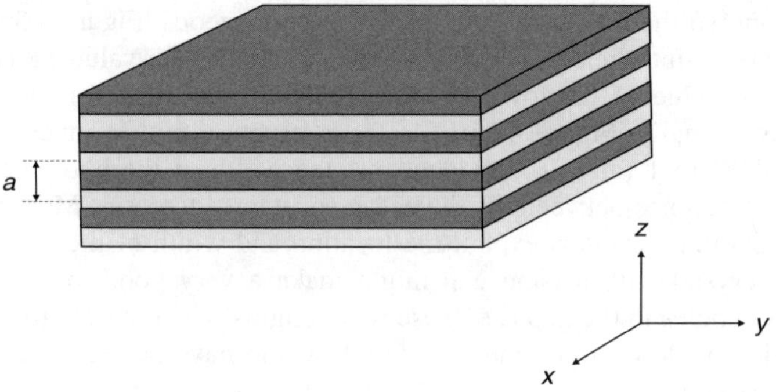

Figure 4.34 One-dimensional photonic crystal. The multilayer film consists of alternating layers with different dielectric constants, periodicity = a.

structure, which became known as yablonovite, was found to prevent microwaves from propagating in any direction, i.e. it exhibited a three-dimensional photonic band gap. An estimate of the distance between voids is given by the wavelength of the light divided by the refractive index of the material. This relationship means that it is even more difficult to create photonic crystals in materials with high refractive index.

The simplest possible photonic crystal, shown in Figure 4.34, consists of alternating layers of material with different dielectric constants. Such a photonic crystal can form an *optical microcavity* and, as will be shown in Chapter 9 (Section 9.6.2) can be used to concentrate the output of light-emitting devices. The traditional approach to an understanding of this system is to allow a plane wave to propagate through the material and to consider the multiple reflections that take place at each interface. Figure 4.35 shows how an optical band gap arises in the frequency (energy) versus wavevector diagram for the multilayer by changing the refractive index difference between the two layers [10]. The figure is shown in the reduced zone representation as discussed for electrons in Chapter 3, Section 3.3.2. In Figure 4.35(a), all of the layers in the multilayer structure have the same

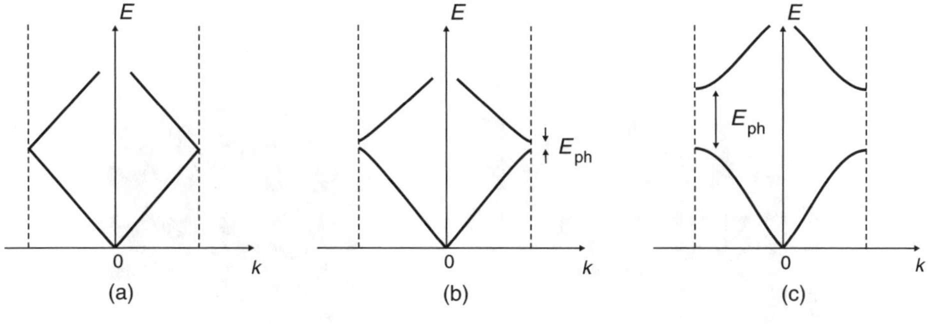

Figure 4.35 Photonic energy band structure showing energy E versus wavevector k for multilayer films as depicted in Figure 4.34. Photonic band gap = E_{ph}. (a) Each layer has same dielectric constant $\varepsilon_r = 13$. (b) Layers alternate between $\varepsilon_r = 13$ and $\varepsilon_r = 12$. (c) Layers alternate between $\varepsilon_r = 13$ and $\varepsilon_r = 1$. From Joannopoulos JD, Meade RD, Winn JD, *Photonic Crystals*, 2nd ed, 1995. Reprinted with permission of Princeton University Press.

dielectric constant, so the medium is completely homogeneous. Figure 4.35(b) represents a structure with alternating dielectric constants of 12 and 13 (the value 13 is chosen as it is approximately the value of the low-frequency dielectric constant for the inorganic semiconductor GaAs). Finally, Figure 4.35(c) is for a structure with a much higher dielectric constant ratio of 13 to 1 (air). It is evident that the width of the photonic band gap, E_{ph}, increases as the difference between the dielectric constants increases. Most of the promising applications of photonic crystals exploit the location and width of the photonic band gaps. For example, a crystal with a band gap might make a very good, narrow-band filter, by rejecting all frequencies in the gap (i.e. these wavelengths would pass through the material). A resonant cavity, made from a photonic crystal, would have perfectly reflecting walls for frequencies in the gap.

Two- and three-dimensional photonic crystals can take on a large number of forms. Examples of such structures are shown in Figure 4.36. Figure 4.36(a) shows a square lattice of dielectric columns. For certain values of the column spacing, this crystal can have a photonic band gap in the *xy* plane. Inside this gap, no extended states are permitted, and the incident ray is reflected. Although a multilayer film, such as depicted in Figure 4.34, only reflects light at normal incidence, this two-dimensional photonic crystal can reflect light from any direction in the plane. The optical analogue of an ordinary crystal is a three-dimensional photonic crystal, a dielectric that is periodic along three mutually orthogonal axes. A simple example is provided in Figure 4.36(b), in which dielectric spheres are arranged in a lattice structure and surrounded by a different dielectric medium. This model is simply that of atoms making up solids, in which case the 'host' medium is air. The dielectric constants can also be reversed, with air bubbles embedded in a regular fashion in a dielectric material. Breaking the periodicity of the voids in the photonic crystal, by either enlarging or reducing the size of a few of the voids, introduces new energy levels within the photonic band gap. This is analogous to the creation of energy levels within the band gap by the addition of dopant atoms in semiconductor crystals. Examples of photonic crystals in organic thin films are described in Chapter 11, Section 11.3.3.

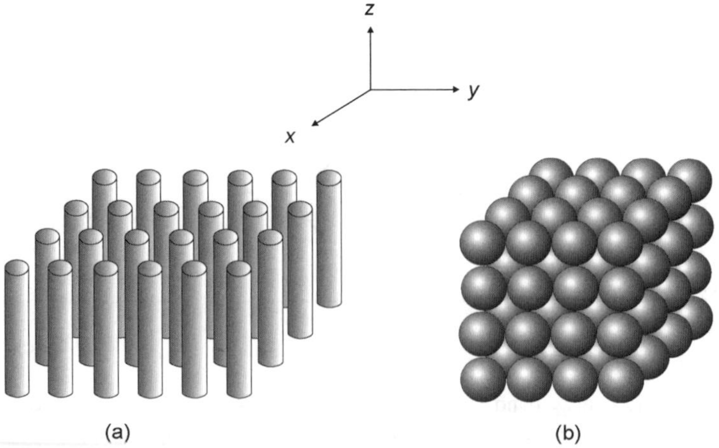

Figure 4.36 (a) A two-dimensional photonic crystal showing a square lattice of dielectric columns. (b) A three-dimensional photonic crystal consisting of a regular array of dielectric spheres.

4.8.1 Subwavelength Optics

The nature of surface plasmons changes when they propagate on metal surfaces that are periodically textured on the scale of the wavelength of light. When the period of the nanostructure is half that of the effective wavelength of the surface plasmon mode, scattering may lead to the formation of surface plasmon standing waves and the opening of a surface plasmon 'band gap' [11]. This is illustrated in Figure 4.37. There are two standing wave solutions, each with the same wavelength but of different frequencies (because of their different field and surface charge distributions). The upper frequency solution ω_+ is of higher energy because of the greater distance between the surface charges and the greater distortion of the field. Surface plasmons with frequencies between ω_- and ω_+ cannot propagate, so this frequency interval is known as the *stop gap*. At the band edges, the density of surface plasmon states is high and there is a significant increase in the associated field enhancement.

Surface plasmon effects can give rise in enhanced transmission through periodic arrays of subwavelength (~100 nm) holes in optically thick metallic films [12]. Not only is the transmission much higher than expected from classic diffraction theory, it can be greater than the percentage area occupied by the holes, implying that even the light impinging on the metal between the holes can be transmitted. The entire periodic structure acts like an antenna in the optical regime. The transmission spectra of hole arrays display peaks that can be tuned by adjusting the period and symmetry. As the holes are too small to act as hollow waveguides, the light cannot get through directly. In fact, the interference pattern produced by the counter-propagating plasmons on the metal surface coincides with the positions of the holes and the amplitude of the field inside is boosted and light is transmitted.

One of the most intriguing developments in this area over recent years has been the development of materials with a negative refractive index, i.e. light refracts in the opposite direction to normal as it enters a material. It has been suggested that if both the permittivity ε and permeability μ of a material were negative, then the refractive index, which determines

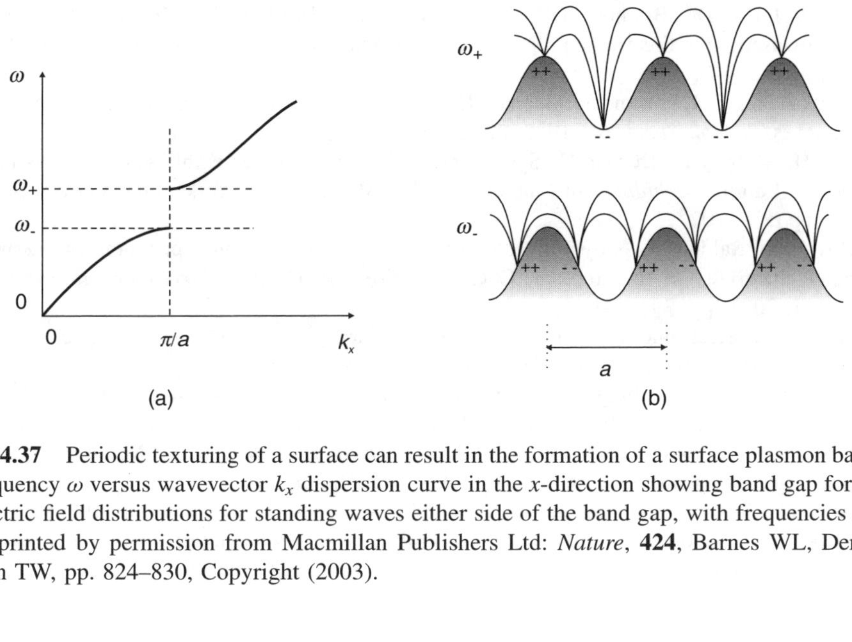

Figure 4.37 Periodic texturing of a surface can result in the formation of a surface plasmon band gap. (a) Frequency ω versus wavevector k_x dispersion curve in the x-direction showing band gap formation. (b) Electric field distributions for standing waves either side of the band gap, with frequencies ω_+ and ω_-. Reprinted by permission from Macmillan Publishers Ltd: *Nature*, **424**, Barnes WL, Dereux A, Ebbesen TW, pp. 824–830, Copyright (2003).

the velocity of light within the material [Equations (4.2) and (4.3)] should also be negative [13]. A thin planar slab of such a 'meta' material could act as a perfect lens [14]. Such materials have been developed for microwave radiation, but it is more difficult to use the same approach for the optical regime. However, it has been shown that gold films patterned with small metal posts may have a negative permeability at visible light frequencies [15]. This led to the observation of optical impedance matching, i.e. the total suppression of reflection from an interface (Section 4.5.2) and paves the way towards components with negative refraction at visible frequencies.

BIBLIOGRAPHY

Barford W. *Electronic and Optical Properties of Conjugated Polymers*. Oxford: Oxford University Press; 2005.

Duffin WJ. *Electricity and Magnetism*. 4th ed. London: McGraw-Hill; 1990.

Eisberg R, Resnick R. *Quantum Physics of Atoms, Molecules, Solids, Nuclei, and Particles*, 2nd ed. New York: John Wiley & Sons, Inc.; 1985.

Joannopoulos JD, Meade RD, Winn JD. *Photonic Crystals*. Princeton, NJ: Princeton University Press; 1995.

Nabook A. *Organic and Inorganic Nanostructures*. Boston: Artech House; 2005.

Raether H. Surface plasma oscillations and their applications. *Phys. Thin Films* 1977; **9**: 145–261.

Schwoerer M, Wolf HC. *Organic Molecular Solids*. Berlin: Wiley-VCH; 2006.

Wilson J, Hawkes J. *Optoelectronics*. London: Prentice Hall; 1998.

REFERENCES

[1] Richardson T, Roberts GG, Polywka MEC, Davies SG. Preparation and characterization of organotransition metal Langmuir–Blodgett films. *Thin Solid Films* 1988; **160**: 231–239.

[2] Friend RH, Bott DC, Bradley DDC, Chai CK, Feast WJ, Foot PJS, Giles JRM, Horton ME, Pereira CM, Townsend PD. Electronic properties of conjugated polymers. *Phys. Trans. R. Soc. London A* 1985; **314**: 37–49.

[3] Kuroda H, Kunii T, Hiroma S, Akamatu H. Charge-transfer bands in crystal spectra of molecular complexes. *J. Mol. Spectrosc.* 1967; **22**: 60–75.

[4] Kuhn H, Möbius D, Bücher H. Spectroscopy of monolayer assemblies. In A Weissberger, B Rossiter (Editor). *Techniques of Chemistry*, Vol. 1, Part 3B. New York: John Wiley & Sons, Inc.; 1972, pp. 577–702.

[5] Bulović VG, Baldo MA, Forrest SR. Excitons and energy transfer in doped luminescent molecular organic materials. In R Farchioni, G Grosso (Editors). *Organic Electronic Materials*. Berlin: Springer; 2001, pp. 391–441.

[6] Tanaka J. The electronic spectra of aromatic molecular crystals. The crystal structure and spectra of perylene. *Bull. Chem. Soc. Jpn.* 1963; **36**: 1237–1249.

[7] Allen S, Ryan TG, Hutchings MG, Devonald DP, Ferguson I, Swart RM, Froggatt ES, Burgess A, Eaglesham A, Cresswell JP, Petty MC. Characterization of nonlinear optical Langmuir–Blodgett oligomers. In GJ Ashwell, D Bloor (Editors). *Organic Materials for Non-linear Optics III*. Cambridge: Royal Society of Chemistry; 1993, pp. 50–60.

[8] Swalen JG. Optical properties of Langmuir–Blodgett films. *J. Mol. Electron.* 1986; **2**: 155–181.

[9] Zizlsperger M, Knoll W. Multispot parallel on-line monitoring of interfacial binding reactions by surface plasmon microscopy. *Prog. Colloid Polym. Sci.* 1998; **109**: 244–253.

[10] Joannopoulos JD, Meade RD, Winn JD. *Photonic Crystals*, 2nd ed. Princeton, NJ: Princeton University Press; 1995.
[11] Barnes WL, Dereux A, Ebbesen TW. Surface plasmon subwavelength optics. *Nature* 2003; **424**: 824–830.
[12] Ebbesen TW, Lezec HJ, Ghaemi HF, Thio T, Wolff PA. Extraordinary optical transmission through sub-wavelength hole arrays. *Nature* 1998; **391**: 667–669.
[13] Pendry J. Positively negative. *Nature* 2003; **423**: 22–23.
[14] Pendry JB. Negative diffraction makes a perfect lens. *Phys. Rev. Lett.* 2000; **85**: 3966–3969.
[15] Grigorenko N, Geim AK, Gleeson HF, Zhang Y, Firsov AA, Khrushchev IY, Petrovic J. Nanofabricated media with negative permeability at visible frequencies. *Nature* 2005; **438**: 335–338.

5 Electroactive Organic Compounds

Fillet of a fenny snake, Into the cauldron boil and bake

5.1 INTRODUCTION

This chapter concerns the nature and properties of some important groups of electroactive compounds used in molecular electronics research. The focus is on organic materials exhibiting interesting electrical and magnetic behaviour. Discussions of other important organic compounds are to be found elsewhere in this book. For example, materials with nonlinear optical behaviour are introduced in Chapter 4 and photochromic compounds are described in Chapter 11. Liquid crystals and biological compounds are the subjects of separate chapters (Chapters 8 and 12, respectively). Synthetic details on all the electroactive compounds are beyond the scope of this book and are covered in some detail elsewhere (e.g. [1]). By way of an introduction, and particularly as an aid to the nonchemist, some notes on basic organic chemistry are first given.

5.2 SELECTED TOPICS IN CHEMISTRY

The selection is purely a personal one. As an engineer, these are simply the topics that I have found that I needed to look up most frequently!

5.2.1 Moles and Molecules

An important principle stated in 1811 by the Italian chemist Amadeo Avogadro was that equal volumes of gases at the same temperature and pressure contain the same number of molecules regardless of their chemical nature and physical properties. This idea was introduced in Chapter 2, Section 2.2.1. Avogadro's number, $N_A = 6.02 \times 10^{23}$ – the number of molecules of any gas present in a volume of 22.41 l, is the same for the lightest gas (hydrogen) as for a heavy gas such as carbon dioxide or bromine. It is one of the fundamental constants of chemistry and allows calculation of the amount of a pure substance, the mole (Chapter 2, Section 2.2.1).

One mole is the amount of substance of a system which contains as many elementary entities as there are atoms in 0.012 kg of the isotope carbon-12; it is often abbreviated simply as 'mol.' When the mole is used, the elementary entities must be stated and may be atoms,

© 2007 John Wiley & Sons, Ltd

molecules, ions, electrons, other particles or specified groups of such particles. A few books prefer to use the unit *kilogram-mole* or *kmol*. When the kilogram-mole is used, Avogadro's constant becomes 6.02×10^{26} kmol^{-1}. For a material of molecular weight M and density ρ, the number of atoms per unit volume n may be calculated as follows:

$$n = \frac{N_A \rho}{M} \tag{5.1}$$

Because of its role as a scaling factor, Avogadro's number provides the link between a number of useful physical constants as we move between the atomic and macroscopic worlds. For example, it provides the relationship between the *universal gas constant*, R, and the Boltzmann constant, k_B (Chapter 3, Section 3.2):

$$R = k_B N_A \tag{5.2}$$

5.2.2 Acids and Bases

In 1884, Svante Arrhenius suggested that salts such as NaCl dissociate when they dissolve in water to give electrically charged particles, which he called *ions*:

$$\mathrm{NaCl(s)} \rightarrow \mathrm{Na^+\,(aq)} + \mathrm{Cl^-\,(aq)} \tag{5.3}$$

where (s) and (aq) refer to the solid and aqueous phases, respectively. Later, Arrhenius extended this theory by suggesting that *acids* are neutral compounds that ionize when they dissolve in water to give H^+ ions and a corresponding negative ion. According to the theory, hydrogen chloride is an acid because it ionizes when it dissolves in water to give hydrogen (H^+) *cations* and chloride (Cl^-) *anions*, according to the following scheme:

$$\mathrm{HCl(g)} \rightarrow \mathrm{H^+(aq)} + \mathrm{Cl^-(aq)} \tag{5.4}$$

where (g) refers to the gaseous phase. In contrast, *bases*, such as NaOH, were considered to produce a hydroxide ion, OH^-. Although a major advance in the understanding of these compounds, the Arrhenius theory ran into difficulties when compounds such as Na_2CO_3 were considered. These exhibited the properties of bases, but clearly could not ionize directly to yield a hydroxide ion. In the 1920s, the Arrhenius theory was independently developed by Thomas Lowry and Johannes Brønsted into a general and more powerful description of acids and bases. The Lowry–Brønsted definition of an acid is a species that tends to donate or lose a hydrogen ion, or *proton*. A Lowry–Brønsted base has a tendency to gain or accept a proton. In the Lowry–Brønsted model, HCl donates a proton to a water molecule (and is therefore an acid).

Under the Brønsted–Lowry definition, both acids and bases are related to the concentration of hydrogen ions present. Acids increase the concentration of hydrogen ion whereas bases decrease the concentration of hydrogen ions (by accepting them). Hence the acidity or basicity of something can be measured by its hydrogen ion concentration. The well-established *pH* scale for measuring acidity was introduced in 1909 by the Danish biochemist Sören Sörensen. This is defined as

$$\mathrm{pH} = -\log[\mathrm{H^+}] \tag{5.5}$$

The concentration is commonly abbreviated by using square brackets, thus $[H^+]$ = hydrogen ion concentration. When measuring pH, $[H^+]$ is in units of moles of H^+ per litre of solution. For example, for pure water the experimentally determined value of $[H^+]$ is 1.0×10^{-7} mol l^{-1} (or 100 nmol l^{-1}). Thus water has a pH equal to 7. A change in $[H^+]$ by a factor of 2 will result in a pH change of 0.3. The pH scale ranges from 0 to 14. Substances with a pH between 0 and 7 are acids whereas those with a pH greater than 7 and up to 14 are bases. In the middle, at pH = 7, are neutral substances, for example, pure water.

According to the Brønsted–Lowry theory, reactions between acids and bases always involve the transfer of an H^+ ion from a proton donor to a proton acceptor. Acids can therefore be neutral molecules:

$$\underset{\text{acid}}{HCl} + \underset{\text{base}}{NH_3} \rightarrow NH_4^+ + Cl^- \tag{5.6}$$

positive ions:

$$\underset{\text{acid}}{NH_4^+} + \underset{\text{base}}{OH^-} \rightarrow NH_3 + H_2O \tag{5.7}$$

and also negative ions:

$$\underset{\text{acid}}{H_2PO_4^-} + \underset{\text{base}}{H_2O} \rightarrow HPO_4^{2-} + H_3O^+ \tag{5.8}$$

The Brønsted–Lowry theory therefore expands the number of potential acids. It also enables the prediction of whether a compound is an acid from its chemical formula.

Gilbert Lewis defined these two terms in even a less restrictive way and included even more examples: an acid was considered to be an electron acceptor while a base was an electron donor. It may be more accurate to say that Lewis acids are substances which are electron deficient (or of low electron density) and Lewis bases are substances which are electron rich (or of high electron density). This approach further expands the number of acids.

5.2.3 Ions

Many substances dissolve in water to produce ions. Water itself contains ions. At 25 °C, the density of water is about 1.0 g cm^{-3} (or 1.0 g ml^{-1}). This corresponds to a concentration of water of about 55.4 mol l^{-1}. As noted in the previous section, the concentration of the H^+ (or OH^-) ions formed by the dissociation of H_2O molecules at 25 °C is only 1.0×10^{-7} mol l^{-1}. Therefore, the ratio of the concentration of H^+ (or OH^-) ions to that of neutral H_2O molecules is about 2×10^{-9}. In other words, only about two parts per 10^9 of the water molecules dissociate into ions at room temperature.

Because the proton is several orders of magnitude smaller than the water molecule, the charge on a particular H^+ ion is distributed over such a small volume that this H^+ ion is attracted towards any source of negative charge that exists in solution. Therefore, the instant that an H^+ ion is created in an aqueous solution, it bonds to a water molecule. In fact, each H^+ ion that an acid donates to water is bound to four neighbouring water molecules, so an appropriate formula for the entity produced when an acid loses a proton is $H(H_2O)_4^+$ or $H_9O_4^+$.

When an ion (positively or negatively charged) is inserted into water, it changes the structure of the hydrogen bond network. A water molecule tends to rotate (reorient) so that its dipole (polarized charge) faces the opposite charge of the ion. As the water molecules orient towards the ion, they break the hydrogen bonds to their nearest neighbours. The group of water molecules thus formed around an ion is called a *hydration shell*. The orientation of the molecules in the hydration shell results in a net charge on the outside of this shell, a charge of the same sign as that of the ion in the centre. The charge on the outside of the hydration shell tends to orient water molecules in the immediate vicinity, leading to a second hydration shell. A further highly diffuse region, the tertiary hydration shell, may well be present before distances are reached that are so far removed from the ion that the water molecules become essentially indistinguishable from those of bulk water. The result of forming hydration shells is to weaken the structure of the hydrogen bond network. This explains the observation that salt water has a lower freezing point than pure water: each ion in the liquid has a hydration shell of oriented water molecules around it, which prevents the water molecules from forming the hexagonal structure of ice.

The degree of hydration depends on a number of factors, in particular the ionic size and its associated charge density. Figure 5.1 shows a schematic diagram of the hydration sphere of a metal cation, M^{n+}, having a primary shell of six water molecules and a secondary shell of about 12 water molecules [2]. Such a situation is probably the case for trivalent metal ions such as Cr^{3+} and Rh^{3+}.

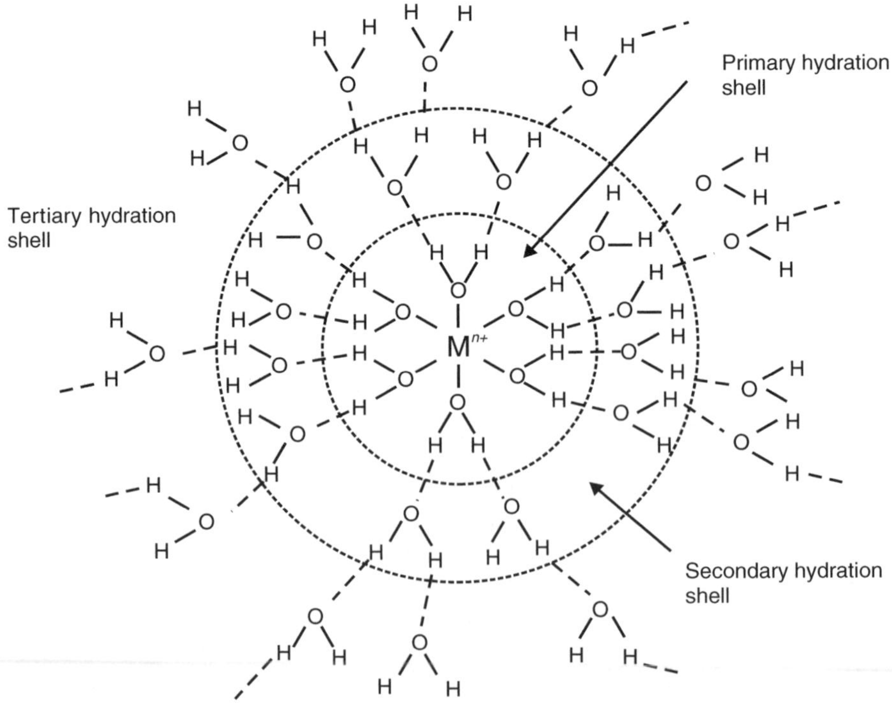

Figure 5.1 The localized structure of a hydrated metal cation in aqueous solution [2].

5.2.4 Solvents

If molecular electronics is to prove a useful technology, the appropriate organic compounds must be able to be easily processed, i.e. they must be capable of being made into the shapes required for devices, probably in the form of thin films. The processing can take many forms but frequently requires the organic electroactive compounds first to be dissolved in a solvent to enable other steps, e.g. spin-coating or self-assembly, to proceed. The majority of chemical reactions are also undertaken in solution. Here, a solvent fulfils several functions. It solvates the reactants and reagents so that they dissolve. This facilitates collisions between the reactant(s) and reagents to transform the reactant(s) to product(s). The solvent also provides a means of temperature control, either to increase the energy of the colliding particles so that they will react more quickly, or to absorb heat that is generated during a reaction. This section briefly looks at the properties of some of the more common organic solvents.

The rule of thumb for choosing a solvent is that 'like dissolves like'. For example, polar solvents will dissolve polar reactants. Generally, there are three measures of the polarity of a solvent: its dipole moment (Chapter 2, Section 2.3.5); its dielectric constant; and its miscibility with water. Molecules with large dipole moments and high dielectric constants are considered polar whereas those with small dipole moments and small dielectric constants are classified as nonpolar. On a practical basis, solvents that are miscible with water are polar and those that are not are nonpolar.

Solvents can be conveniently categorized according to their polarity. A number of examples are shown in Table 5.1. There are three main groupings: polar *protic*, polar *aprotic* and nonpolar. In this context, protic refers to a hydrogen atom attached to an electronegative atom (almost always an oxygen atom). Therefore, protic compounds can generally be represented by the formula ROH, where R is a chemical group. The dipole moment of polar protic solvents originates from the O–H bond. The large difference in electronegativities (Chapter 2, Section 2.2.4) of the oxygen and the hydrogen atom, combined with the small size of the hydrogen atom, justify separating molecules that contain an OH group from those polar compounds that do not. Examples of protic compounds from Table 5.1 are water (HOH), methanol (CH_3OH) and acetic acid (CH_3CO_2H).

The aprotic solvents do not contain an O–H bond, but nevertheless possess large dipole moments (and large dielectric constants). Typically, the dipole originates from a multiple bond between carbon and either oxygen or nitrogen. Most polar aprotic solvents contain a C=O double bond; a good example is acetone, $(CH_3)_2C{=}O$. Nonpolar compounds possess zero (i.e. benzene, C_6H_6) or low dipole moments (diethyl ether, CH_3CH_2OCH) and are not miscible with water.

5.2.5 Functional Groups

The following sections review some of the common functional groups found in organic molecules and outlines their properties. The chemical structures of the groups described are depicted in Table 5.2.

Alcohols

The functional group of an *alcohol* is the *hydroxyl group* (–OH) bonded to an sp^3 hybridized carbon. It can therefore be regarded as a derivative of water, with an alkyl

Table 5.1 Properties of some common organic solvents.

Solvent	Structure	Boiling point (°C)	Dielectric constant	Class	
				Polar	Protic
Hexane	$CH_3(CH_2)_4CH_3$	69	2		
Benzene		80	2.3		
Diethyl ether	$CH_3CH_2OCH_2CH_3$	35	4.3		
Chloroform	$CHCl_3$	61	4.8		
Acetic acid	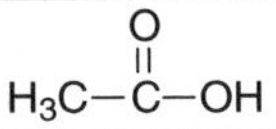	118	6.1	*	*
Tetrahydrofuran (THF)	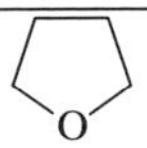	66	7.6		
Acetone	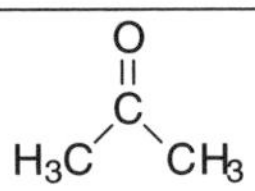	56	21	*	
1-Propanol	$CH_3CH_2CH_2OH$	97	20	*	*
Ethanol	CH_3CH_2OH	78	25	*	*
Methanol	CH_3OH	65	33	*	*
Acetonitrile	$H_3CC{\equiv}N$	82	38	*	
Dimethyl sulfoxide (DMSO)	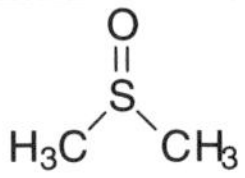	189	47	*	
Water	H—O—H	100	78	*	*
Formamide	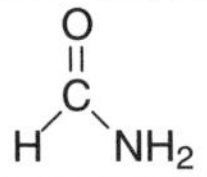	210	111	*	*

group replacing one of the hydrogen atoms. If an aromatic group (Section 5.2.6) is present rather than an alkyl group, the compound is generally called a *phenol* rather than an alcohol. There are three major subsets of alcohols – primary, secondary and tertiary – based on the number of carbons to which the C–OH carbon is bonded. Ethanol or ethyl alcohol, CH_3CH_2OH, is the simplest primary alcohol. The simplest secondary alcohol is propan-2-ol, $CH_3CHOHCH_3$, and an example of a tertiary alcohol is 2-methylpropan-2-ol, $(CH_3)_3COH$.

The hydroxyl group generally makes the alcohol molecule polar. These groups can form hydrogen bonds to one another and to other compounds. Two opposing solubility trends in alcohols are the tendency of the polar OH to promote solubility in water and of the carbon chain to resist it. Thus, methanol, ethanol and propanol are miscible in water because the hydroxyl group dominates over the short carbon chain. Butanol, with a four-carbon chain, is moderately soluble; however, alcohols with five or more carbons (pentanol and higher) are effectively insoluble because of the greater influence of the hydrocarbon chain.

Table 5.2 Chemical structures of common functional groups found in organic compounds.

Compound type	Functional group
Alcohol	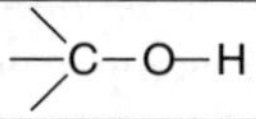
Aldehyde	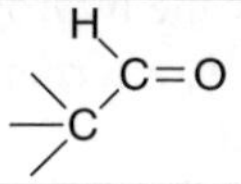
Amine	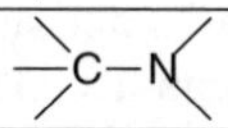
Amide	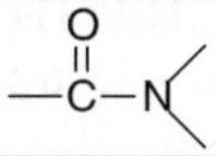
Carboxylic acid	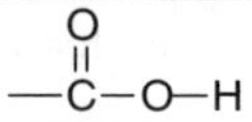
Ester	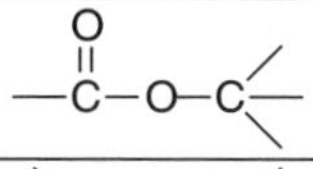
Ether	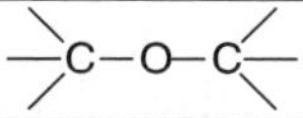
Ketone	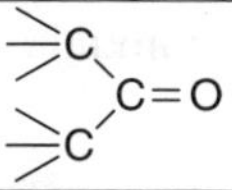
Thiol	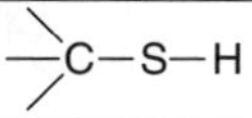
Nitrile	$-C\equiv N$

As a result of hydrogen bonding, alcohols tend to have higher boiling points than comparable hydrocarbons and ethers. All simple alcohols are miscible in organic solvents. The hydrogen bonding means that alcohols can be used as protic solvents. Alcohols, like water, can show either acidic or basic properties at the OH group. They are generally slightly weaker acids than water, but they are still able to react with strong bases such as sodium hydride or reactive metals such as sodium. The salts that result are called *alkoxides*, with the general formula $RO^- M^+$. At the same time, the oxygen atom has lone pairs of nonbonded electrons (Chapter 2, Section 2.4.1) that render it weakly basic in the presence of strong acids such as sulfuric acid.

Aldehydes

An *aldehyde* is either a functional group consisting of a terminal *carbonyl* group (–C=O), or a compound containing a terminal carbonyl group. The carbonyl group is polar; because oxygen is more electronegative than carbon, it results in an electron deficiency at the carbon atom. Aldehydes can be oxidized to form *carboxylic acids* or reduced to primary alcohols.

Amines

Amines are organic compounds containing nitrogen as the key atom in the functional group. Amines have structures resembling ammonia (NH_3), where one or more hydrogen atoms are replaced by alkyl groups or other groups where the nitrogen is bonded to a carbon atom in the group. However, if any of the carbons bonded to the nitrogen is part of a carbonyl group, then the compound is considered an *amide* (see below) rather than an amine. If only one of the hydrogen atoms in ammonia is replaced by a carbon-based group, then the functional group is a primary amine. If two of the hydrogens are replaced by two carbon-based groups, then it is a secondary amine. Finally, a tertiary amine is formed if all three hydrogen atoms are replaced by carbon-based groups.

Like ammonia, amines act as bases and are reasonably strong. The nitrogen atom has a lone electron pair available which can accept an H^+ ion to bond to the nitrogen forming a positive substituted ammonium ion. (NB: The pairs of dots on the N atoms depicted in some chemical reactions represent the lone electron pairs on the nitrogens in the amines.) These lone pairs also contribute to the solubility of simple amines due to hydrogen bonding between water molecules and the lone electron pairs.

Amides

Amides are also members of a group of chemical compounds containing nitrogen. Specifically, an amide is a derivative of a carboxylic acid in which the hydroxyl group has been replaced by an amine or ammonia.

Carboxylic Acids

Carboxylic acids are organic acids characterized by the presence of a carboxyl group, which has the formula –C(=O)–OH, usually written as –COOH or – CO_2H. The salts and anions of carboxylic acids are generally called *carboxylates*. The simplest series of carboxylic acids are the *alkanoic acids*, R–COOH, where R is a hydrogen atom or an alkyl group. Compounds may also have two or more carboxylic acid groups per molecule. The classic compounds for forming insoluble monolayers at the air/water interface are long-chain carboxylic acids (Chapter 7, Section 7.3.1). The acid group confers water solubility whereas the long alkyl chain prevents it: the balance between these opposing tendencies leads to monolayer formation.

Carboxylic acids are typically weak acids that partially dissociate into H^+ cations and $RCOO^-$ anions in aqueous solution. The carboxylate anion R–COO^- is usually named with the suffix '-ate,' so acetic acid, for example, gives rise to the acetate ion. Only about 0.02% of all acetic acid molecules are dissociated at room temperature in solution.

The two electronegative oxygen atoms tend to pull the electron away from the hydrogen of the hydroxyl group, and the remaining proton H^+ can more easily leave. The resulting negative charge is then distributed symmetrically among the two oxygen atoms, and the two carbon–oxygen bonds take on a partial double bond character (i.e. they are delocalized). This is a result of the resonance structure created by the carbonyl component of the carboxylic acid, without which the OH group does not as easily lose its H^+ (see alcohol).

The presence of electronegative groups (such as –OH or –Cl) next to the carboxylic group increases the acidity. Thus, for example, trichloroacetic acid (three –Cl groups) is a stronger acid than lactic acid (one –OH group), which in turn is a stronger acid than acetic acid.

Esters

Esters are compounds where an organic group replaces a hydrogen atom (or more than one) in an *oxygen acid* (i.e. an acid whose molecule has an –OH group from which the hydrogen can dissociate as an H^+ ion). Esters are named in a similar manner to salts. Although these do not really possess cations and anions, the terminology follows the same pattern: a more electropositive part followed by a more electronegative part. The most common esters are the *carboxylate esters*, where the acid in question is a carboxylic acid. For example, if the acid is acetic acid, the ester is called an acetate. The simplest ester is H–COO–CH_3 (methyl formate, also called methyl methanoate).

Esters can participate in hydrogen bonds as hydrogen-bond acceptors, but cannot act as hydrogen-bond donors, unlike their parent alcohols. This ability to participate in hydrogen bonding makes them more water soluble than their parent hydrocarbons. However, the limitations on their hydrogen bonding also make them more hydrophobic than either their parent alcohols or parent acids. Their lack of hydrogen bond donating ability means that ester molecules cannot hydrogen bond to each other. This makes esters generally more volatile than a carboxylic acid of similar molecular weight. Many esters have distinctive odors, which has led to their widespread use as artificial flavourings and fragrances.

Esters may undergo hydrolysis – the breakdown of an ester by water. This process can be catalyzed both by acids and bases; the base-catalyzed process is called saponification (Chapter 12, Section 12.2.5). The hydrolysis yields an alcohol and a carboxylic acid or its carboxylate salt.

Ethers

Ether is the general name for a class of compounds that contain an oxygen atom connected to two (substituted) alkyl groups (an ether group). A typical example is the solvent diethyl ether (ethoxyethane, CH_3CH_2–O–CH_2CH_3). Ethers should not be confused with the compounds with the same general structure R–O–R; for example, aromatic compounds such as furan where the oxygen is part of the aromatic system or compounds where one of the carbon atoms next to the oxygen is connected to oxygen, nitrogen or sulfur such as esters R–C(=O)–O–R or anhydrides R–C(=O)–O–C(=O)–R.

The terms primary ether, secondary ether and tertiary ether are occasionally used and, in common with alcohols and amines, refer to the carbon atom next to the ether oxygen. So, in a primary ether, this carbon is connected to only one other carbon as in diethyl ether CH_3CH_2–O–CH_2CH_3. An example of a secondary ether is diisopropyl ether, $(CH_3)_2$CH–O–CH$(CH_3)_2$, and that of a tertiary ether is di-*tert*-butyl ether, $(CH_3)_3$C–O–C$(CH_3)_3$. *Polyethers* are polymeric compounds with more than one ether group. Examples are the polymers of ethylene oxide such as the crown ethers and polyethylene glycol.

Ether molecules cannot form hydrogen bonds among each other, resulting in a relatively low boiling point comparable to that of the analogous alkanes. Ethers are more hydrophobic than esters or amides of comparable structure.

Ketones

A *ketone* is either the functional group characterized by a carbonyl group linked to two other carbon atoms or a compound that contains this functional group. A ketone can be generally represented by the formula $R^1(CO)R^2$, where R^1 and R^2 are different functional groups. A carbonyl carbon bonded to two carbon atoms distinguishes ketones from carboxylic acids, aldehydes, esters, amides and other oxygen-containing compounds. The double bond of the carbonyl group distinguishes ketones from alcohols and ethers. The simplest ketone is acetone (also called propanone).

As noted above, a carbonyl group is polar. This also makes ketones polar compounds. The carbonyl group interacts with water by hydrogen bonding. It is a hydrogen bond acceptor, but not a hydrogen bond donor, and cannot hydrogen bond to itself. This results in ketones being more volatile than alcohols or carboxylic acids of similar molecular weight.

Thiols

Thiols (formerly known also as *mercaptans*) are those compounds which contain the sulfhydryl group –SH attached to a carbon atom. Thiols are similar to alcohols, with the oxygen of the hydroxyl group –OH being replaced by a sulfur atom. (Oxygen and sulfur have similar chemical properties, as they belong to the same group in the periodic table.) Thiols form *thioethers*, *thioacetals* and *thioesters*, in which the alcohol-derived oxygen atom is replaced by a sulfur atom. Many thiols are colourless liquids having a smell resembling that of garlic. Thiols bind strongly to skin proteins, and are responsible for the intolerable, persistent odour produced by the spraying of skunks! The thiol group is also exploited for chemical self-assembly (Chapter 7, Section 7.3.2) on to surfaces such as gold.

Nitriles

A *nitrile* is a compound which has a –CN functional group (the nitrile group). In the –CN group, the carbon atom and the nitrogen atom are triple bonded together. The prefix *cyano* is used in chemical nomenclature to indicate the presence of a nitrile group in a molecule. A cyanide ion is a negative ion with the formula CN^-. The –CN group is sometimes, less properly, referred to as a *cyanide group* and compounds with them are sometimes referred to as cyanides.

5.2.6 Aromatic Compounds

Aromatic compounds represent a large class of molecules which includes benzene and compounds that resemble benzene in some of their chemical properties, e.g. toluene, naphthalene and anthracene. These compounds contain at least one ring that consists of six carbon atoms, each joined to at least two other carbon atoms, and each joined to adjacent carbon atoms by one single and one double bond. The resulting hexagonal structure is characteristic of many aromatic compounds. Most aromatic compounds have a delocalized π-electron system of alternating single and double bonds and are planar structures. This conveys a particular chemical stability to these systems. The number of π-electrons is $4n + 2$, where n is an integral number. This is known as the *Hückel rule*. Thus the number of delocalized electrons in benzene (6), naphthalene (10) and anthracene (14) is consistent with their aromatic character. In contrast, cyclobutadiene, C_4H_4, is not aromatic, since the number of π-delocalized electrons is 4, which is not satisfied by any n integer value. The cyclobutadienide (2−) ion, however, is aromatic. The eight-membered cyclic compound with four alternating double bonds (cyclooctatetraene) is also not aromatic and shows reactivity similar to that of alkenes. Molecules which are not aromatic are said to be *aliphatic*. Typically, they contain hydrocarbon chains.

Aromatic molecules normally exhibit enhanced chemical stability, compared with similar non-aromatic molecules. The circulating π-electrons in an aromatic molecule generate significant local magnetic fields that can be detected by *nuclear magnetic resonance* (NMR) spectroscopy. The bonds in an aromatic ring are less reactive than ordinary double bonds; aromatic compounds tend to undergo ionic substitution (e.g. replacement of a hydrogen bonded to the ring with some other group) rather than addition (which would involve breaking one of the resonant bonds in the ring).

The presence of the six-membered benzene ring is not essential for aromatic compounds. There are a large number of compounds that contain heteroatoms (O, N, S, etc.) which are also aromatic. Some examples are shown in Table 5.3. The aromicity arises because the heteroatom is either involved in a double bond in the ring or it can make use of a lone pair of electrons (Chapter 2, Section 2.4.1) to interact with the π-electrons to satisfy Hückel's rule. For example, pyridine, C_5H_5N, has a lone pair of electrons at the nitrogen atom that does not participate in the aromatic π-system. This makes pyridine a basic compound with chemical properties similar to these of tertiary amines. In the case of pyrrole, C_4H_5N, which has a very low basicity compared with pyridine, the ring nitrogen is connected to a hydrogen atom. The lone pair of electrons of the nitrogen atom then becomes delocalized in the aromatic ring. It should be noted that the purine molecule depicted in Table 5.3 is a precursor to the DNA bases adenine and guanine (Chapter 12, Section12.3.1).

Table 5.3 Example of aromatic molecules containing heteroatoms.

Molecule	Structure		Molecule	Structure
Pyridine	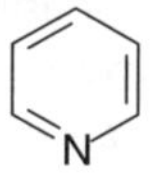		Indole	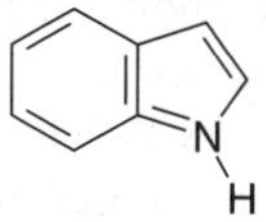
Pyrazine	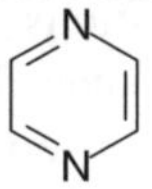		Purine	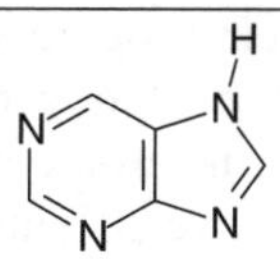
Pyrimidine	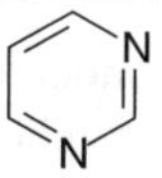		Furan	
Thiophene			Oxazole	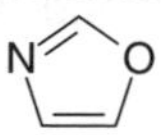
Pyrrole	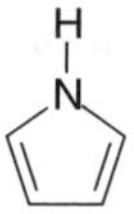		Thiazole	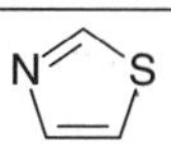

Table 5.4 Structures of pyrene and coronene, two aromatic compounds that do not follow the Hückel rule.

Molecule	Structure
Pyrene	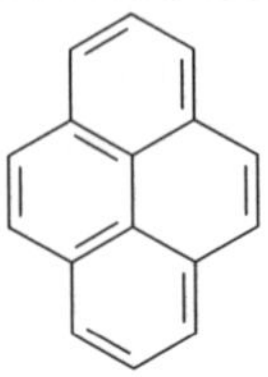
Coronene	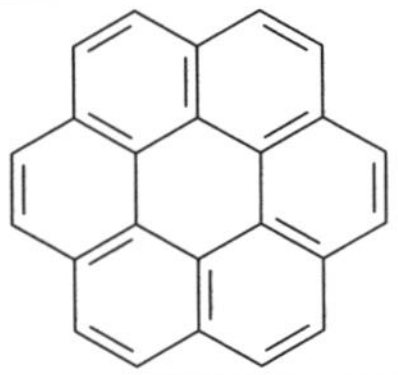

There are also some compounds that are aromatic but which do not follow Hückel's rule. Examples are the large macrocycles pyrene, $C_{16}H_{10}$, and coronene, $C_{24}H_{12}$, shown in Table 5.4. Such structure can be delocalized in 'subcycles' such that every atom participates in some number of resonance structures.

5.3 CONDUCTIVE POLYMERS

The idea of an organic polymeric material being electrically conductive was introduced in Chapter 3, Section 3.4.1. The simplest conductive polymer is *trans*-polyacetylene, consisting of a hydrocarbon chain with alternating single and double bonds. This results in a delocalized electron system, which allows charge carriers to move along the polymer backbone. *trans*-Polyacetylene is a semiconductor with an energy band gap about 1.4 eV. One of the advantages of conjugated polymers is that, to some extent, it is possible to control and vary their properties through appropriate modifications to their structure. The local arrangement of the carbon atoms on the backbone can be altered, for instance by incorporating heteroatoms, or the hydrogen atoms can be replaced with other side groups. Many conductive polymers have been synthesized to provide particular electronic features (band gap, electron affinity).

The monomer repeat units are often based on five-membered or six-membered (benzene) carbon ring systems. These include polyphenylene, polyphenylenevinylene, polypyrrole, polythiophene (and various other polythiophene derivatives) and polyaniline [3]. The chemical structures of some of these materials are shown in Figure 5.2 (the structure of polyaniline has been given previously in Figure 3.35). Generally, the band gap of the conductive polymer decreases with the length of conjugation, in accordance with Equation (3.59). Many conductive polymers have been developed so that they can be easily solution processed and used as thin films in devices. This is accomplished by attaching alkyl chains, which confer solubility in organic solvents. These chains are positioned on the molecules so that they do not disrupt the delocalized electron system. Some examples are given in Chapter 9, Section 9.6.

Heteroatoms are often incorporated into the polymer backbone; polycarbonitrile is the simplest example. Starting with the *trans*-isomer of polyacetylene, polycarbonitrile can be obtained by replacing every second CH group with an N atom [Figure 5.2(b)]. In Chapter 3, Section 3.4.1, we learnt that the occurrence of a bond-length alternation for *trans*-polyacetylene could be considered as a Peierls distortion, i.e. a doubling of the unit cell that accompanies the creation of a band gap at the Fermi level. For polycarbonitrile, the presence of the N atoms leads by itself to a doubling of the unit cell and, therefore, one might not directly expect a C–N bond-length alternation. However, bond length alternation does occur and leads to an increase in the band gap at the Fermi energy. This particular polymer has the unusual property of having a sigma band very close to the Fermi level. This (occupied) band is formed by the nitrogen lone-pair orbitals.

Poly(*p*-phenylenevinylene) (PPV) [depicted in Figure 5.2(d)] is one of the most studied conjugated polymers and was one of the first used in the fabrication of organic light-emitting devices (Chapter 9, Section 9.6); the LUMO and HOMO configuration for this polymer have

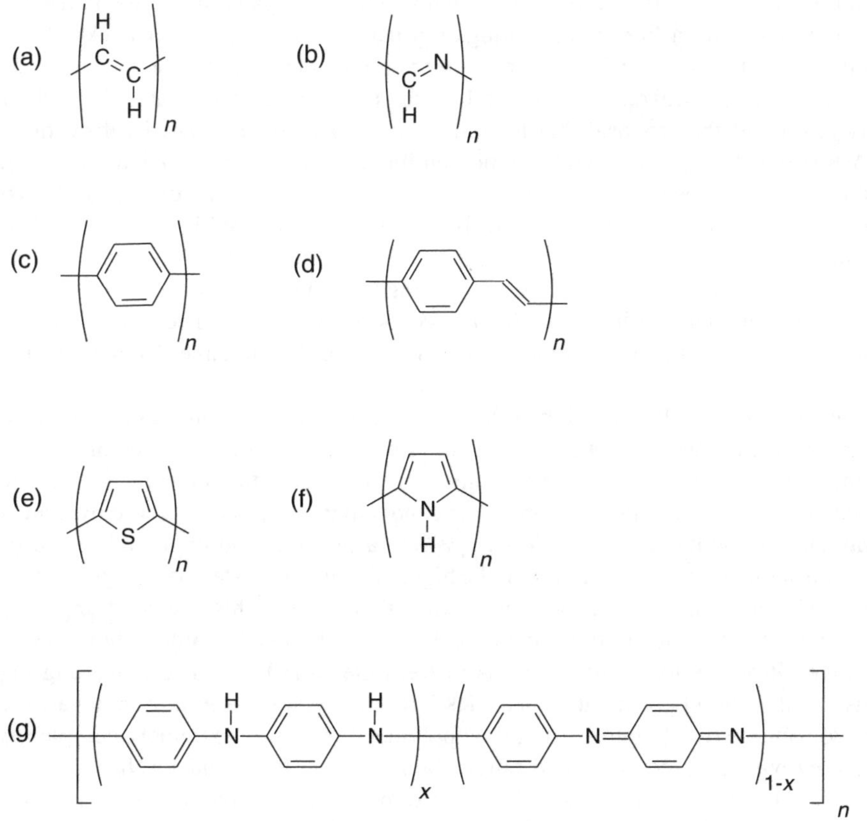

Figure 5.2 Chemical structures of some conductive polymers: (a) polyacetylene;(b) polycarbonitrile (c) polypheneylene; (d) poly(*p*-phenylenevinylene); (e) polythiophene; (f) polypyrrole; and (g) polyaniline.

been described in Chapter 3, Section 3.4.1. Due to the larger unit cell compared with polyacetylene, there are many more bands [4]. The total width of the valence bands is somewhat larger than for polyacetylene, but the position of the Fermi level does not change significantly. The most important result is that both the highest occupied and lowest occupied bands are, as expected, of π symmetry and, in addition, have their major components on the *vinyl* linkages between the phenylene rings (the term vinyl is used generically to refer to a substitution at a carbon atom that is part of an alkene double bond). Therefore, when attempting to modify the electronic properties, the best approach is to replace the hydrogen atoms associated with the vinyl link by other side groups. Substitution of one of the hydrogens with the electron donor NH_2 leads to an overall upward shift of the bands in energy, whereas the electron acceptor CN produces an overall downward shift.

Instead of replacing the hydrogen atoms with other side groups, some of the backbone atoms may be substituted with others to modify the electronic properties, equivalent to passing from polyacetylene to polycarbonitrile. Replacing one of the CH groups of the phenylene rings of PPV with an N atom produces poly(2,5-pyridinevinylene), PPyV. Conversely, replacing the vinyl linkages with either single N atoms or single NH groups results in polyaniline, PANi.

Polyaniline [Figure 5.2(g)], differs from many other conjugated polymers in a number of ways. The N heteroatom lies in the conjugation path along the polymer backbone. Due to steric hindrances, the backbone of this polymer is not planar, but the polymer instead forms helices. Therefore, the separation into σ and π orbitals is only approximate. The lack of strict planarity has led to the proposal that local distortions, so-called ring-twist distortions, may occur. As previously noted in Chapter 3, polyaniline can exist in three oxidation states: fully reduced leucoemeraldine [$x = 1$ in Figure 5.2(g)], which has no quinoidimine units, oxidized emeraldine ($x = 0.5$) and doubly (or fully) oxidized pernigraniline ($x = 0$). Only the pernigraniline form is strictly conjugated. However, the emeraldine form has the property that upon protonation of the imine sites, i.e. placing the emeraldine base in an acidic solution of pH < 4, the conductivity increases by 11 orders of magnitude. The emeraldine form of PANi is also unique in that both the base form and the salt (conductive) form are soluble and can be processed.

The charge carriers in PANi and the role of the lattice distortions are more complex than in other nondegenerate ground-state polymers. First, only protons are required to 'dope' emeraldine. The addition of a proton to imine nitrogen causes the number of π-electrons on the backbone to remain constant; however, the conductivity is seen to increase by many orders of magnitude. Furthermore, *thermoelectric power* (a phenomenon exploited in the thermocouple) measurements reveal that, even in the highly conductive state, the charge carriers in the films have negative charge, i.e. electrons rather than holes. This seems strange as, upon protonation, holes are added to the backbone. An explanation is that protonation removes one of the imine nitrogen's lone pair electrons to form the N–H bond. The remaining unpaired electrons can then hop between the vacancies left at these sites, resulting in negative charge carriers. Polyaniline (or the family of aniline polymers) is thus an extremely complex system.

Compared with polyacetylene and polycarbonitrile, the PPV- and PANi-based polymers contain more or less aromatic rings and also, in some cases, heteroatoms. A less complicated system, and one of the most intensively studied conjugated polymers, is polythiophene, shown in Figure 5.2(e). In its simplest form, it consists of a planar zigzag sequence of C_4H_2S units and of the two forms, shown in Figure 5.3. The aromatic structure [Figure 5.3(a)] has a lower total energy than the *quinoid* structure [Figure 5.3(b)]. The aromatic structure has a

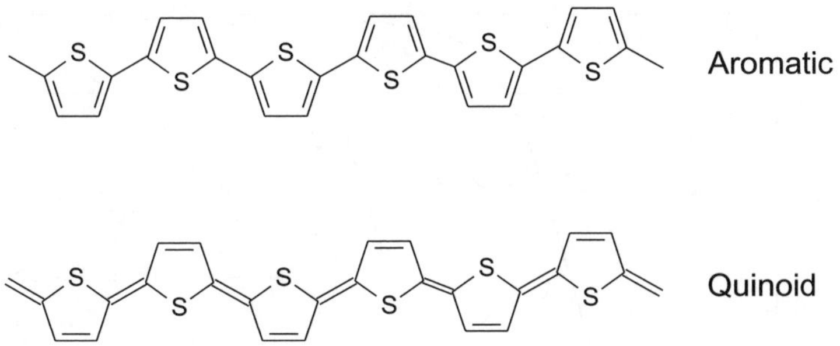

Figure 5.3 Aromatic and quinoid forms of polythiophene.

larger band gap at the Fermi energy than the quinoid structure, but both structures have π bands close to the Fermi level [4].

During the second half of the 1980s, scientists at the Bayer AG research laboratories in Germany developed a new polythiophene derivative, poly(3,4-ethylenedioxythiophene), having the backbone structure shown in Figure 5.4; the polymer is often abbreviated as PEDOT or PEDT [5, 6]. Initially, PEDOT was found to be an insoluble polymer, but it exhibited some interesting properties. For example, it possessed a high conductivity (approximately $300\,\mathrm{S\,cm^{-1}}$), was almost transparent as thin oxidized films and was highly stable in the oxidized state. The solubility problem was subsequently solved by using a water-soluble polyelectrolyte, poly(styrenesulfonic acid), PSS (Figure 5.4), to produce a water-soluble complex, PEDOT:PSS. The role of the PSS is twofold: first to provide a counterbalancing ion and second to keep the PEDOT polymer segments dispersed in an aqueous medium. Figure 5.5 shows the optical transmission versus the surface, or sheet, resistance (in units of Ω per square) (Chapter 3, Section 3.2.2). Possible applications for the PEDOT:PSS complex include capacitors, antistatic coatings, printed circuit boards and as a semi-transparent electrode in electroluminescent devices (Chapter 9, Section 9.6).

The optical and electrical behaviour of PEDOT:PSS is dependent on its electrochemical oxidation state. In its pristine state, the PEDOT component is partly oxidized with the PSS^- counter ions ensuring the overall charge neutrality of the complex. However, under the application of an appropriate bias voltage, it is possible to oxidize or reduce the PEDOT further. Figure 5.6 shows how this can be achieved by the incorporation of PEDOT:PSS in a

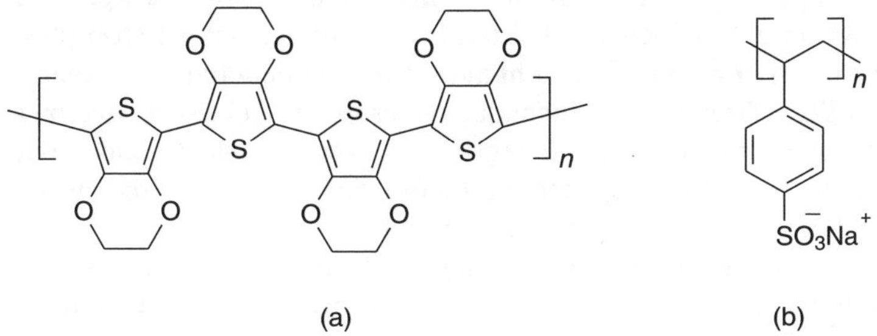

Figure 5.4 Chemical structures of (a) poly(3,4-ethylenedioxythiophene), often abbreviated as PEDOT (or PEDT), and (b) poly(styrenesulfonic acid), PSS. PEDOT is frequently used as a complex with PSS.

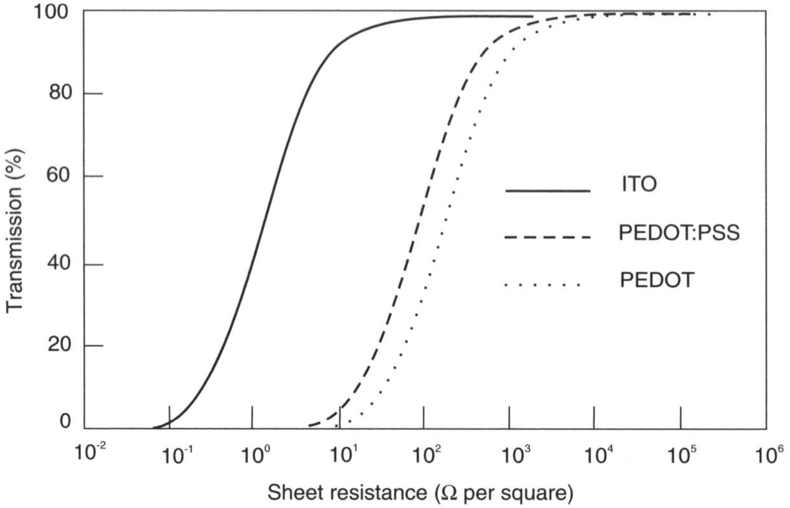

Figure 5.5 Transmission versus sheet resistivity for indium–tin–oxide (ITO) ($\sigma = 5600\,\mathrm{S\,cm^{-1}}$), a proprietary PEDOT:PSS formulation, Baytron® PHC ($\sigma = 450\,\mathrm{S\,cm^{-1}}$), and *in situ* polymerized PEDOT ($\sigma = 600\,\mathrm{S\,cm^{-1}}$). From Kirchmeyer S, Reuter K, 'Scientific importance, properties and growing applications of poly(3,4-ethylenedioxythiophene)', *J. Mater. Chem.*, **15**, pp. 2077–2088, Copyright (2005). Reproduced by permission of The Royal Society of Chemistry.

simple electrochemical cell: two thin adjacent PEDOT:PSS patterns are deposited on a solid substrate and in direct contact with a layer of a solid electrolyte [7]. When a voltage is applied to the adjacent PEDOT:PSS patterns, the PEDOT in the negatively biased pattern becomes reduced, whereas that in the positively biased pattern becomes further oxidized. The electrochemical reaction is described by

$$\mathrm{PEDOT:PSS} + \mathrm{M}^+ + \mathrm{e}^- \rightleftharpoons \mathrm{PEDOT} + \mathrm{M:PSS} \qquad (5.9)$$

where M^+ denotes positively charged metal ions in the electrolyte and e^- denotes the electrons in the PEDOT.

The above reaction is associated with a change of the electrical and optical properties of the PEDOT:PSS. In its oxidized state, PEDOT possesses a high concentration of free charge carriers and the optical properties are determined by transitions between bipolaronic states (Chapter 3, Section 3.4.3). This results in a low degree of optical absorption in the visible wavelength region [Figure 5.6(b)] and high electrical conductivity. In contrast, the reduced, neutral form of PEDOT exhibits semiconductive properties, characterized by a strong optical absorption in the visible wavelength region and low electrical conductivity. It has been suggested that the structure depicted in Figure 5.6 might be used as the basis for an electrochemical active matrix addressed display [7].

Atoms of sulfur or nitrogen possess s and p valence electrons, as in the case of carbon. The chemical bonds that they make in carbon-based polymers are therefore expected to be relatively similar. This situation may change when, instead, metal atoms with d valence electrons (Chapter 2, Section 2.2) are incorporated into the backbone. However, the presence of metal atoms does not necessarily destroy the conjugation along the polymer backbone [4].

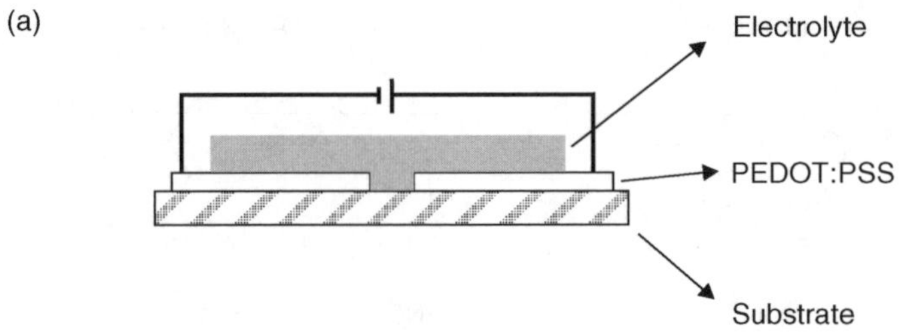

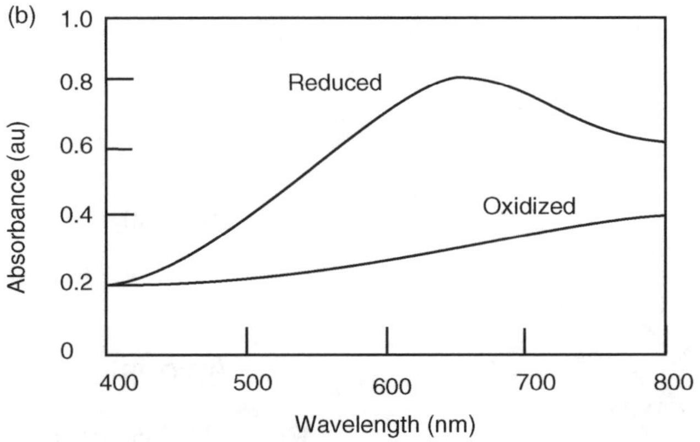

Figure 5.6 (a) Structure of an electrochemical cell incorporating PEDOT:PSS. Application of a voltage between the left and right PEDOT patterns results in the oxidation and reduction of the polymer regions. (b) Optical absorption spectra of the oxidized and reduced PEDOT:PSS patterns. Reprinted from *Advanced Materials*, **14**, Andersson P, Nilsson D, Svensson P-O, Cham M, Malmström A, Remonen T, Kuyler T, Berggren M, 'Active matrix display based on all-organic electrochemical smart pixels printed on paper', pp 1450–1464, Copyright (2002), with permission from Wiley-VCH.

5.4 CHARGE-TRANSFER COMPLEXES

Charge-transfer compounds were introduced in Chapter 3, Section 3.4.1. The best known compounds are materials such as tetrathiafulvalene (TTF) and tetracyanoquinodimethane (TCNQ). The latter was first synthesized at the DuPont laboratories in 1962 and formed the basis of the first stable conductive organic materials. A wide range of other donor and acceptor complexes that exhibit semiconductive behaviour have been synthesized since that time. Some examples are shown in Figure 5.7. Many charge-transfer systems are arranged in separate donor and acceptor stacks (Chapter 3, Section 3.4.1, Figure 3.33), resulting in extensive electron delocalization. The classic complex TTF–TCNQ has room temperature conductivity of about $500\,S\,cm^{-1}$ and, between temperatures of 298 and 54 K, shows metallic behaviour, with decreasing electrical resistance with decreasing temperature. Below 54 K, a Peierls transition opens up an energy gap in the band structure and the complex becomes semiconducting. This characteristic is depicted in Figure 5.8, together with the conductivity versus temperature data for other organic charge-transfer systems [8].

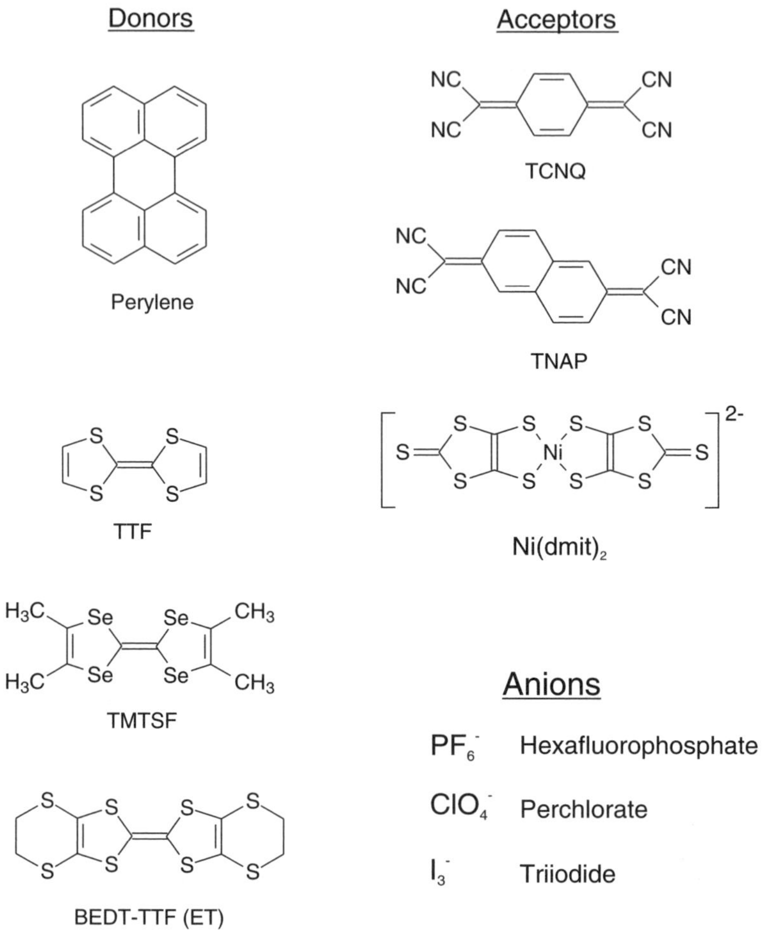

Figure 5.7 Examples of compounds that can form charge-transfer complexes. Electron donors are shown on the left: perylene, tetrathiafulvalene (TTF), tetramethyltetraselenafulvalene (TMTSF) and bis(ethylenedithio)-TTF (BEDT–TTF or ET). Electron acceptors and anions are shown on the right: tetracyanoquinodimethane (TCNQ), tetracyanonaphthalene (TNAP) and $Ni(dmit)_2$ ($H_2dmit = 4,5$-dimercapto-1,3-dithiole-2-thione).

Organic superconductivity (Chapter 3, Section 3.4.4) was first observed in 1979 in salts of tetramethyltetraselenafulvalene (TMTSF) (Figure 5.7). For example, $(TMTSF)_2^{+\cdot}\,PF_6^-$ has a transition temperature of 0.9 K under a pressure of about 9 kbar (which suppresses the Peierls transition); this compound also exhibits the Meissner effect (the exclusion of magnetic field from the material), an important test for superconductivity.

Several other salts of TMTSF are superconducting at temperatures less than 4 K; conductivity versus temperature data for $(TMTSF)_2^{+\cdot}\,ClO_4^-$ are included in Figure 5.8. The molecule bis(ethylenedithio)tetrathiafulvalene (BEDT–TTF or ET) contains eight sulfur atoms and is an important donor. A family of electrically conductive salts has been synthesized with the stoichiometry $(ET)_2X$, where X is the charge-compensating monovalent anion. When the counter ion is $Cu[N(CS)_2]Br$, Figure 5.8 reveals a relatively high

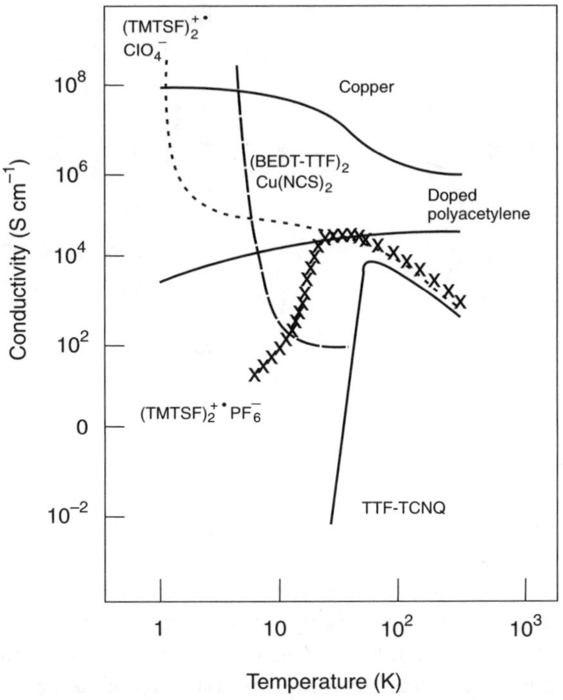

Figure 5.8 Conductivity versus temperature data at ambient pressure for a range of charge-transfer salts. The conductivity for copper is included for comparison. From Petty MC, Bryce MR, Bloor D (Edition), *An Introduction to Molecular Electronics*, Copyright (1995), Edward Arnold. Reproduced with permission of Edward Arnold (Publishers) Ltd.

superconducting transition temperature of about 9 K. Such salts can possess little, or no, columnar stacking, and are characterized by short inter-stack S–S interactions.

Metal complexes of the 1,2-dithiolene *ligand* (a ligand is an atom, ion or functional group that is bonded to one or more central atoms or ions, usually metals generally through a coordinate covalent bond) form the basis of a range of highly conductive materials and, in some cases, superconducting systems. The redox behaviour of these systems can be varied by changing the central metal atom. Figure 5.7 shows one of the most widely studied compounds, $Ni(dmit)_2$ (H_2dmit = 4,5-dimercapto-1,3-dithiole-2-thione).

Phthalocyanines, depicted in Figure 5.9, are examples of metallomacrocyclic compounds. These are flat, ring-shaped molecules that become electrically conductive on oxidation, for example on doping with iodine or bromine. Figure 5.9(a) shows the structure of a metallic phthalocyanine; a variety of transition metals are found to coordinate into the centre of the molecule, e.g. M = Fe, Co, Ni, Cu, Zn and Pt. In many of the charge-transfer systems, the metal retains a +2 oxidation state and does not play a significant role in the conduction process. The charge carriers are associated with the delocalized π orbitals on the macrocyclic ligand. Phthalocyanines are highly stable compounds and many of these can be heated to several hundred degrees before they sublime. This makes them suitable for processing into thin films by physical vapour deposition (thermal evaporation or molecular beam epitaxy – Chapter 7, Section 7.2.2). The phthalocyanine molecule can also be substituted using, for

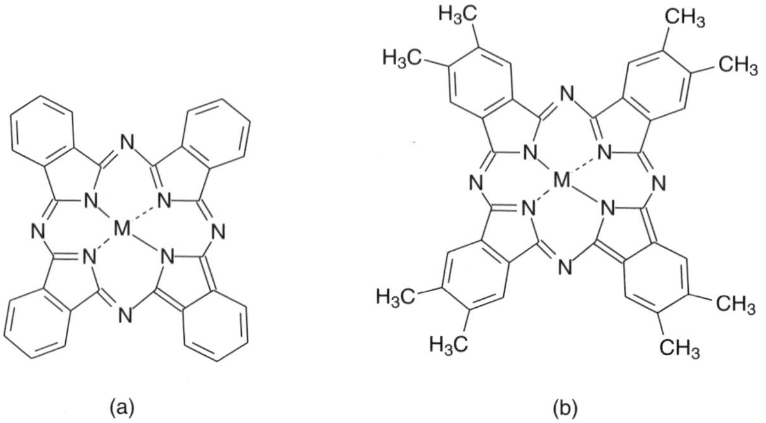

Figure 5.9 (a) Metal phthalocyanine; (b) substituted phthalocyanine compound.

example, methyl groups [Figure 5.9(b)], to make the material soluble in common organic solvents.

Highly conductive organic solids differ from inorganic metals in several respects. First, the overlap between adjacent units is highly anisotropic, leading to the formation of one-dimensional bands rather than the three-dimensional bands of conventional metals. Second, the weak intermolecular forces lead to a high probability of defects even in regular one-dimensional structures, and these severely limit the mean free path of charge carriers in otherwise ideal bands. Finally, the molecules in an organic solid are separated by distances of the order of van der Waals contacts, so that some shortening of these intermolecular spacings is possible without severe consequences for electron repulsions – as would be encountered for species which started at separations already significantly shorter than van der Waals distances. This favours Peierls distortions at low temperatures.

5.5 BUCKYBALLS AND NANOTUBES

5.5.1 Fullerenes

Important electroactive compounds that may find application in molecular electronics are based on forms of carbon. Graphite consists of vast carbon sheets which are stacked one on top of another like a sheaf of papers. In pure graphite, these layers are about 0.335 nm apart, but they can be separated further by intercalating various molecules. The bonding between the carbon atoms in the planes is mainly sp^2 hybridizations consisting of a network of single and double bonds. Weak interactions between the delocalized electron orbitals hold adjacent sheets together. The delocalized electron system in the planes results in semiconductive electrical behaviour.

Under certain conditions, carbon forms regular clusters of 60, 70, 84, etc, atoms. A C_{60} cluster, shown in Figure 5.10, is composed of 20 hexagons and 12 pentagons and resembles a football. The diameter of the ball is about 1 nm. As with graphite, each carbon atom in C_{60} is bonded to three other carbon atoms. Thus C_{60} can be considered as a rolled-up layer of a single graphite or *graphene* sheet. The term *buckminsterfullerene* was given originally to the

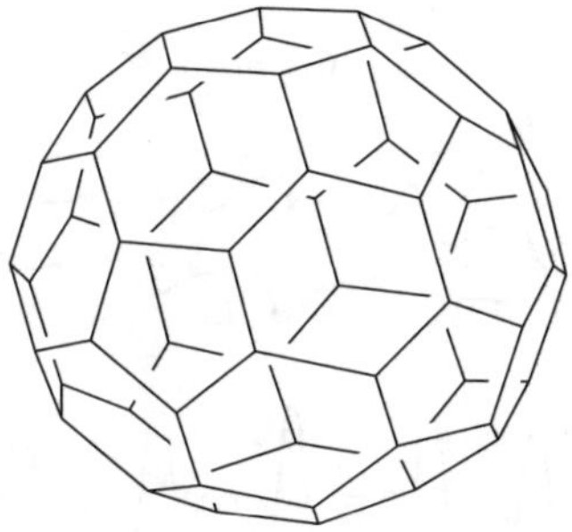

Figure 5.10 Structure of C_{60}.

C_{60} molecule because of the resemblance to the geodesic domes designed and built by Richard Buckminster Fuller. However, this term (or fullerene or buckyball) is used generally to describe C_{60} and related compounds. For example, a molecule with the formula C_{70} can be formed by inserting an extra ring of hexagons around the equator of the sphere, producing an elongated shell more like a rugby ball.

The electronic structure of C_{60} is unique for a π-bonded hydrocarbon, in that the molecule is a strong electron acceptor with an electron affinity of 2.65 eV. This is a consequence of its geometric structure, which influences the electronic energy levels in two simple ways. First, as noted above, the structure may be regarded as 20 six-membered rings with 12 five-membered rings. Conjugated five-membered rings always lead to higher electron affinity as a result of the aromatic stability associated with the $C_5H_5^-$ (cyclopentadienyl) anion. The non-planarity of the structure also means that the π-electrons are no longer pure p in character. The slight pyramidalization of each carbon atom (referring to the downward deflection of the three atoms surrounding each carbon from the plane in which they would all lie in the graphite structure) induces a small rehybridization in which some s-character is introduced into the π orbitals. The molecular orbital energy diagram for C_{60} reveals that the molecule might be expected to accept at least six electrons, and possibly up to 12. Figure 5.11 shows six separate reversible electrochemical redox waves for C_{60} in solution [9].

In the solid state, the C_{60} molecules form a crystal lattice with a face-centered cubic (fcc) structure and a molecular spacing of 1 nm (Chapter 2, Section 2.5.3). Alkali metal atoms can easily fit into the empty spaces in the fcc lattice. Consequently, when C_{60} crystals are heated with potassium, the metal vapour diffuses into the fullerene lattice to form the compound K_3C_{60}. Figure 5.12 shows the location of the alkali metal atoms in the fullerene lattice, where they occupy the two vacant tetrahedral sites (shown by the open spheres) and a larger octahedral site (shaded spheres) per C_{60} molecule [10]. In the tetrahedral site, the potassium ion has four surrounding C_{60} molecules, and in the octahedral site, there are six surrounding fullerene molecules.

The molecule C_{60} is an insulator, but when it becomes doped with an alkali metal it becomes electrically conducting. In the case of K_3C_{60}, the potassium atoms become ionized to form K^+ and their electrons are associated with the C_{60}, which becomes a C_{60}^{3-} triply negative ion. Each C_{60} molecule has three electrons that are loosely bonded to the fullerene molecule and can move through the lattice. K_3C_{60} has a superconducting transition temperature of 18 K. Higher transition temperatures are found with other metal alkali metal atoms, e.g. 33 K for Cs_2RbC_{60}. The transition temperature increases with the radius of the dopant alkali metal ion.

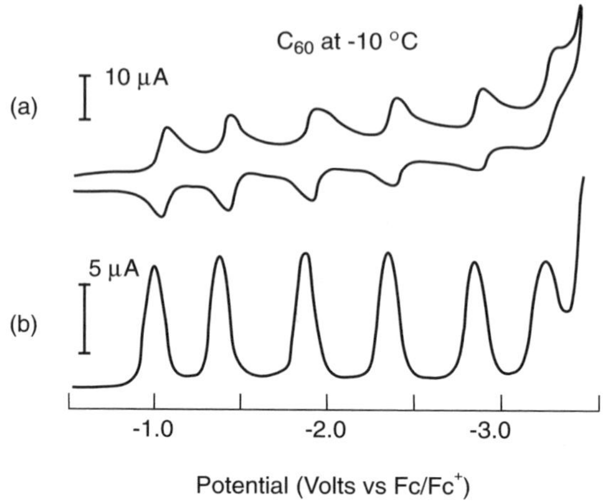

Figure 5.11 Reduction of C_{60} in CH_3CN–toluene solution at $-10\,°C$ using (a) cyclic voltammetry at a $100\,mV\,s^{-1}$ scan rate and (b) differential pulse voltammetry (50 mV pulse width, 300 ms period, $25\,mV\,s^{-1}$ scan rate). Fc = ferrocene. Reprinted with permission from *J. Am. Chem. Soc.*, **114**, Xie Q, Pérez-Corders, Echegoyen, L, 'Electrochemical detection of C_{60}^{-6} andC_{70}^{-6}: enhanced satiability of fullerides in solution', pp. 3978–3980. Copyright (1992) American Chemical Society.

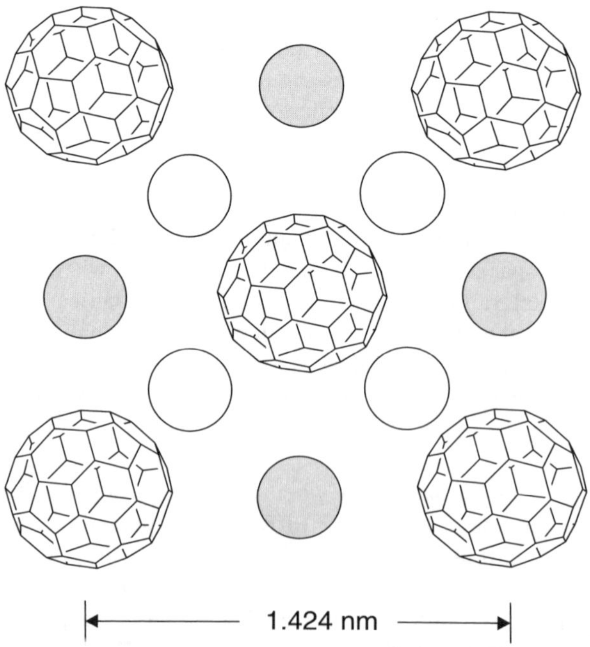

Figure 5.12 Structure of K_3C_{60}. The open and shaded spheres represent the potassium ions at tetrahedral and octahedral sites, respectively [10].

Larger fullerenes such as C_{70}, C_{76}, C_{80} and C_{84} have been found. Some smaller fullerenes also exist. For example, a solid phase of C_{22} has been identified in which the lattice consists of C_{20} molecules bonded together by an intermediate carbon atom. When suitably doped, such smaller fullerenes may exhibit higher superconducting transition temperatures than the C_{60} materials.

The existence of the carbon fullerenes has stimulated some discussion about similar clusters of other atoms, such as silicon or nitrogen. Theory has shown that the N_{20} cluster should be stable. This material has also been predicted to be a powerful explosive! However, it has yet to be synthesized.

5.5.2 Carbon Nanotubes

In addition to the spherical-shaped fullerenes, it is possible to synthesize tubular variations – *carbon nanotubes*. Such tubes are comprised of graphene sheets, curled into a cylinder. Each tube may contain several cylinders nested inside each other. The tubes are capped at the end by cones or faceted hemispheres. Because of their very small diameters (down to around 0.7 nm), carbon nanotubes are prototype one-dimensional nanostructures.

Figure 5.13 depicts a single graphene sheet. The unit vectors of the hexagonal lattice $\mathbf{a}_1$ and $\mathbf{a}_2$ generate the graphene lattice (these unit vectors are not orthogonal in the hexagonal

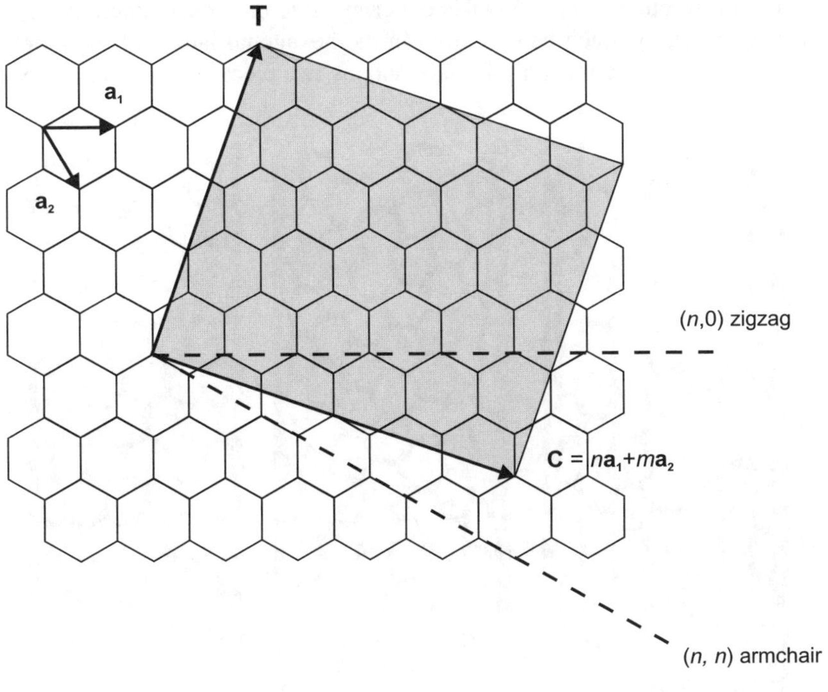

Figure 5.13 Graphene sheet showing the unit vectors $\mathbf{a}_1$ and $\mathbf{a}_2$ of the two-dimensional unit cell. In cutting a rectangular sheet, shown as the shaded region, a circumference vector $\mathbf{C} = n\mathbf{a}_1 + m\mathbf{a}_2$ is defined. The direction of the axis of the nanotube is shown by the vector $\mathbf{T}$. Zigzag and armchair structures are defined by the depicted directions of $\mathbf{C}$.

lattice; Chapter 2, Section 2.5.3). If a (= 0.142 nm) is the carbon–carbon bond length, then $|\mathbf{a_1}| = a(\sqrt{3}, 0)$ and $|\mathbf{a_2}| = a(\sqrt{3/2}, 3/2)$. An important feature of a carbon nanotube is the orientation of the six-membered carbon ring in the honeycomb lattice relative to the axis of the nanotube. In cutting a rectangular sheet, shown as the shaded region in Figure 5.13, a circumference vector $\mathbf{C} = n\mathbf{a_1} + m\mathbf{a_2}$ is defined, where the integers n and m denote the number of unit vectors along the crystallographic axes. The direction of the axis of the nanotube is shown by the vector $\mathbf{T}$ in the figure (in a direction orthogonal to $\mathbf{C}$). The radius of the resulting nanotube R is then given by

$$R = \frac{C}{2\pi} = \left(\frac{\sqrt{3}}{2\pi}\right) a\sqrt{n^2 + m^2 + nm} \tag{5.10}$$

The primary classification of a carbon nanotube is as either being chiral or achiral (Chapter 2, Section 2.4.2). An achiral nanotube is one whose mirror image has a structure identical with the original. There are only two cases of achiral nanotubes: armchair and zigzag (these names arise from the shape of the cross-sectional ring). When the circumference vector lies along one of the two basis vectors, the nanotube is said to be of the zigzag type, for which $m = 0$. For an armchair nanotube, the circumference vector is along the direction exactly between the two basis vectors; in this case, $n = m$. All other (n, m) indexes correspond to chiral nanotubes. Three examples of single-wall nanotubes (SWNTs) are shown in Figure 5.14.

The electronic structure of an SWNT is either metallic or semiconducting, depending on its diameter and chirality. Each carbon atom in the hexagonal lattice of the graphene sheet possesses six electrons. The inner 1s orbital contains two electrons, whereas three electrons

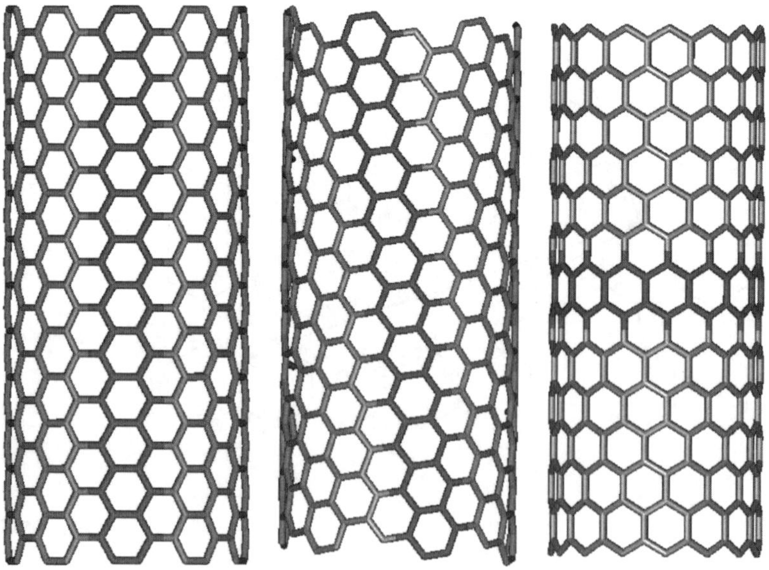

Figure 5.14 Three classes of single-wall carbon nanotube (SWNT): (a) (10,10) armchair SWNT; (b) (12,7) chiral SWNT; and (c) (15,0) zigzag SWNT.

in $2sp^2$ hybridized orbitals form three bonds in the plane of the graphene sheet. This leaves the final electron in a 2p orbital, perpendicular to the graphene sheet and to the nanotube surface. A delocalized π-electron network is therefore formed across the nanotube surface, responsible for its electronic properties. The band structure of a carbon nanotube can be derived from that of graphene, which is *semi-metal* with the valence and conduction bands meeting a several points in the Brillouin zone (Chapter 3, Section 3.3.2). A full analysis leads to the following condition for carbon nanotubes to be metallic [11]:

$$|n - m| = 3I \tag{5.11}$$

where I is an integer. Nanotubes for which this condition does not hold are semiconducting. Furthermore, it can be shown that the band gap of semiconducting nanotubes decreases inversely with an increase in the tube radius, as depicted in Figure 5.15 for zigzag nanotubes. The relationship is approximately that the band gap $E_g = 0.45/R$ eV, where R is the radius of the tube in nanometres.

Some deviations in the electronic properties of nanotubes (from the simple π-electron network model of graphene) arise due to the curvature of the tube. As a result, nanotubes satisfying Equation (5.11) develop a small curvature-induced band gap and hence are semi-metallic. Armchair nanotubes are an exception because of their special symmetry, and remain metallic for all diameters. The band gap of semi-metallic nanotubes is small and varies inversely as the square of the nanotube diameter. For example, a semi-metallic nanotube with a diameter of 1 nm has a band gap of about 40 meV [11].

Nanotubes are found in a variety of forms and shapes other than the single wall tube. Bundles of SWNTs are frequently observed. The individual tubes in the bundle are attracted to their nearest neighbours via van der Waals interactions (Chapter 2, Section 2.3.5), with typical distances between the tubes being comparable to the interplanar spacings in graphite, 0.31 nm. Multiwall nanotubes (noted above) consist of SWNTs nested inside one another,

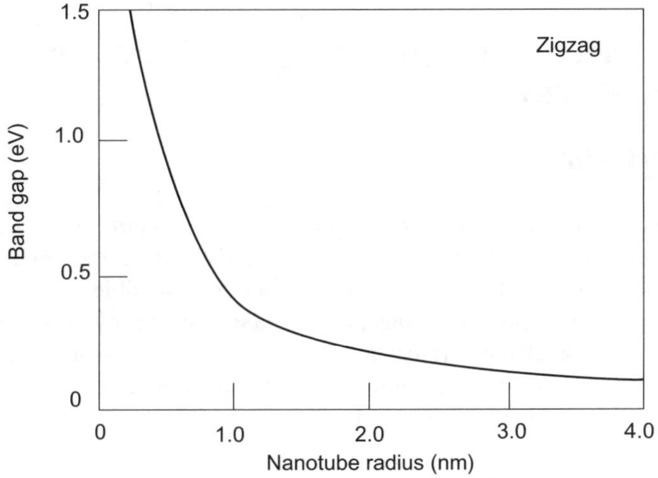

Figure 5.15 Band gap versus radius for zigzag carbon nanotubes. Reprinted from *Rep. Prog. Phys.*, **69**, Anantram MP, Léonard F, 'Physics of carbon nanotube electronic devices', pp. 507–561. Copyright (2006), with permission from IOP Publishing Ltd.

like a Russian doll. Carbon nanotubes also occur in more interesting shapes such as junctions between nanotubes of different chiralities and three terminal junctions. Such junctions are atomically precise in that each carbon atom is bonded primarily to its three nearest neighbours and there are no dangling bonds.

At low temperature, a single-wall carbon nanotube is a quantum wire in which the electrons in the wire move without being scattered (ballistic transport – Chapter 3, Section 3.2.1). Carbon nanotubes hold promise as interconnects in both silicon nanoelectronics and molecular electronics applications because of their low resistance and strong mechanical properties. Carbon nanotubes can also be doped by either electron donors or electron acceptors. After reaction with the host materials, the dopants are intercalated in the intershell spaces of the multiwalled nanotubes and, in the case of single-walled nanotubes, either in between the individual tubes or inside the tubes. The remarkable electronic behaviour of carbon nanotubes is discussed separately in Chapter 11, Section 11.10. In addition to their unique electronic properties, single-wall carbon nanotubes are also some of the strongest materials that are known, exhibiting very high *tensile strengths*, around 45×10^9 Pa (20 times stronger than steel) and large values of *Young's modulus*, $(1.3 - 1.8) \times 10^9$ Pa (almost 10 times that of steel). When carbon nanotubes are bent, they buckle like straws but do not break, and can be straightened back without damage. This results from the sp^2 bonds, which can rehybridize as they are bent. Multiwall nanotubes also possess excellent mechanical properties, but they are not as good in this respect as their single-walled counterparts.

The above confirms carbon's uniqueness as an electronic material. It can be a good conductor in the form of graphite, an insulator in the form of diamond or a flexible polymer (conductive or insulating) when reacted with hydrogen and other species. Carbon differs from other Group 4 elements, such as Si and Ge, which exhibit sp^3 hybridization. Carbon does not have any inner atomic orbitals except for the spherical 1s orbital, and the absence of nearby inner orbitals facilitates hybridizations involving only the valence (outer) s and p orbitals. The fact that sp and sp^2 hybridizations do not readily occur in Si and Ge might be related to the absence of 'organic materials' made from these elements.

5.6 PIEZOELECTRICITY, PYROELECTRICITY AND FERROELECTRICITY

5.6.1 Basic Principles

Piezoelectricity is the ability of certain crystals (for example, quartz) to generate a voltage in response to applied mechanical stress. The word is derived from the Greek *piezein*, which means to squeeze or press. The piezoelectric effect is reversible; piezoelectric crystals, subject to an externally applied voltage, can change shape by a small amount. The deformation, about 0.1% of the original dimension in certain ceramic materials, is of the order of nanometres, but nevertheless finds useful applications such as the generation and detection of sound, high voltages and high frequencies (MHz) and ultrafine manipulation of optical assemblies [12, 13].

On a microscopic scale, piezoelectricity results from a nonuniform charge distribution within the unit cell of a crystal. When the crystal is mechanically deformed, the positive and negative charge centres are displaced by differing amounts. Although the overall crystal remains electrically neutral, the difference in charge centre displacements results in an

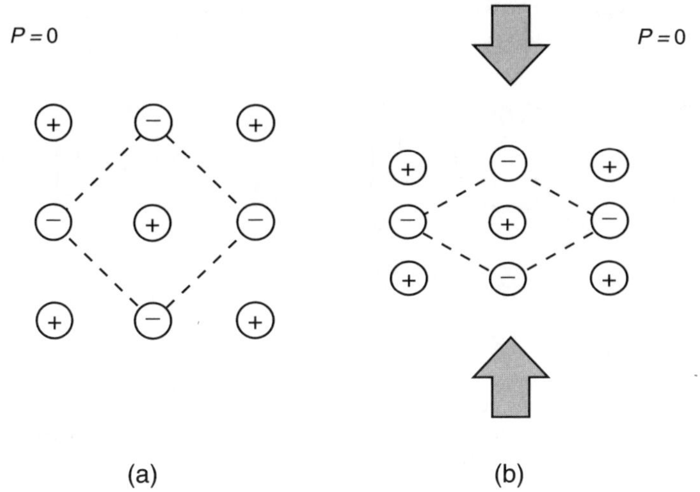

Figure 5.16 Schematic representation of a cubic unit cell. (a) In the absence of an applied force, the centre of mass of the positive ions coincides with that of the negative ions and there is no resulting dipole moment or polarization. (b) Under an applied force, the situation does not change. Reproduced from *Priniciples of Electrical Engineering Materials and Devices, Kasap SO,* Copyright (1997), with permission from McGraw-Hill Education.

electric polarization. Only certain classes of crystals (Chapter 2, Section 2.5.3), those without a centre of symmetry, can be piezoelectric. This is illustrated by the two-dimensional unit cells shown in Figures 5.16 and 5.17, which comprise of positively and negatively charged ions [12]. In Figure 5.16, the cubic unit cell has a centre of symmetry. When unstressed, the centres of mass of the negative and positive ions coincide with the centre of the unit cell. Therefore, there is no net dipole moment or polarization, P (Chapter 2, Section 2.3.5; P is the dipole moment per unit volume). This situation does not change if a force is applied, as depicted in Figure 5.16(b).

If the unit cell has no centre of symmetry, the two centres of mass become displaced on application of a force. The example shown in Figure 5.17 is for a hexagonal unit cell. The direction of the induced polarization depends on the direction of the applied stress. For Figure 5.17(b), a polarization is produced in the same direction as the applied stress. When the same unit cell is stressed along a different direction [Figure 5.17(c)], there is no polarization in this direction, because there is no net displacement of the centres of mass in the direction of the force. However, the direction of the force will produce a polarization in an orthogonal direction. Generally, an applied stress in one direction in a piezoelectric crystal can give rise to a polarization in other crystal directions. If T_j is the applied mechanical stress along some j direction ($j = x, y, z$) and P_i is the induced polarization along some i direction, then the two are linearly related by

$$P_i = d_{ij}T_j \tag{5.12}$$

where d_{ij} are called the *piezoelectric coefficients*. In fact, d_{ij} is a tensor quantity. We have previously come across such quantities in the case of permittivity (Chapter 4, Section 4.3.1). Formally, d_{ij} is a *third-rank tensor* (an array of $3 \times 3 \times 3$ coefficients) as it relates a vector

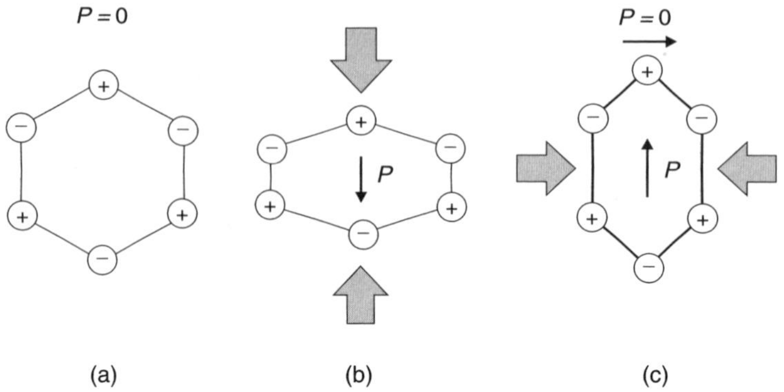

Figure 5.17 The hexagonal unit cell depicted has no centre of symmetry. (a) In the absence of an applied force, the centres of mass for positive and negative ions coincide. (b) Under an applied force, there is a displacement in the centres of mass for the positive and negative ions, resulting in a polarization in the direction of the applied force. (c) When the force is in a different direction, the polarization does not occur in the direction of the force. Reproduced from *Priniciples of Electrical Engineering Materials and Devices, Kasap SO,* Copyright (1997), with permission from McGraw-Hill Education.

quantity (polarization) to a second-rank tensor quantity (stress). In the converse piezoelectric effect, the application of an electric field will produce a strain (change in dimensions) of the crystal, or

$$S_j = d_{ij}\boldsymbol{E}_i \tag{5.13}$$

where S_j is the strain along the j direction and $\boldsymbol{E}_i$ is the electric field along i. The coefficients in Equations (5.12) and (5.13) are the same.

Of the 32 crystal classes, 21 are noncentrosymmetric (not having a centre of symmetry) and, of these, 20 exhibit direct piezoelectricity (the 21st class is a particular cubic class in which the various symmetry elements combine to produce nonpolar behaviour). Ten of these are polar (i.e. spontaneously polarize), having a dipole in their unit cell, and exhibit *pyroelectricity.* If this dipole can be reversed by the application of an electric field, the material is said to be *ferroelectric*. Figure 5.18 illustrates the relationship between the crystal structure and the phenomena of piezoelectricity, pyroelectricity and ferroelectricity.

If a pyroelectric material is heated, a change in the material polarization is produced. This can be measured as a voltage developed across the crystal. The *pyroelectric coefficient p* is given by the rate of change of polarization with temperature

$$p = \frac{\mathrm{d}P}{\mathrm{d}T} \tag{5.14}$$

Very small temperature changes ($\sim 10^{-6}\,°\mathrm{C}$) can give rise to voltages that are easily measurable. Because pyroelectric materials are also piezoelectric, they may also develop an additional polarization when heated due to any temperature-induced stresses (this will depend on if and how the material is mechanically constrained, e.g. as a thin film of material on a substrate with a different thermal expansion coefficient). This is referred to as a secondary pyroelectric effect.

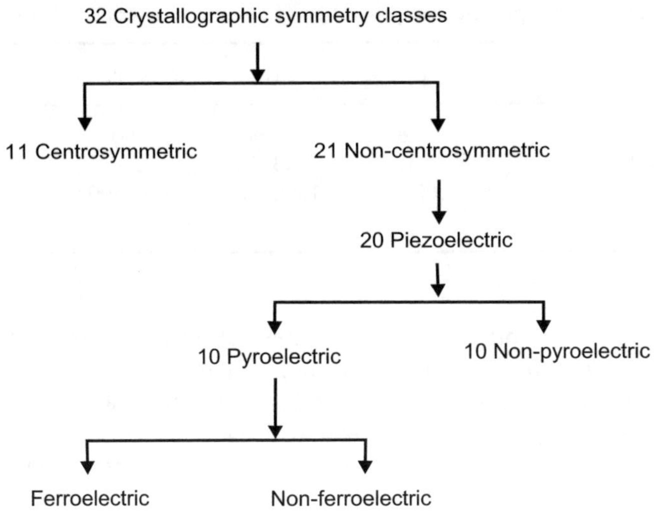

Figure 5.18 Relationship between the crystal symmetry and the properties of piezoelectricity, pyroelectricity and ferroelectricity.

Ferroelectric crystals are permanently polarized even in the absence of an applied field. This effect disappears above a certain temperature, the *Curie temperature*, T_C, and a *paraelectric* phase is formed. There are two main types of ferroelectrics: displacive and order–disorder. The effect in barium titanate, $BaTiO_3$, a typical ferroelectric of the displacive type, is due to a the Ti ion being displaced from equilibrium slightly, the force from the local electric fields due to the ions in the crystal increases faster than the elastic restoring forces. This leads to an asymmetric shift in the equilibrium ion positions and hence to a permanent dipole moment. In an order–disorder ferroelectric, there is a dipole moment in each unit cell, but above T_C these are pointing in random directions. Upon lowering the temperature and going through the Curie point, the dipoles order, all pointing in the same direction within a region or domain.

Ferroelectric crystals often show several Curie points and domain structure hysteresis, much as do ferromagnetic crystals (Section 5.7). By analogy with magnetic core memory, this hysteresis can be used to store information (see below) in memory elements based on ferroelectric capacitors. The nature of the phase transition in some ferroelectric crystals is still not well understood. Ferroelectrics often have very large dielectric constants, and therefore are often found in capacitors. They also often have unusually large nonlinear optical coefficients. The term *electret*, introduced by Oliver Heaviside to suggest the electrical equivalent of a magnet, is often used to describe a ferroelectric material.

5.6.2 Organic Piezoelectric, Pyroelectric and Ferroelectric Compounds

Generally, inorganic piezoelectric and pyroelectric materials (Table 5.5) possess higher values of d and p coefficients than their organic counterparts. However, piezoelectric and

Table 5.5 Piezoelectric and pyroelectric properties of common materials [12–14].

Material	T_C (°C)	ε_r	d (pC N^{-1})	p (μC m^{-2} K^{-1})
$BaTiO_3$, barium titanate	120	1900	190	200
TGS, triglycine sulfate	49	43–50	25	280–350
PZT, lead zirconate titanate, $PbTi_{1-x}Zr_xO_3$	200–400	1000–2000	250–500	60–500

pyroelectric polymeric sensors and actuators offer the advantage of processing flexibility because they are lightweight, tough, readily manufactured into large areas and can be cut and formed into complex shapes. Polymers also exhibit high strength and high impact resistance. Other notable features of polymers are their low dielectric constants, low elastic stiffnesses and low densities, which result in a high voltage sensitivity (an important sensor characteristic). Furthermore, these materials possess low acoustic and mechanical impedances, essential for medical and underwater applications. Polymers also typically have high dielectric breakdowns and high operating field strengths, which means that they can withstand much higher driving fields than ceramics.

An interesting crystalline organic ferroelectric compound is triglycine sulfate, $(NH_2CH_2COOH)_3H_2SO_4$, usually abbreviated as TGS (Table 5.5). This is an order–disorder ferroelectric that is known to exhibit a second-order phase transition (Chapter 2, Section 2.5.2). Above the Curie temperature, $T_C = 49\,°C$, the crystal is centrosymmetric with monoclinic symmetry. The glycine group NH_2CH_2COOH (glycine is the only amino acid that is not chiral – Chapter 12, Section 12.2.1) commonly crystallizes in two different forms. One is a structure in which the two carbon atoms and the two oxygen atoms are approximately coplanar, while the nitrogen is significantly displaced out of this plane; the other is a structure in which all the carbon, nitrogen and oxygen atoms are close to planar. In TGS, two of the glycine groups (denoted II and III) are quasi-planar whereas the other (I) is non-planar. Chemically, two protons from the H_2SO_4 group are more properly associated with the glycines. Glycine I is more correctly a glycinium ion; the other proton can be thought of as an O_{II}–H–O_{III} proton and resonates between glycine II and glycine III, at least in the paraelectric phase above T_C. All three glycines groups participate in the polarization reversal on application of an electric field, but the main reversible dipole is that associated with glycine I.

Table 5.6 compares the piezoelectric and pyroelectric properties of some polymeric compounds, along with data on the glass transition temperatures T_g and melting points T_m (NB: for many polymeric materials, the Curie temperature is greater than the melting point). Probably the best-known organic piezoelectric material is the polymer poly(vinylidene difluoride) (PVDF). The crystal structure of this material has been described in Chapter 2, Section 2.6.3. This material is a semicrystalline polymer with a monomer unit (CH_2CF_2). The dipole moment is 7.0×10^{-30} C m perpendicular to the chain direction. The pyroelectric β-phase of PVDF is obtained from the parent α-phase by a process of stretching and then poling, in which a high electric field is applied to align the dipoles. If all the monomer dipoles were aligned along the chain direction, a maximum microscopic polarization of about 100 mC m^{-2} can be obtained. However, semicrystalline PVDF is approximately 50% crystalline and the observed polarization is around half of this maximum value [13, 14].

Table 5.6 Piezoelectric and pyroelectric properties of polymers [12–14].

Polymer	Mer unit	T_g (°C)	T_m (°C)	ε_r	d (pC N^{-1})	p [μC m^{-2} K^{-1}]
PVC	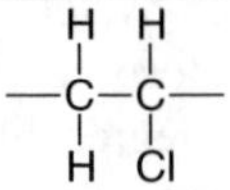	83	212	3.5	0.7	1
PVDF	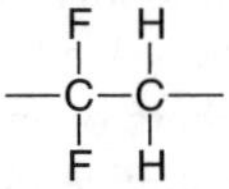	–35	175	12	28	40
Nylon-11	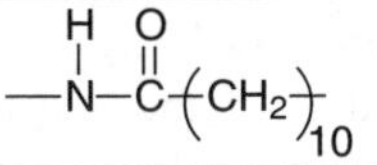	68	195	3.7	0.3	5

PVDF is also a pyroelectric and ferroelectric polymer containing polar crystals in which the direction of polarization can be reversed by the application of an electric field. The material can therefore be taken through a *hysteresis loop*, shown in Figure 5.19 in the form of a plot of the polarization as a function of the applied electric field [13]. The applied electric field required to reduce the polarization to zero, E_c, is called the *coercive field* while the polarization retained by the material with no applied field, P_r, is the *remanent* or *residual polarization*.

Copolymers of poly(vinylidene fluoride) with trifluoroethylene (TrFE) and tetrafluoroethylene (TFE) have also been shown to exhibit strong piezoelectric, pyroelectric and ferroelectric effects. An attractive morphological feature of the comonomers is that they

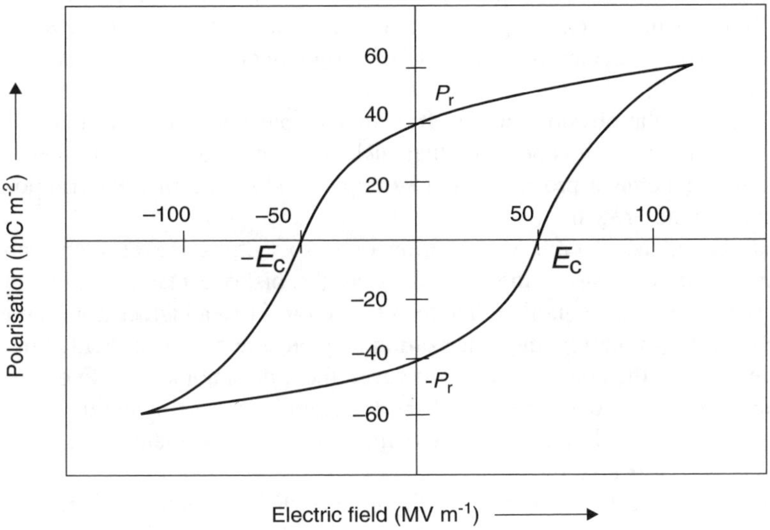

Figure 5.19 Typical polarization versus electric field hysteresis loop for PVDF. P_r = remanent polarisation; E_c = coercive field.

force the polymer into an all-*trans* conformation that has a polar crystalline phase. This eliminates the need for mechanical stretching to yield a polar phase. Poly(VDF–TrFE) crystallizes to a much greater extent than PVDF (up to 90% crystalline), yielding a higher remanent polarization, lower coercive field and much sharper hysteresis loops. TrFE also extends the operational temperature by about 20 °C, to close to 100 °C. Conversely, copolymers with TFE have been shown to exhibit a lower degree of crystallinity and a suppressed melting temperature compared with the PVDF homopolymer.

A low level of piezoelectricity is also observed in polyamides (also known as nylons). The monomer unit of odd-numbered nylons consists of even numbers of methylene groups and one amide group with a large dipole moment. Polyamides crystallize in all-*trans* conformations and are packed so as to maximize hydrogen bonding between adjacent amine and carbonyl groups. The amide dipoles align synergistically for the odd-numbered monomer, resulting in a net dipole moment. The unit dipole density is dependent on the number of methylene groups present and the polarization increases with decreasing number of methylene groups. Other polymer materials exhibiting piezoelectric and pyroelectric effects include polyurea, polytrifluoroethylene and copolymers of vinylidene cyanide and vinyl acetate [13]. However, the electroactive responses of these polymers are inferior to those of PVDF and its copolymer poly(VDF–TrFE). Pyroelectric polymers can also act as a host matrix for other pyroelectric compounds, such as TGS [15].

The phenomena of piezoelectricity and pyroelectricity are also found throughout the natural world. For example, piezoelectricity effects were found in keratin (Chapter 12, Section 12.2.2) in 1941 [14]. When a bundle of hair was immersed in liquid air, a potential of a few volts was generated between the tip and the root. When pressure was applied on the cross-section of the bundle, a voltage was generated. Subsequently, piezoelectricity has been observed in a wide range of other biopolymers, including collagen, polypeptides such as poly(methyl glutamate) and poly(benzyl-L-glutamate), oriented films of DNA, poly(lactic acid) and chitin [14]. Since most natural biopolymers possess D symmetry (Chapter 2, Section 2.4.2), they exhibit 'shear' piezoelectricity. A shear stress in the plane of polarization produces electric displacement perpendicular to the plane of the applied stress, as depicted in Figure 5.17(c). The piezoelectric constants of biopolymers are small relative to synthetic polymers (Table 5.6), ranging in value from $0.01\,pC\,N^{-1}$ for DNA to $2.5\,pC\,N^{-1}$ for collagen. Currently, the physiological significance of piezoelectricity in many biopolymers is not well understood, but it is believed that such electromechanical phenomena may have a distinct role in biochemical processes. For example, it is known that electric polarization in bone influences bone growth.

As noted above, the bistable polarization of ferroelectrics makes them candidates for binary memory applications in the same way as the bistable magnetization of ferromagnetics. The memory is non-volatile (Chapter 11, Section 11.6) and does not require a holding voltage. To record information, the polarization may be reversed or reoriented by application of a field greater than the coercive field. However, there does not appear to be any advantage of this type of memory over the established magnetic and semiconductor technologies. The writing process is relatively slow and most ferroelectric materials show ageing and fatigue effects with repeated cycling.

The topic of ferroelectric behaviour in liquid crystals is covered separately in Chapter 8, Section 8.5. It should be noted that many organic ferroelectric materials possess high second-order nonlinear optical properties and can be used for second-harmonic generation and electro-optic switching. This is discussed in Chapter 4, Section 4.3.2.

5.7 MAGNETIC MATERIALS

5.7.1 Basic Principles

In Chapter 4, Section 4.3, the definition of an electric dipole moment was given. Magnetism has its origin in the *atomic magnetic dipole*. Magnetic dipoles, or magnetic moments, result on the atomic scale from the two kinds of movement of electrons. The first is the orbital motion of the electron around the nucleus. The second, much stronger, source of electronic magnetic moment is due to the electron spin (although current quantum mechanical theory suggests that electrons neither physically spin nor orbit the nucleus).

Consider a current loop carrying a current I, as depicted in Figure 5.20(a). The area enclosed by the current loop (enclosed within a single plane) is A. The magnetic dipole moment, $\mathbf{m_m}$, is then defined:

$$\mathbf{m_m} = IA\mathbf{u_n} \tag{5.15}$$

where $\mathbf{u_n}$ is a unit vector perpendicular to the plane of A. The direction of this unit vector is such that looking along it, the (conventional) current circulates clockwise. When a magnetic moment is placed in a magnetic field, it experiences a torque that tries to rotate the magnetic moment to align with the axis of the magnetic field [Figure 5.20(b)].

An orbiting electron (Figure 5.21) behaves much like a current loop. The circulating electron therefore produces its own *orbital magnetic moment*, $\mathbf{m_{orb}}$. There is also a *spin magnetic moment*, $\mathbf{m_{spin}}$, associated with it due to the electron itself spinning, on its own axis. The overall magnetic moment of an electron consists of $\mathbf{m_{orb}}$ and $\mathbf{m_{spin}}$. However, as these are both vector quantities, they cannot be added numerically. The total magnetic moment of the atom $\mathbf{m_{atom}}$ depends on the orbital motions of all the electrons. However, only unfilled subshells contribute to the overall magnetic moment of an atom (the magnetic moments of pairs of electrons in closed subshells cancel with each other).

When a magnetic field $\mathbf{B}$ is applied to a material, the combined effect of the individual atomic magnets produced is that the material develops a net magnetic moment along the field and becomes magnetized. The *magnetization*, $\mathbf{M}$, is a vector quantity and is defined (in a

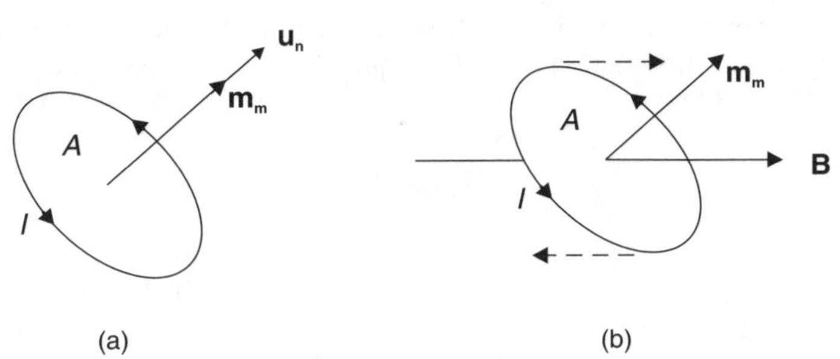

Figure 5.20 (a) Magnetic dipole moment $\mathbf{m_m}$. (b) In a magnetic field, a magnetic dipole moment experiences a torque.

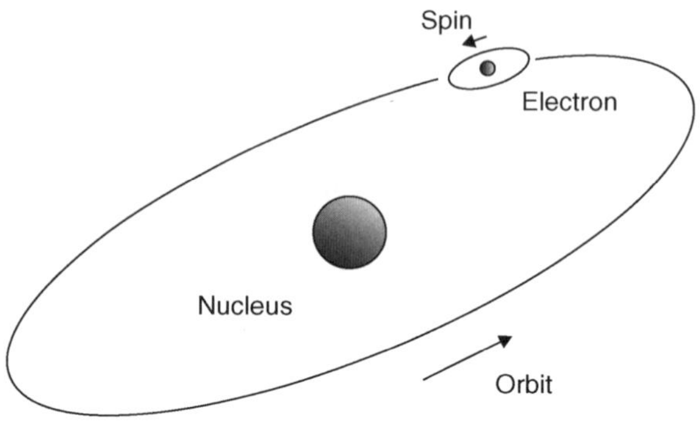

Figure 5.21 Motion of an electron around the nucleus of an atom.

similar way to polarization) as the magnetic dipole moment per unit volume. The magnetizing field (also called the applied field or the magnetic field intensity) is given the symbol **H**, and is the total field that would be present if the field were applied to a vacuum. The magnetic field **B** (sometimes called the magnetic induction) is the total flux of magnetic field lines through a unit cross-sectional area of the material, considering both lines of force from the applied field and from the magnetization of the material. **B**, **H** and **M** are related by

$$\mathbf{B} = \mu_0(\mathbf{H} + \mathbf{M}) \tag{5.16}$$

and

$$\mathbf{B} = \hat{\mu}_r \mu_0 \mathbf{H} \tag{5.17}$$

where $\hat{\mu}_r$ is the relative permeability, formally a tensor quantity since both **B** and **H** are vectors. The constant μ_0 is the *permeability of free space* ($4\pi \times 10^{-7}$ H m^{-1}), which is the ratio **B/H** measured in a vacuum. Equation (5.16) can be contrasted with Equation (4.8) in Chapter 4, which relates the electric displacement **D** to the polarization **P**. The SI units of magnetic field **B** are tesla (T) and those of the magnetizing field **H** are A m^{-1}. The latter is the *cause* (e.g. depending only on the external conduction currents in, say, a coil) whereas **B** is the *effect* and depends on the magnetization of the material.

In the electrical world, the polarization **P** produced in a material is related to the applied electric field **E** (Chapter 4, Equation 4.9). The equivalent expression in magnetism involves the magnetizing field **H**:

$$\mathbf{M} = \hat{\chi}_m \mathbf{H} \tag{5.18}$$

where $\hat{\chi}_m (= \hat{\mu}_r - 1)$ is the *magnetic susceptibility* – a tensor quantity.

Magnetic units are frequently the cause of confusion. While the SI (Système International d'Unités) system is generally preferred by engineers, CGS and Gaussian units are widespread [the defining equation here is $B = H + 4\pi M$, rather than that given in Equation (5.16)]. Table 5.7 shows the conversion factors between these systems for common magnetic units.

Table 5.7 Units for magnetic properties. Multiply a number in the CGS system by the conversion factor to change it to SI units. In the CGS system, $4\pi M$ is usually quoted as it has units of gauss and is numerically equivalent to B and H.

Magnetic quantity	Symbol	CGS unit	Conversion factor	SI unit
Magnetic dipole moment	m_m	emu	10^{-3}	$\mathrm{A\,m^2}$
Magnetic field	B	G	10^{-4}	T
Magnetizing field; magnetic field intensity	H	Oe	$10^3/4\pi$	$\mathrm{A\,m^{-1}}$
Magnetization	M	$\mathrm{emu\,cm^{-3}}$	10^3	$\mathrm{A\,m^{-1}}$
Magnetization	$4\pi M$	G		
Permeability	μ	Dimensionless	$4\pi \times 10^{-7}$	$\mathrm{H\,m^{-1}}$
Magnetic susceptibility	χ_m	$\mathrm{emu\,cm^{-3}\,Oe^{-1}}$	4π	Dimensionless

All materials can be classified in terms of their magnetic behaviour, falling into one of five categories depending on their bulk magnetic susceptibility. The two most common types of magnetism are *diamagnetism* and *paramagnetism*, which account for the magnetic properties of most of the periodic table of elements at room temperature, as shown in Figure 5.22. These elements are usually referred to as nonmagnetic, whereas those which are referred to as magnetic are actually classified as *ferromagnetic*. The only other type of magnetism observed in pure elements at room temperature is *antiferromagnetism*. Finally, magnetic materials can also be classified as *ferrimagnetic*, although this is not observed in any pure element but can only be found in compounds, such as the mixed oxides known as ferrites, from which ferrimagnetism derives its name. The origins of the different forms of magnetism, which are contrasted in Table 5.8, are outlined in the following sections.

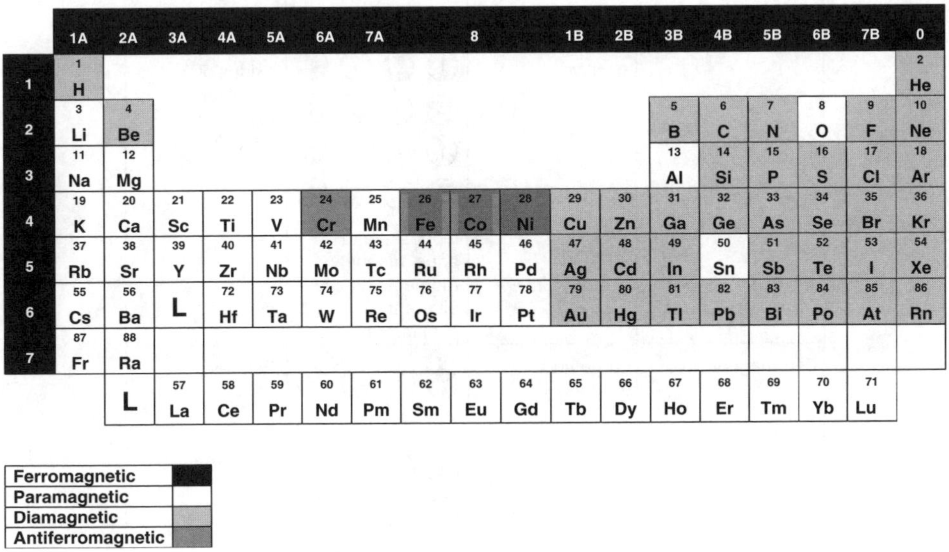

Figure 5.22 Periodic table showing magnetic properties of the elements.

Table 5.8 Summary of different types of magnetic behaviour.

Type of magnetism	Susceptibility	Atomic arrangement	Example
Diamagnetism	Small and negative	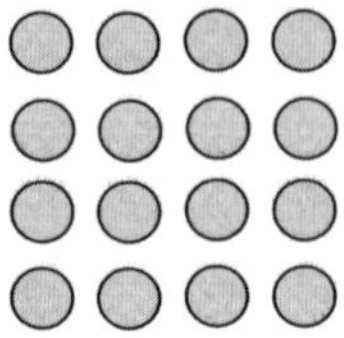 Atoms have no magnetic moment	Au, Ag, Cu, many polymers. Atoms have closed shells
Paramagnetism	Small and positive	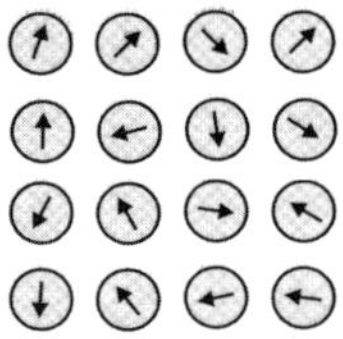 Atoms have randomly oriented magnetic moments	Gaseous and liquid oxygen. Alkali and transition metals
Ferromagnetism	Large and positive	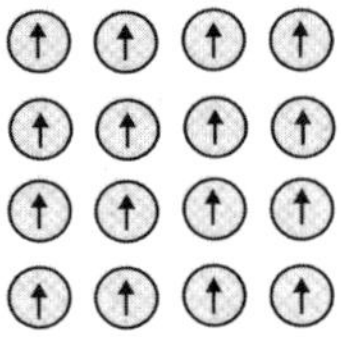 Atoms have parallel aligned magnetic moments	Fe, Co, Ni
Antiferromagnetism	Small and positive	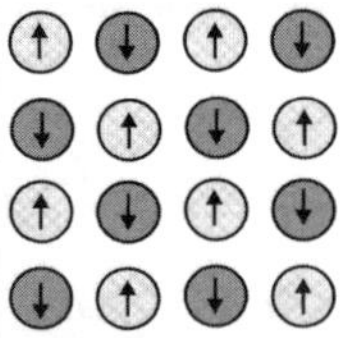 Opposite and equal magnetic moments on two different sublattices	Salts and oxides of transition metals. e.g. MnO, NiO
Ferrimagnetism	Large and positive	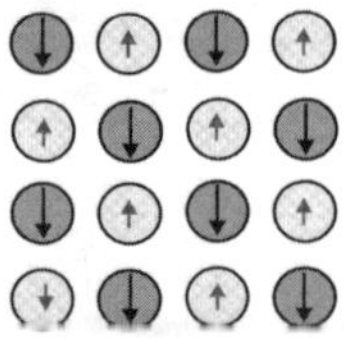 Magnetic moments on different sublattices do not cancel	Ferrites

Diamagnetism

In a diamagnetic material, the atoms have no net magnetic moment when there is no applied field. Under the influence of an applied field, the spinning electrons precess and this motion, which is a type of electric current, produces a magnetization in the opposite direction to that of the applied field (the susceptibility is negative). All materials have a diamagnetic effect; however, it is often the case that the diamagnetic effect is masked by the larger paramagnetic or ferromagnetic term. The value of susceptibility is independent of temperature. As noted in Chapter 3, Section 3.4.4, below its critical temperature a superconductor is a perfectly diamagnetic substance ($\chi_m = -1$).

Paramagnetism

There are several theories of paramagnetism, which are valid for specific types of material. In the *Langevin model*, each atom has a magnetic moment which is randomly oriented as a result of thermal energy. The application of a magnetic field creates a slight alignment of these moments and hence a low magnetization in the same direction as the applied field. As the temperature increases, the thermal agitation will increase and it will become harder to align the atomic magnetic moments. As a consequence, the susceptibility decreases. This behaviour is described by the *Curie law*, which relates magnetic susceptibility to the temperature:

$$\chi_m = \frac{C}{T} \tag{5.19}$$

where C is a constant, in units of temperature, known as the *Curie temperature* (or Curie constant). Materials which obey this law are substances in which the magnetic moments are localized at the atomic or ionic sites and where there is no interaction between neighbouring magnetic moments. The hydrated salts of the transition metals, e.g. $CuSO_4 \cdot 5H_2O$, are examples of this type of behaviour because the transition metal ions, which have a magnetic moment, are surrounded by a number of nonmagnetic ions/atoms. These prevent interaction between neighbouring magnetic moments.

The *Pauli model of paramagnetism* is applicable to materials where the electrons interact to form a conduction band; this is valid for most paramagnetic metals. In this model, the conduction electrons are considered essentially to be free and under an applied field. An imbalance between electrons with opposite spin is set up, leading to a low magnetization in the same direction as the applied field. The susceptibility is independent of temperature, although the electronic band structure may be affected, which will then have an effect on the susceptibility.

Ferromagnetism

Ferromagnetism is only possible when atoms are arranged in a lattice and the atomic magnetic moments can interact to align parallel to each other. This effect is explained in classical theory by the presence of a molecular field within the ferromagnetic material, which was first postulated by Weiss in 1907. This field is sufficient to magnetize the material to

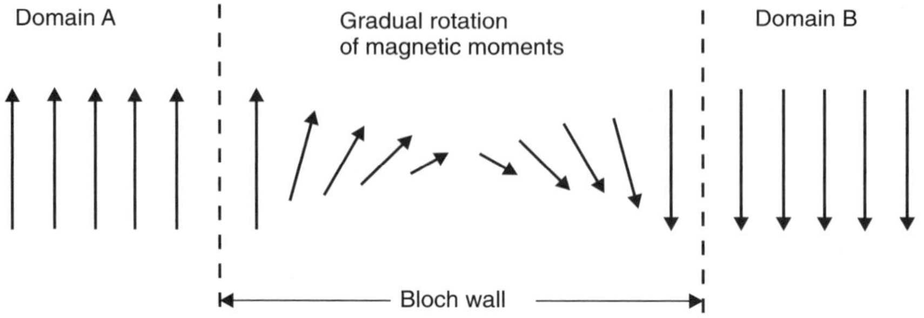

Figure 5.23 A Bloch wall separating two magnetic domains with opposite magnetic polarizations. The magnetic moments change direction gradually within the Bloch wall.

saturation. In quantum mechanics, the *Heisenberg model of ferromagnetism* describes the parallel alignment of magnetic moments in terms of an exchange interaction between neighbouring moments.

On the macroscopic level, ferromagnetic materials contain billions of individual spins, coupled in such a way that the individual moments will respond together to an external magnetic field. Weiss postulated the presence of *magnetic domains* within the material, which are regions where the atomic magnetic moments are aligned. By forming such domains, the magnetic energy is minimized. In the absence of an external field, the orientation of the moments of the domains will be random and the magnetization of the sample will be zero. The transition from one domain to its neighbour occurs through a region where the local magnetic moments are rapidly varying, called *Bloch walls*, illustrated in Figure 5.23. Exchange forces between neighbouring atomic spins tend to keep the moments aligned, and therefore to keep the Bloch walls as thick as possible. However, the magnetic moments within the walls will possess relatively high energy as they are unaligned with the 'easy' directions of magnetization in the two domains. This *anisotropy energy* favours a thin Bloch wall. Consequently, the actual wall thickness is a compromise between the magnetic anisotropy and the exchange energy. In unmagnetized polycrystalline samples, each crystal grain will possess domains, as shown in Figure 5.24. Very small grains may constitute a single domain but, in general, the majority of grains will have many domains.

When a ferromagnetic sample is magnetized, the domains that are favourably aligned with the applied field will grow at the expense of the other domains. Eventually, all the individual moments will be parallel to each other and the magnetization reaches its saturation value. If the field is decreased, the formation of domains will not be reversible. Thus, the magnetization at zero field will not be zero. The finite value of magnetization in zero field (apart from the initial magnetization) is called the *remanent magnetization* M_r (similar to the remanent polarization of a ferroelectric material – Section 5.6.1). In order to demagnetize the sample, it is necessary to apply a negative magnetic field, called the *coercive field* H_c (again, an analogy may be made with ferroelectric materials). A ferromagnetic material will therefore exhibit a hysteresis loop such as that shown in Figure 5.25, where the magnetization M is shown as a function of the applied magnetic field H. The movement of the domains determines how the material responds to a magnetic field. As a consequence, the susceptibility is a function of applied magnetic field.

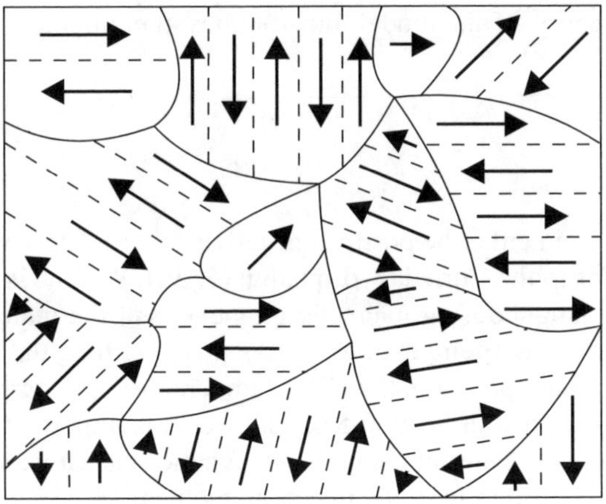

Figure 5.24 Schematic representation of magnetic domains in an unmagnetized polycrystalline material.

In the periodic table, only Fe, Co and Ni are ferromagnetic at and above room temperature. As ferromagnetic materials are heated, the thermal agitation of the atoms results in a decrease in the degree of alignment of the atomic magnetic moments and hence in the saturation magnetization. The thermal agitation finally becomes so great that the material becomes paramagnetic; the temperature of this transition is the Curie temperature, T_C (Fe, $T_C = 770\,°C$; Co, $T_C = 1131\,°C$; Ni, $T_C = 358\,°C$). Above T_C, the susceptibility varies according to the *Curie–Weiss law*. The Curie law, described above for paramagnetism, is a special case of this more general law, which incorporates a temperature constant θ and

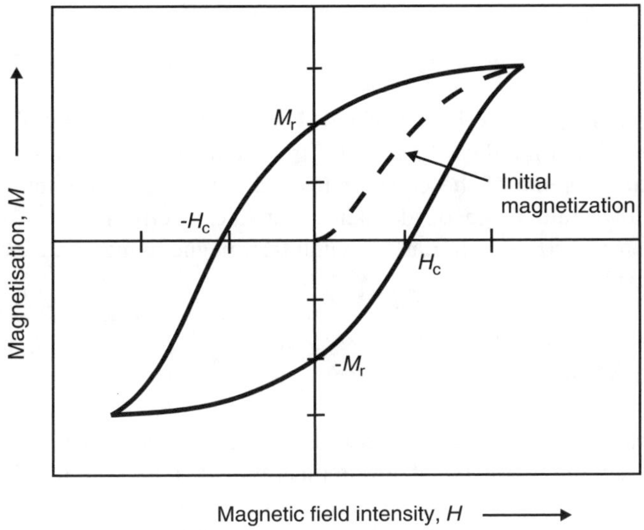

Figure 5.25 Typical magnetization versus magnetic field intensity curve for a ferromagnetic material. H_c = coercive field; M_r = remanent magnetization.

derives from Weiss theory. This model incorporates the interaction between magnetic moments:

$$\chi_m = \frac{C}{T - \theta} \tag{5.20}$$

In the above equation, θ can either be positive, negative or zero. Clearly, when $\theta = 0$ then the Curie–Weiss law equates to the Curie law [Equation (5.19)]. When θ is nonzero, then there is an interaction between neighbouring magnetic moments and the material is only paramagnetic above a certain transition temperature. If θ is positive, then the material is ferromagnetic below the transition temperature, the Curie temperature T_C (compare this with T_C for a ferroelectric material described in Section 5.6.1 above). It is important to note that Equation (5.20) is only valid when the material is in a paramagnetic state. It is also not valid for many metals as the electrons contributing to the magnetic moment are not localized. However, the law does apply to some metals, e.g. the rare earth elements, where the 4f electrons, which create the magnetic moment, are closely bound.

Antiferromagnetism

In the periodic table, the only element exhibiting antiferromagnetism at room temperature is chromium. Antiferromagnetic materials are very similar to ferromagnetic materials but the exchange interaction between neighbouring atoms leads to the anti-parallel alignment of the atomic magnetic moments. The magnetic field cancels out and the material appears to behave in the same way as a paramagnetic material. Like ferromagnetic materials, these materials become paramagnetic above a transition temperature, the *Néel temperature*, T_N. (Cr, $T_N = 37\,°C$) [θ is negative in Equation (5.20); however, the value of θ does not relate to T_N].

Ferrimagnetism

Ferrimagnetism is only observed in compounds which have more complex crystal structures than pure elements. Within these materials, the exchange interactions lead to parallel alignment of atoms in some of the crystal sites and anti-parallel alignment of others. The material breaks down into magnetic domains, just like a ferromagnetic material, and the magnetic behaviour is also very similar, although ferrimagnetic materials usually have lower saturation magnetizations.

5.7.2 Organic Magnets

Molecular magnets are systems where a permanent magnetization and magnetic hysteresis can be achieved (although usually at low temperatures) not through a three-dimensional magnetic ordering, but as a purely single molecule phenomenon. Many organometallic charge-transfer salts exhibit ferromagnetic effects. A well-known example is decamethylferrocene (DMeFc) radical cation (Chapter 3, Section 3.4.1)/tetracyanoethylene (TCNE) radical anion complex, studied in extensively by Miller, Epstein and colleagues in the 1980s,

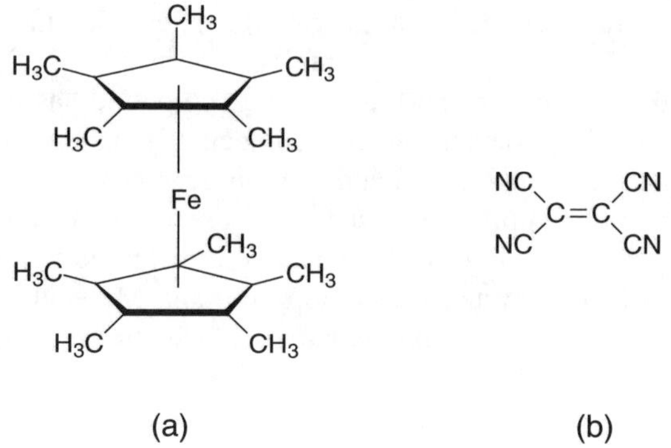

Figure 5.26 (a) Decamethylferrocene; (b) tetracyanoethylene.

and shown in Figure 5.26 [16]. This was the first ferromagnetic system with electron spins residing in p orbitals (in contrast to the more familiar ferromagnets with spins solely residing in d or f orbitals) that exhibited hysteresis and did not have an extended network bonding in one, two or three dimensions. However, the Curie temperature was somewhat low, at 4.8 K. The driving force behind the ferromagnetism is the coupling between $[DMeFc]^{+\bullet}$ and $[TCNE]^{-\bullet}$ both within the stack and between out-of-registry stacks, providing the full three-dimensional coupling of spins required for bulk ferromagnetism. Organic-based magnets now include many diverse examples of materials exhibiting magnetic ordering, including p-orbital-based organic nitroxides, p- or d-orbital-based mixed organic radicals/organometallic or inorganic coordination systems [17–19]. Electron-transfer salts of TCNE and manganese porphyrin compounds form a large family of ferrimagnets ($T_C < 28$ K), demonstrating that metallomacrocycles are viable components of organic magnets [17, 18].

One of the highest Curie temperatures recorded for a molecular magnet is $V(TCNE)_x(CH_2CH_2)_y$, with $x \approx 2$ and $y \approx ½$ ($T_C > 350$ K). This material is produced by reacting bis(benzene)vanadium with tetracyanoethylene in dichloromethane. Because of its extreme air/water reactivity, its insolubility, compositional inhomogeneities within and between samples and its amorphous nature, the structure of this compound remains elusive [18]. A proposed structure has each vanadium atom being coordinated with up to six ligands (nitrogen atoms from six different TCNE molecules) and each TCNE is reduced and is either planar or twisted and bonds as many as four Vs.

The magnetic behaviour of electron-doped C_{60} compounds was first reported in 1991. Buckminsterfullerene has no intrinsic magnetic moment. For a magnetic moment to exist, an electron must be transferred to C_{60} from a donor molecule. The fullerene-based charge-transfer salt, TDAE-C_{60}, has a Curie temperature of 16 K. A further example is a cobaltocene-doped fullerene derivative, which has a T_C of 19 K. In this case, only the fullerene molecules have magnetic spins, and so are solely responsible for the observed magnetism. Pristine C_{60} is a van der Waals crystal, which can be converted to covalently bonded crystalline phases by compression. Depending on the treatment, the molecules interconnect to form one-, two- or three-dimensional polymers. It has been suggested that the two-dimensional polymerized highly oriented rhombohedral C_{60} (Rh-C_{60}), which resembles highly oriented pyrolytic graphite (HOPG), is ferromagnetic [20]. However, the origin of the ferromagnetism, which persisted up to 500 K, is unclear.

There have been several false reports on 'organic magnets' that included trace amounts of ferromagnetic material. Despite some encouraging reports [21, 22], magnetic ordering at room temperature and above in a metal-free organic compound has yet to be unequivocally demonstrated. Due to the high density of iron, molecular magnetic materials are generally inferior to conventional magnets on a 'magnet per unit weight or unit volume' basis [21]. However, for more specialist applications, such as data storage, factors such as the coercive field (i.e. magnetizing field required to demagnetize the sample totally) become important and, in this respect, molecular systems become important. Molecular magnets may also be candidates for quantum computing qubits (Chapter 11, Section 11.14).

BIBLIOGRAPHY

Anantram MP, Léonard F. Physics of carbon nanotube electronic devices. *Rep. Prog. Phys.* 2006; **69**: 507–561.

Atkins P, de Paula J. *Physical Chemistry*, 7th ed. Oxford: Oxford University Press; 2002.

Barford W. *Electronic and Optical Properties of Conjugated Polymers*. Oxford: Oxford University Press; 2005.

Burgess J. *Ions in Solution: Basic Principles of Chemical Interactions*. Chichester: Ellis Horwood; 1988.

Bushby RJ, Paillaud J-L. Molecular magnets. In MC Petty, MR Bryce, D Bloor (Editors). *An Introduction to Molecular Electronics*. London: Edward Arnold; 1995, pp. 72–91.

Gatteschi D, Sessoli R, Villain J. *Molecular Nanomagnets*. Oxford: Oxford University Press; 2006.

Das-Gupta DK. Piezoelectric and pyroelectric materials. In MC Petty, MR Bryce, D Bloor (Editors). *An Introduction to Molecular Electronics*. London: Edward Arnold; 1995, pp. 47–71.

Farchioni R, Grosso G (Editors). *Organic Electronic Materials*. Berlin: Springer; 2001.

Ferraro JR, Williams JM. *Introduction to Synthetic Electrical Conductors*. Orlando, FL: Academic Press; 1987.

Jones RAL. *Soft Machines*. Oxford: Oxford University Press; 2004.

Kroto HW, Walton DRM (Editors). *The Fullerenes*. Cambridge: Cambridge University Press; 1993.

Leznoff CC, Lever ABP (Editors. *Phthalocyanines: Properties and Applications*. New York: VCH; 1989.

Lines ME, Glass AM. *Principles and Applications of Ferroelectric and Related Materials*. Oxford: Clarendon Press; 1977.

Miller JS, Drillon M (Editors). *Magnetism: Molecules to Materials: Molecule-based Materials*, Vol. 2. Weinheim: Wiley-VCH; 2001.

Műller TJJ, Bunz UHF (Editors). *Functional Organic Materials*. Weinheim: Wiley-VCH; 2006.

Neto AC, Guina F, Peres NM. Drawing conclusions from graphene. *Phys. World* 2006; **19**: 33–37.

Richens DT. *The Chemistry of Aqua Ions*. Chichester: John Wiley & Sons, Ltd; 1997.

Saito R, Dresselhaus G, Dresselhaus MS. *Physical Properties of Carbon Nanotubes*. London: Imperial College Press; 1998.

Salaneck WR, Lundström I, Rånby B (Editors). *Conjugated Polymers and Related Materials*. Oxford: Oxford University Press; 1993.

Schwoerer M, Wolf HC. *Organic Molecular Solids*. Berlin: Wiley-VCH; 2006.

Streitwieser A Jr., Heathcock CH. *Introduction to Organic Chemistry*, 3rd ed. New York: Macmillan; 1985.

Wallace GG, Spinks GM, Kane-Maguire LAP, Teasdale PR. *Conductive Electroactive Polymers: Intelligent Materials Systems*. Boca Raton, FL: CRC Press; 2002.

Whatmore RW. Pyroelectric devices and materials. *Rep. Prog. Phys.* 1986; **49**: 1335–1386.

Wright JD. *Molecular Crystals*. Cambridge: Cambridge University Press; 1995.

REFERENCES

[1] Tour JM. *Molecular Electronics*. Singapore: World Scientific; 2003.

[2] Richens DT. *The Chemistry of Aqua Ions*. Chichester: John Wiley & Sons, Ltd; 1997.

[3] Feast WJ, Tsibouklis J, Pouwer KL, Groenendaal L, Meijer EW. Synthesis, processing and material properties of conjugated polymers. *Polymer* 1996; **37**: 5017–5047.

[4] Springborg M, Schmidt K, Meider H, De Maria L. Theoretical studies of electronic properties of conjugated polymers. In R Farchioni, G Grosso (Editors). *Organic Electronic Materials*. Berlin: Springer; 2001, pp. 39–87.

[5] Kirchmeyer S, Reuter K. Scientific importance, properties and growing applications of poly(3,4-ethylenedioxythiophene). *J. Mater. Chem.* 2005; **15**: 2077–2088.

[6] Groenendaal L 'Bert', Jonas F, Freitag D, Pielartzik H, Reynolds JR. Poly(3,4-ethylenedioxythiophene) and its derivatives: past, present, and future. *Adv. Mater.* 2000; **12**: 481–494.

[7] Andersson P, Nilsson D, Svensson P-O, Chen M, Malmström A, Remonen T, Kugler T, Berggren M. Active matrix displays based on all-organic electrochemical smart pixels printed on paper. *Adv. Mater.* 2002; **14**:1460–1464.

[8] Bryce MR. Conductive charge-transfer complexes. In MC Petty, MR Bryce, D Bloor (Editor). *An Introduction to Molecular Electronics*. London: Edward Arnold; 1995, pp. 168–184.

[9] Xie Q, Pérez-Cordero E, Echegoyen L. Electrochemical detection of $C_{60}{}^{6-}$ and $C_{70}{}^{6-}$: enhanced stability of fullerides in solution. *J. Am. Chem. Soc.* 1992; **114**: 3978–3980.

[10] Stephens PW, Mihaly L, Lee PL, Whetten RL, Huang S-M, Kaner R, Deiderich F, Holczer K. Structure of single-phase superconducting K_3C_{60}. *Nature* 1991; **351**: 632–634.

[11] Anantram MP, Léonard F. Physics of carbon nanotube electronic devices. *Rep. Prog. Phys.* 2006; **69**: 507–561.

[12] Kasap SO. *Principles of Electrical Engineering Materials and Devices*. Boston: McGraw-Hill; 1997.

[13] Das-Gupta DK. Piezoelectric and pyroelectric materials. In MC Petty, MR Bryce, D Bloor (Editors). *An Introduction to Molecular Electronics*. London: Edward Arnold; 1995, pp. 47–71.

[14] Harrison JS, Ounaies Z. *Piezoelectric Polymers*. NASA ICASE Report No. 2001-43. Washington, DC: NASA; 2001.

[15] Petty M, Tsibouklis J, Petty MC, Feast WJ. A TGS/70:30 VDF;TrFE copolymer composite: thin film formation and pyroelectric properties. *Ferroelectrics* 1993; **150**: 267–278.

[16] Miller JS, Epstein AJ, Reiff WM. Ferromagnetic molecular charge-transfer complexes. *Chem. Rev.* 1988; **88**: 201–220.

[17] Miller JS, Epstein AJ. Tetracyanoethylene-based organic magnets. *Chem. Commun.* 1998: 1319–1325.

[18] Miller JS. Organic- and molecule-based magnets. *Electrochem. Soc Interface*. 2002; Fall: 22–27.

[19] Mrozińsli J. New trends in molecular magnetism. *Coord. Chem. Rev.* 2005; **249**: 2534–2548.

[20] Makarova TL, Sundqvist B, Höhne R, Esquinazi P, Kopelevich Y, Scharff P, Davydov VA, Kashevarova LS, Rakhmanina AV. Magnetic carbon. *Nature* 2001; **413**: 716–718.

[21] Bushby RJ, Paillaud J-L. Molecular magnets. In MC Petty, MR Bryce, D Bloor (Editors). *An Introduction to Molecular Electronics*. London: Edward Arnold; 1995, pp. 72–91.

[22] Zaidi NA, Giblin SR, Terry I, Monkman AP. Room temperature magnetic order in an organic magnet derived from polyaniline. *Polymer* 2004; **45**: 5683–5689.

6 Tools for Molecular Electronics

Make instruments to plague us

6.1 INTRODUCTION

This chapter gives an overview of some of the techniques that can be used to provide information on the physical structure of organic compounds. The list is by no means exhaustive, but illustrates the physical principles behind the more popular methods. In most instances, the materials are in the form of thin films (e.g. suitable for device fabrication). A detailed understanding of (and ultimate control over) the morphology of these thin layers is essential if the compounds are to be exploited technologically.

The physical properties that underpin the operation of electronic devices ultimately depend on the chemical bonding and crystalline nature of the material. One crystalline form of the same substance (e.g. diamond) can exhibit markedly different properties from another (e.g. graphite). The influence of impurities can be just as dramatic. In Chapter 3, we learned that the intrinsic electrical conductivity depends on the nature and organization of the atoms and molecules that constitute the material (band theory). In practice, chemical impurities (either intentionally or unintentionally added) may ultimately determine the electrical conductivity of a particular semiconductor. It has taken many years (over 50, and work is continuing!) to understand and control the properties of just one important electronic material – silicon. Sadly, the lessons learned from inorganic semiconductor technology are often ignored by researchers in molecular electronics and there are many reports in the literature of irreproducible and poor quality devices based on impure (or even unknown) materials. This has not helped in establishing the credibility of the subject.

Although there are many simple characterization tools (e.g. the humble microscope) that can provide a useful insight into the morphology of organic thin films, many of the techniques that need to be used are sophisticated, and usually very expensive. Part of the reason is that highly sensitive instrumentation is needed for the study of nanoscale amounts of materials. With progress down the Moore's law curve, new equipment has had to be developed. The surface analytical methods noted in Section 6.7 and the scanning microscopies described in Section 6.8 are good examples.

The characterization of organic materials can also be challenging. In contrast to their inorganic counterparts, the materials are usually soft and fragile. In a high-vacuum environment under bombardment by an electron beam, organic substances can be easily damaged. The results from such a physical investigation may therefore not be representative of the material in its natural environment, a particular problem for biological scientists.

© 2007 John Wiley & Sons, Ltd

Consequently, elaborate and careful sample preparation usually accompanies the physical study of organic materials.

6.2 DIRECT IMAGING

6.2.1 Optical Microscopy

The observation of a material under an optical microscope can provide a simple insight into its morphology. Unfortunately, organic compounds, particularly in the form of thin films, do not generally provide high contrast images and therefore special techniques have been devised for their study. Two examples are *phase contrast microscopy* and *confocal microscopy*, which are described below. A further technique, *scanning near-field optical microscopy*, is also introduced. This has much in common with the scanning probe microscopy methods discussed later in this chapter (Section 6.8).

Phase contrast microscopy was developed in the early 20th century by Frits Zernike (who was awarded the Nobel Prize for Physics in 1953). In an optical microscope, light from the object being imaged passes through the centre of the lens as well as the periphery. If the light passing through the edge of the objective lens is retarded by a half wavelength and the light at the centre is not retarded, then the light rays will be out of phase by a half wavelength and they will cancel each other when the objective lens brings the image into focus. A reduction in brightness of the object is then observed (the degree of reduction in brightness will depend on the refractive index of the object). The phase shift is produced by rings etched accurately on to glass plates so that they introduce the required degree of phase shift when inserted into the optical path of the microscope. When in use, this technique allows the phase of the light passing through the object to be inferred from the intensity of the image produced by the microscope. Phase contrast is only useful on specimens that are colourless and transparent and difficult to distinguish from their surroundings. The method is, therefore, suited to the study of organic, particularly biological specimens.

Confocal microscopy also offers several advantages over conventional optical microscopy, including controllable depth of field, the elimination of image degrading out-of-focus information and the ability to collect serial optical sections from thick specimens. The key to the confocal approach is the use of *spatial filtering* to eliminate out-of-focus light or flare in specimens that are thicker than the plane of focus. Current instruments are highly evolved from the earliest versions, but the principle of confocal imaging, which was patented by Marvin Minsky in the 1950s, is employed in all modern confocal microscopes; a schematic diagram is shown in Figure 6.1. In a conventional widefield microscope, the entire specimen is bathed in light from a mercury or xenon source, and the image can be viewed directly by eye or projected on to an image capture device or photographic film. The method of image formation in a confocal microscope is fundamentally different. Illumination is achieved by scanning one or more focused beams of light, usually from a laser or arc-discharge source, across the specimen. This point of illumination is brought to focus in the specimen by the objective lens, and laterally scanned under computer control. The sequences of points of light from the specimen are detected by a photomultiplier tube through a pinhole (or in some cases, a slit) and the output is built into an image and displayed by the computer. Although unstained specimens can be viewed using light reflected back from the specimen, they usually are labelled with one or more fluorescent probes.

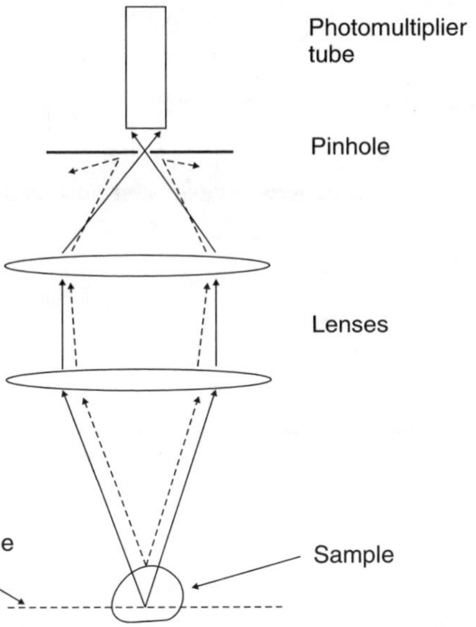

Figure 6.1 A confocal optical microscopy system. The small pinhole permits data collection from a thin optical plane within the sample. Points that are outside the plane of focus will have a different secondary focal plane and therefore most of the light is deflected (dashed lines).

In conventional optical microscopy, or *far-field optical microscopy*, the standard arrangement consists of a system of lenses which focus light from the sample into a virtual, magnified image. In the early 1870s, Ernst Abbé formulated a rigorous criterion for being able to resolve two objects in a light microscope; this is the basis of *diffraction limited optics*:

$$d > \frac{\lambda}{2\theta} \tag{6.1}$$

where $d =$ the distance between the two objects, $\lambda =$ the wavelength of the incident light and $2\theta =$ the angle through which the light is collected. According to this equation, the best resolution achievable with optical light is about 200 nm.

Scanning near-field optical microscopy (SNOM or NSOM) is an imaging technique used to obtain resolution beyond the Abbé diffraction limit. The operational principle, as shown in Figure 6.2, involves illuminating a specimen through a sub-wavelength-sized aperture while keeping the specimen within the near-field regime of the source. The first suggestion that this was possible came from E. H. Synge in 1928. Although not strictly feasible with the technology of the time, his technical criteria form an accurate basis for a super-resolution optical microscope. If the aperture-to-sample separation is kept roughly less than half the diameter of the aperture, the source does not have the opportunity to diffract before it interacts with the sample, and the resolution of the system is determined by the aperture diameter as opposed to the wavelength of light used. In essence, the SNOM uses the evanescent field (Chapter 4, Section 4.7.1), which can be confined on structures much

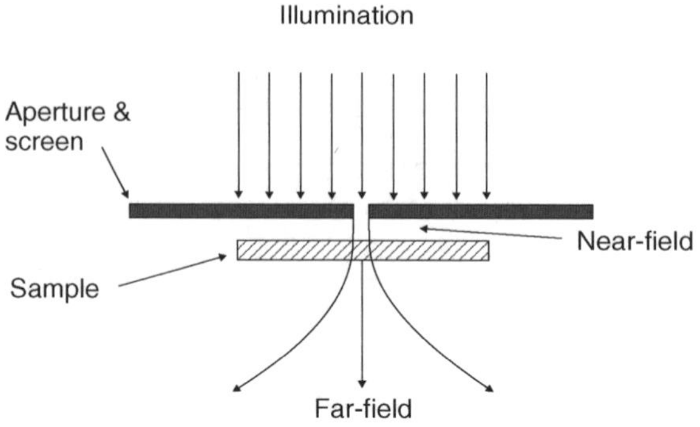

Figure 6.2 Schematic diagram showing the principles of scanning near-field optical microscopy (SNOM or NSOM).

smaller than the wavelength; this field does not propagate into the far-field. A common way to define the aperture and confine the optical field is to stretch an optical fibre, so as to obtain a conical tip, and to coat it with a thin, but opaque, metal layer, usually aluminium. The end of the fibre is uncoated, and is therefore a small pinhole through which an evanescent light wave can pass. The diameter of the hole is often 50–100 nm. An image is built up by raster-scanning the aperture across the sample and recording the optical response of the specimen through a conventional far-field microscope objective. An optical resolution of <50 nm can be achieved using SNOM.

6.2.2 Electron Microscopy

Electrons are generated in an electron microscope by thermionic emission from a metal filament and accelerated through a potential. For an accelerating voltage of 100 kV, the electron wavelength is 3.7×10^{-3} nm. Atomic resolution should therefore be achievable.

The main interactions taking place when the electron beam is incident on matter are shown in Figure 6.3. Unscattered and diffracted electrons form the basis for conventional *transmission electron microscopy* (TEM) and diffraction. Diffracted electrons will form a pattern that can be transformed directly into an image by a magnetic lens. Either the diffraction pattern or the image can be projected onto a viewing screen. Electron diffraction is discussed later in Section 6.5.

Low-energy (<50 eV) secondary electrons emitted from the surface of the sample can be used for *scanning electron microscopy* (SEM); the experimental set-up is depicted in Figure 6.4. A beam of electrons is condensed by a condenser lens and then focused to a very fine point on the sample by the objective lens. The beam can be concentrated in a small probe (2 nm in diameter), which may be deflected across the sample in a raster fashion using scanning coils (which create a magnetic field). Secondary electrons can be detected above the sample and an image showing the intensity of secondary electrons emitted from different parts of the sample can be built up.

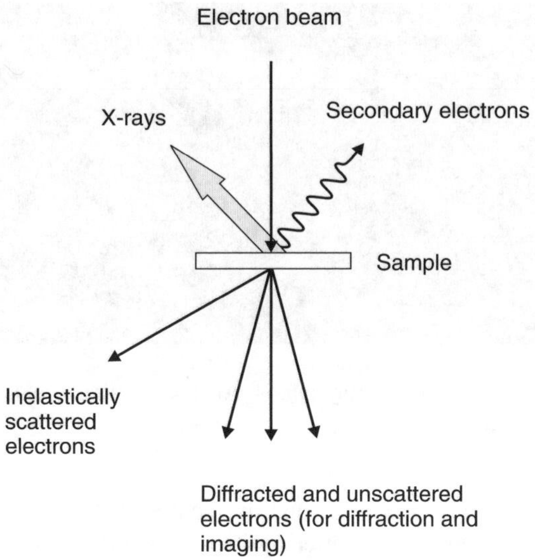

Figure 6.3 Interaction of an electron beam with a sample.

A related technique involves the collection of electrons transmitted through the specimen during scanning. These can be used to produce a scanning transmission (STEM) image, with the advantage, compared with TEM, that radiation damage is reduced because the beam is not stationary. Emitted X-rays are characteristic of the elements in the sample and may be used for elemental analysis.

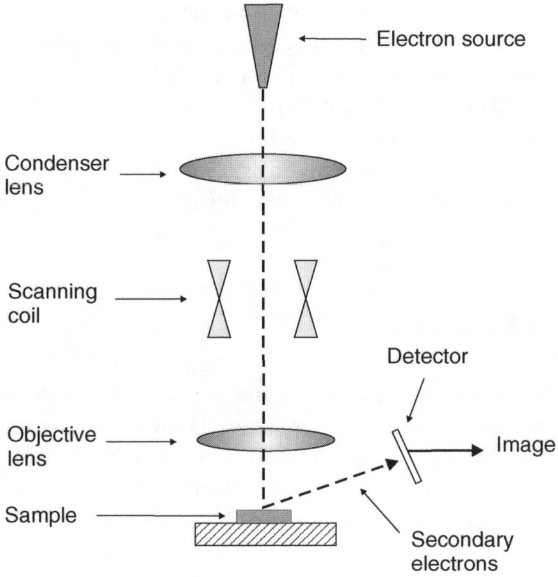

Figure 6.4 Schematic diagram of a scanning electron microscope.

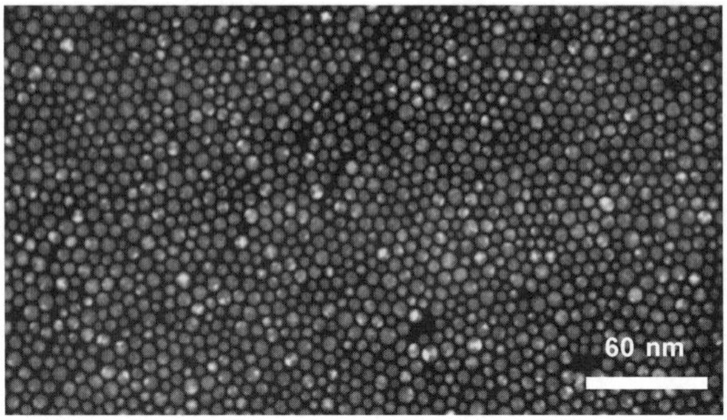

Figure 6.5 Transmission electron micrograph of Au nanoparticles deposited on a carbon-coated microscope grid using the LB technique [1].

In the electron microscope, the electrons are scattered by the atomic potentials of the atoms in the sample. Generally, the scattering increases with atomic number. In contrast to X-rays and neutrons, the scattering of electrons by matter is very strong and diffraction is even feasible with gaseous samples.

Transmission electron microscopy studies of thin organic films, such as those produced by self-assembly or the Langmuir–Blodgett (LB) technique, require the layers either to be removed from the substrate or to be deposited on to amorphous supports. A relatively simple sample preparation technique for LB layers involves the deposition of the film on a previously anodically oxidized aluminium substrate. The organic film on its alumina support is then removed by etching the aluminium layer in a mercuric chloride solution. Alternatively, LB films may be floated from glass substrates in highly diluted hydrofluoric acid and picked up on a carbon-coated copper microscope grid, or transferred to carbon-coated grids by raising the latter through the condensed floating monolayer. Figure 6.5 shows an example of a transmission electron micrograph of an LB film of gold nanoparticles deposited on a carbon-coated grid [1]. The nanoparticles possess a well ordered close-packed arrangement with an average size of about 8 nm.

Reflection measurements are less demanding on sample preparation: thin films of organic materials can be deposited on a variety of substrates and then rapidly studied in the electron microscope.

6.3 X-RAY REFLECTION

When electromagnetic energy is incident on the surface of a material, some of it will be reflected specularly, so that the angle of incidence is equal to the angle of reflection (well known for visible light; Chapter 4, Section 4.5). For crystalline materials, constructive and destructive interference between the radiation reflected from successive crystal planes will occur when the wavelength λ of the incident radiation is of the same order as the lattice spacing (tenths of a nanometre) (Figure 6.6). The condition for maxima in the reflected radiation is provided by *Bragg's law*:

$$n\lambda = 2d_{hkl} \sin\theta \tag{6.2}$$

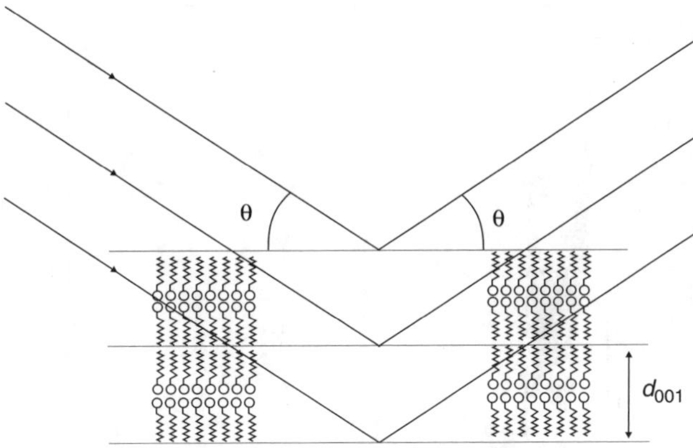

Figure 6.6 Bragg reflection from LB film planes with spacing d_{001}.

where θ is the angle of incidence, conventionally measured from the plane of reflection in X-ray crystallography (in contrast to geometric optics) d_{hkl} is the interplanar spacing (Chapter 2, Section 2.5.6) and the integer n is known as the *order of the reflection*. Clearly, high-order ($n = 2, 3$, etc.) scattering from planes d_{hkl} is indistinguishable from first-order scattering from planes $(d_{hkl})/2$, $(d_{hkl})/3$, etc.

A small correction is required to Equation (6.2) to take into account the fact that the refractive index for X-rays will be slightly less than unity. Bragg's law then becomes

$$n\lambda = 2d \sin\left(1 - \frac{\delta}{\sin^2 \theta}\right) \tag{6.3}$$

where δ is related to the X-ray refractive index by $n = 1 - \delta$. Typically, δ is of the order 10^{-6} [NB: a refractive index of less than unity may be thought to be impossible as it leads to an EM wave velocity greater than the speed of light (Chapter 4, Section 4.3). However, the refractive index is related to the *phase velocity* of the radiation (Chapter 3, Section 3.3.1)].

The simplest interplanar spacing in an LB multilayer film is d_{001}, which for a fatty acid salt (e.g. cadmium octadecanoate) is approximately 5.0 nm. Using Equation (6.2) with an X-ray wavelength of 0.154 nm gives $\theta \approx 1°$ for the first-order Bragg peak. Therefore, low-angle measurements are necessary for X-ray studies of LB and similar films of long-chain compounds.

Figure 6.7 shows X-ray diffraction data (both experimental and theoretical) from a 43-layer fatty acid salt film [2]. As predicted, the first-order Bragg reflection is observed for an angle of incidence close to 1°. In a Y-type LB film, the d_{001} spacing is equal to the distance between the polar planes, i.e. the thickness of two monomolecular layers. The monolayer thicknesses for fatty acid salts, obtained from X-ray experiments, are plotted as a function of the number of carbon atoms in the molecule in Figure 6.8 [3]. The X-ray data are in close agreement with those calculated from the lengths of the molecules, inferring that the hydrocarbon chains in transferred monolayers are oriented almost at right-angles to the substrate surface. For Y-type LB films of other amphiphilic materials (including simple fatty acids), the X-ray d-spacing is often less than twice the molecular length, suggesting some tilting or interdigitation (or both) of the molecules. Unfortunately, the d-spacing value alone cannot be used to distinguish between these situations.

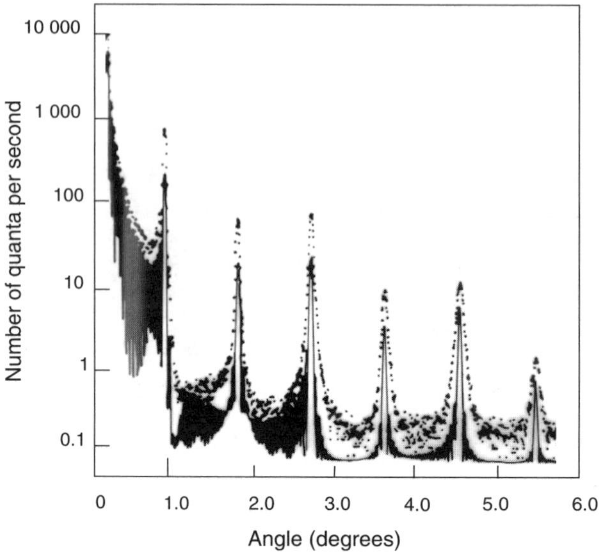

Figure 6.7 X-ray diffraction data from a 43-layer LB film of perdeuterated manganese octadecanoate on a silicon substrate. Experimental values are shown as points. The solid curve is based on calculation and is displaced from the data points [2]. Reprinted with permission from *Phys. Rev. B*, **23**, Nicklow RM, Pomerantz M, Segmüller A, 'Neutron diffraction from small numbers of Langmuir–Blodgett monolayers of manganese stearate', pp. 1081–1087, Copyright (1981), American Physical Society.

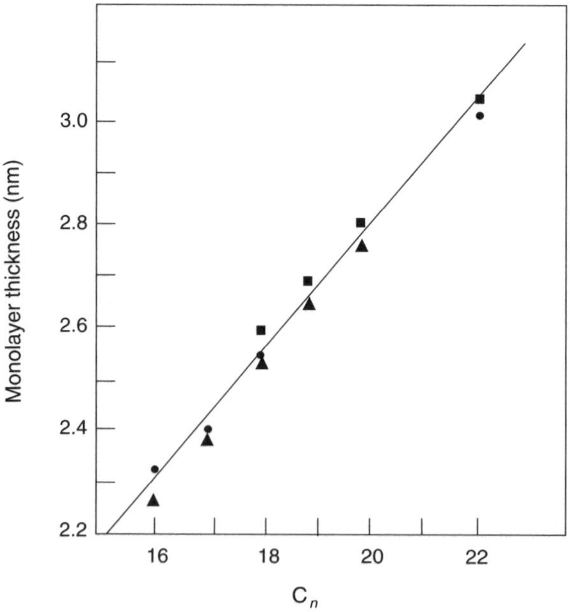

Figure 6.8 Monolayer thickness, obtained from X-ray diffraction experiments, versus number of carbon atoms in the molecule for salts of long-chain fatty acids [3]. Reprinted from Petty MC, *An Introduction to Langmuir–Blodgett Films*, Cambridge: Cambridge University Press, 1996, p. 100.

X-rays are scattered by their interaction with atomic electrons and interference takes place between X-rays scattered from different parts of an atom. The scattering power or scattering factor decreases with increasing scattering angle 2θ, resulting in a decrease in the intensity of the Bragg peaks. However, it is evident that the height of the third Bragg reflection in Figure 6.7 is greater than that of the second. This is because the intensity of a particular Bragg peak is related not only to the scattering power of each atom in the lattice, but also to the position of the atom in the unit cell. In summing the individual waves to give the resultant diffracted beam, both the amplitude and phase of each wave scattered by the individual atoms are important. The intensity of the scattered radiation I_{hkl} from a set of planes $\{hkl\}$ may be written as

$$I_{hkl} \propto F_{hkl}^2 \tag{6.4}$$

where F_{hkl} is called the *structure factor.*

The structure factor of the unit cell depends on the constituent atoms and their individual scattering factors. For a particular d-spacing, the only variable in the cell structure is the nature of the atom. Since F depends directly on the effective number of scattering electrons per atom, a large F value is given by a high atomic number. Hence organic materials containing heavy metals are generally superior for X-ray diffraction experiments.

The angular width of the Bragg peaks at half-height, Δ, is inversely proportional to the length of the ordered region L according to the Scherrer equation [4]:

$$\Delta = \frac{\lambda}{L\cos\theta} \tag{6.5}$$

The parameter L can be interpreted either as an apparent size of the crystalline grain, assuming a grain structure with perfectly ordered domains, or, in the case of an imperfect crystalline order, L is related to the correlation length of the order.

6.3.1 Electron Density Profile

If the structure factors F_{hjk} for a complete set of X-ray reflections are known, the electron density ρ at any position xyz in the unit cell may be calculated using

$$\rho(xyz) = \frac{1}{V}\sum_{\mathrm{h}}\sum_{\mathrm{k}}\sum_{\mathrm{l}} F_{hkl}\cos 2\pi(hx + ky + lz) \tag{6.6}$$

where V is the unit cell volume. This is an example of a *Fourier summation*; the electron density profile is the Fourier transform of the X-ray diffraction pattern. To calculate $\rho(xyz)$, values of F_{hkl} are needed. Unfortunately, only F_{hkl}^2 can be measured (the + and − signs of F_{hkl} correspond to phases of 0° and 180°). It is therefore possible to obtain $|F_{hkl}|$, but not the sign of the structure factor. This dilemma is known as the *phase problem.*

For a set of data containing m structure factors, each of which may be positive or negative, there are $2m$ possible combinations of phases; e.g. for 300 reflections there are 2×10^{90} possibilities! Clearly, the solution of the phase problem by trial-and-error methods is not very practicable! A physical criterion is often used to evaluate the electron density profile of multilayer films perpendicular to the substrate. For example, the optimization of a plateau in the electron density profile (representing the region of constant electron density corresponding to the alkyl chain) can be used to simplify these computations in many LB film systems.

6.3.2 Kiessig Fringes

The observation of X-ray interferences, first described by Kiessig, is one of the most precise methods to measure the thickness of a thin film. Specular reflection of X-rays by a flat surface is observed at glancing angles close to the critical angle θ_c (Chapter 4, Section 4.5) for total reflection. For all materials, θ_c is a few tenths of a degree. In a thin-film system, each interface of two layers with different refractive indices gives rise to a specularly reflected X-ray beam. With monochromatic X-radiation, interference maxima and minima can be observed near the critical angle. Kiessig showed that is was possible to determine the thickness of the film from the angular spacing of these intensity maxima. If deviation of the X-ray refractive index from unity is neglected, the fringe maxima θ_m are related to the total thickness t by

$$\sin\theta_m = m\frac{\lambda}{2t} \tag{6.7}$$

where m is an integer. The film thickness can therefore be obtained from the slope of the $\sin\theta_m$ versus m plot. Figure 6.9 reveals a series of Kiessig fringes obtained for layer-by-layer

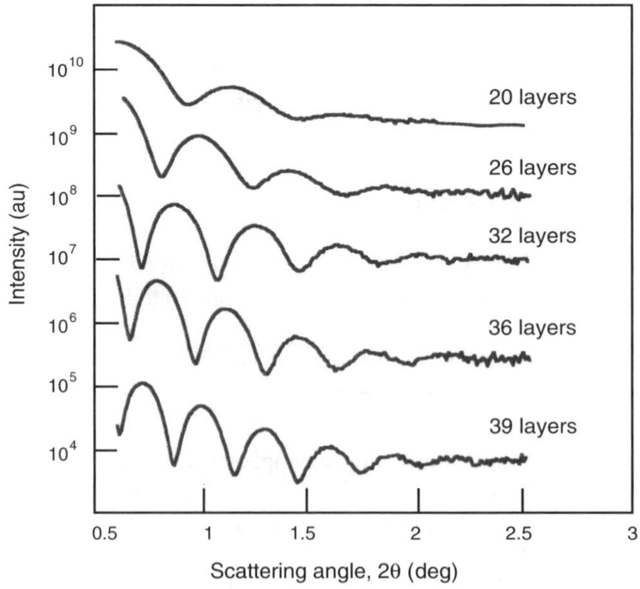

Figure 6.9 X-ray diffraction data for different numbers of layers of alternating anionic and cationic polyelectrolyte films deposited by the layer-by-layer technique. Kiessig interference fringes are evident. Reprinted from *Protein Architecture*, Lvov Y, p. 133, Copyright (1999), with permission from Taylor & Francis Group LLC.

assemblies of the polyelectrolytes [5]. There are no Bragg peaks, as might be expected for a multilayer assembly, suggesting a poor degree of layer-by-layer ordering. The number of Kiessig fringes increases with the increase in the number of polyelectrolyte bilayers.

6.3.3 In-Plane Measurements

X-ray diffraction can also be used to obtain information about the in-plane structure of multilayer molecular assemblies. However, since most substrates supporting the films (e.g. glass slides) will almost completely attenuate an X-ray beam, diffraction in transmission either requires the organic layer to be removed from the substrates or special substrate materials to be used. The technique of Prakash *et al.* [6] consists of tilting the plane of the multilayer to small angles with respect to the horizontal (with the incident and diffracted X-ray beams in the horizontal plane). This allows the observation of (11l) diffraction peaks.

6.4 NEUTRON REFLECTION

The use of a neutron beam gives rise to interference effects in a similar way to an X-ray beam. The major difference between the two types of radiation lies in the factors governing the intensity of diffraction. For X-rays, this depends on the electron density variation across the layers. In the case of neutrons, it is the variation of nuclear scattering length density that determines the Bragg intensity. Because neutron scattering is a nuclear property, it may vary considerably from one element to the next and is different for different isotopes, e.g. hydrogen and deuterium. Neutron absorption is usually negligible and the interference effects that cause X-ray scattering to diminish with increasing angle are absent with neutrons and the scattering is isotropic.

6.5 ELECTRON DIFFRACTION

Electron diffraction also provides useful structural information on organic films. To obtain the structure normal to the film plane, the electron beam impinges at grazing angles and the reflected beam is observed. For fatty acid-type LB films, electron diffraction experiments reveal the packing of the C_2H_4 subcells in the aliphatic chains. Consequently, two levels of organization must be considered for long-chain molecules in layered structures. These are illustrated in Figure 6.10. In addition to the nature of the subcell, the packing of the individual molecules in the multilayer will define the main cell. (The X-ray diffraction data shown in Figure 6.7 reveal the packing associated with the main cell.) In the solid state, long-chain fatty acid-type compounds exhibit polymorphism, i.e. they can exist in a variety of crystallographic states, depending on the packing arrangements of the main and subcells. In all, there are 10 different packing arrangements for long-chain fatty acids depending on the relative displacements of neighbouring chains [7]. These can conveniently be distinguished by the Miller indices of the interface plane between layers of molecules and the type of subcell symmetry. There are three possible close-packed structures with similar packing densities: orthorhombic (R), monoclinic (M) and triclinic (T) (Chapter 2, Section 2.5.4). In

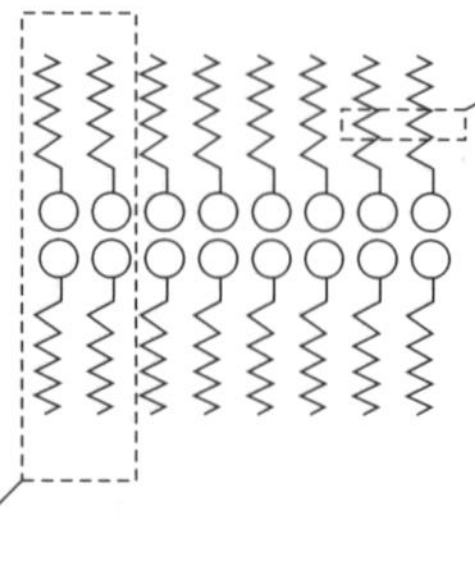

Figure 6.10 Main cell and subcells for fatty acid multilayers.

the T and M arrangements, the zigzags in the chain are parallel to one another. However, this is not so for the R structure.

The transmission geometry is used to obtain in-plane structural information. As noted above, an advantage of the electron beam compared with X-rays is that the interaction of electrons with matter is much stronger. Consequently, a diffraction pattern can be obtained with a beam of smaller diameter. Both transmission electron diffraction (TED) and reflection high-energy electron diffraction (RHEED) techniques may be used. Figure 6.11 shows a transmission electron diffraction pattern for an LB film built up from 22-tricosenoic acid [8]. The diffraction pattern can be indexed as arising from the orthorhombic R(001) packing of the subcells. In-plane X-ray diffraction data (Section 6.3.3) can provide similar information on the packing of the subcells in LB arrays of molecules [9].

If the LB multilayer structure were a simple crystalline system, the above identification system would allow only a few possible orientations for the molecular chains. In practice, it is found that the tilt angle of the alkyl chains in fatty acid materials (e.g. 22-tricosenoic acid) varies continuously with the deposition pressure; the greater the pressure, the smaller is the angle of tilt from the substrate normal [10, 11]. The structure of these LB layers of tilted molecules may therefore be fairly complex, consisting of regions of crystallinity in which grains are inclined to the substrate with a distribution of tilt azimuths and permeated by holes [12].

The angle of inclination to the shadow edge of the line of the spots in RHEED patterns of fatty acid LB films also varies with the deposition pressure [10, 11]. Typical data are shown in Figure 6.12 [10, 12]. As the deposition pressure is increased, the molecular tilt elevation also increases (i.e. the molecules become more upright). However, this is a continuous process and is not what would be expected from the molecules taking up one of a few fixed orientations.

6.6 INFRARED SPECTROSCOPY

Infrared (IR) spectroscopy (wavelength in the range 1–100 μm) probes the vibrational features of an organic molecule. A particular bond must have a permanent dipole moment associated with it to interact with the IR radiation.

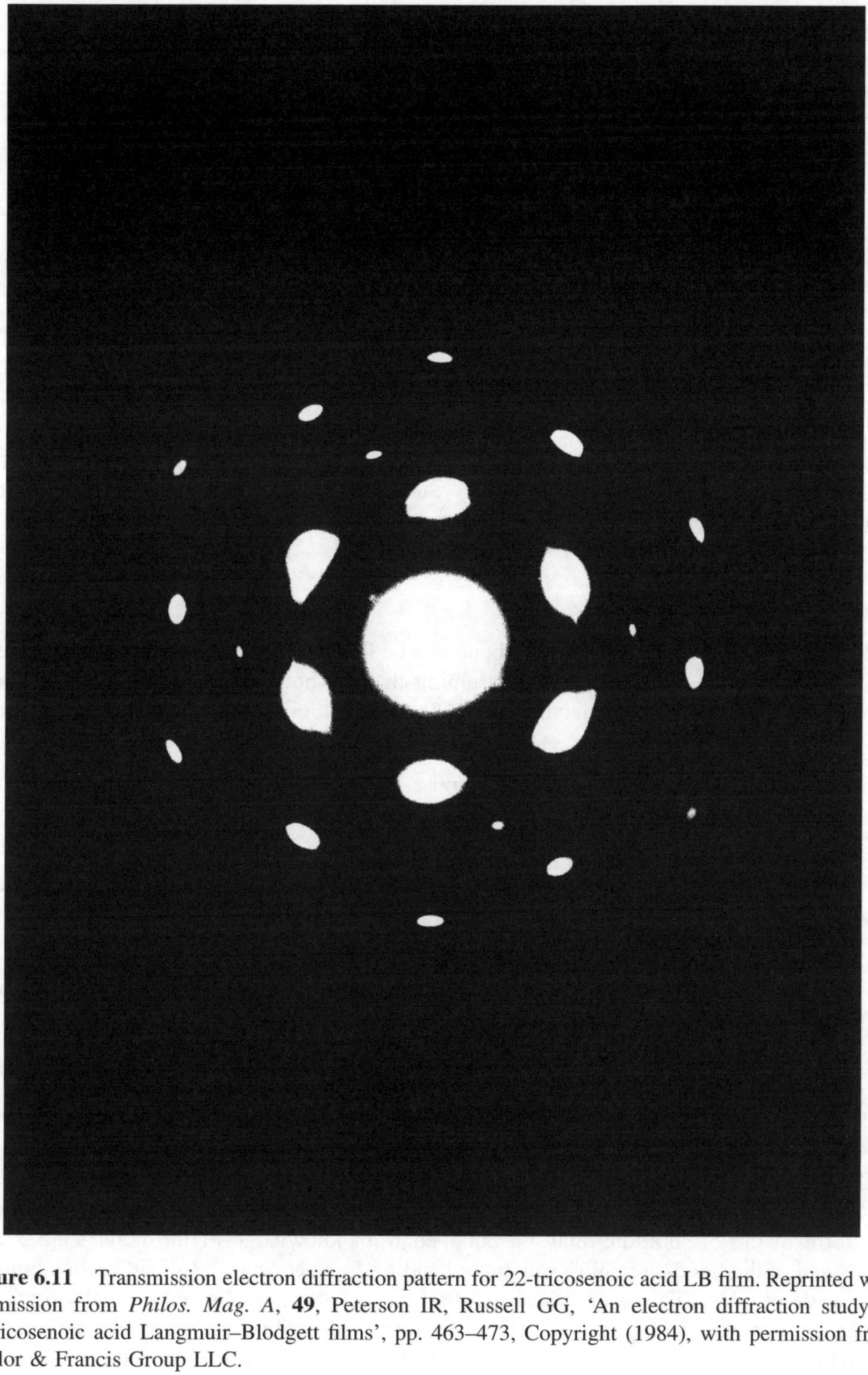

Figure 6.11 Transmission electron diffraction pattern for 22-tricosenoic acid LB film. Reprinted with permission from *Philos. Mag. A*, **49**, Peterson IR, Russell GG, 'An electron diffraction study of ω-tricosenoic acid Langmuir–Blodgett films', pp. 463–473, Copyright (1984), with permission from Taylor & Francis Group LLC.

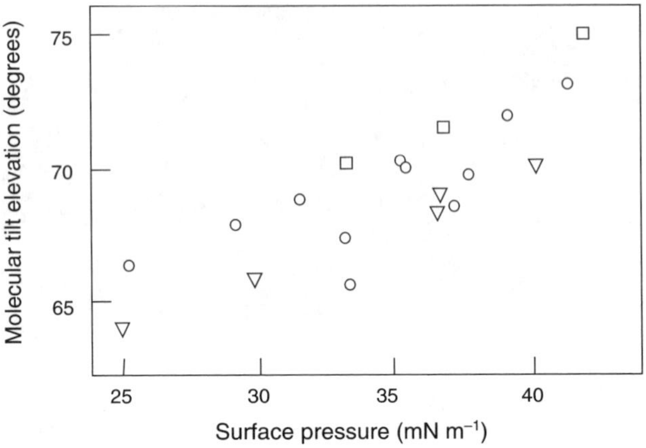

Figure 6.12 Molecular tilt (measured from substrate plane) versus deposition surface pressure for 22-tricosenoic acid LB films. □, deposition pH 7; ○, pH 3. Reprinted from *Thin Solid Films*, **161**, Peterson IR, Russell GJ, Earls JD, Girling IR. 'Surface pressure dependence of molecular tilt in Langmuir–Blodgett films of 22-tricosenoic acid', pp. 325–331, Copyright (1988), with permission from Elsevier. ∇, pH 7. Data from Barnes and Sambles [12].

The compression and extension of a bond can be likened to the behaviour of a spring and this analogy may be taken further by assuming that the bond, like a simple spring, obeys Hooke's law. For a diatomic molecule (e.g. HCl), the vibrational energy levels E_v are given by

$$E_v = h\nu\left(\epsilon + \frac{1}{2}\right) \tag{6.8}$$

where h is Planck's constant and the vibrational quantum number $\epsilon = 0,\ 1,\ 2\ldots$.

In IR spectroscopy, it is usual for the energy of the electromagnetic radiation to be quoted in terms of a *wavenumber*, $\bar{\nu}$. This is the reciprocal of the wavelength, in centimetres, i.e.

$$\bar{\nu} = \frac{1}{\lambda}\ \text{cm}^{-1} \tag{6.9}$$

The wavenumber expresses the number of waves or cycles contained in each centimetre length of the radiation and is a useful concept in spectroscopy.

The number of vibrational modes associated with polyatomic molecules can be very large. An N-atomic molecule has $3N - 5$ normal modes of vibration if it is linear and $3N - 6$ if it is nonlinear. A normal mode is one in which all the nuclei undergo harmonic motion, have the same frequency of oscillation and move in phase (but with different amplitudes). The form of these vibrations may be obtained from knowledge of the bond lengths and angles and of the bond-stretching and angle-bending force constants. Consider, for example, a molecule of water. This is nonlinear and triatomic, with three allowed normal vibrational modes, as depicted in Figure 6.13. Each motion is described as stretching or bending, depending on the nature of the change in molecular shape. Furthermore, the motions are designated either symmetric or antisymmetric.

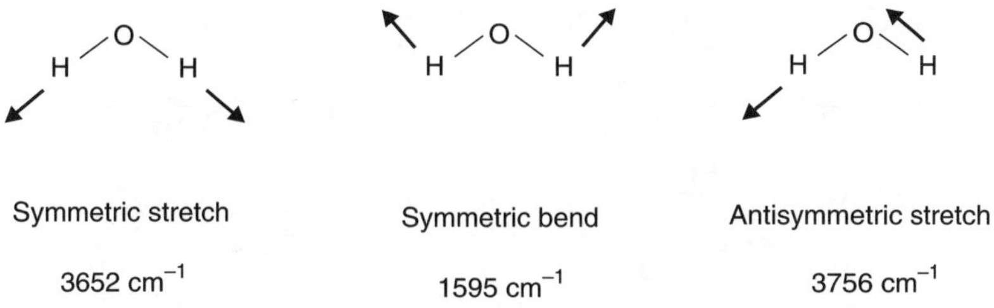

Figure 6.13 Three fundamental vibrations of the water molecule.

Although a normal mode of vibration involves movement of all the atoms in a molecule, there are circumstances in which movement is almost localized in one part of the molecule. If the vibration involves the stretching or bending of a terminal –*XY* group, where *X* is heavy compared to *Y* (e.g. an OH group in a fatty acid) the corresponding vibration wavenumbers are almost independent of the rest of the molecule to which –*XY* is attached. A typical wavenumber for the *XY* stretching vibration may therefore be referred to. For example, the (OH) stretching frequency is normally in the region of 3600 cm^{-1}. Many group vibrations occur in the region 1500–3700 cm^{-1}; stretching and bending vibrations of some well-known groups are listed in Table 6.1.

Not all parts of a molecule are characterized by group vibrations. Many normal modes involve strong coupling between stretching or bending motions of atoms in a straight chain, a branched chain or a ring. Such vibrations are called skeletal vibrations and tend to be specific to a particular molecule. For this reason, the region where skeletal vibrations mostly occur, from about 1400 cm^{-1} to low wavenumbers, is sometimes called the *fingerprint region.*

Table 6.1 Characteristic stretching and bending frequencies of molecular groups.

Group	Approximate wavenumber (cm^{-1})
–OH	3600
$–NH_2$	3400
$=CH_2$	3030
$–CH_3$	2960 (antisym. stretch)
	2870 (sym. stretch)
	1460 (antisym. bend)
	1375 (sym. bend)
$–CH_2–$	2920 (antisym. stretch)
	2850 (sym. stretch)
	1470 (bend)
–C≡C–	2220
>C=O	1750–1600
>C=C<	1650

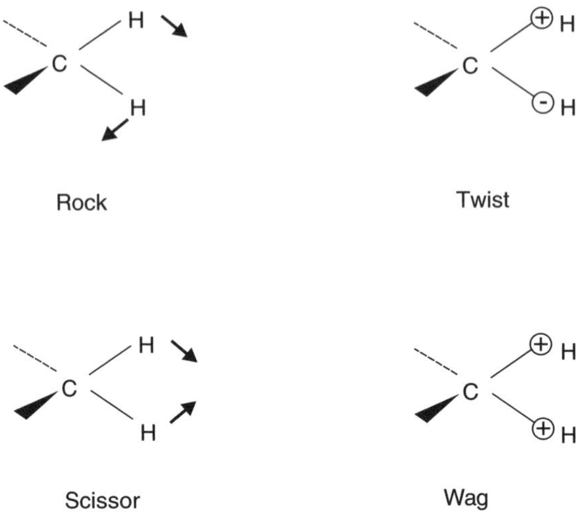

Figure 6.14 Rocking, twisting, scissoring and wagging vibrations in a CH_2 group. The + and − signify movement in and out of the plane of the diagram; the arrows indicate movement in this plane.

In addition to the description of group vibrations as stretch and bend (or deformation), the terms rock, twist, scissor, wag, torsion, ring breathing and inversion (umbrella) are used frequently. Some of these are illustrated for the CH_2 group in Figure 6.14.

Only the component of radiation polarized in the direction of the transition dipole moment (Chapter 4, Section 4.4.1) will induce a change from one vibrational energy level to another. The absorption intensity of the radiation may be written as

$$I \propto |\mathbf{p} \cdot \mathbf{E}|^2 \tag{6.10}$$

where $\mathbf{p}$ is the transition dipole moment and $\mathbf{E}$ is the electric field vector. If $\mathbf{E}$ is parallel to the transition dipole moment, the probability of absorption is high, whereas if $\mathbf{E}$ is perpendicular to it, no radiation is absorbed.

The absorption or reflection intensities resulting from the interaction of IR radiation with monolayer samples are very low, a direct result of the relatively small numbers of molecules being sampled. In a transmission experiment with normal incidence of the IR beam, the electric field vector is oriented parallel to the layer plane. In this geometry [Figure 6.15(a)], the projection of the transition moments on the layer plane is probed. An increase in surface sensitivity may be obtained by using a method based on *attenuated total reflection* (ATR) [Figure 6.15(b)]. In this technique, the organic compound is deposited on either side of an IR-transmitting crystal (e.g. silicon or germanium). The radiation is incident at an angle greater than the critical angle and undergoes multiple reflections inside the crystal. On each reflection, the evanescent field of the IR beam penetrates the organic material under study and may be absorbed by it (the evanescent field decays over micrometre distances – Chapter 4, Section 4.7.1). Both the simple transmission and ATR experiments may be used with either polarized or unpolarized radiation.

Reflection absorption infrared spectroscopy, RAIRS, requires IR radiation to be incident at a grazing angle (85–88°) to a metal surface, on which an organic layer has been deposited [Figure 6.15(c)]. Incident s-polarized radiation (Chapter 4, Section 4.2) undergoes a phase

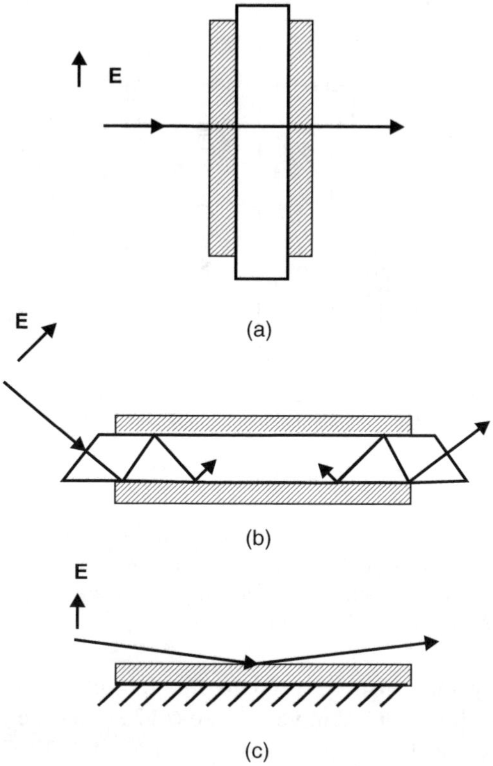

Figure 6.15 Comparison of (a) simple transmission, (b) ATR (attenuated total reflection) and (c) RAIRS (reflection absorption infrared spectroscopy) sampling techniques.

shift of 180° on reflection from the metal surface and so the electric field components of the incident and reflected radiation cancel at the metal/LB film interface. In contrast, the incident and reflected components of p-polarized IR radiation differ by only about 90° at grazing incidence. This is the origin of the *surface selection rule*, which results in the very useful ability to distinguish vibrations that possess a transition dipole moment with a large component perpendicular to the surface.

The orientation of the molecules in self-assembled or LB films may be investigated using different polarizations of incident radiation or by comparing ATR and RAIRS measurements. Figure 6.16 contrasts the ATR spectrum with an approximate RAIRS spectrum (taken in ATR mode with a metal overlayer on the organic film) in the CH stretching region for 21 layers of cadmium docosenoate [13]; both spectra were recorded using p-polarized radiation. The bands that are observed are 2960 cm^{-1} (antisymmetric CH_3 stretch), 2920 cm^{-1} (antisymmetric CH_2 stretch), 2870 cm^{-1} (symmetric CH_3 stretch) and 2850 cm^{-1} (symmetric CH_2 stretch). The symmetric and antisymmetric CH_2 stretches are both intense in the ATR spectrum. In the RAIRS mode, the CH_3 bands increase in intensity. The transition dipole moments of the CH_2 vibrations are perpendicular to the long axis of the fatty acid molecule, whereas the symmetric and antisymmetric CH_3 stretches have components along the chains axis. Therefore, the experimental data shown in Figure 6.16 are consistent with the long axes of the fatty acid molecules being almost perpendicular to the substrate plane.

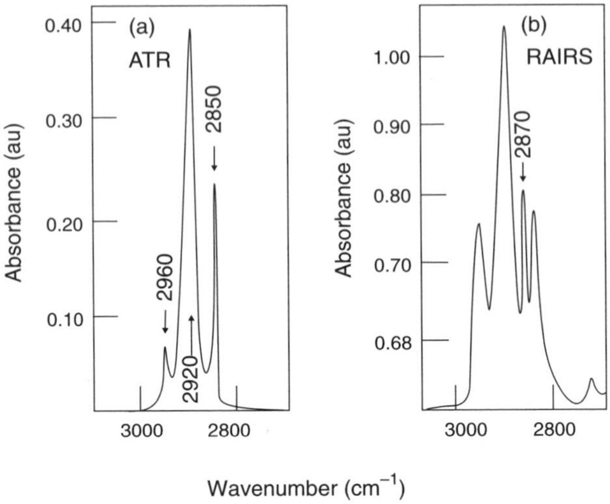

Figure 6.16 Infrared spectra of 21 LB layers of cadmium docosanoate on a silicon ATR crystal. (a) ATR mode; (b) RAIRS mode. Reprinted with permission from *Langmuir*, **5**, Davies GH, Yarwood J, 'Infrared intensity enhancement for Langmuir–Blodgett monolayers using thick metal overlayers', pp. 229–232, Copyright (1989) American Chemical Society.

Infrared spectroscopy may also be used to monitor chemical and structural changes occurring in multilayer films. For example, Figure 6.17 shows the results of an IR study of LB layers of a derivative of the charge-transfer compound TTF (Chapter 3, Section 3.4.1) before and after doping with iodine vapour [14]. The molecules of this compound, known as

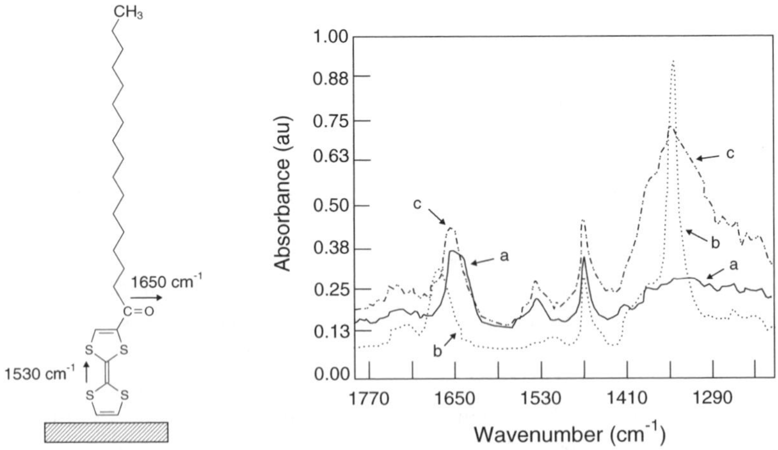

Figure 6.17 ATR infrared spectra of HDTTF LB film as a function of doping with Br_2 vapour. a, Before doping; b, immediately after doping; c, 1 h after doping. The orientation of the HDTTF molecule on the substrate is shown on the left. Reprinted with permission from *Langmuir*, **6**, Dhindsa AS, Bryce MR, Ancelin H, Petty MC, 'Infrared spectroscopic studies on the structure and ordering hexadecanolyltetrathiafulvalene conducting Langmuir–Blodgett multilayers', pp. 1680–1682, Copyright (1990) American Chemical Society.

HDTTF, transfer to the substrate with their long axes roughly perpendicular to the surface. As depicted in Figure 6.17, the C=O stretching vibration, at 1650 cm^{-1}, has its transition dipole oriented perpendicular to the long axis of the molecule whereas the C=C stretching mode of the fulvalene ring, at approximately 1530 cm^{-1}, is parallel to this direction. The intensity of these bands reflects strongly the chemistry of the initial doping (oxidation) and subsequent loss of I_2 to produce an organic conductor. Immediately after doping, the C=O stretching band moves to a higher frequency (curve b in Figure 6.17) because of electronic changes associated with the oxidation of the TTF derivative HDTTF to produce a radical cation $HDTTF^{+\bullet}$. The C=C band (1530 cm^{-1}) almost disappears, showing that the radical cation carries a largely C–C single bond. The band reappears when iodine molecules leave the film to form a mixed-valence complex, with some fulvalene rings in their ground (not charge-transfer) state. The charge transfer in such systems often leads to vibronic excitation of originally IR-forbidden vibrations. Figure 6.17 shows a very intense band near 1350 cm^{-1} immediately after doping with I_2. This arises from the coupling of the motion of electrons with vibrational modes in the fulvalene ring, referred to as electron–molecule vibronic coupling.

6.6.1 Raman Scattering

When electromagnetic energy falls on an atomic or molecular sample, it may be absorbed if the energy of the radiation corresponds to the separation of two energy levels in the atoms or molecules. If it does not, the radiation will be either transmitted or scattered. Of the scattered radiation, most is of unchanged wavelength λ and is called *Rayleigh scattering*. The intensity I_s is related to the wavelength by

$$I_s \propto \lambda^4 \tag{6.11}$$

However, a small amount of the scattered radiation is at a slightly increased or decreased wavelength. This is *Raman scattering*; the radiation scattered with an energy lower than the incident beam is referred to as *Stokes' radiation*, whereas that at higher frequency is called *anti-Stokes' radiation*. To be Raman active, a molecular rotation or vibration must cause some change in a component of the molecular polarizability. The process can be easily understood in the context of the simplified energy-level diagram shown in Figure 6.18. Incident light of energy $h\nu_0$ impinges on a molecule, originally in its ground vibrational level, causing a momentary excitation to what is termed a *virtual state*. During this brief

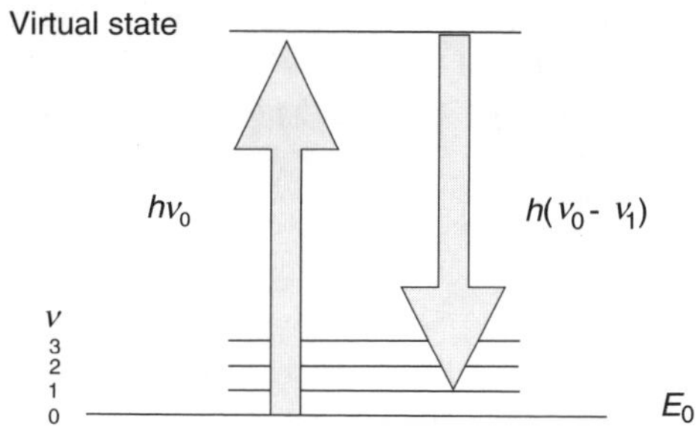

Figure 6.18 Energy level diagram illustrating the Raman scattering process.

interaction, a quantum of vibrational energy is acquired by the molecule from the light. In this case, the scattered light is Stokes' radiation with energy, $h(\nu_0 - \nu_n)$, less than the original light by an amount equal to that necessary for excitation of a particular vibrational mode, ν_n, of the molecule, and the molecule is left vibrationally excited.

Raman scattering is an inherently weak process, so an enhancement mechanism is usually needed to organic monolayer samples. An increase in the signal can usually be achieved by depositing films on the surfaces of noble metals such as Au or Ag or by the interaction with surface plasmons. These methods are generally referred to as *surface-enhanced Raman spectroscopy*. For materials with electronic transitions in the visible, the resonance Raman effect can be used to enhance certain bands. The Raman process may be used as a basis for a microscopy technique. The mode of operation is similar to a fluorescence microscope. Images are recorded with a sensitive (e.g. CCD) camera in light that has been scattered in a particular Raman band by using optical filters.

6.7 SURFACE ANALYTICAL TECHNIQUES

Surface analytical techniques such as *Auger electron spectroscopy* (AES), *X-ray photoelectron spectroscopy* (XPS) and *secondary-ion mass spectrometry* (SIMS) may be used to provide useful information about the structure and/or chemical composition of the surfaces of organic materials. These methods are contrasted in Table 6.2.

Auger electron spectroscopy is a two-step process and is depicted schematically in Figure 6.19. First, a high energy photon ejects an electron from the core orbital of an atom, E_A, in the material under investigation. As this vacancy is immediately filled by an electron from a higher energy, E_B, X-ray photon emission can occur [Figure 6.19(a)]. However, the energy may also be transferred to a second electron, an Auger electron, releasing it from one of the higher energy orbitals, E_C, as shown in Figure 6.19(b). The energies of these Auger electrons are low (20–1000 eV) so that, although they may be generated from as far within the sample as the original electron beam penetrates, only those produced within the first few atomic layers of the surface can escape. Therefore, the technique has immense surface sensitivity.

Table 6.2 Important surface analytical methods.

Technique	Acronym	Probe beam	Depth probed	Sample beam	Comments
Auger electron spectroscopy	AES	Low-energy (1–10 keV) electrons	0.5–10 nm	Electron energy (20–1000 eV)	Very high surface sensitivity
X-ray photoelectron spectroscopy	XPS	Low-energy X-rays	0.5–5 nm	Photoelectrons	Little surface damage; chemical composition can be determined
Secondary ion mass spectrometry	SIMS	Pulsed ion (Ar^+) beam (few keV)	2 nm–100 μm	Secondary ions	Identification of chemical compounds

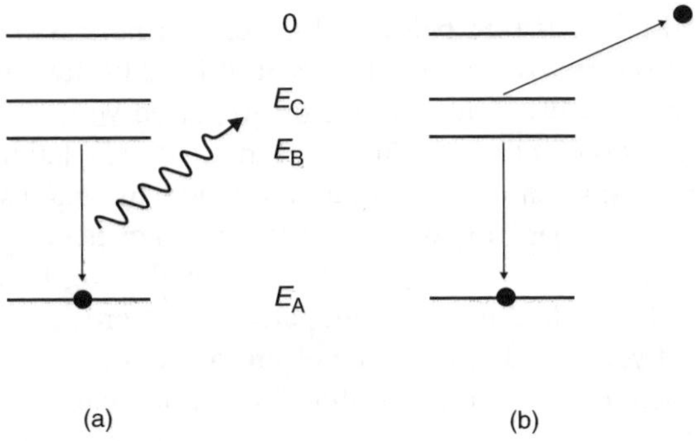

Figure 6.19 Possible processes occurring following the removal of an inner core electron by an incident high energy electron. This vacancy is immediately filled by an electron from a higher energy. (a) X-ray photon emission occurs as the vacancy is filled; (b) Auger effect – the energy is transferred to a second electron, which is subsequently emitted.

In the XPS experiment, the sample surface is irradiated by a source of low-energy X-rays. Photoionization takes place in the sample producing photoelectrons of a characteristic energy distribution. Because X-rays do not normally cause appreciable surface damage, XPS is usually preferred as an analytical technique for organic materials. Analysis of the kinetic energy of the photoelectrons permits the elemental composition of the surface layers (up to 5 nm) to be determined quantitatively. Information concerning the bonding environments of the elements may also be obtained.

Secondary ion mass spectrometry involves bombarding a sample surface with a pulsed primary ion beam (usually Ar^+) with energy of a few keV. This results in the emission of both positively and negatively charged secondary ions from the uppermost surface layers (a few nm). The identification of chemical compounds is possible by the detection of either molecular ions with masses up to 104 amu or by the detection of characteristic fragments.

The relative position of a particular atom with respect to its nearest neighbours can sometimes be determined with great precision from the *extended X-ray absorption fine structure* (EXAFS). Above an X-ray absorption edge, the absorption by the atom is slightly affected by the waves being scattered by neighbouring atoms. This is manifested by weak oscillations in the absorption as a function of frequency for a considerable range above the edge. Just above the absorption edge (around 5 eV beyond the absorption threshold) there are stronger oscillations, called *X-ray absorption near-edge structure* (XANES), which derive from the same process of scattering from neighbouring atoms. *Near-edge X-ray absorption fine structure* (NEXAFS) is another method using synchrotron radiation that can be used to obtain orientation information on the molecules in thin organic films.

6.8 SCANNING PROBE MICROSCOPIES

Gerd Binnig and Heinrich Rohrer introduced the *scanning tunnelling microscope* (STM) in 1981. For this, they were awarded half of the Nobel Prize for Physics in 1986 (the other half went to Ernst Ruska for the design of the first electron microscope). Like *atomic force microscopy* (AFM – invented in 1986 by Binnig, Quate and Gerber), the STM technique may be used to provide direct images of metal surfaces with nanometre resolution. There are now very many other related methods (see below) which can all be classified as *scanning probe microscopes* (SPMs). These microscopes work by measuring a local property – such as height, optical absorption, or magnetism – with a probe or tip placed very close to the sample. The small probe–sample separation (of the order of the instrument's resolution) makes it possible to take measurements over a small area. To acquire an image, the microscope raster-scans the probe over the sample while measuring the local property in question.

Scanning tunnelling microscopy measures the current flowing between the tip and the surface of the sample. The method may provide lateral and vertical resolutions of less than 0.3 and 0.02 nm, respectively. Furthermore, the electron energies are usually less than 3 eV, thus avoiding the degradation of an organic thin film, which can be a problem with other imaging methods.

Whereas STM records the overlap of the local electron density of states between a tip and a surface (or its modulation by adsorbate molecules in the gap), AFM measures the interatomic forces between a cantilevered spring tip and the sample surface. The image contrast in AFM is achieved by probing the elastic response of the molecules to the force exerted by the scanning tip. Figure 6.20 shows the principles of the technique. The microscope, which essentially consists of a tip attached to a cantilever [Figure 6.20(a)], can operate in three different modes: contact, noncontact and tapping. Figure 6.20(b) shows the dependence on the interaction force on the distance between the tip and the surface (Chapter 2, Section 2.3.1). At a relatively large distance from the surface, the force of attraction between the tip and the surface dominates, whereas the force of repulsion is the most significant at very small distances.

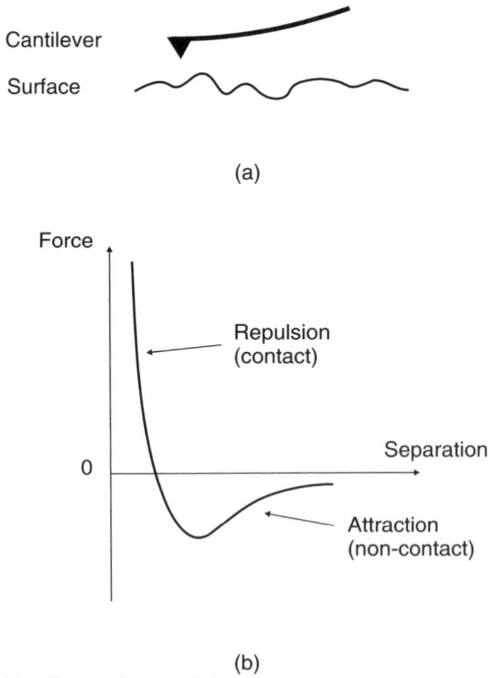

Figure 6.20 (a) Cantilever interaction with a surface in atomic force microscopy. (b) Force versus separation between the AFM tip and the substrate.

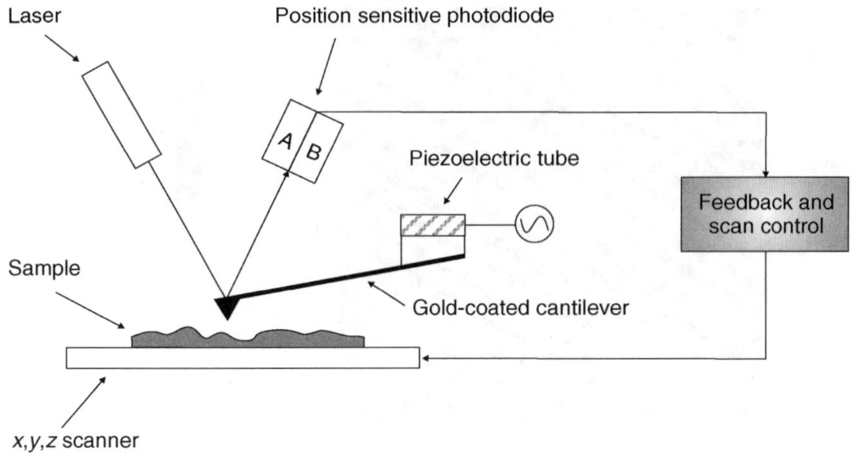

Figure 6.21 Schematic diagram of an atomic force microscope.

In the contact mode, the tip is brought into close contact with the sample surface, so that the force between the tip and the sample becomes repulsive. The resulting deflection by the cantilever is then measured. For organic materials, one problem with this mode of operation is that it can easily scratch the soft surface of the sample. If the tip is moved away from the surface by a small distance (5–10 nm), the tip will be attracted to the surface, causing the cantilever to bend towards it. This is the noncontact mode of operation. However, since the force of attraction is much weaker than that of repulsion, the resolution is normally lower than for the contact mode of operation. The tapping mode combines the advantages of the contact and noncontact techniques. The cantilever is oscillated at a frequency in the range 100–500 kHz, with amplitude of approximately 20 nm, so that the tip is just touching (tapping) the surface. The resolution of this tapping mode is almost the same as that of the contact mode, but it is faster and much less damaging.

Figure 6.21 shows a schematic diagram of an AFM. The instrument can generally measure the vertical deflection of the cantilever with picometre resolution. To achieve this, most AFMs use an optical lever, a device that achieves resolution comparable to that of an interferometer while remaining inexpensive and easy to use. It works by reflecting a laser beam off the cantilever. Angular deflection of the cantilever causes a twofold larger angular deflection of the laser beam. The reflected laser beam then strikes a position-sensitive photodetector, consisting of two side-by-side photodiodes. The difference between the two photodiode signals indicates the position of the laser spot on the detector and thus the angular deflection of the cantilever. Because the cantilever-to-detector distance generally measures thousands of times the length of the cantilever, the optical lever greatly magnifies motions of the tip.

Most scanned-probe microscopes use piezoceramic tubes (usually made of ceramic materials) to position the tip, as these combine a simple one-piece construction with high stability and large scan range. Four electrodes cover the outer surface of the tube, while a single electrode covers the inner surface. Application of voltages to one or more of the electrodes causes the tube to bend or stretch, moving the sample in three dimensions. If the tip were scanned at constant height, there would be a risk that the tip would collide with the surface, causing damage. Hence in most cases a feedback mechanism is employed to adjust the tip-to-sample distance to keep the force between the tip and the sample constant.

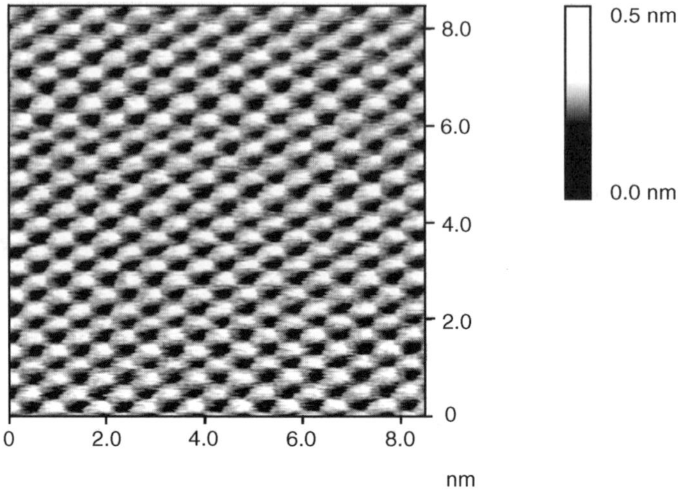

Figure 6.22 Atomic force miscroscope image of a 12-layer 22-tricosenoic acid LB film deposited on silicon. Unfiltered data obtained with a Digital Instruments Nanoscope III. Reprinted with permission from *J. Phys. Chem.* **100**, Evanson SA, Badyal JPS, Pearson C, Petty MC, 'Variation in intermolecular spacing with dipping pressure for arachidic acid LB films', pp. 11672–11674, Copyright (1996) American Chemical Society.

Early experiments using STM on organic samples exhibited poor reproducibility. Artifacts such as steps, domain walls or superstructures due to multiple tip effects and *Moiré patterns* made the images difficult to interpret. A major problem with thin-film samples, such as self-assembled films, is that they possess highly insulating regions (due to the hydrocarbon tails) through which the tunnelling current must pass. Such difficulties are circumvented using AFM. Figure 6.22 shows an AFM image of the surface of a 12-layer 22-tricosenoic acid LB film; lines of individual molecules are evident at the magnification shown [15].

Other measurements can be made using modifications of the SPM. These include variations in surface microfriction with a *lateral force microscope* (LFM), orientation of magnetic domains with a *magnetic force microscope* (MFM), and differences in elastic modulii on the micro-scale with a *force modulation microscope* (FMM). *Scanning capacitance microscopy* (SCM) is sensitive to electrical charges that develop on the surface and on the probe. This is most useful for determining the level of doping on semiconductor chips. These current microscopes can detect electronic charges of attofarads (10^{-18}F).

A further adaptation of the SPM has been developed to probe differences in chemical forces across a surface at the molecular scale. This technique has been called the *chemical force microscope* (CFM). The AFM and STM can also be used to perform electrochemistry on the microscale.

6.9 FILM THICKNESS MEASUREMENTS

Film thickness is an essential parameter in the characterization of thin organic films. Many different measurement techniques have been used and some of the most popular are listed in Table 6.3. However, some of these methods do not provide an independent measure of film

Table 6.3 Summary of common methods to determine the thickness of thin organic films.

Technique	Comments
X-ray diffraction	Provides d_{001} lattice spacing. Section 6.3
Ellipsometry	Commercial instruments available; sample refractive index should be known for high accuracy. Chapter 4, Section 4.5.3
Surface plasmon resonance	Sample refractive index should be known. Chapter 4, Section 4.7.2
Capacitance versus number of layers	Provides dielectric thickness. Requires insulating samples. Possibility of damage to organic layer during metallization. Chapter 7, Section 7.3.1
Mechanical probe	Commercial instruments available. Provides metric thickness directly. Organic film may be damaged. Not suitable for monolayer sensitivity

thickness; other physical parameters must be first determined. For example, some optical techniques (e.g. surface plasmon resonance, Chapter 4, Section 4.7.2) measure the optical thickness (or optical path length), which is equal to the metric thickness multiplied by the refractive index. Figure 6.23 shows a schematic diagram of an ellipsometer. A range of such instruments are available commercially and are used extensively for thin-film (e.g. silicon dioxide) thickness measurements in the semiconductor industry. However, as noted in Chapter 4, Section 4.5.3, the method needed to obtain accurate values of film thickness is complex. The problem of obtaining values of film thickness and refractive index from the ellipsometric parameters Ψ and Δ involves a least-squares routine in which theoretical values of Ψ and Δ are calculated and the subsequent error function is minimized. However, the solutions are not unique (the same values of Ψ and Δ correspond to different refractive index and thickness combinations) and some experience is needed for accurate results.

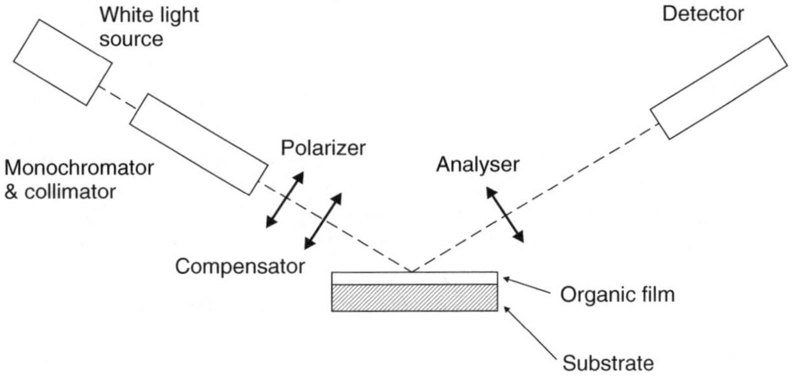

Figure 6.23 Schematic diagram of an ellipsometer.

A further problem with ellipsometric measurements on ultra-thin organic films concerns the nature of the thin-film system under investigation. The organic layer is invariably deposited on a solid surface: this may be a metal (or metallized glass microscope slides) or semiconductor. For such substrates, there is almost certainly a surface ('oxide') layer between the bulk substrate and the organic film. This may result from exposing an evaporated metal film or a freshly etched semiconductor to the atmosphere. It may also be augmented during the organic film deposition process (e.g. for LB deposition, the substrate is lowered into and out of the aqueous subphase). Such an interfacial film may be several nanometres in thickness and can introduce considerable errors if its presence is ignored. The usual solution is to modify the optical constants of the underlying substrate (which are used in the computer iteration) by taking measurements on an uncoated portion of the substrate.

Electrical measurements based on measuring the capacitance of a metal/LB film/metal structure as a function of the number of monolayers (Chapter 7, Section 7.3.1) yield the dielectric thickness (metric thickness/relative permittivity). Direct measurement of the thickness of an LB film is conveniently accomplished using a mechanical probe; however, care must be taken to avoid damage to the relatively soft organic layer.

BIBLIOGRAPHY

Bellamy LJ. *The Infrared Spectra of Complex Molecules*, 3rd edn. London: Chapman and Hall; 1986.

Cheetham AK. Diffraction methods. In AK Cheetham, P Day (Editors). *Solid State Chemistry Techniques*. Oxford: Oxford Scientific Publications; 1988, pp. 39–51.

Feigin LA, Lvov YM, Troitsky VI. X-ray and electron diffraction study of Langmuir–Blodgett films. *Sov. Sci. Rev. A Phys.* 1989; **11**: 285–377.

Hollas JM. *Modern Spectroscopy*, 2nd ed. Chichester: John Wiley & Sons, Ltd; 1992.

Kitaigorodskii AI. *Organic Chemical Crystallography*. New York: Consultants Bureau; 1961.

Petty MC. *An Introduction to Langmuir–Blodgett Films*. Cambridge: Cambridge University Press; 1996.

Roberts GG (Editor). *Langmuir–Blodgett Films*. New York: Plenum Press; 1990.

Russell GJ. Practical reflection electron diffraction. *Prog. Cryst. Growth Charact.* 1982; **5**: 291–321.

Tredgold RH. *Order in Thin Organic Films*. Cambridge: Cambridge University Press; 1994.

Ulman A. *Ultrathin Organic Films*. San Dieg, CA: Academic Press; 1991.

Ulman A (Editor). *Characterization of Organic Thin Films*. Boston: Butterworth-Heinemann; 1995.

Woodruff DR, Delchar TA. *Modern Techniques of Surface Science*, 2nd ed. Cambridge: Cambridge University Press; 1994.

REFERENCES

[1] Paul S, Pearson C, Molloy A, Cousins MA, Green M, Kolliopoulou S, Dimitrakis P, Normand P, Tsoukalas D, Petty MC. Langmuir–Blodgett film deposition of metallic nanoparticles and their application to electronic memory structures. *Nano Lett.* 2003; **3**: 533–536.

[2] Nicklow RM, Pomerantz M, Segmüller A. Neutron diffraction from small numbers of Langmuir–Blodgett monolayers of manganese stearate. *Phys. Rev. B* 1981; **23**: 1081–1087.

[3] Petty MC. *An Introduction to Langmuir–Blodgett Films*. Cambridge: Cambridge University Press; 1996.

[4] Leuthe A, Riegler H. Thermal behaviour of Langmuir–Blodgett films. II. X-ray and polarized reflection microscopy studies on coexisting polymorphism, thermal annealing and epitaxial layer growth of behenic acid multilayers. *J. Phys. D: Appl. Phys.* 1992; **25**: 1786–1797.

[5] Lvov Y. Electrostatic layer-by-layer of proteins and polyions. In Y Lvov, H Mohwald (Editors). *Protein Architecture, Interfacing Molecular Assemblies and Immobilisation Biotechnology.* New York: Marcel Dekker; 2000, pp. 125–167.
[6] Prakash M, Ketterson JB, Dutta P. Study of in-plane structure in lead–fatty acid LB films using X-ray diffraction. *Thin Solid Films* 1985; **134**: 1–4.
[7] Kitaigorodskii AI. *Organic Chemical Crystallography.* New York: Consultants Bureau; 1961.
[8] Peterson IR, Russell GJ. An electron diffraction study of ω-tricosenoic acid Langmuir–Blodgett films. *Philos. Mag. A* 1984; **49**: 463–473.
[9] Kumar NP, Major S, Vitta S, Talwar SS, Dubeck P, Amenitsch H, Bernstorff S, Ganesan V, Gupta A, Dasannacharya BA. Molecular packing in cadmium and zinc arachidate LB multilayers. *Colloids Surf. A* 2002; **198–200**: 75–81.
[10] Peterson IR, Russell GJ, Earls JD, Girling IR. Surface pressure dependence of molecular tilt in Langmuir–Blodgett films of 22-tricosenoic acid. *Thin Solid Films* 1988; **161**: 325–331.
[11] Robinson I, Sambles JR, Peterson IR. A reflection high energy electron diffraction analysis of the orientation of the monoclinic subcell of 22-tricosenoic acid Langmuir–Blodgett bilayers as a function of the deposition pressure. *Thin Solid Films* 1989; **172**: 149–158.
[12] Barnes WL, Sambles JR. Surface pressure effects on Langmuir–Blodgett multilayers of 22-tricosenoic acid. *Surf. Sci.* 1987; **187**: 144–152.
[13] Davies GH, Yarwood J. Infrared intensity enhancement for Langmuir–Blodgett monolayers using thick metal overlayers. *Langmuir* 1989; **5**: 229–232.
[14] Dhindsa AS, Bryce MR, Ancelin H, Petty MC, Yarwood J. Infrared spectroscopic studies on the structure and ordering of hexadecanolyletrathiafulvalene conducting Langmuir–Blodgett multilayers. *Langmuir* 1990; **6**:1680–1682.
[15] Evanson SA, Badyal JPS, Pearson C, Petty MC. Variation in intermolecular spacing with dipping pressure for arachidic acid LB films. *J. Phys. Chem.* 1996; **100**: 11672–11674.

7 Thin Film Processing and Device Fabrication

When I consider everything that grows

7.1 INTRODUCTION

For most current (e.g. liquid crystal displays) and future (e.g. plastic transistors) applications of molecular electronics, the organic materials are required in the form of thin films (in this context, 'thin' is generally taken to mean 1 nm to 10 μm). This presents a considerable challenge to materials scientists, as organic compounds, in their bulk form, can be fragile and difficult to handle. The physical properties of a thin film may be very different from those of the bulk, particularly if the film thickness is very small. This behaviour can be related to the morphology of the layer, which is determined by the processes that occur during the film formation.

In this chapter, an overview of the more popular methods that may be used to fabricate thin layers of organic compounds will be described. A distinction will be drawn between established deposition technologies, in many cases developed for use with inorganic materials, and those techniques that actually allow molecular-scale architectures to be built up. Each deposition method differs in complexity and may be more suited to provide films in a particular thickness range. Specific types of organic compound are necessary for certain processes, e.g. self-assembly exploits the attraction between certain chemical groups. Some methods are 'wet' (spinning) whereas others are inherently 'dry' (plasma deposition). There are also implications for the degree of order and contamination levels in the deposited film.

Film thickness uniformity may also be an important parameter. For instance, thin film optical interference filters can demand a thickness uniformity of $\pm 1\%$. A related issue is conformal coverage, which refers to the ability to coat both the vertical and horizontal surfaces of substrates. The problem arises primarily in the fabrication of integrated circuits as the semiconductor contacts and device interconnection metallization are needed to cover complex topography, with micro-steps, grooves and raised stripes. Film coverage will not be uniform when geometric shadowing effects cause unequal deposition on the top and sidewalls of steps.

In all forms of thin film growth, it is important to know whether the deposition is *epitaxial*, i.e. is the orientation of the atoms/molecules in the thin film the same as that of the atoms/molecules in the underlying substrate? Molecular beam epitaxy achieves this and allows the growth of extremely pure crystalline films. Finally, important considerations for commercialization are factors such as the cost of the thin film process, the time taken for coating and/or whether the method can easily be adapted to continuous coating.

Molecular Electronics: From Principles to Practice M. C. Petty
© 2007 John Wiley & Sons, Ltd

7.2 ESTABLISHED DEPOSITION METHODS

7.2.1 Spin-Coating

Spin-coating is exploited extensively by the microelectronics industry for depositing layers of photoresist films, generally polymers such as polyimides, on silicon wafers. The various steps involved are illustrated in Figure 7.1. A quantity of a polymer solution is first placed on the semiconductor wafer, which is then rotated at a fixed speed of several thousand revolutions per minute (or the solution can be applied while the wafer is slowly rotating). The resist solution flows radially outwards, reducing the fluid layer thickness. Evaporation of the solvent results in a film of uniform thickness.

A quantity of solution is first applied to the substrate surface; this may be pretreated with an adhesion promoter to improve wetting. The initial volume of fluid dispensed on the rotating disk and the rate of fluid delivery have a negligible effect on the final film thickness. In contrast, the resist viscosity (dependent on the concentration of the starting solution) and final film speed are both important process parameters. An increase in angular velocity decreases the film thickness; an inverse power-law relationship usually holds for the thickness dependence on the final spin speed. For a given speed, the film thickness decreases rapidly at first, but then slows at longer times. A simple theory predicts the following relationship between the thickness of the spun film, d, the viscosity coefficient of the solution, η, its density, ρ, the angular velocity of the spinning, ω, and the spinning time, t [1]:

$$d = \left(\frac{\eta}{4\pi\rho\omega^2}\right)^{\frac{1}{2}} t^{-\frac{1}{2}} \tag{7.1}$$

More sophisticated models have been developed to allow for changes in the resist resulting from solvent evaporation and the non-Newtonian character of the rheological behaviour of photoresist [2–4]. These work reasonably well for practical concentrations of spinning solutions. However, very dilute solutions often lead to pinholes in the final film and failure of the mathematical predictions.

Organic compounds that have been successfully deposited by spin-coating include electrically insulating polymers such as poly(vinylidene fluoride), polymers and dyes

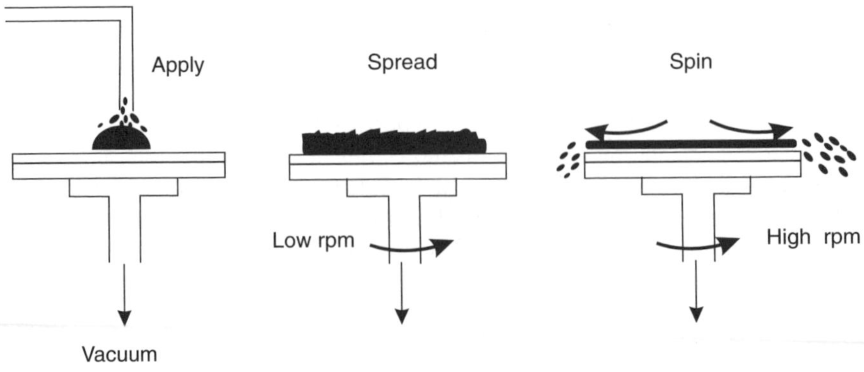

Figure 7.1 Schematic diagram of spin-coating.

developed for electroluminescent displays and certain phthalocyanine materials. Although spin-coating is expected to produce films in which individual molecules are relatively disordered, this is not always the case. For instance, organized phthalocyanine layers have been deposited [5]. A study using X-ray diffraction revealed a crystalline order similar to that observed for Langmuir–Blodgett films (Section 7.3.1) of analogous compounds. The way in which this order is achieved is not fully understood, but may result from the centrifugal forces acting upon the individual molecules during spinning. In other cases, order can be achieved by a post-deposition treatment. Application of heat and an electric field normal to the film plane (the process of poling; Chapter 2, Section 2.6.3) can be used to align the C–F dipoles in poly(vinylidene fluoride) films. The result is a polar film possessing piezoelectric and pyroelectric properties (Chapter 5, Section 5.7). A similar method can be used to induce second-order nonlinear optical behaviour (e.g. second-harmonic generation; Chapter 4, Section 4.3.2) in films of appropriate molecules.

7.2.2 Physical Vapour Deposition

Thermal Evaporation

Solid materials vaporize when heated to sufficiently high temperatures and this process may proceed through the liquid phase. A thin film is then obtained by the condensation of the vapour on a colder substrate. This method has traditionally been used for the deposition of films of inorganic materials, such as metals and their alloys. However, the technique is now being exploited for the formation of layers of low molecular weight organic compounds. The rate of evaporation Γ (in kg m^{-2} s^{-1}) from a surface is given by the Langmuir expression [6]:

$$\Gamma = P\left(\frac{M}{2\pi RT}\right)^{\frac{1}{2}} \tag{7.2}$$

where P is the vapour pressure (in N m^{-2}) of the material at temperature T (in K), M is the molecular weight and R is the gas constant.

Because of collisions with ambient gas atoms, a fraction of the vapour atoms will be scattered. For a straight-line path between the evaporating material (source) and the substrate, it is necessary to use low pressures ($<10^{-4}$ mbar), where the mean free path of the gas atoms λ (Chapter 3, Section 3.2.1) is much greater than the source–substrate distance. From kinetic theory, λ is given by

$$\lambda = \frac{k_B T}{P\pi d^2\sqrt{2}} \tag{7.3}$$

where d is the diameter of the molecules. The mean free path is therefore inversely proportional to gas pressure. For the common gases, He, H_2, O_2, N_2, Ar, CH_4, CO_2 and H_2O, the effective cross-sectional diameters range from 0.2 nm (for He) to 0.5 nm (for H_2O). This gives average mean free path values of about 10 cm at a pressure of 10^{-3} mbar and about 10^4 cm at 10^{-6} mbar [6]. Therefore, at low pressures, a shadow mask, as depicted in Figure 7.2, can be used immediately in front of the substrate to define patterns. The low pressure also prevents contamination of the source material (e.g. by oxidation).

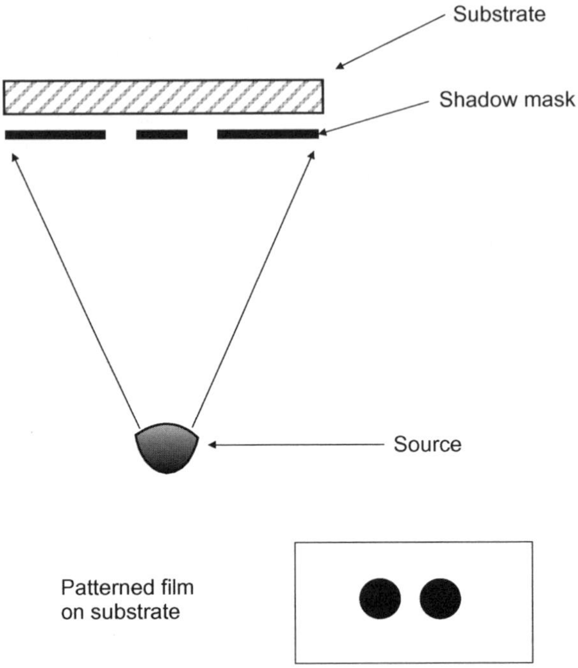

Figure 7.2 Use of a shadow mask to pattern an evaporated thin film.

Figure 7.3 shows a schematic diagram of a typical evaporation system. The system chamber, which can be made out of glass or metal, is evacuated to a pressure of 10^{-4}–10^{-6} mbar, normally with two types of vacuum pump operating in series (a rotary and diffusion pump). It is common to introduce a shutter (either mechanically or electronically

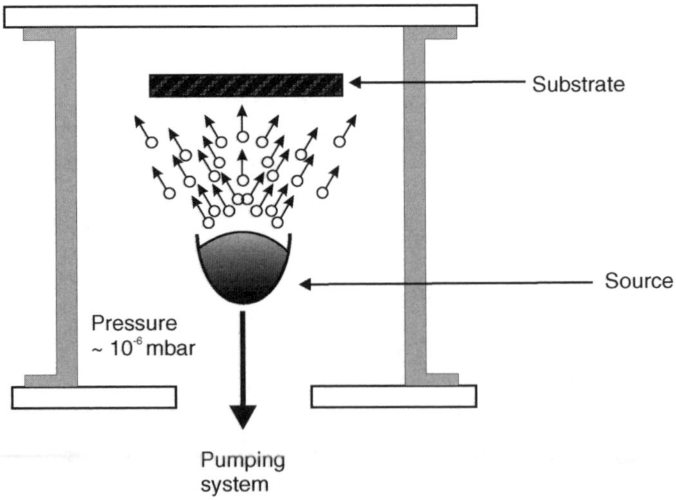

Figure 7.3 Vacuum evaporation system. Evaporating atoms or molecules traverse the space between the source and the substrate at reduced pressure.

operated) between the source and the substrate. This can be used to prevent the material that evaporates initially from the source surface (which may be contaminated) from depositing on the substrate. A film thickness sensor is usually included in the evaporation chamber. This is based on an oscillating quartz crystal. As the evaporating species condenses on this crystal, its vibration frequency is altered. The change in frequency can then be used as an indication of the film thickness. (This technique is also exploited in certain chemical sensors; Chapter 10, Section 10.4.5.)

Although commonly thought of as a single process, the deposition of thin films by thermal evaporation consists of several distinguishable steps:

(i) transformation of the source material, in solid or liquid form, into the gaseous state;

(ii) vapour molecules traversing the space between the source and the substrate; and

(iii) condensation of the vapour upon arrival on the substrate.

The first step, governed by Equation (7.2), requires the conversion of thermal to mechanical energy, and is achieved by a variety of physical methods. Resistive heating has been used to deposit low molecular weight dyes, charge-transfer salts and large macromolecules, such as the phthalocyanines. Typical evaporation rates are 1–10 nm min^{-1}. Other techniques include arc evaporation, RF heating and heating by electron bombardment. Deposition of polymer films by laser ablation is an area that offers some promise [7]. Laser pulsed methods have been used successfully for polyethylene, polycarbonate, polyimide and poly(methyl methacrylate). Typical film growth rates are 0.02–0.1 nm per laser pulse.

Materials that dissociate in the vapour phase may provide solid films with a stoichiometry (i.e. composition) that differs from that of the source. Therefore, special techniques have been devised. One approach is to use the method of *flash evaporation*. Rapid evaporation is achieved by continuously dropping fine particles of the materials on to a hot surface. Although fractionation occurs during the evaporation of each particle (the more volatile component evaporating first), at any time there will be several particles at different stages of fractionation. Consequently, the vapour phase will possess a similar composition to that of the source material. A further technique is to evaporate from two or more sources and control the flux from each to obtain a vapour with the required composition (Figure 7.4). This has been used effectively to deposit thin films of doped organic charge-transfer salts [8]: one source is the charge-transfer salt, e.g. tetrathiafulvalene (TTF), and the other is the dopant, e.g. iodine. Laser co-ablation techniques can be used to deposit films of metal–polymer composites [7].

Whatever method is used to heat the material, the vapour will be in an atomic or molecular form when it arrives at the substrate. Figure 7.5 illustrates the various stages that occur during the formation of a solid film on the surface of the substrate [9]. The surface of the substrate is covered with a large number of adsorption sites and a molecule becomes bound to one of these with a characteristic energy. The adsorbed atoms do not remain stationary [Figure 7.5(a)] but may re-evaporate, requiring an energy equal to the adsorption energy, or migrate or hop to an adjacent adsorption site [Figure 7.5(b)]. Other processes are collision and recombination [Figure 7.5(c)]. Each group of atoms eventually reaches a size that is more likely to grow than to decay. The formation of such stable islands of material is known as nucleation [Figure 7.5(d)]. Individual islands continue to grow by the addition of more single hopping atoms [Figure 7.5(e) and (f)].

Eventually, the islands coalescence with neighbours to form an interlinking network [Figure 7.5(g)]. The uncoated areas gradually diminish until a single continuous film is

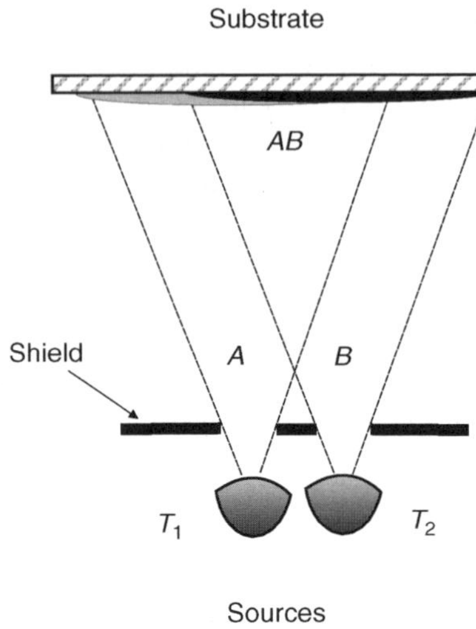

Figure 7.4 Two-source evaporation arrangement. Two independent sources, held at temperatures T_1 and T_2 are used to evaporate independently materials A and B. The two vapour fluxes combine at the substrate.

formed [Figure 7.5(h)]. This may not occur until a great deal of material has been deposited. For example, an evaporated film of gold on glass at room temperature becomes continuous at an average thickness of about 30 nm, or 100 atoms thick!

The microstructure of an evaporated thin film depends on the evaporation rate, substrate temperature and the chemical and physical nature of the substrate surface. The size of the grains in a polycrystalline film will generally be larger for high source and substrate temperatures. However, if the kinetic energy of the incoming molecules is too high, the surface mobility of the adsorbed species is reduced because the vapour molecules will penetrate the condensed film. The effect of substrate temperature on grain size is greater for relatively thick films. For a given material–substrate combination and under a fixed set of deposition conditions, the grain size increases initially as the film thickness increases. Beyond a certain thickness, the grain size remains constant, suggesting that coherent growth with the underlying grains does not go on forever. Figure 7.6 shows an atomic force microscope (AFM) image (see Chapter 6, Section 6.8) of an evaporated film of the organic semiconductor pentacene on a glass substrate [10]. The film thickness is approximately 40 nm. A granular structure, with large (μm in size) dendritic grains is evident.

The physical nature of evaporated films can be changed by post-deposition heat treatment (annealing). This is usually associated with a change in the crystallinity of the film. For example, following thermal evaporation, the DC in-plane conductivity of thin films of the electron donor bis(ethylenedithio)tetrathiafulvalene (BEDT-TTF) can be increased by many orders of magnitude by doping with iodine and then annealing at 60 °C [11].

One important aspect, often neglected, is the thickness distribution in evaporated films. This will depend on the geometry of the evaporation system. For a point source and plane substrate, parallel to the plane of the emitting surface, depicted in Figure 7.7, the rate of deposition varies as $(\cos\theta)/r^2$, where r is the radial distance of the substrate from the source

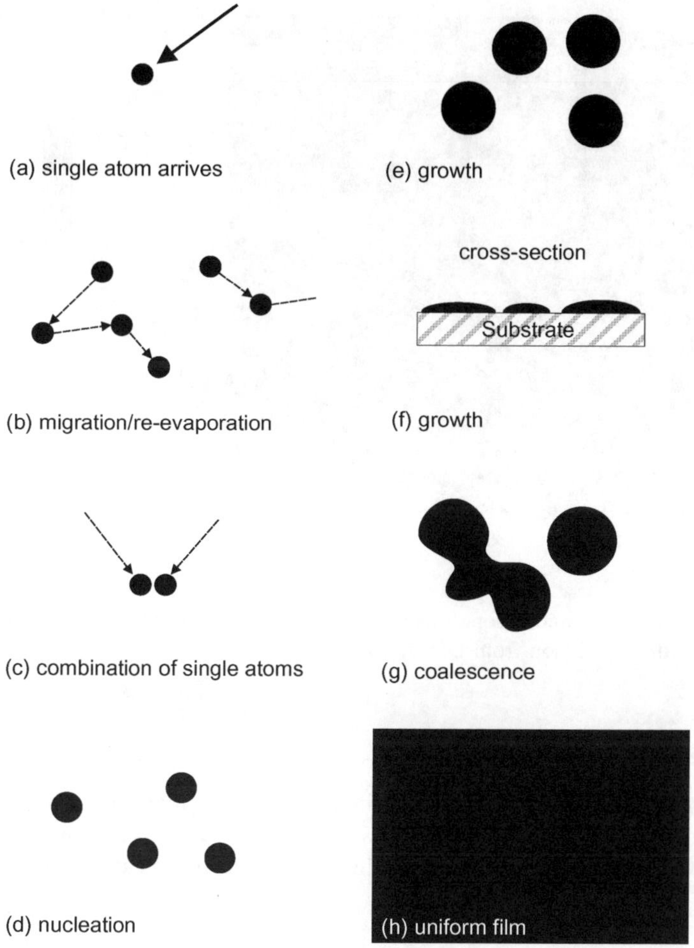

Figure 7.5 Stages of growth of an evaporated thin film. Reprinted with permission from Leaver KD, Chapman BN, *Thin Films.* Copyright (1971), Taylor and Francis.

and θ is the angle between the radial vector and the normal to the substrate. The following expression can be derived for the film thickness d as a function of l, the distance along the substrate measured from its central position [6]:

$$\frac{d}{d_0} = \left[1 + \left(\frac{l}{h}\right)^2\right]^{-\frac{3}{2}} \tag{7.4}$$

where d_0 is the film thickness at the central position on the substrate (where the vapour condenses normally) and h is the normal distance of the point source to the substrate. Equation (7.4) reveals that the film thickness at a distance along the substrate surface equal to h will be about 35% of its maximum value, measured in the centre of the substrate. This effect may be reduced by using a large area source.

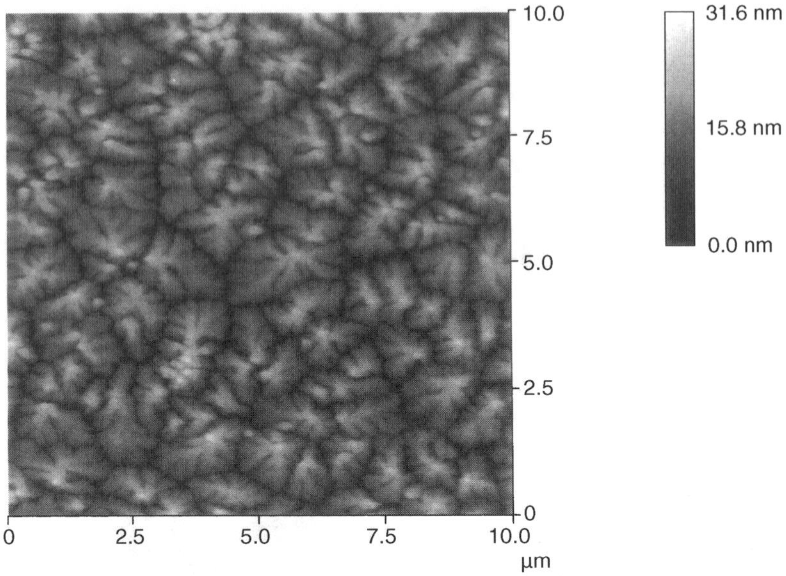

Figure 7.6 Atomic force microscope image of an evaporated film of pentacene, thickness ~40 nm [10]. Reprinted with permission from Dan Kolb.

Molecular Beam Epitaxy

Molecular beam epitaxy (MBE) is, in principle, very similar to the method of vacuum evaporation described in the previous section [12]. However, an ultrahigh vacuum ($<10^{-9}$ mbar) is required to eliminate the scattering by residual gas molecules. The

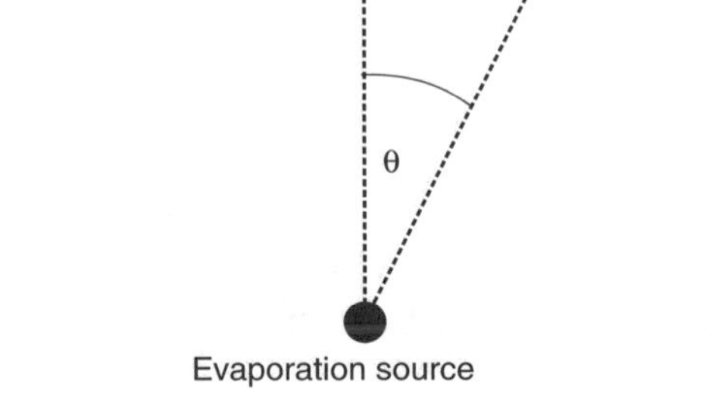

Figure 7.7 Evaporation geometry from a point thermal evaporation source on to a plane substrate.

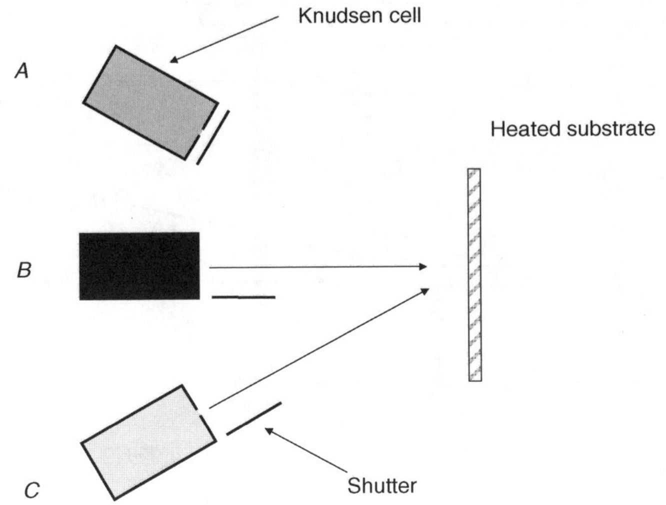

Figure 7.8 The use of Knudsen cells for molecular beam epitaxy. The three cells shown are use to provide molecular beams of materials *A*, *B* and *C*.

technique consists of directing controlled 'beams' of the required molecules towards a heated substrate. The experimental arrangement is shown in Figure 7.8. Multiple sources, *Knudsen cells* (for source materials *A*, *B* and *C* in Figure 7.8) can be shuttered (usually computer-controlled) and used to create a superlattice structure on the substrate, with precise control of the molecular composition, orientation and packing in two dimensions. Each Knudsen cell encloses an evaporating surface that is large compared with the orifice. The diameter of the orifice must be about one-tenth or less of the mean free path of the gas molecules, and the wall around the orifice must be vanishingly thin so that gas molecules leaving the enclosure are not scattered, adsorbed and desorbed by the orifice wall.

Some of the first trial observations of real-time epitaxial growth in organic molecular systems have involved the growth of phthalocyanine monolayers on cleaved surfaces of MoS_2 and on highly oriented pyrolyzed graphite and alkali metal halides. *In situ* reflection high-energy diffraction (RHEED; Chapter 6, Section 6.5) can be used to monitor the actual film growth. Multiple quantum well structures based on organic molecular materials, such as perylene and naphthalene derivatives and copper phthalocyanine, can be fabricated using MBE. The method has also been used successfully to deposit electrically conductive organic semiconductors for the fabrication of organic field effect transistors.

Molecular beam epitaxy is a very slow process. The typical rate of growth is a single monolayer per second, or approximately 1 μm per hour. However, this allows for abrupt changes in material composition.

Sputtering

Sputtering is based on the momentum exchange of accelerated ions incident on a target of source material [6]. At very low kinetic energies (<5 eV) the interaction is confined to the outermost surface layer of the target material. Ions, such as Ar and Xe, rather than neutral

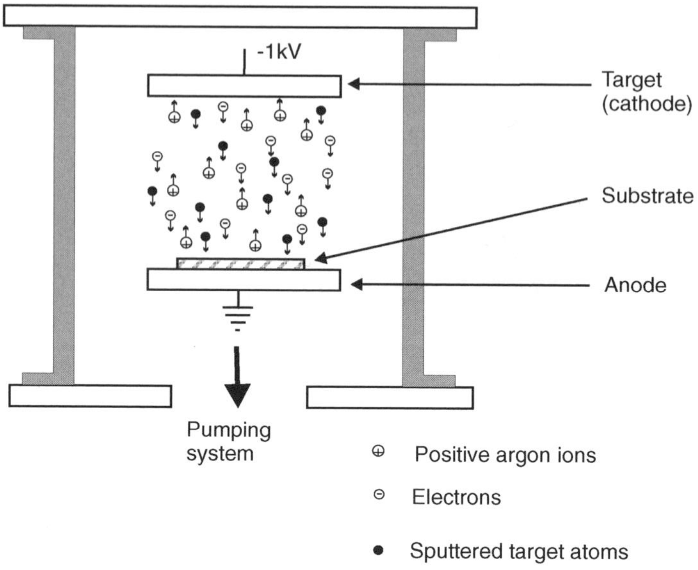

Figure 7.9 Sputtering system. A target, held at a negative potential, is bombarded with positively charged ions.

atoms are used for bombardment as these can be accelerated to any desired kinetic energy with applied electric fields.

A source of ions is provided by a glow discharge created by an electric field between two electrodes in a gas at low pressure. Figure 7.9 shows a schematic diagram of a simple sputtering system. The material to be sputtered, the target, is the cathode. The vacuum chamber is evacuated and then filled with the inert gas at a low pressure. A DC potential difference of several kilovolts is applied to the electrodes, which causes the gas to become ionized and form a plasma. The positive gas ions are then accelerated by the electric field so that they arrive at the cathode with considerable energy and sputter the target atoms. Some secondary electrons are also produced at the cathode and these accelerate towards the anode and help to maintain the plasma.

Effective sputtering is possible only when both the number of ions and their energy are large and controllable. As the gas pressure is increased, the discharge current increases and the number of ions increases (approximately proportional to pressure squared). There is, however, an upper limit to the pressure since ejected atoms suffer more collisions and are prevented from reaching the anode. Optimum pressures for glow discharge sputtering are in the range 25–75 mbar.

There are several variations on the basic sputtering arrangement described above. If the substrate is held at a negative potential, it will be subjected to steady ion bombardment which effectively 'cleans' the film of adsorbed gases otherwise trapped in as impurities. This is called bias sputtering. A magnetic field may also be applied to improve the ionization efficiency (by increasing the path length of the ionizing electrons). For insulating source materials, DC sputtering is inappropriate because of the accumulation of positive surface charges on the target; RF (radiofrequency) sputtering is therefore used.

The principal advantage of sputtering is that almost any material can be deposited. Since no heating is required, materials that are difficult to melt (and therefore to evaporate) are easily sputtered, as are compounds that would dissociate in an evaporation source. The method has been used for some organic polymers, e.g. polytetrafluoroethylene (PTFE). However, a relatively large amount of the material is needed as a target. This is not always practical for the new organic compounds, as these may only be available in very small quantities.

Sputtering takes place from the whole of the target surface, and this means a uniformly thick deposit over larger areas can be obtained, in contrast to thermal evaporation. Furthermore, the sputtering rate is proportional to the current flowing between the electrodes, hence the process is more controllable than evaporation.

7.2.3 Chemical Vapour Deposition

Chemical vapour deposition (CVD) is a process in which one or more gaseous species react on a solid surface and one or more of the reaction products is a solid phase material [6, 13]. The CVD process is based on the decomposition and/or radical generation of chemical species by stimulating vapour with heat, plasma (discharge) or light (laser). The method is used in the microelectronics industry for the fabrication of inorganic semiconducting and insulating films. Chemical reactions that are commonly exploited include pyrolysis (thermal decomposition), oxidation, reduction, disproportionation and various transfer reactions. During the pyrolysis of silane (SiH_4), the molecules strike a hot surface and decompose into Si and H_2, with the latter going back into the gas phase. The silicon left behind can build up as a solid film. Pyrolysis can be used to produce layers of insulating polymer films. For example, solid poly-*p*-xylene films have been formed by pyrolysis of *p*-xylene at 600 °C [14]. Polymers can also be formed by the high-temperature electron bombardment of monomers. This approach has been used to form polymer films of siloxane, styrene, butadiene and methacrylate [1, 14].

For plasma-enhanced CVD (PECVD), glow discharge plasmas are sustained within chambers where simultaneous CVD reactions occur [15, 16]. The discharge is normally excited by an RF field. One example is the formation of polystyrene films for use as the dielectric in a nuclear battery [1, 15]. The surface polymer may be simultaneously removed by reaction with the residual gases (ablation); the balance between ablation and polymerization may be controlled by adding gases that promote one or other effect. For example, CF_4 is an effective plasma etch material under normal discharge conditions; however, if a small amount of hydrogen is added, a polymer is deposited.

Laser or, more generally, optical chemical processing involves the use of monochromatic photons to enhance and control reactions at substrates [17]. Two reactions may be involved during laser-assisted deposition. In the pyrolytic mechanism, the laser heats the substrate to decompose gases above it and enhance rates of chemical reactions there. Pyrolytic deposition requires substrates that melt above the temperatures necessary for gas decomposition. Photolytic processes, on the other hand, involve direct dissociation of molecules by energetic photons.

Metallorganic chemical vapour deposition (MOCVD), the other common technique for device growth, is illustrated in Figure 7.10. The example shown is that for the formation of the inorganic semiconducting material gallium aluminium arsenide. As shown, different gases containing the molecular species of interest (in this case, Ga, Al and As) pass through a

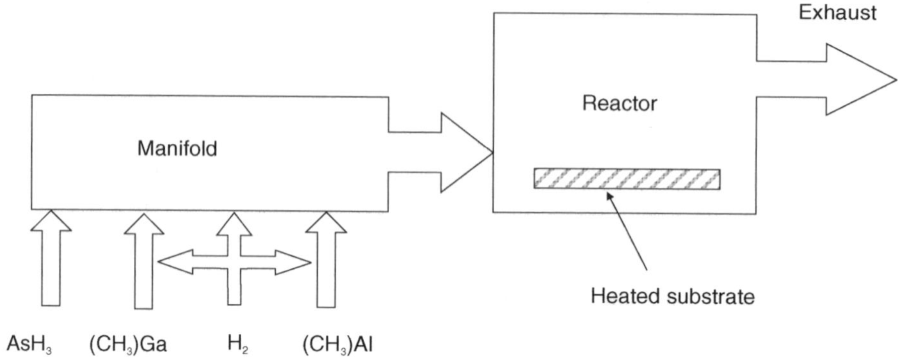

Figure 7.10 Schematic diagram for metallorganic chemical vapour deposition. The example depicts the arrangement to form thin films of the compound inorganic semiconductor alloy AlGaAs.

reactor that incorporates a heated substrate. The composition of the deposited material is controlled by the respective composition of the gases and the temperature of the substrate. The gases pass through the chamber by forced convection; these then participate in a chemical reaction, which encourages the desired growth material to precipitate on to the substrate. The following equation shows the basic reaction between the organometallic compound trimethylgallium and arsine to form gallium arsenide and methane:

$$(CH_3)_3Ga + AsH_3 \rightarrow GaAs + 3CH_4 \tag{7.5}$$

This reaction takes place at about 700 °C and epitaxial growth of high-quality GaAs layers can be obtained. The addition of a trimethylaluminium source, shown in Figure 7.10, allows the alloy AlGaAs to be grown. The flexibility of gas mixing arrangements permits the growth of multiple thin layers similar to those discussed for MBE. The gas flow rates, reactor pressure and reactor temperature are all important parameters in determining the composition and quality of material deposited using MOCVD.

7.2.4 Electrochemical Methods

Electrochemical deposition, or electroplating, has been known for at least 100 years. Approximately half of the 70 or so metals can be electrodeposited, either singly or as alloys. The equipment required consists of an anode and a cathode immersed in a suitable electrolyte. Metal is deposited onto the cathode and the relationship between the weight of the material deposited and the various parameters can be expressed by Faraday's first and second laws of electrolysis. These state:

(i) The weight of the deposit is proportional to the amount of electric charge that is passed.

(ii) The weight of material deposited by the same charge is proportional to the *equivalent weight E* (defined by the atomic weight of an element or radical divided by the valence it assumes in compounds), provided that there are no other electrolysis reactions occurring that consume a fraction of the current.

Expressed as an equation, the weight deposited m is given by

$$m = q\alpha E = q\alpha\frac{M}{z} = \frac{1}{F}\frac{qM}{z} \tag{7.6}$$

where q is the total charge passed through the electrolyte, M is the molar mass, z is the valence number of the substance as an ion in solution (electrons per ion) and α is known as the current efficiency, i.e. the fraction of the total current associated with deposition of the material concerned. In Equation (7.6), F is *Faraday's constant*, a common unit of charge in electrochemistry. This is equal to the charge of 6.02×10^{26} electrons (or 1 kmol; see Chapter 2, Section 2.2.1), i.e.

$$F = N_A e \tag{7.7}$$

Therefore, $F = 9.65 \times 10^7$ C kmol^{-1}. A charge of 1 F passing through an electrochemical cell will liberate 1 kg equivalent of a substance at an electrode. A deposition speed of about 1 nm s^{-1} for an electrode current density of 100 A m^{-2} is typical for an electrochemical process.

Anodization is a particular kind of electrolysis. In a copper electroplating bath, the copper anode goes into the electrolytic solution as a positive ion. With some metals this does not happen, but instead the surface of the anode slowly oxidizes by reaction with water in the electrolyte; the increasing resistance of the growing anodic oxide layer causes a continuous decrease in the electrolytic current. This anodization process is limited to a few metals such as aluminium and tantalum, which can be oxidized to form excellent thin film capacitors. Many polymeric films can be prepared by the anodic oxidation of suitable monomer species, such as pyrrole, thiophene and aniline [18]. The full details of electropolymerization are not fully understood. In the very first step, the neutral monomer is oxidized to a radical cation. It must therefore have an oxidation potential that is accessible via a suitable solvent–electrolyte system and should react more quickly with other monomers than with other nucleophiles in the electrolyte solution.

Electroless deposition does not require electrodes or an external energy source. Ions in the plating solution are simply converted to neutral atoms by a reducing agent in the presence of a catalyst. The process is restricted to particular chemical reactions, i.e. each depositing layer of the film is must be catalyzed by the preceding layer. In some cases, such as silver, electroless deposition begins and proceeds readily, but most surfaces have to be specifically treated ('sensitized') before a film will deposit. A general method for the electroless metallization of polyamide has been described [19]. Here, the polymer itself is used as either an oxidizing or reducing agent to deposit a metal species from solution. The equipment required for electroless deposition is very simple and both conductive and non-conductive substrates can be used. Furthermore, as there are no requirements for uniform current density, irregular objects and inaccessible surfaces can be coated.

7.2.5 Inkjet Printing

The need to combine large area coatings with device patterning has resulted in the development of direct-write fabrication methods, such as inkjet printing. This particular method is attracting the interest of many scientists and technologists and is currently a rapidly moving area of research and development, producing many research papers and patents [20, 21]. A number of different inkjet technologies have been developed. Continuous

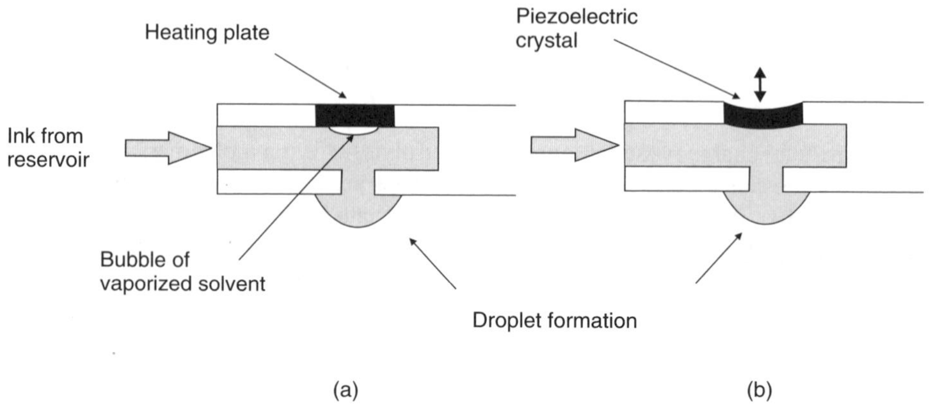

Figure 7.11 Inkjet print heads: (a) thermal (bubble-jet) operation; (b) piezoelectric operation.

inkjet deposition is based on inducing an electric charge in the liquid by ejecting a jet of ink from an orifice through a region with an external electric field. The drops can then be deflected to a substrate or to a reservoir for recycling. In the case of drop-on-demand printers, droplet ejection can be achieved with thermal (bubble-jet) and piezoelectric modes of operation, both shown schematically in Figure 7.11. The majority of published literature on inkjet printing as a tool for manufacturing organic devices has been the result of using piezoelectric actuated printers. This technology is favoured primarily because it applies no thermal load to the organic 'inks.' The combination of solution-processable emissive polymers with inkjet printing offers some promise in the development of low-cost, high-resolution displays (Chapter 9, Section 9.6). The technique has also been applied to the manufacture of all-polymer transistor circuits (Chapter 9, Section 9.4).

The sizes of nozzles used by inkjet printers are in the range 20–70 μm, giving droplet sizes of 4–180 pl. However, some printers can produce 1 pl droplets. Figure 7.12 shows a sequence of photographs taken during the formation and ejection of a droplet from a piezoelectric inkjet head with a 50 μm nozzle [22]. The formation of the satellite drop (which eventually recombines with the main droplet) is a typical phenomenon. The physical properties of the ink must be matched to the performance of a specific printer; the ink viscosity and surface tension are both important parameters. The viscosity must be low enough to allow for rapid refilling (a typical ink has a viscosity of about 2 cP) and the surface tension (typically 40 mN m^{-1}) must be sufficiently high

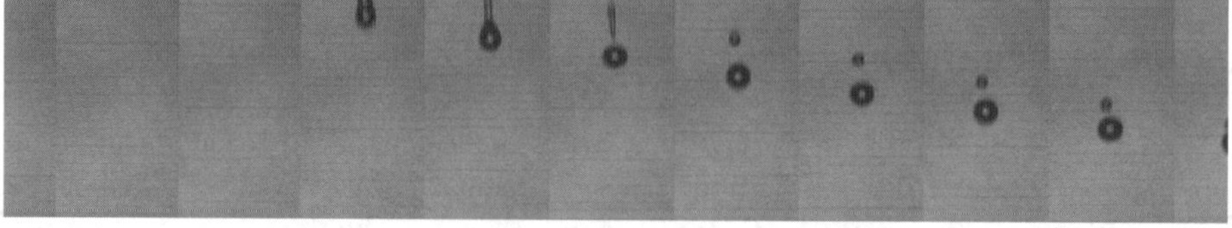

Figure 7.12 Ejection of drops from print head [22]. Nozzle size = 50 μm. Reprinted with permission from David Morris.

to hold the ink in the nozzle without dripping. The clogging of the nozzle with partly dried ink is a major problem for this method of thin film deposition. Low-volatility, water-miscible liquids such as ethylene glycol are added at 10–20% to prevent this.

When the droplet strikes the substrate, it may be adsorbed. However, this will depend on the nature of the surface. For example, overhead transparency sheets used for conventional printing are non-absorbing and have to be coated with a hydrophilic layer, or may be patterned to achieve a rough surface. 'Phase-change' inks can be used on hard surfaces [21]. These materials freeze rapidly on deposition and are cold-rolled to produce a strongly bonded layer. The rate of expulsion of the droplets from the nozzle and the relative movement of the inkjet head and substrate will determine if sequential drops coalesce on the surface. Figure 7.13 shows the effect of depositing droplets of a PEDOT:PSS (Chapter 5, Figure 5.3) on or near to other droplets [22]. When successive droplets are deposited so that part of the wet droplet overlaps with a previous, dry droplet, the wet droplet interacts with it in such a way as to cover a greater area than if there had been no overlap. The morphology of very thin films of inkjet printed conductive polymers can be exploited in chemical sensing devices (Chapter 10, Section 10.4.3) [23].

The resolution of the inkjet printing process can be improved by using various surface pattering techniques [20]. This is illustrated in Figure 7.14. The substrate surface is pretreated to form hydrophilic and hydrophobic regions. This can be accomplished with a variety of methods, including laser printing, deposition of self-assembled monolayers using a rubber stamp (see Section 7.4.3) or ultraviolet light exposure. Water-based ink droplets are then confined to the hydrophilic pattern by the surrounding hydrophobic regions.

7.2.6 Sol–Gel Processing

Sol–gel processing involves the suspension of a solid in a liquid (the *sol*), the removal of the liquid (the *gel*) and finally the densification of the solid, e.g. by sintering [24, 25]. The method is used to form highly dense films of ceramics and glasses. Thin layers can be prepared using much lower temperatures than would be needed in the 'conventional' processing of the same material.

The sol–gel technique covers a wide range of processing of colloidal and macromolecular systems. The staring point is the formation of a stable colloid (Chapter 2, Section 2.7). This can be achieved by breaking a material down from a macroscopic to a microscopic state, perhaps by mechanical means such as milling. Alternatively, chemical processes may be used to build up appropriately sized particles from smaller ones. The chemistry of the sol–gel process is largely based on an alkoxide solution route. Alkoxides are traditional organometallic precursors for silica, alumina, titania and other metal oxides. The sol–gel reaction, driven by a catalyst, starts with the hydrolysis of alkoxides in a water–alcohol mixed solution, followed by polycondensation reactions.

The crucial second stage is the conversion of the sol to a gel. In the latter state, the particles have cross-linked and formed a structure that is capable of immobilizing the remaining solvent. The sol–gel reaction is often performed in the dip-coating regime, in which the solid substrate is slowly pulled out of the solution containing the required chemicals. This reaction, known as *gelation*, takes place in a thin layer wetting the substrate.

Drying by evaporation under normal conditions gives rise to capillary pressure that causes shrinkage of the gel network. The resulting dried gel, called a *xerogel* ('xero' means dry), is

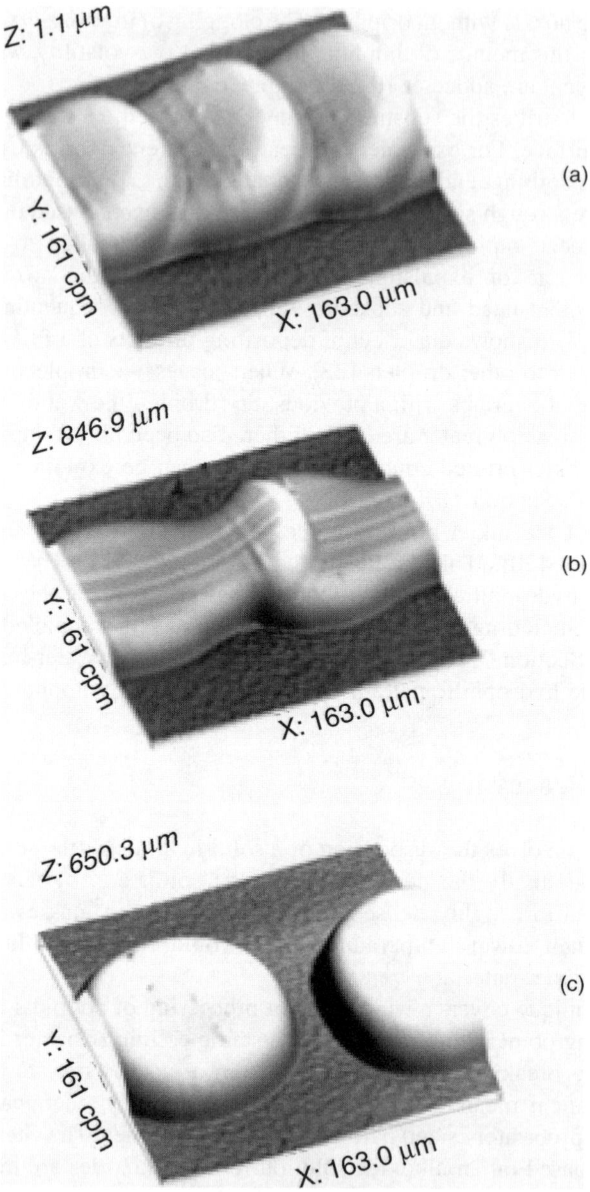

Figure 7.13 Inkjet-printed dots of PEDOT:PSS on substrate [22]. (a) 60 μm line pitch. (b) 100 μm line pitch. (c) 120 μm line pitch. Reprinted with permission from David Morris.

often reduced in volume by a factor of 5–10 compared with the original wet gel. Most gels are amorphous, even after drying, but many recrystallize when heated. A relatively dense product may be formed by heating a xerogel to high temperatures.

If the wet gel is placed in an autoclave and dried under *supercritical* conditions, there is no interface between the liquid and vapour, so there is no capillary pressure and relatively little shrinkage. The process is called *supercritical* (or *hypercritical*) *drying* and the product

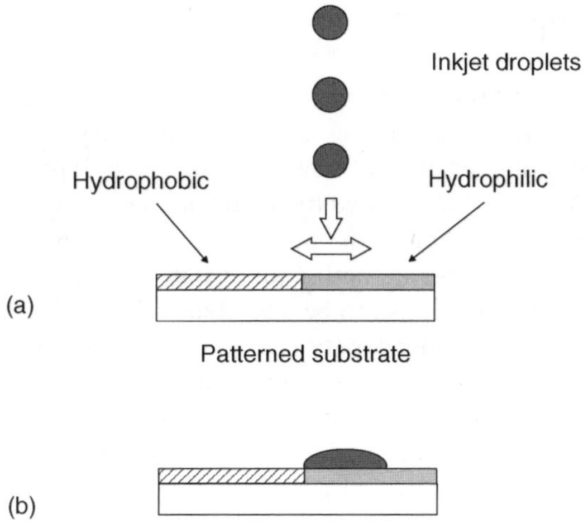

Figure 7.14 Use of surface treatments to define patterns for ink-jet printing.

is called an *aerogel.* This may be extremely porous, made mostly of air. The classification of sol–gel technology and products is given in Figure 7.15. Although used mainly to process inorganic materials, sol–gel technology can also be exploited to make thin layers of organic compounds and organic–inorganic polymer hybrids [26, 27].

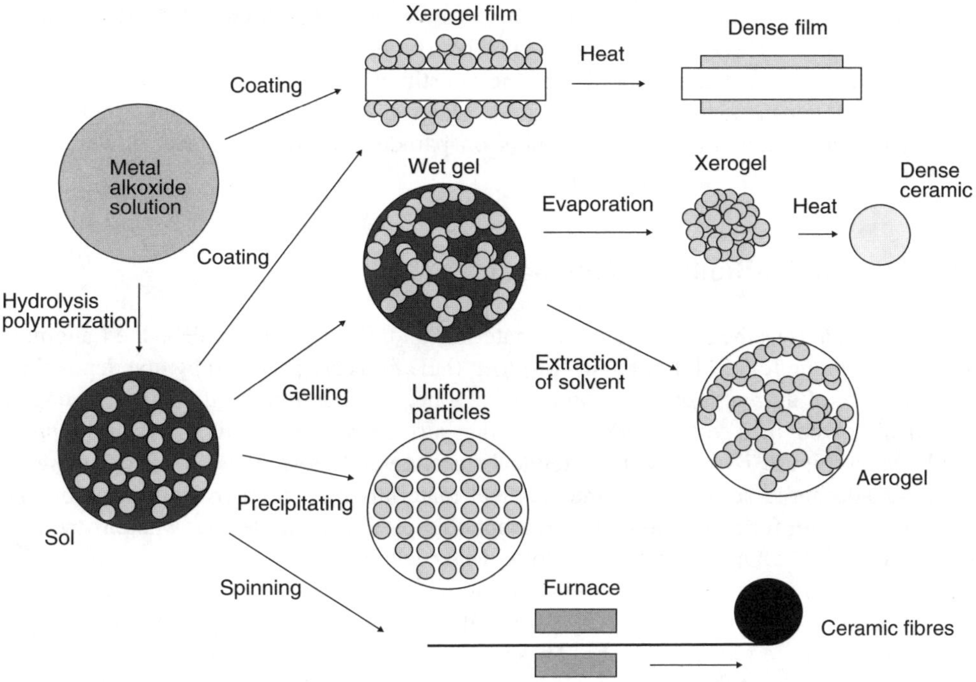

Figure 7.15 Sol–gel technologies and products. Reproduced from *Sol–Gel Science: the Physics and Chemistry of Sol–Gel Processing*, Brinker CJ, Scherer GW, p. 1 Copyright (1990), with permission from Academic Press.

7.2.7 Other Techniques

Other traditional methods of thin film formation include dip-coating, spraying, painting and screen printing. Most of these techniques are relatively easy to perform and require a minimum of equipment. Polymer films of materials such as polypropylene, polystyrene and poly(vinyl chloride) (PVC) can be obtained by the technique of direct isothermal immersion of a substrate into a suitable solution of the polymer (e.g. PVC in cyclohexanone). Material will deposit on the immersed substrate until equilibrium is reached between the deposition and re-solution rates. Satisfactory films can be also obtained by solution casting: allowing the evaporation of a polymer-containing solution placed on a substrate (e.g. polystyrene in chloroform). Both of these methods will give relatively thick films.

Screen printing (sometimes called silk screen printing) offers a further inexpensive method of preparation of films. The process consists of dispersing a paste (the ink) of material on a mesh-type screen on which a desired pattern may be defined photolithographically. The substrate is placed a short distance beneath the screen. A flexible wiper (squeegee) then moves across the screen surface, deflecting it vertically and bringing it into contact with the substrate. This forces the paste through the open mesh areas. The substrate is allowed to stand at ambient temperature for some time to enable the paste to coalesce to form a coherent film.

7.3 MOLECULAR ARCHITECTURES

The techniques outlined in the following sections are generally restricted to the deposition of organic materials. These methods can, in theory, provide monolayer and multilayer structures with a high degree of order of the constituent molecules. However, meticulous attention to experimental detail is required. Diagrams of such thin film architectures, showing the molecules aligned like soldiers on parade, are commonplace; however, they can be very deceptive.

7.3.1 Langmuir–Blodgett Technique

A method that allows the manipulation of materials on the nanometre scale is the Langmuir–Blodgett technique [28–31]. Langmuir–Blodgett (LB) films are prepared by first depositing a small quantity of an *amphiphilic* compound (i.e. one incorporating both polar and nonpolar groups) dissolved in a volatile solvent, on the surface of a carefully purified *subphase*, usually water. Figure 7.16 shows the chemical structure and molecular dimensions of stearic acid (*n*-octadecanoic acid), one of the classical materials used for LB film formation. The hydrocarbon chain forms the nonpolar, hydrophobic part of the molecule, whereas the polar carboxylic acid –COOH group confers some water solubility.

When the solvent has evaporated, the organic molecules may be carefully compressed to form a floating two-dimensional solid. During this process, the floating film will undergo a multiplicity of phase transformations. These are similar to three-dimensional gases, liquids and solids, but perhaps the closest analogy is with the mesophases shown by liquid crystals (Chapter 8, Section 8.2). Phase changes may be readily identified by monitoring the surface pressure, Π, as a function of the area occupied by the film, a. This is the two-dimensional

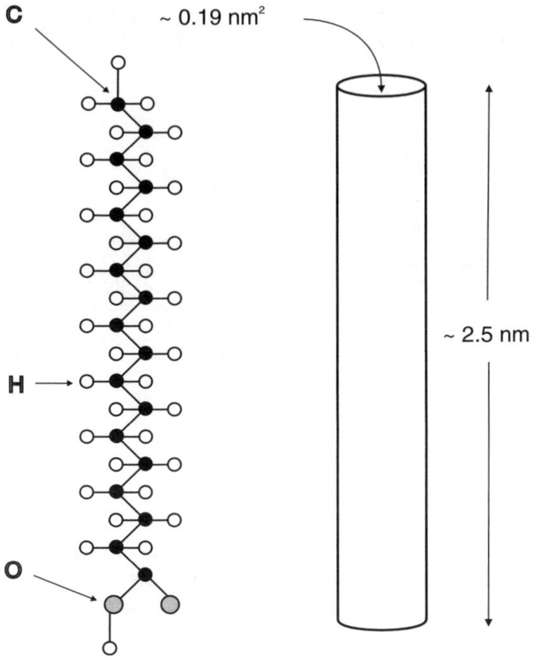

Figure 7.16 Chemical structure for *n*-octadecanoic acid (stearic acid). The approximate geometric shape and dimensions of the molecule are shown on the right.

equivalent of the pressure versus volume isotherm for a gas/liquid/solid. Figure 7.17 shows such a plot for a hypothetical long-chain organic monolayer material (e.g. a long-chain fatty acid). This diagram is not meant to represent that observed for any particular substance, but displays most of the features observed for long-chain compounds.

In the 'gaseous' state (*G* in Figure 7.17), the molecules are far enough apart on the water surface that they exert little force on one another. As the surface area of the monolayer is reduced, the hydrocarbon chains will begin to interact. The 'liquid' state that is formed is generally called the expanded monolayer phase (*E*). The hydrocarbon chains of the molecules in such a film are in a random, rather than a regular orientation, with their polar groups in contact with the subphase. As the molecular area is progressively reduced, condensed (*C*) phases may appear. There may be more than one of these and their emergence can be accompanied by constant pressure regions of the isotherm, as observed in the cases of a gas condensing to a liquid and a liquid solidifying. These regions will be associated with enthalpy changes (Chapter 2, Section 2.5.2) in the monolayer. In the condensed monolayer states, the molecules are closely packed and are oriented with the hydrocarbon chain pointing away from the water surface. The area per molecule in such a state will be similar to the cross-sectional area of the hydrocarbon chain, i.e. about $0.19\ \text{nm}^2\ \text{molecule}^{-1}$.

If the surface pressure versus area measurements are undertaken at several temperatures, and the points corresponding to the same phase transitions are plotted on a pressure versus temperature graph, the resulting diagram will show the range of temperatures and pressures

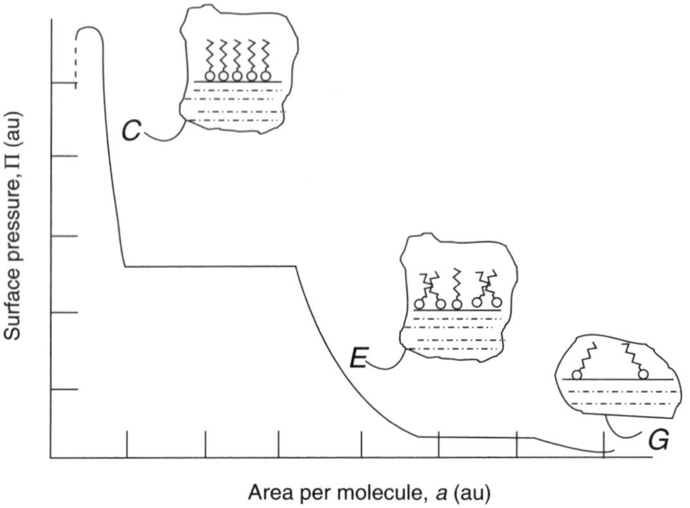

Figure 7.17 Surface pressure, Π, versus area per molecule, *a*, isotherm for a long-chain organic compound. The surface pressure and area are in arbitrary units (au). Reprinted from Petty MC, *An Introduction to Langmuir–Blodgett Films*, p. 18. Copyright (1996), Cambridge University Press.

over which the various phases exist. Such a phase diagram for *n*-docosanoic acid is given in Figure 7.18 [32]. The nomenclature of the various phases follows the method adopted by Harkins [33]. The condensed phases include L_2, L_2' (liquid condensed), *LS* (super-liquid), *S* (solid) and *CS* (close-packed solid). The use of the term 'liquid' for some of these monolayer states simply reflects the historical assumptions that the phases were liquid-like. However,

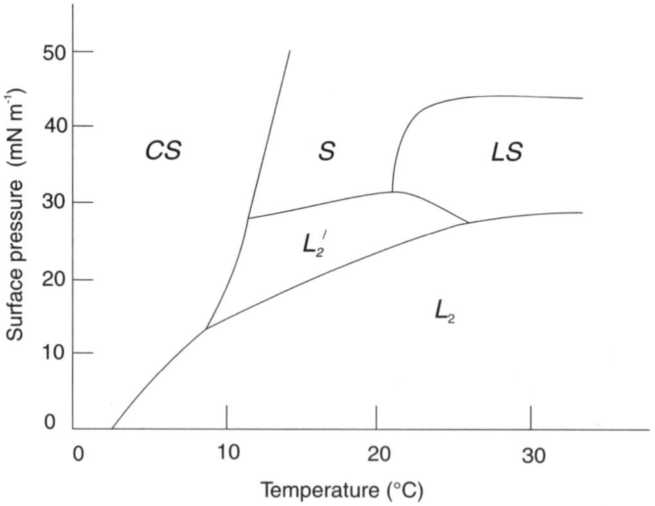

Figure 7.18 Surface pressure versus temperature phase diagram for *n*-docosanoic acid. The different phases (see text for details) are depicted as *CS*, *S*, *LS*, L_2 and L_2'. From Peterson IR, 'Langmuir–Blodgett films', in GJ Ashwell (Editor), *Molecular Electronics*. Taunton: Research Studies Press, 1992, pp. 117–206. Reproduced by permission of Research Studies Press Ltd.

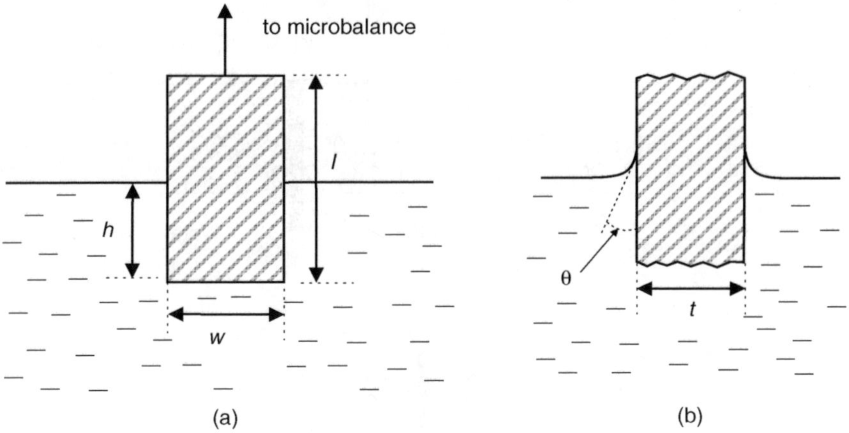

Figure 7.19 A Wilhelmy plate: (a) front view; (b) side view.

it is now known that all the condensed phases have well-defined in-plane structures and exhibit X-ray diffraction peaks (Chapter 6, Section 6.3). The monolayer characteristics of many organic materials may be found in the comprehensive handbook edited by Mingotaud *et al.* [34].

The two most common methods for monitoring the surface pressure are the Langmuir balance and the Wilhelmy plate. Both have similar sensitivities ($\sim 10^{-3}$ mN m^{-1}), but the use of the Wilhelmy plate technique is perhaps the more popular. An absolute measurement of Π is made by suspending a plate from a sensitive balance in the monolayer. Figure 7.19 shows the experimental arrangement. The forces acting on the plate are due to gravity and surface tension downwards and buoyancy, due to displayed water, upwards. For a rectangular plate of dimensions l, w and t and of material of density ρ_W immersed to a depth h in a liquid of density ρ_L, the net downward force F is given by

$$F = \rho_W glwt + 2\gamma(t + w)\cos\theta - \rho_L gtwh \tag{7.8}$$

where γ is the surface tension of the liquid, θ is the contact angle on the solid plate and g is the acceleration due to gravity. The usual procedure is to choose a plate that is completely wetted by the liquid (i.e. $\theta = 0$) and measure the change in F for a stationary plate. This change in force ΔF is then related to the change in surface tension $\Delta\gamma$ by

$$\Delta\gamma = \frac{\Delta F(t + w)}{2} \tag{7.9}$$

If the plate is thin enough so that $t \ll w$:

$$\Delta\gamma = \frac{\Delta F}{2w} \tag{7.10}$$

The Langmuir balance is a differential technique. A clean water surface is separated from the monolayer-covered area by a partition (usually a movable PTFE float connected to a conventional balance) and the force acting on this partition is measured [29].

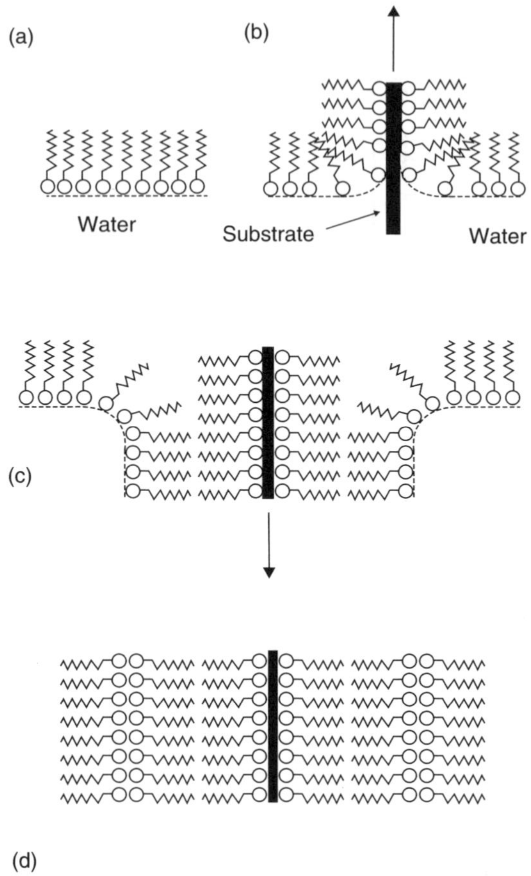

Figure 7.20 Y-type Langmuir–Blodgett film deposition. Film transfer on both the upward and downward movements of the substrate.

Figure 7.20 shows the commonest form of LB film deposition. The substrate is hydrophilic and the first monolayer is transferred, like a carpet, as the substrate is raised through the water. The substrate may therefore be placed in the subphase before the monolayer is spread. Subsequently, a monolayer is deposited on each traversal of the monolayer/air interface. As shown, these stack in a head-to-head and tail-to-tail pattern; this deposition mode is called *Y-type.* Although this is the most frequently encountered situation, instances in which the floating monolayer is only transferred to the substrate as it is being inserted into the subphase, or only as it is being removed, are observed. These deposition modes are called *X-type* (monolayer transfer on the downstroke only) and *Z-type* (transfer on the upstroke only) and are illustrated in Figure 7.21. Mixed deposition modes are sometimes encountered and, for some materials, the deposition type can change as the LB film is built up.

LB films of long-chain fatty acids are often prepared by adding divalent cations to the subphase to improve the deposition characteristics of the monolayer material. The floating layer will be a mixture of the fatty acid and the fatty acid salt. The salt concentration in the

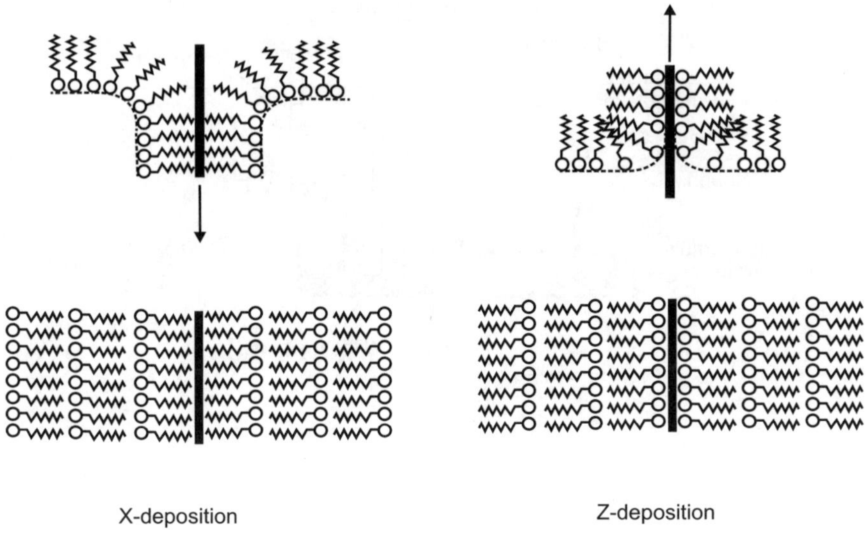

Figure 7.21 X-type and Z-type Langmuir–Blodgett film deposition.

monolayer will depend on the pH. In a subphase containing, say, cadmium chloride, the following reaction will take place:

$$2C_nH_{2n+1}COOH + CdCl_2 \rightleftharpoons (C_nH_{2n+1}CO_2)_2^- Cd^{2+} + 2HCl \tag{7.11}$$

The salt formation is favoured by a high subphase pH. In the specific case of a floating monolayer of *n*-eicosanoic acid (arachidic acid) on a water subphase at room at room temperature containing a cadmium salt at a concentration of 10^{-4} M and having a pH of 5.7, a monolayer comprising about 50% cadmium eicosanoate and 50% eicosanoic acid will be formed [29].

Film transfer is characterized by measurement of the *deposition ratio*, τ (also called the *transfer ratio*). This is the decrease in the area occupied by the monolayer (held at constant pressure) on the water surface divided by the coated area of the solid substrate, i.e.

$$\tau = \frac{A_L}{A_S} \tag{7.12}$$

where A_L is the decrease in the area occupied by the monolayer on the water surface and A_S is the coated area of the solid substrate. If asymmetric substrates are used (e.g. a glass slide metallized on one surface), then it is unlikely that τ will be identical for both surfaces. Transfer ratios significantly outside the range 0.95–1.05 suggest poor film homogeneity. The accurate measurement of τ for alternate-layer films (see later in this section) can present a particular problem as, in most alternate-layer deposition systems, the substrate holder is passed through the floating monolayer [28].

Many different approaches have been made to the design of equipment for the deposition of LB films. Figure 7.22 shows a schematic diagram of a simple experimental arrangement. The subphase container – the *Langmuir trough* – is made from PTFE and a working area is defined

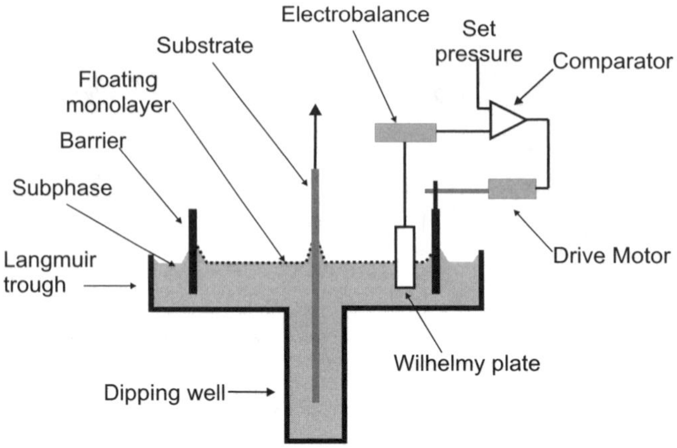

Figure 7.22 Schematic diagram of equipment for Langmuir–Blodgett film deposition.

by a PTFE-coated glass fibre barrier, which can be moved using a low-geared electric motor. The barrier motor is coupled to a sensitive electronic balance which continuously monitors, using a Wilhelmy plate, the surface pressure of the monolayer. Using a feedback arrangement, this pressure can be maintained at a predetermined value. The physical dimensions of the Langmuir trough arrangement are not critical (systems with dimensions ranging from centimetres to metres have been demonstrated) and are governed by the size of the substrate used.

The diagrams shown in Figures 7.20 and 7.21 are simple cartoons and may not represent accurately the real arrangements of molecules on a solid surface. For LB films of *fatty acid salts*, X-ray diffraction experiments show that the long axes of the molecules in the LB film are indeed orthogonal to the substrate plane, as shown in most 'molecular stick' drawings [28]. The molecular arrangement in the deposited film is similar to that of the *LS*, *S* and *CS* condensed monolayer phases on the water surface. However, the long molecular axes in many simple *fatty acid* LB layers are inclined at an angle to the substrate normal, the tilt angle depending on the precise deposition conditions. For example, the molecular tilt elevation of fatty acid films deposited from the L_2' phase can vary with the deposition pressure. This tilt elevation may also change from layer to layer.

During LB film deposition, the first monomolecular layer will be transferred on to a solid substrate of a different material. This is an example of *heterogeneous crystal growth* [35]. For subsequent monolayers, transferring onto an existing film, the deposition will be *homogeneous.* It is likely that the chemical and physical structure of the first monolayer will be different from that of subsequent layers. There is now much evidence to support this. For example, infrared investigations of fatty acid films show that the first monolayer possesses a hexagonal packing of the CH_2 subcells in the alkyl chain, with the chain axis oriented perpendicular to the substrate surface [28]. As the film thickness increases, a transition occurs to a structure in which the subcells have an orthorhombic packing and the main cells are packed in a monoclinic crystallographic form, with the axes of the alkyl chains inclined at an angle of 20–30° to the substrate normal.

For fatty acid salt materials deposited on metallic substrates, there may be an ion exchange between the fatty acid salt and the thin oxide layer on the substrate surface [28]; for example, when a calcium stearate film is transferred on to an aluminium plate, a layer of aluminium stearate is formed. Consequently, a strong chemical bond can anchor the polar head of the first LB layer to the substrate surface. Under certain conditions, epitaxial deposition is observed for LB films. For fatty acids, evidence from electron diffraction shows that each monolayer has the same local orientation of its molecular lattice as that of the underlying monolayer; however, this does not mean that translational order also extends from layer to layer [36]. LB films of fatty acids generally consist of domains with in-plane dimensions ranging from several hundred micrometres to a few tenths of a millimetre. If the long axes of the molecules are tilted with respect to the substrate normal, then the films will be birefringent (Chapter 4, Section 4.3.1) and the domain structure can be seen by observation under a polarizing microscope. Figure 7.23 shows such an image of a 170-layer LB film of 22-tricosenoic acid deposited on an etched silicon substrate [37]. Most of the area depicted in the photograph was coated initially with a bilayer of *n*-octadecanoic (stearic) acid. However, the region in the top left corner of the photograph was not coated, providing clear evidence that the 22-tricosenoic acid had developed the domain features of the underlying *n*-octadecanoic acid bilayer.

The reproducibility of the LB deposition process is monitored by measuring a suitable physical characteristic of the organic film as a function of time. Measurements of the film thickness, its optical density or the frequency shift of a quartz crystal oscillator are all straightforward. A further useful method, particularly if the multilayer structure is to be used for some electrical application, is to sandwich the multilayer film between two metal electrodes and to monitor the capacitance of this structure as a function of the number of dipping cycles. Since capacitance varies as the reciprocal of thickness, a plot of reciprocal capacitance of the

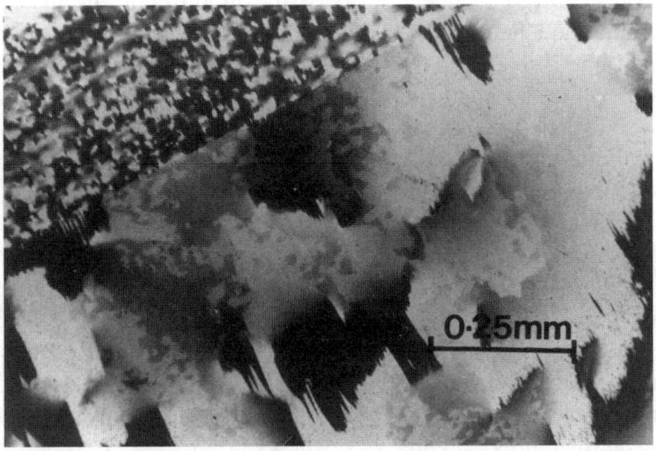

Figure 7.23 Photomicrograph of 22-tricosenoic acid (170 bilayers) deposited on single crystal silicon. The substrate, apart from that shown in the top left-hand corner, was initially covered with a bilayer of *n*-octadecanoic acid. From *B. Polym. J.*, **17**, Peterson IR, Russell GJ, 'Deposition mechanisms in Langmuir–Blodgett Films', pp. 364–367, 1985. Copyright Society of Chemical Industry. Reproduced with permission. Permission is granted by John Wiley & Sons, Ltd, on behalf of the SCI.

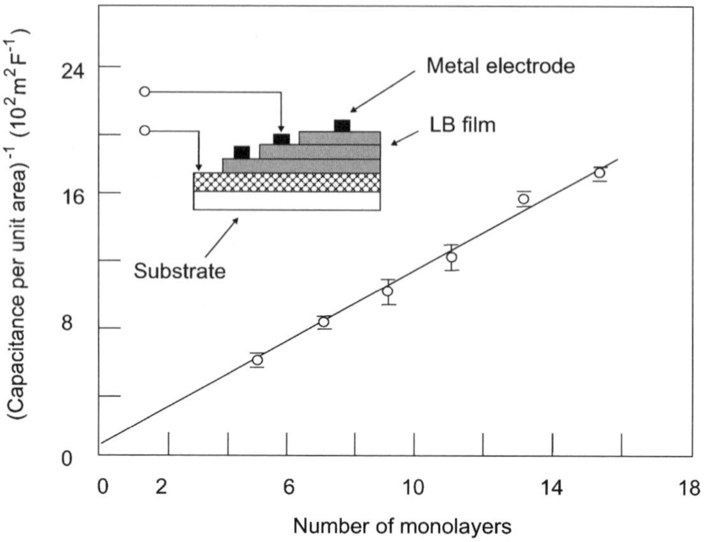

Figure 7.24 Reciprocal capacitance per unit area versus number of monolayers for Au/cadmium stearate/Al structures. The inset shows the structure of the metal/LB film/metal capacitors [38]. Reprinted with permission from John Batey.

metal/LB film/metal structure versus the number of dipping cycles should be a straight line. Figure 7.24 shows such a plot for cadmium stearate layers sandwiched between evaporated aluminium electrodes [38]. The linear form of the graph confirms the reproducibility of monolayer capacitance and therefore of the film deposition for one monolayer to the next. The slope of the straight line is related to the permittivity (Chapter 3, Section 3.6.1) and thickness of each monolayer, while the intercept on the ordinate yields similar information about the surface (e.g. oxide) layers on the metal electrodes.

It is possible to build up films containing more than one type of monomolecular layer. In the simplest case, condensed floating monolayers of two different amphiphilic materials are confined (using mechanical barriers) to different regions of the water surface. By lowering the solid substrate through the first layer of, say material *A*, and raising it up through the other, material *B*, alternate layers of structure *ABABAB*... may be built up, as shown in Figure 7.25. This permits the fabrication of organic superlattices with precisely defined symmetry properties. Such molecular assemblies can exhibit pyroelectric, piezoelectric and nonlinear optical phenomena (Chapter 11, Section 11.3).

The LB technique may also be combined with solid-state chemistry methods to produce novel molecular architectures. Figure 7.26 shows a method to produce a network of conductive polypyrrole in a fatty acid matrix [39]. First, monolayers of the iron salt of a long-chain fatty acid (e.g. ferric palmitate) are assembled on an appropriate substrate. The multilayer film is then exposed to saturated HCl vapour at room temperature for several minutes. During this process, a chemical reaction transforms the fatty acid salt into layers of ferric chloride separated by layers of fatty acid. In the third and final step, the film is exposed to pyrrole vapour and a reaction occurs between the pyrrole and the ferric chloride producing polypyrrole within the multilayer assembly.

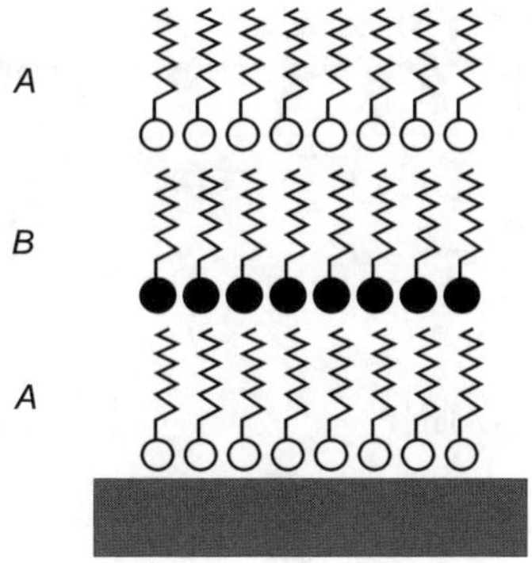

Figure 7.25 Schematic diagram of an alternate-layer Langmuir–Blodgett film built up from monolayers of compound *A* and monolayers of compound *B*.

As described above, long-chain fatty acid materials are often deposited as a mixture of a fatty acid and a salt. Following deposition, the free acid may be removed from the film by soaking the LB layers in a suitable solvent. This *skeletonization* process reduces the refractive index of the multilayer structure, making it suitable for use as an antireflection coating for glass.

It should be noted that the LB process is very different to many other thin film techniques. During film deposition, a monomolecular layer will be transferred, like a carpet, from the surface of the subphase to the substrate. The resulting LB film is therefore continuous at one monolayer coverage, in contrast to vacuum evaporation (Figure 7.5).

Many biological molecules form condensed monolayers on a water surface. Phospholipids, chlorophyll a, the green pigment in higher plants, and vitamins A, E and K are all examples. Biochemists and biophysicists have also been long aware that monomolecular films bear a close resemblance to naturally occurring biological membranes (Chapter 12, Section 12.6) and many revealing experiments may be undertaken with floating and transferred layers of biological compounds. The structurally similar proteins avidin (a toxic protein found in egg white) and streptavidin and have been the model system for many protein binding studies. Each avidin molecule tightly binds biotin, a member of the

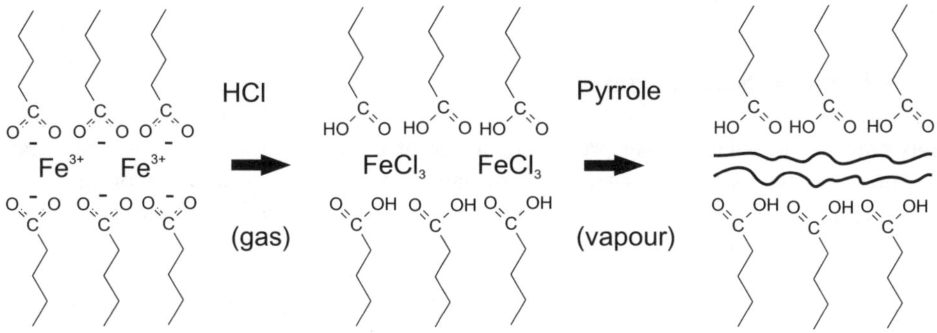

Figure 7.26 Idealized solid-state reactions of ferric stearate multilayers with HCl and pyrrole vapour to form a conductive polymer film within the fatty acid matrix [39].

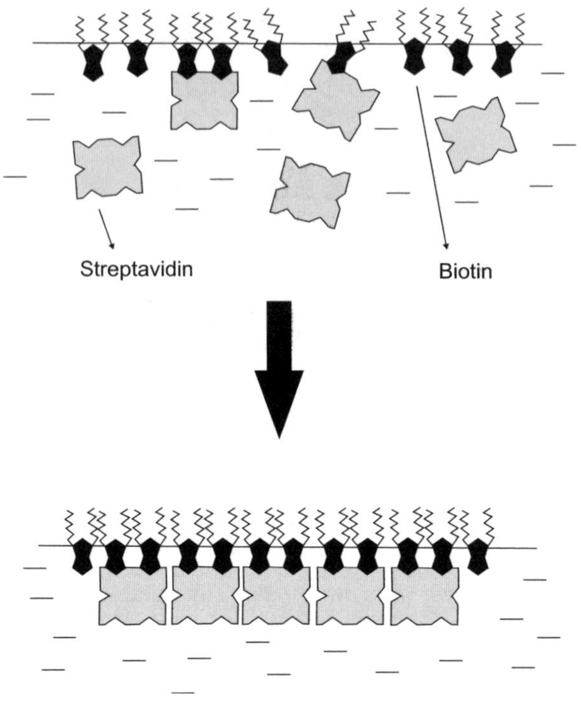

Figure 7.27 Illustration of the specific binding of streptavidin (in solution) to a biotin-derivatized lipid at the air/water interface.

B-vitamin family (found in egg yolk), at four symmetrically located sites. Streptavidin in an aqueous subphase will bind to a biotin lipid at the air/water interface, as shown in Figure 7.27. The resulting complex forms two-dimensional crystalline domains.

The vertical dipping LB process is not the only way to transfer a floating molecular film to a solid substrate or to build up multilayer films. Other methods are based on touching one edge of a hydrophilic substrate with the monolayer-covered subphase or lowering the substrate horizontally so that it contacts the ends of the floating molecules, the *Langmuir–Schaefer* technique [28]. The latter approach is useful for the transfer of highly rigid monolayers to solid supports.

7.3.2 Chemical Self-Assembly

Self-assembly is a much simpler process than that of LB deposition. Monomolecular layers are formed by the immersion of an appropriate substrate into a solution of the organic material (Figure 7.28) [31]. The best known examples of self-assembled systems are organosilicon compounds on hydroxylated surfaces (SiO_2, Al_2O_3, glass, etc.) and alkanethiols on gold, silver and copper. However, other combinations include dialkyl sulfides or disulfides on gold, alcohols and amines on platinum and carboxylic acids on aluminium oxide and silver.

The self-assembly process is determined predominantly by strong interactions between the head group of the self-assembling molecule and the substrate, resulting in a chemical

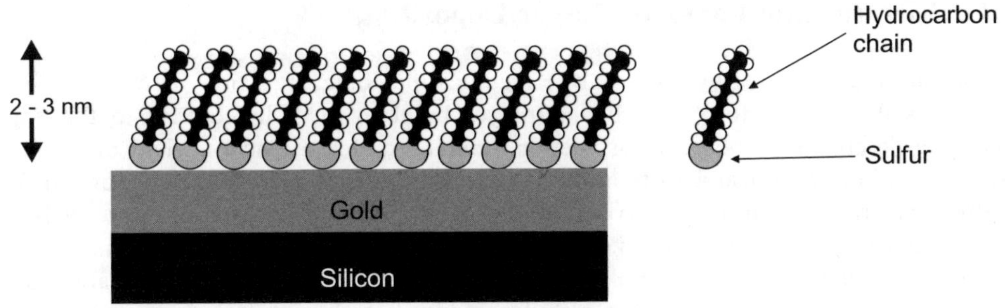

Figure 7.28 Self-assembled monolayer film of an alkanethiol on an Au-coated substrate.

bond between the head group and a specific surface site. This can be a covalent Si–O bond for alkyltrichlorosilanes on hydroxylated surfaces, a covalent, but slightly polar, Au–S bond in the case of alkanethiols on gold or an ionic $-CO_2^- Ag^+$ bond for carboxylic acids on AgO/Ag. Secondary considerations are the interactions between the alkyl chains and those between the terminal functionalities, a methyl (CH_3) group in the case of a simple alkyl chain.

Chemical means can be used to build up multilayer organic films. A method pioneered by Sagiv's group is based on the successive absorption and reaction of appropriate molecules [40]. As shown in Figure 7.29, the head groups react with the substrate to give a permanent chemical attachment and each subsequent layer is chemically bonded to the one before in a very similar way to that used in systems for supported synthesis of proteins.

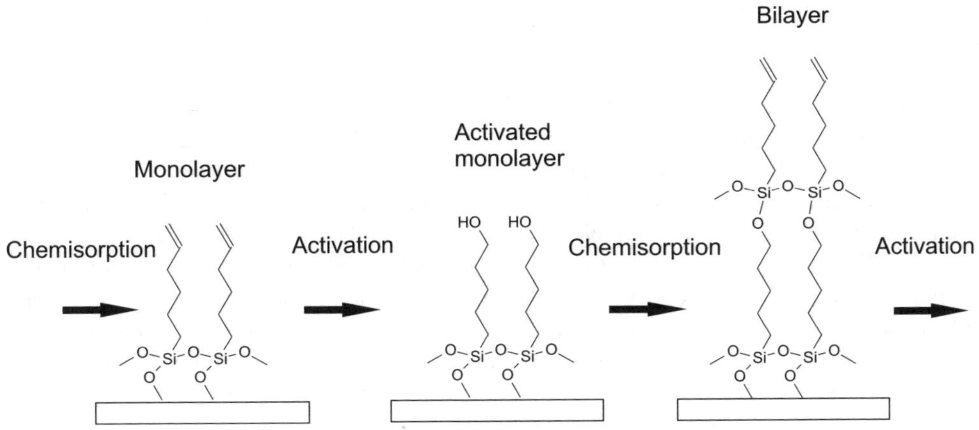

Figure 7.29 Preparation of a chemically attached polymeric multilayer. Reprinted from *Thin Solid Films*, **99**, Netzer L, Iscovici R, Sagiv J, 'Adsorbed monolayers versus Langmuir–Blodgett monolayers–why and how, I: from monolayer to multilayer, by adsorption', pp. 235–241, Copyright (1983), with permission from Elsevier.

7.3.3 Electrostatic Layer-by-Layer Deposition

In the electrostatic layer-by-layer (LbL) method, the ionic attraction between opposite charges is the driving force for the multilayer build-up [41]. In contrast to the Sagiv technique described in the previous section, which requires a reaction yield of 100% to maintain surface functional density in each layer, no covalent bonds need to be formed. The starting materials are usually *polyelectrolytes* (PEs) [42]. This term refers to polymer systems consisting of a macroion, i.e. a macromolecule carrying covalently bound anionic or cationic groups, and a low molecular weight counter ion for electroneutrality. Such materials possess a wide range of molecular and supramolecular structures and a strong dependence of their properties, in solution or as a dispersion, on the surrounding medium. The interest in polyelectrolytes stems mainly from their ability to form organized structures in, or from, solution.

The PE family can be divided into three main subgroups, shown in Figure 7.30. The first group includes all the strong polyelectrolytes. Such polyions possess a well-defined and constant charge in solution, e.g. sodium poly(styrene sulfonate) (PSS) [Figure 7.30(a)] or poly(diallyldimethylammoium chloride) (poly-DADMAC) [Figure 7.30(b)]. These materials are dissociated into macroions and counterions in aqueous solutions over the entire pH range 0 to 14. All weak polyelectrolytes belong to the second class. These charged polymers are able to form a polyion–counter ion system in solution only over a limited range of pH (being dissociated outside this range), e.g. polyethyleneimine (PEI) [Figure 7.30(c)] or poly(ethylene-*co*-maleic acid) (PMAE) [Figure 7.30(d)]. Finally, polyampholytes carry both anionic and cationic groups that are activated in alkaline or acidic media, respectively, e.g. proteins or copolymers [Figure 7.30(e)]. Therefore, by varying the pH of the solution in which they are immersed, it is possible to reverse the sign of their electrostatic charge.

One of the most interesting properties is the ability of PEs to dissolve in water (aqueous media are generally the most common environments) even if, as for polystyrene, they possess a hydrophobic backbone. The physical chemistry and processing characteristics of PEs are dependent on their behaviour in a solution or dispersion. These materials combine properties derived from long-chain molecules with those that result from charge interactions.

The electrostatic LbL deposition equipment simply consists of the process steps shown in Figure 7.31. A suitably charged substrate is immersed sequentially in the polyanion and polycation solutions; an intermediate washing step is included to remove excess polyelectroyte. A schematic molecular illustration of the processing sequence is shown in Figure 7.32. The solid substrate with a positively charged planar surface is placed in the solution containing the anionic polyelectrolyte and a monolayer of the polyanion is adsorbed [Figure 7.32(a)]. Since the adsorption is carried out at relatively high concentrations of the polyelectrolyte, most ionic groups remain exposed to the interface with the solution and the surface charge is reversed. After rinsing in pure water, the substrate is immersed in the solution containing the cationic polyelectrolyte. Again, a monolayer is adsorbed but now the original surface charge is restored [Figure 7.32(b)], resulting in the formation of multilayer assemblies of both polymers [Figure 7.32(c)].

The range of substrates that can be used as templates for the multilayer self-assembly is wide, with little restriction on their type or shape. The only feature required by the surface of the substrate is for it to possess an electric charge. This can be achieved in different ways. For example, it is possible to use freshly cleaved mica, a glass or quartz slide or a silicon

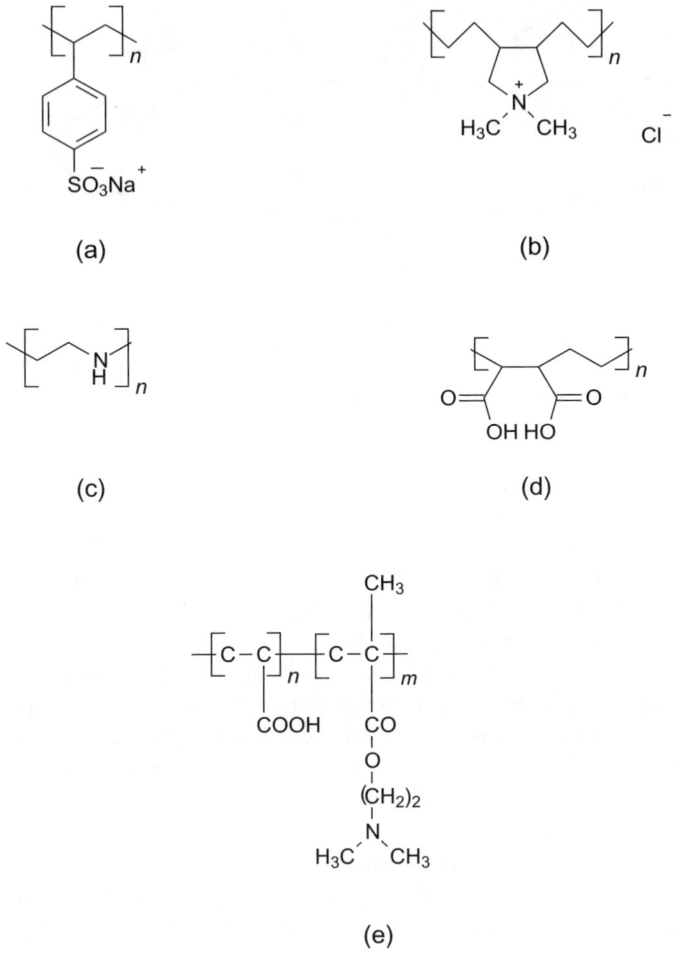

Figure 7.30 Types of polyelectrolyte: (a) sodium poly(styrene sulfonate) (PSS); (b) poly(diallyldimethylammoium chloride) (poly-DADMAC); (c) polyethylenimine (PEI); (d) poly(ethylene-*co*-maleic acid) (PMAE); (e) a copolymer formed from acrylic acid and dimethylaminoethyl methacrylate. (a), (b) are examples of strong polyelectrolytes; (c), (d) are weak polyelectrolytes; and (e) is a polyampholyte.

wafer covered by a layer of cationic polyethylenimine. Good-quality surfaces are obtainable by aminosilanization procedures, by plasma treatment and by sonification in an appropriate solution or by the deposition of a charged amphiphile monolayer using the LB method. Freshly prepared metal substrates, usually slightly negatively charged, can also be used. After a few polyelectrolyte adsorption cycles, the top surface will develop a strong electric charge. The choice of substrate is not limited to planar surfaces; microcapsules, colloids, tubules and biological cells are all common templates.

The advantages of LbL deposition over other thin film deposition methods include: (a) a high versatility (being applicable to almost every solvent accessible surface); (b) a wide range of substances may be deposited (not just polymers, but colloids, proteins, DNA and inorganic compounds [41]); and (c) tailoring of surface properties and therefore of the

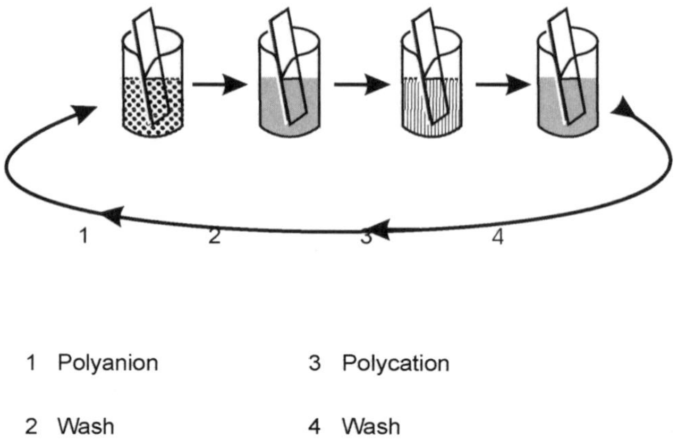

Figure 7.31 Layer-by-layer electrostatic deposition in a series of beakers.

interaction between the assembled object and its environment (e.g. surfaces specially designed for biocompatibility, corrosion protection, anti-static coatings, sensing, surfaces with specific adhesion or wetting properties, improved surface conductivity or ion transport).

Many studies reveal that the LbL assembly process exhibits a nonlinear film growth in its early stages. During the first several minutes of deposition, the charged macromolecules tend to adsorb preferentially on selected defect sites of the oppositely charged support (scratches, micro-particles and edges) and form islands composed of polymer coils, as shown by the atomic force microscopy image in Figure 7.33 [43]. As the number of deposition steps is increased, the islands coalesce until complete coverage of the underlying support is achieved. In this respect, LbL deposition shows some similarities to the nucleation and growth of thermally evaporated thin films (Figure 7.5).

Weak and strong polyelectrolytes can generate LbL thin films with different surface morphologies. Such behaviour can be attributed to their different charge densities in

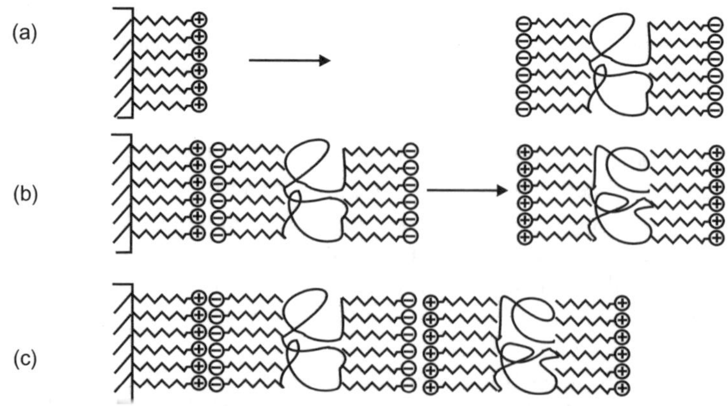

Figure 7.32 Build-up of multilayer assemblies by consecutive adsorption of anionic and cationic polyelectrolytes [41].

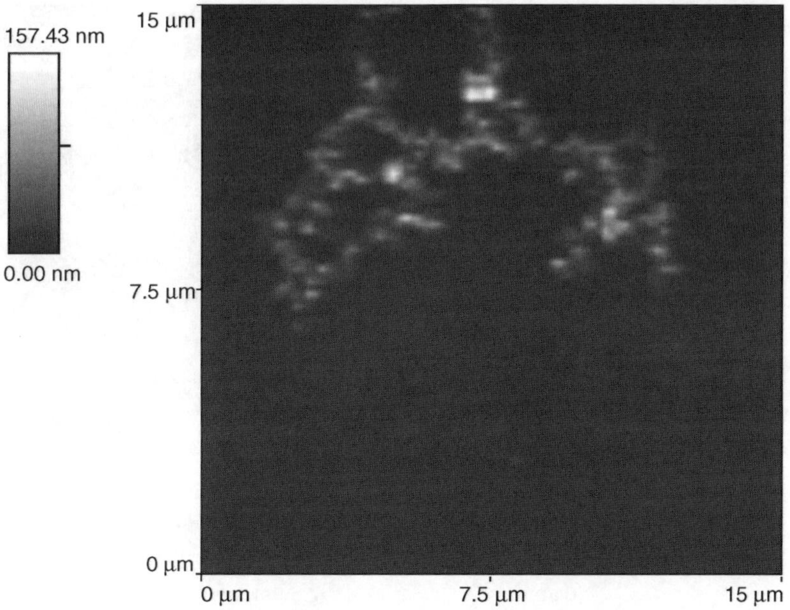

Figure 7.33 Atomic microscope scan in air using contact mode of PEI/PSS LbL film where the electrostatic self-assembly has only been partial. A group of polymer coils is visible in the upper half of the scan. Reprinted from *Thin Solid Films*, **483**, Palumbo M, Pearson C, Petty MC, 'Atomic force microscope characterization of poly(ethyleneimine)/poly(ethylene-*co*-maleic acid) and poly(ethyleneimine)/poly(styrene sulfonate) multilayers', pp. 114–121, Copyright (2005), with permission from Elsevier.

solution. For example, polymeric chains can assume the form of a coil (strong polyelectrolytes) or a globule (weak polyelectrolytes at low charge density). Consequently, during the process of adsorption, the former would produce a more uniform surface whereas the latter would generate a rougher and more porous surface. Atomic force microscope images of two different LbL architectures are shown in Figure 7.34 [43]. The image in Figure 7.34(a) is for three bilayers of PEI/PMAE whereas that in Figure 7.34(b) depicts the morphology of a three bilayer film of PEI/PSS (the chemical structures have been given in Figure 7.29). The much smoother surface for the latter structure is probably related to the strong polyelectrolyte character of the PSS. In solution, the chains of this polymer will be rigid. In contrast, PMAE is a relatively weak polyacid and, in solution, its chains will be influenced by lower repulsion forces along their length. The molecular conformation in solution will more globular as the chains will not be a rigid as those of the PSS.

A related, but alternative, approach to the electrostatic LbL assembly outlined above uses LbL adsorption driven by hydrogen-bonding interactions [44]. This has been accomplished with polyvinylpyrrolidone, poly(vinyl alcohol), polyacrylamide and poly(ethylene oxide). In the case of polyaniline, comparisons with films assembled via the electrostatic mechanism, using sulfonated polystyrene, indicate that the nonionic polymers adsorb on polyaniline with a greater density of loops and tails and form highly interpenetrated bilayers with a high polyaniline content.

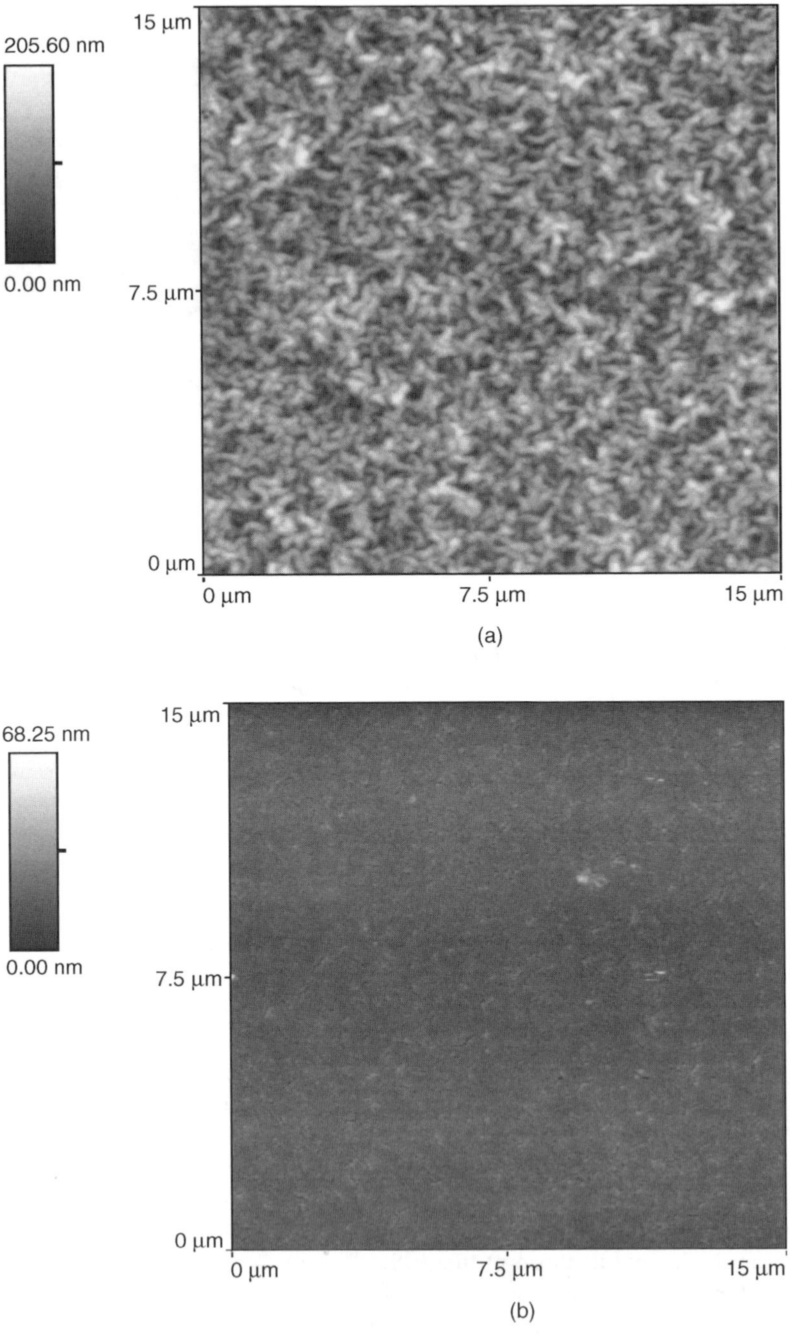

Figure 7.34 Atomic force microscopy scans in air using contact mode of LbL films. (a) Three bilayers of PEI/PMAE; (b) three bilayers of PEI/PSS. Reprinted from *Thin Solid Films*, **483**, Palumbo M, Pearson C, Petty MC, 'Atomic force microscope characterization of poly(ethyleneimine)/poly(-ethylene-*co*-maleic acid) and poly(ethyleneimine)/poly(styrene sulfonate) multilayers', pp. 114–121, Copyright (2005), with permission from Elsevier.

7.4 NANOFABRICATION

In this section, some methods that may prove useful for defining nanometre-scale patterns in organic thin films are described. First, the well-established principles of photolithography are outlined.

7.4.1 Photolithography

Planar microelectronic components are patterned using photolithography. A surface is covered with a light-sensitive photoresist and then exposed to ultraviolet light through a mask. Either the exposed photoresist (positive resist) or the unexposed regions (negative resist) can subsequently be washed away to leave positive or negative image of the mask on the surface. Positive resists are typically workable to finer feature sizes and are more generally used. Figure 7.35 illustrates the processing steps that are needed [from (a) to (h)] to define a pattern on a silicon substrate using a positive resist. A thin oxide layer is first grown on the silicon in a furnace [Figure 7.35(b)]. Subsequently, the oxide is coated with the resist

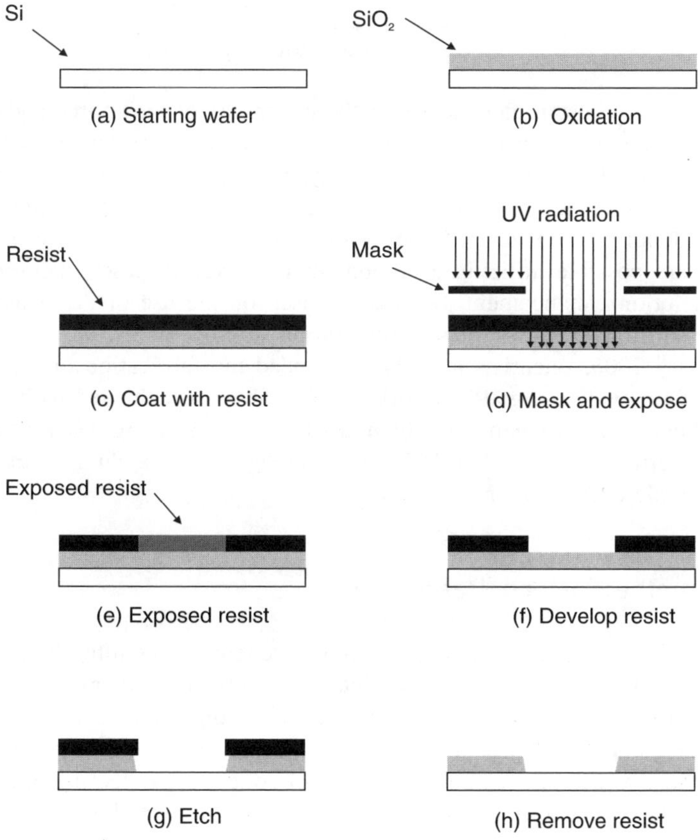

Figure 7.35 Schematic diagram of the process of photolithography.

layer (for example, by spin-coating) and the resist is exposed through a contact mask to UV radiation [Figure 7.35 (d)]. After removing the exposed areas of photoresist, the silicon oxide defined by the patterned mask is etched away [Figure 7.35(g)]. The resist is dissolved in a solvent, and the resulting pattern in the silicon dioxide [Figure 7.35(h)] can be used as a window to diffuse or implant dopant atoms in the silicon in the predefined regions.

The lateral resolution of diffraction-limited optics (Chapter 6, Section 6.2.1) can be expressed by

$$\text{minimum feature size} = \frac{k_1 \times \text{wavelength}}{\text{numerical aperture}} \tag{7.13}$$

where

$$\text{numerical aperture} = \frac{1}{f\text{number}} = \frac{\text{aperture diameter}}{\text{focal length}}$$

and k_1 is a system-dependent constant in the range 0.5–0.8.

A reasonable depth of focus is also needed to expose wafers with a particular topography with height variations patterned in previous lithography steps. This is given by

$$\text{depth of focus} = \frac{k_2 \times \text{wavelength}}{(\text{numerical aperture})^2} \tag{7.14}$$

where k_2 is a further constant depending on the coherence of the source and on the resist.

To define ever smaller patterns, radiation of shorter wavelength can be used. The mercury light source (365 nm line) used extensively for photolithography is being replaced by a krypton fluoride laser radiating at 248 nm (deep ultraviolet). This improves both the resolution and depth of focus and enables features down to 250 nm to be defined. Further improvements are possible utilizing an argon fluoride excimer laser operating at 193 nm. Additional resolution enhancement techniques based on the use of the phase of the light, such as phase-shifting masks or mask illumination improvements, are increasingly being used. In February 2006, scientists at IBM announced the fabrication of uniformly spaced ridges only 29.9 nm wide using 193 nm lithography. This was achieved with an interference immersion lithography arrangement which uses two intersecting laser beams to create interference patterns. The use of a 157 nm wavelength from a fluorine excimer laser is expected to provide even smaller features.

7.4.2 Nanometre Pattern Definition

In order to extend the ongoing technologies to wavelengths in the ultraviolet and soft X-ray range (collectively known as *extreme ultraviolet*, XUV) below 100 nm, some effort is being directed at the use of reduction optics. This uses reflecting optics and masks with special multilayer coatings. The preferred wavelength is 13 nm, just below the absorption edge of silicon. The reflective coatings consist of a stack of coherently reflecting bilayer systems of Si and Mo. The XUV radiation requires optical elements fabricated with a surface accuracy and roughness in the subnanometre range. This specification is beyond values achieved so far, yet is being rapidly approached.

X-ray lithography involves the use of even shorter wavelengths of about 1 nm. For this wavelength, no suitable optics are available and the imaging scheme reduces to a one-to-one shadow printing process. The source of X-rays is a *synchrotron* (a very expensive capital investment). One synchrotron can have several exposure beam lines. Structures well below 100 nm can be reproduced (limited by the finite gap between the mask and the wafer). However, X-ray mask fabrication is a nontrivial task and has not yet been successfully commercialized.

Electron-beam lithography is a method widely used in research and is the primary technique for mask fabrication. However, serial writing is too slow for integrated circuit manufacture. The *SCALPEL* (SCattering with Angular Limitation Projection Electron-beam Lithography) circumvents the serial limitations. The ultimate resolution of e-beam lithography is not determined by the wavelength (in contrast to optical lithography) as wavelengths of around 10^{-13} m (smaller than atomic dimensions) are achievable. An e-beam can be focused to a spot size of approximately 1 nm and the resolution is then determined by the resist properties. Patterns of 80 nm wide holes have been demonstrated, although the ultimate resolution is probably much less than this.

Ion-beam projection lithography utilizes a beam of hydrogen or helium ions, which are focused electrostatically. The resolution is currently about 180 nm, but the method is able to produce structures well below 100 nm.

Structures of less than 10 nm may also be fabricated using techniques of shadowing and edge-step deposition [1]. Substrate steps with a square profile are first formed by ion-beam etching, as shown in Figure 7.36. Wires of triangular cross-section are then produced by ion-etching the film-coated substrate at an angle, so that the wire is formed in the shadow of the step. Metal wires as narrow as 30 nm and as long as 0.5 mm have been fabricated in this way.

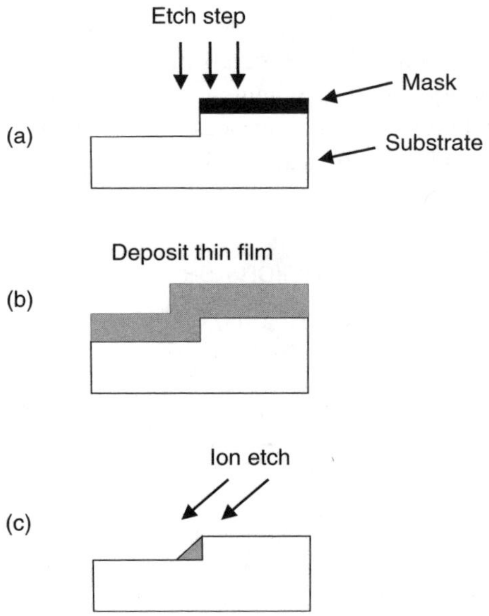

Figure 7.36 Fabrication process for the production of fine wires by the step-edge technique.

7.4.3 Soft Lithography Techniques

Chemical approaches to the deposition of ultra-thin organic films also offer some control over the composition and structure of the surface. An example is shown in Figure 7.37 [45]. A substrate is patterned with gold and aluminium strips using conventional photolithography. The aluminium oxidizes spontaneously in air and provides a layer of aluminium oxide at the solid/vapour interface; in contrast, the gold remains clean. Two adsorbates, L_1 and L_2, are chosen so that they adsorb strongly and selectively on gold and alumina. In the example, L_1 is an alkanethiol and L_2 a fluorine-labelled carboxylic acid. Exposure of the substrate to a solution containing L_1 and L_2 results in replicating the gold pattern with a self-assembled monolayer derived from the alkanethiol and the aluminium pattern with a self-assembled monolayer derived from the carboxylic acid.

An example of pattern generation using a rubber stamp, so-called *soft lithography*, is shown in Figure 7.38 [46]. A pattern-transfer element is formed by pouring a liquid polymer, such as polydimethylsiloxane (PDMS), on to a 'master' made from silicon. The polymer is allowed to cure to form an elastomer, which can then be removed from the master. This replica can be used as a stamp to transfer a chemical ink, such as a solution of an alkanethiol, to a surface.

A relatively high throughput method for lithographically processing one-dimensional nanowires is *on-wire lithography* [47]. This procedure combines template-directed synthesis of nanowires with electrochemical deposition and wet-chemical etching and allows routine fabrication of face-to-face disk arrays and gap structures in the range from five to several hundred nanometres.

7.4.4 Scanning Probe Manipulation

The tip of a scanning tunnelling or an atomic force microscope (Chapter 6, Section 6.8) may be used to manipulate atoms and molecules on a surface. Careful control of the tip can allow patterns to be drawn in an organic film. Scanning microscopy can also be used to reposition molecules, such as the fullerene C_{60}, on surfaces and to break up an individual molecule [48]. A wide range of scanning probe lithographic techniques is available to pattern self-assembled monolayers [49].

$$L_1 = Cl(CH_2)_{11}SH$$

$$L_2 = CF_3(CF_2)_8COOH$$

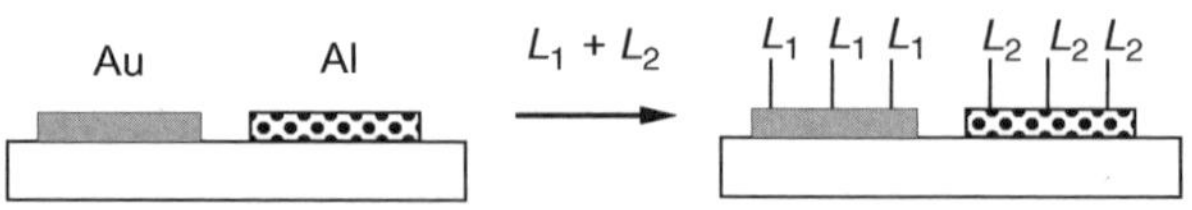

Figure 7.37 Schematic diagram showing the simultaneous formation of two independent self-assembled monolayers. Reprinted with permission from *Langmuir*, **6**, Whitesides GM, Laibinis PE, 'Wet chemical approaches to the characterization of organic surfaces: self-assembled monolayers, wetting, and the physical-organic chemistry of the solid–liquid interface', pp. 87–96. Copyright (1990) American Chemical Society.

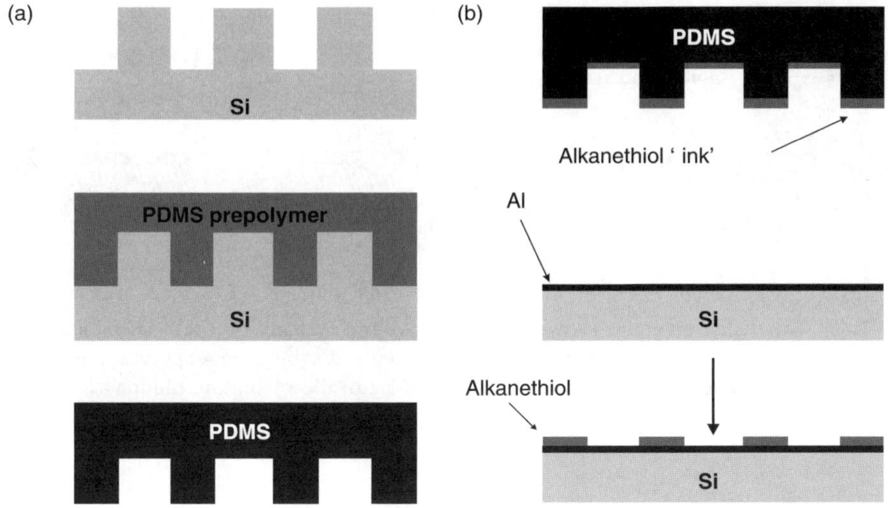

Figure 7.38 Soft lithography method to apply a pattern to a surface using a rubber stamp. Reprinted from *Physics World*, **11** (5), Brittain S, Paul K, Zhao X-M, Whitesides GM, 'Soft lithography and microfabrication', pp. 31–36, Copyright (1998), with permission from IOP Publishing Limited.

Figure 7.39 shows the IBM company logo written using a scanning tunnelling microscope with Xe atoms on a (110) Ni surface [50]. This experiment was undertaken in an ultra-high vacuum system and the entire chamber housing the microscope was cooled to 4 K. Each of the letters is 5 nm from top to bottom.

The photolithographic process may also be used to create nanometre-scale patterns in self-assembled monolayers by utilizing a UV laser coupled to a scanning near-field optical microscope (SNOM; Chapter 6, Section 6.2.1) as a light source [51–53]. The alkanethiolates

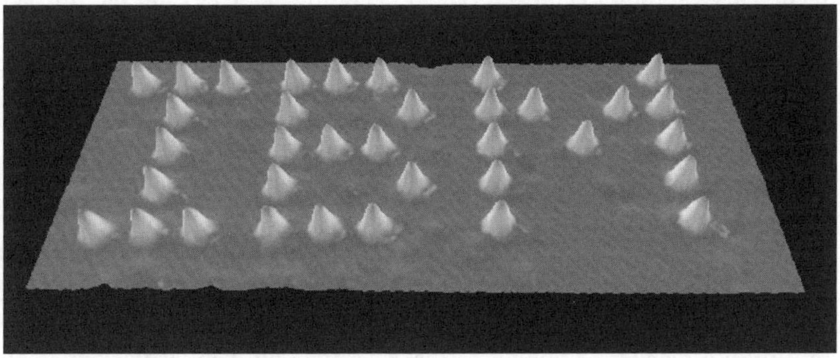

Figure 7.39 IBM company logo written using a scanning tunnelling microscope with Xe atoms on a (110) Ni surface. Each of the letters is 5 nm from top to bottom. Reprinted from *Nature*, **344**, Eigler DM, Schweizer Ek, 'Positioning single atoms with a scanning tunnelling microscope, pp. 524–526. Copyright (1990), with permission from Macmillan Publishers Ltd.

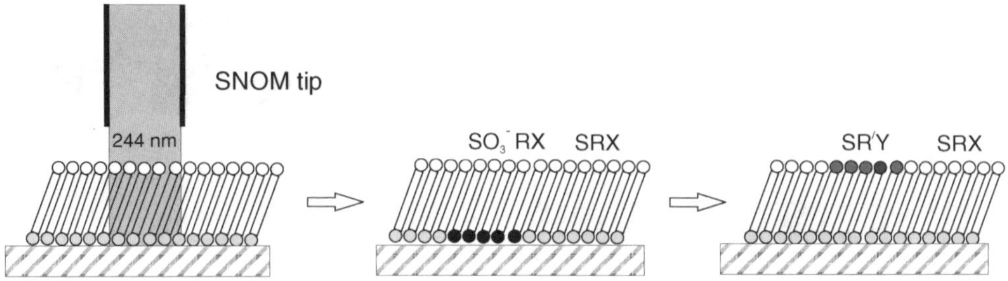

Figure 7.40 Schematic diagrams showing the selective replacement of a self-assembled layer of an X-terminated thiol following oxidation using a UV source and subsequent exposure to a solution of a contrasting Y-terminated thiol. The use of the SNOM tip allows high-resolution structures to be defined.

that are formed by the adsorption of alkanethiols on a gold surface (Section 7.3.2) can be oxidized by UV light in the presence of air to alkylsulfonates. The reaction is

$$Au{-}SR(CH_2)_nX + \frac{3}{2}O_2 \rightarrow Au^+ + X(CH_2)_nSO_3^- \qquad (7.15)$$

The process is depicted schematically in Figure 7.40. The coupling of the UV source to the SNOM allows features smaller than the diffraction limit to be fabricated. The alkylsulfonates are weakly bound to the underlying gold surface and may be displaced by immersing the substrate into a solution of a contrasting thiol. Figure 7.41 shows a lateral force microscopy (LFM; Chapter 6, Section 6.8) image that has been created using this technique [53]. The lines represent $C_{15}CH_3$ written into C_2COOH self-assembled monolayers on polycrystalline Au with a grain size of approximately 20 nm. Parallel lines as small as 20 nm, with a separation of about 20 nm, are evident. The contrast between regions of different terminal group functionality is very clear, and lines written close together can be differentiated. The resolution achieved ($\lambda/12$) is significantly beyond the diffraction limit and rivals the performance of electron beam lithography for materials of this type. Furthermore, the method is an ambient technique compatible, in principle, with operation in a fluid medium.

7.4.5 Dip-Pen Nanolithography

A further approach that has been developed at Northwestern University is called *dip-pen nanolithography* (DPN) [54, 55]. This technique, illustrated in Figure 7.42, is able to deliver organic molecules in a positive printing mode. An AFM tip is used to 'write' alkanethiols on a gold thin film in a manner analogous to that of a fountain pen. Molecules flow from the AFM tip to a solid substrate (the 'paper') via capillary transport, making DPN a potentially useful tool for assembling nanoscale devices.

The chemisorption of the 'ink' is the driving force that moves the organic molecules from the AFM tip through the water to the substrate as the tip is scanned across this

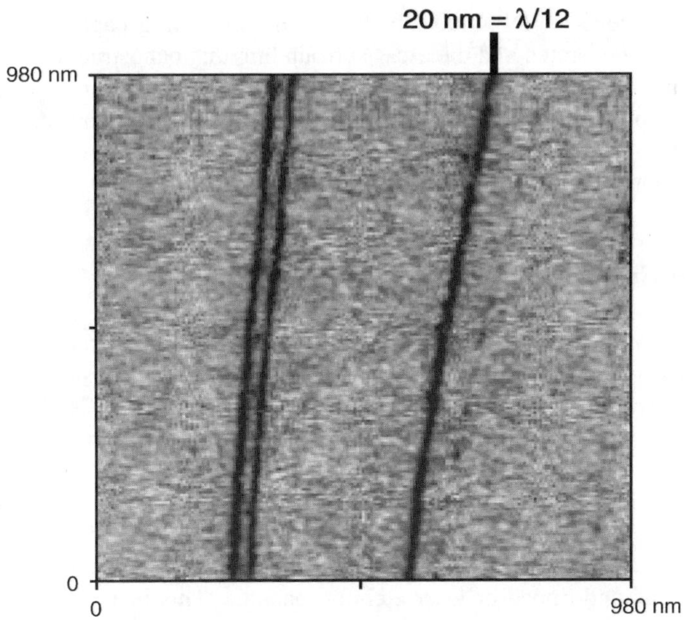

Figure 7.41 Lateral force microscopy image showing lines of $C_{15}CH_3$ written into a C_2COOH self-assembled monolayer on polycrystalline gold (grain size ~20 nm). Reprinted from *Nano Lett.*, **4**, Sun S, Leggett GJ, 'Matching the resolution of electron beam lithography by scanning near-field photolithography', pp. 1381–1384. Copyright (2004) American Chemical Society.

surface. Adjusting the scan rate and relative humidity can control line widths. The basic DPN method has some similarity to LB film deposition, described above in Section 7.3.1. For example, if the DPN experimental arrangement shown in Figure 7.42 is turned by 90°, a 'nano-LB trough' (similar to those that use a moving subphase to compress the monolayer film) is evident. As the amphiphilic molecules flow down the AFM tip, supported by a thin layer of water, their surface pressure will rise. The resulting condensed monomolecular film is then transferred to a solid substrate. However, for DPN, the 'substrate' is stationary while the 'nano-trough' moves in relation to it.

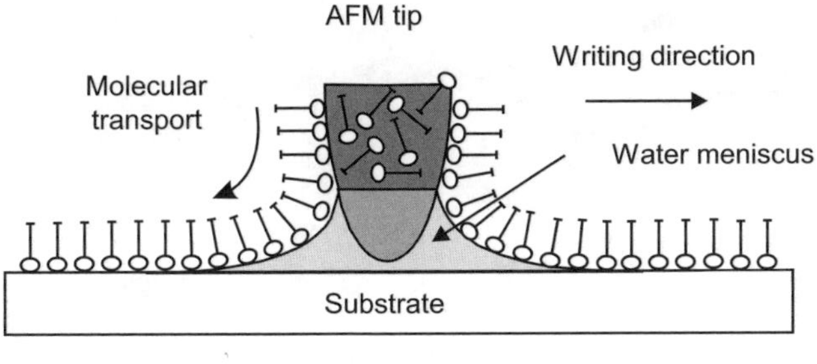

Figure 7.42 Dip-pen nanolithography.

Recent developments of DPN have included an overwriting capability that allows one nanostructure to be generated and the areas surrounding that nanostructure to be filled with a second type of ink. Perhaps the greatest limitation in using scanning probe methodologies for ultra-high resolution nanolithography over large areas derives from the serial nature of most techniques. However, the use of multiple scanning-probe microscopy cantilever probes as pens has been shown to increase patterning speeds.

7.4.6 Other Methods

In addition to patterning during deposition using the techniques described above, methods have also been developed to pattern the organic material following deposition [56]. These include nano-imprinting of the polymer, lithographically induced self-assembly, micro-cutting, photo-cross-linking of polymer regions to form conductive contacts and cold welding.

Finally, for the large-scale production of molecular electronic devices, methods such as roll-to-roll processing will become important. In the *gravure* (or rotogravure) process, depicted schematically in Figure 7.43, a pattern or image is engraved on a metal cylinder, usually with a diamond-tipped or laser etching machine. This image is composed of small recessed cells (or dots) that act as tiny wells. Their depth and size control the amount of ink that is transferred to the substrate. Such a process can be easily adapted as a continuous deposition process to fabricate organic devices [56]. Starting with a roll of plastic (or, perhaps, fabric) substrate material, the organic materials forming the device active regions are first deposited from the liquid or vapour phase, and then a metal 'strike' layer is deposited. The film is then passed between rollers with the embossed pattern representing the ultimate electrode scheme required on their surfaces. These rollers directly pattern the surface (possibly using one of the soft-lithography techniques described above). In the final step, the strike layer metal is removed by dry etching.

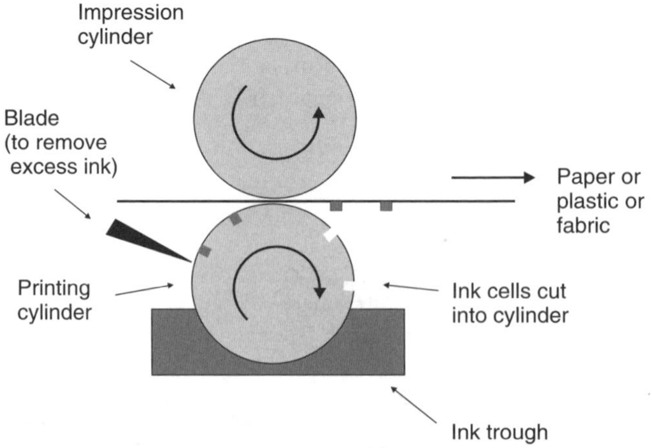

Figure 7.43 Schematic diagram of gravure printing where the ink is held in a recess etched or cut into the printing roller.

BIBLIOGRAPHY

Adamson AW. *Physical Chemistry of Surfaces*, 5th ed. Chichester: Wiley-Interscience; 1990.

Brinker CJ, Scherer GW. *Sol–Gel Science: the Physics and Chemistry of Sol–Gel Processing.* San Diego, CA: Academic Press; 1990.

Chapman B. *Glow Discharge Processes: Sputtering and Plasma Etching.* New York: John Wiley & Sons, Inc.; 1980.

Forrest SR. The path to ubiquitous and low-cost organic electronic appliances on plastic. *Nature* 2004; **428**; 911–918.

Hodes G (Editor). *Electrochemistry of Nanomaterials.* Weinheim: Wiley-VCH; 2001.

Jones RW. *Fundamental Principles of Sol–Gel Technology.* London: Institute of Metals; 1989.

Maissel LI, Glang R (Editors). *Handbook of Thin Film Technology.* New York: McGraw-Hill; 1970.

McShane MJ, Lvov YM. Layer-by-layer electrostatic self-assembly. In J Schwartz, C Contescu (Editors). *Dekker Encyclopedia of Nanoscience and Technology.* New York: Marcel Dekker; 2004, pp. 1–20.

Mitsuishi M, Matsui J, Miyashita T. Functional organized molecular assemblies based on polymer nanosheets. *Polym. J.* 2006; **38**: 877–896.

Nabook A. *Organic and Inorganic Nanostructures.* Boston: Artech House; 2005.

Paunovic M. *Fundamentals of Electrochemical Deposition.* New York: John Wiley & Sons, Inc.; 1998.

Petty MC. *An Introduction to Langmuir–Blodgett Films.* Cambridge: Cambridge University Press; 1996.

Pierre AC. *Introduction to Sol–Gel Processing.* Boston: Kluwer; 1998.

Roberts GG (Editor). *Langmuir–Blodgett Films.* New York: Plenum Press; 1990.

Talham DR. Conducting and magnetic Langmuir–Blodgett films. *Chem. Rev.* 2004; **104**: 5479–5501.

Tredgold RH. *Order in Thin Organic Films.* Cambridge: Cambridge University Press; 1994.

Ulman A. *An Introduction to Organic Thin Films.* San Diego, CA: Academic Press; 1991.

Wallace GG, Spinks GM, Kane-Maguire LAP, Teasdale PR. *Conductive Electroactive Polymers: Intelligent Materials Systems.* Boca Raton, FL: CRC Press; 2002.

REFERENCES

[1] Brodie I, Muray JJ. *The Physics of Micro/Nano-Fabrication.* New York: Plenum Press; 1992.

[2] Flack WW, Soong DS, Bell AT, Hess DW. A mathematical model for spin-coating of polymer resists. *J. Appl. Phys.* 1984; **56**: 1199–1206.

[3] Bornside DE, Macosko CW, Scriven LE. On the modelling of spin-coating. *J. Imaging Technol.* 1987; **13**: 122–130.

[4] Bornside DE, Macosko CW, Scriven LE. Spin-coating: one-dimensional model. *J. Appl. Phys.* 1989; **66**: 5185–5193.

[5] Critchley SM, Willis MR, Cook MJ, McMurdo J, Maruyama Y. Deposition of ordered phthalocyanine films by spin-coating. *J. Mater. Chem.* 1992; **2**: 157–159.

[6] Glang R. Vacuum evaporation. In LI Maissel, R Glang (Editors). *Handbook of Thin Film Technology.* New York: McGraw-Hill; 1970, pp. 1-3–1-330.

[7] Chrisey DB, Hubler GK (Editors). *Pulsed Laser Deposition of Thin Films.* New York: Wiley-Interscience; 1994.

[8] Breen JJ, Tolman JS, G W Flynn GW. Scanning tunnelling microscopy studies of vapour deposited films of tetrathiafulvalene with iodine. *Appl. Phys. Lett.* 1993; **62**: 1074–1076.

[9] Leaver KD, Chapman BN. *Thin Films.* London: Wykeham; 1971.

[10] Kolb D, University of Durham, unpublished work.

[11] Kilitziraki M, Moore AJ, Petty MC, Bryce MR. Evaporated thin films of tetrathiafulvalene derivatives and their charge-transfer complexes. *Thin Solid Films* 1998; **335**: 209–213.

[12] Hara M, Sasabe H. Organic molecular beam epitaxy. In MC Petty, MR Bryce, D Bloor (Editors). *An Introduction to Molecular Electronics.* London: Edward Arnold; 1995, pp. 243–260.
[13] Sherman A. *Chemical Vapour Deposition for Microelectronics.* Park Ridge: Noyes; 1987.
[14] Chopra KL. *Thin Film Phenomena.* New York: McGraw-Hill; 1969.
[15] Yasuda H. *Plasma Polymerization.* New York: Academic Press; 1985.
[16] Lucovsky G, Ibbotson DE, Hess DW (Editors). *Characterization of Plasma-Enhanced CVD Processes.* Pittsburgh: Materials Research Society; 1990.
[17] Eden JG. *Photochemical vapour deposition.* New York: Wiley-Interscience; 1992.
[18] Heinze J. Electrochemistry of conducting polymers. *Synth. Met.* 1991; **41–43**: 2805–2823.
[19] Haushalter RC, Krause LJ. Electroless metallization of organic polymers using the polymer as a redox reagent: reaction of polyamide with zintl anions. *Thin Solid Films* 1983; **102**: 161–171.
[20] Sirringhaus H, Shimoda T (Editors). Inkjet printing of functional materials. *MRS Bull.* 2003; **28**: 802–803.
[21] Calvert P. Inkjet printing for materials and devices. *Chem. Mater.* 2001; **13**: 3299–3305.
[22] Morris DJ. Electrical and spectral characterization of inkjet printed poly(3,4-ethylenedioxythiophene): poly(4-styrenesulphonate). *PhD Thesis.* Bangor: University of Wales; 2004.
[23] Mohammed MF, Pearson C, Petty MC. An inkjet-printed chemical fuse. *Appl. Phys. Lett.* 2005; **86**: 013507.
[24] Jones RW. *Fundamental Principles of Sol–Gel Technology.* London: Institute of Metals; 1989.
[25] Brinker CJ, Scherer GW. *Sol–Gel Science: the Physics and Chemistry of Sol–Gel Processing.* San Diego, CA: Academic Press; 1990.
[26] Sugiyasu K, Fujita N, Shinkai S. Design of novel composite materials by functional low molecular-weight organogels. *J. Synth. Org. Chem. Jpn.* 2005; **63**: 359–369.
[27] Ogoshi T, Chujo Y. Organic–inorganic polymer hybrids prepared by the sol–gel method. *Composite Interfaces* 2005; **11**: 539–566.
[28] Petty MC. *An Introduction to Langmuir–Blodgett Films.* Cambridge: Cambridge University Press; 1996.
[29] Roberts GG (Editor). *Langmuir–Blodgett Films.* New York: Plenum Press; 1990.
[30] Tredgold RH. *Order In Thin Organic Films.* Cambridge: Cambridge University Press; 1994.
[31] Ulman A. *An Introduction to Organic Thin Films.* San Diego, CA: Academic Press; 1991.
[32] Peterson IR. Langmuir–Blodgett films. In GJ Ashwell (Editor). *Molecular Electronics.* Taunton: Research Studies Press; 1992, pp. 117–206.
[33] Harkins WD. *Physical Chemistry of Surface Films.* New York: Reinhold; 1952.
[34] Mingotaud A-F, Mingotaud C, Patterson LK (Editors). *Handbook of Monolayers.* Orlando, FL: Academic Press; 1993.
[35] Neugebauer CA. Condensation, nucleation and growth of thin films. In LI Maissel, R Glang (Editors). *Handbook of Thin Film Technology.* New York: McGraw-Hill; 1970, pp. 8-3–8-44.
[36] Peterson IR. Langmuir–Blodgett films. *J. Phys. D: Appl. Phys.* 1990; **23**: 379–395.
[37] Peterson IR, Russell GJ. Deposition mechanisms in Langmuir–Blodgett films. *B. Polym. J.* 1985; **17**: 364–367.
[38] Batey J. Electroluminescent MIS structures incorporating Langmuir–Blodgett films. *PhD Thesis.* Durham: University of Durham; 1983.
[39] Rosner RB, Rubner MF. Solid state polymerization of pyrrole within a Langmuir–Blodgett film of ferric stearate. *Chem. Mater.* 1994; **6**: 581–586.
[40] Netzer L, Iscovici R, Sagiv J. Adsorbed monolayers versus Langmuir–Blodgett monolayers – why and how? I: from monolayer to multilayer, by adsorption. *Thin Solid Films* 1983; **99**: 235–241.
[41] Decher G, Schlenoff JB (Editors). *Multilayer Thin Films: Sequential Assembly of Nanocomposite Materials.* Weinheim: Wiley-VCH; 2003.
[42] Dautzenberg H, Jaeger W, Kőtz, Philipp B, Seidel Ch, Stscherbina D. *Polyelectrolytes: Formation, Characterization and Application.* Munich: Hanser; 1994.

[43] Palumbo M, Pearson C, Petty MC. Atomic force microscope characterization of poly(ethyleneimine)/poly(ethylene-*co*-maleic acid) and poly(ethyleneimine)/poly(styrene sulfonate) multilayers. *Thin Solid Films* 2005; **483**: 114–121.
[44] Stockton WB, Rubner MF. Molecular-level processing of conjugated polymers. 4. Layer-by-layer manipulation of polyaniline via hydrogen-bonding interactions. *Macromolecules* 1997; **30**: 2717–2725.
[45] Whitesides GM, Laibinis PE. Wet chemical approaches to the characterization of organic surfaces: self-assembled monolayers, wetting, and the physical-organic chemistry of the solid–liquid interface. *Langmuir* 1990: **6**: 87–96.
[46] Brittain S, Paul K, Zhao X-M, Whitesides G. Soft lithography and microfabrication. *Phys. World* 1998; **11** (5): 31–36.
[47] Qin L, Park S, Huang L, Mirkin CA. On-wire lithography. *Science* 2005; **309**: 113–115.
[48] Gimzewski J. Molecules, nanophysics and nanoelectronics. *Phys. World* 1998; **11** (6): 29–33.
[49] Krämer S, Fuierer RR, Gorman CB. Scanning probe lithography using self-assembled monolayers. *Chem. Rev.* 2003; **103**: 4367–4418.
[50] Eigler DM, Schweizer EK. Positioning single atoms with a scanning tunnelling microscope. *Nature* 1990; **344**: 524–526.
[51] Sun S, Chong KSL, Leggett GJ. Nanoscale molecular patterns fabricated by scanning near-field optical lithography. *J. Am. Chem. Soc. Commun.* 2002; **124**: 2414–2415.
[52] Sun S, Chong KSL, Leggett GJ. Photopatterning of self-assembled monolayers at 244 nm and applications to the fabrication of functional microstructures and nanostructures. *Nanotechnology* 2005; **16**: 1798–1808.
[53] Sun S, Leggett GJ. Matching the resolution of electron beam lithography by scanning near-field photolithography. *Nano Lett.* 2004; **4**: 1381–1384.
[54] Piner RD, Zhu J, Xu F, Hong S, Mirkin CA. 'Dip-pen' nanolithography. *Science* 1999; **283**: 661–663.
[55] Lee SW, Sanedrin RG, Oh B-K, Mirkin CA. Nanostructured polyelectrolyte multilayer organic thin films generated via parallel dip-pen nanolithography. *Adv. Mater.* 2005; **17**: 2749–2753.
[56] Forrest SR. The path to ubiquitous and low-cost organic electronic appliances on plastic. *Nature* 2004; **428**; 911–918.

8 Liquid Crystals and Devices

But, soft! what nymphs are these?

8.1 INTRODUCTION

The melting of a solid is normally a sharp transition. As the material is heated, the arrangement of the atoms or molecules abruptly changes from that of the three-dimensional order of the solid to the zero order associated with the liquid, as described in Chapter 2, Section 2.5. However, it is not uncommon for an organic solid to pass through intermediate phases as it heated from a solid to a liquid, depicted in Figure 8.1. The Austrian botanist Friedrich Reinitzer is often credited with discovering this new phase of matter, the *liquid crystal phase*, in 1888. Reinitzer observed that, on heating a sample of cholesteryl benzoate, the solid crystal changed into a hazy liquid. As he increased the temperature further, the material changed into a clear, transparent liquid. The compound seemed to possess two melting points!

8.2 LIQUID CRYSTAL PHASES

Perhaps one in every few hundred organic compounds exhibits liquid crystalline behaviour. These phases are known as *mesophases*. When such a state is formed, the translational order of the solid phase may be lost, but the orientational order remains. Sometimes a mesophase may display translational order but no orientational order; this is termed a *plastic crystal*. A very wide range of liquid crystal phases is now known. These are identified by the degree of short-range translational order and by the shape of the molecules.

8.2.1 Thermotropic Liquid Crystals

Liquid crystals compounds can be *thermotropic* or *lyotropic*. In the former, a phase transition into a liquid crystal occurs as the temperature is changed. In contrast, lyotropic liquid crystals exhibit phase transitions as a function of concentration. Thermotropic liquid crystals can be further subdivided into two main classes: high molecular weight compounds, including both main-chain and side-chain polymers, and low molecular weight materials, which encompass *calamitic* (rod-like molecules) and *discotic* (disc-like molecules). The classes and their relationships to one another are shown in Figure 8.2. The following section provides an overview of these liquid crystalline mesophases.

Molecular Electronics: From Principles to Practice M. C. Petty
© 2007 John Wiley & Sons, Ltd

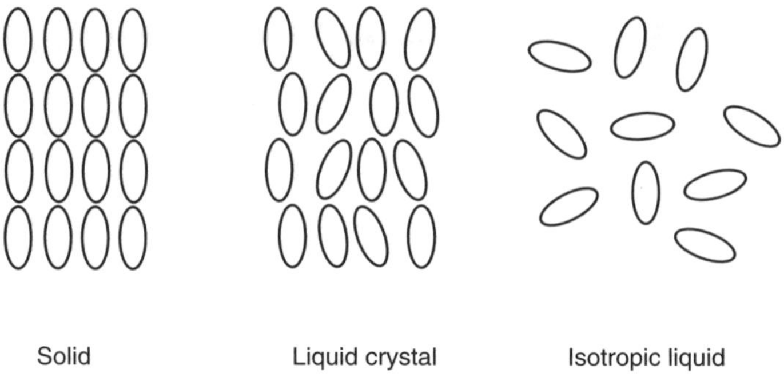

Figure 8.1 Representation of solid, liquid and liquid crystal phases of rod-shaped molecules.

Nematic Phases

The *nematic* (from the Greek word for thread) phase is the least ordered liquid crystal phase and is exploited extensively in electro-optic applications. This phase has no long-range translational order and only orientational order. There is only one nematic phase and, on heating, this will eventually become an isotropic liquid. A schematic diagram, showing the arrangement of rod-shaped molecules in a nematic phase, is given in Figure 8.3. The molecules in the liquid crystal are free to move about in much the same fashion as a liquid; as they do so, they tend to remain orientated in a certain direction. The direction of preferred orientation is called the *director* of the liquid crystal. Each molecule is oriented at some

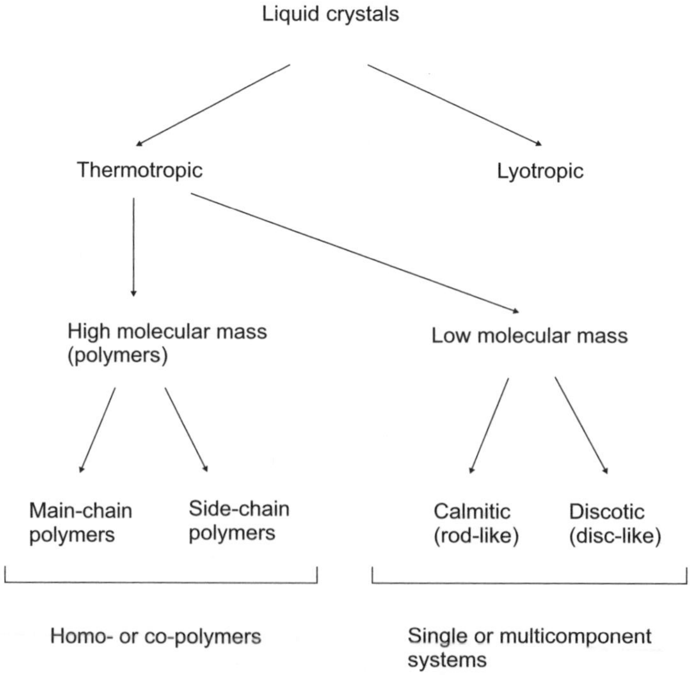

Figure 8.2 Liquid crystal categories.

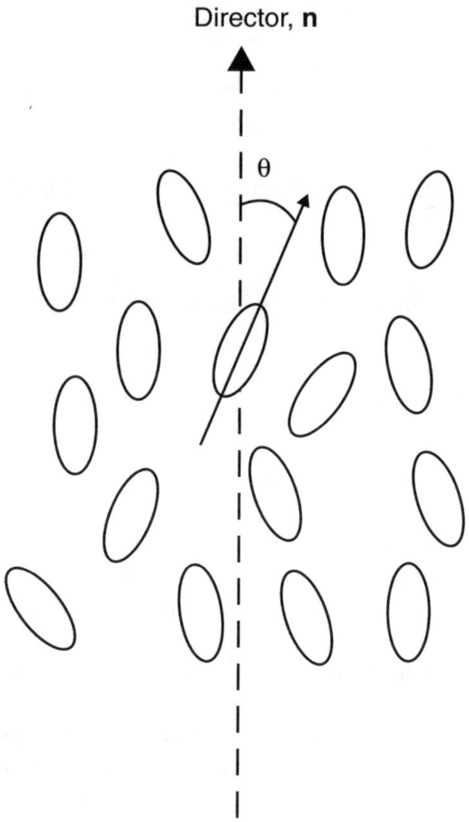

Figure 8.3 Arrangement of rod-shaped molecules in a liquid crystalline phase. The long axis of each molecule makes an angle θ with the director **n**.

angle θ about the director. The degree of orientational order is then given by the *order parameter S*, which is defined as

$$S = \tfrac{1}{2}\langle 3\cos^2\theta - 1\rangle \qquad (8.1)$$

where the angular brackets denote a statistical average. A value of $S = 1$ indicates perfect orientational order whereas no orientational order results in $S = 0$. Typical values for the order parameter of a liquid crystal range between 0.3 and 0.9, with the exact value a function of temperature, as a result of kinetic molecular motion. Liquid crystals are anisotropic materials, and the physical properties of the system vary with the average alignment with the director. If the alignment is large, the material is highly anisotropic. Similarly, if the alignment is small, the material is almost isotropic. Figure 8.4 shows an example of a nematic liquid crystal material based on a cyanobiphenyl group – the compound 4′-*n*-pentyl-4-cyanobiphenyl. The CN group provides the molecule with a strong electric dipole moment (Chapter 2, Section 2.3.5). The transitions from solid to nematic to liquid take place reversibly on heating and cooling.

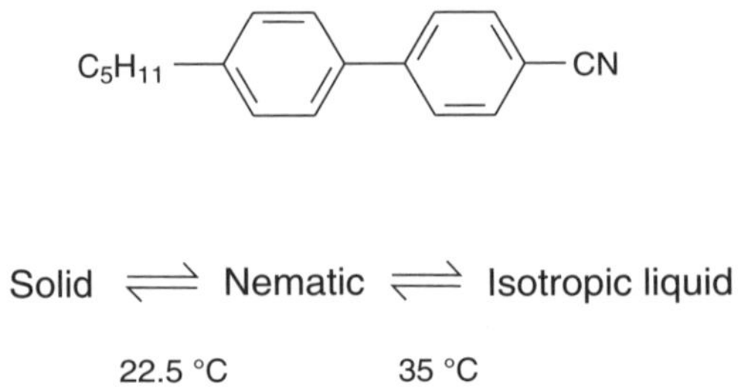

Figure 8.4 Example of an alkylcyanobiphenyl molecule, 4′-*n*-pentyl-4- cyanobiphenyl, that exhibits a nematic liquid crystal phase.

Smectic Phases

Smectic (from the Greek word for soap) phases are usually formed by thermotropic liquid crystals at lower temperature than the nematic phase. In addition to the orientational order, smectic phases possess one-dimensional translational ordering into layers. The smectic phases can be further subdivided and, at present, 12 different types have been identified. These are designated S_A, S_B, etc., up to S_K; there are two smectic S_B phases – the crystal *B* and *hexatic* S_B phases [1]. Some of these mesophases (crystal *B*, S_E, S_G, S_H and S_J) have very long-range correlation of position over many layers and are more similar to crystalline solids.

The S_A phase is the least ordered of the thermotropic smectic phases. The molecules are arranged in disordered layers, each layer having a liquid-like freedom of motion of its constituent molecules in two dimensions, with the director perpendicular to the layer planes. By contrast, in the S_C phase the molecules are tilted from this direction by about 35°. This tilt is correlated between molecules within each layer and from one layer to another. In the hexatic S_B phase, there is again ordering of orientationally aligned molecules into layers. The molecules are arranged in an hexagonal array but the translational order is short-range only. The orientation of the hexagonal net is, however, maintained over a long-range and, unlike the ordering of molecular positions, is correlated between layers. Two tilted variations of the S_B phase exist in which the tilt direction is constrained to point either towards one face of the hexagonal lattice (S_F) or towards one apex (S_I). The S_A, S_C, hexatic S_B and crystal *B* arrangements of rod-shaped molecules are contrasted in Figure 8.5.

Several liquid crystalline phases may be shown by a material as it is heated from the solid state to an isotropic liquid. Figure 8.6 shows an example. Most of the phase transitions are observed on both heating and cooling, although the crystal *H* phase is only seen on cooling the isotropic liquid below the melting point of the compound.

Chiral Phases

The final distinct type of liquid crystalline mesophase is the *cholesteric* or *chiral nematic*. The molecules in such a phase are optically active (Chapter 2, Section 2.4.2) and the optical chiral centre results in a unique helical structure in which the director gradually rotates from one plane of molecules to the next. An important characteristic of the cholesteric mesophase

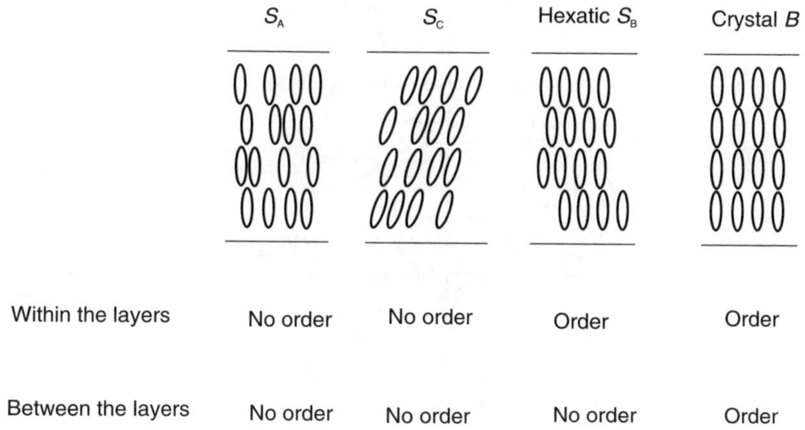

Figure 8.5 Different smectic liquid crystalline arrangements of rod-shaped molecules. S_C is a tilted version of S_A.

is the pitch. This is defined as the distance it takes for the director to rotate one full turn in the helix, as illustrated in Figure 8.7. A by-product of the helical structure of the chiral nematic phase is its ability to reflect selectively light of wavelengths equal to the pitch length, so that a colour will be reflected when the pitch is equal to the corresponding wavelength of light in the visible spectrum. An increase in temperature tightens the pitch. This effect is exploited in the liquid crystal thermometer, which displays the temperature of its environment by the reflected colour. Mixtures of various types of these liquid crystals are often used to create sensors with different responses to temperature change. The wavelength of the reflected light can also be controlled by adjusting the chemical composition, since cholesterics can consist either of exclusively chiral molecules or of nematic molecules with a chiral dopant dispersed throughout. In this case, the dopant concentration is used to adjust the chirality and thus the pitch.

The importance of the chiral centre for cholesteric behaviour is illustrated by the two molecules depicted in Figure 8.8. The compound shown in Figure 8.8(a), 4-methoxylbenzylidene-4′-butylaniline, transforms from crystalline to nematic liquid crystal at 20°C, and from nematic to an isotropic liquid at 74°C. In contrast, the very similar molecule shown in Figure 8.8(b) forms a cholesteric liquid crystal. The chiral centre is marked with an asterisk.

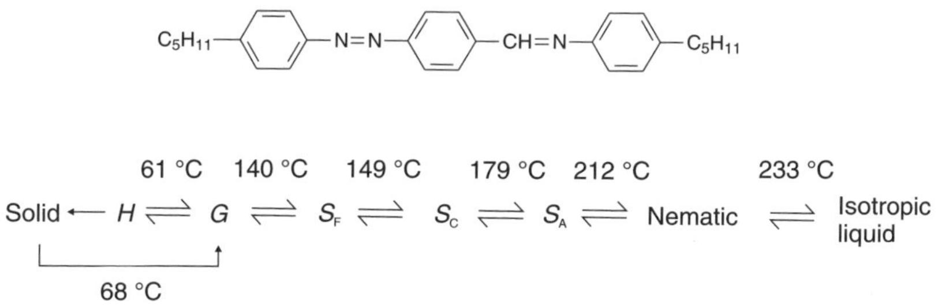

Figure 8.6 Multiphase liquid crystal behaviour exhibited by the organic molecule depicted.

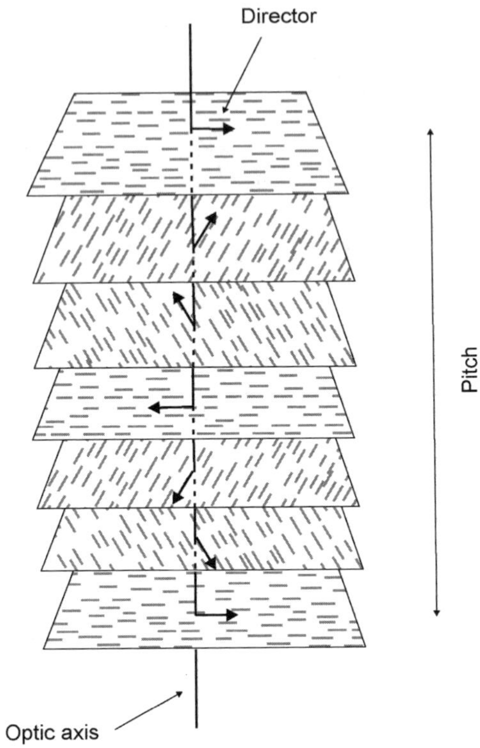

Figure 8.7 The helical structure of the cholesteric (chiral nematic) phase. The director direction (shown by an arrow) gradually rotates from one layer to the next, providing a unique helical structure.

Discotic Phases

As their name suggests, discotic liquid crystals consists of disc-shaped molecules. Structurally, most of the mesophases fall into two distinct categories: the *columnar* and the *nematic*, contrasted in Figure 8.9. The columnar phase, in its simplest form shown in Figure 8.9(a), consists of discs stacked one on top of the other to form liquid-like columns, whereas the

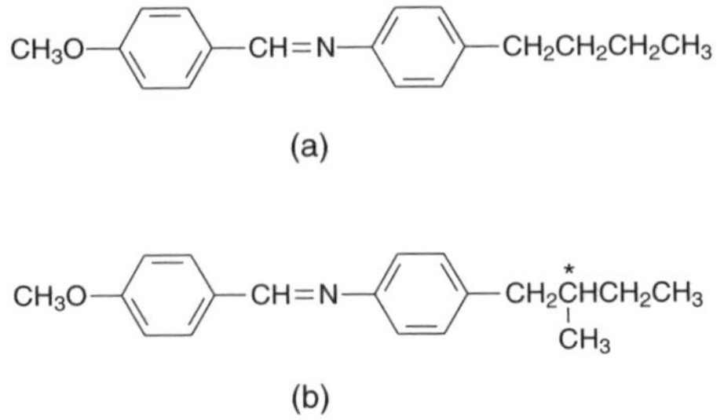

Figure 8.8 (a) An organic compound, 4-methoxylbenzylidene-4′-butylaniline, that exhibits a nematic liquid crystal phase. (b) A similar compound, now incorporating a chiral centre, which shows chiral nematic (cholesteric) behaviour. The chiral centre is denoted by an asterisk.

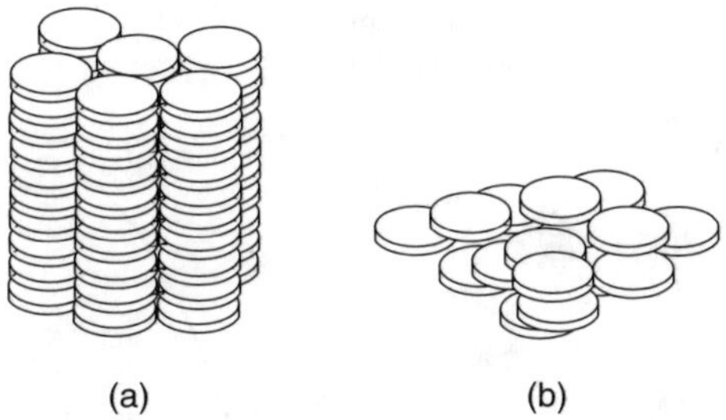

Figure 8.9 Discotic liquid crystal phases. (a) Columnar; (b) nematic.

nematic phase in Figure 8.9(b) has an orientationally ordered arrangement of the discs without any long-range translational order. Two examples of the types of compounds forming discotic liquid crystal phases are shown in Figure 8.10. The molecules possess a fairly rigid, planar centre with hydrocarbon chains emanating in all directions. Such features are fairly common in discotic liquid crystals.

8.2.2 Lyotropic Liquid Crystals

In contrast to thermotropic mesophases, lyotropic liquid crystal transitions occur with the influence of solvents, and not by a change in temperature. Lyotropic mesogens are typically amphiphilic, meaning that they are composed of both *lyophilic* (solvent-attracting) and *lyophobic* (solvent-repelling) parts (in the case of water being the solvent, the terms hydrophilic and hydrophobic are used; Chapter 7, Section 7.3.1). This causes them to form into *micellar* structures (Chapter 12, Section 12.2.5) in the presence of a solvent, since the lyophobic ends will stay together as the lyophilic ends extend outwards toward the solution. As the concentration of the solution is increased and the solution is cooled, the micelles increase in size and eventually coalesce. This separates the newly formed liquid crystalline state from the solvent.

The molecules that make up lyotropic liquid crystals are *surfactants*, similar to compounds that form LB films (Chapter 7, Section 7.3.1) or biological membranes (Chapter 12, Section 12.6). Not all surfactants, however, form lyotropic liquid crystals. When

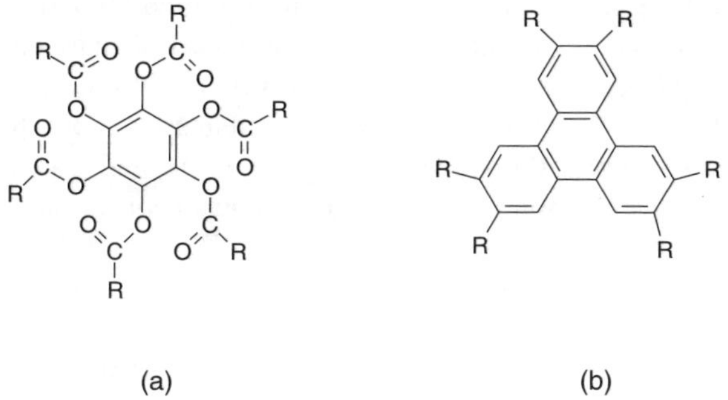

Figure 8.10 Examples of discotic liquid crystalline compounds. (a) Hexa-substituted benzene; (b) triphenylene.

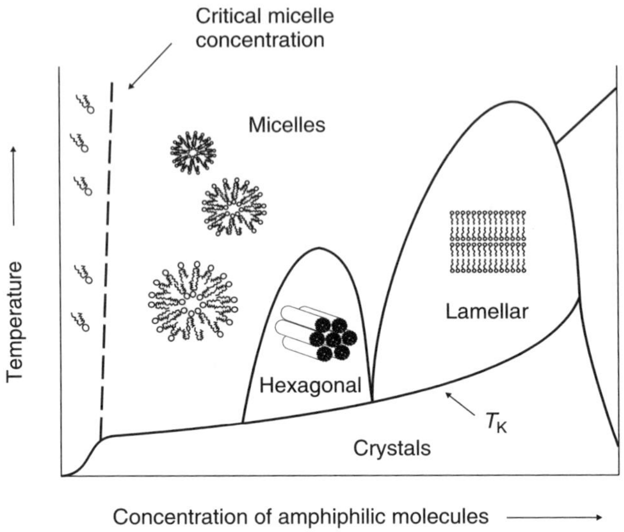

Figure 8.11 Phase diagram of a typical lyotropic liquid crystal. The nearly vertical dashed line on the left shows the minimum concentration for micelle formation. T_K is the Kraft temperature. Various liquid crystal phases occur in the region close to the 100% concentration axis.

dissolved in high enough concentrations, the molecules arrange themselves so that the polar heads are in contact with a polar solvent and/or the nonpolar tails are in contact with a nonpolar solvent.

Figure 8.11 shows a generic phase diagram of a typical lyotropic compound, illustrating the changes in structure that occur as the concentration of amphiphilic molecules increases. In very dilute solutions (far left of the phase diagram), the individual molecules are dispersed in the solvent, which is assumed to be water for the purposes of this discussion. However, as the concentration of the molecules in solution increases, they take on different micelle structures. These are dictated by the packing efficiencies of different shaped objects. For example, the molecules can begin to arrange themselves in hollow spheres, rods, and sheets. The concentration at which micelles form in solution, called the *critical micelle concentration*, is shown as the nearly vertical dashed line in Figure 8.11. If the temperature is too low, the molecules tend to form rigid crystalline structures. The temperature above which crystals do not form, but liquid crystalline structures do form, is called the *Kraft* temperature – T_K in Figure 8.11. Micelles come in various sizes, but the smallest ones have a diameter about twice the length of a hydrocarbon chain with all-*trans* bonds. As the weight concentration of the amphiphile increases, the micelles become increasingly able to dissolve nonpolar substances. When this occurs, the micelles become large and swollen. If they reach a sufficiently large size, the solution becomes cloudy and becomes an emulsion (Chapter 2, Section 2.7). At lower concentrations, the swollen micelles are not large enough to interfere with light, but they are still extremely stable and exist in equilibrium. This is an example of a microemulsion.

Rod-shaped micelles often form into hexagonal arrays made out of six rods grouped around a central micelle, depicted in Figure 8.11. This hexagonal arrangement offers more efficient packing than can be achieved by using spheres (closely packed spheres can fill no more than 74% of space, but closely packed cylinders can fill up to 91% of space).

Hexagonal liquid crystals generally exist in solutions that are 40–70% amphiphile. The liquid crystals may come apart if too much water or salt is added to the solution, but many varieties can absorb oil by expanding the diameter of the rod-shaped micelle.

At even higher concentrations, the molecules form another liquid crystalline phase – the lyotropic liquid crystal bilayer (such sheet-like structures can fill up 100% of space). The molecular arrangement is similar to that of smectic liquid crystals in the thermotropic category. Because the sheet-like layers can easily slide past each other, this phase is less viscous than the hexagonal phase, at least in the direction of the sliding, despite its lower water content. Other behaviour can occur when the situation is something other than a simple water solution. If the molecules are placed on the surface of water without actually being dissolved in it, they form a monolayer in which the polar heads are in contact with the water and the hydrophobic tails point into the air (Chapter 7, Section 7.3.1).

If the concentration by weight of amphiphilic molecules is higher than that of water, the molecules form a sort of matrix with water droplets scattered inside, in contact with the polar heads. In the case of molecules dissolved in a nonpolar solvent, their behaviour is similar to that when dissolved in water, except that now the nonpolar tails are in contact with the solvent and the polar heads are isolated in the centres of the micelles and bilayers. If the solution contains both water and a higher concentration of nonpolar solvent, similar *inverse micelles* form with water droplets trapped inside the micelle and nonpolar solvent on the outside (Chapter 12, Section 12.2.5). Finally, if weaker amphiphilic molecules and simple salts are dissolved together in water, they form lyotropic nematic phases.

One very common example of lyotropic liquid crystalline behaviour is household soap. Soaps work better than pure water at removing dirt and grease because the nonpolar insides of the micelles are capable of dissolving nonpolar substances that will not dissolve in water. (This also works in reverse if the solvent is nonpolar and some of the substance to be removed is polar.) Soaps also help water dissolve more dirt and grease because the molecules tend to remain at the surface, hydrocarbon tail away from the water, thus lowering the surface tension of the water and allowing more material to enter it and be dissolved.

8.3 LIQUID CRYSTAL POLYMERS

Polymer liquid crystals are a class of materials that combine the properties of polymers with those of liquid crystals. Such hybrids show the same mesophases which are characteristic of ordinary liquid crystals, yet retain many of the useful and versatile properties of polymers.

In order for normally flexible polymers to display liquid crystal characteristics, the mesogens must be incorporated into their chains. The placement of the mesogens plays a large role in determining the type of polymer liquid crystal that is formed. *Main-chain polymer liquid crystals* are formed when the mesogens are themselves part of the main chain of a polymer. In contrast, *side-chain polymer liquid crystals* are formed when the mesogens are connected as side chains to the polymer by a flexible 'bridge', called the *spacer*. Figure 8.12 shows examples of main-chain, (a), and side-chain, (b) and (c), polymer liquid crystals.

Main-chain polymer liquid crystals are formed when rigid elements are incorporated into the backbone of normally flexible polymers. These stiff regions along the chain allow the polymer to orient in a manner similar to ordinary liquid crystals, and thus display liquid crystal characteristics. There are two distinct groups of materials, differentiated by the

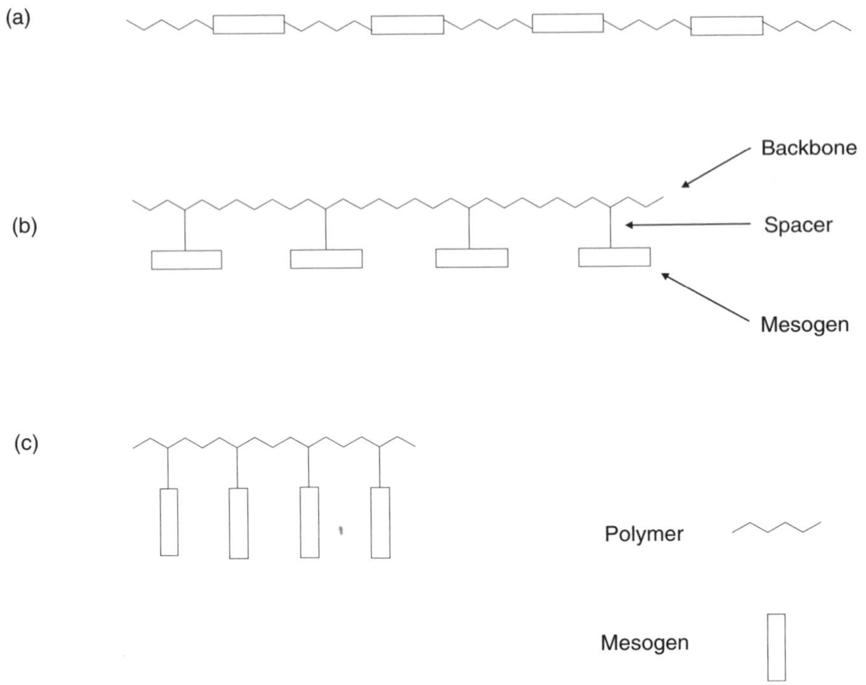

Figure 8.12 Examples of polymer liquid crystals: (a) represents a main-chain polymer liquid crystal and (b) and (c) are examples of side-chain liquid crystals.

manner in which the stiff regions are formed. The first group is characterized by stiff, rod-like monomers; these monomers are typically made up of several aromatic rings. In the second and more prevalent group of main-chain polymer liquid crystals, the mesogens are separated by a spacer group. Decoupling of the mesogens provides for independent movement of the molecules, thereby facilitating proper alignment. Generally, the mesogenic units are made up of two or more aromatic rings which give the necessary restriction on movement to allow the polymer to display liquid crystal properties. The stiffness necessary for liquid crystallinity results from restrictions on rotation caused by *steric hindrance* (arising from the fact that each atom within a molecule occupies a certain amount of space) and resonance (Chapter 2, Section 2.4.3).

Another characteristic of the mesogen is its *axial ratio*, defined as the length of the molecule divided by its diameter. Experimental results have indicated that these molecules must be at least three times long as they are wide, otherwise they are not sufficiently 'rod-like' to display the characteristics of liquid crystals.

Side-chain polymer liquid crystals have three major components: the backbone, the spacer and the mesogen [Figure 8.12(b) and (c)]. The versatility of such compounds arises because these structures can be varied in a large number of ways. For example, the nature of the backbone can be crucial in determining if the polymer shows liquid crystal behaviour. Polymers with rigid backbones normally have high glass transition temperatures, and therefore liquid crystal behaviour is often difficult to observe. Perhaps the most important part of a side-chain polymer liquid crystal is the mesogen. It is the alignment of these groups that gives rise to the liquid crystal behaviour. Figure 8.13 shows the chemical structure of a

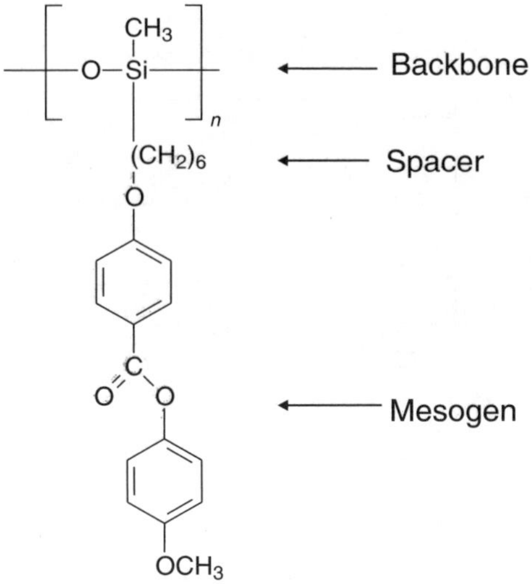

Figure 8.13 Molecular structure of a side-chain polymer liquid crystal based on a polysiloxane backbone.

side-chain liquid crystalline polymer based on a polysiloxane backbone. The spacer unit comprises methylene units whereas the mesogen is composed of aromatic rings. Like their main-chain counterparts, mesogens attached as side groups on the backbone of side-chain polymer liquid crystals are able to orient because the spacer allows for independent movement. Even though the polymer may be in a tangled conformation, orientation of the mesogens is still possible because of the decoupling action of the spacer. The spacer length has a profound effect on the temperature and type of phase transitions. Usually, the glass transition temperature decreases with increasing spacer length. Short spacers tend to lead to nematic phases, whereas longer spacers lead to smectic phases.

8.4 DISPLAY DEVICES

In the 1930s, it was suggested that display devices could be based on liquid crystals, where their optical behaviour would be controlled by temperature. However, the early liquid crystals were unstable with small temperature changes, and their behaviour was too erratic for practical use. The breakthrough finally came 40 years later, in the chemistry laboratories of the University of Hull, UK, where George Gray and his colleagues succeeded in making the first really stable liquid crystals. These were the alkylcyanobiphenyls (Figure 8.4), which were both cheap and stable. Furthermore, their optical properties could be precisely controlled using an electric field rather than heat.

8.4.1 Bifrefingence

Due to their anisotropic nature, liquid crystals are birefringent (Chapter 4, Section 4.3.1). They possess two refractive indices, one for light polarized parallel to the director, $n_{\parallel}$, and

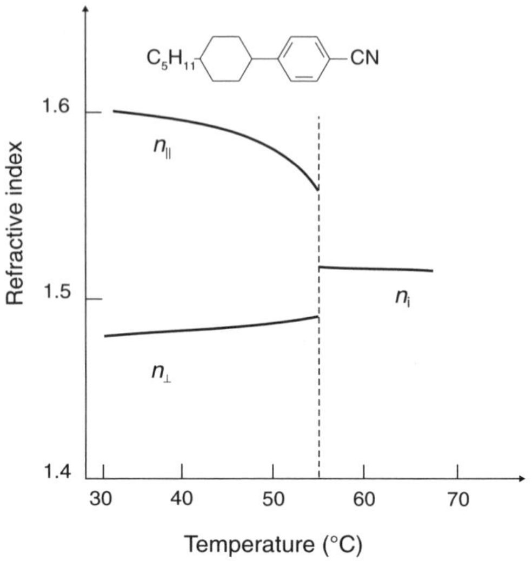

Figure 8.14 Temperature dependence of the birefringence of the liquid crystal molecule shown. The upper curve shows the extraordinary refractive index $n_{\|}$ and the lower curve the ordinary refractive index $n_{\perp}$. At the nematic liquid transition of around 54°C, both refractive indices merge into that of the isotropic liquid n_i. Reprinted from *Nanoelectrics and Information Technology*, 2nd ed., R. Waser (Editor), pp. 889–909. Copyright (2005), with permission from Wiley–VCH.

the other, for light polarized perpendicular to the director, $n_{\perp}$. The former gives rise to the extraordinary ray whereas the latter is responsible for the ordinary ray propagating through the sample. The birefringence is characterized by the difference in the refractive indices for the ordinary and extraordinary rays. Figure 8.14 shows the temperature dependence of the birefringence of a nematic liquid crystal. At the nematic–liquid transition around 54°C, the two refractive indices merge into one value, n_i, associated with the isotropic liquid phase.

If an isotropic, transparent material is placed between two polarizers, oriented at 90° to each other (known as *crossed polarizers*), no light will normally emerge because the light from the first polarizer is completely absorbed by the second. Insertion of an isotropic material does not change this situation because the polarization of light is unchanged as it travels through an isotropic material. The polarized light propagating through a liquid crystal may be considered to comprise of two components – light polarized along the director and light polarized perpendicular to the director. As the radiation propagates through the material, these two polarizations get out of phase and emerge from the material as elliptically polarized light. The electric vector of such light is constantly rotating during each cycle of the electromagnatic radiation. Therefore, it will be parallel to the polarization axis of the second polarizer twice during each cycle and some light will emerge from it. The introduction of a liquid crystal between two polarizers will therefore cause the field of view to appear bright (unless the incident polarized light has its polarization direction either parallel or perpendicular to the director). Devices that exploit this effect are called *phase retarders* and are found in many optical applications. If the thickness of the birefringent material is carefully adjusted to cause a 90° change in the phase difference, light that is linearly polarized on entering the material emerges as circularly polarized light. Similarly, incident circularly polarized light emerges as linearly polarized light.

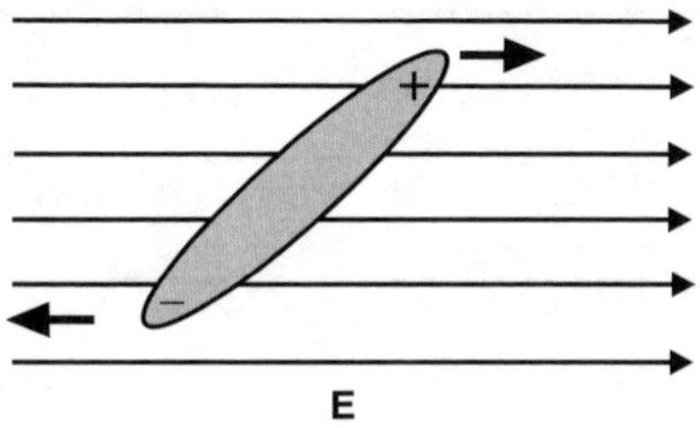

Figure 8.15 Effect of applied electric field on a dipolar molecule.

8.4.2 Freedericksz Transition

The response of liquid crystal molecules to an electric field is widely used in display applications. The ability of the director to align along an external field is caused by the electric nature of the molecules. Permanent electric dipoles result when one end of a molecule has a net positive charge and the other end a net negative charge (Chapter 2, Section 2.3.5). When an external electric field is applied to the liquid crystal, as depicted in Figure 8.15, the dipole molecules attempt to orient themselves along the direction of the field. This tendency is, of course, opposed by thermal motion.

Even if a molecule does not form a permanent dipole, it can still be influenced by an electric field. In some cases, the field produces slight rearrangement of electrons and protons in molecules such that an induced electric dipole results. Although not as strong as permanent dipoles, orientation with the external field still occurs. The effects of magnetic fields on liquid crystal molecules are analogous to those of electric fields. When a magnetic field is applied, the molecules will tend to align with or against the field. Liquid crystals are very sensitive to magnetic fields, achieving complete alignment of the director for magnetic fields of relatively low strength.

In the absence of an external field, the director of a liquid crystal is free to point in any direction. It is possible, however, to force the director to point in a specific direction by introducing an external influence other than an electric or magnetic field. For example, when a thin polymer coating (usually a polyimide) is spread on a glass substrate and rubbed in a single direction with a cloth, it is observed that the liquid crystal molecules in contact with that surface align with the rubbing direction. The mechanism for this is believed to be an epitaxial growth of the liquid crystal layers on the partially aligned polymer chains in the near-surface layers of the polyimide.

If mesogenic materials are confined between closely spaced plates with rubbed surfaces (as described above) and oriented with rubbing directions parallel, the entire liquid crystal sample can be oriented in a *planar* or *homogeneous texture*, as shown in Figure 8.16(a). Mesogens can also be oriented normal to a surface with the use of appropriate surface layers (e.g. trichlorosilanes incorporating a long aliphatic chain) or in the presence of an electric field applied normal to the surface, giving rise to the *homeotropic texture*, as illustrated in Figure 8.16(b).

The competition between orientation produced by surface anchoring and by electric field effects is exploited in many liquid crystal devices. Consider the case in which liquid crystal molecules are aligned parallel to the surface and an electric field is applied perpendicular to the cell. At first, as the electric field increases in magnitude, no change in alignment occurs.

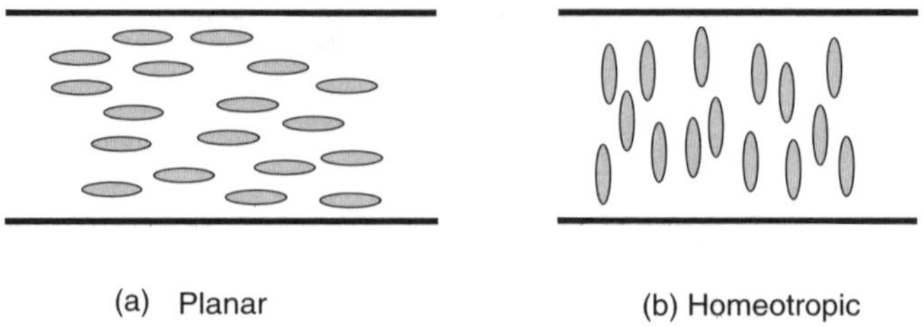

Figure 8.16 (a) Planar (or homogeneous) alignment and (b) homeotropic alignment of liquid crystal molecules between parallel plates.

However, at a threshold magnitude of electric field, deformation occurs where the director changes its orientation from one molecule to the next. The occurrence of such a change from an aligned to a deformed state is called a *Freedericksz transition* and can also be produced by the application of a magnetic field of sufficient strength. An important aspect of the Freedericksz transition is that typical values for the threshold field (electric or magnetic) are quite modest, approximately 4×10^4 V m^{-1} for the electric field (which can be achieved by the application of 1 V to a sample 25 μm in thickness) and about 0.2 T in the case of the magnetic field. This is the result of the relative freedom that liquid crystal molecules possess as they diffuse throughout the sample. The Freedericksz transition is fundamental to the operation of many liquid crystal displays because the director orientation (and thus the properties) can be controlled easily by the application of external fields.

8.4.3 Twisted Nematic Display

The simplest liquid crystal display device is the *twisted nematic display*, developed in the early 1970s. Figure 8.17 shows a schematic diagram of its construction. The liquid crystal is sandwiched between two glass plates, which have each been coated with a transparent conductive coating such as indium tin oxide. These form the conductive electrodes across which a voltage may be applied. Polymer spacer beads maintain a uniform gap between the sheets of glass and the edges are sealed with an epoxy resin.

In a twisted nematic display, alignment layers are positioned with their rubbing directions perpendicular to each other and polarizers are applied to match the orientation of the

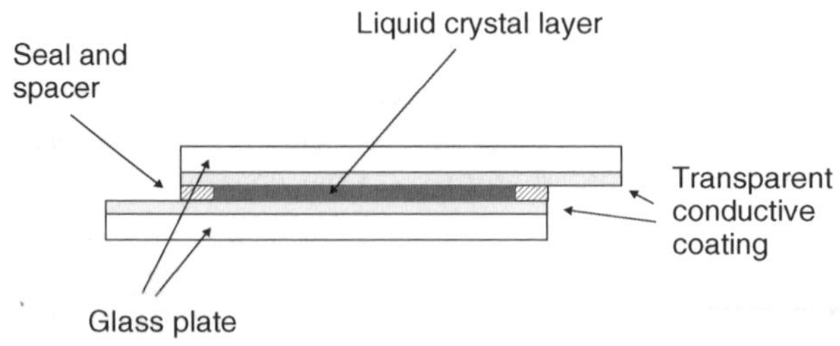

Figure 8.17 Schematic diagram of the construction of a twisted nematic liquid crystal cell.

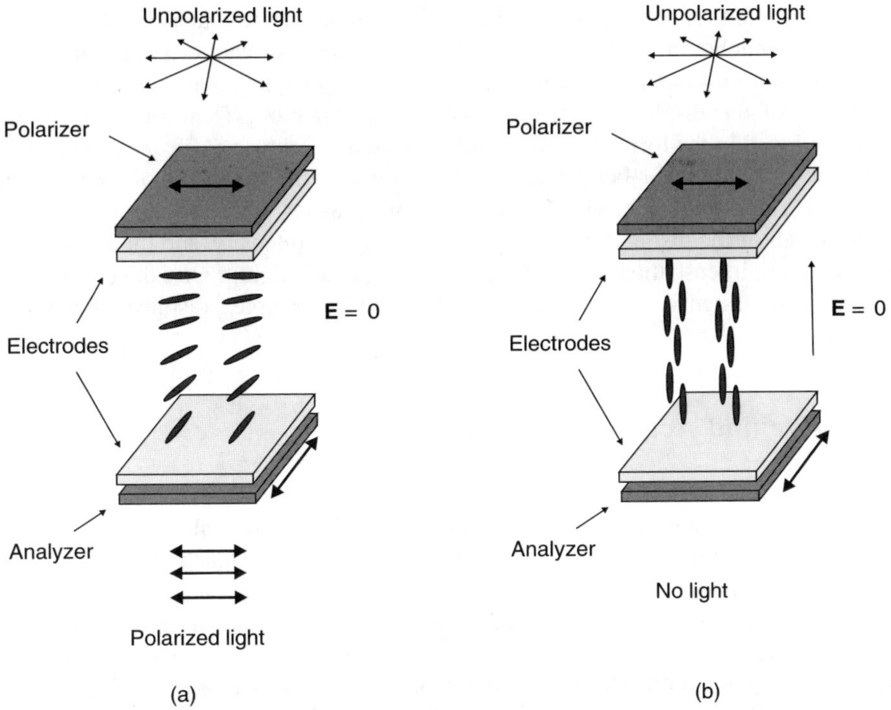

Figure 8.18 Twisted nematic cell. (a) With no applied field. (b) Electric field **E** applied perpendicular to the cell plates.

alignment layers. The arrangement is shown in Figure 8.18. With no electric field applied between the electrodes, there will be light transmitted as the twist in the alignment of the liquid crystal molecules will rotate the plane of polarization of the incoming polarized light by 90°. However, when a voltage above the threshold value for the Freedericksz transition is applied to the cell, there will be a torque on the liquid crystal molecules tending to align them parallel to the field. Although the molecules immediately adjacent to the glass surfaces will retain their parallel order (this feature is not shown in Figure 8.18), the majority of the liquid crystal molecules will align with the electric field. The polarization direction of the light crossing the cell will now be rotated only very slightly and almost no light will be transmitted by the cell.

A twisted nematic polymer liquid crystal cell can also be used to make energy-efficient displays. However, the response times of these devices can be relatively slow. Where a fast response is not needed, a laser may be used to melt selectively portions of the display into the liquid crystal phase. The orientation of the cell is then chosen by applying a field across it, just as in an ordinary twisted nematic liquid crystal cell. When the polymer cools and hardens into a glass, the mesogens will be locked in that configuration and the field can be turned off.

In order for a display to function, it must have a light source. A twisted nematic display can be operated in either a reflection or a transmission mode. In the former, a reflective layer forms the base layer of the cell and the amount of ambient light that is reflected back to the viewer is controlled by the applied voltage. The liquid crystal displays found in most calculators and watches are of this type. These displays are not very bright because the light must pass through multiple polarizers, which severely reduce its intensity, in addition to the various layers of the display which are only semi-transparent. Therefore, a more intense source is employed in the form of a *back lighting* system. Lights (e.g. incandescent bulbs or a solid-state source) mounted behind and at the edges of the display replace the reflected ambient light. This results in brighter displays for two reasons: the light does not have to pass through so much of the display and therefore does not lose intensity, and the lighting system can be made more intense than ambient light. Although back lighting has the disadvantage of being very power intensive, it is found widely in more complex displays such as laptop computer screens.

8.4.4 Passive and Active Addressing

The information on a twisted nematic device can be displayed either as a series of dots or in the form of a *seven-segment array* (very common in clocks and calculators). In the latter case, the seven segments can be used to show numeric and alphabetical characters (by choosing which segments to turn on). *Addressing* is the process by which individual display elements, known as *pixels*, are turned on and off in order to create an image. There are two main types of addressing, direct and multiplexing. Direct addressing is convenient for displays where there are only a few elements that have to be activated, such as the seven-segment display. With direct addressing, each pixel in the display has its own drive circuit. A microprocessor must individually apply a voltage to each element.

For a large laptop computer display with $768 \times 1024 \approx 8 \times 10^5$ pixels (*XGA* or eXtended Graphics Array – an IBM display standard introduced in 1990), it is impossible to realize such a large number of drive circuits. The individual pixels are therefore addressed as a matrix through rows and columns. For example, for a 100×100 matrix of pixels, 10^4 drivers are required for direct addressing in contrast to only 200 for multiplex addressing. This *passive matrix display* is addressed by a set of multiplexed transparent electrodes, perpendicular to one another, above and below the liquid crystal layer in a row and column formation (similar to the cross-bar arrangement used for memory devices and described in Chapter 11, Section 11.4). A passive pixel is addressed when there is a sufficient voltage across it to cause the liquid crystal molecules to align parallel to the electric field. A display can have more than one pixel on at any one time because of the response time of the liquid crystal material. When addressed, a pixel has a short turn-on time during which the liquid crystal molecules align in such a way as to make the pixel opaque. When the voltage is removed, the pixel behaves in a similar fashion to the discharge of a capacitor, slowly turning off as charge dissipates and the molecules return to their undeformed orientation. Because of this response time, a display can scan across the matrix of pixels, turning on the appropriate ones to form an image. As long as the time to scan the entire matrix is shorter than the turn-off time, a multiple pixel image can be displayed.

Active matrix displays are available in some laptop computers. In this type of display, the addressing takes place completely behind the liquid crystal film. The front surface of the display is coated with a continuous electrode whereas the rear surface electrode is patterned into individual pixels. As shown in Figure 8.19, a transistor (a thin film transistor) acts as a switch for each pixel. Electrical contacts to the transistor are made using a set of narrow

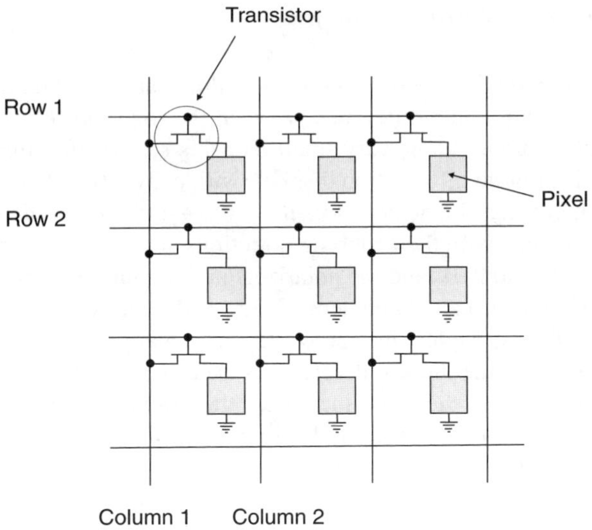

Figure 8.19 Active matrix display diagram. Each liquid crystal cell, or pixel, is turned on or off by a transistor.

multiplexed electrodes (gate lines and source lines) running along the gaps between pixels. A pixel is addressed by applying a voltage to a gate line which switches the transistor on and allows charge from the source line to flow to the rear electrode. A voltage is thereby established across the pixel, which turns on. An image is created in a similar fashion to the passive display as the addressing circuitry scans across the matrix. An active matrix display does not suffer from many of the limitations of the passive display. It can be viewed at an angle of up to 45° and can have a contrast of 40:1, meaning that the brightness of an 'on' pixel is 40 times greater than that of an 'off' pixel. It does, however, require a more intense back lighting system because the transistors and the gate and source lines are not very transparent.

8.4.5 Full-Colour Displays

To achieve a colour display, it is first necessary to have a device that is black in one state and white in the other. In a white display, all wavelengths pass through and, therefore, all wavelengths can be manipulated to create the desired colour. To achieve full colour, each individual pixel is divided into three subpixels: red, green and blue. So, for each full colour pixel, three distinct pixels are used. These subpixels are formed by using colour filters which only allow certain wavelengths to pass through them while absorbing the rest. With a combination of red, blue and green subpixels of various intensities, a pixel can be made to produce any number of different colours. This is analogous to a colour cathode ray tube, such as used in a television or computer monitor, in which different phosphors glow red, green or blue when excited by an electron beam. The number of colours that can be made by mixing red, green and blue subpixels depends on the number of distinct *grayscales* (intensities) that can be achieved by the display.

8.4.6 Super-Twisted Nematic Display

The simple twisted nematic display has a number of drawbacks for high-density information display devices. For example, the difference between the 'on' and 'off' voltages in displays with many rows and columns can be very small. For this reason, the simple twisted nematic device is impractical for large information displays with conventional addressing schemes. A significant improvement may be achieved using a *super-twisted nematic display.* Here, the alignment layers are placed with their rubbing directions at a variety of angles to one another to set up a twist from 180 to 270° and the polarizers are not applied parallel to the alignment layers. Figure 8.20 shows the calculated dependence of the mid-cell tilt angle on the applied voltage [3]. For small twist angles, the molecules all remain approximately parallel to the glass plates of the cell until the threshold voltage is exceeded; the tilt then rises slowly as the applied voltage increases. As the twist increases, the voltage dependence of the mid-cell angle on voltage becomes much steeper, leading to improved multiplexing capabilities. Beyond about 240°, a bistable situation even occurs (where the tilt increases with decreasing voltage). However, one disadvantage of the super-twisted display arises from the wavelength dependence of the transmitted light. The result is that most displays of this type are coloured, e.g. a pale blue or pale yellow/green background. As noted in the previous section, black and white operation, i.e. no colour dependence, is a necessary condition if a display is to achieve true full-colour operation. The best, but most expensive, solution is to use optical compensation in the form of a double super-twisted nematic cell. Here, two layers of a super-twisted nematic are used, but with opposite twists. In the 'off' state, the phase shift resulting from the first layer is compensated by the second layer. This pixel appears black. The 'on' state is not affected by the second super-twisted nematic layer, and white light emerges. Since the two layers consist of the same liquid crystal material, the behaviour is constant over the entire temperature range.

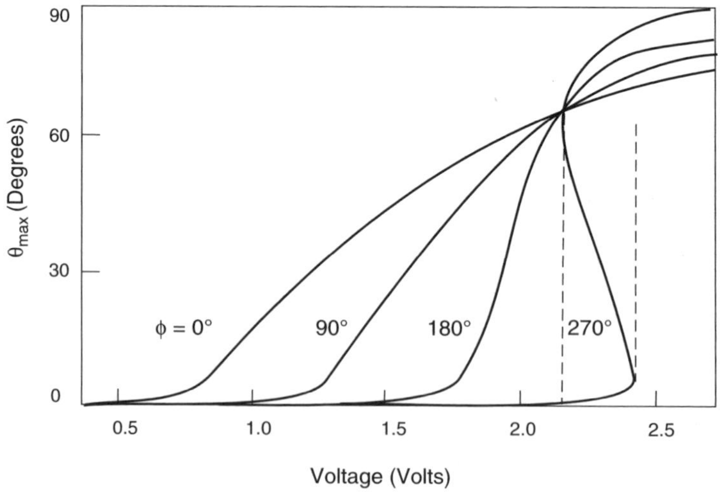

Figure 8.20 Theoretical dependence of the mid-cell angle θ_{max} of a twisted nematic liquid crystal cell on the applied voltage. The different curves represent different total twist angles ϕ. Reprinted from *Mol. Cryst. Liq. Cryst. Lett.*, **4**, Raynes EP, 'The theory of supertwist transitions', pp. 1–8. Copyright (1986) with permission from Taylor & Francis/ Gordon & Breach (US).

8.5 FERROELECTRIC LIQUID CRYSTALS

The requirements for a material to exhibit ferroelectricity are that it must possess a permanent polarization and that the polarization direction can be changed using an applied electric field (Chapter 5, Section 5.6). The smectic *C* phase of liquid crystals first appeared in literature in 1933. However, it was not until 1974 that it was realized that the phase ought to be ferroelectric. This was discovered by Robert Meyer, who later demonstrated it in a synthesized chiral smectic C material DOBAMBC, *p*-decyloxybenzylidene-*p*′-amino-2-methylbutyl cinnamate, the chemical structure of which is shown in Figure 8.21.

In the S_A phase (Figure 8.5), the molecules are upright and, since there is no head-to-tail ordering, there is no polarization normal to the layers. Moreover, even if the molecules themselves are chiral, there is equal probability of their assuming any orientation about their long axes. Hence the transverse component of the dipole moment is averaged out and there is also no net polarization parallel to the layers.

In the chiral S_C phase (S_C^*), the molecules are arranged as for an ordinary S_C phase However, if the molecules possess a lateral dipole, then this will lie parallel to the layer plane but perpendicular to the tilt plane. Thus, in the S_C^* phase, each layer is spontaneously polarized. Since the structure has a twist about the layer normal, the polarization direction will rotate from one layer to the next, resulting in an average dipole of zero. A bulk S_C^* sample, free to develop its helical structure, will therefore not show ferroelectric behaviour since the spontaneous polarization will average to zero over one pitch (since polarization vectors go around an entire circle and cancel each other out). This is often referred to as the *helielectric phase*.

In 1980, Clark and Lagerwall proposed a way to suppress the helix and developed the surface stabilized ferroelectric liquid crystal [4]. The helix is constrained by using a cell gap that is less than the helical pitch. The smectic layers are oriented approximately perpendicular to the glass bounding plate. Interaction forces between the liquid crystal and the bounding plates unwind the intrinsic helix. Symmetry arguments show that this boundary condition also causes the molecular orientation for each layer to be the same and the material exhibits ferroelectric behaviour. The director likes to lie in the plane of the bounding plates. Because of this condition and the fact that the director is constrained to be at a certain angle from the normal to the layer (i.e. to lie on the intersection of a cone and the bounding plate), there are two stable states, as shown in Figure 8.22. The polarization vector, therefore, must be normal to the bounding plates and its two states are in opposite directions. The values of the spontaneous polarization in these materials is fairly small, usually between 10 and 1000 nC cm^{-2}, i.e. one to two orders of magnitude less than that for a solid inorganic ferroelectric such as KH_2PO_4. The switching times for ferroelectric liquid crystal devices can be just a few tens of microseconds, about 1000 times faster than the twisted nematic device. This is because (a) the ferroelectric liquid crystal possesses ordered permanent electric dipoles and (b) the electric field forces both the on and the off transitions.

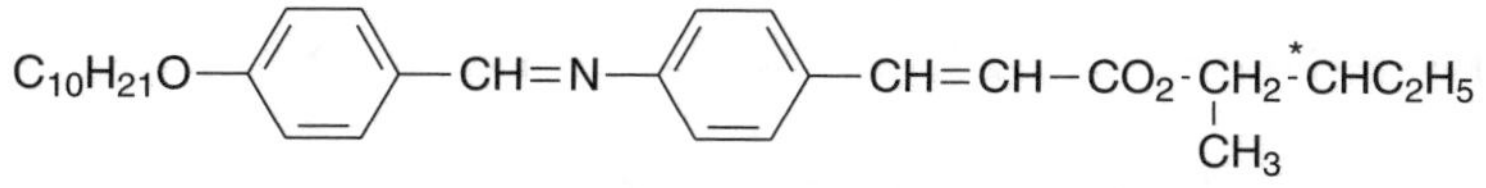

Figure 8.21 The structure of DOBAMBC, *p*-decyloxybenzylidene-*p*′-amino-2-methylbutyl cinnamate, a ferroelectric liquid crystal. The chiral centre is marked with an asterisk.

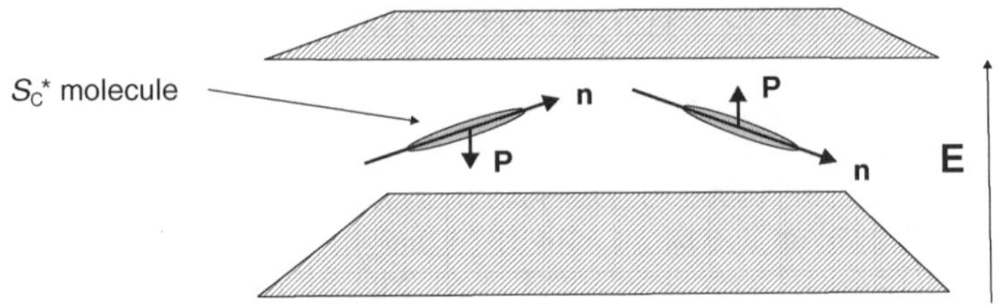

Figure 8.22 Surface-stabilized ferroelectric liquid crystal. Between closely spaced plates only two orientations are possible, with the directors **n** oriented into or out of the page. The corresponding polarizations **P** are down and up, respectively.

The bistability offered by surface-stabilized ferroelectric liquid crystal devices makes them ideal where low energy consumption is important for still images. This is because additional power is not required once the image has been created. The switching time is still fast enough to provide high frame rates required for video and also allows full colour on each pixel, giving a higher quality on a display of a given size. A further advantage of ferroelectric liquid crystal displays is the small pixel size, made possible by the thin cells. With a size of approximately 5 μm, the pixels can be made a factor of 10 smaller than those in nematic displays. This is significantly better than required for computer displays, e.g. in an XGA 768 × 1024 display of 14 inch diagonal, the subpixel representing a single colour is 90 μm wide [2]. Ferroelectric displays are therefore very suitable for microdisplay applications, e.g. projection or head-mounted displays.

Perhaps the main drawback of the liquid crystal display is that the molecular orientation may be lost by mechanical stress during the use of the display. Experience with an ordinary nematic display reveals that if the display is gently pressed, it will lose its function momentarily because the twisted structure is distorted. This may lead to unwanted colours or even a completely black display. However, within a short time the display will again function correctly. Unfortunately, this is not true for a ferroelectric display. As soon as regions, or domains, are formed which do not have a correct orientation, they will remain stable.

In the late 1980s, a different arrangement of the molecules in their layer plane was discovered. The phase is known as the *antiferroelectric liquid crystal* (AFLC) phase. This phase occurs in some materials at a temperature below the FLC phase. These materials, like FLCs, are chiral and possess a spontaneous polarization. The difference is that, in the AFLC phase, the director is tilted in opposite direction in alternate layers (see antiferromagnetic behaviour in Chapter 5, Section 5.7). For AFLCs, the pitch is the distance for the director to precess, 180° instead of 360° as for FLCs – due to the opposite tilt in adjacent layers the director also has gone around half of the cone. As for the FLC, the AFLC helix must be unwound through a boundary constraint for the material to be used in displays.

8.6 POLYMER-DISPERSED LIQUID CRYSTALS

Polymer-dispersed liquid crystals represent a class of materials which may have applications ranging from switchable windows to projection displays. They consist of liquid crystal droplets that are dispersed in a solid polymer matrix. The liquid crystal director inside each

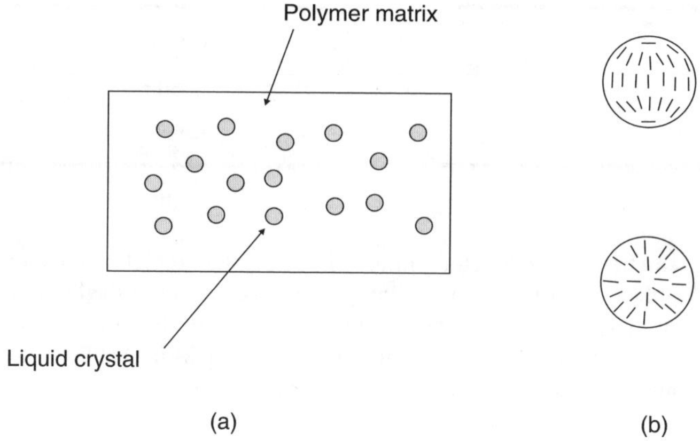

Figure 8.23 (a) Polymer dispersed liquid crystal with droplets of liquid crystal distributed throughout the polymer. (b) Two possible configurations for the directors, shown as lines, with respect to the polymer surface: top, parallel alignment; bottom, perpendicular alignment.

droplet adopts one of two configurations, depending on whether the liquid crystal prefers to align parallel or perpendicular to the polymer surface; the two arrangements are illustrated in Figure 8.23. In the case of parallel alignment, two points where the director is undefined occur; these are called *disclinations*, by analogy with dislocations in crystals (Chapter 2, Section 2.5.7). Perpendicular alignment causes only one disclination to form at the centre of the droplet.

For the case of parallel alignment, a random orientation is expected from the directors in the liquid crystal droplets. Light passing through each droplet will travel at one velocity for light polarized parallel to the director (determined by the extraordinary refractive index $n_{\parallel}$) and at a different velocity for light polarized perpendicular to the director (determined by the ordinary refractive index $n_{\perp}$). The solid polymer is isotropic, possessing a single refractive index n_s, which, in general, will be different from either refractive index of the liquid crystal. This situation is depicted in Figure 8.24(a). The difference in refractive indices between the polymer and the liquid crystal results in a structure that scatters light very effectively.

If an electric field is now applied to the polymer dispersed liquid crystal, the directors will try to orient with the field, as shown in Figure 8.24(b). Light passing through the cell now has its electric field nearly perpendicular to the director, so if $n_{\perp}$, the ordinary refractive index of the liquid crystal, is made equal to the polymer refractive index n_s, little reflection occurs at the boundary of each droplet (Chapter 4, Section 4.5.2). The polymer dispersed liquid crystal then appears clear.

It is relatively easy to produce polymer dispersed liquid crystals in large areas. As a consequence, their cost is low. Applications for these materials range from switchable windows to displays. The main drawback to their widespread application originates from incomplete alignment of the liquid crystal molecules under an applied electric field. Figure 8.23 reveals that the liquid crystal molecules close to the droplet surfaces tend to orient parallel to the surface and not always to the field. This results in a residual scattering which is manifest by cloudiness in window applications.

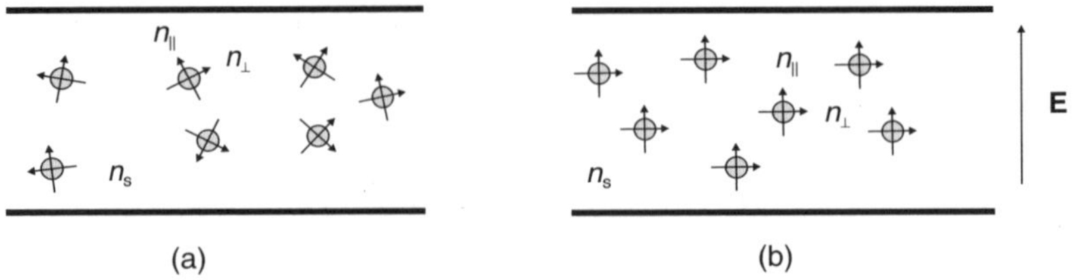

Figure 8.24 Reorientation of the director within a liquid crystal droplet in a polymer dispersed liquid crystal. $n_{\parallel}$ and $n_{\perp}$ are the extraordinary and ordinary refractive indices of the liquid crystal and n_s is the refractive index of the polymer. (a) With no applied electric field, the directors within the individual liquid crystal droplets are unaligned and incident light is scattered. (b) Application of an electric field aligns the directors and the cell appears clear if $n_{\perp}$ and n_s are equal.

8.7 LIQUID CRYSTAL LENSES

The ability of liquid crystals to control the phase of a light beam, in addition to its intensity, has led to their exploitation in electrically switchable optical elements such as lenses. A conventional fixed lens can be thought of a device that produces a phase shift across the incident beam, as a function of distance. It achieves this by a medium of constant refractive index (usually glass) with a variable thickness across the lens. A liquid crystal lens can produce the same result by keeping the physical thickness constant and by tuning the refractive index.

Most liquid crystal lenses are made up from a number of individual pixels, as in the case of displays. These can be in the form of a rectangular grid, or as concentric rings, in order to realize a circular lens. The desired shape of the wavefront is then approximated by controlling the phase of the individual pixels, illustrated in Figure 8.25. Alternatively, modal addressing may be used [5]. This removes the need for individual pixels and exploits the fact that a liquid crystal cell can be modelled as a distributed electrical circuit, similar to a transmission line. By using a relatively high-resistance transparent top electrode, and applying a voltage via a low-resistance contact electrode surrounding it, the voltage across the liquid crystal cell, and thereby the phase profile, will vary with position. Such lenses are

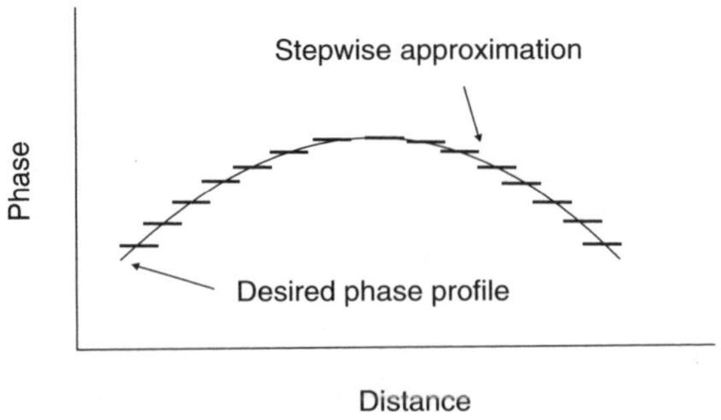

Figure 8.25 Approximation of a continuous phase profile (full line) with zonal (pixelated) liquid crystal elements.

(a) (b) (c)

Figure 8.26 Example of a liquid crystal lens in operation. (a) Lens is off and the optical imaging system is adjusted to provide a good focus. (b) Optical system is mechanically adjusted to defocus the image. (c) Liquid crystal lens is turned on to refocus the image. Reprinted from *Liq. Cryst. Today*, **10**, Love GD, Maumov AF, 'Modal liquid crystal lenses', pp. 1–4. Copyright (2001), with permission from Taylor and Francis Ltd. http://www.informaworld.com

simple to make and can produce a smooth change in focal length. The lens will respond to an applied AC field. The frequency of this field defines the magnitude of the voltage across the liquid crystal, which in turn shapes the electro-optic response. Figure 8.26 shows the effect of focusing such a lens [6]. A simple imaging system is set up using fixed lenses to focus some text on to a camera [Figure 8.26(a)]. The liquid crystal lens is placed in front of the system, but turned off. One of the fixed lenses is then displaced to defocus the image [Figure 8.26(b)] and this is subsequently corrected using the liquid crystal lens [Figure 8.26(c)].

Smart lenses based on hydrogels (Chapter 2, Section 2.7) which adapt to the environment have also been demonstrated [7]. For example, the focal length of a lens can be made to change as the result of a biological reaction.

8.8 OTHER APPLICATION AREAS

The most common application of liquid crystal technology is liquid crystal displays, which have been discussed in some detail above. Other applications noted above have included liquid crystal thermometers, phase retarders, electro-optic switches and lenses. Side-chain polymer liquid crystals exhibit good properties for applications in optically nonlinear devices, including optical waveguides and electro-optic modulators in poled polymeric slab waveguides. Other devices that may be fabricated from polymer liquid crystals include optically addressed spatial light modulators, tuneable notch filters, optical amplifiers and laser beam deflectors. The properties of ferroelectric chiral smectic *C* phases make this material useful for films with applications in nonlinear optics.

One area that is being explored is optical imaging and recording. In this technology, a liquid crystal cell is placed between two layers of photoconductor. Light is applied to the photoconductor, which increases its conductivity. This causes an electric field to develop in the liquid crystal corresponding to the intensity of the light. The electric pattern can be transmitted by an electrode, which enables the image to be recorded.

Liquid crystals have been put to many other uses. They are used for non-destructive mechanical testing of materials under stress. This technique is also used for the visualization of radiofrequency (RF) waves in waveguides. They are exploited in medical applications where, for example, the transient pressure transmitted by a foot walking on the ground is measured. Low molar mass liquid crystals have applications including erasable optical discs, full colour 'electronic slides' for computer-aided drawing (CAD), and light modulators for colour electronic imaging.

A few application areas focus on the alignment of molecules in polymer liquid crystals. Ordinary polymers have never been able to demonstrate the stiffness necessary to compete against traditional materials such as steel. It has been observed that polymers with long, straight chains are significantly stronger than their tangled counterparts. Main-chain liquid crystal polymers are well suited to ordering processes. For example, the polymer can be oriented in the desired liquid crystal phase and then quenched to create a highly ordered, strong solid. One success story is the development of *Kevlar*, poly(*p*-phenylene terephthalamide). This polymer is a synthetic fibre that is five times stronger than steel, weight for weight. Kevlar is very heat resistant and decomposes above 400°C without melting. It is usually used in bulletproof vests, in extreme sports equipment and for composite aircraft construction. It is also used as a replacement for steel cords in car tyres, in fire suits and as an asbestos replacement.

Kevlar was invented by the DuPont Corporation in the early 1960s, following the work of Stephanie Kwolek. Chemically, Kevlar is an *aramid* liquid crystalline polymer, which means an aromatic nylon (a nylon containing benzene rings). The chemical structure of Kevlar is shown in Figure 8.27. It derives its strength from intramolecular hydrogen bonds and aromatic stacking interactions between aromatic groups in neighboring strands. These interactions are much stronger than the van der Waals interactions found in other synthetic polymers and fibers. Kevlar consists of relatively rigid molecules, which form a planar sheet-like structure similar to silk protein. These properties result in its high mechanical strength and its remarkable heat resistance. Because it is highly unsaturated (Chapter 2, Section 2.4.3), it has a low flammability.

Kevlar molecules have polar groups accessible for hydrogen bonding. Water that enters the interior of the fibre can take the place of bonding between molecules and reduce the

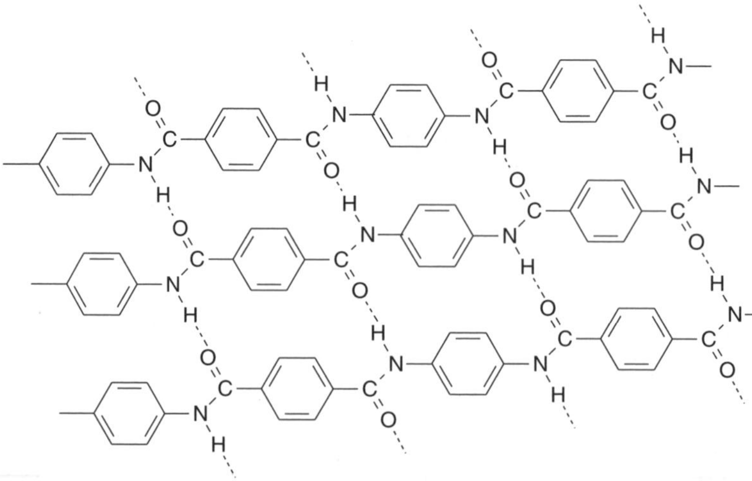

Figure 8.27 Chemical structure of Kevlar, poly(*p*-phenylene terephthalamide). Hydrogen bonding is depicted by dashed lines.

material's strength. The groups exposed at the surface lead to good wetting properties. This is important for bonding the fibres to other types of polymer, forming a fibre-reinforced plastic. This property also makes the fibres feel more natural and 'sticky' compared with nonpolar polymers such as polyethylene. In structural applications, Kevlar fibres can be bonded to one another or to other materials to form a composite.

BIBLIOGRAPHY

Chandrasekhar S. Liquid Crystals. 2nd ed. Cambridge: Cambridge University Press; 1992.

Collins PJ. *Liquid Crystals: Nature's Delicate Phase of Matter.* Bristol: Adam Hilger; 1990.

Collins PJ, Hird M. *Introduction to Liquid Crystals: Chemistry and Physics.* London: Taylor and Francis; 1997.

De Gennes PG, Prost J, Prost J. *The Physics of Liquid Crystals.* Oxford: Clarendon Press; 1995.

Demus D, Goodby J, Gray GW, Spiess H-W, Vill V (Editors). *Physical Properties of Liquid Crystals.* Weinheim: Wiley-VCH; 1999.

Khoo I-C, Simoni F (Editors). *Physics of Liquid Crystalline Materials.* Philadelphia: Taylor and Francis; 1991.

Lacey D. Liquid crystals and devices In MC Petty, MR Bryce, D Bloor (Editors). *An Introduction to Molecular Electronics.* London: Edward Arnold; 1995, pp. 185–219.

Singh S. *Liquid Crystals: Fundamentals.* Singapore: World Scientific; 2003.

Lagerwell ST. *Ferroelectric and Antiferroelectric Liquid Crystals.* Weinheim: Wiley-VCH; 1999.

REFERENCES

[1] Tredgold RH. *Order in Thin Organic Films.* Cambridge: Cambridge University Press; 1994.

[2] Zorn R, Wu S-T. Liquid crystal displays. In R Waser (Editor). *Nanoelectronics and Information Technology.* 2nd ed. Weinheim: Wiley-VCH; 2005, pp. 889–909.

[3] Raynes EP. The theory of supertwist transitions. *Mol. Cryst. Liq. Cryst. Lett.* 1986; **4**: 1–8.

[4] Clark NA, Lagerwall ST. Submicrosecond bistable electro-optic switching in liquid crystals. *Appl. Phys. Lett.* 1980; **36**; 899–901.

[5] Naumov AF, Love G. Control optimization of spherical modal liquid crystal lenses. *Opt. Express* 1999; **4**: 344–352.

[6] Love GD, Maumov AF. Modal liquid crystal lenses. *Liq. Cryst. Today* 2001; **10**: 1–4.

[7] Jiang H, Dong L. Smart lenses. *Phys. World* 2006; **19**: 29–31.

9 Plastic Electronics

Sell when you can, you are not for all markets

9.1 INTRODUCTION

In Chapter 1, we noted that the subject of molecular electronics could be divided broadly into two themes: *molecular materials for electronics* and *molecular-scale electronics.* This chapter is concerned with the former. Since the discovery of semiconducting behaviour in organic materials, there has been a considerable research effort aimed at exploiting this property in electronic and opto-electronic devices. The term *plastic electronics* is commonly used to describe this activity. Although this strictly refers to electronic devices incorporating polymeric compounds, the term is frequently used to include devices incorporating other semiconducting organic materials.

Organic semiconductors can have significant advantages over their inorganic counterparts. For example, thin layers of polymers can easily be made by low-cost methods such as spin-coating. Appropriate processing allows organic thin films to be produced in large areas, and sometimes in the form of free-standing films. High-temperature deposition from vapour reactants is generally needed for inorganic semiconductors. Synthetic organic chemistry also offers the possibility of designing new materials with different band gaps. As noted in Chapter 3, Section 3.4.1, the mobilities of the charge carriers in organic field effect transistors are low when compared with inorganic semiconductors such as Si and GaAs. Nevertheless, the simple fabrication techniques for organic materials have attracted several companies to work on transistor applications, such as data storage and thin film arrays to address liquid crystal displays.

9.2 ORGANIC DIODES

Semiconducting organic compounds may be used in a similar fashion to inorganic semiconductors in metal/semiconductor/metal structures. Perhaps the simplest example is that of a diode, or rectifying device, which can be made by sandwiching a semiconductor between different metals. Such devices are examined in the following section. The special case of 'molecular diodes' is discussed separately in Chapter 11, Section 11.5.

9.2.1 Schottky Diode

A *Schottky barrier* is formed when a metal and semiconductor, with appropriate *work functions*, are brought into contact. In practice, this process is frequently carried out by the

Molecular Electronics: From Principles to Practice M. C. Petty
© 2007 John Wiley & Sons, Ltd

thermal evaporation of a metal on to the surface of the semiconductor in a vacuum (Chapter 7, Section 7.2.2). Consider the case for an n-type semiconductor where the work function of the metal is greater than the work function of the semiconductor. The energy band diagrams before and after contact is made are shown in Figure 9.1. The work function Φ is the energy difference between the *vacuum level* and the Fermi level. The vacuum level defines the energy at which the electron is free from that particular solid, and where the electron has zero kinetic energy.

For the metal, the work function Φ_m is the minimum energy required to remove an electron from the solid. Typical values of Φ_m for very clean surfaces are 4.3 eV for Al and 5.1 eV for Au. However, these figures are changed if the surfaces become contaminated and also differ for the same substance in different morphological forms (e.g. single crystal and polycrystalline). In the metal, there are electrons at the Fermi level E_{Fm}, but in the semiconductor there are usually none at E_{Fs} (i.e. the semiconductor Fermi level is probably located within the band gap). Nevertheless, the semiconductor work function Φ_s still represents the energy required to remove an electron from the semiconductor. In Figure 9.1, the energy difference between the lowest lying state in the conduction band and the vacuum level is the electron affinity X (Chapter 3, Section 3.4.1). For the particular case depicted, $\Phi_m > \Phi_s$. Therefore, when the two solids come into contact, the more energetic electrons in the conduction band of the semiconductor can transfer to the metal,

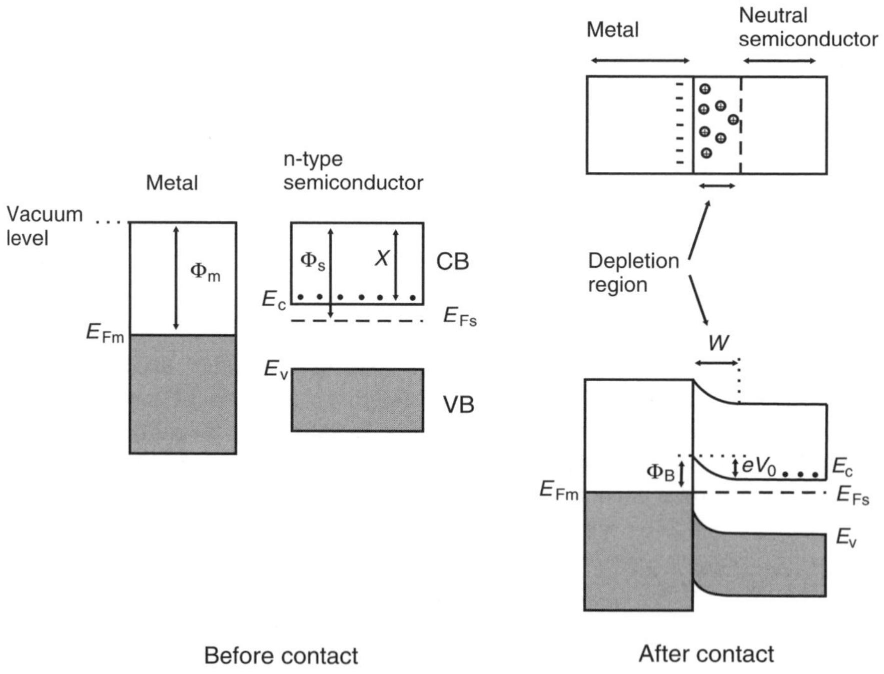

Figure 9.1 Formation of a Schottky barrier between a metal and an n-type semiconductor. CB and VB are the conduction and valence bands, respectively; E_c and E_v are the edges of the conduction and valence bands, respectively; E_{Fs} and E_{Fm} are the Fermi levels of the semiconductor and metal, respectively; Φ_s and Φ_m are the work functions of the semiconductor and metal, respectively; X is the electron affinity of the semiconductor; Φ_B is the Schottky barrier height; and V_0 is the contact potential.

into lower empty energy levels (just above E_{Fm}). These transferred electrons accumulate at the surface of the metal and leave behind an electron-depleted region of width W in which there are uncompensated positively-charged donor atoms, i.e. a net positive space-charge. The positive charge within this so-called *depletion region* matches the negative charge on the metal. A *contact potential* V_0 is established between the metal and semiconductor, which prevents further electron diffusion from the semiconductor conduction band to the metal. In equilibrium, the value of V_0 is given by

$$V_0 = (\Phi_m - \Phi_s)/e \tag{9.1}$$

Associated with the contact potential is an internal electric field directed from the semiconductor to the metal surface. This field is nonuniform and will have its maximum value at the metal–semiconductor interface (where there are a maximum number of field lines from positive to negative charge).

The Fermi level throughout the metal and semiconductor must be uniform in equilibrium (a Fermi level difference across the metal/semiconductor junction would result in an electric current flowing in an external circuit). In order for this to occur, the energy bands in the semiconductor must bend, as depicted in Figure 9.1. At the semiconductor edge of the depletion region, the semiconductor is still n-type. However, the carrier concentration in the conduction band n decreases towards the metal–semiconductor interface as $E_c - E_{Fs}$ increases (for intrinsic or undoped material, E_{Fs} is located close to the middle of the band gap; Chapter 3, Section 3.3.5). The potential energy required for an electron to move from the semiconductor bulk to the metal is called the *Schottky barrier height* Φ_B, given by

$$\Phi_B = \Phi_m - \mathrm{X} = eV_0 + (E_c - E_{Fs}) \tag{9.2}$$

which is greater than eV_0.

Provided that no external connections are made to the Schottky barrier (open-circuit conditions), no net current will flow through the metal–semiconductor interface. The number of electrons thermally emitted over the potential energy barrier Φ_B from the metal to the semiconductor is equal to the number of electrons thermally emitted over eV_0 from the semiconductor to the metal. However, when the semiconductor side of the junction is connected to the negative terminal of an external DC power source (e.g. a battery) and the metal to the positive terminal, the effect will be to reduce the contact potential from V_0 to $V_0 - V$, where V is the magnitude of the external voltage. It is now easier for the electrons to overcome the potential energy barrier into the metal. This situation is called *forward bias* and the resulting forward current is given by

$$I = I_0\left[\exp\left(\frac{eV}{k_B T}\right) - 1\right] \tag{9.3}$$

where I_0 is a constant that is related to the barrier height of the metal–semiconductor junction.

When the polarity of the external power source is reversed, the Schottky barrier becomes *reverse biased.* Now, the contact potential is increased by an amount V and it is more difficult for a current to flow. Under these conditions, the current is very low and is primarily due to thermal emission of electrons over the barrier Φ_B from the metal to the semiconductor.

Figure 9.2 shows the band structures of the Schottky barrier under (a) forward and (b) reverse bias, together with (c) the predicted current versus voltage behaviour of the device. The *I–V* characteristics exhibit rectifying properties, and the device is called a *Schottky diode.*

A Schottky diode can also be made on a p-type semiconductor (many organic semiconductors exhibit p-type conductivity). In this case, the requirement is that the work function of the metal is less than that of the semiconductor, $\Phi_m < \Phi_s$. The band diagram is shown in Figure 9.3. At equilibrium, the alignment of the Fermi levels requires a positive charge on the metal and an equal negative charge distributed in the depletion region of the semiconductor; this is provided by ionized acceptors left uncompensated by holes. For both the p- and n-type Schottky diodes, the forward current is due to injection of *majority carriers* (i.e. electrons in the n-type structure and holes in the case of the p-type device) from the semiconductor into the metal.

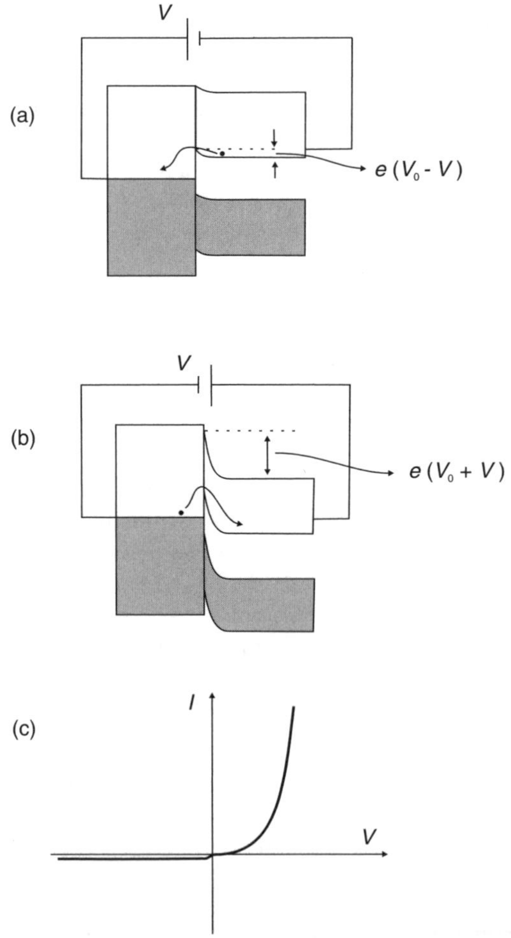

Figure 9.2 Schottky diode. (a) Bands in forward bias. (b) Bands in reverse bias. V is the applied voltage and V_0 is the contact potential. (c) Current I versus voltage V characteristics.

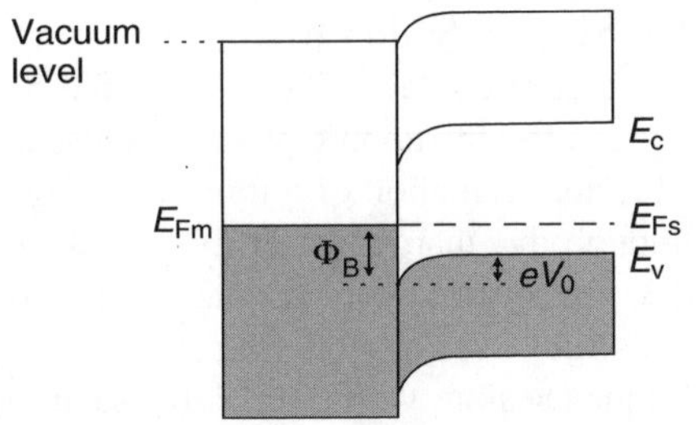

Figure 9.3 Band diagram for a Schottky barrier formed on a p-type semiconductor. E_c and E_v are the edges of the conduction and valence bands, respectively; E_{Fs} and E_{Fm} are the Fermi levels of the semiconductor and metal, respectively; Φ_B is the Schottky barrier height; and V_0 is the contact potential.

Real Schottky diodes can exhibit different current voltage characteristics to those predicted for the ideal structure [e.g. Equation (9.3) for the forward current]. For example, for forward bias voltages greater than k_BT/e, the current often takes the form

$$I = I_0 \exp\left(\frac{eV}{nk_BT}\right) \tag{9.4}$$

where n is called the *ideality factor* of the diode and usually has a value between 1 and 2 for diodes based on Si. Significantly higher values are often encountered for organic semiconductors. An example of the current versus voltage behaviour of organic diode based on the polymer poy(3-hexylthiophene) is shown in Figure 9.4 [1]. The polymer was spin-coated

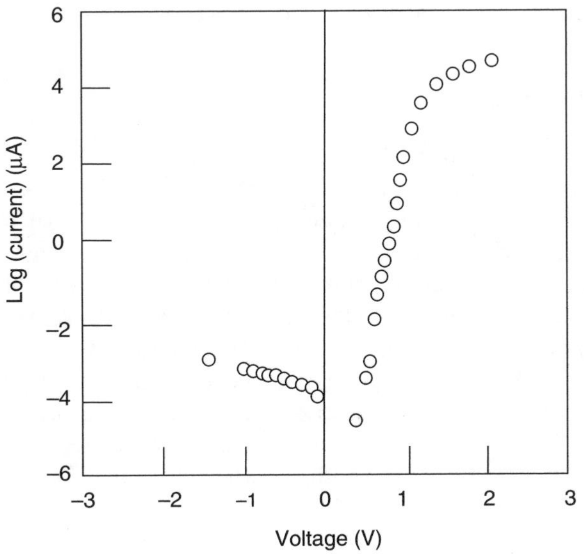

Figure 9.4 Current versus voltage characteristics for an In/polythiophene/Au diode. Reprinted from *Jpn. J. Appl. Phys.*, **33**, Kuo CS, Wakin FG, Sengupta SK, Tripathy, SK, 'Schottky and metal–insulator–semiconductor diodes using poly(3-hexylthiophene)', pp. 2629–2632, Copyright (1994), with permission from the Japanese Society of Applied Physics.

on to a gold substrate and subsequently doped with $FeCl_3$; a metallic top electrode was established using an indium pressure contact. The ideality factor was 3.8 for measurements taken in a short a time as possible after the doping process. Subsequent changes to the electrical characteristics were attributed to the formation of a thin insulating layer between the top metal contact and the polymer. Although the charge carrier mobilities in organic semiconductors are relatively low, organic diodes can operate to reasonably high frequencies. For example, a pentacene diode has been successfully used to rectifying an AC signal at 50 MHz and it has been argued that GHz operating frequencies are within reach for such devices [2].

The simple analysis of a Schottky barrier assumes that the metal is in intimate contact with a semiconductor surface, with no interfacial layer between. This is not usually the case in practice [as suggested for the poly(3-hexylthiophene) diode described above]. A semiconductor surface will contain defects due to incomplete covalent bonds (Chapter 2, Section 2.5.7) and other effects. These are so-called *surface states*, which can acts as traps for electrons or holes. Furthermore, the contact is seldom an atomically sharp discontinuity between the semiconductor and the metal. There is typically a thin interfacial layer, which is neither semiconductor nor metal. For example, silicon crystals are covered by a thin (1–2 nm) oxide layer even after etching or cleaving in atmospheric conditions. Because of such effects, it is difficult to fabricate Schottky diodes with barrier heights predicted from work function differences [i.e. Equation (9.2)]. In some cases, for example inorganic compound semiconductors such as GaAs, the density of surface states within the band gap is sufficient to produce an effect called *Fermi level pinning*, resulting in barrier heights that are independent of the work function of the metal contact.

An ultra-thin insulating layer may deliberately be inserted between a metal and a semiconductor to produce a so-called *metal–insulator–semiconductor* (MIS) structure (see Section 9.3). The effect of the insulator is to control transport and to shape the electric field in the device. The application to electroluminescent display devices is discussed in the Section 9.6.

9.2.2 Ohmic Contacts

An *ohmic contact* is a junction between a metal and a semiconductor that does not restrict the current flow. The current is essentially limited by the resistance of the semiconductor outside the contact region rather than the thermal emission rate of carriers across a potential barrier at the contact. Ohmic contacts exhibit linear *I–V* characteristics in both biasing directions. Very often such contacts are more difficult to fabricate than Schottky barriers! Ideal metal–semiconductor contacts are ohmic when the charge induced in the semiconductor in aligning the Fermi levels is provided by majority carriers. This requires the use of low work function metals (e.g. Al, $\Phi_m \approx 4.3\,eV$) for n-type semiconductors and high work function metals (e.g. Au, $\Phi_m \approx 5.1\,eV$) for p-type semiconductors. An approach that is often used to make ohmic contacts is to dope the semiconductor heavily in the contact region. If a barrier is then formed at the interface, the depletion width is sufficiently small to allow carriers to tunnel through the barrier (Chapter 3, Section 3.5.2).

9.3 METAL–INSULATOR–SEMICONDUCTOR STRUCTURES

The operation of many switching devices, such as the *metal–insulator–semiconductor transistor* or MISFET, is based on a phenomenon called the *field effect* (NB: the term

MOSFET refers to the situation in which the insulator is an oxide, typically silicon dioxide). This can best be understood by first considering the operating principles of a simple metal–insulator–semiconductor (MIS) device.

9.3.1 Idealized MIS Devices

The band structure for an MIS device, based on a p-type semiconductor, is shown in Figure 9.5. This device differs from the MIS structure noted in the Section 9.2.1 only in the thickness of the insulating layer. For the MIS device shown, the insulator is assumed to be perfect and relatively thick so that charge transport between the metal and semiconductor does not occur. The structure is essentially that of a parallel plate capacitor in which one plate is a semiconductor. For the idealized case (Figure 9.5), it is assumed that the work function of the metal is equal to that of the semiconductor ($\Phi_m = \Phi_s$). With no external bias, the Fermi levels of the metal and semiconductor align and the various bands are flat throughout the MIS structure – *flat-band conditions.*

Figure 9.6 shows the effect of applied an external voltage to the MIS structure. If a negative voltage is applied to the metal contact (with respect to the semiconductor), a negative charge will be produced on the metal. Using the capacitor analogy, an equal and opposite charge must therefore be produced at the semiconductor surface. In the case under consideration, i.e. that of a p-type semiconductor, this occurs by hole (majority carrier) *accumulation* at the semiconductor–insulator interface [Figure 9.6(a)]. The detailed explanation for this is as follows. Since the applied negative potential reduces the electrostatic potential of the metal relative to the semiconductor, the electron energies are raised in the metal relative to the semiconductor. As a result, the metal Fermi level E_{Fm} lies above its equilibrium position by an amount eV, where V is the applied voltage. As no current flows through the ideal MIS structure, there can be no variation in the semiconductor Fermi level E_{Fs} within the semiconductor. The conduction and valence bands in the semiconductor are displaced upwards with respect to E_{Fs}. The carrier density in the valence band depends exponentially on the energy difference between the valence band edge E_v and E_{Fs} [3].

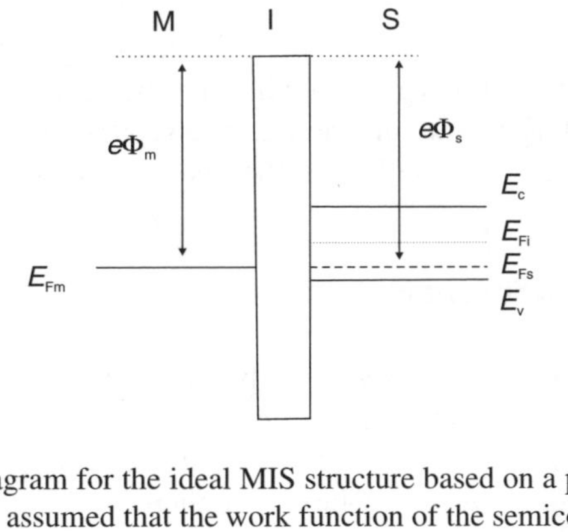

Figure 9.5 Energy band diagram for the ideal MIS structure based on a p-type semiconductor at zero applied bias. In this case, it is assumed that the work function of the semiconductor is equal to the work function of the metal ($\Phi_s = \Phi_m$). E_c and E_v are the edges of the conduction and valence bands, respectively; E_{Fs} and E_{Fm} are the Fermi levels of the semiconductor and metal, respectively; and E_{Fi} represents the position of the Fermi level for intrinsic material.

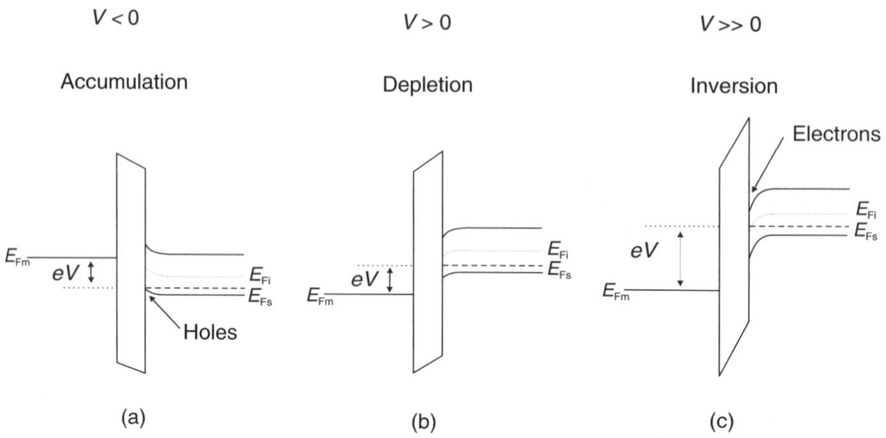

Figure 9.6 Energy band diagrams for an MIS structure based on a p-type semiconductor with different voltages applied to the gate electrode. (a) Negative voltage applied to gate, holes accumulate at semiconductor surface; (b) positive voltage applied to gate, depletion layer forms at semiconductor surface; (c) large positive voltage applied to gate, layer of negative charge (inversion layer) forms at semiconductor surface. V = applied voltage; E_{Fm} = metal Fermi level; E_{Fs} = semiconductor Fermi level; E_{Fi} = intrinsic Fermi level.

Therefore, the band bending causes an accumulation of majority carriers (holes) near the semiconductor surface. This makes the semiconductor surface layer more p-type, leading to a hole accumulation layer.

If a positive voltage is now applied to the metal, a negative charge appears in the semiconductor. In the p-type material, this arises from the depletion of holes from the region near the surface, leaving behind uncompensated ionized acceptors. The situation is called *depletion* and is similar to that for the Schottky diode discussed in Section 9.2.1. In the depleted region, shown in Figure 9.6(b), the hole concentration decreases, bending down the bands at the semiconductor–insulator interface.

If the positive voltage on the gate is increased further, the semiconductor bands bend down more strongly. At some point E_{Fi}, the Fermi level position for intrinsic material, becomes lower than E_{Fs} at the surface. The semiconductor region adjacent to the surface then has the conduction properties of n-type material. This situation is referred to as *inversion* and is illustrated in Figure 9.6(c). The inverted layer, separated from the underlying p-type material by a depletion region, is the key to the operation of many MIS transistor devices.

9.3.2 Organic MIS Structures

The different regimes in an MIS structure may be monitored conveniently by measuring the *differential capacitance* (i.e. $\partial C/\partial V$) of the device. Figure 9.7 shows the capacitance and conductance at a measurement frequency of 111 Hz and a temperature of 270 K as a function of bias for an Au/silicon dioxide/poly(3-hexylthiophene) MIS structure [4]. The accumulation A, depletion D and inversion I regimes are marked on the figure. The decrease in the capacitance in the depletion region is due to the capacitance of the depletion layer (which decreases as the applied bias moves toward positive voltages and the width of the

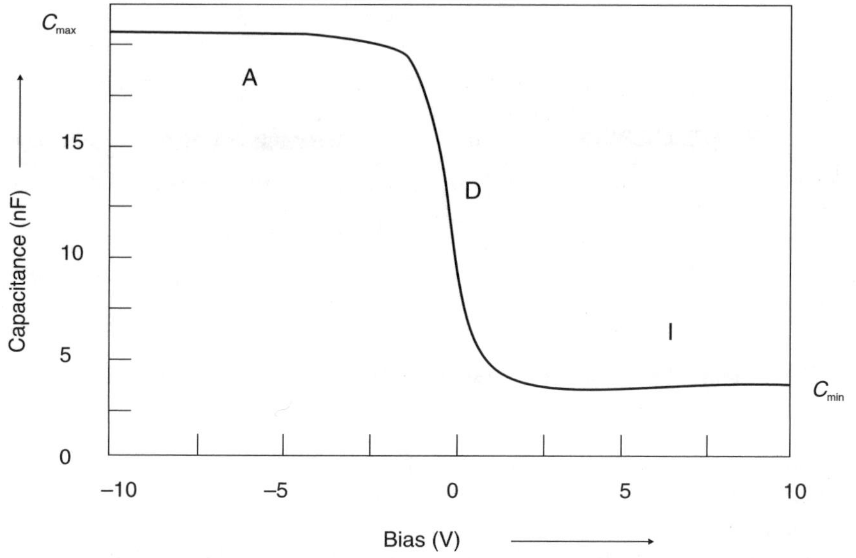

Figure 9.7 Capacitance versus voltage behaviour for an MIS structure comprising Si/SiO_2/polythiophene/Au measured at 111 Hz and at 270 K. The accumulation A, depletion D and inversion I regions are indicated. Reprinted from *Synthetic Metals*, **146**, Grecu S, Bronner M, Opitz A, Brütting W,' Characterization of polymeric metal–insulation–semiconductor diodes; pp. 359–363, Copyright (2004), with permission from Elsevier.

depletion layer grows) appearing in series with the SiO_2 capacitance. When minority carriers (electrons in this case) are generated in the inversion regime, the applied electric field cannot further penetrate the depletion region, which remains a constant thickness. The capacitance therefore remains constant, as shown in Figure 9.7. At very low frequencies, the capacitance in inversion would be expected to rise to the value in accumulation, as the minority carriers respond to the AC measuring voltage [3]. However, this situation is not usually observed for MIS devices based on organic semiconductors. The capacitance versus voltage scan shown in Figure 9.7 showed little dependence on the direction of the voltage sweep [4]. However, MIS structures incorporating organic materials (as either the insulator, semiconductor layer or both) often show large hysteresis effects. These can be attributed to mobile ions in the organic layer, to trapping of charge at the interface between the semiconductor and insulator and/or to polarization of the insulating layer.

9.4 FIELD EFFECT TRANSISTORS

Field effect transistors are three-terminal structures: a voltage applied to a metallic *gate* affects an electric current flowing between *source* and *drain* electrodes. Figure 9.8 shows how the FET device is related to that of the MIS structure described in the previous section. Application of a voltage to the gate electrode V_g alters the charge density at the semiconductor surface. The mode of operation of organic FETs is invariably in the accumulation region. Hence, the majority carrier current flowing between the source and drain electrodes I_d is changed by the gate voltage. This control feature allows the

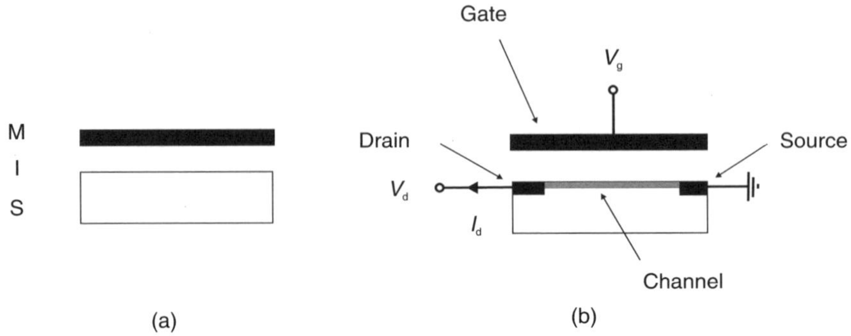

Figure 9.8 Contrast between (a) MIS structure and (b) MISFET.

amplification of small AC signals or the device to be switched from an on state to an off state and back again. These two important operations, amplification and switching, are the basis of many electronic functions.

The drain current–voltage characteristics for a typical MISFET based on a p-type semiconductor are shown in Figure 9.9. The electrical characteristics can be adequately described by models developed for inorganic semiconductors [3, 5]. When the gate voltage is biased negatively with respect to the grounded source electrode, the accumulation of holes occurs at the semiconductor–insulator interface. For low drain voltage V_d, the channel acts as a resistance; the drain current I_d increases linearly with V_d (the linear regime) and is given approximately by the equation

$$I_d = \frac{WC_i}{L}\mu\left(V_g - V_t - \frac{V_d}{2}\right)V_d \tag{9.5}$$

where L and W are the length and width, respectively of the conducting channel, C_i is the capacitance per unit area of the insulating layer, V_t is a constant (for a particular device)

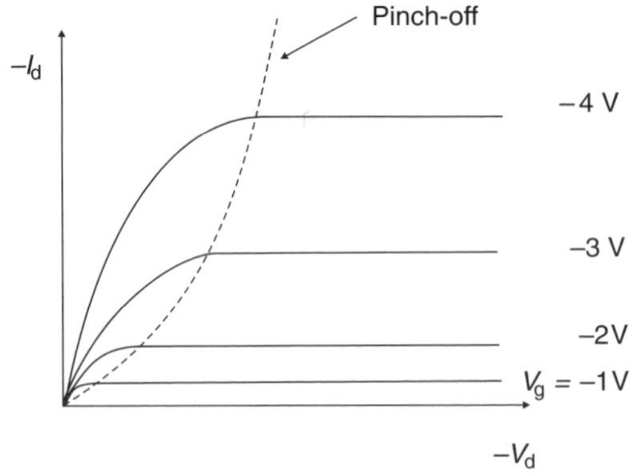

Figure 9.9 Drain current I_d versus drain voltage V_d characteristics for a MISFET based on a p-type semiconductor. The curves are shown for different applied gate voltages. The dashed line indicates where the channel of the FET becomes pinched off as a result of increasing V_d.

called the *turn-on* or *threshold voltage* and μ is the field effect charge carrier mobility. The latter can be calculated in the linear regime from the *transconductance*, g_m:

$$g_m = \left(\frac{\partial I_d}{\partial V_g}\right)_{V_d=\text{constant}} = \frac{WC_i}{L}\mu V_d \tag{9.6}$$

As the magnitude of V_d increases, I_d tends to saturate, as shown in Figure 9.9. This is the result of *pinch-off* off the accumulation layer. The effect is illustrated in Figure 9.10. The combination of the voltages applied to the gate and drain will produce a wedge-shaped accumulation layer. At pinch-off, the effective cross-sectional area of the conduction channel becomes zero at the drain. The drain current beyond the pinch-off point remains essentially the same and can be modelled using

$$I_d = \frac{WC_i}{2L}\mu(V_g - V_t)^2 \tag{9.7}$$

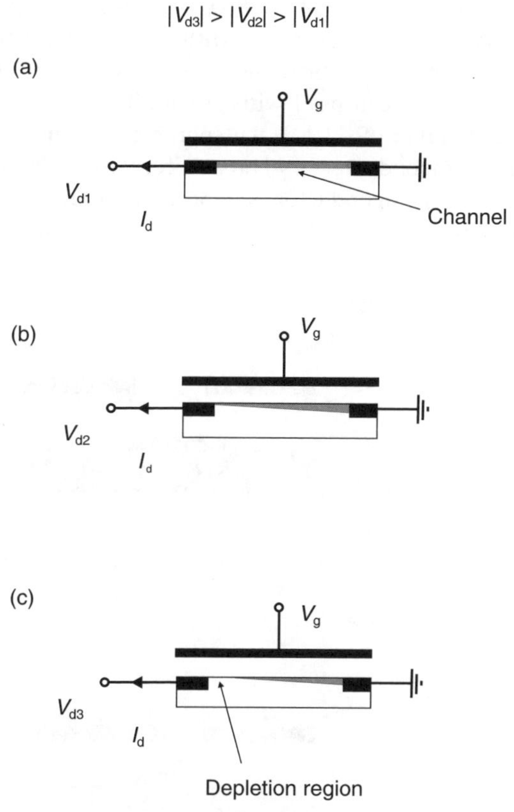

Figure 9.10 Schematic diagrams showing the effect of applying an increasing drain voltage to a MISFET (constant gate bias). (a) Linear regime: channel extends between source and drain electrodes. (b) Pinch-off: cross-sectional area of channel becomes zero at the drain electrode. (c) Saturation region: carriers must cross the depletion layer to reach the drain electrode.

In the saturation regime, the carrier mobility can be calculated from the slope of the plot of $|I_d|^{0.5}$ versus V_g. The transition between the saturated and the linear region appears approximately where $V_d = V_g - V_t$ and is indicated in Figure 9.9 with a dashed line. When the gate electrode for the FET device discussed above is positively biased, the channel is depleted of carriers and little current flows between the source and the drain. The ratio of the current flowing in the accumulation regime to that in depletion is often referred to as the I_{on}/I_{off} ratio of the device.

Looking at the expression for the drain current in the saturation regime given by Equation (9.7), it appears that I_d becomes zero when V_g is reduced to V_t. In reality, there is still some drain current conduction below threshold, and this is known as the *subthreshold conduction.* In polymer FETs, this current is related to the conductivity of the organic semiconductor. For state-of-the-art MOSFETs, the slope of the plot of $\log(I_d)$ versus V_g in the subthreshold region is typically 70 mV per decade. This means that a change in the input voltage of 70 mV will change the output I_d by an order of magnitude. Clearly, the smaller the value of this subthreshold slope, the better the transistor is as a switch.

For efficient transistor operation, charge must be injected easily from the source electrode into the organic semiconductor and the carrier mobility should be high enough to allow useful quantities of source–drain current to flow. The organic semiconductor and other materials with which it is in contact must also withstand the operating conditions without thermal, electrochemical or photochemical degradation. Figure 9.11 shows two common device configurations that have been used with organic FETs. In the first, referred to as a 'top-contact' configuration [Figure 9.11(a)], the source and drain electrodes are formed on top of the organic semiconductor, whereas for the 'bottom-contact' structure [Figure 9.11(b)], the organic semiconductor is deposited on prefabricated source and drain electrodes. The latter structure has the advantage of the organic layer being deposited in the

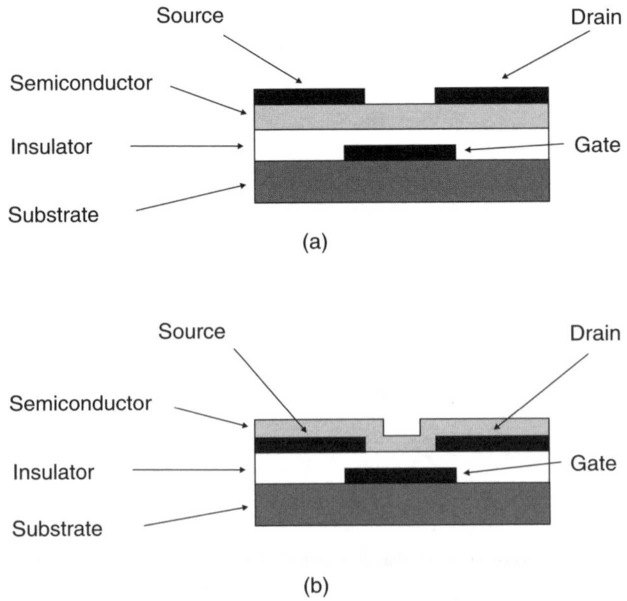

Figure 9.11 Schematic diagrams of thin film MISFETs. (a) Top and (b) bottom electrode structure.

final stage of device processing; it will therefore not be exposed to subsequent thermal or chemical processing steps. It is important to note that the electronic nature of the metal/organic semiconductor junction can depend on whether the semiconductor is deposited on the metal or vice versa [6].

The semiconducting and insulating layers in organic FETs can be produced using a variety of different thin film techniques, including spin-coating, thermal evaporation and Langmuir–Blodgett assembly (Chapter 7). Figure 9.12 illustrates a series of processing steps that can be used to fabricate a polymer FET using the method of inkjet printing [7]. In the first stage, a substrate (either rigid or flexible) is patterned (Chapter 7, Section 7.4). The conductive polymer is inkjet printed to form the transistor source and drain electrodes. A layer of semiconducting polymer is then printed over the channel defined by the source and drain and a layer of dielectric is deposited by spin-coating. Finally, a conducting gate electrode is inkjet printed.

The operating characteristics of organic transistors have improved markedly since they were first demonstrated in the 1980s [5]. This has been brought about both by improvements

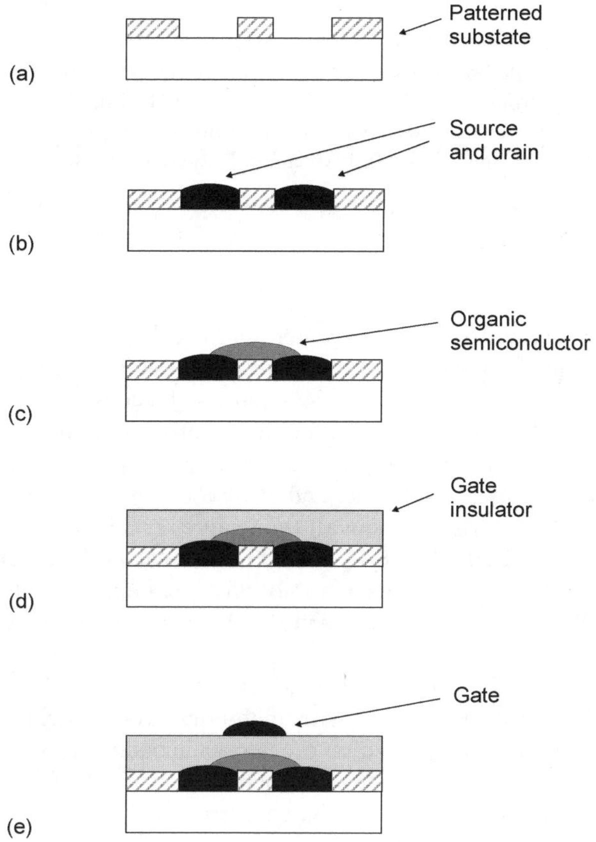

Figure 9.12 Processing steps to form an organic MISFET using the technique of inkjet printing. (a) Prepatterned substrate; (b) organic source and drain electrodes printed; (c) organic semiconducting layer printed; (d) gate insulator deposited by spin-coating; (e) organic gate electrode printed. Reproduced from Plastic Logic [7].

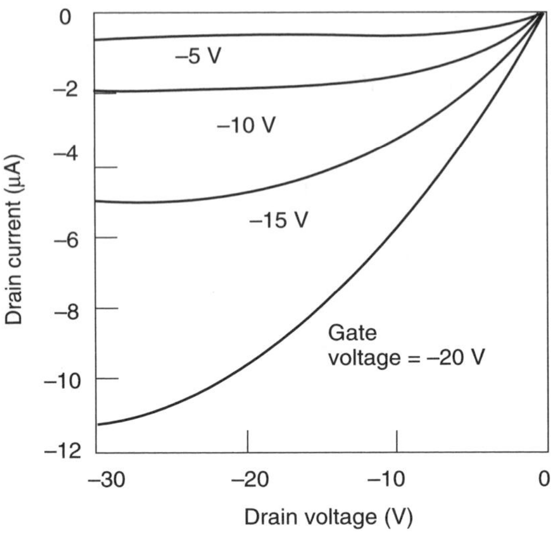

Figure 9.13 Drain current versus drain voltage characteristics of a MISFET based on pentacene. Device fabricated on a flexible polymer substrate. Gate dielectric: polyvinylphenol, thickness 270 nm. Reprinted with permission from Klauk H, Halik M, Zschieschang U, Eder F, Schmid G, Dehm C, *Appl. Phys. Lett.*, **82**, 'Pentacene organic transistors and ring oscillators on glass and on flexible polymeric substrates', pp. 4175–4177. Copyright (2003) American Institute of Physics.

in the material synthesis and in the thin film processing techniques and by a realization that organic compounds can be degraded by water and/or oxygen in the atmosphere. Consequently, the processing of devices and their subsequent storage usually take place in an inert ambient. The lifetimes of organic devices can also be considerably extended by some kind of encapsulation. Figure 9.13 shows the drain current–voltage characteristics for an organic FET based on thermally evaporated pentacene [8]. This particular device was fabricated on a flexible polymer substrate and using a polymer gate insulator deposited by spin-coating. These characteristics show all the features of Figure 9.9. Figure 9.14 shows the dependence of I_d on V_g and $I_d^{0.5}$ on V_g for the same device. The latter plot is reasonably linear for high values of V_g and reveals a mobility figure of approximately 0.3 cm^2 V^{-1} s^{-1}. The pentacene FET has a subthreshold swing of 1.6 V per decade and an on/off current ratio of 10^6.

State-of-the-art organic FETs possess characteristics similar to those of devices prepared from hydrogenated amorphous silicon, with mobilities around 1 cm^2 V^{-1} s^{-1} and on/off ratios $> 10^6$. Thin film transistors based on organic semiconductors are unlikely to compete with single crystal inorganic semiconductors such as Si or GaAs for fabricating very fast switching devices. This is because the charge carrier mobilities in organic compounds (Chapter 3, Table 3.1) are very much less than found in their inorganic counterparts. However, organic FETs may find a role in niche areas, such as components of plastic circuitry for use as display drivers in portable computers and pagers, and as memory elements in transaction cards and identification tags. Such prospects are discussed in the next section.

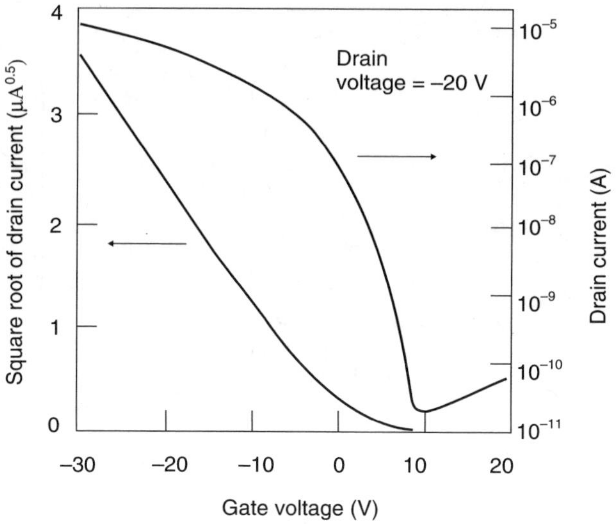

Figure 9.14 Drain current and square root of drain current versus gate voltage characteristics for pentacene MISFET, the I_d versus V_d characteristics of which are given in Figure 9.13. Reprinted with permission from Klauk H, Halik M, Zschieschang U, Eder F, Schmid G, Dehm C, *Appl. Phys. Lett.*, **82**, 'Pentacene organic transistors and ring oscillators on glass and on flexible polymeric substrates', pp. 4175–4177. Copyright (2003) American Institute of Physics.

9.5 INTEGRATED ORGANIC CIRCUITS

A significant step in the microelectronics industry was the ability to manufacture a complete circuit, an *integrated circuit*, on the same substrate. Simple integrated circuits can be fabricated using organic materials. Figure 9.15 shows the characteristics of an *inverter* device (i.e. a circuit that produces logic '1' for logic '0' input and vice versa) based on two interconnected pentacene MISFETs [8]. One of the transistors acts as the load for the switching device. The resulting inverter has sufficiently large gain but, due to the positive switch-on voltage, the input and output levels do not match. Also evident is a hysteresis of a few hundred millivolts in the inverter characteristic, which is possibly due to mobile charges in the gate dielectric. This situation can be improved by the incorporation of a voltage-level-shifting network in the integrated circuit. These simple inverter structures can form the basis of more complex oscillator circuits.

A particularly useful silicon device for digital applications is a combination of n-channel and p-channel MOS transistors on adjacent regions of an integrated circuit. This *complementary MOS* (referred to as CMOS) circuitry dominates the market for applications such as microprocessors (Chapter 1, Section 1.3.1). Complementary circuits are also of interest for less sophisticated applications, especially those that are battery-powered, since they can provide very low static power dissipation and therefore extend battery life. Although a number of p-channel organic MISFETs are available with performance comparable to that of amorphous silicon, n-channel devices have not received the same attention [9–11]. Notable examples are FETs based on organometallic charge-transfer complexes [9] and the fullerene

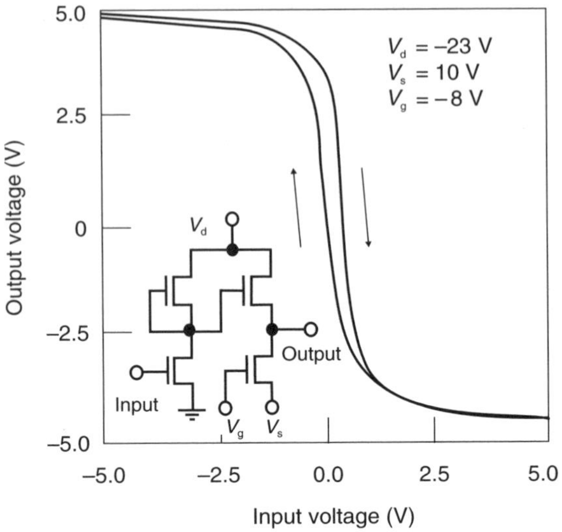

Figure 9.15 Output versus input voltage of an inverter device based on pentacene. The circuit of the structure is shown in the inset. Reprinted with permission from Klauk H, Halik M, Zschieschang U, Eder F, Schmid G, Dehm C, *Appl. Phys. Lett.*, **82**, 'Pentacene organic transistors and ring oscillators on glass and on flexible polymeric substrates', pp. 4175–4177. Copyright (2003) American Institute of Physics.

C_{60}, an electron acceptor [11]; these can possess high n-channel mobilities (Chapter 3, Table 3.1). By combining low-voltage p-channel and n-channel organic thin film transistors in a complementary circuit, a static power consumption of less than 1 nW per logic gate may be achieved [12].

9.5.1 Radiofrequency Identification Tags

Radiofrequency identification (RFID) is a method of remotely storing and retrieving data using devices called RFID tags. An RFID tag is a small object, such as an adhesive sticker, which can be attached to, or incorporated into, a product. The tags contain antennas to enable them to receive and respond to radiofrequency (RF) queries from an RFID transceiver. A typical system is depicted in Figure 9.16. These tags can be either *active* or *passive.* Passive RFID tags do not have their own power supply: the minute electrical current induced in the antenna by the incoming RF scan provides enough power for the tag to send a response. Due to power and cost concerns, the response of a passive RFID tag is necessarily brief: typically just an ID number. Lack of its own power supply makes the device fairly small, and commercially available products exist that can be embedded under the skin. As of 2004, the smallest such devices commercially available measured 0.4×0.4 mm, and thinner than a sheet of paper; such devices are practically invisible. Passive tags have practical read ranges that vary from about 10 mm up to about 5 m.

Active RFID tags, on the other hand, must have a power source, and may have longer ranges and larger memories than passive tags, in addition to the ability to store additional

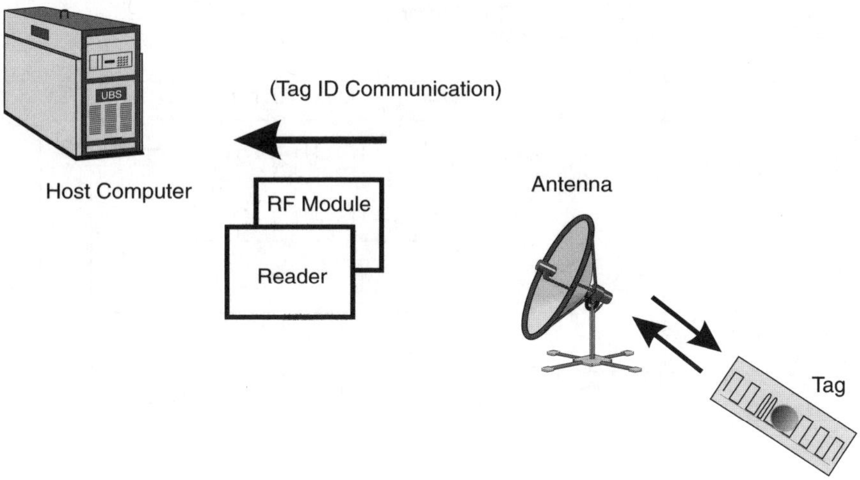

Figure 9.16 Schematic diagram of an RFID system. The antenna captures the tag ID number, first as analogue radiofrequency waves, then it is converted to digital information.

information sent by the transceiver. At present, the smallest active tags are about the size of a coin. Many active tags have practical ranges of tens of metres and a battery life of up to several years. Identification tags can operate at different frequencies, ranging from about 100 kHz to the microwave region (around 2 GHz).

Figure 9.17 shows (a) the circuit and (b) a photograph of an RFID system based on pentacene thin film transistors [13]. This comprises a seven-stage ring-oscillator and a five-stage output buffer. The gate length for all the transistors was nominally 20 μm, and the gate width for each of the ring-oscillator inverters was 60 and 300 μm for the load and driver transistors, respectively. The oscillator, at 73 V DC power, had a propagation delay of 1.7 μs. The output buffer consists of an inverter equal in size to the ring-oscillator inverters, followed by successively larger inverters. The largest inverter has a drive transistor gate width of about 6 mm, made using interdigitated source and drain electrodes. Circuit simulations indicate that such an inverter provides adequate load modulation (amplitude modulation) of the RF carrier.

The circuit shown in Figure 9.17(a) uses the AC power from a tuned amplifier to power directly the logic circuitry, without an intermediate rectification stage. This approach greatly simplifies fabrication. Furthermore, using AC power allows for much higher RF frequencies than would be practical using organic-semiconductor-based diodes in a rectification stage. During the forward biased cycle of the AC power (in which a negative voltage is applied to the drain of the load thin film transistor), the logic circuitry operates as expected. In the reverse bias portion, the load transistor is turned off, and the inverter output is either floating or driven towards ground in the case of a logic 0 or logic 1 input, respectively. The circuit also benefits from a relatively large logic gate input capacitance to filter out the RF and to maintain the output signal.

The reader uses a circular two-turn antenna, approximately 3 inches in diameter, with an inductance of about 1.5 μH. A commercial silicon diode feeding into a preamplifier makes up the detection circuit, which was shown to operate at 125 kHz, with an upper limit of around 6.5 MHz.

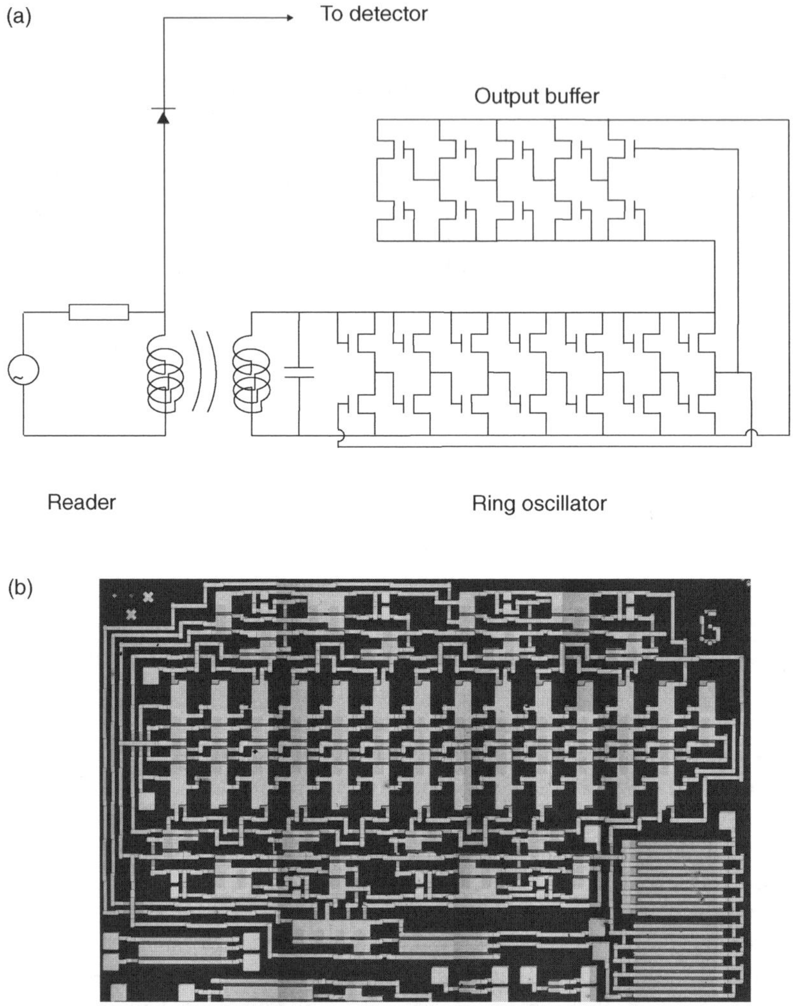

Figure 9.17 (a) Circuit diagram of a ring oscillator, fabricated using pentacene FETs. (b) Photograph of the organic integrated circuit. Reprinted with permission from Baude PF, Ender DA, Haase MA, Kelley TW, Muyres DV, Theiss SD, *Appl. Phys. Lett.*, **82**, 'Pentacene-based radio-frequency identification circuitry', pp. 3964–3966. Copyright (2003) American Institute of Physics.

9.6 ORGANIC LIGHT-EMITTING DISPLAYS

Reports of light emission from organic materials on the application of an electric field (electroluminescence) have been around for many years. In 1987, workers at Kodak reported efficient low-voltage electroluminescence in an organic thin film device incorporating dye molecules [14]. Interest intensified in the 1990s following the report of organic light-emitting devices (OLEDs) incorporating the conjugated polymer polyphenylenevinylene (PPV) (Chapter 5, Section 5.3) [15]. The simplest OLED is an electroluminescent (EL) compound, such as a polymer or dye, sandwiched between metals of high and low work function, as

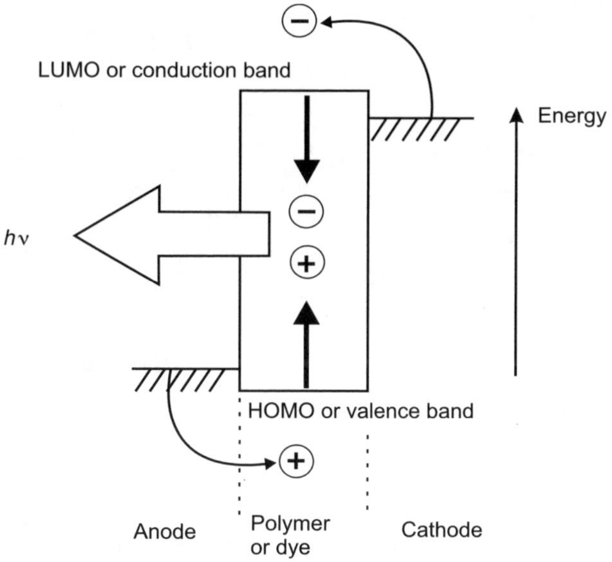

Figure 9.18 Schematic energy band structure of an organic light emitting device (OLED). The recombination of electrons and holes results in the emission of light of frequency ν and energy $h\nu$.

depicted in Figure 9.18. In practice, just like all other electronic devices based on organic compounds, the devices are encapsulated to prevent the inclusion of water and oxygen, both of which can lead to short operating lifetimes.

On application of a voltage, electrons are injected from the low work function electrode into the LUMO level (or conduction π^* band in the case of an organic compound possessing a delocalized electron system) of the organic compound and holes from the high work function electrode into the HOMO level (or valence π band in the case of an organic compound possessing a delocalized electron system). The recombination of these oppositely charged carriers then results in the emission of light. Work is focused on the use of both low molecular weight organic molecules and polymers as EL materials and there is considerable industrial interest in the application of such materials to various display technologies [16, 17]. The chemical formulae of some of the materials that have been used as the EL compounds in OLED devices are shown in Figure 9.19. The ability to make thin films of such compounds is an important practical issue. For example, the addition of the side chain to the PPV monomer, to form MEH-PPV, allows the latter compound to be easily spin-coated on to a variety of substrates. Polymers such as poly(9,9-dialkylfluorene) have a significant advantage over soluble polyphenylenes in that solubilizing side chains can be introduced without inhibiting the intrachain conjugation with the aromatic π system [18].

The nonpolymeric compounds, such as tris(8-hydroxyquinoline)aluminium (Alq_3), shown in Figure 9.19 are usually formed into thin films by the process of thermal evaporation. The fabrication of OLEDs is invariably undertaken in an inert atmosphere such as nitrogen to prevent degradation of the organic layers. Figure 9.20 shows the forward current versus voltage and EL versus voltage characteristics of an OLED based on indium tin oxide (ITO)/PPV/Al [19]. The ITO electrode is positively biased to obtain these characteristics. Negligible current and light emission result with the opposite polarity of applied voltage.

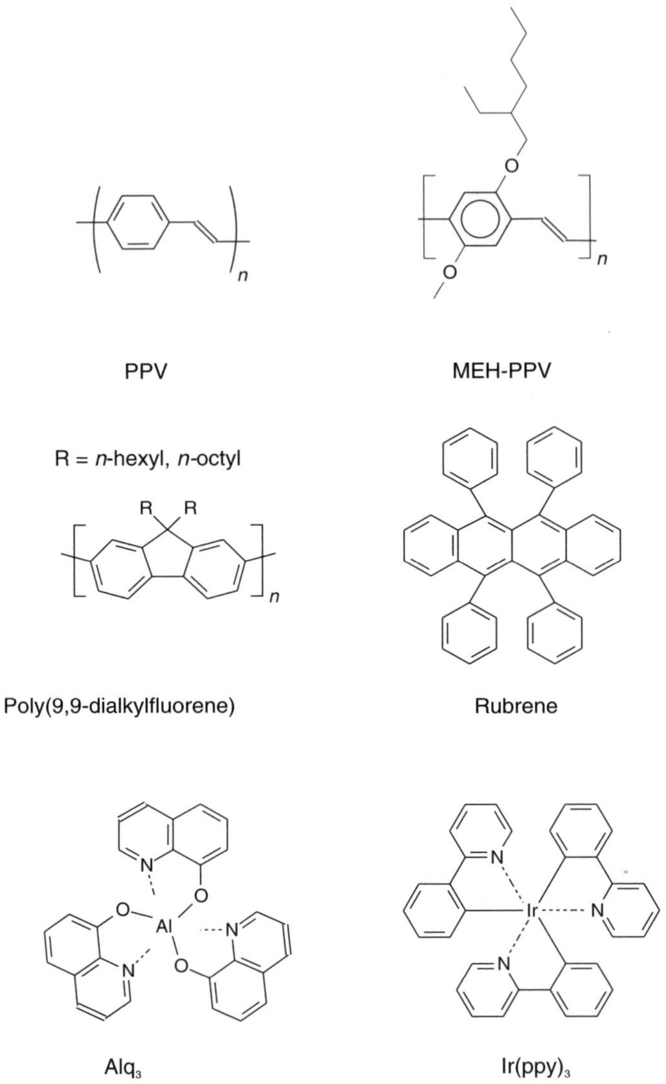

Figure 9.19 Organic compounds used as the emissive layer in OLEDs.

The OLEDs therefore act as rectifying devices. The steep rise in the forward current between 1 and 2 V can be modelled with the Schottky diode Equation (9.4), with an ideality factor typically in the range 1.6–2.4 for different devices. The deviation from this exponential behaviour above 2 V results from the limiting series resistance associated with the device. The EL from the OLED is observed as the forward voltage is increased beyond about 1.6 V. Generally, OLEDs based on other materials exhibit rectifying electrical characteristics, but the results sometimes suggest other conductivity mechanisms. For example, space-charge limited conduction [20] and Fowler–Nordheim conduction (Chapter 3, Sections 3.5.4 and 3.5.5) are often reported [21].

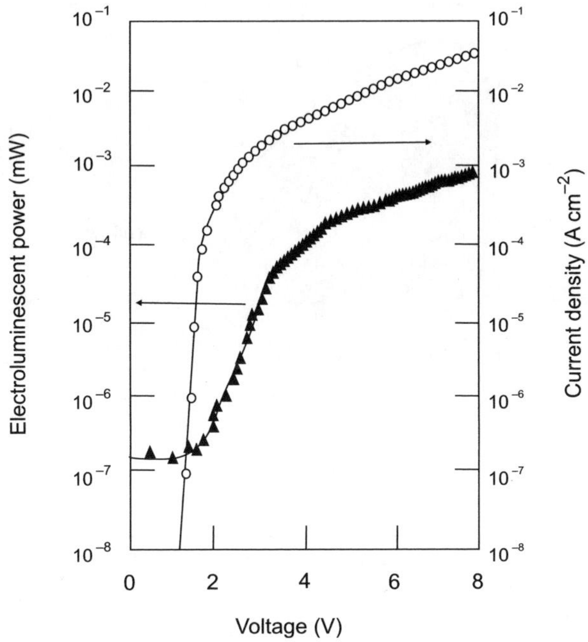

Figure 9.20 Current versus voltage and electroluminescence output versus voltage characteristics for an ITO/MEH-PPV/Al OLED. Reprinted with permission from Karg S, Meier M, Riess W, *J. Appl. Phys.*, **82**, 'Light-emitting diodes based on poly-*p*-phenylene-vinylene: I charge-carrier injection and transport', pp. 1951–1966. Copyright (1997) American Institute of Physics.

By judicious choice of compound, EL emission at different wavelengths across the visible spectrum is possible. For example, fluorene-containing polymers can emit across the entire visible spectrum. A convenient way of representing the colour output of an OLED is to use the CIE (Commission International de l'Eclairage) *chromaticity diagram* (Figure 9.21) [22]. Although the human eye has three different types of colour-sensitive cones, it is found that any colour can be expressed in terms of just two colour coordinates x and y. The boundary represents maximum saturation for the spectral colours, and the diagram forms the limit of all perceivable hues. The calculation of the CIE chromaticity coordinates for an OLED requires the multiplication of its spectral power at each wavelength times the weighting factor from each of the three colour matching functions.

Red, green and blue emission with good colour purity is required for display applications, so narrow emission bands are required. However, an inherent problem with organic compounds is that they tend to possess relatively broad emission spectra, as a result of the coupling to vibronic states in the molecules. Colour filters can be used to improve the colour purity. An alternative approach is to use materials which have very narrow emission bands, such as rare earth complexes. Emission in these materials is from atomic states in the metal ion so there is little interaction with the organic ligands and the emission bands are very narrow. For example, terbium and europium compounds are known to be excellent phosphors with extremely narrow emission bands (half-widths of around 5 nm).

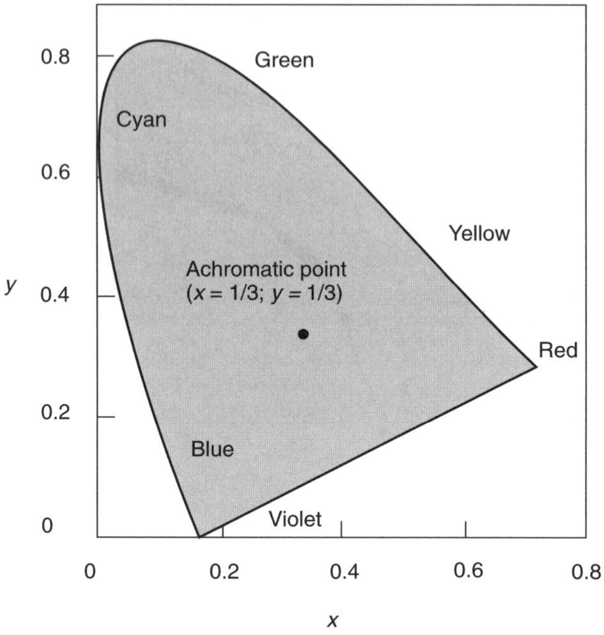

Figure 9.21 CIE chromaticity diagram.

There is also a keen interest in developing white light organic displays (CIE coordinates approximately 0.31, 0.32 – the interpretation of 'white' does vary throughout the world). For example, a large-area white light-emitting device could provide a solid-state light source that might compete with conventional lighting technologies. Different methods of making an intrinsically white-emitting OLED by blending emissive species, in either single or multiple layers, have been demonstrated [23–25]. Alternatively, a blue OLED can be used with one or more down-conversion layers [26]

9.6.1 Device Efficiency

There is a proliferation of units associated with the characterization of light-emitting devices and much confusion abounds. A very brief introduction is therefore provided in this section.

First, it is necessary to make a distinction between *photometric* and *radiometric* units [27]. Radiometry is the measurement of optical radiation, which is electromagnetic radiation within the frequency range $3 \times 10^{11} - 3 \times 10^{16}$ Hz. This corresponds to wavelengths between 0.01 and 1000 μm, and includes the ultraviolet, visible and infrared regions. Photometry, on the other hand, is the measurement of light, defined as electromagnetic radiation which is detectable by the human eye. It is therefore restricted to the wavelength range from about 380 to 780 nm. Photometry is essentially radiometry except that everything is normalized to the spectral response of the eye, known as the *photoptic* response. Radiometry uses the familiar unit of the watt as a measure of power (rate of energy emission) or radiant flux. The photometric equivalent of the watt is the *luminous flux*,

Table 9.1 Common photometric units.

Property	Description	Units
Energy	Total amount of light emitted from source	Joules (J)
Luminous flux	Rate of energy emitted from source	Lumen (lm)
Luminous intensity	Flux emitted from a point source per unit solid angle	Candela (cd) ≡ lumens per steradian (sr)
Luminance	Flux emitted per unit surface area of extended source per unit solid angle	$cd\,m^{-2}$

measured in *lumens* (lm). The *luminous intensity* measured in *candelas* (cd) refers to the flux emitted into a unit solid angle in space or *steradian* (sr). Since there are 4π sr in a sphere, in the case of a light source emitting equally in all directions (isotropic) the relationship between lumens and candelas is $1\,cd = 4\pi\,lm$. The amount of light leaving a surface is the *luminance*, measured in $cd\,m^{-2}$; these units are often called *nits* (from the Latin *nitere* = to shine). A typical computer screen has a luminance of about $100\,cd\,m^{-2}$, and the luminance of an average clear sky will be around $8000\,cd\,m^{-2}$.

The CIE has adopted as a standard an average eye with a predictable response to light at various frequencies. As a result, one light watt consists of 683 lm. This represents the photosensitivity of the 'standard' human eye; this response has a peak value of $683\,lm\,W^{-1}$ at a wavelength of 555 nm. According to this function, 700 nm red light is only about 4% as efficient as 555 nm green light (i.e. 1 W of 700 nm red light is only 'worth' $683 \times 0.04 \approx 27$ lm). Table 9.1 provide a list of the photometric quantities encountered in OLED work.

Quantum efficiencies are often quoted for OLEDs [27]. An important distinction must be made between the definitions of the external and internal efficiencies. For display applications, the commonly accepted definition for the *external quantum efficiency*, η_{ext}, is the ratio of the number of photons emitted by the OLED into the viewing direction to the number of electrons injected. Although this external quantum efficiency could also be defined as the ratio of the total number of photons emitted from the device (in all directions) to the number of electrons injected, this definition is not useful for display devices. A large fraction of the light can be waveguided by the organic layer(s)–substrate combination, ultimately emerging out at the edge of the device. Thus, the total amount of light emitted from the device will be significantly higher than the light emitted in the viewing direction, leading to an efficiency based on total light emitted which can be up to four times larger than η_{ext}. The *internal quantum efficiency* is the ratio of the total number of photons generated within the structure to the number of electrons injected. The internal and external efficiencies therefore differ by the fraction of light coupled out of the structure into the viewing direction.

A convenient measure of the properties of an OLED is the *luminous efficiency*, η_l, in $cd\,A^{-1}$. In many respects, η_l is equivalent to η_{ext}, with the exception that η_l weights all incident photons according to the photopic response of the eye. In this case

$$\eta_l = \frac{AL}{I} \tag{9.8}$$

Table 9.2 Performance of polymer-based organic light emitting devices (Sumation Co. Ltd, 2006) [28].

Colour	Voltage (V)	Luminous efficiency at 400 cd m^{-2} (cd A^{-1})	Luminous power efficiency at 400 cd m^{-2} (lm W^{-1})	Measured lifetime at room temperature at fixed luminance (h) at [cd m^{-2}]	Extrapolated lifetime at 100 cd m^{-2} at room temperature (h)
Phosphor red	6.5	10.6	5.1	512 at 4029	800 000
Fluorescent red	3.6	2.3	2.0	936 at 3000	300 000
Green	4.9	15.8	10.1	320 at 6000	500 000
Blue	5.5	9.6	5.5	619 at 1800	200 000
White	6.8	4.5	2.1	474 at 1440	65 000

All data taken using common cathode and may include an interlayer.

where L is the luminance of the OLED (in cd m^{-2}), A is the device active area (not necessarily equal to the area of light emission) and I is the current.

Another frequently used display efficiency unit is the *luminous power efficiency* or *luminosity*, η_p, measured in lm W^{-1}. This is the ratio of luminous power emitted in the forward direction (in lm) to the total electrical power required to drive the OLED at a particular voltage. The luminous power efficiencies for incandescent light bulbs are typically less than 20 lm W^{-1}, and the figures for fluorescent tubes are in the 50 – 100 lm W^{-1} range. Table 9.2 gives the luminous efficiency and the luminous power efficiency and lifetimes of some commercial polymer-based OLEDs [28]. These data are taken from a commercial source in 2006; the same company announced extrapolated lifetime figures in March 2007 (at 100 cd A^{-1}) of phosphoresecent red (2.4×10^6 h), green (2.8×10^6 h), blue (10^6 h) and white (3.3×10^5 h). The performance of OLEDs is continuing to improve rapidly and the reader should check for the latest information.

9.6.2 Methods of Efficiency Improvement

The external quantum efficiencies of the early OLEDs were around 10^{-2}%. This value has now increased to over 10%. Some of the methods used to achieve this dramatic increase are outlined below.

Electrodes

The anode in OLEDs is typically indium tin oxide (ITO), since this is transparent and highly conductive. However, the material has a fairly rough surface and is treated with a thin layer of a hole conductor (see below) such as PEDOT (Chapter 5, Section 5.3), phthalocyanine or polyaniline to improve the contact to the emissive layer and the device stability. Another advantage of these layers is that they smooth out the relatively rough surface of the ITO, thereby preventing any local short-circuiting that would otherwise cause the device to fail. In many OLED devices, the hole current can far exceed the electron current. This results in significant energy wastage, since the excess holes cannot combine with electrons to generate light. Efficiency improvements can be achieved by coating the anode with a thin layer that reduces the hole current (hole blocking layer).

A low work function metal or alloy is used for the cathode in OLEDs. An important requirement is to allow the efficient injection of electrons into the organic material. Unfortunately, metals with very low work functions, such as Li and Ca, are reactive and require careful encapsulation. A thin inorganic insulating layer such as LiF, or an organic monolayer, may be inserted between the cathode and the emissive material. The introduction of such an interfacial layer can modify the electronic band structure in a number of ways. For example, the layer may introduce fixed charge, affecting the height of the barrier to electron injection (i.e. modifying the work function of the metal). An important difference between MIS and MS structures is that, in the former case, the metal Fermi level is no longer be 'tied' to the energy band structure of the semiconductor (see earlier discussion in Section 9.3). This means that, as the forward applied voltage is increased, the cathode Fermi level can move with respect to the LUMO level in the organic emissive layer, aligning filled electron states in the metal with vacant states in the organic semiconductor. As long as the interfacial layer is 'transparent' to electrons, the EL will increase. This, of course, imposes some restrictions on the nature of the interfacial layer: it must support a DC voltage, but at the same time be sufficiently thin that the minority electron current can pass through, e.g. by tunnelling.

Multilayer Structures

A breakthrough in terms of improved OLED performance was achieved when the functions of charge transport and light emission were separated. For example, electron- and hole-transporting layers can be introduced between the cathode and the emitting layer and between the anode and the emitting layer, respectively, to improve and balance the injection of charge carriers. There are three requirements for good charge-transport materials for use in OLEDs: they must have a high mobility for either electrons or holes, demonstrate good injection efficiency from the chosen electrode and possess suitable band offsets with other organic layers used within the device.

Figure 9.22 shows a range of the organic compounds that have been used as hole transport materials and to modify the anode electrode (see above); some of these were originally developed for use in xerography (Section 9.8.3). One of the most common classes of compound that are used are the arylamines, of which *N*,*N*′-diphenyl-*N*,*N*′-bis(3-methylphenyl)-[1,1′-biphenyl]-4,4′-diamine (TPD) is one example. However, this particular compound has a glass transition temperature, T_g (Chapter 2, Section 2.6), of only 63 °C, which can result in recrystallization and delamination from the ITO anode.

Some of the electron transport materials that have been used in OLEDs are depicted in Figure 9.23. Many of these contain oxadiazole groups (i.e. the five-membered rings containing one oxygen and two nitrogens); these are very electron deficient, which allows them to block holes and transport electrons very well.

The overall structure of an OLED can be fairly complex (Figure 9.24). The two electrodes can each be modified to improve the carrier injection (electron and hole injection layers – EIL and HIL); furthermore, carrier transporting layers (ETL and HTL) can provide efficient transfer of electrons and holes to the layer that emits the light (EML). In many cases, the

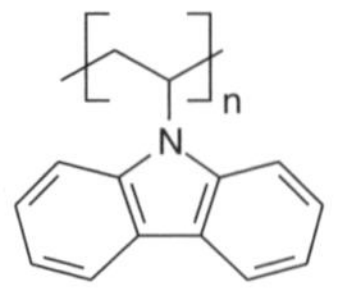

PVK

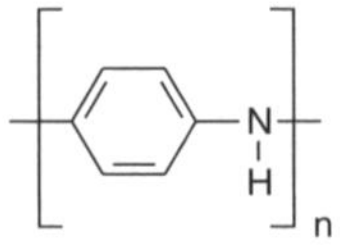

Polyaniline

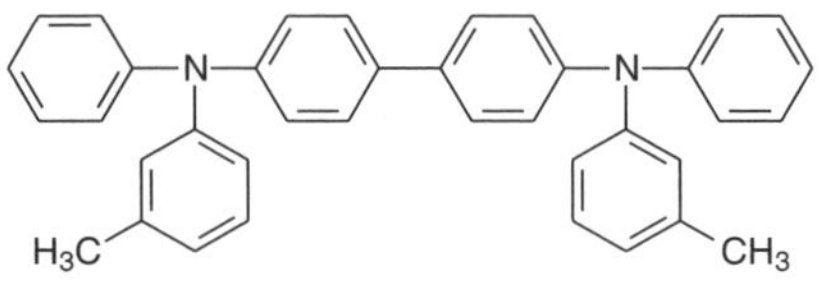

TPD

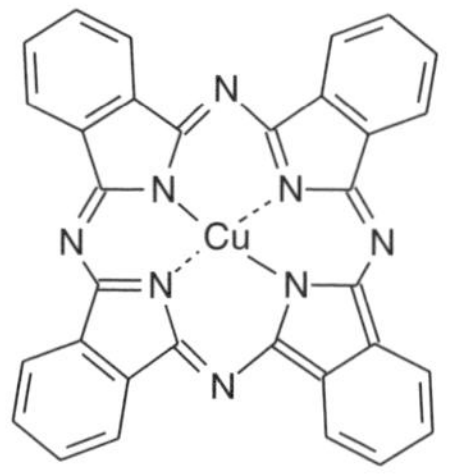

Copper phthalocyanine

Figure 9.22 Hole-transporting molecules used in OLEDs.

carrier transport layers are doped (Chapter 3, Section 3.4.2) to provide low-resistance contact regions. Common p-type dopants are strong electron acceptors such as F_4-TCNQ, whereas alkali metals such as Li or Cs can be effective as n-type dopants. The resulting *p–i–n* devices can exhibit high efficiencies and operating voltages that are close to the thermodynamic limit, i.e. the photon energy divided by the electronic charge [29–31]. The efficiency gain from the multilayer structure depicted in Figure 9.24 must, of course, be offset against the increased fabrication costs.

Triplet Emission

When an electron and a hole recombine to form an excited molecular state in an organic material, the spins of the electrons in the excited and ground levels can point either in the same direction (triplet state) or in the opposite directions (singlet state) (Chapter 4, Section 4.4.5). For quantum mechanical reasons, 75% of recombination events are associated with triplet states, which, in most cases, do not emit photons when they decay to the ground state. Hence the production of emission from the triplet state of organic materials is a further means to improve the device efficiency [16, 17]. It has been shown theoretically that systems

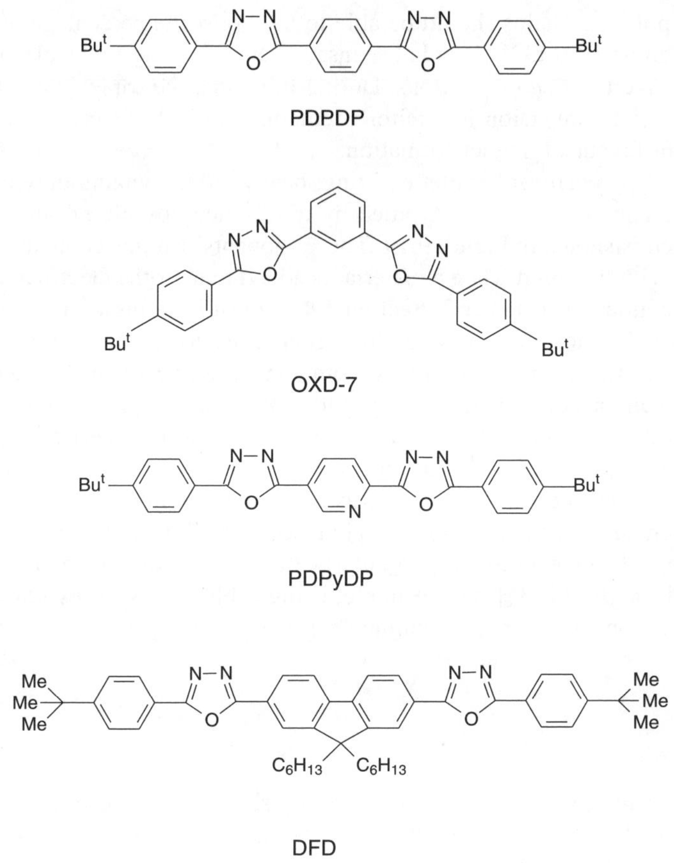

Figure 9.23 Electron-transporting molecules used in OLEDs.

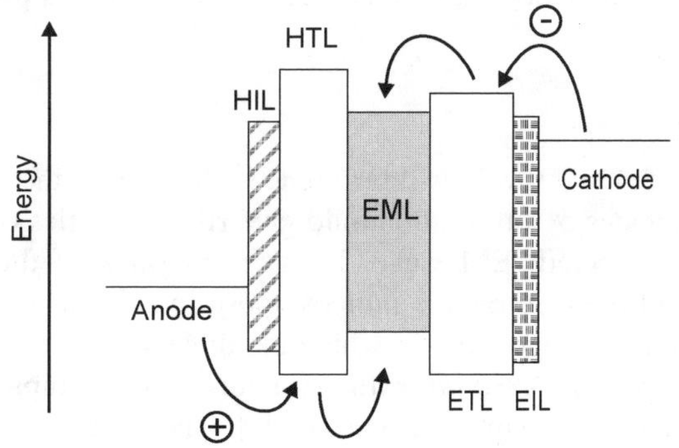

Figure 9.24 Multilayer OLED. HIL, hole injection layer; HTL, hole transport layer; EML, emissive layer; ETL, electron transport layer; EIL, electron injection layer.

built from long polymer chains should be able to the boost the percentage of light-emitting singlets to as high as 50% [32]. This is because, with increasing molecular weight, triplets take longer to convert to neutral excitons. During this time, the triplet state can convert to a singlet, while singlet conversion to excitons remains rapid. As a result, the spin statistics become biased in favour of singlet formation.

An alternative approach uses triplet emitting, heavy atom containing, (electro)phosphorescent metal–organic complexes, as typified by tris(2-phenylpyridine)iridium, $Ir(ppy)_3$, the structure of which is shown in Figure 9.19. These 'dopants' (in this context, the term should not be confused with that used when a material is added to alter the electrical conductivity of an organic semiconductor; Chapter 3, Section 3.4.2) must have high luminescence quantum yields and relatively short lifetimes, of the order 1 μs, to avoid exciton-ion quenching. Unfortunately, the triplet emission suffers from long lifetime of the excited states in the device, which opens a non-radiative decay path and makes a proper device design more difficult. Yet another method utilizes a phosphorescent 'guest' material in a light-emitting polymer 'host.' It is possible to use all the excited states, both singlet and triplet, for light emission provided that the triplet energy gap of the host is higher than that of the guest. Luminous efficiencies can be high for phosphorescent OLEDs, e.g. $>50\,\mathrm{cd\,A^{-1}}$ [33].

The efficiency of white OLEDs, particularly those incorporating phosphorescent dyes, can also be fairly high [34, 35]. For example, using a blue phosphorescent OLED with an appropriate down-conversion layer, a luminous efficiency reaching almost $40\,\mathrm{cd\,A^{-1}}$ can be achieved [34].

Shaped Substrates

A further increase in efficiency can be achieved by shaping the substrate to ensure that the maximum possible fraction of light emitted from the OLED leaves the device in the forward-scattered direction. Due to internal reflections at interfaces between materials of different refractive indices in the OLED, the light output from the OLED exceeding the glass–air critical angle ($\sim 19°$) is waveguided in the substrate, and is thus wasted. Patterning of the substrate can enhance the effective emission angle, thereby scattering additional light energy into the viewing direction. One particularly useful structure is a preformed glass mesa, an example of which is shown schematically in Figure 9.25 [36]. A factor of four enhancement in the external quantum efficiency may be achievable using this approach.

Microcavity Structures

In Chapter 4, Section 4.5.4, we saw how interference effects from the reflection of light from the upper and lower surfaces of a thin film could give rise to a series of maxima and minima in the transmitted (or reflected) EM wave. By using a stack of thin films, formed from alternate materials of different refractive indices, these interference fringe maxima become very sharp. The particular arrangement, in which the dielectric stack is formed between two mirrors, is called a *Fabry–Perot resonator*; the structure is an example of a one-dimensional photonic crystal described in Chapter 4, Section 4.8. The term optical microcavity is also used widely to refer to such structure with dimensions of the order of light. These have been used successfully with OLEDs to both increase their EL output and to tune their emission wavelength. The schematic structure of such an arrangement is shown in Figure 9.26. The

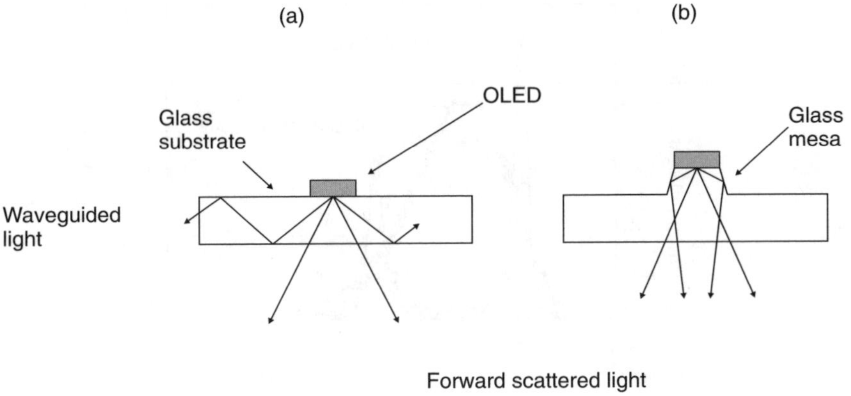

Figure 9.25 Schematic cross-section (not to scale) of an OLED fabricated on a shaped substrate (mesa) designed to increase the proportion of light emitted by the OLED in the forward direction. (a) Much of the emitted light is waveguided in the glass substrate and lost. (b) Most of the waveguided light is directed into the viewing direction by internal reflection from the walls of the glass mesa. From *IEEE Trans. Electron Dev.* **44**, Burrows PE, Gu G, Bulović Shen Z, Forrest SR, Thomson ME, 'Achieving full-colour organic light-emitting devices for lightweight, flat-panel displays', pp. 1188–1202, Copyright (1997) Reproduced by permission of IEEE.

SiO_2 and TiO_2 have different refractive indices (1.4 and 2.3, respectively) and are alternated to provide a total thickness of one quarter of a wavelength. The OLED is then fabricated on top of this stack. The EL output of a microcavity OLED based on Alq_3 is contrasted with that from a reference device in Figure 9.27 [37]. The form of the EL output for the OLED fabricated on the microcavity is of narrow and intense peaks. The wavelengths of these can be predicted using calculations based on classical optics [37]. By varying the stack, the emission colour of the OLED can be altered, providing a means to produce a range of colours from the same OLED [36, 37].

Blended and Doped Layer OLEDs

As an alternative to multilayer OLEDs, such as those based on the structure shown in Figure 9.24, devices based on blended single layers of emissive and charge transport materials may be fabricated [38, 39]. These generally possess lower efficiencies than

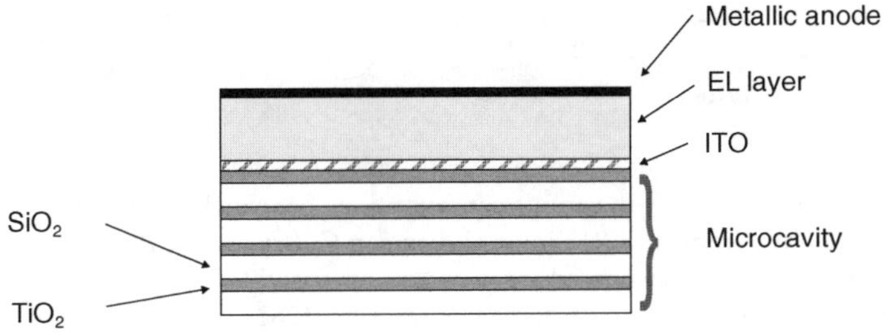

Figure 9.26 Organic light-emitting device fabricated on an optical microcavity.

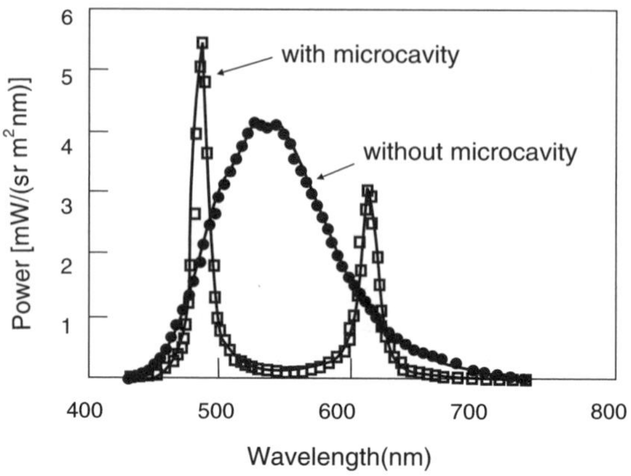

Figure 9.27 Electroluminescent output for an Alq_3 OLED fabricated on a microcavity compared to a reference device. Reproduced from *Organic Electroluminescent Materials and Devices*, T Nakayama, p. 365, Copyright (1997), with permission from Taylor & Francis Group LLC.

multilayer structures but they have the considerable advantage of ease of manufacture. Figure 9.28(a) shows how the external quantum efficiency of a blended layer OLED based on a polymer, poly[2-(2-ethylhexyloxy)-5-methoxy-1,4-phenylenevinylene] (MEH-PPV), and an electron transport compound, 2,7-bis[2-(4-*tert*-butylphenyl)-1,3,5-oxadiazol-5-yl]-9,9-dihexylfluorene (DFD), varies as the MEH-PPV:DFD ratio is changed in a spin-coated film [38]. The device efficiency increases with the amount of DFD so that the OLED fabricated with a 95% blend is about 100 times more efficient than that of pure MEH-PPV.

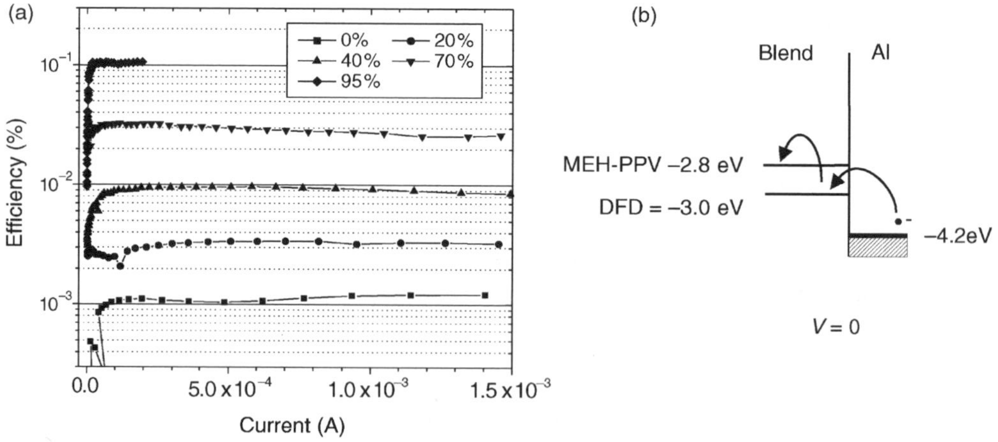

Figure 9.28 (a) External quantum efficiency versus current for OLEDs fabricated with MEH-PPV/DFD blended layers using blends with various concentrations (by weight) of DFD, an electron acceptor. Reprinted with permission from Ahn JH, Wang C, Pearson C, Bryce MR, Petty MC, *Appl. Phys. Lett.*, **85** 'Organic light-emitting diodes based on a blend of poly[2-(2-ethylhexyloxy)-5-methoxy-1,4-phenylenevinylene] and an electron transport material', pp. 1283–1285. Copyright (2004) American Institute of Physics. (b) Energy band diagram showing charge transfer from DFD to the MEH-PPV emissive polymer.

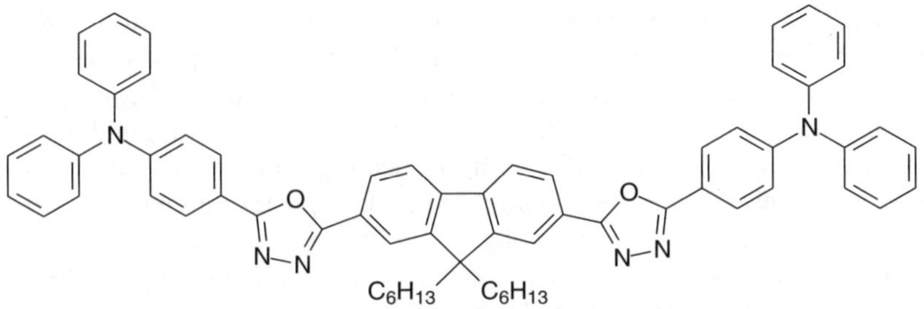

Figure 9.29 2,7-Bis[2-(4-diphenylaminophenyl)-1,3,4-oxadiazol-5-yl]-9,9-dihexylfluorene: a triad molecule for OLEDs containing electron transport, hole transport and emissive groups [43].

The electrons are injected from the metal (aluminium) cathode into the LUMO level of the DFD, and then into the LUMO level of the emissive polymer [Figure 9.28(b)]. For all the devices, the electroluminescence originates from the MEH-PPV material, indicating that the energy, or charge, transfer between the electron transport compound and the polymer is very efficient. Electron transport and hole transport moieties can also be covalently attached to the emissive polymer [40–43]. Figure 9.29 shows an example of such a triad molecule [43]. The oxadiazole, triarylamine and fluorene units serve as the electron transport, hole transport and emitter segments of the molecule.

Many of the fluorescent dyes that have been incorporated into OLEDs were originally developed as laser dyes, since these materials were designed to possess high photoluminescence quantum efficiencies in dilute solution, along with good stability. However, in general, their solid-state fluorescence is extremely weak due to concentration quenching, and this reduces the EL efficiency. One solution to this problem is to add bulky side groups to the molecule to prevent aggregation by increasing the steric hindrance. Unfortunately, this often leads to poor charge transport through the material. A further remedy is to dope the emissive dye into an organic matrix, such as Alq_3. This effectively dilutes the concentration of the emissive dye (the dopant) and thus prevents aggregation. Provided that the dopant is red shifted (i.e. has its emission at a lower energy) compared with the host, excitons formed in the host material will tend to migrate to the dopant prior to relaxation. This results in emission which is predominantly from the dopant. Red emitters are fairly polar, and it is therefore difficult to find noncrystallizing materials. Thus many luminescent systems emitting at long wavelengths are realized by molecular doping. Popular dopants for Alq_3 are the yellow-emitting 5,6,11,12-tetraphenylnaphthacene (rubrene; Figure 9.19) and the red-emitting derivatives of 4-(dicyanomethylene)-2-methyl-6-(*p*-dimethylaminostyryl)-4*H*-pyran (DCM) [17, 42]. If there is a good overlap of the absorption band of the dye with the emission band of the host, Förster transfer (Chapter 4, Section 4.4.6) is very efficient, with quantum yields that are sometimes higher than that of the undoped host. For example, quantum yields are almost unity in the Alq_3–rubrene system [44]. If Förster transfer is not so efficient, as for Alq_3–DCM, a mixed colour from the host and guest is observed.

9.6.3 Full-Colour Displays

An important application area for organic light-emitting devices is in flat panel displays – OLED technology that might eventually replace active-matrix LCDs (Chapter 8,

Section 8.4.4). A display consists of a matrix of contacts made to the bottom and top surfaces of each organic light-emitting element, or pixel. The individual pixels may be addressed either actively or passively. In the former case, the display is addressed one line at a time, so that if a display has 480 lines, then an individual pixel can only be emitting for 1/480th of the time. High drive currents are needed, leading to heating problems and to expensive driver circuits. However, theses issues are to some extent offset by the simplicity of the technique.

Active addressing schemes involve using an active device, such as a thin film transistor, attached to each pixel. As a consequence, the pixel can remain emitting for the entire frame rather than for a small fraction of it. Organic EL technology is suited to active matrix addressing since it is a low-voltage technology. Organic FETs may eventually provide the drive circuitry.

To generate a full-colour image, it is necessary to vary the relative intensities of three closely spaced, independently addressed pixels, each emitting one of the three primary colours of red, green or blue. Some of the different approaches that have been used are shown in Figure 9.30 [36]. Perhaps the simplest scheme [Figure 9.30(a)] is the side-by-side-positioned red, green and blue (R, G, B) subpixels. However, this arrangement requires each of the three

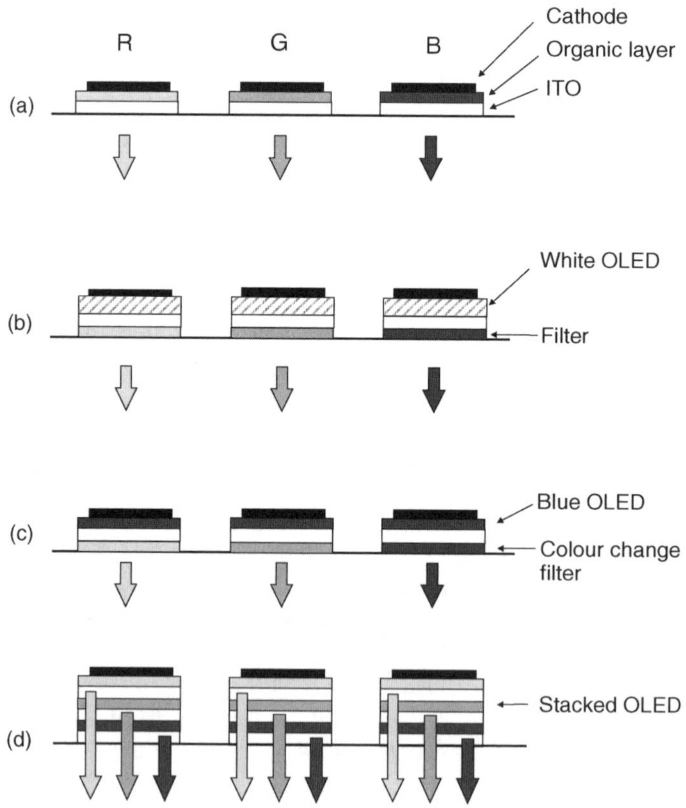

Figure 9.30 Schemes for generating full-colour displays. (a) Separate red, green and blue emitters (R, G and B) providing pixels side-by-side. (b) The light from white-emitting OLEDs is filtered to provide R, G and B emission. (c) The light from blue-emitting OLEDs is used to generate R, G and B emission through colour changing filters. (c) Stacked OLEDs emit R, G and B [36].

closely spaced OLEDs to be sequentially grown and patterned. Alternatively, optical filtering of white OLEDs can produce red, green and blue emission [Figure 9.30(b)]. This method is not particularly efficient as a significant amount of light is absorbed in the filters. An alternative approach [Figure 9.30(c)] exploits a single blue or ultraviolet OLED to pump organic fluorescent wavelength down-converters. Each of these 'filters' consists of a material that efficiently absorbs the blue light and re-emits the energy as either green or red light.

An elegant way of achieving fill-colour displays is to stack the red, green and blue pixels on top of each other, as shown in Figure 9.30(d). Since there are no sub-pixels, this has the advantage of increasing the resolution of the display by a factor of three. However, the method requires semi-transparent electrodes that are compatible with high current densities. One problem encountered with stacking numerous transparent organic layers is the formation of unwanted optical cavities whose resonances alter the emission spectra of OLEDs (see the section above on microcavities). Such effects can be eliminated by careful control of layer thickness and composition.

The challenge to making full-colour polymer-based displays is very different from that for making such displays using small-molecule OLEDs. Solution chemistry makes it difficult to deposit and pattern a polymer pixel of one colour, and then repeat the process using a second colour emitter, because the solvents employed may dissolve or attack the devices already on the substrate.

Several schemes have been suggested to circumvent this problem. One particularly promising method involves depositing a single blue-emitting polymer and then selectively diffusing green and red dyes into adjacent regions. A further approach involves depositing the various polymer constituents of a full-colour display using inkjet printing. Here, control of the thickness and shape of the droplet, which eventually sets into a high-resolution pixel, remains an as yet unsolved problem

9.6.4 Electronic Paper

Perhaps the ultimate goal for portable displays in terms of readability, ease of access and use, ruggedness and ultra-low power consumption is the electronic emulation of paper – *e-paper* [45]. Paper is a reflective 'device' with no energy consumption. There are some suggestions that e-paper should be in a similar format, modulating the ambient light rather than generating it. *Electrophoretic* display technology offers some advantages as the display technology has a wide angular viewing range and a high contrast ratio over a wide range of illumination. The principle of operation exploits the movement of charged particles in an applied electric field. These might be combined with an organic transistor active matrix backplane.

9.7 PHOTOVOLTAIC CELLS

Concerns over global climate change, local air pollution and resource depletion are making photovoltaics (PVs) an increasingly attractive method of energy supply. The current technology is based on single crystal silicon solar cells. These have been developed since the 1940s and offer conversion efficiencies of around 15% for off-the-shelf commercial devices (although figures around 25% are reported in the laboratory). However, the technology is more expensive than conventional power generation. Options for reducing

PV generating costs include working on incremental advances to improve silicon efficiencies, but also developing fundamentally different devices based on alternative materials. PVs based on organic compounds, such as polymers or dyes, offer the possibility of large-scale manufacture at low temperature coupled with low cost. However, until the end of the 20th Century, little progress had been made and energy conversion efficiencies of up to only about 1% were achieved. The availability of new conductive organic materials and different PV designs has significantly improved on this figure. To 2004, several laboratories have reported conversion efficiencies of 4–5%, with lifetimes of around 10^4 hours [46].

9.7.1 Organic Semiconductor Solar Cell

The structure of a simple organic solar cell is depicted in Figure 9.31. In essence, the device is very similar to the OLED described in the previous section. If the incoming photons have energy greater than the band gap of the polymer (or greater than the HOMO–LUMO separation in the case of organic molecular materials), then these will be absorbed, creating electrons and holes. In an inorganic PV cell, these electrons and holes would be generated within, or close to, a depletion region in the semiconductor and they would be free to migrate to opposite electrodes, where they can do useful work in an external electrical load. However, in the organic material the electrons and holes are bound together in excitons (Chapter 4, Section 4.4.3). An immediate problem in organic PV cells is to split these excitons. This can be conveniently done at an interface, the simplest interface being the junction between the electrodes and the organic material. Under open-circuit conditions, holes are collected at the high work function electrode (e.g. ITO) and electrons at the low work function electrode (e.g. Al). The open-circuit output voltage of the PV device depends on the work function difference between the electrodes.

The current versus voltage behaviour of the PV cell is shown in Figure 9.32. In the dark [Figure 9.32(a)], the device behaves as a diode, e.g. with the I–V relationship given by Equation (9.4). Under illumination, the I–V characteristic is moved down the current axis as a photocurrent is generated. Two important device parameters, shown in Figure 9.32(a), are V_{oc}, the open-circuit voltage, and I_{sc}, the short-circuit current. The former represents the

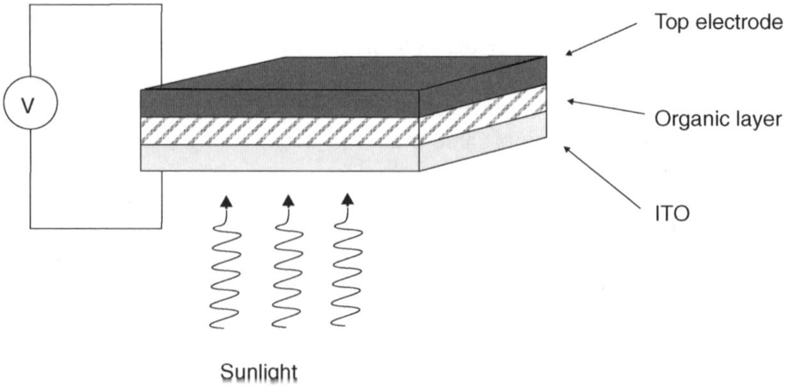

Figure 9.31 Schematic diagram showing the structure of a photovoltaic cell based on a conductive organic compound.

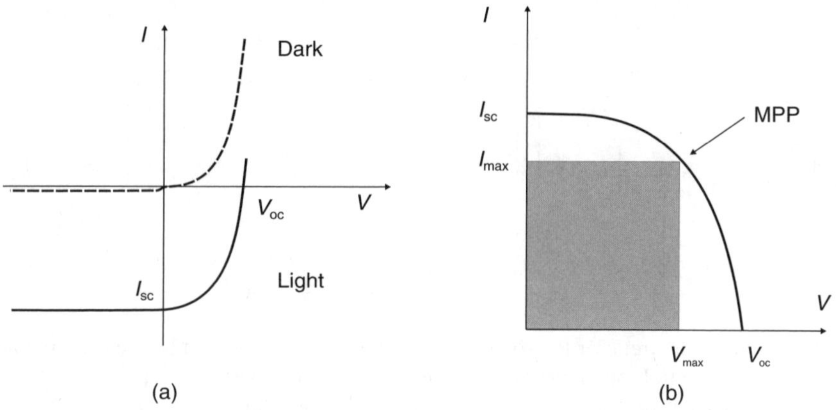

Figure 9.32 (a) Current I versus voltage V characteristics for an organic PV structure in the dark and in the light. V_{oc} = open-circuit voltage; I_{sc} = short-circuit current. (b) Expanded region of I–V curve showing the maximum voltage V_{max} and the maximum current I_{max} available from the solar cell. The shaded area represents the maximum power available from the cell. MPP = maximum power point.

voltage appearing across the terminals of the solar cell when no current is drawn from the device; this is the maximum attainable voltage that the PV cell can provide. When the device terminals are short-circuited, I_{sc} represents the maximum current that the solar cell can provide. The I–V characteristics under illumination are shown in more detail in Figure 9.32(b). The actual operating point on a given solar-cell I–V characteristic is determined by the load resistance that is connected between its terminals. To maximize the output power, it is desirable to choose the value of load resistance such that the operating point is at the maximum power point, MPP in Figure 9.32(b). In this case, the current and voltage outputs of the PV cell are I_{max} and V_{max}, respectively. The energy conversion efficiency η is given by

$$\eta = \frac{I_{max} V_{max}}{P_i A} = \frac{I_{sc} V_{oc} F}{P_i A} \tag{9.9}$$

where P_i is the incident solar power density and A is the area of the solar cell. The ratio of $I_{max} V_{max}$ to $I_{sc} V_{oc}$ is defined as the fill-factor, F, of the solar cell. It is clearly desirable to produce devices with F values as close to unity as possible.

Figure 9.33 shows the equivalent circuit of a simple organic PV structure, such as that based on a Schottky barrier. Imperfections in the device leading to current leakage are represented by a shunt resistance, R_{shunt}, and parasitic series resistance effects are represented by R_{series}. The current generated by the incoming EM radiation is taken into account by the incorporation of a constant current generator of value I_l. Note that a forward dark current I_d will flow through the rectifying junction. The load resistance, voltage and current are given by R_{load}, V_{load} and I_{load} in Figure 9.33. Using simple circuit theory, it can be shown that [47]

$$I_{load}\left(1 + \frac{R_{series}}{R_{shunt}}\right) = I_l - I_d - \frac{V_{load}}{R_{shunt}} \tag{9.10}$$

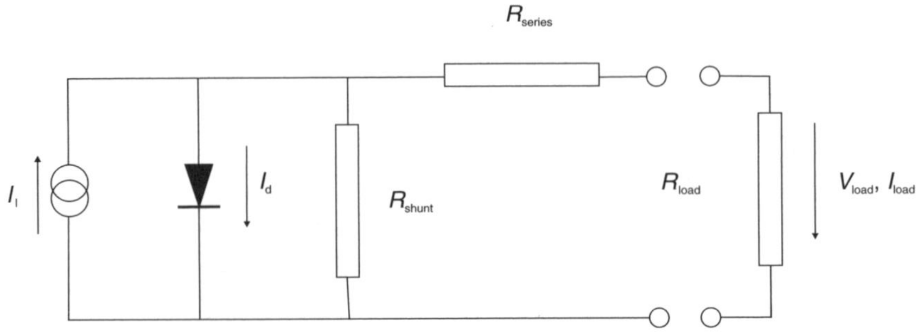

Figure 9.33 Equivalent circuit of PV solar cell. I_l is the current generated by the solar radiation; I_d is the forward biased current through rectifying barrier; R_{series}, R_{shunt} and R_{load} represent the series, shunt and load resistors, respectively; and V_{load} and I_{load} are the voltage and current, respectively, associated with the load resistor.

This equation can be solved numerically to ascertain the effects of R_{series} and R_{shunt} on the cell output characteristics. However, it is evident that, in general, it is important for R_{series} to be small and for R_{shunt} to be large. To absorb most of the incoming light, the organic layer needs to be as thick as possible. However, the low charge carrier mobilities found in organic semiconductors results in an unacceptable series resistance. This limits the thickness of the photoactive layer to about 100 nm.

The optical absorption of the photoactive layer must also be 'tuned' to match the incoming photon flux from the solar irradiation. The peak in the solar spectrum corresponds to a wavelength of approximately 0.6 μm (depending on location on the Earth's surface) and it has a useful range of about 0.3–2 μm. For PV devices, the most effective conversion of this radiation to electrical energy is achieved using semiconductors with a band gap of around 1 μm (i.e. $E_g \approx 1.2\,\text{eV}$).

A number of different polymeric semiconductors have been used in PV devices. These include polyacetylene, polythiophene, polypyrrole and polyphenylenevinylene. The incorporation of additional electron-donating and electron-accepting species can be used to create interfaces among the polymer molecules of differing electron affinities. As a consequence, it is possible to enhance the probability of electron transfer between the molecules. This photoexcited charge transfer causes the bound excitons to separate. Figure 9.34 shows the current density versus behaviour of a blended layer organic PV device based on poly (2-methoxy-5-(3′,7′-dimethyloctyloxy)-1,4-phenylenevinylene) (MDMO-PPV) copolymer blended with [6,6]-phenyl-C_{61} butyric acid methyl ester (PCBM). The solar cell possesses a low series resistance and a high fill factor [46].

9.7.2 Dye-Sensitized Solar Cell

An artificial photosystem containing a light-harvesting antenna was manufactured by Grätzel's group in 1991 using a dye-sensitized titanium dioxide thin film with a redox electrolyte [48]. Grätzel's dye-sensitized photovoltaic device operates in the following manner, depicted schematically in Figure 9.35. Upon photon absorption, a dye electron is elevated to an excited state. The dye then injects the excited electron directly into the conduction band of a semiconductor, which subsequently creates a current in an external

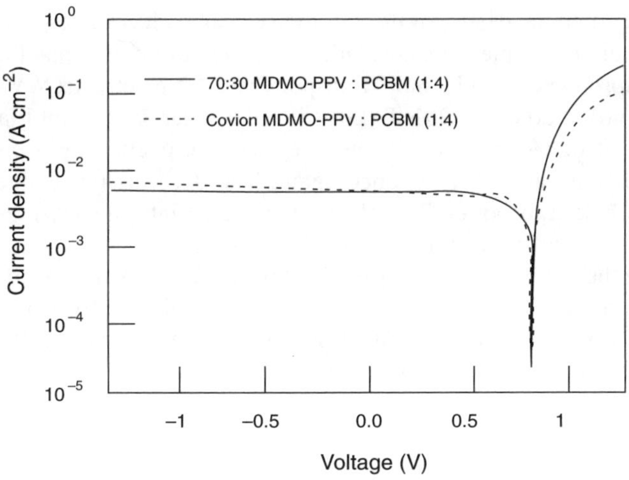

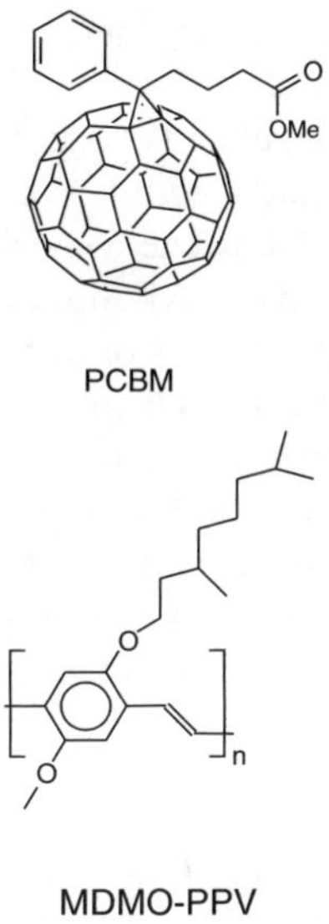

Figure 9.34 Current versus voltage characteristic for a PV cell based on a 1:4 mixture of MDMO-PPV and PCBM. The chemical structures of MDMO-PPV and PCBM are shown on the right. Reprinted from *Materials Today*, **7**, Sarifici NS, 'Plastic photovoltaics', pp. 36–40, Copyright (2004), with permission from Elsevier.

circuit. An electrolyte then reduces the dye, regenerating it. Grätzel used a thin film of nanoparticles of titanium dioxide (a wide band gap semiconductor) and a lithium ion liquid electrolyte. This choice of materials offers a high interfacial surface area, which causes rapid and efficient charge transfer.

Two factors combined to give a high (7%) conversion efficiency for Grätzel's cell. First, the dye, Ru(II)$L_2(SCN)_2$ (where L is 4,49-dicarboxy-2,29-bipyridyl), and the titanium oxide film were able to absorb 46% of the incoming solar flux. By using a light-harvesting antenna, Grätzel was also able to separate light absorption and charge transport. This provided the cell with a superior fill factor than a conventional silicon solar cell, since exciton recombination losses were minimized due to the fact that there was no exciton travelling through the

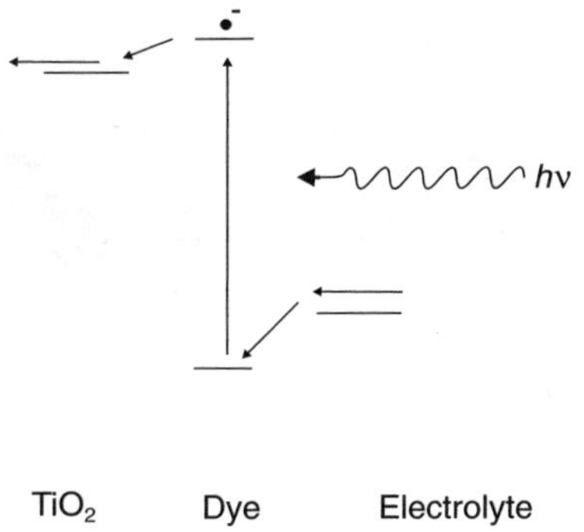

Figure 9.35 Carrier transport in dye-sensitized solar cell.

semiconductor defects. Conversion efficiencies of around 10% have now (2006) been achieved for such liquid electrolyte cells.

Grätzel and co-workers modified the design of the cell to use a solid electrolyte [49] since the liquid electrolyte cells required complete encapsulation and were difficult to handle. The new electrolyte was the amorphous organic hole transporter 2,2′,7,7′-tetrakis(*N*,*N*′-di-*p*-methoxyphenyl-amine)9,9′-spirobifluorene (OMeTAD) with the solar cell exhibiting an energy conversion efficiency of 0.7%. In this cell, the dye electron, excited by a photon, was transferred to the conduction band of TiO_2 once again. The dye was regenerated by injecting an electron into the hole transporter. The electrons injected into the TiO_2 and the holes injected into OMeTAD then diffused to the electrodes.

Grätzel continued to refine the devices and increased the device efficiency to around 2.5% by blending 4-*tert*-butylpyridine (tBP) and $Li(CF_3SO_2)_2N$ into the solid electrolyte, which minimized charge recombination losses at the TiO_2/spiro-MeOTAD interface [50]. The power conversion efficiency for this structure was about 2.6%.

9.7.3 Luminescent Concentrator

The luminescent concentrator has been under study since the 1970s [51]. The concept is based on incorporating, in a transparent matrix, materials that are both absorbing and luminescent. The refractive index of the transparent matrix is chosen larger than that of the surrounding medium (air). As the luminescent materials emit light in all directions, part of the emitted light will be internally reflected at the matrix/air interface, as shown schematically in Figure 9.36. By designing geometries for which the length and width of the transparent matrix are larger than its thickness (as for a plate configuration), geometric concentration is accomplished. The emitted light is totally internally reflected and will be concentrated within the matrix, where it can be harvested at the edges. There it can be

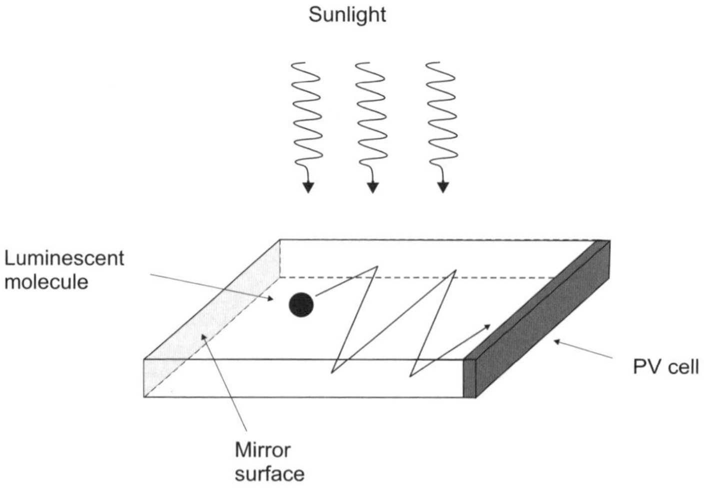

Figure 9.36 Schematic diagram of a luminescent flat plate concentrator. Direct and diffuse sunlight are absorbed by luminescent molecules. The emission is confined to the thin plate by total internal reflection. The radiation is collected by a PV cell at one end of the plate.

transformed into electrical energy by a PV cell. Although increases in photon flux of 100, compared with exposing the same photocell to direct sunlight, have been predicted, only modest overall system efficiencies (∼5%) have been achieved. This is partly due to the fact that for the organic luminescent materials used, the luminescent wavelengths were only slightly higher than the absorption wavelengths. Under these conditions, a significant part of the emitted light is lost by reabsorption. However, the design of newer materials, based on quantum dots and photonic crystals, is offering renewed interest in this area.

9.8 OTHER APPLICATION AREAS

9.8.1 Conductive Coatings

Conductive coatings have applications as diverse as electromagnetic shielding and corrosion control. Traditionally, the simplest systems have consisted of fine metallic particles dispersed in an insulating polymer matrix. However, as noted earlier in this chapter, the development of intrinsically conductive polymers, which can be processed by methods such as spin-coating and solution casting, has opened up further opportunities.

Static electrical charge can accumulate on the surfaces of insulating plastic (and other) surfaces by build up on insulating surfaces by the *triboelectric effect*. These lead to the attraction of dust, the difficulty in handling very thin films and in some circumstances to the generation of electric shocks and sparks. Static charging be reduced significantly or eliminated by the addition of *antistatic agents*. These compounds work either by making the surface of the material itself slightly conductive or by absorbing moisture from the air. In the former case, conductive polymer such as PEDOT:PSS (Chapter 5, Section 5.3) can be conveniently used. Surfactants, possessing both hydrophilic and hydrophobic groups (Chapter 8, Section 8.2.2), can be used to promote the uptake of water on the surface of an insulator. The hydrophobic moiety interacts with the surface of the material, while the hydrophilic part of the molecule interacts with the air moisture and binds the water molecules. Common antistatic agents are based on long-chain aliphatic amines and amides and quaternary ammonium salts. Antistatic agents are also added to some aircraft jet fuels, to impart electrical conductivity and to avoid the buildup of static charges, which could lead to sparking igniting the fuel vapors. The development of textiles that can dissipate static electricity, and can also screen the microwave radiation from mobile telephones, has led to the incorporation of conductive polymers into fibres.

9.8.2 Batteries and Fuel Cells

Lithium ion batteries (sometimes abbreviated Li-Ion) are a type of rechargeable battery commonly used in consumer electronics. These are currently one of the most popular types of battery, with one of the best energy-to-weight ratios, no memory effect and a slow loss of charge when not in use. However, one drawback of the Li-Ion battery is that its life span is dependent upon ageing from time of manufacture (shelf-life) regardless of whether it was charged, and not just on the number of charge/discharge cycles.

Lithium ion polymer batteries, or more commonly *lithium polymer batteries* (abbreviated Li-Poly or LiPo) are rechargeable batteries which have technologically evolved from Li-Ion batteries. Ultimately, the lithium salt electrolyte is not held in an organic solvent as in the proven Li-Ion design, but in a solid polymer composite such as polyacrylonitrile. There are many advantages of this design over the classic Li-Ion design, including the fact that the solid polymer electrolyte is not flammable (unlike the organic solvent that the Li-Ion cell uses). Since no metal battery cell casing is needed, the battery can be lighter and it can be specifically shaped to fit the device it will power. Because of the denser packaging without intercell spacing between cylindrical cells and the lack of metal casing, the energy density of Li-Poly batteries is over 20% higher than that of a classical Li-Ion battery and approximately three times better than that of nickel–cadmium (NiCd) and nickel metal hydride (NiMH) batteries. Commercialized technologies use an ion-conducting polymer instead of the traditional combination of a microporous separator and a liquid electrolyte. This promises not only better safety, as polymer electrolyte does not burn as easily, but also the possibility of making battery cells very thin, as they do not require pressure applied to 'sandwich' the cathode and anode together. The polymer electrolyte seals both electrodes like glue.

Other organic batteries that have been developed use one metallic electrode, for example lithium, and one polymer electrode, such as polypyrrole or polyaniline. However, the problems of self-discharge, low current drain and energy storage capacity mean that Li-Poly batteries are not competitive with Ni–Cd and other established batteries.

A *fuel cell* is an electrochemical energy conversion device similar to a battery, but designed for continuous replenishment of the reactants consumed. This device produces electricity from an external supply of fuel and oxygen as opposed to the limited internal energy storage capacity of a battery. Additionally, the electrodes within a battery react and change as a battery is charged, or discharged, whereas a fuel cell's electrodes are catalytic and relatively stable. Perhaps the most important device is the hydrogen fuel cell, shown schematically in Figure 9.37. The anode and cathode reactions are as follows:

$$\begin{aligned} &H_2 \rightarrow 2H + 2e^- \\ &O_2 + 4H^+ + 4e^- \rightarrow 2H_2O \end{aligned} \quad (9.11)$$

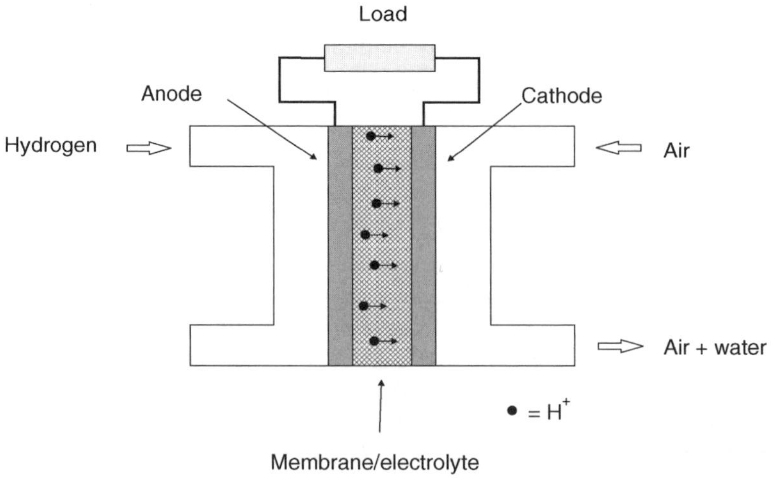

Figure 9.37 Schematic diagram of a hydrogen fuel cell.

On the anode side, hydrogen diffuses to the anode catalyst, where it dissociates into protons and electrons. The protons are conducted through the membrane to the cathode, but the electrons are forced to travel in an external circuit (supplying power) because the membrane is electrically insulating. On the cathode catalyst, oxygen molecules react with the electrons (which have travelled through the external circuit) and protons to form water. In this example, the only waste product is water vapor and/or liquid water. In addition to pure hydrogen, there are hydrogen-carrying fuels for fuel cells, including diesel, methanol and chemical hydrides, the waste product with these types of fuel is carbon dioxide.

An important requirement for an electrochemical fuel cell is a good solid-state ionic conductor. Here the membrane which separates the two electrodes must be able to transfer protons efficiently. A material that combines the flexibility and toughness of a plastic with high protonic activity would be an ideal candidate. Prototype cells have been successfully operated with membranes made from poly(styrenesulfonate) (Chapter 7, Section 7.3.3, Figure 7.30). In the presence of water, this polyelectrolyte is an excellent protonic conductor, although it suffers some degradation in a working fuel cell. Probably the most successful polyelectrolytes developed are based on perfluorinated polymers, the first of which was Nafion®, with the structure depicted in Figure 9.38(a). Nafion's unique ionic properties are a result of incorporating perfluorovinyl ether groups terminated with sulfonate groups on a tetrafluoroethylene backbone. There is a range of similar polyelectrolytes with the same generic structure, differing only in the values of x, y, m and n and either with or without $-[CF(CF_3)]-$ moieties in the pendant. Such material readily absorbs water to achieve conductivities as high as $1\,S\,m^{-1}$. Other, lower cost proton-conducting electrolytes using aromatic polymers, which exhibit high thermal and chemical stability, have been developed. One example, poly(ether ether ketone) (PEEK), is shown in Figure 9.38(b); this compound may be sulfonated to produce a polyelectrolyte. PEEK is a thermoplastic with extraordinary mechanical properties. The Young's Modulus is 3.6 GPa and its tensile strength 170 MPa. PEEK is partially crystalline, and has a glass transition temperature of 143 °C and a melting temperature of 334 °C and is highly resistant to

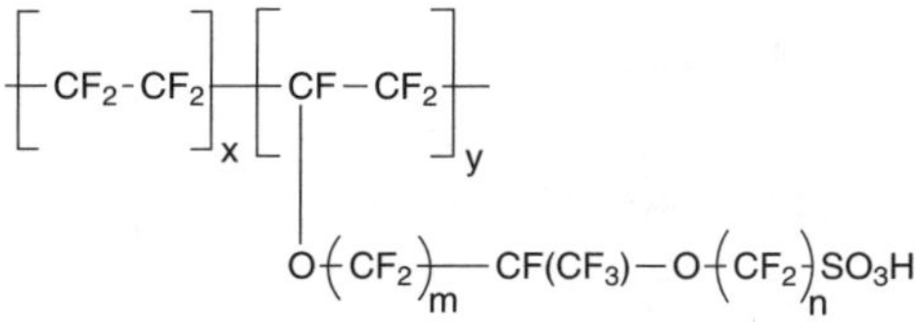

(a)

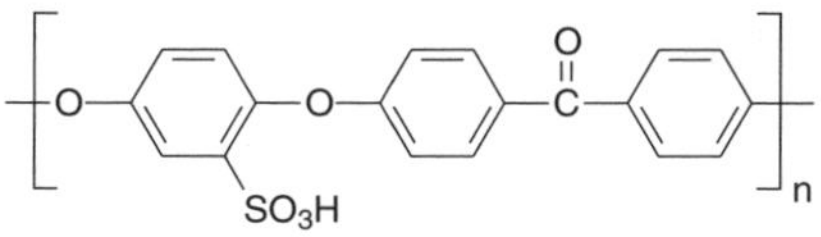

(b)

Figure 9.38 Ionic conductors: (a) Nafion® and (b) poly(ether ether ketone) (PEEK).

thermal degradation. The material is also resistant to both organic and aqueous environments, and is used in bearings, piston parts, pumps, compressor plate valves and cable insulation applications.

9.8.3 Xerography

The most common realization of xerography (from the Greek *xeros* for dry and *graphos* for writing) is the office photocopier. The process was invented in 1938 and patented in 1940 by Chester F. Carlson. Xerography exploits electrostatics and photoconductivity. There are a number of variations of the process, but the most common is based on the development of a latent electrostatic image formed on the surface of a precharged insulating photoconductor. The five 'fundamental' stages are shown in Figure 9.39.

Initially (Step I. Charging), an electrostatic charge is uniformly distributed over the surface of the photoconductor by a corona discharge. The photconductor is invariably deposited on the surface of a drum that rotates as the image of the document is scanned. The charge is deposited on the surface by the discharge from a fine wire held at a high potential close to the surface of the drum. An image of the document to be copied is projected on to the photoconductor. Electron–hole pairs are generated in the photoconductor at points where the light is incident. For a positively charged surface, electrons will drift to the surface, neutralizing the positive charge. Hence the initial surface potential is reduced locally and the

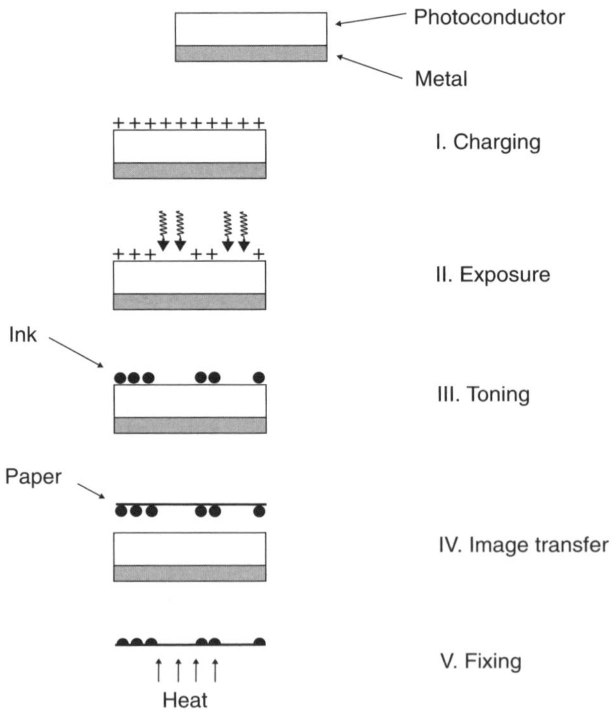

Figure 9.39 The important processing steps in xerography.

optical image is converted into a latent electrostatic image (Step II. Exposure). The ideal photoconductor will have a high dark resistivity and a *panchromatic* photoconductive response (i.e. sensitive to all wavelengths of visible light) with high quantum efficiency. The latent image is converted into a visible image by bringing charged toner particles (the 'ink') close to the surface of the photoconductor (Step III. Toning). The toner particles are charged by agitated contact with carrier particles, which also serve to transport the toner to the surface of the photoconductor. The toner neutralizes the charge on the photoconductor, so that the image can be transferred to a charged piece of paper (Step IV. Image transfer). Finally, the paper is heated to fix the image onto the surface of the paper (Step V. Fixing). In two further processes, cleaning and erase, the drum is first cleaned of any remaining toner that did not transfer to the paper by a rotating brush or a wiper blade under suction, and then any residual electrostatic image is removed by flooding the entire photoreceptor with uniform illumination.

The properties required for the photoreceptor are a high dark surface resistivity, a reasonable level of photoconductivity when illuminated with either incandescent lamps or semiconductor laser diodes and good abrasion resistance. The latter is needed to ensure an adequate photoreceptor lifetime by minimizing damage during the image transfer, cleaning and discharging. The first practical copiers used photoreceptors fabricated by the vacuum deposition of selenium and selenium alloys. Amorphous silicon has also been employed by plamsa decomposition of volatile silicon and dopant compounds. Today, most commercial photocopiers use semiconducting polymers. These offer simple fabrication techniques such as dip or spray coating of an aluminium drum or solution coating of metallized polyethylene terephthalate fabric to make photoreceptor belts for high-speed copiers.

Typically, as illustrated in Figure 9.40, polymeric photoreceptors have multilayer structures comprising an outer charge transport later (CTL) adjacent to a charge-generation layer (CGL). An additional layer between the CGL and the metal substrate can be used to prevent charge injection from the metal. The CTL must have a high carrier mobility and be transparent at wavelengths that generate electrons and holes in the underlying CGL. For this layer, much work has focused on polymers with pendant conjugated groups, such as PVK (polyvinylcarbazole; Figure 9.22). However, most commercial photocopiers use molecularly-doped, insulating polymers [52]. As noted in Section 9.6.2, many of the organic compounds developed as the CTL in xerography are now used as transport layers in organic light emitting devices. The CGL usually comprises fine crystals of pigment dispersed in a

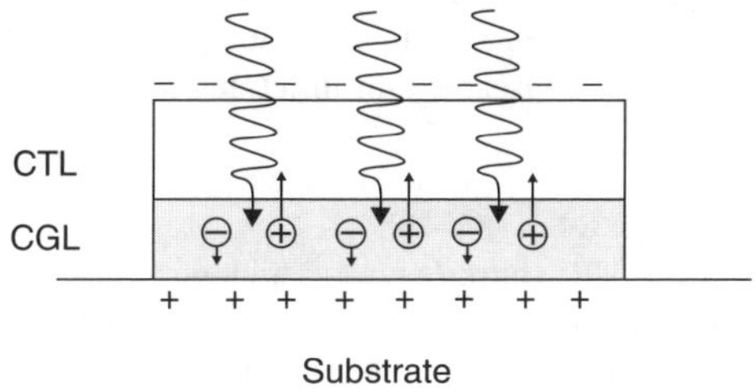

Figure 9.40 A two-layer xerographic material. CTL = charge-transport layer; CGL = charge generation layer.

polymer binder. Materials with suitable properties include azo dyes, perylene (Chapter 4, Section 4.4.5, Figure 4.15) and phthalocyanines (Chapter 5, Section 5.4, Figure 5.9).

BIBLIOGRAPHY

Barford W. *Electronic and Optical Properties of Conjugated Polymers.* Oxford: Oxford University Press; 2005.

Blythe T, Bloor D. *Electrical Properties of Polymers.* Cambridge: Cambridge University Press; 2005.

Borsenberger PM, Weiss DS. *Organic Photoreceptors for Xerography.* New York: Marcel Dekker; 1998.

Cuniberti G, Fagas G, Richer K (Editors). *Introducing Molecular Electronics.* Berlin: Springer; 2005.

Eisberg R, Resnick R. *Quantum Physics*, 2nd ed. New York: John Wiley & Sons, Inc.; 1985.

Forrest SR. The path to ubiquitous and low-cost organic electronic appliances on plastic. *Nature* 2004; **428**: 911–918.

Klauk H (Editor). *Organic Electronics: Materials, Manufacturing, and Applications.* Weinhein: Wiley-VCH; 2006.

Miyata S, Nalwa HS (Editors). *Organic Electroluminescent Materials and Devices.* Amsterdam: Gordon and Breach; 1997.

Petty MC, Bryce MR, Bloor D (Editors). *An Introduction to Molecular Electronics.* London: Edward Arnold; 1995.

Richardson TH (Editor). *Functional Organic and Polymeric Materials.* Chichester: John Wiley & Sons, Ltd; 2000.

Schwoerer M, Wolf HC. *Organic Molecular Solids.* Berlin: Wiley-VCH; 2006.

Sze SM. *Physics of Semiconductor Devices.* New York: John Wiley & Sons, Inc.; 1969.

Wallace GG, Spinks GM, Kane-Maguire LAP, Teasdale PR. *Conductive Electroactive Polymers: Intelligent Materials Systems.* Boca Raton, FL: CRC Press; 2002.

Waser R (Editor). *Nanoelectronics and Information Technology*, 2nd ed. Weinhein: Wiley-VCH; 2005.

Zhenan B, Locklin J (Editors). Organic Field Effect Transitors. Boca Rafon, FL: CRC Press; 2007.

REFERENCES

[1] Kuo CS, Wakin FG, Sengupta SK, Tripathy SK. Schottky and metal–insulator–semiconductor diodes using poly(3-hexylthiophene). *Jpn. J. Appl. Phys.* 1994; **33**: 2629–2632.

[2] Steudel S, Myny K, Arkhipov V, Deibel C, De Vusser S, Genoe J, Heremans P. 50 MHz rectifier based on an organic diode. *Nature Mater.* 2005; **4**: 597–600.

[3] Sze SM. *Physics of Semiconductor Devices.* New York: John Wiley & Sons, Inc.; 1969.

[4] Grecu S, Bronner M, Opitz A, Brütting W. Characterization of polymeric metal–insulator–semiconductor diodes. *Synth. Met.* 2004; **146**: 359–363.

[5] Dimitrakopoulos CD, Mascaro DJ. Organic thin-film transistors: a review of recent advances. *IBM J. Res. Dev.* 2001; **45**: 11–27.

[6] Watkins NJ, Yan L, Gao Y. Electronic structure symmetry of interfaces between pentacene and metals. *Appl. Phys. Lett.* 2002; **80**: 4384–4386.

[7] Plastic Logic. Available from URL http://www.plasticlogic.com/.

[8] Klauk H, Halik M, Zschieschang U, Eder F, Schmid G, Dehm C. Pentacene organic transistors and ring oscillators on glass and on flexible polymeric substrates. *Appl. Phys.Lett.* 2003; **82**: 4175–4177.

[9] Pearson C, Moore AJ, Petty MC, Bryce MR. A field effect transistor based on a Langmuir–Blodgett film of an $Ni(dmit)_2$ charge transfer complex. *Thin Solid Films* 1992; **210–211**: 257–260.

[10] Pappenfus TM, Chesterfield RJ, Frisbie CD, Mann KR, Casado J, Raff JD, Miller LL. A π-stacking tertiophene-based quinodimethane is an n-channel conductor in a thin film transistor. *J. Am. Chem. Soc.* 2002; **124**: 4184–4185.

[11] Katz HE. Organic molecular solids as thin film transistor semiconductors. *J. Mater. Chem.* 1997; **7**: 369–376.

[12] Klauk H, Zschieschang U, Pflaum J, Halik M. Ultralow-power organic complementary circuits. *Nature* 2007; **445**: 745–748.

[13] Baude PF, Ender DA, Haase MA, Kelley TW, Muyres DV, Theiss SD. Pentacene-based radio-frequency identification circuitry. *Appl. Phys. Lett.* 2003; **82**: 3964–3966.

[14] Tsang CW, VanSlyke SA. Organic electroluminescent diodes. *Appl. Phys. Lett.* 1987; **51**: 913–915.

[15] Burroughes JH, Bradley DDC, Brown AR, Marks RN, Mackay K, Friend RH, Burns PL, Holmes AB. Light-emitting diodes based on conjugated polymers. *Nature* 1990; **347**: 539–541.

[16] Miyata S, Nalwa HS (Editors). *Organic Electroluminescent Materials and Devices.* Amsterdam: Gordon and Breach; 1997.

[17] Hudson AJ, Weaver MS. In TH Richardson (Editor). *Functional Organic and Polymeric Materials.* Chichester: John Wiley & Sons, Ltd; 2000, pp. 365–394.

[18] Rees ID, Robinson KL, Holmes AB, Towns CR, O'Dell R. Recent developments in light-emitting polymers. *MRS Bull.* 2002; **27**: 451–455.

[19] Karg S, Meier M, Riess W. Light-emitting diodes based on poly-*p*-phenylene-vinylene: I charge-carrier injection and transport. *J. Appl. Phys.* 1997; **82**:1951–1966.

[20] Campbell AJ, Weaver MS, Lidzey DG, Bradley DDC. Bulk limited conduction in electroluminescent polymer devices. *J. Appl. Phys.* 1998; **48**: 6737–6746.

[21] Parker ID. Carrier tunnelling and device characteristics in polymer light-emitting devices. *J. Appl. Phys.* 1994; **75**: 1656–1666.

[22] Coaton JR, Marsden AM (Editors). *Lamps and Lighting*, 4th ed. London: Edward Arnold; 1997.

[23] Kido J, Kimura M, Nagai K. Multilayer white light-emitting organic electroluminescent device. *Science* 1995; **267**: 1332–1334.

[24] Kido J, Shionoya H, Nagai, K. Single-layer white light-emitting organic electroluminescent devices based on dye-dispersed poly(*N*-vinylcarbazole). *Appl. Phys. Lett.* 1995; **67**: 2281–2283.

[25] Thompson J, Blyth RIR, Mazzeo M, Anni M, Gigli G, Cingolani R. White light emission from blends of blue-emitting organic molecules: a general route to the white organic light-emitting diode? *Appl. Phys. Lett.* 2001; **79**: 560–562.

[26] Duggal AR, Shiang JJ, Heller CM, Foust DF. Organic light-emitting devices for illumination quality white light. *Appl. Phys.Lett.* 2002; **80**: 3470–3472.

[27] Forrest SR, Bradley DDC, Thompson ME. Measuring the efficiency of organic light-emitting devices. *Adv. Mater.* 2003; **15**: 1043–1048.

[28] Sumation Co. Ltd. Available from URL http://www.sumation.co.uk/.

[29] Pfeiffer M, Leo K, Zhou X, Huang JS, Hofmann M, Werner A, Blochwitz-Nimoth J. Doped organic semiconductors: physics and application in light emitting diodes. *Organic Electron.* 2003; **4**: 89–103.

[30] Huang J, Pfeiffer, Werner, Blochwitz J, Leo K, Liu S. Low-voltage organic electroluminescent devices using pin structures. *Appl. Phys. Lett.* 2002; **80**: 139–141.

[31] Gebeyehu D, Walzer K, He G, Pfeiffer M, Leo K, Brandt J, Gerhard A, Stössel, P, Vestweber H. Highly efficient deep-blue light-emitting diodes with doped transport layers. *Synth. Met.* 2005; **148**: 205–211.

[32] Borchardt JK. Developments in organic displays. *Mater. Today* 2004; **7**: 43–46.

[33] Universal Display Corporation. Available from URL http://www.universaldisplay.com/.

[34] Krummacher BC, Choong V-E, Mathai MK, Choulis SA, So F, Jermann F, Fielder T, Zachau M. Highly efficient white organic light-emitting diode. *Appl. Phys. Lett.* 2006; **88**: 113506.

[35] Kanno H, Sun Y, Forrest SR. White organic light-emitting device based on a compound fluorescent-phosphor-sensitized-fluorescent emission layer. *Appl. Phys. Lett.* 2006; **89**: 143516.

[36] Burrows PE, Gu G, Bulović Shen Z, Forrest SR, Thompson ME. Achieving full-colour organic light-emitting devices for lightweight, flat-panel displays. *IEEE Trans. Electron Dev.* 1997; **44**: 1188–1202.

[37] Nakayama T. Organic luminescent devices with a microcavity structure. In S Miyata, HS Nalwa (Editors). *Organic Electroluminescent Materials and Devices.* Amsterdam: Gordon and Breach; 1997, pp. 359–389.

[38] Ahn JH, Wang C, Pearson C, Bryce MR, Petty MC. Organic light-emitting diodes based on a blend of poly[2-(2-ethylhexyloxy)-5-methoxy-1,4-phenylenevinylene] and an electron transport material. *Appl. Phys. Lett.* 2004; **85**: 1283–1285.

[39] Oyston S, Wang C, Hughes G, Batsanov AS, Perepichka IF, Bryce MR, Ahn JH, Pearson C, Petty MC. New 2,5-diaryl-1,3,4-oxadiazole–fluorene hybrids as electron transporting materials for blended-layer organic light emitting diodes. *J. Mater. Chem.* 2005; **15**: 194–203.

[40] Zang Z, Hu Y, Li H, Wang L, Jing X, Wang F, Ma D. Polymer light-emitting diodes based on a bipolar transporting luminescent polymer. *J. Mater. Chem.* 2003; **13**: 773–777.

[41] Thomas KRJ, Lin JT, Tao Y-T, Chuen C-H. New carbazole–oxadiazole dyads for electroluminescent devices: influence of acceptor substituents on luminescent and thermal properties. *Chem. Mater.* 2004; **16**: 5437–5444.

[42] Peng Q, Kang ET, Neoh KG, Xiao D, Zou D. Conjugated alternating copolymers of fluorene and 2-pyridine-4-ylidenemalononitrile: synthesis, characterization and electroluminescent properties. *J. Mater. Chem.* 2006; **16**: 376–383.

[43] Kamtekar KT, Wang C, Bettington S, Batsanov AS, Perepichka IF, Bryce MR, Ahn JH, Rabinal M, Petty MC. New electroluminescent bipolar compounds for balanced charge-transport and tuneable colour in organic light emitting diodes: triphenylamine–oxadiazole–fluorene triad molecules. *J. Mater. Chem.* 2006; **16**: 3823–3835.

[44] Fuhrmann T, Salbeck J. Organic materials for photonic devices. *MRS Bull.* 2003; **28**: 354–359.

[45] Jung S, Theis D. Electronic paper. In R Waser (Editor). *Nanoelectronics and Information Technology*, 2nd ed. Weinheim: Wiley-VCH; 2005, pp. 953–965.

[46] Saricifti NS. Plastic photovoltaics. *Mater. Today* 2004; **7**: 36–40.

[47] Pulfrey DL. *Photovolatic Power Generation.* New York: Van Nostrand; 1978.

[48] O'Regan B, Grätzel M. A low-cost, high-efficiency solar cell based on dye sensitized colloidal TiO_2 films. *Nature* 1991; **353**: 737–740.

[49] Bach U, Lupo D, Comte P, Moser JE, Weissörtel F, Salbeck J, Spreitzer H, Grätzel M. Solid-state-dye-sensitised mesoporous TiO_2 solar cells with high photon-to-electron conversion efficiencies. *Nature* 1998; **395**: 583–585.

[50] Krüger J, Plass R, Cevey L, Piccirelli M, Grätzel M, Bach U. High efficiency solid-state photovoltaic device due to inhibition of interface charge recombination. *Appl. Phys. Lett.* 2001; **79**: 2085–2087.

[51] Weber WH, Lambe J. Luminescent greenhouse collector for solar radiation. *Appl. Opt.* 1976; **15**: 2299–2300.

[52] Blythe T, Bloor D. *Electrical Properties of Polymers.* Cambridge: Cambridge University Press; 2005.

10 Chemical Sensors and Actuators

Above the sense of sense

10.1 INTRODUCTION

This chapter will focus on chemical sensors and physical actuators that exploit organic materials. Some examples of devices (e.g. electronic nose) already exist as commercial products whereas others (e.g. lab-on-a-chip, smart textiles) are fast developing technologies that will certainly make an impact in the marketplace over the next decades.

The development of effective devices for the identification and quantification of chemical and biochemical substances for process control and environmental monitoring is a growing need. Many sensors do not possess the specifications to conform to existing or forthcoming legislation; some systems are too bulky or expensive for use in the field. Inorganic materials such as the oxides of tin and zinc have traditionally been favoured as the sensing element in gas and vapour sensors. However, one disadvantage of devices based on metallic oxides is that they usually have to be operated at elevated temperatures, limiting some applications. As an alternative, there has been considerable interest in trying to exploit the properties of organic materials. Many such substances, in particular phthalocyanine derivatives, are known to exhibit a high sensitivity to gases. Lessons can also be taken from the biological world; for example, a household carbon monoxide detector can be designed to simulate the reaction between CO and haemoglobin. A significant advantage of organic compounds is that their sensitivity and selectivity can be tailored to a particular application by modifications to their chemical structure. Moreover, thin film technologies, such as self-assembly or layer-by-layer electrostatic deposition (Chapter 7), enable ultra-thin layers of organic materials to be engineered at the molecular level.

There are many physical principles upon which sensing systems might be based; changes in electrical resistance (chemiresistors), refractive index (fibre optic sensors) and mass (quartz microbalance) have all been exploited in chemical sensing. The main challenges in the development of new sensors are in the production of cheap, reproducible and reliable devices with adequate sensitivities and selectivities.

In this chapter, a brief survey of the different sensor technologies and their applications will be given. The concept of a sensor or actuator will first be described. Then, some of the most important parameters that define a sensor will be introduced. Sensing systems will be classified through the means of transduction exploited and on the basis of the type of measured quantity.

Molecular Electronics: From Principles to Practice M. C. Petty
© 2007 John Wiley & Sons, Ltd

10.2 SENSING SYSTEMS

One of the first challenges encountered in describing a sensing system is to provide a satisfactory definition of the term sensor. A *sensor* (the term *transducer* is also used) is a device that is able to convert a variation of any quantity, property or physical state – called the input signal or *measurand* – into a 'useful' (i.e. something we are able to read and interpret) output signal. A sensor, therefore, is an element sensitive to the input and able to communicate with a measuring or control system. Sensors generally have their output in an electrical form, whereas the term *actuator* is used to describe sensors with a non-electrical output. Natural sensors, such as those found in living organisms, usually respond to electrochemical signals; their physical nature is based on ion transport. In artificial sensors, information is usually transmitted through the transport of electrons or photons.

It has been proposed by Middlehoek and Noorlag that a sensor or actuator can be classified according to the energy domain of its primary input–output [1]. Six classes of input and output signals can be distinguished: mechanical, thermal, electrical, magnetic, radiation and (bio)chemical.

Sensors are also one of two types: *active* or *passive*. Active sensors directly generate a signal in response to an external stimulus. The input stimulus energy is converted into output energy without the need for an additional power source. Examples are thermocouples, pyroelectric radiation detectors and piezoelectric pressure sensors. In contrast, passive sensors require an external energy source for their operation. The signal is then modified by the sensor to produce a useful output. Passive sensors are sometimes called *parametric* because their properties change in response to an external stimulus, and these changes can subsequently be converted into an output signal. For instance, a *thermistor* is a temperature-sensitive resistor. This device does not directly generate any energy signal, but by passing an electric current through it (excitation signal) the resistance can be measured by detecting variations in current and/or voltage across the thermistor. These variations (measured in ohms) are directly related to the temperature change.

Figure 10.1 shows a schematic diagram of a sensing system. Modern devices can be highly miniaturized and the sensing element can be integrated with circuitry for signal

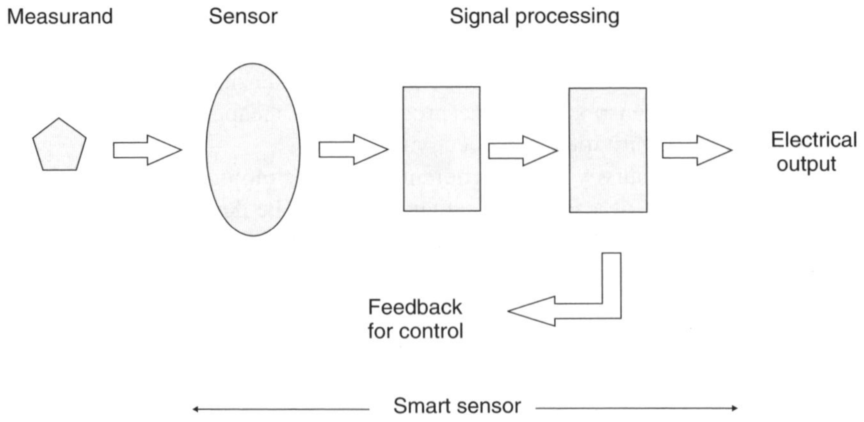

Figure 10.1 Schematic diagram of an electronic sensor system.

processing (e.g. amplification, filtering). A feedback path for control purposes may also be provided. Such a system may be referred to as an *integrated* or *smart sensor.*

Issues related to the degradation of the device, to thermal drift and to nonlinearity in the response can be addressed by a smart sensor in an automatic manner. Moreover, self-diagnosis of the working state and self-calibration are very attractive features in such a measuring system, making it able to adapt to different environments.

10.3 DEFINITIONS

A correct definition of the parameters of a sensing system is fundamental for a better understanding of the sensor behaviour and for establishing uniform standards. It is then possible to have a consistent comparison between the performances of different sensing systems. The descriptions also need to be attributed to the correct part of the sensing system. For example, some parameters only characterize the active element(s), whereas others define the performance of the entire sensing system.

An ideal sensor should have characteristics such as a high *selectivity* and *sensitivity,* complete *reversibility,* long-term *reliability* and *stability,* short *response* and *recovery times* and a good *signal-to-noise ratio.* Other factors might include multiplexing capabilities, low cost, a minimal complexity for use and portability for *in situ* applications. Compatibility with integrated circuits and microprocessors, flexibility in the assembly of different platforms and of several sensors within a single package are also highly desirable features.

Reliability, sensitivity and selectivity are characteristics that depend directly on the material used as a sensor. Features such as response time and noise are more complex, since several components of the sensing system play a role. In particular, it is important to note that the main source of noise in a sensing system is, very often, the system itself, with all its electronic components making a contribution. Shielding of the multiple noise sources within the system is, therefore, essential.

The response of a sensor is calculated from the value of the output before the measurement starts, R_0 and the value measured in the presence of a measurand, R_{meas}:

$$\left|\frac{\Delta R}{R_0}\right| = \left|\frac{R_{\text{meas}} - R_0}{R_0}\right| \tag{10.1}$$

The sensitivity of a sensor is defined as the derivative of the parameter R_{meas} with respect to the measurand, m (the input). If there is a linear variation between the input and the sensor response, as shown in Figure 10.2, then the sensitivity, S, is given by

$$S = \frac{\Delta R}{\Delta m} \tag{10.2}$$

When the response curve of the sensing system is not linear, then the sensitivity can be expressed as

$$S = S(m) = \frac{\mathrm{d}R_{\text{meas}}}{\mathrm{d}m} \tag{10.3}$$

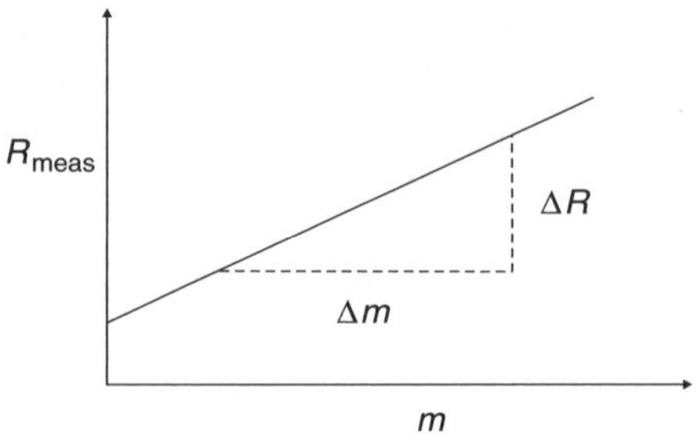

Figure 10.2 Output versus input for a sensor operating in a linear regime. R_{meas} is the output or response of the sensor and m is the input or measurand.

As shown in Figure 10.3, the sensitivity S_{m_0} for a practical value of the measurand m_0 is

$$S = \left.\frac{dR_{meas}}{dm}\right|_{m=m_0} = S_{m_0} \tag{10.4}$$

The response time of a sensor can be defined in a number of ways. For example, it may define the time interval for the response signal R_{meas} to change from 10 to 90% of its final saturation value, R_f, after applying the stimulation as a step-function, as shown in Figure 10.4. If the system is reversible, then it is also possible to define a recovery time, as the time for the response signal to change from 90 to 10% of R_f. In other cases, the response and recovery times describe the time until the sensor output signal has reached $1-1/e$ (i.e. approximately 63%) of its final value. The reciprocal of the response time is related to the *bandwidth* of the sensor; the quicker the response, the larger is the bandwidth.

The signal-to-noise (S/N) ratio of a measuring system is defined as the ratio of the strength of the signal plus noise carrying information to the unwanted interference (noise). It is measured using the logarithmic decibel (dB) scale and it is equal to 20 times the base-10 logarithm of the amplitude ratio (i.e. voltage of the signal divided by the voltage of the noise, V_s/V_n), or 10 times the logarithm of the power ratio. If $V_s = V_n$, then the S/N ratio is equal to zero. In this situation, the signal is difficult to measure directly and sophisticated signal processing techniques must be used to recover the signal.

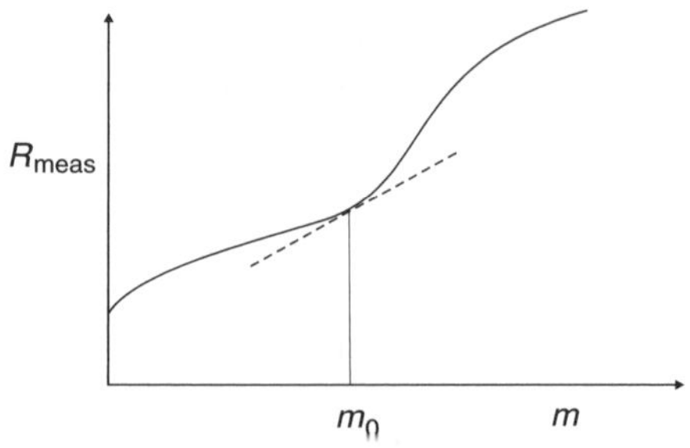

Figure 10.3 Output versus input for a sensor operating in a nonlinear regime. R_{meas} is the output or response of the sensor and m is the input or measurand.

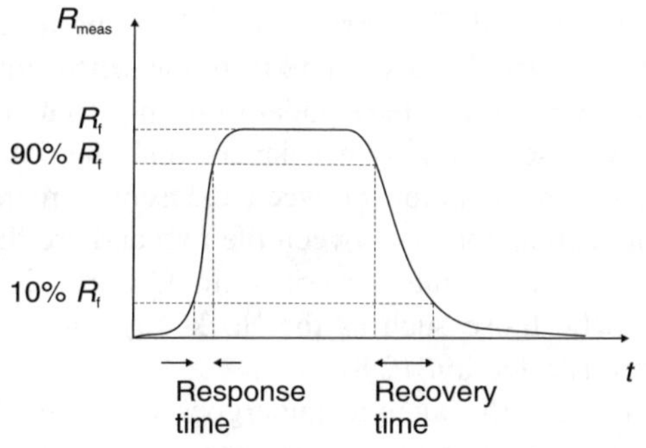

Figure 10.4 Response and recovery times of a sensor following a step-input stimulus.

For biochemical sensors, selectivity is very important. This is a measurement of the ability of the sensor to respond to only one input in the presence of interferences. In most biological systems, specificity is achieved by shape recognition, which involves a comparison with a reference. High selectivity means that the contribution from the primary species dominates, and that the interference from other species is minimal. However, an absolutely selective sensor does not really exist and there is always some interference present.

Sensor *resolution* is the minimum change in the measurand which a sensing device can resolve. In other words, it is the size of the incremental steps in the response obtained for a linear increase of the input signal. Thus the resolution is not necessarily constant over the whole measuring range. A further important parameter of a sensing system is its operative or *dynamic range*, i.e. the range of values of the measurand, which can be detected by the sensor.

A number of different sensor classification systems can be found in the literature. One method is to consider all of its properties, such as what it measures (measurand or stimulus), its specifications, the conversion mechanism, the material(s) from which it is fabricated and its field of application. Categorization schemes based on the physical or chemical means of transduction have been used, but these can be difficult to follow. Sensors are often grouped according to what they measure. It is then possible to look in detail at the different options possible and to consider the characteristics and features of each sensing system. According to this approach, there are three sensor families or groups: physical sensors, chemical sensors, and biological sensors. The latter will be covered in some detail in Chapter 12, Section 12.8. Here, we shall be concerned with chemical and physical sensors and, of course, the focus will be on the exploitation of organic materials as the sensing elements.

10.4 CHEMICAL SENSORS

Chemical sensors are devices sensitive to stimuli produced by various chemical compounds or elements. The most important property of these sensors is selectivity. Changes in the chemical environment are transformed into sensor properties such as electrical conduction, thermal conductivity or optical effects. Chemical sensors can be subdivided into gas and liquid sensors. The operation of all these devices depends on the interaction of a chemically sensitive material with the *analyte* (i.e. the measurand). This reaction subsequently produces a physical change in the sensor, which is converted into a measurable electronic signal. The interaction between the analyte and the sensor material can be reversible or irreversible. In the former case, the analyte molecules dissociate from the sensor material when the external concentration is removed and, overall, they undergo no net change. Examples of this type include the adsorption of gases on polymer layers and the interactions of gases with conductive polymers. Here, the reaction between the sensor material and the analyte is determined by the intermolecular forces between the two and are the results of one or more of the chemical bonding arrangements described in Chapter 2, Section 2.3. The most selective reactions tend to be those such as the 'lock–key' mechanism that operates in a biological sensor (Chapter 12, Section 12.8).

For irreversible interactions, the analyte undergoes a chemical reaction at the sensor surface catalyzed by the sensor material; the analyte is consumed in the sensing process, although the numbers of molecules reacting represent a small proportion of the total number within the sample. The term 'irreversible' in this context can be confusing. Removal of the analyte concentration will still reverse the sensing process, but the associated time constant may be longer than for the 'reversible' sensors. The sensitivity and selectivity of this type of irreversible sensor are determined by the choice of the catalytic surface. A good example of this type of sensor is the amperometric electrochemical sensor described in Section 10.4.2.

10.4.1 Calorimetric Gas Sensors

Calorimetric sensors, also called *pellistors*, operate by detecting the heat liberated when a combustible gas reacts with the atmosphere. In the case of a detector for hydrocarbons, C_xH_y, the reaction proceeds according to the following scheme

$$C_xH_y + \left(x + \frac{y}{4}\right)O_2 \rightarrow x\,CO_2 + \frac{y}{2}H_2O \tag{10.5}$$

A typical pellistor consists of a fine (25–50 μm) platinum wire spiral supported within a bead of porous alumina, as shown in Figure 10.5. The porous bead is about 2 mm in diameter and contains particles of finely divided, high surface area, precious metal catalyst, such as Pd or Pt. The platinum spiral is used both to heat the catalyst bead to around 500 °C, in order to speed up the combustion reaction, and to measure the temperature of the catalyst bead. In the presence of a combustible gas, the heat liberated by reaction with oxygen on the catalyst surface is detected as a temperature change of the platinum wire. Platinum coil pellistors have been commercially available for many years. The devices are non-specific in their response to combustible gases and respond reasonably rapidly (around 20 s) to gas concentrations down to 500 ppm.

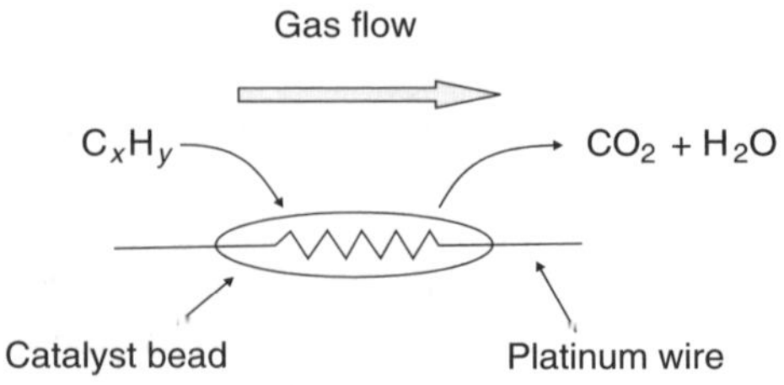

Figure 10.5 Pellistor or calorimetric gas sensor.

10.4.2 Electrochemical Cells

Electrochemical sensors are probably the most versatile chemical sensors. Some of these measure voltage (*potentiometric*); others monitor electric current (*amperometric*). Such detectors are used where a chemical reaction takes place or when the charge transport is modulated by the interaction of the chemical species under observation. Electrochemical gas sensors ionize gas molecules at a three-phase boundary layer (atmosphere, electrode of a catalytically active material, electrolyte). One of the ions, e.g. O^{2-}, H^+, Cl^-, which is involved in the reaction on the surface, can be preferentially conducted in the electrolyte.

In the simplest form of amperometric sensor (sometimes called a micro fuel cell; Chapter 9, Section 9.8.2) two electrodes – a sensing and a counter electrode – are separated by a thin layer of electrolyte. Gas diffusing to the sensing electrode reacts at the surface of the electrode either by oxidation (removal of electrons) or reduction (addition of electrons). This reaction causes the potential of the electrode to increase or decrease with respect to the counter electrode. With a resistor connected across the electrodes, a current is generated which can be detected and used to determine the concentration of gas present.

One of the conditions required for amperometric sensors to work accurately is that the potential of the counter electrode should remain constant. In practice, the surface reactions at each electrode causes these to polarize (become charged). This process may be small initially, but it increases with the level of reactant gas and effectively limits the concentration range to which the sensor can respond. The effect can be counteracted by the introduction of a reference electrode of stable potential. This is placed within the electrolyte in close proximity to the sensing electrode. As the electrode must maintain a constant potential for correct operation, it is important that no current is drawn from it. A *potentiostatic* feedback circuit is therefore used to measure the potential difference between sensing and reference electrodes. Certain amperometric sensors, e.g. oxygen sensors, do not require a reference electrode, as the reaction involves a chemical change in the counter electrode.

Figure 10.6 shows a schematic diagram of a three-electrode amperometric electrochemical sensor. Gas that comes into contact with the sensor first passes through a small capillary-type opening. A gas-permeable membrane (also called a hydrophobic membrane and typically made from PTFE) is used to cover the sensor's surface. Besides offering a mechanical protection to the sensor, the membrane fulfils the additional function of filtering out unwanted particulates. Electrochemical amperometric cells are available for a range of simple gases and are reasonably selective. The detection limits are down to 0.1 ppm with response times of less than 1 min.

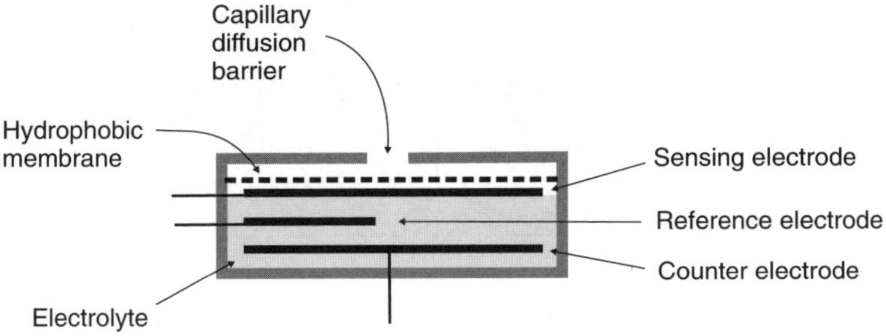

Figure 10.6 Schematic diagram of a typical electrochemical gas sensor.

In potentiometric cells, the potential difference generated between the electrodes by different partial pressures on either side of the cell provides the output. An example is the *Gas–FET*, a chemically sensitive field effect transistor. These devices use a FET (Chapter 9, Section 9.4) with a gate metallization exposed to the surrounding atmosphere. In the case of Pd-gate devices, hydrogen gas becomes adsorbed on the surface and dissociates into atoms, which subsequently diffuse to the oxide–metal interface where a dipole layer is created. This effective change in the work function of the gate electrode results in a shift in the threshold voltage, with a consequent change in the output characteristics of the transistor, as shown in Figure 10.7.

For solution work, *ion-selective electrodes* are extensively used. These devices consist of membranes that respond to one particular ionic species in the presence of others. A well-known example is the pH electrode. The output is in the form of a voltage V that is related to the concentration of an ion by the *Nernst equation*:

$$V = V_0 + 2.303\frac{RT}{nF}\log\left(\frac{a_{\text{red}}}{a_{\text{ox}}}\right) \tag{10.6}$$

where V_0 is a constant to account for all the other potentials in the system, R is the gas constant, n is the number of electrons transferred and F is Faraday's constant (Chapter 7, Section 7.2.4: F = electronic charge × Avogadro's number = 9.65×10^7 C kmol^{-1}) and a_{red}

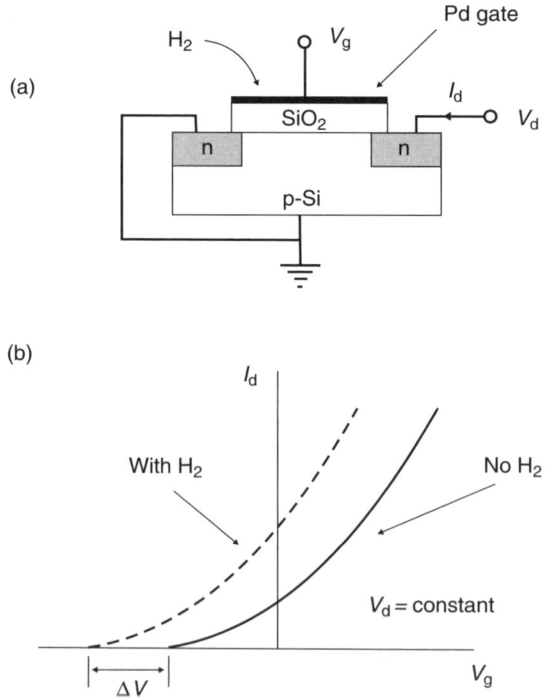

Figure 10.7 (a) Pd-gate gas-sensitive field effect transistor. (b) The effect of exposure to hydrogen gas on the drain current I_d versus gate voltage V_g characteristic.

and a_{ox} are the *activities* of the reduced and oxidized species, respectively. The activity is related to the ion concentration c by the relationship

$$a = \gamma c \tag{10.7}$$

where γ is the *activity coefficient*. This parameter is essentially a correction factor, introduced to take into account nonideal behaviour. In dilute solutions, γ approaches unity and the activity can then be identified with concentration.

For a solid electrolyte in equilibrium with an ion in solution, a plot of the measured potential versus the logarithm of the ion concentration will give a straight line, with a slope of about $58\,\text{mV}(= 2.303RT/F)$ per decade at room temperature for a reaction involving the transfer of one electron, e.g. a pH sensor.

The complete measurement system, depicted in Figure 10.8, consists of an ion-selective electrode, an internal reference electrode and an external reference electrode. Commercial ion-selective electrode systems often combine the two electrodes into one unit. The need for a reference electrode has been outlined above, for the case of an amperometric electrochemical gas sensor. There are three main requirements which such a reference potential must satisfy: it must be stable, reversible and reproducible [2]. The condition for reversibility (good ohmic contact) is satisfied by selecting an electrochemical reaction that is very fast, i.e. has a very high exchange-current density ($> 1\,\text{mA}\,\text{cm}^{-2}$). An example of such a reaction is that utilized by the silver/silver chloride electrode, which comprises silver metal coated with a few microns of silver chloride. The Ag/AgCl reference electrode must be stored in a separate compartment with defined and constant activity of the chloride ion. However, the electrical circuit must be completed. This contact is realized by the liquid junction shown in Figure 10.8. The junction is an open channel, but the outflow of the internal reference solution is very low, typically $2\,\text{nl}\,\text{h}^{-1}$.

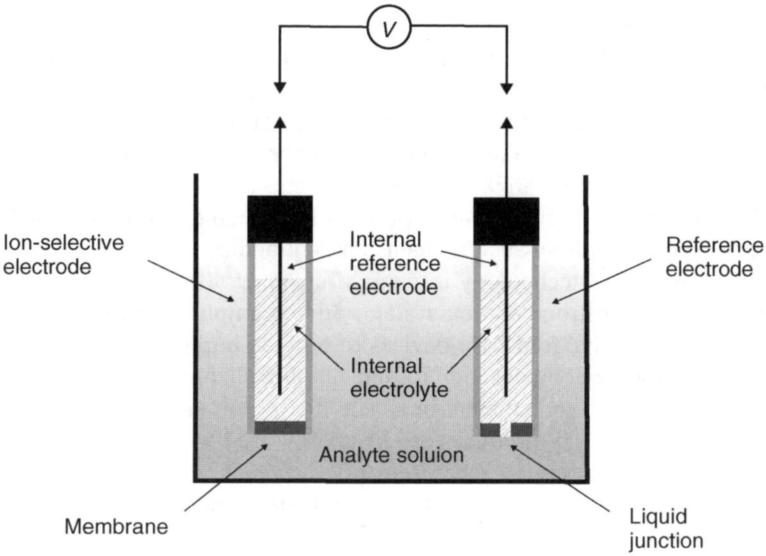

Figure 10.8 Measurement arrangement using an ion-selective electrode and a reference electrode.

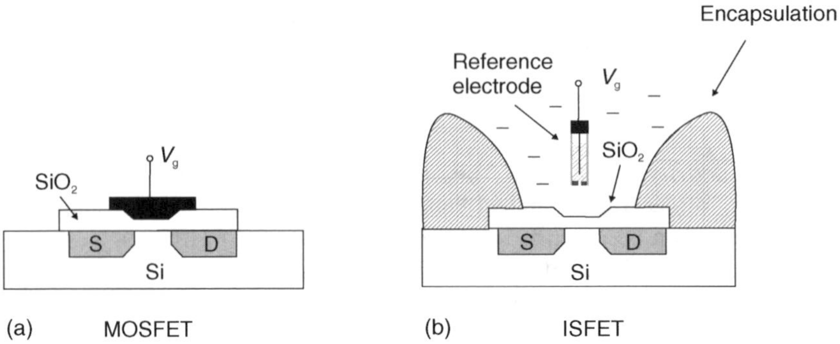

Figure 10.9 Comparison of the structure of (a) MOSFET (metal oxide semiconductor field effect transistor) and (b) ISFET (ion-sensitive field effect transistor) devices.

Ion-sensitive field effect transistors (ISFETs) have been developed since the 1970s [3, 4]. In essence, the ISFET is a MOSFET in which the gate connection is separated from the silicon substrate in the form of a reference electrode. This is inserted in an aqueous solution that is in contact with the gate insulator, as shown in Figure 10.9. A change in the interfacial potential at the liquid/gate insulator interface will produce a change in the threshold voltage of the transistor. Gate insulators such as SiO_2, Si_3N_4, Al_2O_3 and Ta_2O_5 are responsive to hydrogen ions. The mechanism of operation is associated with the formation of a thin hydrated surface layer (4–5 nm). In the case of quartz, it is known that surface silanol groups dissociate:

$$\equiv SiOH \rightleftharpoons SiO^- + H^+ \tag{10.8}$$

Sensitivity to ions other than H^+ can be achieved by the deposition of suitable gate materials. Polymeric membranes containing the ionophore valinomycin (Chapter 12, Section 12.6.1) can provide a response to potassium ions. Layers of immobilized enzymes (Chapter 12, Section 12.2.3) are often used to convert a substance for which no sensor is available into a substance for which a chemical sensor exists. A well-known example is the urea sensor, which makes use of immobilized urease. This converts urea into ammonium ions, carboxyl ions and hydroxide ions, and provides the option of monitoring the pH as an indirect method for detecting area. Such devices are called enzymeFETs or ENFETs. ImmunoFETs or IMFETs make use of highly specific immunological reactions (Chapter 12, Section 12.2.2) between antigens and antibodies.

One advantage of ISFET technology is that it offers integration with the signal processing required for a complete instrumentation system. Silicon amplifiers can be easily combined with the chemically sensitive transistor devices to provide improved signal-to-noise ratio, a reduction in drift and some compensation for temperature changes. However, one problem is the need for an external reference electrode. The highly miniaturized solid-state chemical sensors that ISFET technology produces are invariably used with conventional liquid-filled reference electrodes. Differential measurements between an ISFET and an identical reference FET which does not respond to the ions under measurement (i.e. a REFET) can introduce additional problems [3]. A number of companies sell ISFET and related technologies. Their products are generally aimed at markets where the vulnerability of glass membrane electrodes presents measurement difficulties, such as the food industry.

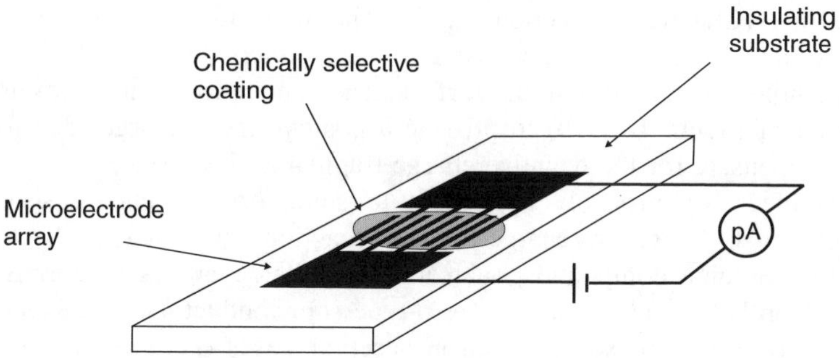

Figure 10.10 Schematic diagram of a chemiresistor device, fabricated by coating a thin film of gas or vapour-sensitive semiconductive material on to a planar microelectrode array.

10.4.3 Resistive Gas Sensors

A simple *chemiresistor* sensor exploits the resistance change of a thin layer of a gas-sensitive material. Although the most common type of resistive gas sensor uses semiconducting metal oxides, such as tin oxide, as the sensing element (so-called Taguchi sensors), devices based on organic materials are under development. The latter respond to a broad range of gases and vapours and may operate at room temperature (chemiresistors based on inorganic semiconductors usually have to be heated). Furthermore, a wide range of compounds can be synthesized, which offers the prospect of tailoring devices to respond to particular analytes. A simple chemiresistor device can be fabricated by depositing a thin film of the active (gas-sensitive) material on an interdigitated microelectrode structure, such as that shown in Figure 10.10. This structure allows a large surface area of the sample to be exposed to the analyte gas or vapour.

In the case of metal oxide-type devices, a model that has been derived involves the gas affecting the space-charge region (depletion region; Chapter 9, Section 9.2) adjacent to the surface of the material or at grain boundaries within the material. Figure 10.11 illustrates the process that can occur on exposure to oxygen. The gas will first become chemisorbed

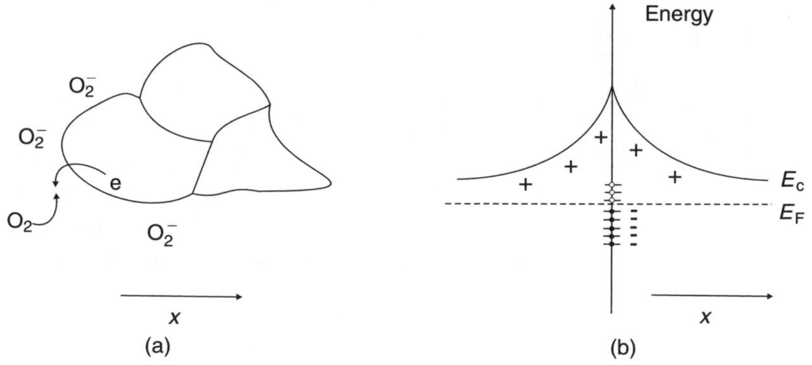

Figure 10.11 (a) Grain structure in a polycrystalline material showing the chemisorption of oxygen at the surface of the grains. (b) The negatively charged oxygen molecules produce an upward curvature of the electronic energy bands at the grain boundary, providing a depletion region.

[Figure 10.11(a)] (Chapter 2, Section 2.5.7). The molecules act as surface acceptors, removing electrons from the valence band of the semiconductor and giving rise to an increase in the hole concentration at the surface. The conduction band bends upwards in the energy diagram of Figure 10.11(b) to provide a positive fixed charge (due to fixed donor atoms) that compensate for the negative charge trapped at the surface.

The above ideas are probably applicable to some organic chemiresistors based on molecular materials. However, charge-transfer reactions (Chapter 3, Section 3.4) can occur in certain organic compounds, such as the phthalocyanines or porphyrins [5]. For the majority of phthalocyanines, which are p-type semiconductors, an increase in conductivity is observed when exposed to electron acceptor gases (i.e. oxidizing gases) such as oxygen or NO_2, whereas a decrease is seen on exposure to electron donor gases (reducing gases) such as ammonia. The conductivity changes are often accompanied by the appearance of charge-transfer bands in the visible spectra. A typical example for a chemiresistor response is illustrated in Figure 10.12 for exposure of an evaporated lead phthalocyanine film to 50 ppb of nitrogen dioxide [6]. The charge-transfer reaction also depends on the nature of the central metal ion in the phthalocyanine molecule and is dependent on the morphology of the thin film. The latter must accommodate both the charge-transfer interaction and charge-carrier transport. It is believed that the charge transfer interaction can occur more readily if the face of the phthalocyanine or porphyrins ring rather than an edge is available for electron transfer through complex formation. On the other hand, charge transport is facilitated by a long-range stacking of cofacially oriented phthalocyanine rings.

The extreme thinness of the Langmuir–Blodgett layers (Chapter 7, Section 7.3.1) provides a high surface to volume ratio for the sensor and, therefore, a potential for high sensitivity. Such films of phthalocyanine derivatives can show significantly different responses to vapours as compared with evaporated layers. For example, increases in conductivity are observed on exposure to ammonia [5]. Asymmetrically substituted materials

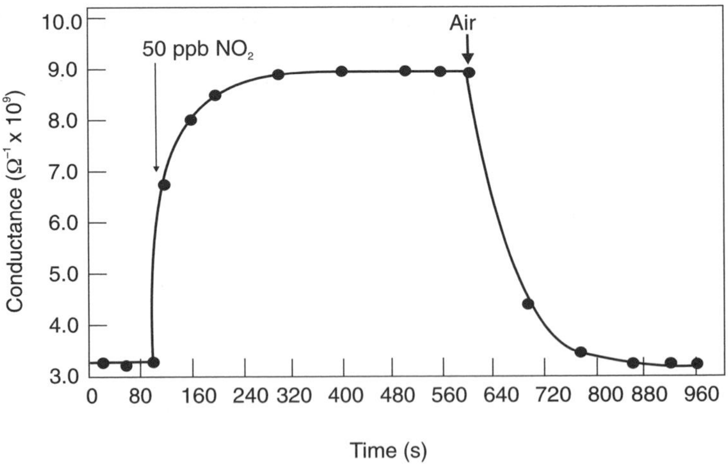

Figure 10.12 Response and recovery of a chemiresistor coated with a vacuum-evaporated lead phthalocyanine films and exposed to 50 ppb of nitrogen dioxide in air at 150 °C. Reprinted from *Sensors Actuators*, **5**, Bott B, Jones TA, 'A highly sensitive NO_2 sensor based on electrical conductivity changes in phthalocyamine film', pp. 43–53. Copyright (1984), with permission from Elsevier.

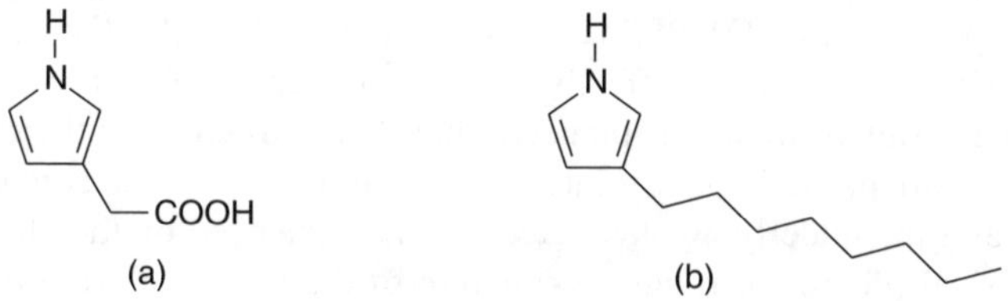

Figure 10.13 Examples of two substituted pyrrole monomers used to make conductive polymers [7]. (a) The incorporation of the carboxylic acid group gives a hydrophilic polymer. (b) The incorporation of the alkyl chain substituent produces a hydrophobic polymer.

may show enhanced response and recovery times in comparison with LB films of unsubstituted compounds. Such results indicate the importance of the peripheral groups (required to provide the solubility for LB processing) in the phthalocyanine molecules.

Many conductive polymers have been used as gas or vapour sensors. The most common materials exploited are polypyrrole, polyaniline and polythiophene. Exposure to gases leads to oxidation or reduction of the polymer chains, resulting in a variation of conductivity. Substitution of the basic polymer can also lead to modifications to the sensing response. For example, Figure 10.13 shows two examples of substituted pyrrole monomers [7]. While the attachment of a carboxylic acid group [Figure 10.13(a)] leads to a more hydrophilic polymer (likely to interact with similar molecular species), the substitution with the hydrocarbon chain [Figure 10.13(b)] leads to a more hydrophobic material. One of the attractions of using conductive polymers is their ease of preparation. Techniques such as solution casting, spin-coating and electrochemical deposition are relatively straightforward methods to fabricate sensing devices (Chapter 7).

Inkjet printing (Chapter 7, Section 7.2.5) is a further method that is attracting attention as the polymer is deposited only where it is needed. The morphology of such films can be important for very thin chemiresistors. Figure 10.14 shows electrical data for a single inkjet-printed layer

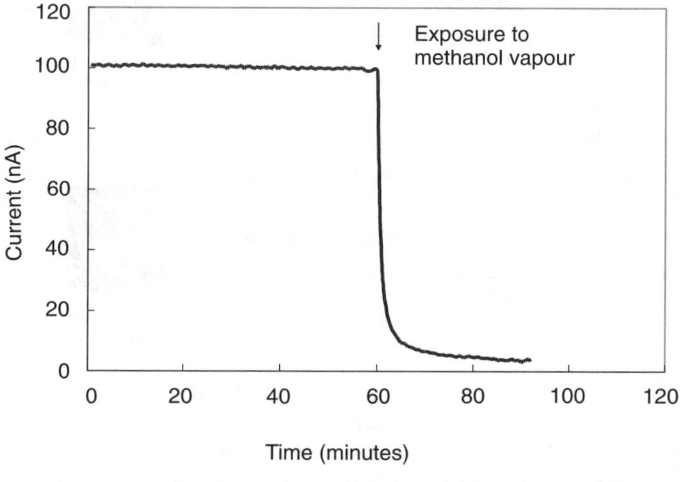

Figure 10.14 Current response of thin PEDOT-PSS layer to the exposure of 5000 ppm methanol. Reprinted permission from *Appl. Phys. Lett.*, **86**, Mabrook MF, Pearson C, Petty MC, 'An inkjet-printed chemical fuse', p. 013507. Copyright (2005) American Institute of Physics.

of the conductive polymer PEDOT-PSS (Chapter 5, Section 5.3) [8]. On exposure to alcohol vapour, the electrical conductivity effectively falls to zero. This irreversible decrease in conductivity is the result of a vapour-induced change in structure of the organic film. The morphology of the as-deposited inkjet printed layer will depend on the contact angle between the PEDOT-PSS and the underlying substrate. This is influenced by the chemical (degree of polarity) and physical (degree of roughness) nature of the substrate. Very thin inkjet-printed PEDOT-PSS layers will be in the form of a series of small islands, as depicted in Figure 10.15. If such a film is exposed to the vapour of an alcohol, the vapour can redissolve the organic film in the regions where it is very thin, i.e. between the islands of organic film. Therefore, the electrical connection between the PEDOT-PSS islands is lost and the current passing through the organic film falls effectively to zero. When the vapour is turned off, the polar PEDOT-PSS molecules are preferentially adsorbed on the existing islands of the polymer rather than on the substrate. The morphology of the inkjet printed film, shown in the AFM images in Figure 10.15, has thereby been permanently altered. This type of sensing response has been called a 'chemical fuse'.

Thicker inkjet-printed layers of PEDOT-PSS reveal a more 'normal' response. For example Figure 10.16 shows the effect of conductivity change for a five-layer film on exposure to pulses of methanol [9]. The gas-sensing mechanism(s) for conductive polymers is complex. Five possible sites of interaction between the vapour and the polymer have been identified [10]. First, the vapour molecules could affect charge transfer between the polymer

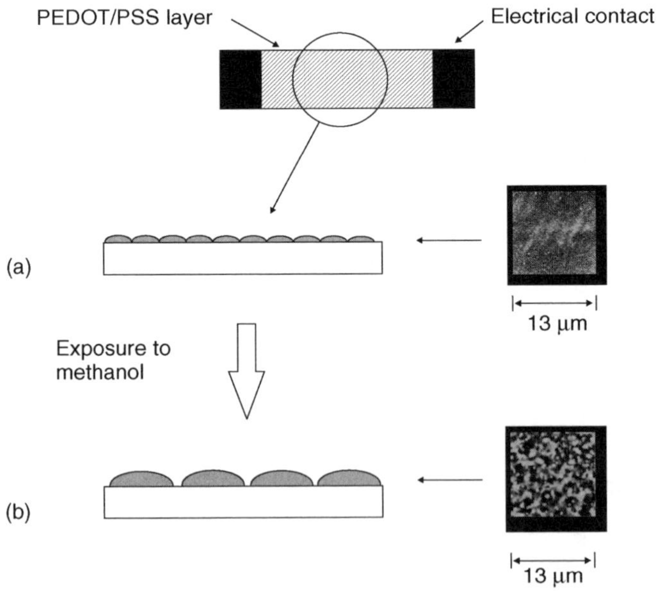

Figure 10.15 Schematic diagrams of the inkjet-printed PEDOT-PSS 'chemical fuse.' (a) Fresh PEDOT-PSS thin film with an island distribution providing an electrically conductive path through the film. (b) Proposed reorganization of the PEDOT-PSS material where almost all the conductive paths in the film surface are broken due to the presence of alcohol. The atomic force micrographs on the right reveal the film morphology before and after exposure to 5000 ppm of methanol for 30 min. From *IEEE Sens. J.*, **6**, Mabrook MF, Pearson C, Petty MC, 'Inkjet-printed chemical sensors for the detection of organic vapours', pp. 1435–1444. Copyright (2006). Reproduced by permission of IEEE.

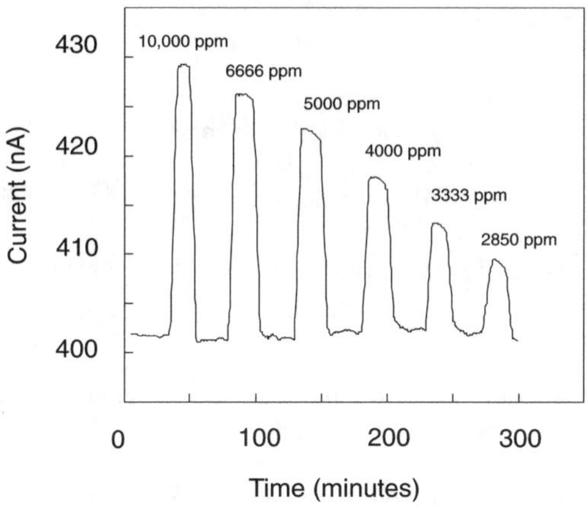

Figure 10.16 Current response of a five-layer inkjet-printed PEDOT-PSS chemiresistor on exposure to step changes in the concentration of methanol vapour in a nitrogen carrier gas at room temperature. From *IEEE Sens. J.*, **6**, Mabrook MF, Pearson C, Petty MC, 'Inkjet-printed chemical sensors for the detection of organic vapours', pp. 1435–1444. Copyright (2006). Reproduced by permission of IEEE.

and the electrode contact. Second, the vapour could lead to oxidation, or reduction, of the polymer chains, influencing the number of charge carriers. Although this mechanism could be important for reactive gases, such as H_2S or ammonia, it is less likely for most organic vapours, such as simple alcohols, which are not strong oxidizing or reducing agents. In the third case, the vapour could interact with the mobile charge carriers on the polymer chains, thereby changing their mobility. Fourth, the vapour molecules could interact with the counter ions within the film. Finally, the vapour could influence the interchain hopping of the charge carriers. The reversible increase in current for the PEDOT-PSS chemiresistor shown in Figure 10.16 on exposure to alcohol vapour can be attributed to the dielectric properties of the vapour molecules [9]. Polar solvents with high dielectric constants can induce a screening effect between counter ions and charge carriers. This reduces the Coulomb interaction between positively charged PEDOT and negatively charged PSS dopant, thereby enhancing the hopping rate and DC conductivity.

The response of organic and inorganic resistive gas detectors is often described mathematically by the *Langmuir isotherm*, developed by Irving Langmuir in 1916. For molecules in contact with a solid surface at a fixed temperature, this describes the partitioning between the gas phase and adsorbed species as a function of applied pressure. The isotherm is based on three assumptions: that adsorption cannot proceed beyond monolayer coverage; that all sites are equivalent and that the surface is uniform; and, finally, that the ability of a molecule to absorb at a given site is independent of the occupation of neighbouring sites. The model is depicted in Figure 10.17.

The resulting isotherm relates the fraction of surface sites occupied $\theta(0 < \theta < 1)$ to the gas pressure P

$$\theta = \frac{bP}{1 + bP} \tag{10.9}$$

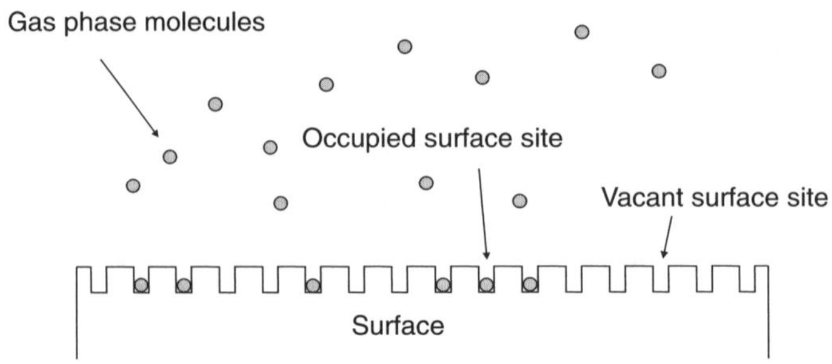

Figure 10.17 Schematic diagram that is the basis for the Langmuir isotherm. Molecules in the gas phase can occupy fixed sites on the surface of a material.

where b is the equilibrium constant. Note that $\theta \rightarrow bP$ at low pressures but $\theta \rightarrow 1$ at high pressures. This is illustrated by the graph shown in Figure 10.18(a). Assuming that the conductivity is directly related to the number of molecules of the active gas adsorbed on the sample surface, then under a constant voltage, the current as a function of the partial gas

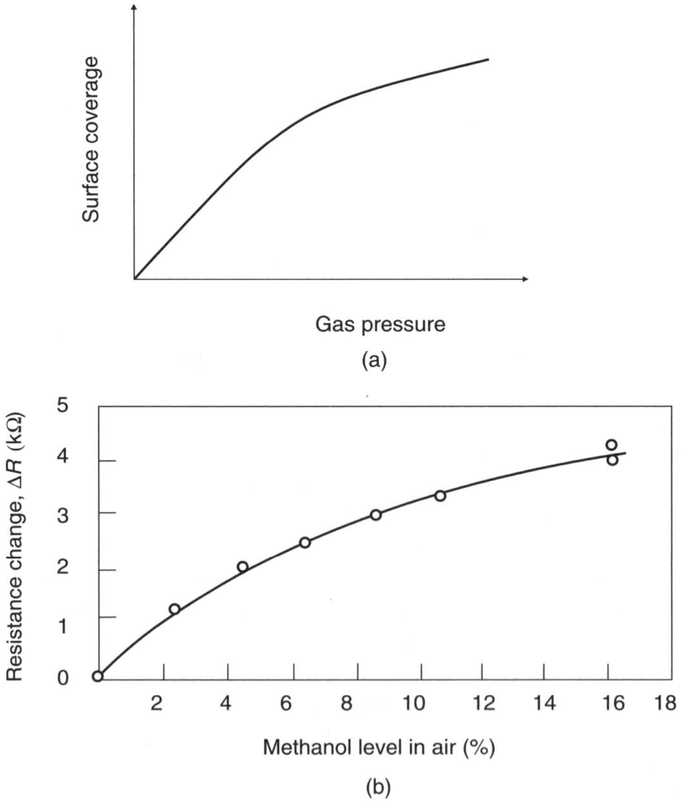

Figure 10.18 (a) Theoretical form of the Langmuir isotherm. (b) Typical isotherm for a conductive polymer gas sensor showing its response to methanol vapour. Reprinted from *Sens. Actuators*, **5**, Bartlett PN, Ling-Chung SK, 'Conducting polymer gas sensors. Part V. Response of polypyrrole to methanol vapour', pp. 141–150. Copyright (1989), with permission from Elsevier.

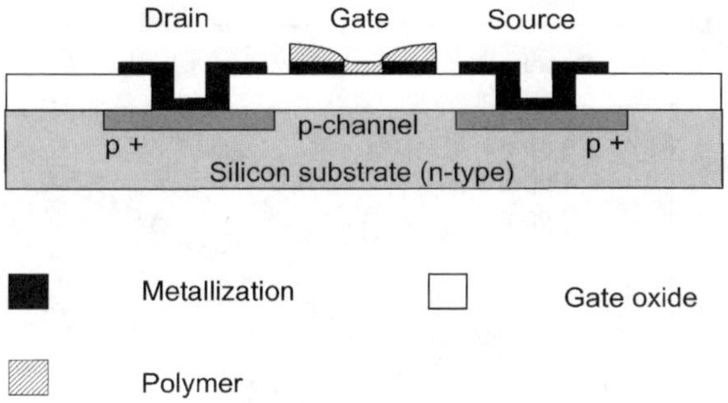

Figure 10.19 Schematic diagram of a charge-flow transistor [11]. Part of the gate metallization in the silicon MOSFET is replaced by a semiconductive organic material.

pressure can be described by the Equation (10.9). The change in resistance of a polypyrrole chemiresistor to increasing concentrations of methanol in air is shown in Figure 10.18(b) [10]. The shape of the response curve is similar to that of the Langmuir isotherm [Figure 10.18(a)], showing a linear response at low concentrations of vapour and a saturation of the response at high concentrations.

One problem associated with many chemiresistor devices described above is that the current outputs are low (typically picoamperes) requiring elaborate detection electronics and careful shielding and guarding of components. This difficulty may be overcome by incorporating the organic sensing layer into a silicon field effect transistor. A schematic diagram of such a structure is shown in Figure 10.19 [11, 12]. Note that configuration is different from that depicted for the Gas-FET shown in Figure 10.7, which is a potentiometric device based on the change in work function of a gate metal on exposure to a gas. The FET device shown in Figure 10.19 has a 'hole' in its gate metal which is filled with a conductive polymer. The principle of operation of this so-called *charge-flow* transistor is as follows [11]. When a voltage is applied to the gate of the FET, the capacitor that is formed between the gate and the silicon substrate charges in two stages. First, the metallic part of the gate charges very rapidly to the applied voltage. Charge then gradually flows through the conductive polymer (which will have a lower conductivity than the gate metal) until the organic part of the capacitor is uniformly charged to the applied voltage. The time required for this charging process to be complete will depend on the sheet resistance (the resistivity divided by the film thickness; Chapter 3, Section 3.2.2) of the polymer. Therefore, exposure to gas will affect the turn-on time of the FET. Figure 10.20 shows data for such a FET incorporating a film of a spin-coated phthalocyanine derivative in the gate (gate 'hole' = 35 μm) [13]. Increasing the concentration of NO_2 increases the conductivity of the phthalocyanine and reduces the turn-on time of the transistor. Because of the amplification provided by the transistor configuration, the current flowing in this FET device is significantly higher than that for the chemiresistor structure shown in Figure 10.10. Of course, it is also possible to use a chemically sensitive organic film as the semiconductive layer in a diode or a transistor. However, in these cases the device must be fabricated so that the gas can interact readily with the organic material.

A further example of a resistive gas sensor uses a polymeric matrix with conductive filler. The matrix swells in the presence of the hydrocarbons resulting in the separation of the conductive particles. Thus, the conductivity of the system falls. When the hydrocarbon source is removed, the polymer returns to its original volume and the sensor is conductive again [14].

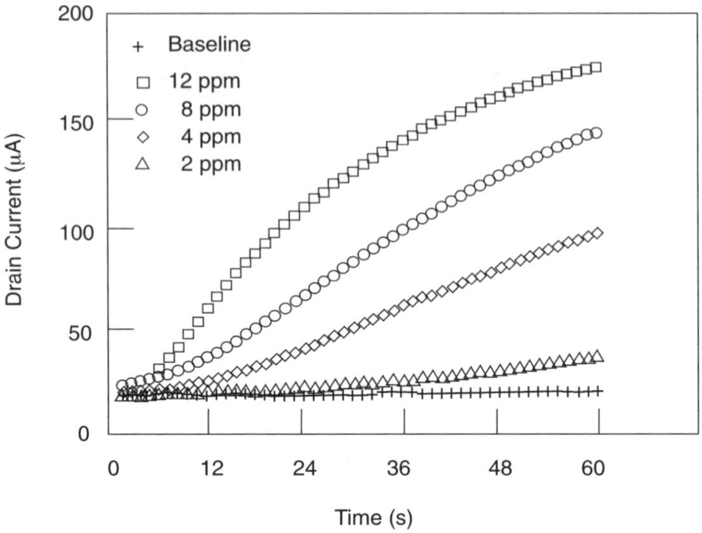

Figure 10.20 Drain current I_d response for a charge flow transistor when a voltage step of −5 V is applied to the gate electrode. Data are shown for the device exposed to different concentrations of nitrogen dioxide. Width of hole in the gate metallization = 35 μm. A phthalocyanine derivative is spin-coated to fill the gate hole. Reprinted from *Thin Solid Films*, **284–285**, Barker PS, Petty MC, Monkman AP, McMurdo J, Cook MJ, Pride R, 'A hybrid phthalocyam/silicon field effect transistor sensor for NO_2', pp. 94–97. Copyright (1996), with permission from Elsevier.

10.4.4 Dielectric Sensors

For chemical sensors, the issue of reversibility is closely related to that of chemical selectivity. Weak interactions between the vapour and sensor coating will produce sensors with good reversibility and little hysteresis. However, such sensors will not have sufficient sensitivity and selectivity to be widely useful. Very strong interactions may improve the sensitivity and selectivity, but can result in sensors that are irreversible or only slowly reversible. Hence, if reversibility is important, a compromise must be reached between the selectivity and the reversibility. This balance can be achieved by sensor coatings which interact with vapours via solubility interactions [15]. When a polymer film is exposed to a vapour, as shown in Figure 10.21, the equilibrium distribution of solute molecules between the gas phase and the polymeric phase is measured by the *partition coefficient*, K_p, given by

$$K_p = \frac{c_p}{c_v} \tag{10.10}$$

where c_p is the concentration of solute in the polymer and c_v is the concentration of solute in the vapour phase. Values of $\log K_p$ can be as high as 6–7 for some vapour–polymer combinations [15]. These numbers demonstrate that the concentration of vapour molecules in the sensing material can be 1–10 million times more than in the vapour phase. Thus absorption is very effective at collecting and concentrating analyte molecules on the sensor's

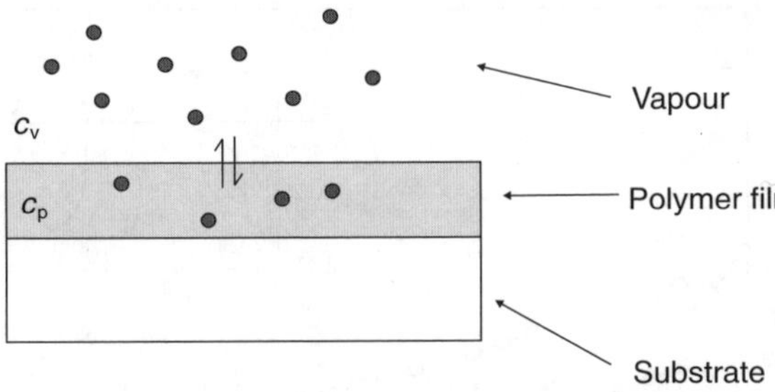

Figure 10.21 Partition of vapour molecules between the gas phase and a polymer film. c_v and c_p are the concentrations of solute in the vapour phase and polymer, respectively.

surface. Partition coefficients are strongly temperature dependent. Absorption normally decreases with increasing temperature. As a result, sensors will become less sensitive and less selective as the temperature increases.

Using the principle outlined above, nonconductive polymers (i.e. those not containing (π-bonds) can be used as dielectric-or capacitance-based sensors. This class of polymer is more wide-ranging than that of conductive polymers, offering more scope for chemical modification to tailor the gas–polymer interaction. The detection principle is based on the dissolution of the gas or vapour into the polymer film with a consequent change of the permittivity. To monitor such changes, the sensor takes the form of a capacitor structure.

Amorphous elastomeric polymers represent an important category of nonconductive sensor coatings. In the elastomeric phase (i.e. above the glass transition temperature; Chapter 2, Section 2.6) constant thermal motion of the polymer chains allows rapid vapour diffusion. Adsorption and desorption of a vapour leave the material in the same state. The softness of elastomeric materials has an additional advantage for piezoelectric sensors such as surface acoustic wave devices (Section 10.4.5) that are sensitive to changes in material stiffness. An example of such materials is the family of polysiloxanes characterized by a repeat

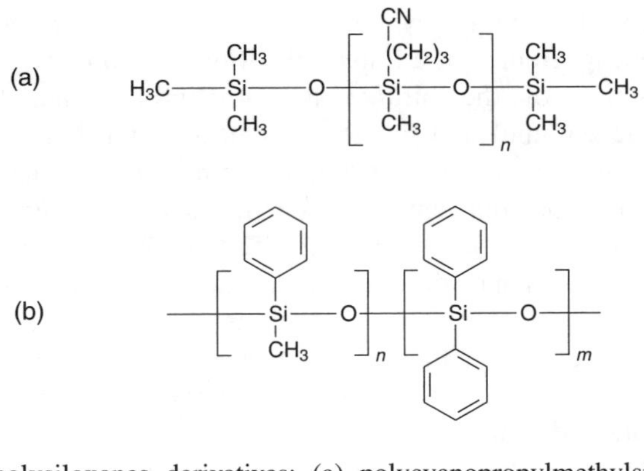

Figure 10.22 Two polysiloxanes derivatives: (a) polycyanopropylmethylsiloxane; (b) a phenyl-methyldiphenylsiloxane copolymer.

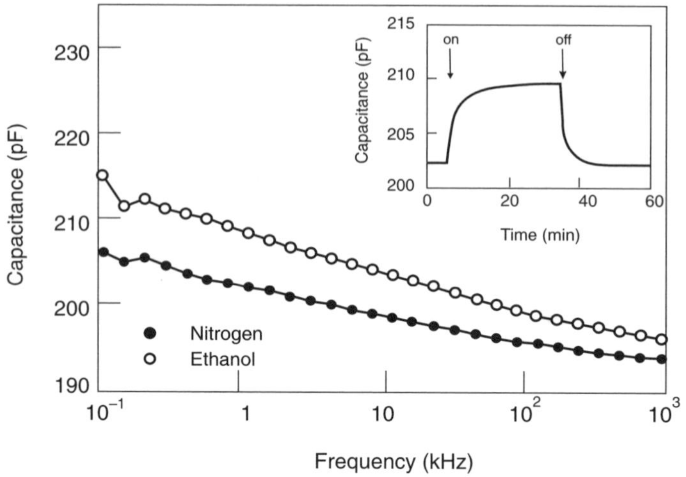

Figure 10.23 Capacitance versus frequency for a coordination polymer LB film in nitrogen and exposed to ethanol vapour (3.3%). The inset shows the transient behaviour measured at a fixed frequency (1 kHz) when the ethanol vapour was turned on and off (indicated by the arrows). Reprinted from *Sens. Actuators B*, **57**, Casalini R, Wilde JN, Nagel J, Oertel U, Petty MC, 'Organic vapour sensing using thin films of a co-ordination polymer: comparison of electrical and optical properties', pp. 28–34. Copyright (1999), with permission from Elsevier.

unit –RR′–Si–O–, where R and R′ are generic functional groups. By modifying the side chains, nonpolar, polar and polarizable polymers can be obtained. Two examples are shown in Figure 10.22: (a) polycyanopropylmethylsiloxane and (b) a phenylmethyldiphenylsiloxane copolymer. The former compound is highly polar (the nitrile or –CN bond is associated with a large dipole moment) and is expected to interact well with polar vapours [e.g. acetonitrile (CH_3CN) or 1,1,1-trifluoroethane (CF_3CH_3)] whereas the copolymer structure depicted in Figure 10.22(b) has a large polarizability (i.e. a large ease of dipole formation in an electric field) and should interact well with vapours that are similarly polarizable (e.g. benzene).

Figure 10.23 shows the frequency behaviour at room temperature, over the range 10^{-1}–10^3 kHz, of the capacitance of a Langmuir–Blodgett film of a coordination polymer formed by the reaction of the bifunctional amphiphilic ligand 5,5′-methylenebis-(*N*-hexadecylsalicylideneamine) and copper ions in an interfacial reaction at the water surface [16]. The measurements were taken in both nitrogen gas and ethanol vapour (3.3%). The inset shows the transient behaviour measured at a fixed frequency (1 kHz) when the ethanol vapour was turned on and off. The conductance of the dielectric sensor is often measured at the same time as its capacitance. Measurements at different frequencies form the basis of the technique of impedance (or admittance) spectroscopy, a powerful method for the investigation of the dynamic electrical properties of dielectrics (Chapter 3, Section 3.6.2). Discrimination between different vapours may be achieved by monitoring the complex admittance at several frequencies.

The percentage changes in the capacitance and conductance for the coordination polymer described above, on exposure to a number of different vapours, and normalized to the vapour

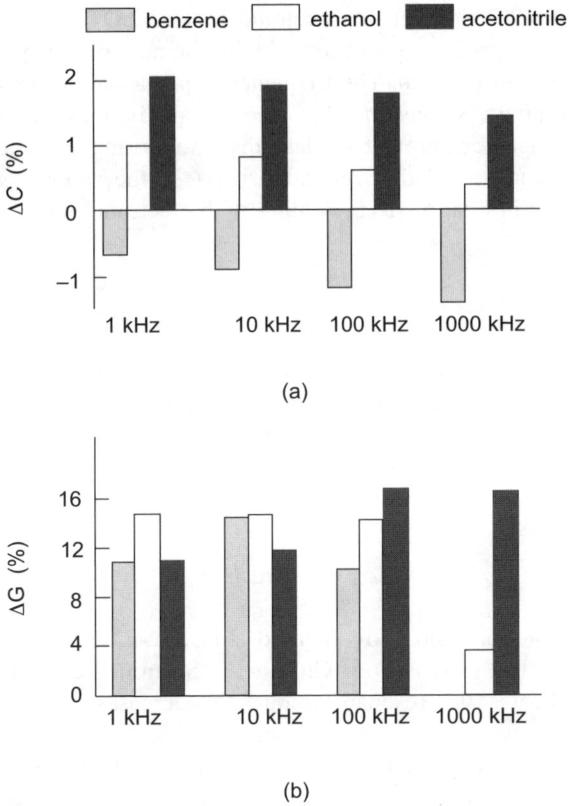

Figure 10.24 Average changes in (a) the capacitance ΔC and (b) the conductance ΔG for a coordination polymer LB film, measured at four frequencies for three different vapours. Reprinted from *Sens. Actuators B*, **57**, Casalini R, Wilde JN, Nagel J, Oertel U, Petty MC, 'Organic vapour sensing using thin films of a co-ordination polymer: comparison of electrical and optical properties', pp. 28–34. Copyright (1999), with permission from Elsevier.

concentration, are depicted in Figure 10.24. Such changes in may result from a variety of processes, including:

(i) a change in the permittivity due to bulk dissolution of the vapour into the LB film;

(ii) a change in the polymer permittivity induced by the interaction between the polymer and the vapour molecules; and

(iii) a change in the film thickness (i.e. swelling).

As the dielectric relaxation for the organic solvent is expected to occur at a very high frequency (beyond the maximum used in the experiment), mechanism (i) should produce changes in capacitance and conductance that are constant over the frequency range used. This should also be the case for mechanism (iii). From Figure 10.24, it is evident that the capacitance increases when the dielectric sensor is exposed to ethanol and acetonitrile, with a greater fractional increase for acetonitrile. This is consistent with the strongly polar nature of the latter solvent (dipole moment for acetonitrile = 3.92 D compared with 1.69 D for ethanol). A decrease in capacitance is observed with the nonpolar benzene, almost certainly associated with swelling of the film. The frequency dependence of the admittance is not the same for the three vapours, suggesting that mechanism (ii) may be significant. However, from a practical sensing viewpoint, it is clear that monitoring the admittance at different frequencies provides a means of discrimination between the vapours studied. This idea is explored in more detail in Section 10.5, dealing with electronic noses.

10.4.5 Acoustic Devices

Highly sensitive chemical sensing devices can be made by measuring mass. Many such *gravimetric* sensors operate at ultrasonic frequencies and register the shift in the resonant frequency $\Delta\nu$ of a piezoelectric crystal oscillating a frequency ν when an additional mass is deposited on its surface, as shown schematically in Figure 10.25 The relationship between the frequency shift and additional mass Δm is given by the *Sauerbrey equation*, one form of which may be written as

$$\Delta\nu = -(B\Delta m)\nu^2 \tag{10.11}$$

where B is a calibration constant. (This method is also used as the basis of film thickness measurements for thermal evaporation; Chapter 7, Section 7.2.2.) The negative sign in Equation (10.11) indicates the resonant frequency decreases as the mass of additional material increases.

This class of sensors is extremely sensitive; for a typical quartz crystal operating at a frequency of 10 MHz, an additional mass of the order of a nanogram can produce a shift in frequency of 1 Hz. To improve the selectivity, a chemically active layer (e.g. a polymer film)

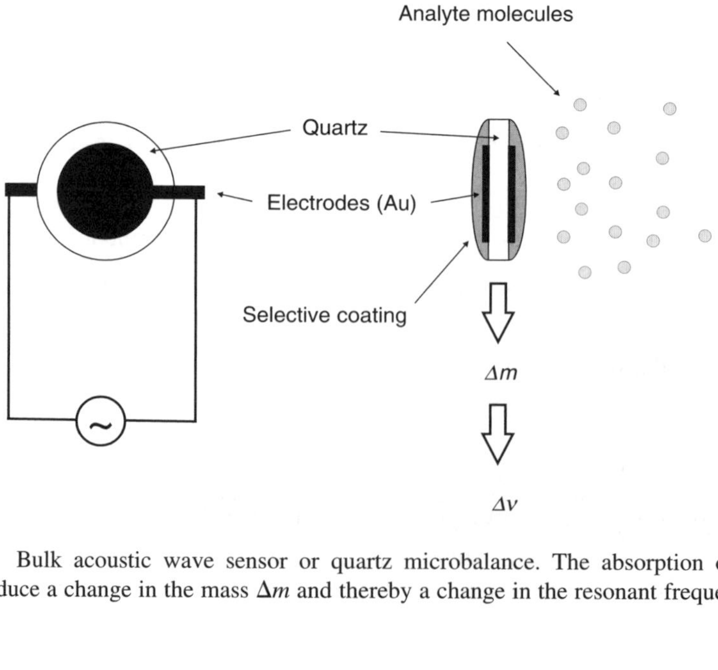

Figure 10.25 Bulk acoustic wave sensor or quartz microbalance. The absorption of the analyte molecules produce a change in the mass Δm and thereby a change in the resonant frequency $\Delta\nu$ of the quartz crystal.

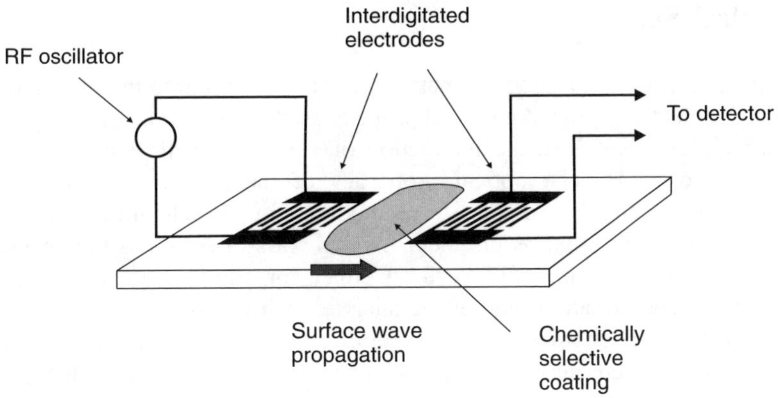

Figure 10.26 Surface acoustic wave (SAW) device for chemical sensing.

can be used to coat the resonant crystal. However, this can complicate the analysis (as the vapour will affect the viscoelastic properties of the polymer in addition to the mass loading) and the sensing layer/quartz structure should be treated as a multiple resonator in which the reflection and/or refraction of acoustic energy occurs at each interface [2].

Another class of gravimetric sensors is represented by *surface acoustic wave* (SAW) detectors. Their working principle is based on the propagation of mechanical waves along a solid surface which is in contact with a medium of lower density (e.g. air). Figure 10.26 shows a schematic diagram of an SAW sensor. The surface acoustic wave is propagated from one set of interdigitated electrodes on a quartz substrate by the application of a high frequency (RF signal). The energy of the SAW is confined to a region close to the substrate surface, which is just a few wavelengths thick. A similar electrode arrangement is used to detect the signal. Interaction between a gas or vapour and a selective coating, which is positioned between the two sets of interdigitated electrodes, results in a perturbation of the acoustic wave. This can be detected at the receiving electrodes by a change in the phase or amplitude of the signal. Surface acoustic wave devices typically operate in the frequency range 30 MHz to 1 GHz, i.e. much higher than bulk oscillators. As the SAW devices also have a similar $(\Delta v)/v^2 \propto$ mass relationship to Equation (10.11), a significant increase in sensitivity might be expected. Problems with drift and noise mean that large increases in sensitivity are not always achieved. However, other advantages, such as the ability of fabricate more than one SAW device on the same substrate and the fact that the devices can be made smaller than bulk devices, have led to much interest in this form of chemical sensor.

There are different types of surface wave that can be exploited in SAW devices. The most widely known is the *Raleigh wave*, which consists of wave components both parallel (longitudinal) and perpendicular (transverse) to the direction of propagation. As the SAW wave propagates along the substrate, surface particles move in an elliptical path (an analogy may be made with the ripples on a water surface caused by a physical disturbance). A different type of surface acoustic wave is the *Love wave*. Such waves are purely transverse and are only associated with shear stresses and are generally exploited for liquid sensing.

Piezoelectric materials other than quartz can be used in gravimetric sensing. For example, the piezoelectric polymer poly(vinylidene difluoride) (PVDF) has been introduced in Chapter 2, Section 2.6.3. Such materials can possess low *acoustic impedances* in comparison with inorganic compounds, providing an excellent mechanical match to water and biological systems.

10.4.6 Optical Sensors

Optical chemical sensors translate information from the chemical to the optical domain, and then finally to an electric signal. An optical probe is used to interrogate the chemically active layer; the principle is based on the interaction of electromagnetic waves with a material (Chapter 4, Section 4.4). The optical properties of the sensors (e.g. refractive index, absorption, scattering and fluorescence) are modified by the chemical reaction with the measurand. In general, the first requirement for this class of sensor is the generation of an optical excitation beam. This could be used either for chemically modifying the active sensors (e.g. photoisomerization) and/or the analyte, or it could be used simply as a probe. For sensors exploiting phenomena such as chemiluminescence or bioluminescence, optical excitation is not required because it is the chemical reaction itself that generates an electromagnetic signal. For other optical sensing systems, it is necessary to convert the chemical reaction and the molecular selectivity to a second optical signal, through, for example, indicators immobilized on a support. Alternatively, secondary reaction products such as oxygen, carbon monoxide and ammonia might be detected using a bioreactive layer. Finally, waveguides or optical fibres transmit and propagate the optical response to light sensors, which convert it to an electric signal.

Simple optical gas or vapour sensors exploit changes in the optical absorption spectra generated by the interaction of the sensing layer with the analytes (e.g. CO_2, O_2). For example, Figure 10.27 shows the time evolution of the UV–visible spectrum of an LB film of a porphyrin on exposure to 4.4 ppm NO_2 [17]. The optical absorption spectra of porphyrins consists of an intense absorption band around 425 nm, known as the *Soret band*. Four bands of lower intensity and longer wavelength, referred to as *Q bands*, are also observed. The relative position and intensity of these Q bands gives information on the position and type of

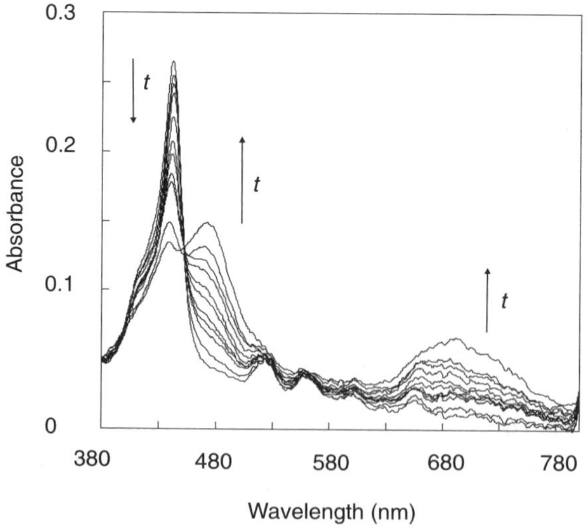

Figure 10.27 Time evolution of the UV–visible spectrum of an LB film of a porphyrin derivative following exposure to 4.4 ppm NO_2 gas. Reprinted from *Adv. Colloid Interface Sci.*, **116**, Richardson TH, Dooling CM, Jones LT, Brook RA, 'Development and optimization of porphyrin gas sensing LB films', pp. 81–96. Copyright (2005), with permission from Elsevier.

substituent groups present. Figure 10.27 shows that immediately following exposure to the NO_2 gas, there is a rapid decrease in the intensity of the Soret band and the appearance of new bands at around 480 and 700 nm. The stationary point at around 450 nm, indicating zero absorbance change at this wavelength throughout the exposure cycle, is called an *isosbestic point*. This indicates a two-state reaction process:

$$P_A + NO_2 \rightarrow P_B \qquad (10.12)$$

The addition of NO_2 causes a change in state for some of the porphyrin (P) molecules from state A to state B, and the intensities of the Soret and 480 nm bands indicate the relative proportions of A and B, respectively.

The rapid decrease in the intensity of the Soret band on exposure to NO_2 shown in Figure 10.27 is followed by a slower decay in the absorbance before full saturation occurs. This is believed to be a result of the difference between surface adsorption and bulk diffusion of the toxic species. The changes in the optical spectra shown in Figure 10.27 can be reversed by heating the sample in nitrogen.

The manifestation of a sensor based on optical changes such as those described above can take a number of forms. For example, the use of optical fibres results in the development of compact, miniaturized and low-cost sensors for laboratory or *in situ* applications. One of the main advantages of this category of chemical sensor is that they do not need to have an electric current passing through the sensing material and consequently the issue of power dissipation is absent.

Every gas, liquid or solid, with a covalent bond and strong dipolar characteristics interacts with infrared radiation at a specific frequency (Chapter 6, Section 6.6). On the basis of this principle, infrared absorption sensors have been developed. Whenever a target molecule interacts with the chemically active layer of the sensor, new absorption bands are generated and the original spectrum is modified. Comparison of a series of spectra allows the identification of the analyte and the determination of its concentration.

Fluorescence-based sensors are able to detect very low concentrations of chemicals with good selectivity. A polymeric matrix may host the receptors or the matrix itself may be a receptor for the analyte. When the sensor probe is exposed to the analyte, the polarity of the sensing receptor changes and, with it, its emission spectrum. Using an optical fibre coated at one end with the sensing material, the same propagation channel (i.e. the fibre) is used first to interrogate the sensor (exciting signal or input) and then to collect the response (output). A multiwavelength spectral detector, positioned at the end of the fibre, then compares the two optical signals. Optical fibre biosensors are described in Chapter 12, Section 12.8.2.

Specificity may be introduced into chemical sensors for the detection of metal ions in liquids by using metal-binding macrocycles. An important class of these *ionophores* is the crown ether, a macrocyclic polyether containing $(\text{–O–CH}_2\text{–CH}_2\text{–})_n$ repeat units. In Chapter 11, Section 11.4, we see that the self-assembly of one layer of such material can be exploited, using electrochemical principles, for metal ion detection. Optical sensors can use *chromoionophores* [18]. These are more sophisticated versions of simple crown ether rings and are constructed from two functionally different chemical groups: an ionophore, recognizing specific ions, and a chromophore, transducing the chemical information produced by the ionophore–ion interaction into an optical signal. Figure 10.28 shows the chemical structure for an amphiphilic chromoionophore based on a benzothiazoliumstyryl dye containing a 1,10-dithia-18-crown-6 ether group. The presence of the hydrocarbon chain

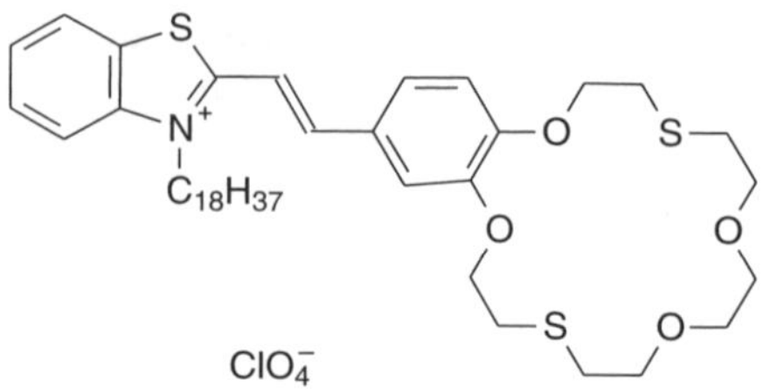

Figure 10.28 An amphiphilic chromoionophore based on a benzothiazoliumstyryl dye containing a 1,10-dithia-18-crown-6 ether group [19].

allows this compound to form an insoluble layer at the air/water interface. The optical absorbance of Langmuir–Blodgett films can be affected by complexation with metal ions such as $Ag.^+$ The changes in absorbance at 440 nm of a two-bilayer LB film on exposure to different silver concentrations is shown in Figure 10.29 [19]. The response to metal cations was found to depend on whether the LB films were used as-deposited, or kept in acidic water for a few hours prior to complexation study. This treatment was thought to influence the formation of aggregates in the LB film. Complexation with Ag^+ cations then results in the break-up of the aggregates. The reaction is fully reversible if the complexed film is retreated in acidic water for 20 h.

Optical sensors exploiting evanescent waves (Chapter 4, Section 4.7.1) utilize a light beam passing through a waveguide that propagates as an evanescent wave in the medium surrounding the waveguide. For example, evanescent waves are formed when sinusoidal waves travelling in a medium of refractive index n_1 are (internally) reflected from an

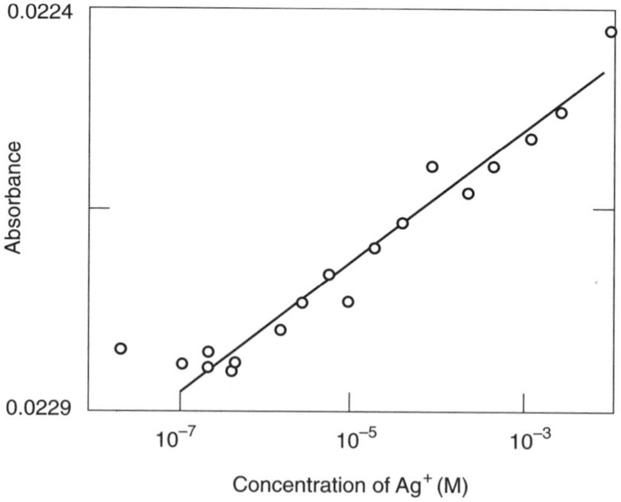

Figure 10.29 Change in absorbance of an LB film of the chromoionophore depicted in Figure 10.28 (two bilayers) at 440 nm as a function of $AgClO_4$ concentration in acidic water. The LB film was deposited from an acidic subphase and held in acidic water for 20 h. *Langmuir*, **10**, Lednev IK, Petty MC, 'Aggregate formation in Langmuir–Blodgett films of an amphiphilic benzothiazolium styryl chromophore', pp. 4185–4189. Copyright (1994) American Chemical Society.

interface with a second medium of refractive index n_2 at an angle greater than a specific value (critical angle). Internal reflection means that no refracted wave is generated in the second medium, but the incident wave is fully reflected within the first medium. Although no net energy is transferred from one medium to the other, an optical disturbance occurs in the second medium, which takes the form of an evanescent wave. The intensity of such radiation decays exponentially with the distance from the interface at which they are formed. If in the second medium (the sample) a chemical species is able to absorb electromagnetic radiation at the wavelength of the probe, then there will be a reduction in the intensity of the reflected light passing through the waveguide. Such a variation will be registered by a detector at the end of the optical support. Reduction in the intensity profile can therefore indicate the presence of the analyte and provide information about its concentration.

Particular types of evanescent wave sensors exploit surface plasmon resonance (Chapter 4, Section 4.7.2). This has been the subject of theoretical and experimental research for several decades. However, only at the beginning of the 1980s was it demonstrated that SPR could be exploited in gas and biological species detection [20, 21]. Since then, growing interest in the SPR principle and technology has been shown by both the scientific community and commercial companies

The electromagnetic fields of a surface plasmon wave are distributed in a highly asymmetric fashion and most of the field is concentrated in the dielectric. A surface plasmon propagating along the surface of silver is less attenuated and exhibits higher localization of the electromagnetic field in the dielectric than a surface plasmon supported by gold. However, gold is more suitable for sensing application in liquids because it is relatively stable (e.g. does not oxidize). As the excitation of surface plasmons by an optical wave results in resonant transfer of energy into the surface plasmons, there will be a resonant absorption of the energy of the optical wave. Because of the strong concentration of the electromagnetic field in the dielectric (an order of magnitude higher than that in typical evanescent field sensors using dielectric waveguides), the propagation constant of the surface plasmon, and consequently the resonance condition, is very sensitive to variations in the optical properties of the dielectric adjacent to the metal layer supporting the wave. Therefore, changes in the optical parameters of the transducing medium can be detected by monitoring the interaction between the surface plasmon and the optical wave.

As noted in Chapter 4, Section 4.7.2, several methods are available to couple optical energy into surface plasmons. Two of the most common techniques are attenuated total reflection (ATR) in prism or semi-cylinder couplers and diffraction at the surface of a grating. It is also possible to excite SPR by an optical fibre, thereby providing a high degree of miniaturization. In the ATR Kretschmann configuration, a simple SPR system for chemical sensing can operate at a fixed angle. This is chosen to be close to the resonant condition, at a point where the reflectivity varies rapidly with angle. The reflected light intensity is then monitored as a chemical species interacts with a coating deposited on the metal. Figure 10.30(a) shows the reflectivity response to copper ions for a fixed-angle SPR sensing arrangement [22]. The sensing film consists of a polyelectrolyte bilayer (PEI/PMAE) deposited by the electrostatic layer-by-layer technique (Chapter 7, Section 7.3.3). On exposure to copper, it is thought that a chelate is formed with polyethylene-co-maleic acid, one of the components of the organic film. The reaction, depicted in Figure 10.30(b), could be fully reversed by exposing the complexed film to hydrochloric acid for 10 min.

In scanning angle SPR, the wavelength is fixed and the incident angle is varied. It is thereby possible to have a qualitative picture for the resonance curve following interactions

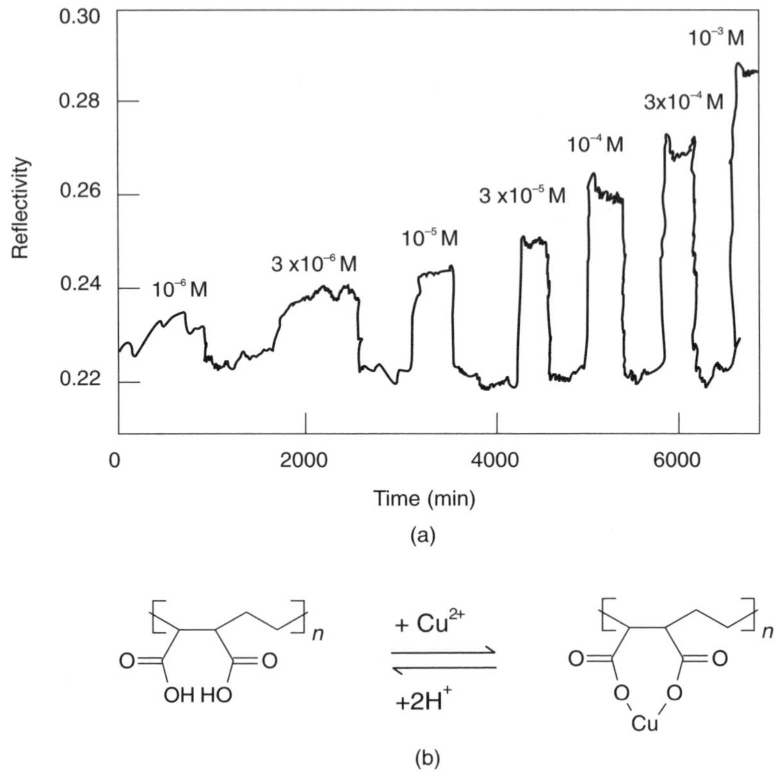

Figure 10.30 (a) Reflectivity (in an SPR experiment at a fixed incident angle) versus time for a polyelectrolyte bilayer (PEI/PMAE) exposed to increasing concentrations of copper acetate. The sample was immersed in hydrochloric acid and purged with pure water between the different copper ion concentrations. Reprinted from *J. Phys. D: Appl. Phys.*, **34**, Pearson C, Nagel J, Petty MC, 'Metal ion sensing using ultrathin organic films prepared by the layer-by-layer adsorption technique', pp. 285–291, Copyright (2001), with permission from IOP Publishing Limited. (b) Reaction of copper ions and maleic acid to form a chelate.

of the vapour with the sensing material. Alternatively, the incident angle can be fixed while the wavelength is varied. Only one wavelength will excite the surface plasmons, resulting in a minimum in intensity of the reflected light. Hence a plot of reflectivity versus wavelength will be similar to that of reflectivity versus incident angle.

SPR systems have now been successfully commercialized and are extensively applied to biochemical analysis [23]. Most of these use prism coupling for the momentum enhancement. The Kretschmann configuration is usually used in preference to the Otto arrangement as the analyte can easily come into contact with the sensing layer. A third hybrid configuration has also been used. Here, the coupling of the light to plasmons is accomplished by the resonant mirror principle [24]. A small layer of silica ($\sim$1 mm) is deposited on a prism base. On top of the silica, there is a layer with a high refractive index, e.g. titania. The silica layer is thin enough to allow the evanescent field generated on the prism/silica interface to couple into the titania. This allows the titania layer to function as an optical waveguide. Repeated total internal reflection of the guided mode within the waveguiding titania layer results in the production of an evanescent field at the titania/adsorbed layer interface. The exact angle of the incident light

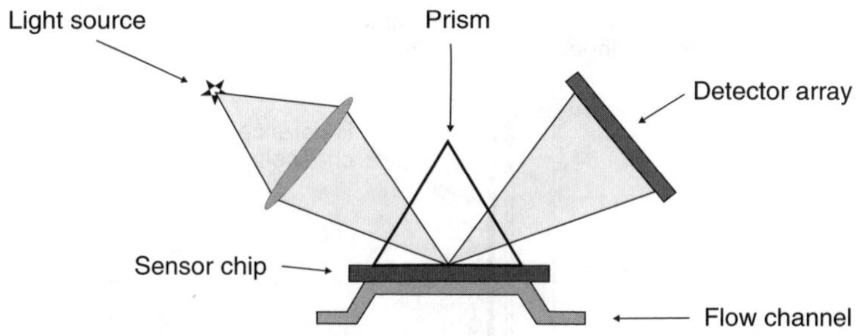

Figure 10.31 Schematic diagram of commercial SPR system with no moving parts produced by Biacore. A wedge-shaped light source and a diode array are used to measure the intensity of the reflected light. Reproduced from [25].

at which there is a resonance between the waveguided mode and the resulting evanescently coupled light is directly dependent on the refractive index of the adsorbed layer. In such a device, there is virtually no loss of the reflected light intensity associated with the resonance condition. Instead, resonance is accompanied by a change in phase of the reflected light, which is recorded interferometrically.

The commercial SPR system produced by Biacore is equipped with a continuous flow arrangement in which four channels are coupled in series [25]. One of the channels can be used to provide an in-line reference. An automatic sample needle delivers buffer solution and analyte to the surface of the sensing 'chip'. The continuous flow ensures that no changes in the analyte concentration occur during the measurement. A wedge-shaped light source and a diode array are used to measure the relative intensities of the reflected light as shown in Figure 10.31. In this way, the optical interface has no moving parts.

With an increased interest in optical sensors for use in remote and portable control instrumentation, much work has been focused on the development of low-cost SPR systems. Figure 10.32 shows a schematic diagram of such a system developed by Texas Instruments

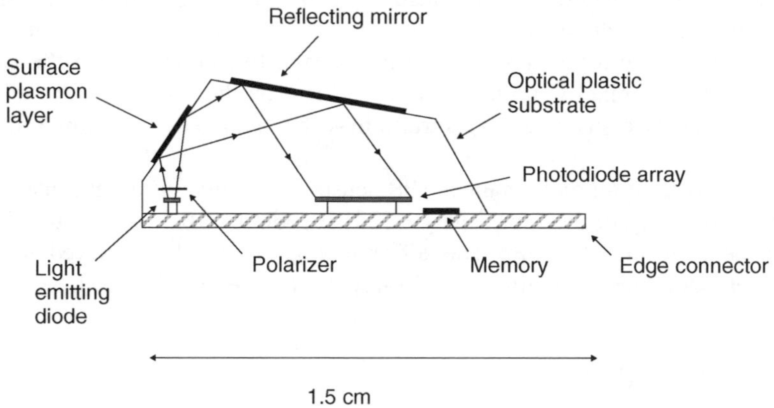

Figure 10.32 Cross-section of SPR sensor developed by Texas Instruments showing the components of the device and the path followed by the light inside the sensor. Reprinted from *Sens. Actuators B*, **91**, Chinowsky TM, Quinn JG, Bartholomew DU, Kaiser R, Elkind JL, 'Performance of the Spreeta 2000 integrated surface plasmon resonance affinity sensor', pp. 266–274, Copyright (2003), with permission from Elsevier.

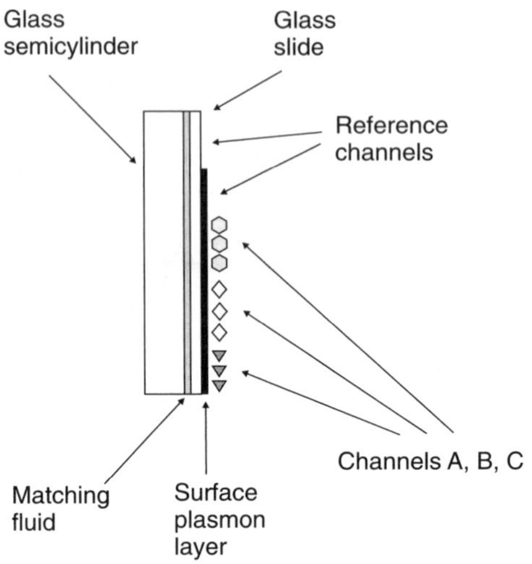

Figure 10.33 Schematic diagram of an arrangement for a multi-channel SPR sensing chip.

[26, 27]. The sensor consists of a plastic prism moulded to a microelectronic circuit contained on a printed circuit board. The circuit incorporates an infrared light-emitting diode (LED), a linear diode-array detector and a non-volatile memory chip for recording identification and calibration information. A diverging beam is emitted from the LED, which passes through a polarizer and strikes an elliptical angle of the sensor surface at a range of angles above a critical angle. The angle at which light is incident on this surface will vary with the location on the surface: smaller incident angles occur closer to the base of the sensor.

The sensor surface is formed by a glass chip coated with a layer of gold, approximately 50 nm in thickness, bonded to the plastic prism. The angle at which SPR occurs will vary with the refractive index of the analyte. The light reflected from the sensor surface then reflects from the sensor's top mirror and back down on to the diode array. Because each detector pixel will collect light incident on the sensor surface at a different angle, a reflectivity versus angle spectrum may be obtained by reading the detector array.

Work is also focused on multichannel SPR sensing, in which different sensing layers are exposed simultaneously to the same analyte [28, 29]. Figure 10.33 shows the definition of several channels on an SPR chip. The different regions can be achieved using various patterning methodologies, including layer-by-layer self-assembly [30–32].

10.5 BIOLOGICAL OLFACTION

The human *olfactory system* has many receptors cells (sensors) which are individually nonspecific; signals from these are fed to the brain via a network of primary and secondary neurons for processing. It is generally believed that the selectivity of the olfactory system is a result of a high degree of parallel processing in the neural architecture. In humans, the

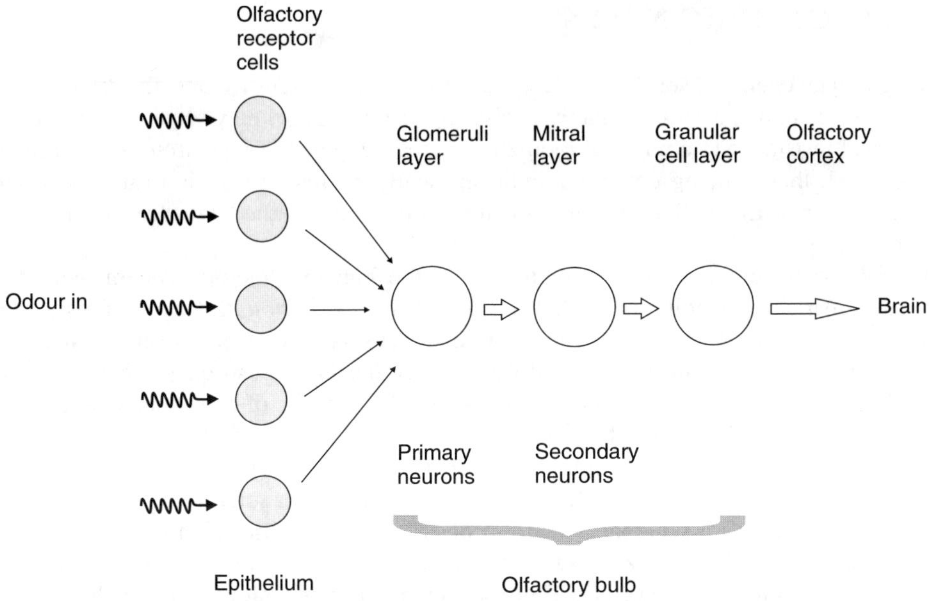

Figure 10.34 Schematic representation of the human olfactory system.

process of olfaction begins with the inhalation of molecules and their transport to the olfactory receptor cells, located within a specialized membrane, of approximate dimensions 2×5 cm, called the *epithelium*, in the nose. The subsequent interactions generate electrical signals, which propagate down the axon (Chapter 12, Section 12.7) of the olfactory receptor cells to the olfactory bulb, as depicted in Figure 10.34. The signals are processed by further layers referred to as *glomeruli*. Here, the nature of the odorous stimulus is encoded as a specific combination of activated glomeruli, which takes the form of a two-dimensional map. Particular odorants are associated with specific topographic patterns of activity. Although mammals possess in the order of 1000 different receptor proteins, it is established that they can detect more than this number of different odorants. This suggests that any given odorant is able to interact with more than one receptor protein and, conversely, that any single receptor is able to interact with more than one odorant. It is worth noting that olfactory receptor cells are replaced every 30–60 days, so at any one time the population of olfactory receptor cells contributing to a particular glomerus will possess a range of ages. This will provide the system with a mechanism to overcome the problem of a loss of sensitivity of the receptor cells with time. Although the sensitivity of any given receptor is likely to decrease with time, the summed response from a population of cells of all ages will be more stable [7].

The map of glomerular activity, as detected by the *mitral* cells, then feeds into the next level of information processing, the *granular cell* level, before passing via the *olfactory cortex* to the brain. Olfactory information travels not only to the *limbic system*, primitive brain structures that govern emotions, behaviour, and memory storage, but also to the brain's *cortex*, or outer layer, where conscious thought occurs. In addition, it combines with taste information in the brain to create the sensation of flavour.

10.6 ELECTRONIC NOSES

Although many chemical sensing devices can show adequate sensitivities, the selectivity can be poor. For example, a semiconductive polymer (chemiresistor) may show a similar change in electrical resistance to a range of oxidizing (reducing) gases. To get around this difficulty, one approach that is being embraced enthusiastically by researchers is to use an array of sensing elements, rather than a single device. This is the method favoured by nature, as described above.

The *electronic* nose is an attempt to mimic the human olfactory system and several companies market such equipment [7, 33]. Emulation of *gustation*, the sense of taste, would similarly lead to an *electronic tongue*. Individual sensors can be based on polymer films. Each element is treated in a slightly different way during deposition so that it responds uniquely on exposure to a particular gas or vapour. The pattern of resistance changes in the sensor array can then be used to fingerprint the vapour. Alternatively, data for a single sensor but measured under different conditions, such as the conductivity of a chemiresistor at different frequencies (Figure 10.24), may be used to provide a suitable fingerprint.

Figure 10.35 shows a schematic diagram of an electronic nose. If more than two linear sensors are used to characterize a sample, a *multivariate* evaluation is usually applied to visualize and evaluate the data [7]. There are many such techniques in use. In electronic noses, pattern recognition approaches are generally used. Unsupervized pattern recognition algorithms try to cluster the results, which are usually visualized in a two-dimensional plot. A common technique that is applied is *principal component analysis*. This involves the use of a mathematical procedure to transform a number of (possibly) correlated variables into a (smaller) number of uncorrelated variables called principal components. Another approach to multivariate analysis uses an *artificial neural network* (ANN). An ANN is a highly parallel information processing system that has been inspired from our understanding of the biological nervous system. The network consists of a lattice of information processing

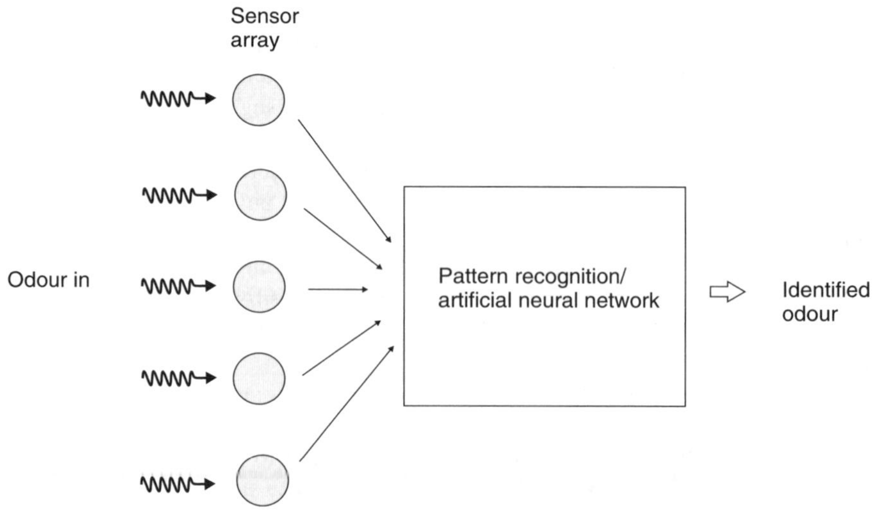

Figure 10.35 Components of an electronic nose.

elements (neurons; Chapter 12, Section 12.7) which are connected together in a certain way. The strengths of these connections are called *synaptic weights* and are determined either during a training (or learning) phase for supervized neural networks, or by an algorithm for unsupervized neural networks.

10.7 PHYSICAL SENSORS AND ACTUATORS

Physical sensors are selective and sensitive to physical phenomenon, such as pressure, force, acceleration, specific heat, current, electric charge, magnetic field, electrical resistance and light. These form a major part of the sensor industry, with an increasing number of applications. Micro-electro-mechanical systems (MEMS or the 'nano' equivalent NEMS) is the integration of mechanical elements, sensors, actuators and electronics on a common silicon substrate through microfabrication technology. This section will be confined to a description of pressure sensors and polymer actuators as far as such structures are being developed with organic materials. A comprehensive review of other sensors is provided in the textbook by Gardner, Varadan and Awadelkarim [34].

10.7.1 Touch Sensors

Touch, or tactile, sensors detect not only the presence of a touch, but also its position. A common touch sensor, exploited in handheld computing devices, is the touchscreen. This can operate on either resistive or capacitive principles. Figure 10.36 shows the principle of the former arrangement. The device consists of a two resistive layers separated by an insulating

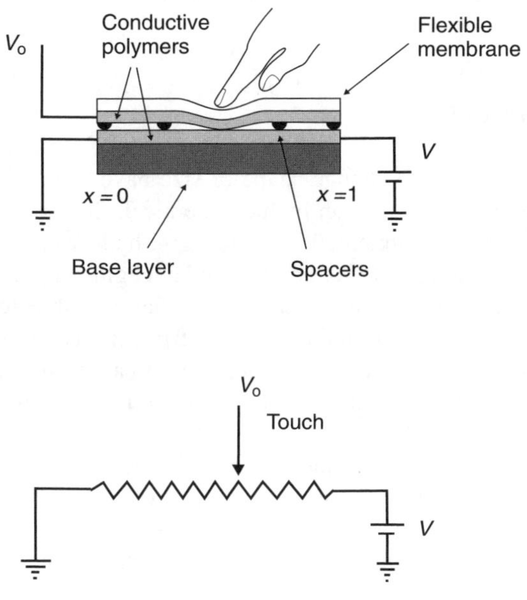

Figure 10.36 Principles of a resistive touch sensor based on a conductive polymer.

(usually air) spacer layer. Some systems use conductive indium tin oxide (ITO) as the resistive layer, although many touchscreens are now being developed based on electrically conductive polymers, such as polythiophene derivatives. If a flexible nonconductive polymer base plate is used, the touchscreen assembly becomes fully flexible. The touchscreen is essentially a variable resistor, where V is the applied voltage and V_o is the output voltage – typically fed, via an analogue-to-digital converter, to a computer. The touch position x is related to the measurement voltage by $x = V_o/V$. Even if the electrical contact is made between the flexible membrane and the base plate at several locations, there will only be one measured voltage and hence only one intermediate touch location measured.

Capacitive touchscreens are another example of tactile sensors. A capacitor is formed when a user's finger touches a dielectric coating deposited on a resistive base (e.g. a conductive polymer). One electrode is this resistive layer and the other electrode is the user's finger. This capacitance is typically detected by observing a shunt to ground of current at frequencies in the kHz range. Position information is obtained by monitoring the magnitude of this current.

Tactile sensors are very important in the development of automated industrial processing equipment. For example, a robot hand must be able to detect and subsequently grip a work piece with sufficient force to hold it firmly, but not to use excessive force that may cause damage. The distributed sensing we have in our hands allows us to hold on to an egg with enough force that it does not slip, but without so much force that it breaks. One of the goals in robotics is to develop an array of sensors that provides the systems with a similar source of tactile feedback, i.e. 'smart skin.' The majority of tactile sensors that have been developed for robots have been based on the resistance changes associated with compression of a conductive elastomer (e.g. a carbon-loaded rubber) or foam. There are also examples of sensors that exploit conductive polymers. When a force is applied to the sensor element, the resistance changes as a result of the *piezoresistive* effect (which is different from the piezoelectric effect observed in insulating polymers such as PVDF; Chapter 2, Section 2.6.3).

10.7.2 Polymer Actuators

Polymers that can be activated to change shape or size have been available for many years [35–37]. The activation mechanisms can include chemical, thermal, pneumatic, optical and magnetic processes. The first documented experiment with electroactive polymeric materials is attributed to Roentgen where, in 1880, he observed a length change in a rubber-band (with one fixed end and a mass attached to the free end) that he subjected to an electric field. Following the observation of a substantial piezoelectric activity in poly(vinylidene difluoride) (PVDF), other polymer systems have been investigated, and a number of effective materials have been developed. Materials that can create linear strains up to 380% are now available.

Generally, electroactive polymer actuators can be divided into two major groups based on their activation mechanism: ionic (involving mobility or diffusion of ions) and electronic (driven by electric field). The former category is often referred to as 'wet', because the actuators use aqueous component materials. The electronic polymers (electrostrictive, electrostatic, piezoelectric and ferroelectric) are driven by electric fields and can be made to hold the induced displacement under activation of a DC voltage, allowing them to be considered for robotic applications. These materials have a greater mechanical energy density and they can be operated in air with no major constraints, but they require a high activation field ($> 100\,\mathrm{V\,\mu m^{-1}}$) close to the breakdown level. Ionic materials (gels, polymer–metal composites, conductive polymers and carbon nanotubes) are driven by diffusion of ions and these require an electrolyte for the actuation mechanism. Their major advantage is their operation at drive voltages as low as 1–2 V. However, there is a need to maintain their wetness and, apart from conductive polymers and carbon nanotubes, it is difficult to sustain DC-induced displacements. The induced displacement of both the electronic and ionic polymer can be geometrically designed to bend, stretch or contract.

Polymeric gels are biphasic systems composed of a solid phase or elastic matrix permeated by a fluid, generally water, and a number of different types have been developed and studied. An example is shown in Figure 10.37. These systems exhibit plastic contraction with changes in temperature, pH, magnetic or electrical field, and have a vast number of applications, for example, soft actuators in the biomedical field and for controlled drug release. Their application in smart lenses has already been noted in Chapter 8, Section 8.7. Gel actuators are characterized by large strains (50% or more) and lower forces than conducting polymers, and several prototype actuating systems have been realized for biomedical use. In the mid-1990s, a group at MIT developed an 'artificial muscle' gel that was able to exhibit the strain and force characteristics of natural muscle [38]. This gel contained polyacrylonitrile fibres which reacted to a change in pH. Contraction in acid solution was fairly slow, whereas expansion in basic solution was much faster; a full cycle took about 40 s.

The motion of ions in and out of a conjugated polymer during oxidation and reduction will be accompanied by changes in the dimensions of the polymer sample. These changes will depend on the size of the ions used (and their associated hydration shells; Chapter 5, Section 5.2.3) and the structure of the polymer, i.e. whether it is a high-density film or a low-density gel, the degree of crystallinity and any preference for orientation of the polymer

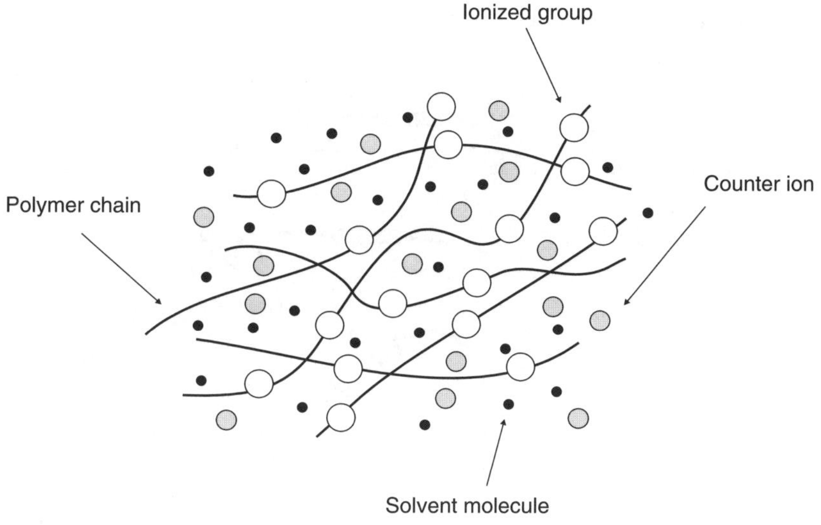

Figure 10.37 Composition of a polyelectrolyte gel.

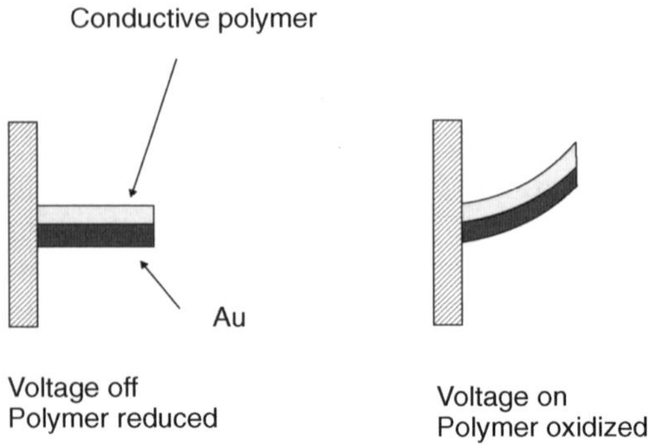

Figure 10.38 Conductive polymer/Au bimorph actuator. (a) No applied voltage; (b) voltage applied (e.g. polymer is oxidized).

chains. Values for dimensional changes reported in the literature vary widely [39]. Measured expansion of polypyrrole and polyaniline films parallel to the film surface varies between approximately 0.5% and a few percent, while estimates of the volume change vary between about 7% for polyacetylene to 35–120% for polypyrrole and polyaniline.

The poor mechanical properties often found in extension restrict the application of electropolymerized films and gels. Therefore, many of the devices described in the literature are *bimorphs*, with a stronger supporting layer. An example is depicted in Figure 10.38, in which a conductive polymer is deposited on a gold layer. Such microactuators have been used in fairly sophisticated applications. Figure 10.39 shows the use of a microrobot to move and position a small object in the laboratory [37]. The microrobot arm is made of polypyrrole microactuators and consists of an 'elbow', a 'wrist' and a 'hand' with two to four 'fingers'. Each joint consists of two separately controllable micromuscles. Using this arrangement, a 100 μm glass bead could be lifted and moved over a system of polyurethane tracks. This micro-actuator might be considered as the 'big brother' version of the molecular actuators described in Chapter 11, Section 11.11.

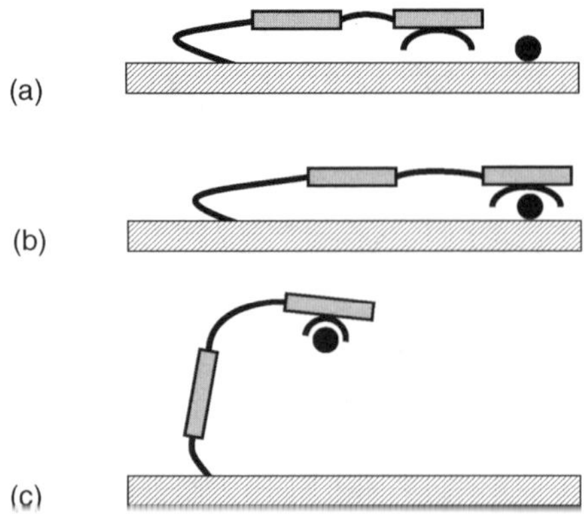

Figure 10.39 (a) A microrobot arm made of polypyrrole microactuators. The microrobot can be used to (b) locate and (c) lift a 100 μm glass bead [37].

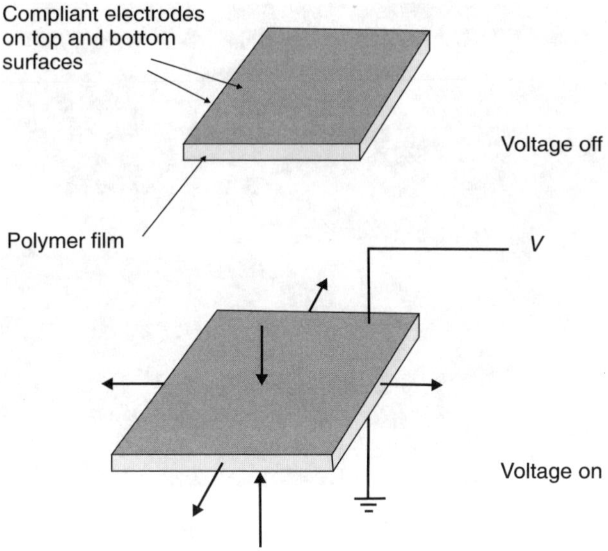

Figure 10.40 A dry polymer actuator. (a) No applied voltage; (b) An applied voltage changes the dimensions of the polymer element.

The electronic (or so-called dry) actuators usually respond to applied electric fields (as opposed to charge or mass-transport). The basic functional element is shown in Figure 10.40. The resulting strain is approximately proportional to the square of the electric field. If the response is determined by the field-induced reorientation of the crystalline or semi-crystalline structure, then the polymer is said to be *electrostrictive*. However, if the response is dominated by the interaction of the electrostatic charges on the electrodes (often called the *Maxwell stress*) then the polymer is called a *dielectric elastomer.* It should be noted that the phenomenon of electrostriction is different to that of piezoelectricity, introduced in Chapter 2, Section 2.6.3, which is exploited by certain acoustic sensors (Section 10.4.5). The latter only occurs in certain types of crystal structure (those without a centre of symmetry) and results in a strain that is proportional to the applied electric field. Unlike piezoelectricity, there is also no inverse electrostrictive effect, i.e. a deformation does not produce an electric field.

As noted above, one problem with micro- or nano-actuators is that of producing free-standing elements that might respond to external stimuli. One approach is to incorporate carbon nanotubes as structural components in the thin films. Some success in this respect has been achieved using the electrostatic layer-by-layer process, which has been used to create composites using polyelectrolytes 'reinforced' with the carbon nanotubes [40–42]. Figure 10.41 shows a carbon nanotube–polyelectrolyte bridge (20 μm wide and 60 μm long; left) with a thickness of 1 μm. The rough edges (±2 μm) are due to the tendency for the LbL deposit to tear away unevenly under sonication [42].

10.7.3 Lab-on-a-Chip

A promising tool for analysing proteins and protein complexes in the biology laboratory of the future is a microfluidic device commonly called a *Lab-on-a-chip* [43, 44]. These

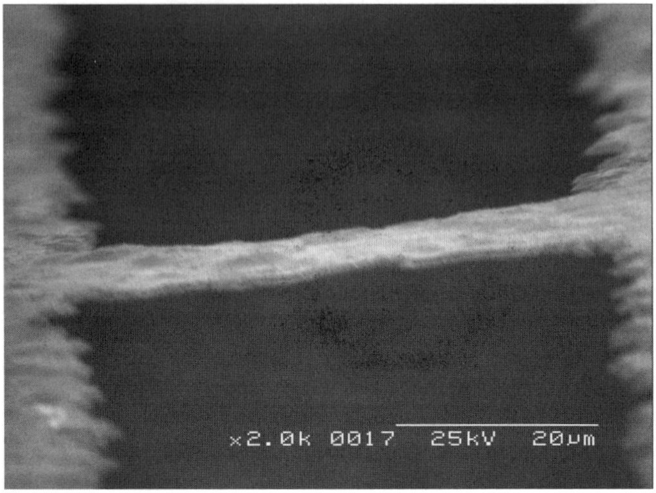

Figure 10.41 A composite carbon nanotube/polyelectrolyte free-standing bridge [42].

'laboratories' are fabricated using photolithographic processes developed in the microelectronics industry to create circuits of miniature chambers and channels in a quartz, silica, or glass chip. They direct the flow of liquid chemical reagents just as semiconductors direct the flow of electrons. These reagents can be diluted, mixed, reacted with other reagents, or separated – all on a single chip.

The flow of a fluid through a microfluidic channel can be characterized by the *Reynolds number*, *Re*, defined as

$$Re = \frac{Lv\rho}{\mu} \tag{10.13}$$

where L is the most relevant length scale, μ is the viscosity, ρ is the fluid density and v is the average velocity of the flow. For many microchannels, L is equal to $4A/P$, where A is the cross-sectional area of the channel and P is the wetted perimeter of the channel. Due to the small dimensions of microchannels, *Re* is usually much less than 100, often less than 1.0. In this *Re* regime, the flow is completely laminar and no turbulence occurs. The transition to turbulent flow generally occurs at *Re* in the region of 2000. Laminar flow provides a means by which molecules can be transported in a relatively predictable manner through microchannels. However, even at $Re < 100$, it is possible to have momentum-based phenomena such as flow separation.

There are two common methods by which fluid actuation through microchannels can be achieved. In pressure-driven flow, the fluid is pumped through the device via positive displacement pumps, such as syringe pumps. One of the basic laws of fluid mechanics for pressure-driven laminar flow, the so-called no-slip boundary condition, states that the fluid velocity at the walls must be zero. This produces a parabolic velocity profile within the channel. Pressure-driven motion, termed *Poiseuille flow*, is well understood but designing and fabricating reliable mechanical pumps in the traditional materials of silicon and glass has been difficult. These devices require multiple levels of fabrication, and are easily damaged by particles of dust and contaminants in the fluid.

Another common technique for pumping fluids is that of electro-osmotic pumping. If the walls of a microchannel have an electric charge, as most surfaces do, an electric double layer of counter ions will form at the walls. When an electric field is applied across the channel, the ions in the double layer move towards the electrode of opposite polarity. This creates motion of the fluid near the walls and transfers, via viscous forces, into convective motion of the bulk fluid. If the channel is open at the electrodes, as is most often the case, the velocity profile is uniform across the entire width of the channel. However, if the electric field is applied across a closed channel (or a back-pressure exists that just counters that produced by the pump), a recirculation pattern forms in which fluid along the centre of the channel moves in a direction opposite to that at the walls. In closed channels, the velocity along the centreline of the channel is 50% of the velocity at the walls. Electrically driven flow has a number of drawbacks: sensitivity to impurities that adsorb on the wall of the channel, ohmic generation of heat in the fluid and the need for high voltages (on the order of kilovolts).

Microfluidic devices can be fabricated from a variety of materials. Based on fabrication techniques developed in the microelectronics industry, silicon has been used extensively to create microfluidic devices. Polymers offer a lot of advantages for microsensors, microactuators and microfluidics. These are relatively low-cost materials, fabrication techniques are simple and there is no need for special clean-room and/or high-temperature processes. Polymers can be deposited on various types of substrates, and there is a wide choice of molecular structures. This enables films to be produced with various physical and chemical properties, including sensing and actuation behaviour. A popular material for fabrication of microfluidic device is the silicone polymer polydimethylsiloxane (PDMS) (Section 10.4.4). Microfluidic devices can be fabricated from PDMS by pouring the liquid over a mould (usually silicon or photoresist) and curing to cross-link the polymer. The result is an optically clear, relatively flexible material that can be stacked on to other cured polymer slabs to form complex three-dimensional geometries.

Common fluids used in microfluidic devices include whole blood samples, bacterial cell suspensions, protein or antibody solutions and various buffers. The use of microfluidic devices for biomedical research and clinically important technologies has a number of significant advantages. First, because the volume of fluids within these channels is very small, usually a few nanolitres, the amount of reagents and analytes used is fairly small. This is especially significant for expensive reagents. The fabrication techniques used to construct microfluidic devices are relatively cheap and are very amenable both to highly elaborate, multiplexed devices and also to mass production. In a manner similar to that for microelectronics, microfluidic technologies permit the fabrication of highly integrated devices, useful for performing several different functions on the same substrate chip. One of the long-term goals in the field of microfluidics is to create integrated, portable clinical diagnostic devices for home and bedside use, thereby eliminating time-consuming laboratory analysis procedures [45].

Methods of manipulating particles, such as living cells, on the chip surface include *electrophoresis* and *dielectrophoresis* [46, 47]. Electrophoretic forces arise from the interaction of a particle's charge in an electric field, whereas dielectrophoresis originates from the polarizability (Chapter 4, Section 4.3.2) and occurs with uncharged particles. Figure 10.42 shows a schematic diagram illustrating the principle of dielectrophoresis. A particle is suspended within an asymmetric electrode arrangement. Under the influence of the field, the particle will become polarized. The interaction between the induced charges and the local electric field will then result in a force on then particle. Because of the inhomogeneous

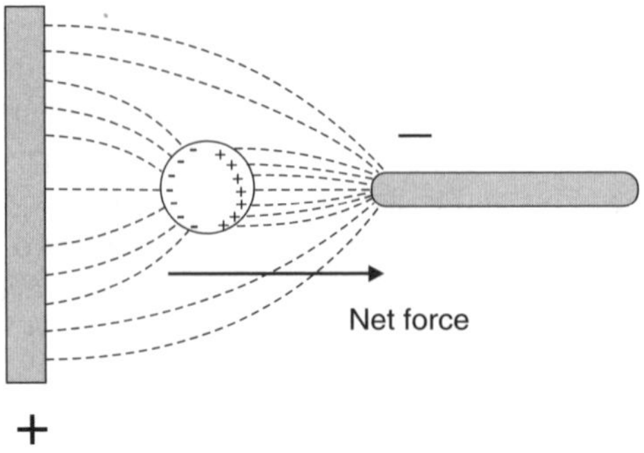

Figure 10.42 Origin of dielectrophoresis. The application of a non-uniform electric field **E** to an uncharged particle produces a polarization. The interaction of the induced dipoles and the external field causes a force to act on the particles.

nature of the electric field in Figure 10.42, this force is greater on the side facing towards the point electrode than that on the side facing the plane electrode, and there is net motion towards the point electrode. This effect is called positive dielectrophoresis. However, if the particle is less polarizable than the surrounding medium, the induced dipole will align in the opposite direction to the field and the particle will be repelled from the high-field regions – negative dielectrophoresis. The force that is created in dielectrophoresis is dependent on the induced dipole and is unaffected by the direction of the electric field, responding only to the field gradient. Since the alignment field is irrelevant, the force can also be generated by AC fields. This has the advantage of eliminating any electrophoretic force (due to any net particle charge). The dielectrophoretic force F_{DEP} acting on a spherical body of radius r is given by [47]

$$F_{\mathrm{DEP}} = 2\pi r^3 \varepsilon'_{\mathrm{m}} / \mathrm{Re}[\mathrm{K}(\omega)] \nabla \boldsymbol{E}^2 \tag{10.14}$$

where $\varepsilon'_{\mathrm{m}}$ is the real part of the complex permittivity (i.e. the dielectric constant; Chapter 3, Section 3.6.1) of the surrounding medium, $\boldsymbol{E}$ is the electric field (rms or root-mean-square value), ∇ is the mathematical *Del vector operator* $\left(\nabla = \frac{\partial}{\partial x} + \frac{\partial}{\partial y} + \frac{\partial}{\partial z}\right)$ and $\mathrm{Re}[K(\omega)]$ is the real part of the complex permittivity factor, given by

$$K(\omega) = \frac{\varepsilon_{\mathrm{p}} - \varepsilon_{\mathrm{m}}}{\varepsilon_{\mathrm{p}} + 2\varepsilon_{\mathrm{m}}} \tag{10.15}$$

where ε_{p} and ε_{m} are the complex permittivities of the particle and medium, respectively.

By careful construction of the electrode geometry that is used to generate the field, it is possible to create electric field morphologies so that potential energy minima are bounded by regions of increasing field strengths. In such electrode arrangements, particles experiencing positive dielectrophoresis are attracted to the regions of maximum field (e.g. the electrode edges) whereas particles experiencing negative dielectrophoresis are trapped in isolated field minima.

Noncharged particles can also rotate under the influence of applied fields via *electro-orientation* and *electro-rotation effects* [47]. The former results from non-spherical particles, whereas the latter exploits a rotating electric field. *Travelling-wave dielectrophoresis* [48] is a linear analogue of electro-rotation. The electrodes are arranged in a track and the phase of the electric field is advanced along the track. This produces an electric field wave that travels along the track. Particles can then be moved along these tracks.

10.8 SMART TEXTILES AND CLOTHING

We are all carrying around more and more electronic products, e.g. mobile phones, laptops, personal hi-fis, personal digital assistants (PDAs), with more such gadgets on the way. It makes sense, therefore, to integrate such products into our clothes. Smart or intelligent clothing is a combination of electronics and clothing textiles. The first step is to integrate existing products into clothing simply by sewing these in. For example, orientation, communications and geographic positioning electronics can all be incorporated into outdoor clothing. The functional architecture is implemented using *GPS* (global positioning system) for navigation and *GSM* (global system for mobile communications) for communication. Power supplies and the user interface can be incorporated in the form of a supporting vest.

New fibre and textile materials and miniaturized electronic components make it possible to create truly usable smart clothes. These intelligent clothes are worn like ordinary clothing, providing help in various situations according to the designed application. Simple designs might include an ability to vary the colour of the garment or 'active' drying. The latter could be achieved using a humidity sensor, continuously monitoring the humidity and controlling a heating element. One example is smart fabric derived from the properties of pinecones, developed at the University of Bath, UK [49]. The fabric adapts to changing temperatures by opening up when warm and shutting tight when cold, just like a pine cone bract does in nature in order to release its seeds.

BIBLIOGRAPHY

Blythe T, Bloor D. *Electrical Properties of Polymers*. Cambridge: Cambridge University Press; 2005.

Gardner JW, Bartlett PN. *Electronic Noses: Principles and Applications*. Oxford: Oxford University Press; 1999.

Gardner JW, Varadan VK, Awadelkarim OO. *Microsensors, MEMS and Smart Devices*. Chichester: John Wiley & Son Ltd; 2001.

Janata J. *Principles of Chemical Sensors*. New York: Plenum Press; 1989.

Jiang H, Dong L. Smart lenses. *Phys. World* 2006; **19**: 29–31.

Koryta J. *Ion-Selective Electrodes*. Cambridge: Cambridge University Press; 1975.

Kortya J. *Ions, Electrodes and Membranes*. Chichester: John Wiley & Sons, Ltd; 1982.

Kumar CSSR (Editor) *Nanomaterials for Biosensors*. Weinheim: Wiley-VCH; 2006.

Moseley PT, Crocker AJ. *Sensor Materials*. Bristol: Institute of Physics; 1996.

Nabook A. *Organic and Inorganic Nanostructures*. Boston: Artech House; 2005.

Voldman J. Electrical forces for microscale cell manipulation. *Annu. Rev. Biomed. Eng*. 2006; **8**: 425–454.

Waiser R (Editor) *Nanoelectronics and Information Technology*. Weinheim: Wiley-VCH; 2005.

Wright RH. *The Sense of Smell*. Boca Raton, FL: CRC Press; 1982.

REFERENCES

[1] Middelhoek S, Noorlag DW. Three-dimensional representation of input and output transducers. *Sens. Actuators* 1981–82; **2**: 29–41.

[2] Janata J. *Principles of Chemical Sensors*. New York: Plenum Press; 1989.

[3] Bergveld P. Thirty years of ISFETOLOGY. What happened in the past 30 years and what may happen in the next 30 years. *Sens. Actuators B* 2003; **88**: 1–20.

[4] Janata J. Thirty years of CHEMFETs – a personal view. *Electroanalysis* 2004; **16**: 1831–1835.

[5] Snow AW, Barger WR. Phthalocyanine films in chemical sensors. In CC Leznoff and ABP Lever (Editors) *Phthalocyanines, Properties and Applications*. New York: VCH; 1989, pp. 341–391.

[6] Bott B, Jones TA. A highly sensitive NO_2 sensor based on electrical conductivity changes in phthalocyanine film. *Sens. Actuators* 1984; **5**: 43–53.

[7] Gardner JW, Bartlett PN. *Electronic Noses: Principles and Applications*. Oxford: Oxford University Press; 1999.

[8] Mabrook MF, Pearson C, Petty MC. An inkjet-printed chemical fuse. *Appl. Phys. Lett.* 2005; **86**: 013507.

[9] Mabrook MF, Pearson C, Petty MC. Inkjet-printed chemical sensors for the detection of organic vapours. *IEEE Sens. J.* 2006; **6**: 1435–1444.

[10] Bartlett PN, Ling-Chung SK. Conducting polymer gas sensors. Part II. Response of polypyrrole to methanol vapour. *Sens. Actuators* 1989; **19**: 141–150.

[11] Senturia SD, Sechen CM, Wishneusky JA. The charge-flow transistor: a new MOS device. *Appl. Phys. Lett.* 1977; **30**: 106–108.

[12] Barker PS, Di Bartolomeo C, Monkman AP, Petty MC, Pride R. Gas sensing using a charge-flow transistor. *Sens. Actuators B* 1995; **25**: 451–453.

[13] Barker PS, Petty MC, Monkman AP, McMurdo J, Cook MJ, Pride R. A hybrid phthalocyanine/silicon field effect transistor sensor for NO_2. *Thin Solid Films* 1996; **284–285**: 94–97.

[14] Carrillo A, Martin-DomÀnguez IR, Glossman D, Márquez. Study of the effect of solvent induced swelling on the resistivity of butadiene based elastomers filled with carbon particles. *Sens. Actuators A* 2005; **119**: 157–168.

[15] Grate JW, Abraham MH. Solubility interactions and the design of chemically selective sorbent coatings for chemical sensors and arrays. *Sens. Actuators B* 1991; **3**: 85–111.

[16] Casalini R, Wilde JN, Nagel J, Oertel U, Petty MC. Organic vapour sensing using thin films of a co-ordination polymer: comparison of electrical and optical techniques. *Sens. Actuators B* 1999; **57**: 28–34.

[17] Richardson TH, Dooling CM, Jones LT, Brook RA. Development and optimization of porphyrin gas sensing LB films. *Adv. Colloid Interface Sci.* 2005; **116**: 81–96.

[18] Lednev IK, Petty MC. Langmuir monolayers and Langmuir–Blodgett multilayers containing macrocyclic ionophores. *Adv. Mater.* 1996; **8**: 615–629.

[19] Lednev IK, Petty MC. Aggregate formation in Langmuir–Blodgett films of an amphiphilic benzothiazolium styryl chromoionophore. *Langmuir* 1994; **10**: 4185–4189.

[20] Nylander C, Ljedeberg B, Lind T. Gas detection by means of surface plasmon resonance. *Sens. Actuators* 1982; **3**: 79–88.

[21] Nylander C, Ljedeberg B, Lundstrom I. Surface plasmon resonance for gas detection and biosensing. *Sens. Acuators* 1983; **4**: 299–304.

[22] Pearson C, Nagel J, Petty MC. Metal ion sensing using ultrathin organic films prepared by the layer-by-layer adsorption technique. *J. Phys. D: Appl. Phys.* 2001; **34**: 285–291.

[23] McDonnell JM. Surface plasmon resonance: towards an understanding of the mechanisms of biological molecular recognition. *Curr. Opin. Chem. Biol.* 2001; **5**: 572–577.

[24] Hall D. Use of optical biosensors for the study of mechanistically concerted surface adsorption processes. *Anal. Biochem.* 2001; **288**: 109–125.

[25] Biacore. Available from URL http://www.biacore.com/.
[26] Texas Instruments. Available from URL http://www.ti.com/.
[27] Chinowsky TM, Quinn JG, Bartholomew DU, Kaiser R, Elkind JL. Performance of the Spreeta 2000 integrated surface plasmon resonance affinity sensor. *Sens. Actuators B* 2003; **91**: 266–274.
[28] Homola J, Lu HB, Nenninger GG, Dostalek J, Yee SS. A novel multichannel surface plasmon resonance biosensor. *Sens. Actuators B* 2001; **76**: 403–410.
[29] Homola J, Lu HB, Lee SS. Dual-channel surface plasmon resonance sensor with spectral discrimination of sensing channels using dielectric overlayer. *Electron. Lett.* 1999; **35**: 1105–1106.
[30] Palumbo M, Nagel J, Petty MC. Surface plasmon resonance detection of metal ions: layer-by-layer assembly of polyelectrolyte sensing layers on a multichannel chip. *IEEE Sens. J.* 2005; **5**: 1159–1164.
[31] Palumbo M, Petty MC. Polyelectrolytes-based SPR sensors. In CA Grimes, EC Dickey, MV Pishko (Editors). *Encyclopaedia of Sensors*, Vol. 8. California: American Scientific Publishers; 2006, pp. 33–52.
[32] Hammond PT. Chemistry directed deposition via electrostatic and secondary interactions: a nonlithographic approach to patterned polyelectrolyte multilayer systems. In G Decher and JB Schlenoff (Editors). *Multilayer Thin Films*. Weinheim: Wiley-VCH; 2003, pp. 271–299.
[33] Electronic Nose Network. Available from URL http://www.nose-network.org/review/.
[34] Gardner JW, Varadan VK, Awadelkarim OO. Microsensors, *MEMS and Smart Devices*. Chichester: John Wiley & Sons, Ltd; 2001.
[35] Bar-Cohen Y (Editor). *Electroactive Polymer [EAP] Actuators as Artificial Muscles – Reality, Potential, and Challenges*. Jet Propulsion Laboratory (NASA). Washington, DC SPIE Press; 2001.
[36] NASA. *Electroactive Polymer Actuators*. Available from URL http://eap.jpl.nasa.gov/.
[37] Jager EWH, Smela E, Inganäs O. Microfabricating conjugated polymer actuators. *Science* 2000; **290**: 1540–1545.
[38] Brock D, Lee WJ, Segalman D, Witkowski M. A dynamic-model of a linear-actuator based on polymer hydrogel. *J. Intelli. Materi. Syst. Struct.* 1994; **5**: 764–771.
[39] Blythe T, Bloor D. *Electrical Properties of Polymers*. Cambridge: Cambridge University Press; 2005.
[40] Hua F, Cui T, Lvov YM. Ultrathin cantilevers based on polymer-ceramic nanocomposite assembled through layer-by-layer deposition. *Nano Lette.* 2004; **4**: 823–825.
[41] Mamedov AA, Kotov NA, Prato M, Guldi DM, Wicksted JP, Hirsch A. Molecular design of strong single-wall carbon nanotube/polyelectrolyte multilayer composites *Nature Mater.* 2002; **1**: 190–194.
[42] Cadd D, Wood D, Petty MC. Free standing polymer cantilevers and bridges reinforced with carbon nanotubes. Micro Nano Lett., in press.
[43] Whitesides GM. The origins and future of microfluidics. *Nature* 2006; **442**: 368–373.
[44] Janasek D, Franzke J, Manz A. Scaling and the design of miniaturised chemical-analysis systems. *Nature* 2006; **442**: 374–380.
[45] Yager P, Edwards T, Fu E, Helton K, Nelson K, Tam MR, Weig BH. Microfluidic diagnostic technologies for global public health. *Nature* 2006; **442**: 412–418.
[46] Voldman J. Electrical forces for microscale cell manipulation. *Annu. Rev. Biomed. Eng.* 2006; **8**: 425–454.
[47] Hughes MP. AC electrokinetics: applications for nanotechnology. *Nanotechnology* 2000; **11**; 124–132.
[48] Hughes MP, Pethig R, Wang X-B. Forces on particles in travelling electric fields: computer-aided simulations. *J. Phys. D: Appl. Phys.* 1996; **29**: 474–482.
[49] University of Bath. Centre for Biomimetric and Natural Technologies. Available from URL http://www.bath.ac.uk/mech-eng//biomimetics/home.htm.

11 Molecular-Scale Electronics

O brave new world!

11.1 INTRODUCTION

The second strand to molecular electronics (molecular-scale electronics) recognizes the spectacular size reduction in the individual processing elements in integrated circuits over recent years (Chapter 1). This 'bottom-up' approach offers many intriguing prospects for manipulating materials on the nanometre scale, thereby providing opportunities to build up architectures with predetermined and unique physical and/or chemical properties. The literature is packed with examples of these activities, far too many to condense into a single chapter of a book. A selection of these ideas has therefore been chosen to provide an illustration of what might be possible as we make progress in the 21st century.

11.2 NANOSYSTEMS

The development of molecular electronics, as with all other areas of scientific endeavour, is constrained by the laws of physics. In most cases, these are the laws of classical or Newtonian physics. However, on the nanoscale, other principles become more important, even crucial. Some of the ideas of quantum mechanics, responsible for example for the chemical bonding that holds solids together, have been introduced in Chapters 2 and 3. Quantum mechanical tunnelling (Chapter 3, Section 3.5.2) becomes a significant electrical conduction process at dimensions less than 5 nm and therefore will have a key role as molecular-scale electronics develops. In the following sections, some other ideas are introduced which could be significant for electronic and/or optoelectronic devices operating at nanometre dimensions.

11.2.1 Scaling Laws

The magnitudes of physical quantities characterizing nanoscale systems differ considerably from those familiar from the macroscale world. Some of these quantities can be estimated by applying scaling laws to the values for macroscale configurations [1]. For example, the strength of a structure and the force it exerts can be assumed, in the first instance, to scale with its cross-sectional area. Nanoscale devices accordingly exert only small forces: a stress of $10^{10}\,\mathrm{N\,m^{-2}}$ equates to $10^{-8}\,\mathrm{N\,nm^{-2}}$ or $10\,\mathrm{nN\,nm^{-2}}$.

Of particular relevance to this chapter is the scaling of classical (macroscopic) electromagnetic systems. Here, it is convenient to assume that electrostatic field strengths (hence

Molecular Electronics: From Principles to Practice M. C. Petty
© 2007 John Wiley & Sons, Ltd

electrostatic stresses) are independent of scale. The onset of strong field-emission currents (Chapter 3, Section 3.5.5) from conductors limits the electrostatic field strength permissible at the electrodes of nanoscale systems; values of $10^9\ \mathrm{V\ m^{-1}}$ can readily be tolerated. At this field strength, one nanometre corresponds to a potential difference of 1 V.

If all the dimensions of a material are reduced by a constant K, then the effects on various important electrical parameters can be calculated. For example, electrical resistance R will scale with K whereas the capacitance C (for a parallel plate capacitor) will scale as K^{-1} (assuming, of course, that the resistivity and permittivity of the material remain unchanged as its dimensions are reduced). An important consequence is that the *time constant* of a resistor capacitor combination, i.e. the RC product, will remain unchanged. This is also the case for the current density $J(J = \mathrm{current/area}$, and both the current and area decrease as K^{-2} with scaling). Current densities in aluminium interconnections in microelectronic circuitry are limited to $10^{10}\ \mathrm{A\ m^{-2}}$ or less by electromigration, which is a diffusive process that redistributes metal atoms under the influence of electrical forces and eventually interrupts circuit continuity. This current density equates to $10\ \mathrm{nA\ nm^{-2}}$.

The above represent general scaling laws. Similar arguments can be applied to particular electronic devices, such as the field effect transistor, as described in Chapter 1, Section 1.3.1 [2]. In some models, as the device dimensions are reduced by K, the doping density is *increased* by K in order to keep the resistance constant. In this case, the RC time constant varies as K^{-1}, thereby providing faster switching times. A disadvantage, however, is that the current density now varies as K. The assumption of a constant electric field also provides problems as this leads to a reduction in operating voltages, which means that compatibility with standard voltage levels used in other parts of the electronic system will be lost.

The diffusion of matter (i.e. the net transport of particles from a region of higher concentration to a region of lower concentration) is a macroscopic manifestation of Brownian motion (Chapter 2, Section 2.7). Brownian motion can be described by a *random walk*, in which a particle moves a fixed distance, hits something and then sets out in a totally different (random) direction. The consequence is that the distance travelled is proportional to the square root of the number of steps (in contrast to a normal walk in which, for travel in a straight line, the distance travelled is proportional to the number of steps). This has important consequences on the nanoscale. The diffusion coefficient of oxygen in water is $18 \times 10^{-10}\ \mathrm{m^2\ s^{-1}}$. The time taken for a molecule to diffuse a given distance is approximately the (distance)^2 divided by the diffusion coefficient. Hence an oxygen molecule will move 10 nm in about 50 ns, and 1 cm in around 90 min. Bigger molecules diffuse somewhat more slowly than smaller ones. The diffusion coefficient for a sugar molecule (sucrose) in water is about $5 \times 10^{-10}\ \mathrm{m^2\ s^{-1}}$, whereas for a macromolecule such as a protein the diffusion coefficient is likely to be less than $10^{-10}\ \mathrm{m^2\ s^{-1}}$.

11.2.2 Interatomic Forces

In the macroworld, solids are held together by strong ionic or covalent forces (Chapter 2, Section 2.3). Other bonds, such the van der Waals bond and the hydrogen bond, are much weaker, but these can become important as nano-dimensions are approached. The classical van der Waals bond is a very short-range interaction with the energy associated with it varying as r^{-6}, where r is the distance between the interacting molecules. However, if a single molecule or atom is interacting with a large body, such as a plane or a large sphere, the dependence of the interaction on the spacing from the point to the plane or sphere will become r^{-3}. Other possibilities are shown in Figure 11.1 [1, 3].

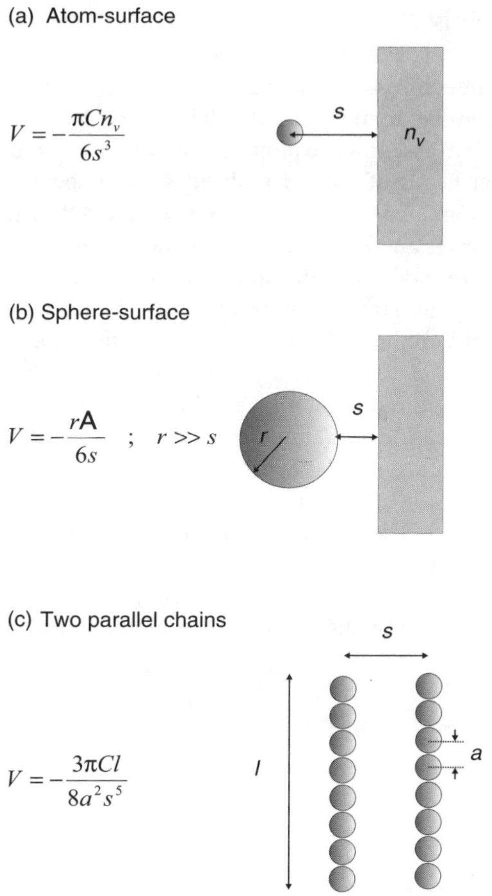

Figure 11.1 Potential energy of the van der Waals attractions V for three different extended geometries. n_v is the number of atoms per unit volume, **A** is the Hamaker constant, a tabulated material parameter, and C is the van der Waals constant [1, 3].

A further force, the *Casimir* force, operates between metallic surfaces, forcing them together. This is associated with the modes of oscillation of the electromagnetic field in an enclosed region. The force is vanishingly small except for extremely closely spaced surfaces, those that are relevant to molecular-scale devices. For example, for two metal surfaces spaced by 10 nm, there is an attractive force of about 1 atm (100 kPa) arising from the Casimir effect.

11.3 ENGINEERING MATERIALS AT THE MOLECULAR LEVEL

In this section, some ideas are explored for introducing functionality into materials by molecular engineering. This type of endeavour might be considered as a sort of half-way house between the two extreme definitions for molecular electronics. We are using molecular-scale processing (LB or LbL deposition, for example) to confer a particular functionality into a macroscopic sample (e.g. an organic film of 100 nm dimensions).

11.3.1 Polar Materials

A simple molecular architecture – an organic 'superlattice' – is that of the alternate-layer molecular film. This can be realized in the laboratory by LB deposition (Chapter 7, Section 7.3.1) or using self-assembly exploiting electrostatic forces (LbL films; Chapter 7, Section 7.3.3). The polar heads of the LB molecules are associated with an electric dipole moment (Chapter 2, Section 2.3.5). In an alternating Y-type *ABABABA*... film, the dipoles of the two different molecules *A* and *B* will not cancel and the multilayer will exhibit an overall polarization in the direction normal to the substrate plane. This is illustrated in Figure 11.2 for an LB film comprising an alternating arrangement of monolayers of a long-chain fatty acid and a long-chain fatty amine. The figure indicates that the deposition can result in a proton transfer from the acid to the amine head group, thereby creating a large dipole moment. The polarization within the multilayer film can be measured using a *Kelvin probe*; data are shown in Figure 11.3 for acid/amine assemblies in the form of surface potential, i.e. the effective potential difference across the film, as a function of the film thickness (number of monolayers) [4]. The only difference between the two sets of data is the order in which the two materials were deposited. The surface potential values for data set (a) were obtained for a structure in which the latter of fatty amine was deposited first on the substrate (aluminium-coated glass); for the data set (b), the acid was transferred first. Thus, the surface potential appears to reflect the dipole orientations in the two multilayer assemblies.

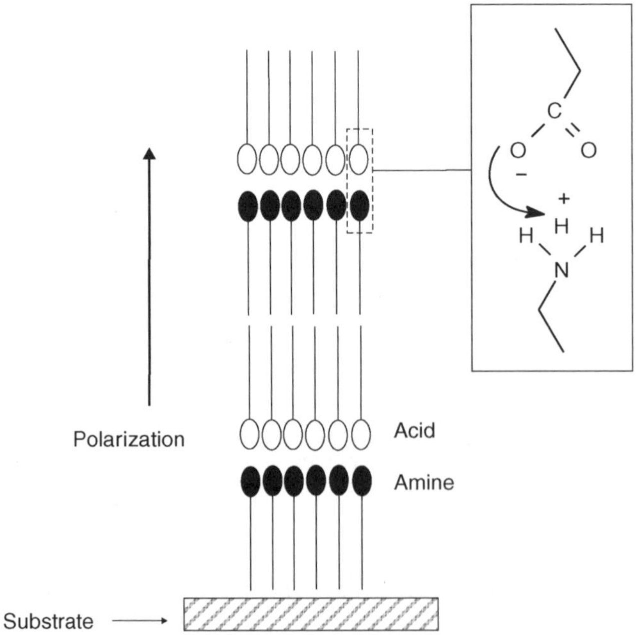

Figure 11.2 Schematic diagram of an organic superlattice formed by the alternate layer Langmuir–Blodgett deposition of a long-chain acid and a long-chain amine. The orientations of the polar head groups are shown on the right.

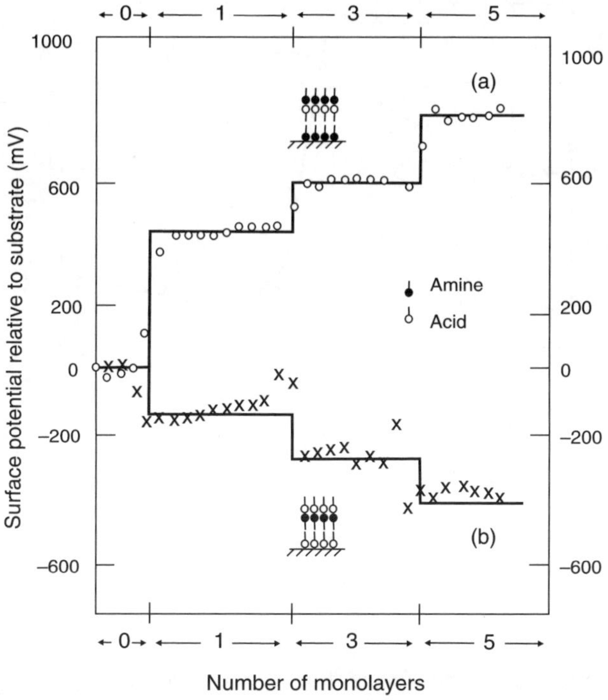

Figure 11.3 Surface potential of acid/amine alternate layer LB arrays as a function of the number of layers. Data are shown for two complementary orientations of the polar axis. Reprinted from Christie P, Roberts GG, Petty MC, *Appl. Phys. Lett.*, **48**,'Spontaneous polarization in organic superlattices', pp. 1101–1103. Copyright (1986) American Institute of Physics.

If the temperature of an acentric thin film structure, such as that shown in Figure 11.2, is varied, then the polarization across the film will change. This is the basis of the pyroelectric effect, which is only observed in materials that do not possess a centre of symmetry (Chapter 5, Section 5.6). If a metallic contact is provided on top of the alternate-layer film, the magnitude of the current I flowing through the film (i.e. perpendicular to the substrate plane) is given by

$$I = pA\frac{\mathrm{d}T}{\mathrm{d}t} \tag{11.1}$$

where p is the pyroelectric coefficient of the film, A is the sample area and $\mathrm{d}T/\mathrm{d}t$ is the rate of temperature change. This equation can be derived easily from the definition of p given by Equation (5.14), Chapter 5, Section 5.6.1. Note that a current will only be produced while the temperature is varied. Figure 11.4 shows the pyroelectric current obtained by heating and cooling an LB acid/amine assembly [5]. A positive current is measured on heating, whereas if the multilayer film is cooled, the direction of current is reversed. The pyroelectric coefficients of LB films are of the order 1–10 $\mu\mathrm{C\,m^{-2}\,K^{-1}}$. These are considerably less than seen in inorganic single crystals and ceramics and slightly less than observed for

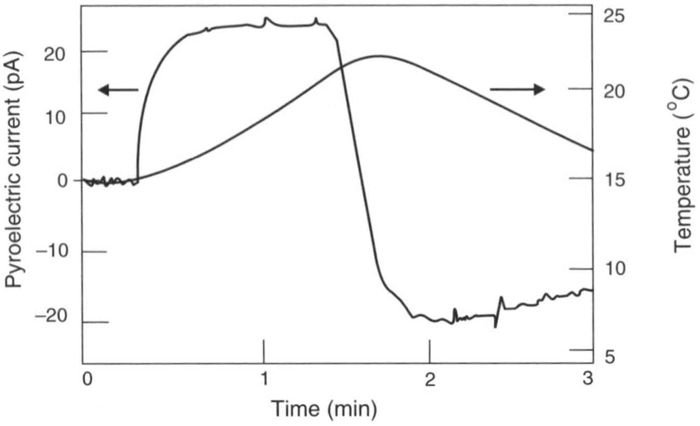

Figure 11.4 Time variation of temperature and pyroelectric current for a 99-layer acid/amine LB film. The arrows show which axis should be consulted for each curve. *From IEE Trans. Ultrasonics Ferroelectrics Frequency Control* **35**, Jones CA, Petty MC, Roberts GG, 'Langmuir–Blodgett films: a new class of pyroelectric materials', pp. 736–740. Copyright (1988). Reproduced by permission of IEEE.

organic materials produced in other forms. For example, poly(vinylidene fluoride) (PVDF) (Chapter 5, Section 5.6.2) possesses a pyroelectric coefficient of 40 $\mu C\, m^{-2}\, K^{-1}$. However, the LB film structure has the advantage that it is relatively easy to fabricate – PVDF must be poled in order (Chapter 2, Section 2.6.3) to produce a noncentrosymmetric structure. The LB film also possesses a relatively small dielectric constant ε_r: an important figure of merit for pyroelectric materials is p/ε_r [5]. Asymmetric LB structures, such as that depicted in Figure 11.2, are also expected to be piezoelectric, i.e. they will become polarized if they are subjected to a mechanical stress.

11.3.2 Nonlinear Optical Materials

The alternating molecular architecture described in the previous section may also form the basis of thin organic layers possessing second-order nonlinear optical properties (Chapter 4, Section 4.3.2). Such films will exhibit phenomena such as second-harmonic generation and the linear electro-optic (Pockels) effect. Conjugated materials possess π-electrons that are loosely bound to the molecules and can contribute considerably to the molecular polarizabilities. As a consequence, some organic compounds exhibit large nonlinear susceptibility coefficients, often considerably larger than those of conventional inorganic dielectrics. However, the measurable nonlinear optical effects are generally fairly small. The second-order coefficients of organic films can be obtained by exciting the film with a source of intense radiation (e.g. a laser) and monitoring the amount of second-harmonic radiation that is produced. An example of the result of such an experiment has already been given in Chapter 4, Section 4.5.4. The SPR technique, also described in Chapter 4, Section 4.7, can be adapted for the measurement of the electrooptic coefficient of an LB film. An electric field, $\boldsymbol{E}$, is applied to the organic film as the angle of incidence of the p-polarized light is changed.

A small change in permittivity, due to the Pockels effect, is produced. The second-order nonlinear susceptibility is then given by [6]

$$\chi^{(2)}(-\omega;\omega,0) = \frac{\Delta\varepsilon' + j\Delta\varepsilon''}{2\boldsymbol{E}} \tag{11.2}$$

The frequency argument in parentheses and to the right of the semicolon represent the applied fields (an optical frequency ω and a DC field in this case), whereas the field resulting from the interaction (which may be regarded as emitted, hence the negative sign) is given to the left of the semicolon.

SPR and Pockels data are shown in Figure 11.5 for monolayers of a hemicyanine and a nitrostilbene dye (both mixed with a fatty acid) [7]. In contrast to second-harmonic measurements, the Pockels experiment gives both the magnitude and the sign of the molecular hyperpolarizability. Comparison of Figure 11.5(a) and (b) reveals that the β coefficient for the hemicyanine layer is in the opposite sense to that of the nitrostilbene layer, i.e. the differential reflectivity curves for the two dyes are inverted.

The second-order nonlinear optical effects noted for monolayers may be retained in thicker films if the molecules are arranged in a noncentrosymmetric manner. This may be achieved by alternating an 'active' compound with a 'passive (e.g. a fatty acid) spacer layer, alternating two active materials that have been chosen so that their β coefficients are additive, or exploiting compounds that deposit as X-type or Z-type films (Chapter 7, Section 7.3.1). If the molecular alignment is preserved as the LB film is built up, then a quadratic

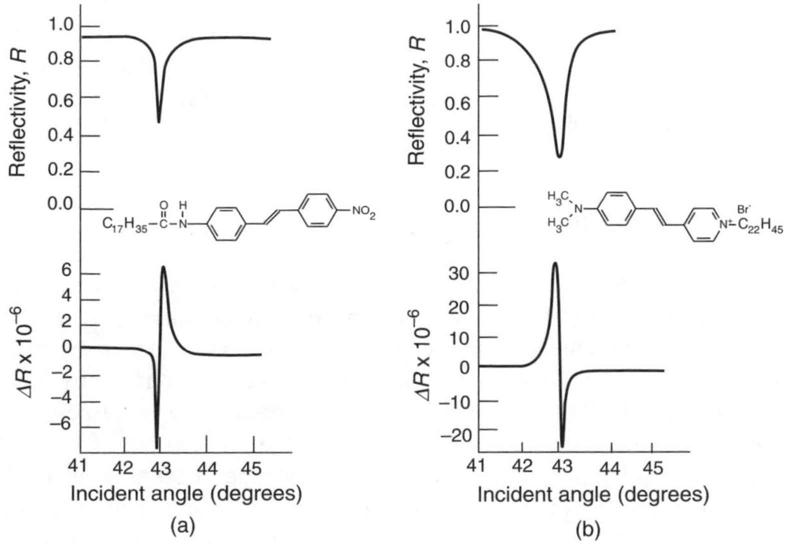

Figure 11.5 Angular dependence of the reflectivity R and the differential reflectivity ΔR for monolayer films of a fatty acid mixed with (a) a nitrostilbene compound and (b) a hemicyanine compound; the molecular structures are depicted. The experimental data and theoretical fits are indistinguishable on the scales shown. Reprinted from Andrea D, Lapiccirella A, Marletta G, Viticoli A (Editors), *Materials for Photonic Devices*. pp. 259–269. Copyright (1991), World Scientific.

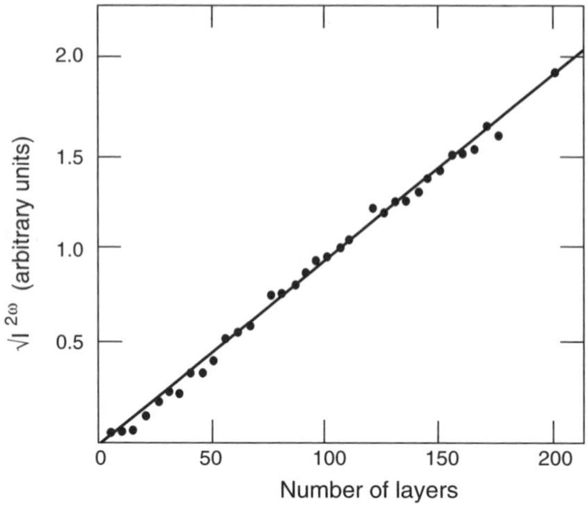

Figure 11.6 Variation of the square root of the second-harmonic intensity, $I^{2\omega}$, with the number of deposited layers for Z-type LB films of a cationic hemicyanine dye. After Ashwell GJ, Ranjan R, Whittam AJ, Gandolfo DS, 'Second-harmonic generation from alternate-layer and Z-type Langmuir–Blodgett films: optimization of the transparency/efficiency trade-off. *J. Mater. Chem.* **10**, pp. 63–68 Copyright (2000). Reproduced by permission of the Royal Society of Chemistry.

relationship between the second-harmonic intensity and the film thickness should result. Figure 11.6 shows an example of this, obtained for Z-type LB films of a cationic hemicyanine dye [8]. The quadratic dependence on the second-harmonic intensity is preserved to in excess of 200 Z-type layers. These films possess high nonlinear susceptibilities and have the additional advantage that they are transparent at the fundamental wavelength (1.06 μm in this case).

11.3.3 Photonic Crystals

Self-assembly approaches can also be used to build photonic crystal architectures (Chapter 4, Section 4.8) [9, 10]. Simple structures can be fabricated using arrays of SiO_2 spheres. These synthetic *opals* (the gemstone opal is made up of tiny spherical particles of quartz) can be assembled from colloidal suspensions or by using the LB technique. In this colloidal self-assembly, spheres of equal size in suspension are used. These spheres have a diameter that is comparable to the wavelength of light. When they are in an appropriate solvent they are charge stabilized; they have a repulsive interaction at short range which prevents them from sticking together. Because these particles are so small and their density is not too different from that of the solvent, they settle very slowly under the influence of gravity and their movement is mostly Brownian motion (Chapter 2, Section 2.7). When the colloidal suspension settles down, the spheres naturally arrange themselves in a face-centred cubic (fcc) structure (Chapter 2, Section 2.5.4). This immediately provides a three-dimensional dielectric that has a length scale comparable to the wavelength of light. The colloidal structure can then be impregnated with a high

Figure 11.7 Scanning electron microscope image of a synthetic opal formed by the Langmuir–Blodgett technique. The particles have a diameter of approximately 300 nm. After Bardosova M, Tredgold RH, 'Ordered layers of monodispersive colloids,' *J. Mater. Chem.*, **12**, pp. 2835–2842. Copyright (2002). Reproduced by permission of the Royal Society of Chemistry.

refractive index material. Once the original SiO_2 template has been removed, an *inverse opal* structure is obtained in which a regular array of air spheres is embedded in a substance which has a high refractive index.

Figure 11.7 shows an example of a synthetic opal created by the LB technique [11]. In this case, the silicon dioxide particles were first made hydrophopbic using a surface treatment before being spread on the surface of a water subphase. The initial layer was formed by the Langmuir–Schaefer method (Chapter 7, Section 7.3.1) in which the glass substrate was slowly moved upward in a horizontal position. Subsequent layers were built-up by the normal LB method. A highly ordered, close-packed arrangement of the individual opal particles is evident in Figure 11.7.

11.4 MOLECULAR DEVICE ARCHITECTURES

As noted in Chapter 1, the evolution of silicon microelectronics, the major enabling technology of the information revolution, is threatened not only by technical issues that may restrict the further miniaturization of integrated circuits, but also from the increased fabrication cost that accompanies shrinking the feature size below 100 nm. To overcome these limitations, a worldwide research effort is under way, which is focused on electronic device structures that go beyond conventional device architectures. The aim is to find materials and concepts that lead to devices which are scalable for at least several generations below 50 nm and fast (ns and less). The technology and materials must also be compatible with present-day and future generations of CMOS.

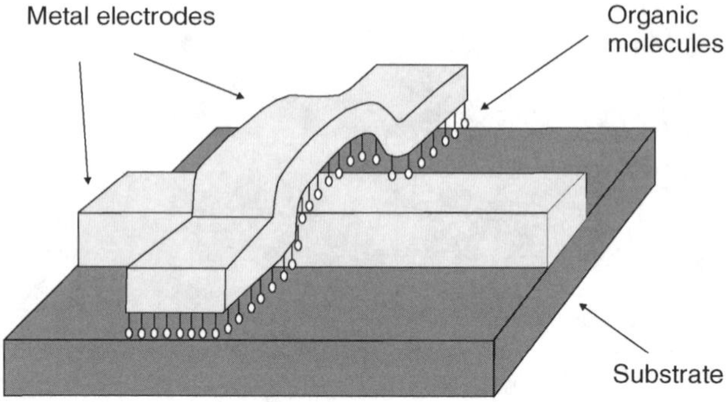

Figure 11.8 Schematic diagram of cross-bar structure for electrical conductivity studies.

Many scenarios for future electronic devices envisage organic molecules as the 'switches' in computational architectures. Different approaches have been used to exploit the electrical properties of individual molecules or assemblies of molecules. Perhaps the simplest method is that of a cross-bar arrangement (Figure 11.8), in which a layer of molecules (deposited by, say, chemical self-assembly) is sandwiched between metallic stripes. One immediate problem is that of depositing the top electrode without damaging the organic layer. However, there are other issues to be faced by the experimentalist using this deceptively simple device structure. For example, the electrodes may possess surface oxide layers that have similar thicknesses to the organic layer (e.g. aluminium oxide on aluminium). The electrical resistances of such surface layers can be high, leading to large contact resistances. If dissimilar electrodes are used to provide the top and bottom contacts (e.g. gold for the bottom contact and evaporated aluminium for the top), then the differences in work functions of the metals will lead to some asymmetry in the current versus voltage characteristics of the devices. Even if the same electrode material is used for both the top and bottom electrodes, it cannot be assumed that work function differences will be eliminated. The work function of a metal film will depend on the method of preparation (Chapter 9, Section 9.2.1). For example, the work function of gold deposited on glass may differ from that of the same metal deposited on an organic material [12].

The electrical conductivity of many organic compounds is also affected by the presence of moisture and/or oxygen in the atmosphere. Certain chemical sensors (Chapter 10) exploit this fact. Stringent precautions are taken to encapsulate plastic electronics devices, such as field effect transistors and displays, to increase their lifetimes. Similar approaches are essential when exploring the electrical behaviour of assemblies of organic molecules. The samples must be stored and measured in dry and, in some cases, dark environments.

A further problem, common to all areas of microelectronics, is that of structural defects in the organic film. Although atomic force microscopy can reveal a very high degree of molecular order over distances of many nanometres (for example, Chapter 6, Section 6.8) the order may not be preserved as the scale is increased. Defects, such as pinholes and grain boundaries, may act as nucleation points for the growth of metallic filaments.

Many reports of 'unusual' electric behaviour in organic thin films have appeared in the literature over the second half of the 20th century. Some of these are almost certainly the result of the poor quality of the organic layer and/or the lack of attention to the issues

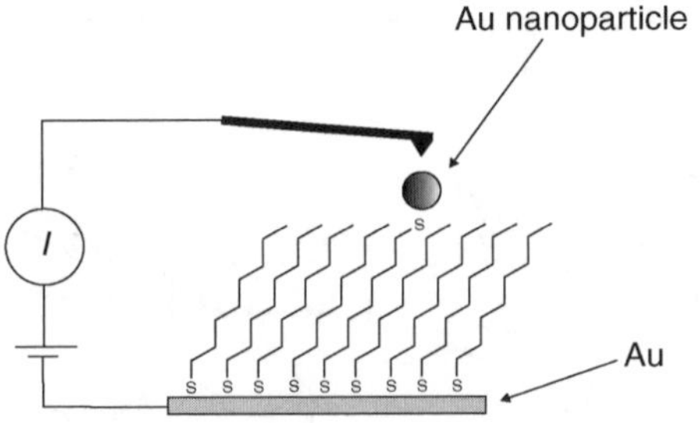

Figure 11.9 Representation of an experimental arrangement for the measurement of the conductivity of a single molecule. After Cui *et al.* [13].

discussed above. Before any electrical data on assemblies of organic molecules are interpreted, a fundamental question that should always be asked is whether these electrical characteristics are a property of the organic molecules, of the metallic electrode or of the presence of any interfacial (e.g. oxide) layer (or a combination of these).

Other methods of measuring the electrical properties of molecular structures use the tip of a scanning probe microscope to provide the top contact. This circumvents some of the difficulties outlined above. Figure 11.9 shows a method in which a monolayer on a gold surface is used in conjunction with gold nanoparticles [13]. A self-assembled monolayer of octanethiol on a gold surface is first obtained. Molecules of 1,8-octanedithiol are then inserted into the octanethiol monolayer using a replacement reaction whereby one of the two thiol groups becomes chemically bound to the gold surface. The octanethiol monolayer acts as a molecular insulator, isolating the dithiol molecules from one another. The thiol groups at the top of the film are then derivatized by incubating the monolayer with a suspension of gold nanoparticles. A gold-coated conducting AFM probe is finally used to locate and contact individual particles bonded to the monolayer. Measurements on over 4000 nanoparticles produced only five distinct families of curves. Figure 11.10 shows representative curves from each family. The curves correspond to multiples of a fundamental curve, which is ascribed to a situation in which a single dithiol molecule links the gold nanoparticle to the underlying gold substrate. In the low-voltage region (between ± 0.1 V) the single molecule has a resistance of 900 ± 50 MΩ.

Alternative measurement techniques use molecular assemblies in contact with liquid electrodes (e.g. Hg or liquid electrolytes) and take advantage of electrochemical principles. The combination of the self-assembly process with molecular recognition offers a powerful route to the development of nanoscale systems that may have technological applications as sensors, devices and switches. For example, the complexation of a neutral or ionic guest at one site in a molecule may induce a change in the redox properties of the system. Figure 11.11 shows a schematic arrangement of a self-assembled layer that includes a derivative of the charge-transfer molecule TTF (Chapter 3, Section 3.4.1). The incorporation of the metal-binding macrocycle is to enable the molecule to function as a metal cation sensor. Monolayers assembled onto platinum have been shown to exhibit electrochemical recognition to Ag^+ ions [14].

One important property of any switching element is that of the switching speed. This is determined by the *RC* time constant of the device. The limiting factors, restricting the switching speed and bandwidth of electronic components, are the carrier mobilities in the

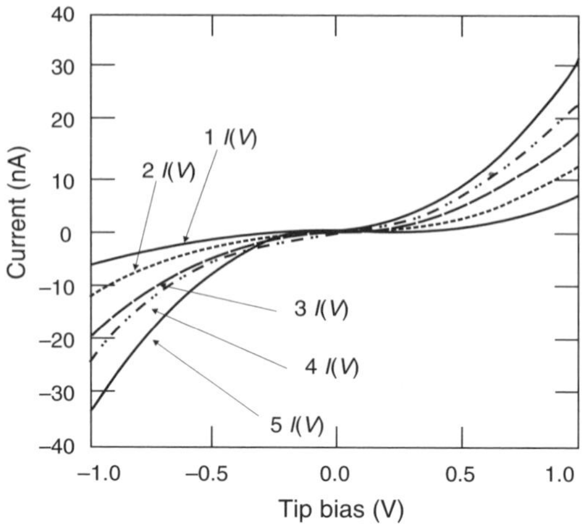

Figure 11.10 Current versus AFM tip bias, *I*(*V*), curves measured with the equipment shown in Fig. 11.9. The five curves shown are representative of distinct families, *NI*(*V*), that are integer multiples of a fundamental curve $I(V)$ $(N = 1, 2, 3, 4$ and $5)$. From *Science*, 2001, **294**, Cui XD, Primak A, Zarate X, Tomfohr J, Sankey OF, Moore AL, Moore TA, Gust D, Harris G, Lindsay SM, 'Reproducible measurement of single-molecule conductivity', pp. 571–574. Reprinted with permission from The American Association for the Advancement of Science.

materials and the parasitic capacitances (requiring constant charging and discharging) associated with the device structure. In inorganic semiconductor devices, these parasitics are not small, as a result of the high dielectric constants and very thin junction depletion regions. However, silicon transistors can still switch fast because the relatively high charge carrier mobilities and the high current densities can be achieved (exceeding 100 MA m^{-2}). For organic semiconductors, time constants are not expected to be high as the charge carrier mobilities in organic materials are relatively small (Chapter 3, Section 3.4.1).

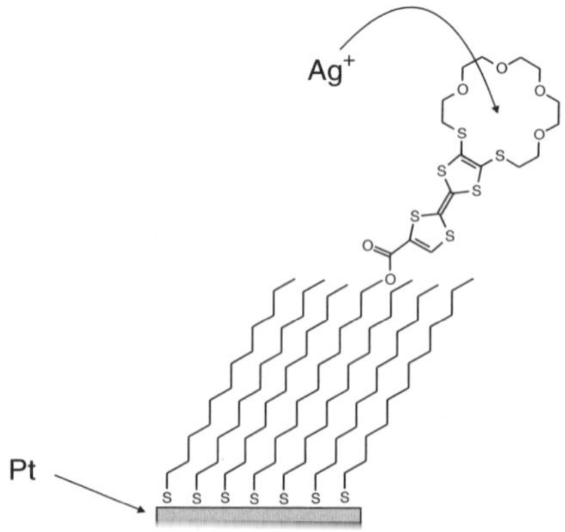

Figure 11.11 A self-assembled film containing a cation-senstitive, redox-active molecule [14].

11.5 MOLECULAR RECTIFICATION

The concept of molecular rectification, i.e. asymmetric current versus voltage behaviour within a molecule, was predicted in 1974 by Aviram and Ratner [15]. The proposal was that an asymmetric organic molecule containing a donor and an acceptor group separated by a short σ-bonded bridge, allowing quantum mechanical tunnelling, should exhibit diode characteristics. An example of such a molecule, together with the energy band structure, is given in Figure 11.12. The ground (HOMO) and excited (LUMO) states of the molecules are E_{D1}, E_{D2} (for the donor) and E_{A1}, E_{A2} (acceptor), respectively. The presence of the σ-bridge is essential to the device operation. If the electronic systems of the donor and acceptor are allowed to interact strongly with one another, a single donor level will exist on the timescale of the experiment. The donor and acceptor sites should effectively be insulated from one another in order for the device to function.

If the molecule is sandwiched between two metallic electrodes, the passage of electrons through the device can be considered as a three-step process: from metal cathode to acceptor, from acceptor to donor and finally from the donor to the anode. As the applied voltage is increased (the acceptor side of the molecule is biased negatively with respect to the donor side), electrons will be transferred from the cathode to the acceptor molecule (to energy state E_{A2}) and from the donor (from E_{D1}) to the anode. Electrons will move from the acceptor molecule LUMO level (E_{A2}) to the donor HOMO level (E_{D1}) by the process of quantum mechanical tunnelling (Chapter 3, Section 3.5.2) if the σ bridge is sufficiently short. If the direction of the applied bias is reversed, conduction can only take place at much higher voltages, i.e. the HOMO level of the donor would have to be higher than the LUMO level of the acceptor. This leads to predicted diode characteristics.

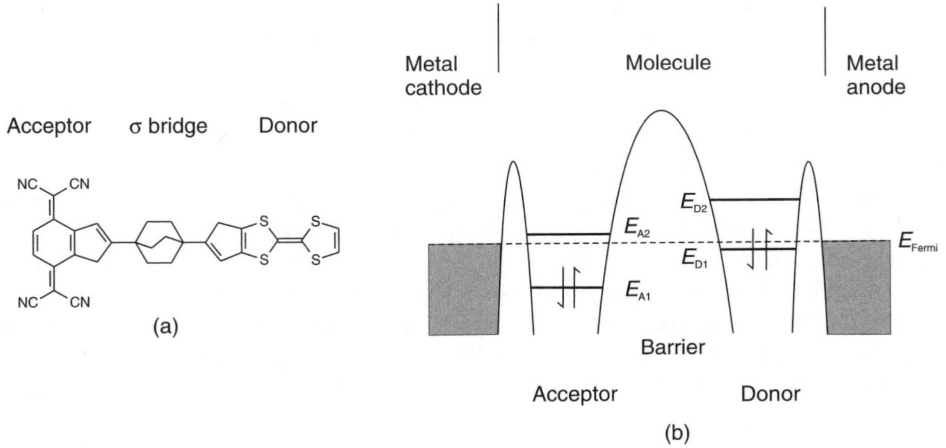

Figure 11.12 (a) An example of a molecular rectifier molecule. (b) Energy band structure of molecular rectifier. Reprinted from *Chem. Phys. Lett.*, **29**, Aviram A, Ratner MA, 'Molecular rectifiers', pp. 277–283, Copyright (1974), with permission from Elsevier.

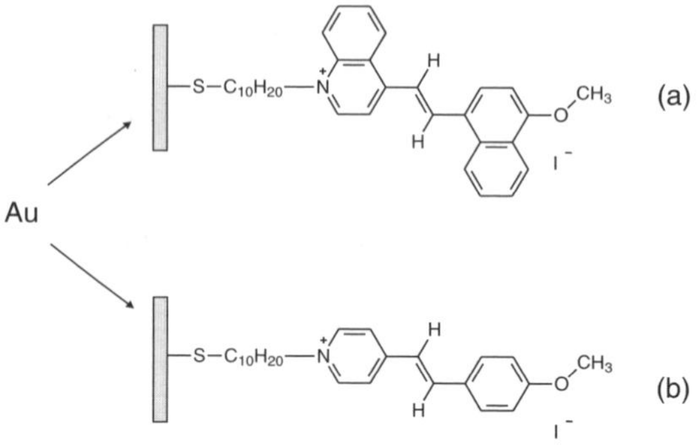

Figure 11.13 Chemical structures of self-assembled monolayers of (a) rectifying and (b) nonrectifying molecules. The counter ion in each case is iodide. From Ashwell GJ, Mohib A, Miller JR, 'Induced rectification from self-assembled monolayers of sterically hindered π-bridged chromophores, '*J. Mater. Chem.*, **15**, 1160–1166. Copyright (2005). Reproduced by permission of the Royal Society of Chemistry.

There have been many attempts to demonstrate this effect in the laboratory, particularly in organic thin films [16–19]. Asymmetric current versus voltage behaviour has certainly been recorded for many metal/insulator/metal structures, although these results are often open to several interpretations as a result of the experimental issues highlighted in the previous section. For example, the asymmetry in the electrical characteristics may originate from work function differences or from Schottky barrier (Chapter 9, Section 9.2.1) formation between the organic material and one of the electrodes. A key test is to relate the observed rectification to a property of the molecule under test. Figure 11.13 shows the chemical structure of two similar self-assembling monolayers consisting of a bulky 4-methoxy-naphthyl donor that is linked via a –CHCH– π-bridge to a bulky quinolinium acceptor [19]. The steric hindrance provided by the donor and acceptor of the molecule depicted in Figure 13(a) enforces the required nonplanarity, which, in turn provides an effective electron tunnelling barrier between the electroactive ends of the molecule. Self-assembled monolayers of this molecule exhibit rectification ratios of about 30 at $\pm$ 1 V, as shown in Figure 11.14. Significantly, the electrical asymmetry is suppressed in self-assembled monolayers of a less bulky analogue [Figure 11.13(b)].

The forward current in a molecular rectifier is a result of quantum mechanical tunnelling through the σ-bonded barrier, and the response of this current will be essentially instantaneous. However, rectifiers not only must conduct when they become forward biased, but also cease to conduct when they are reverse biased. The device current will include a component flowing through the parasitic capacitance which is inherently present. This will not be a problem at low switching frequencies, but will limit the maximum frequency for which the device has useful characteristics. An estimate of the limiting frequency of the device proposed by Aviram and Ratner – about 1.5 kHz – can be obtained from the *RC* time constant of the device and by using some of the original parameters [20]. The key factor is the current density through the device. Even if the parasitic capacitance is large, it can be charged and discharged faster if the device current is increased. Further increases lead to further reductions in switching time, up to as point where transit delays due to charge carrier movement start to dominate.

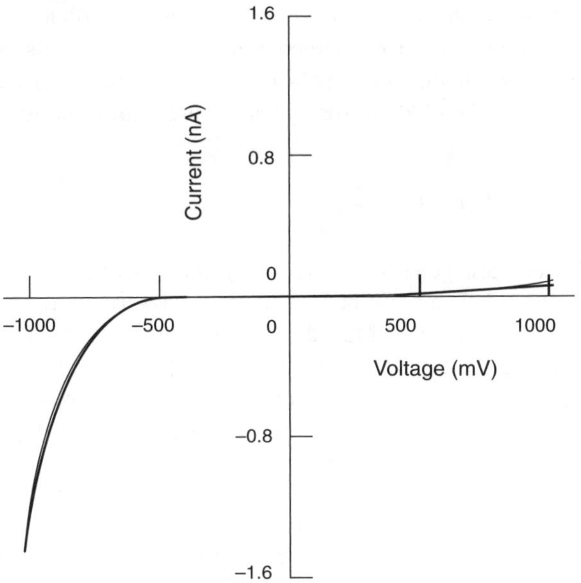

Figure 11.14 Current versus voltage characteristics of the iodide salt of the molecule depicted in Fig. 11.13(a). The polarity relates to the substrate electrode and the higher forward bias current corresponds to electron flow from the gold- coated substrate to the contacting tip. From Ashwell GJ, Mohib A, Miller JR, 'Induced rectification from self-assembled monolayers of sterically hindered π-bridged chromophores, '*J. Mater. Chem.*, **15**, 1160–1166. Copyright (2005). Reproduced by permission of the Royal Society of Chemistry.

11.6 ELECTRONIC SWITCHING AND MEMORY DEVICES

Memories represent by far the largest part of electronic systems; for example, memories in Pentium chips occupy more that 70% of the chip area. Memory devices are inherently *volatile* or *nonvolatile*. In the former case, the information is lost if the power supply is removed. In silicon technology, examples of volatile memory are the *static random access memory* (SRAM) and the *dynamic random access memory* (DRAM). Data stored in the cells of volatile memory must periodically be refreshed. In contrast, *flash memory* (the term 'flash' refers to the erase process, in which blocks of memory are erased in one operation) is non-volatile, storing information, in the ideal case indefinitely, if the power is turned off. In addition, flash memory offers fast read access times (although not as fast as volatile DRAM memory) and better shock resistance than hard disk memories.

It is expected that the majority of nonvolatile memories in the early part of the 21st century will be based on flash technologies. Moore's law (Chapter 1, Section 1.3.2) will continue to drive transistor-based memory technology scaling but technology complexity will increase. For example, as flash memory devices begin to move into the sub-100 nm lithography regime, continued scaling will become difficult due to the high electric fields required for the programming and erase operations and the stringent requirements for long-term charge storage. These demands are imposing fundamental scaling limitations on the memory cell operating voltages and on the physical thickness of the tunnelling dielectric. Overcoming such restrictions will require innovations in cell structure and device materials. A future scenario can be envisaged where logic is implemented in CMOS (or in polymer FETs, in the case of plastic electronics) and the memory is added to the system using a different technology.

11.6.1 Resistive Bistable Devices

Organic memory devices are generally formed by interposing thin layers organic layers between two electrodes [21]. The cross-bar arrangement described in Section 11.4 is popular as it is highly scalable. The data storage density provided by this architecture is limited simply by the cross-point area. A variation uses the tip of a scanning probe microscope instead of a (thermally evaporated) metallic top electrode, leading to ultrahigh data storage density. Write-once/read-many-times organic memory can be fabricated by burning polymer 'fuses'. For example, a thin conductive polymer layer, or fuse, of poly(3,4-ethylenedioxythiophene) (PEDOT; Chapter 5, Section 5.3) is sandwiched between two electrodes [22]. The as-deposited device shows high conductance due to the conductive polymer, but when a burning-voltage pulse is applied (the write process), the PEDOT fuses will 'blow' and cause the device to be in an open-circuit condition. The write process can be as short as microseconds, depending on the thickness of the PEDOT layer and the amplitude of the voltage pulse.

An AFM-based data storage concept (called the 'Millipede') that has potentially ultrahigh density has been developed by IBM [23, 24]. Indentations, 30–40 nm, are made in a thin (50 nm) poly(methyl methacrylate) (PMMA) layer, resulting in a data storage density of 2500–3000 Gbits cm^{-2}. The very large-scale integration of micro-nanomechanical devices on a single chip leads to an impressive two-dimensional arrangement of AFM cantilevers (100 mm^{-2}) with integrated read and write functionality.

Obtaining bistable memory devices poses is a greater challenge. There are numerous examples of functional organic molecules exhibited switching behaviour when sandwiched between metallic electrodes but, as noted in Section 11.4, the physical origin of such effects are not always clear. The creation and dissolution of metal filaments between the electrodes was probably responsible for some of the reported switching data [25]. Bistable rotaxane molecules have been used in a number of these studies; an example is shown in Figure 11.15 [26, 27]. This molecule is amphiphilic and the ring component can move between the polar

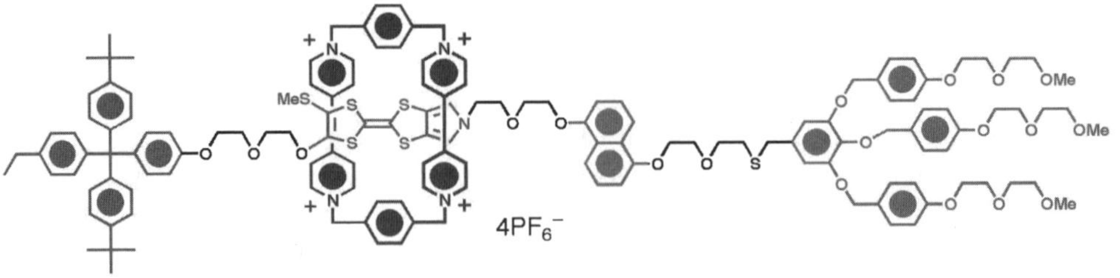

Figure 11.15 Bistable rotaxane structure. The 'ring' component can move between the hydrophobic sulfur-rich region shown centrally to the hydrophilic 'stopper' region on the far right of the molecule [26].

(right-hand side of the molecule) and nonpolar (central sulfur-rich group) regions of the main part of the molecule. The molecules can be assembled on an electrode using the LB approach and a top electrode then deposited to form a cross-bar structure. If the molecules are deposited on nonmetallic substrates (e.g. Si or carbon nanotubes), the absence of switching using 'control' compounds suggests that the rotaxane molecule itself is responsible for the bistability [28]. A 160 000-bit molecular electronic memory circuit, fabricated at a density of 10^{11} bits cm^{-2}, has been fabricated from such bistable rotaxanes [29]. Although the circuit possesses a large number of defects, these could be readily identified through electronic testing and isolated using software coding. The working bits were then configured to form a fully functional random access memory.

Charge-transfer storage at metal/organic interfaces [21] and charge transfer between organic molecules and metallic nanoparticles [30, 31] have also been suggested as operating principles of organic nonvolatile memory. Figure 11.16 shows an example of the latter. The device structure, depicted in Figure 11.16(a), consists of three layers (organic semiconductor/ultra-thin Al layer/organic semiconductor) sandwiched between aluminium. The particular organic semiconductive compound used was tris(8-hydroxyquinoline)

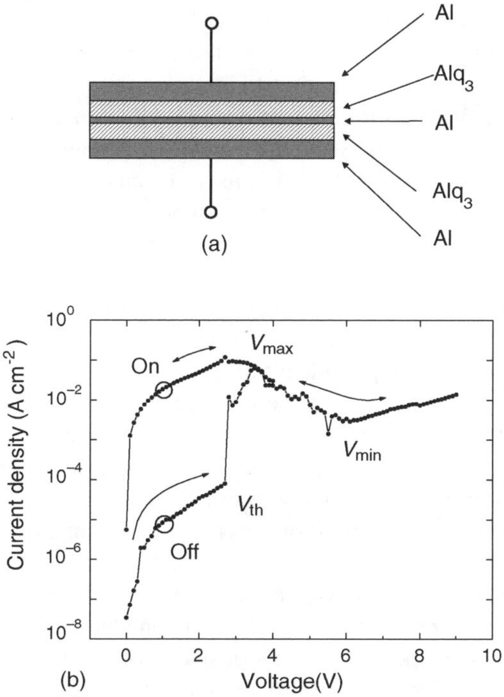

Figure 11.16 Electrical behaviour of an organic electrical bistable device. (a) The structure is based on the organic semiconducting compound Alq_3 and consists of Al (50 nm)/Alq_3 (50 nm)/Al (5 nm)/ Alq_3 (50 nm) /Al (50 nm). (b) Current density versus voltage characteristics. The 'on' and 'off' states are set at voltages close to V_{max} and V_{min}, respectively. The transition from the off to the on state occurs at the threshold voltage V_{th}. Reprinted from Bozano LD, Kean BW, Beinhoff M, Carter KR, Rice PM, Scott JC, *Appl. Phys. Lett.* **84**, 'Mechanism for bistability in organic memory elements', pp. 607–609. Copyright (2004) American Institute of Physics.

aluminium (Alq_3) (a compound used on organic display devices; Chapter 9, Section 9.6). Starting in the 'on' state, the current versus voltage curve is N-shaped, so that there is a local maximum in the current at V_{max} and a local minimum at V_{min}. Beyond this value, the current density increases with an increase in voltage. Between V_{max} and V_{min} is a (rather noisy) regime of *negative differential resistance* (NDR). It is suggested [30] that the mechanism responsible for switching is similar to that reported for electroformed metal–insulator–metal (MIM) diodes [32] and may also be similar for other organic sandwich structures [33–34]. For the 'on' state of the device, the transport levels in the Alq_3 and the trapping levels associated with the sandwich structure are uncharged. At low voltages, charge is injected primarily from one of the electrodes and moves through the transport states. In the NDR regime, a space charge builds up in the device as charges are trapped in deep levels. This opposes the field at the injecting electrodes and reduces the current. If the voltage is reduced rapidly to zero, charge is left in the traps. However, alternative ideas suggest that the device operation is based on two-dimensional single-electron tunnelling between nanometre-sized metal islands formed by the penetration of the top metal electrode during thermal evaporation [35]. The fact that similar electrical switching characteristics can be obtained for organic thin structures that have no intentionally added nanoparticles indicates that the device fabrication process may hold a clue to its operation [35–37]. Bistable devices that do not include nanoparticles (i.e. based on a single layer of an organic compound) are preferable as their presence will restrict the device scaling (when the cell dimensions become comparable to those of the nanoparticles).

Memory devices can also exploit the movement of atoms. For example, it has been proposed to fabricate a high-density memory using the electrostatic attraction between suspended, crossed carbon nanotubes [38]; this is discussed further in Section 11.10. A further device, based on silver sulfide (Ag_2S), works by controlling the formation and annihilation of an atomic bridge at the crossing point between two electrodes [39, 40]. Figure 11.17 shows a schematic diagram of one such memory cell. A 1 nm thick silver layer deposited on top of the Ag_2S layer disappears into the sulfide layer when a current flows from the platinum electrode to the silver electrode [Figure 11.17(a)]. This results in loss of contact between the two electrodes and initializes the device. A bridge of silver atoms is locally formed by applying a voltage of the opposite sign, re-establishing contact between the silver sulphide and the platinum [Figure 11.17(b)]. The conductance through the device can be as small as one quantum unit of conductance, suggesting that the silver bridge can touch the platinum electrode with just one atom. Although this structure is not strictly a molecular electronic device as it is based on inorganic compounds, the principle of operation may have wider applicability.

If molecular components are to be used as functional elements in place of semiconductor-based devices, they must compete with inorganic semiconductors under the extreme conditions required for processing and operating a practical device. There is some concern whether organic materials possess adequate stability to meet the extreme performance conditions required for any type of practical device, e.g. high-temperature processing steps during manufacture ($\sim$400 °C); relatively high-temperature operating conditions (up to 140 °C); and very large numbers of operational cycles over a lifetime ($\sim 10^{12}$). However, as discussed in Chapter 5, there are organic compounds that exhibit good chemical and thermal stability. In this respect, results exploiting the redox behaviour of self-assembled porphyrin molecules are encouraging [41]. Here, information is stored in the discrete oxidation states of the molecules. (See also the discussion in Section 11.8.3 on chemical switching.)

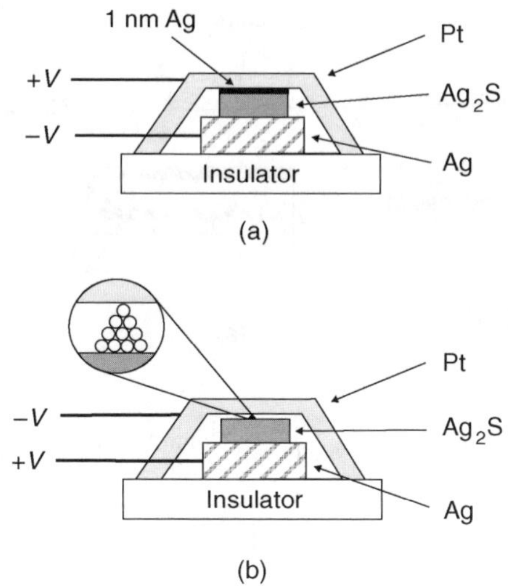

Figure 11.17 A rewritable memory bit based on the properties of a silver sulfide mixed ionic conductor. (a) A 1 nm thick silver layer deposited on top of the Ag_2S layer is incorporated into the sulfide layer when the Pt electrode is positively biased. (b) A bridge of Ag atoms is formed locally when the Pt electrode is negatively biased. Reprinted by permission from Macmillan Publishers Ltd: *Nature*, **433**, J van Ruitenbeek, 'Silver nanoswitch', pp. 21–22, Copyright (2005).

11.6.2 Flash Memories

The section above has shown that charge storage at interfaces can be exploited in a resistive switching device. Charge storage is also the basis of the flash memory [42]. This device is similar in structure to a MOS field effect transistor (MOSFET), except that it has two gate electrodes, one on top of the other. The top electrode forms the control gate, below which a 'floating gate' is capacitatively coupled to the control gate and the underlying silicon. The memory cell operation involves putting charge on the floating gate or removing it, corresponding to two logic levels. Nanoflash devices utilize single or multiple nanoparticles as the charge storage elements. These are usually embedded in the gate oxide of a field effect transistor and located in close proximity to the transistor channel. Figure 11.18(a) depicts a simple version of this device – a metal/insulator/semiconductor or MIS structure (Chapter 9, Section 9.3); the potential energy variation (for either electrons or holes) across the structure is depicted in Figure 11.18(b) [43]. A very thin (<3 nm) barrier between the substrate and the nanoparticles allows charge (electrons or holes) to move between the substrate and the nanoparticles by the process of quantum mechanical tunnelling. This charge is stored on the nanoparticles, resulting in a shift in the capacitance versus voltage characteristics (Chapter 9, Section 9.3.2) of the metal/insulator/semiconductor device [44].

Figure 11.19 shows details of a hybrid silicon/organic MISFET device incorporating a self-assembled layer of gold nanoparticles [45]: Figure 11.19(a) shows a schematic diagram of the self-assembled layer and Figure 11.19(b) is a photograph of the final device. In the self-assembly process, an oxidized silicon surface is first functionalized with a long-chain

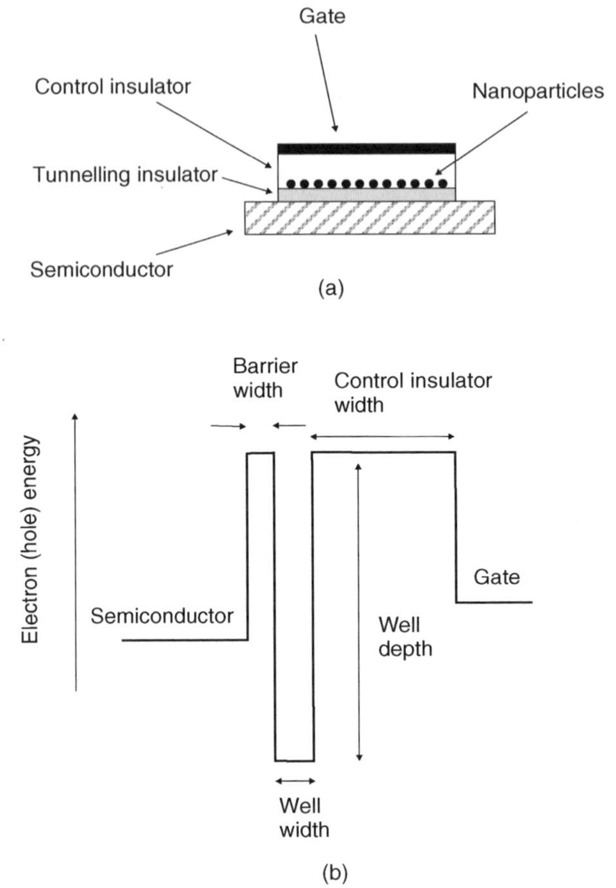

Figure 11.18 (a) Schematic diagram showing the structure of a Flash memory. (b) Potential energy band diagram for electrons (or holes) across the device. From *J. Electron. Mater.*, **34**, 'Self-assembly of metal nanocrystals on ultrathin oxide for non-volatile memory applications', Lee C, Meteer J, Narayan V, Kan EC, p.2. Copyright (2005), with kind permission of Springer Science and Business Media.

amine derivative. This surface is exposed to a solution of carboxylic acid-derivatized gold nanoparticles. Langmuir–Blodgett layers of cadmium arachidate are then used to provide the gate insulator. Testing the memory characteristics of the devices consists of applying positive or negative voltage pulses successively on the gate of a previously unstressed device, keeping the source and drain electrodes grounded (i.e. at 0 V). The voltage pulse height is progressively increased while the pulse duration is kept constant. The injected (rejected) charges into (out of) the gold nanoparticles causes a shift of the transistor threshold voltage, V_t (Chapter 9, Section 9.4), to higher or lower voltages. The high-V_t state is usually called the 'write' state and the low-V_t is the 'read' state. The final device exhibits a clear memory 'window' in Figure 11.20, which is a plot of the threshold voltage shift ΔV_t as a function of the write/erase (programming) voltage $V_{W/E}$. No significant programming window is evident with a pulse period below 1 s. Since no device hysteresis was observed for a gate-voltage

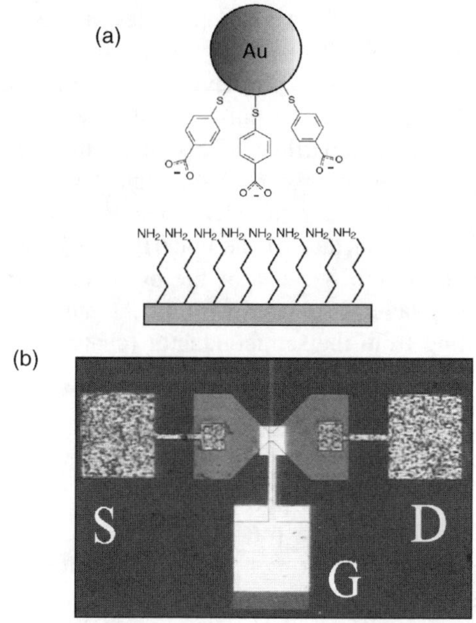

Figure 11.19 (a) Self-assembly of gold nanoparticles on to the surface of functionalized Si/SiO_2. (b) Optical image of FET memory device. Reprinted from *Microelectron. Eng.* **73–74**, Kolliopoulou S, Dimitrakis P, Normand P, Zhang HL, Cant N, Evans SD, Paul S, Pearson C, Molloy A, Petty MC, Tsoukalas D, & 'Integration of organic insulator and self-assembled gold nanoparticles on Si MOSFET for novel non-volatile memory cells', pp. 725–729, Copyright (2004), with permission from Elsevier.

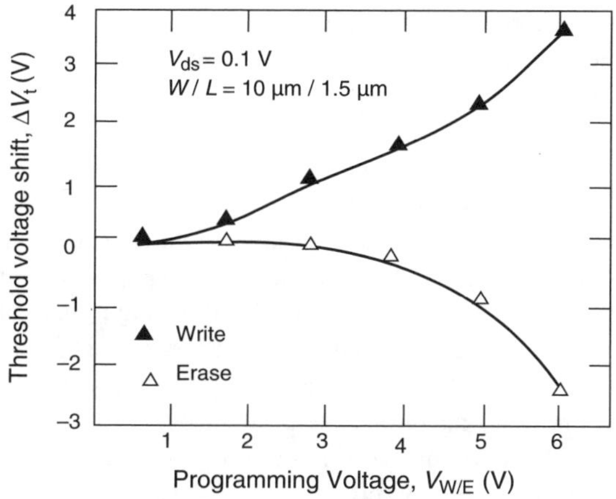

Figure 11.20 Write/erase memory window after application of gate voltage pulses with 1 s duration. V_{ds} = drain/source voltage; W = gate width; L = gate length. $V_{W/E}$ = write/erase (programming) voltage. Reprinted from *Microelectron. Eng.* **73–74**, Kolliopoulou S, Dimitrakis P, Normand P, Zhang HL, Cant N, Evans SD, Paul S, Pearson C, Molloy A, Petty MC, Tsoukalas D, 'Integration of organic insulator and self-assembled gold nanoparticles on Si MOSFET for novel non-volatile memory cells', pp. 725–729, Copyright (2004), with permission from Elsevier.

cycle for reference samples without the Au nanoparticles, it was concluded that these were responsible for the charge storage effects. The application of positive pulses to the gate lowers the threshold voltage of an unstressed device, indicating that electrons are extracted from the nanoparticles to the gate. Negative pulses increase V_t due to electron injection from the gate metal on to the gold nanoparticles. This mechanism reveals that the transistor channel does not contribute to the charge exchange and hence cannot affect the memory properties of the hybrid device. Although the SiO_2 layer used in the work was relatively thin (5 nm), the distance between the surface of the underlying silicon and the gold particles is effectively increased to over 10 nm because of the presence of the amine functionalizing layer and the organic capping layer associated with the Au nanoparticles. This prevents easy charge transfer via tunnelling from the semiconductor (channel) to the Au.

11.6.3 Spintronics

Spintronics (an acronym for SPIN TRansport electrONICS) exploits the spin of an electron to process and store digital information [46–49]; the subject is also called *magnetoelectronics.* The simplest method of generating a spin-polarized current is to inject the current through a ferromagnetic material. Much of the interest in the area started with IBM exploiting magnetically induced resistance or magnetoresistance in the early 1990s. The most successful spintronic device to date is the *spin valve.* This uses a layered structure of thin films of magnetic materials, which changes electrical resistance depending on applied magnetic field direction. When the two magnetization vectors (Chapter 5, Section 5.7.1) of the ferromagnetic layers are aligned, then a relatively high current will flow, whereas if the magnetization vectors are antiparallel, then the resistance of the system is much higher. The magnitude of the change [(antiparallel resistance − parallel resistance)/parallel resistance] is called the giant magnetoresistance (GMR) ratio. Devices have been demonstrated with GMR ratios as high as 200%, with typical values greater than 10%. This is a vast improvement over the anisotropic magnetoresistance effect in single-layer materials, which is usually less than 3%.

Figure 11.21 shows a schematic diagram of a non-volatile memory device based on a magnetic tunnel junction. This has been developed by IBM and other companies and exploits a spin-dependent tunnelling phenomenon. The devices have two ferromagnetic layers separated by a thin insulating barrier. The first layer polarizes the spins of current-carrying electrons, which cross the barrier to the second layer by quantum mechanical tunnelling when both layers are magnetically aligned. When the magnetism of the second layer is reversed, the tunnelling is reduced. The magnitude of the current in the memory cell can be used to indicate a '0' or a '1'. Magnetic random access memory (*MRAM*) based on such spintronics principles is now going into mass production. Early devices (2006) were based on 180 nm process technology. Although this memory is significantly more expensive than Flash or DRAM, it compensates by combining the functions of both. There are also developments that suggest that MRAM can be scaled to 60 nm and below. The most notable is the discovery of the *spin-momentum-transfer effect* [47]. This exploits the net angular momentum that is carried by a spin-polarized current and the transfer of this momentum to the magnetization of the second layer. It offers the potential of orders of magnitude lower switching currents and therefore a much lower energy for writing. The spin-momentum-transfer effect becomes important when the minimum dimension of the memory cell is less

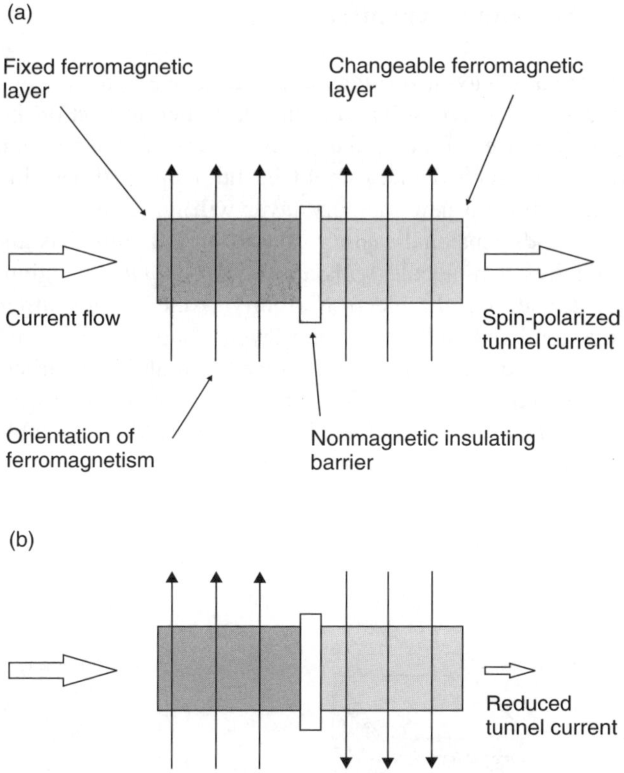

Figure 11.21 Schematic diagram of a magnetic tunnel junction. (a) The fixed ferromagnetic layer polarizes the spins of current-carrying electrons, which cross the barrier to the second layer by quantum mechanical tunnelling when the magnetism of both layers is aligned. (b) When the direction of magnetism in the second layer is reversed, the tunnelling is reduced.

than 100 nm and becomes more efficient as the cell size is reduced (the opposite to what occurs with the use of conventional magnetic field switching).

There are also many proposals for a transistor that works by switching electronics between spin states. In one option, a gate controls whether most of the electrons have the same spin state. It is also possible to exploit magnetic domain wall motion (Chapter 5, Section 5.7.1) to form computational logic elements. For this case, information is stored through the presence or absence of a domain wall in a linear array of domain walls in a magnetic thin film loop confined to a channel in a silicon chip. Such a memory might eventually provide a solid-state replacement for the magnetic hard drive used in computers.

Although most of the materials developed for spintronic devices are inorganic, a case has been made for exploiting organic compounds [48]. This is based on the weak spin–orbit and hyperfine interactions in organic molecules, which leads to the possibility of preserving spin coherence over times and distances much longer than in conventional metals or semiconductors. Spin valves using a tris(8-hydroxyquiniline)aluminium (Alq_3) (Chapter 9, Section 9.6) layer sandwiched between ferromagnetic electrodes have been shown to possess a GMR value of about 5% at low temperatures [49].

11.6.4 Three-Dimensional Architectures

Molecular-scale electronics may also offer increased device densities by fabricating three-dimensional architectures. The cross-bar structure described in Section 11.4 offers a simple means of realizing this. For example, if the 'active' material can be deposited in thin film form at low temperatures, such devices could be built up in three dimensions since the additional processing to form a new memory layer will not affect the underlying layer(s). The principle of a three-dimensional memory based on LB films has also been suggested [50]. The device requires a molecule with a central conjugated region of high electron affinity (for an n-type material, the electron affinity is the energy difference between the bottom of the conduction band and the vacuum level; Chapter 3, Section 3.4.1) surrounded by aliphatic subsitituents of low electron affinity. A multilayer structure, as shown in Figure 11.22, might be used to store one *N*-bit word, the presence or absence of charge on the *n*th layer representing a 0 or 1 of the *n*th bit. The LB film could be assembled on the gate of a FET and, on application of an electric field, transport of bits across the layers may be detected as induced charge on the gate.

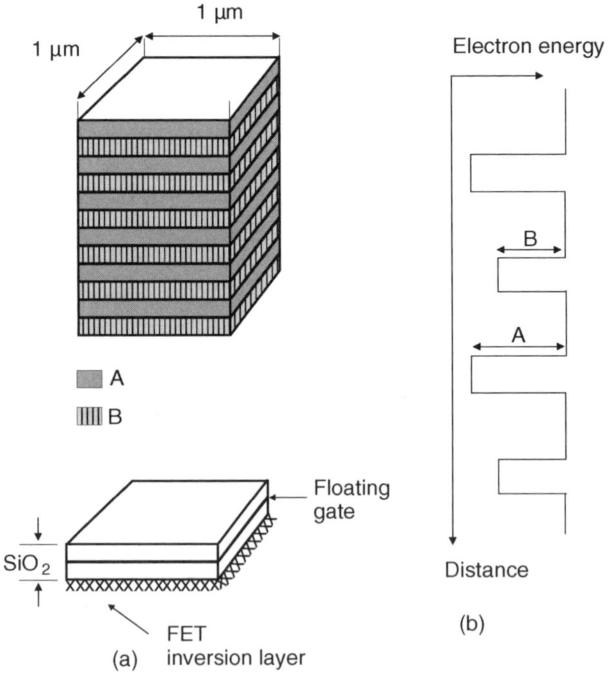

Figure 11.22 (a) Molecular memory holding *N* bits (top) compared with a conventional silicon memory holding 1 bit (bottom). (b) Electron energy versus perpendicular distance for molecular memory with no applied electric field. Reprinted from *Thin Solid Films*, **179**, Burrows PE, Donovan KJ, Wilson EG, 'Electron motion perpendicular to Langmuir–Blodgett multilayers of conjugated macrocyclic compounds: the organic quantum well', pp. 129–136, (1989), with permission from Elsevier.

11.7 SINGLE-ELECTRON DEVICES

Single-electron devices are those that can control the motion of even a single electron and consist of quantum dots or nanoparticles associated with tunnel junctions [51]. However, the terminology can be confusing [52]. The name suggests that the devices work with only one electron which is not completely correct. Even a single atom contains a number of electrons and a metal- or semiconductor-based 'single-electron' device will have a huge number of electrons which have relevance to its operation. The principle of *Coulomb blockade* enables electrons to be localized on an isolated island or transported to one so that a precise number are transferred. In the single-electron transistor based on this effect, a capacitatively coupled gate electrode is used to control this transfer. The simplest device is the arrangement depicted in Figure 11.23. Electrons can tunnel between a reservoir and an island. This electron motion is controlled by the gate voltage. Moving an electron from the reservoir to the island is accomplished by applying a voltage to the gate V_g, where

$$V_g = \frac{e}{C} \tag{11.3}$$

where e is the electronic charge and C is the capacitance of the island due to the gate electrode. The charging energy E is given by the expression for the energy associated with a capacitor, i.e.

$$E = \frac{e^2}{2C} \tag{11.4}$$

The capacitance of a sphere scales with its radius, r:

$$C = 4\pi\varepsilon_r\varepsilon_0 r \tag{11.5}$$

(Students of electrostatics might think it unusual that an isolated sphere has a finite value of capacitance; however, the above equation can be easily demonstrated by deriving the capacitance between two concentric spheres and then allowing the radius of the outer sphere to become infinitely large.)

For a 1 μm island (sphere) in free space, $C \approx 5 \times 10^{-17}$ F and $E \approx 2 \times 10^{-22}$ J or about 1.4 meV. At 300 K, the thermal energy k_BT is about 25 meV, an order of magnitude larger.

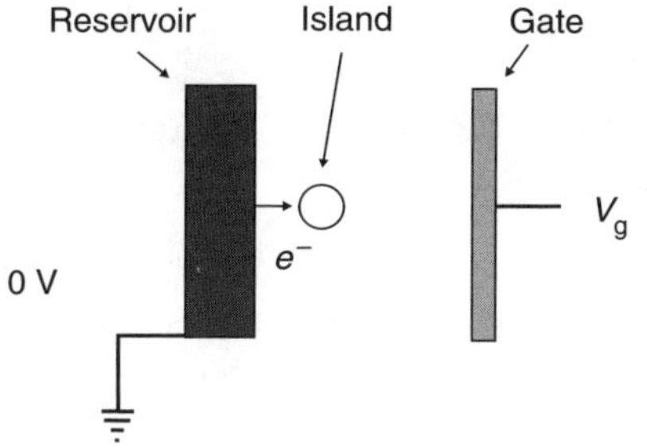

Figure 11.23 Principle of Coulomb blockade. Electronic charge is isolated on an 'island.'

However, if the island is very small (< 10 nm) or the temperature is very low, the charging energy becomes greater than the thermal energy and single electrons may be isolated on the island. Strong Coulomb repulsion – Coulomb blockade – will block the transfer of a second electron.

A second requirement necessary for the observation of Coulomb blockade is that the quantum fluctuations in the electron number n on the island are sufficiently small that the charge is localized on the island. This translates to a lower limit for the resistance of the tunnelling barrier, R_t.

If we consider the typical time to charge or discharge the island, Δt:

$$\Delta t = R_t C \tag{11.6}$$

The Heisenberg uncertainty principle (Chapter 3, Section 3.3.1):

$$\Delta E \Delta t \sim \frac{e^2}{C} R_t C \geq h \tag{11.7}$$

implies that R_t should be much larger than the resistance quantum (von Klitzing constant; Chapter 3, Section 3.3.8):

$$R_t \geq \frac{h}{e^2} \tag{11.8}$$

in order for the energy uncertainty to be much smaller than the charging energy.

By depositing small metal particles on an insulating self-assembled monolayer on a metal substrate, the tunnelling of single electrons through a metal–molecule–metal junction can be studied [53]. Using current versus voltage measurements on Pd clusters on a decanethiol monolayer on Au(111) at 77 K, Coulomb blockade could be observed; the experimental data are shown in Figure 11.24. The tunnelling of an electron to the metal particle is blocked until

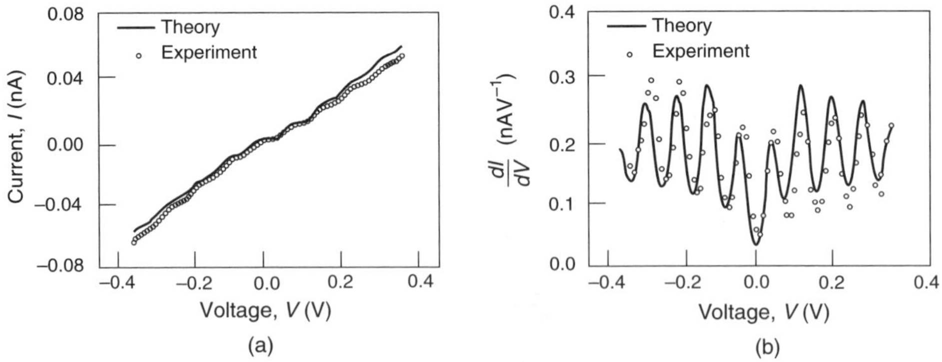

Figure 11.24 (a) Current versus voltage characteristic showing Coulomb staircase. (b) Derivative $\mathrm{d}I/\mathrm{d}V$ for the data presented in (a). In each case, the experimental points are shown as open circles, and the full curves represent the theoretical fits. The experimental I versus V curve is taken on top of a Pd cluster at 80 K. Reprinted from Oncel N, Hallbäck A-S, Zandvliet HJW, Speets EA, Ravoo BJ, Reinhoudt DN, Poelsema B, *J. Chem. Phys.*, **123**, 'Coulomb blockade of small Pd clusters', 004703. Copyright (2005) American Institute of Physics.

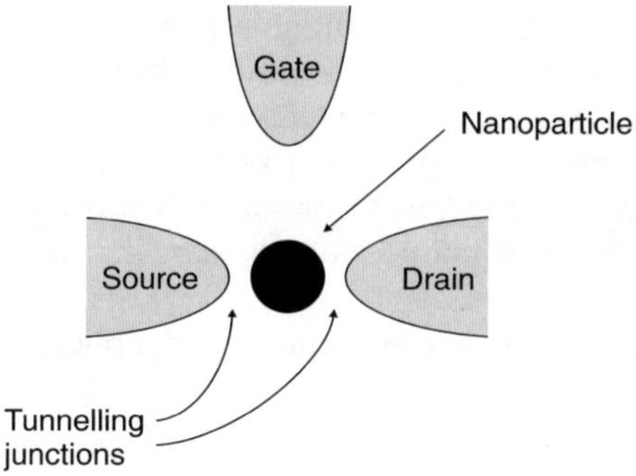

Figure 11.25 Schematic diagram of a single-electron transistor.

the applied voltage reaches a value equal to or higher than the charging energy of the particle. As a consequence, the current will increase in a stepwise manner, and the current versus voltage curve will display the shape of a staircase, the so-called *Coulomb staircase* depicted in Figure 11.24(a). Measurements taken with uncoated 'reference' monolayer exhibit no such features. From the derivative of the I–V data, dI/dV [Figure 11.24(b)], the fractional charge on the cluster can be determined. As can be seen from the curves, the experiment and the theory are in good agreement.

Although a single-electron box can control the number of electrons in the island, it does not have the properties of a switching device. *Single-electron transistors* (SETs) are therefore three-terminal switching devices, which can transfer single electrons from source to drain; a schematic diagram of such a device is shown in Figure 11.25. To make a memory cell, the SET is coupled to a MOSFET and used to transfer precise electron numbers to its gate or to remove them. The MOSFET current is a function of the presence or absence of these electrons on the gate and has two levels that define memory states. It is possible to observe this memory effect with just one extra electron present on the memory node, but in practical cases 10–100 electrons may be used.

11.8 OPTICAL AND CHEMICAL SWITCHES

The three most important stimuli that can be used to switch a chemical compound are electrical energy (electrons or holes), light energy (photons) and chemical energy (in the form of protons, metal ions, specific molecules, etc.). We have already covered some examples from the first category in Section 11.6. Any molecular-level system that can be reversibly switched between two different states by use of an external stimulus can be exploited for storing information. In the case of a photochemical input, the most common switching processes are related to photoisomerization or photoinduced redox reactions. If electrochemistry is exploited, then the induced processes are, of course, redox reactions.

Switching processes can take place under thermodynamic or kinetic control. In the former, the molecule responding to the stimulus is in thermodynamic equilibrium with its surroundings, so that when the stimulus is removed the molecule reverts to its initial state.

An example is a fluorescence sensor, the emission intensity of which is modulated by the presence of a particular substrate. Kinetic control, which means that the two states are separated by some kinetic barrier, is usually found in systems responding to photonic stimulation. The kinetic control can operate over very different timescales, from picoseconds (for some electron excited states) to years (for some electrochromic systems). For systems under thermodynamic control, it is not possible to address a single molecule because of the rapid equilibration between the states.

The following sections contain some examples of optical and chemical switches in molecular materials.

11.8.1 Fluorescence Switching

Sensors based on the control of fluorescence have been described in Chapter 10, Section 10.4.6. Figure 11.26 illustrates the operating principle [54]. A potentially fluorescent unit, e.g. an anthracene moiety, is linked to an electron-donating group, e.g. an amine. Photo-induced electron transfer from the HOMO orbital of the donor will quench the fluorescence [Figure 11.26(a)]. However, when the HOMO level of the donor is bound to an appropriate molecule or ion (in the case of an amine, by protonation, for example), fluorescence can be observed because the HOMO level of the donor is reduced in energy and electron transfer can no longer occur [Figure 11.26(b)].

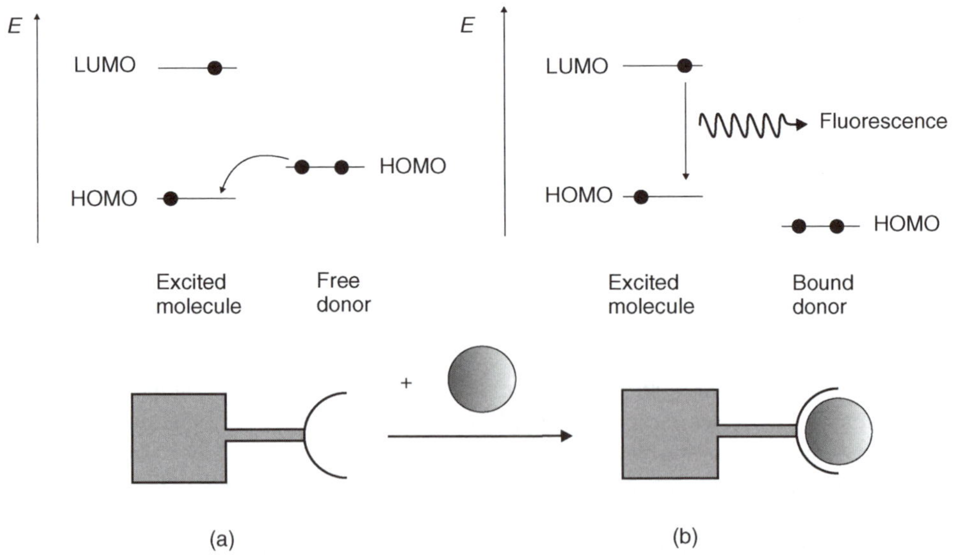

Figure 11.26 Representation of fluorescence quenching. (a) An electron from the HOMO level of a donor group quenches the fluorescence from an excited molecule. (b) The energy of the donor group is lowered as a result of a chemical interaction. The group can no longer donate electrons to the HOMO level of the excited molecule and fluorescence can be observed. Reprinted from *Molecular Devices and Machines*, Balzani V, Credi A, Venturi M, p. 239. Copyright (2003), with permission from Wiley-VCH.

11.8.2 Photochromic Systems

Photochromic materials reversibly change their absorption spectra, i.e. their colour, in response to incident light [55, 56]. In photochromic systems, the interconverting species are isomers, because the photoreaction simply causes rearrangement of the electronic and nuclear structure. A general photochromic reaction is depicted in Figure 11.27. Light of energy $h\nu_1$ causes switching of a stable isomer A to a higher energy isomer B. After photochemical conversion (a process that can be performed in a few femtoseconds), a spontaneous back reaction is expected to occur. This can be fast or slow, depending on the system. The reverse reaction can also occur by illuminating with a different frequency of light, ν_2, or by means of a thermal process, providing energy ΔE. Photochromic materials may also display *thermochromism* and *electrochromism.* In the former case, colour change is induced by heat that causes reorganization of the electronic structure (e.g. in spiropyrans) or dissociation (organometallic compounds). In electrochromism, the colour is modified or generated by an electric field; this is discussed in more detail further on in this section.

The above describes ideal photochromic behaviour. In practice, photochromic systems all exhibit 'fatigue' to a greater or lesser extent. The time required to switch between colourless and coloured species gradually increases and the maximum colouration is reduced after repeated cycling between A and B. Fatigue is caused by unwanted side reactions under the influence of light and heat. These effects become noticeable after any number of switching cycles, but some organic systems can survive more than 10^4 cycles without damage.

A large number of photochromic materials exist and these fall, broadly, into six separate classes, depending on their switching mechanism [55]: hydrogen tautomerism, dissociation, dimerization, *cis–trans* isomomerization, cyclization and charge transfer. Examples of two systems are shown in Figure 11.28. Figure 11.28(a) illustrates *cis–trans* isomerization (Chapter 2, Section 2.4.2) in the azobenzene family. This process is also responsible for light-induced development processes in higher plants and for vision in animals. The photochromic behaviour of bacteriorhodopsin, a natural light-harvesting protein, and the possible applications in memory devices are described in some detail the Chapter 12, Section 12.10.1.

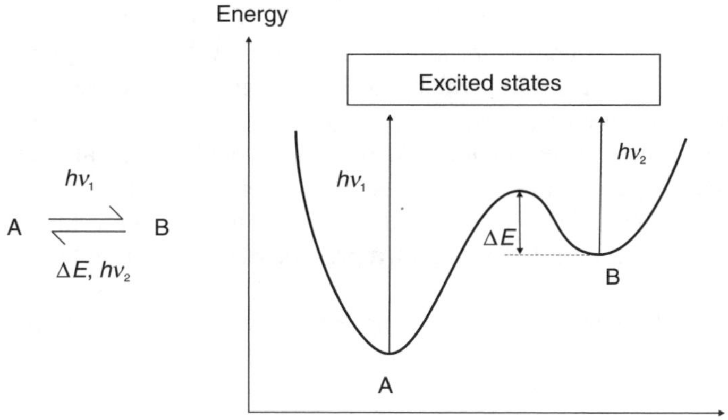

Figure 11.27 Left: a photochromic reaction where A represents the initial unswitched state of the photochromic material and B the switched material. The relative energy levels of the states A and B are depicted on the right.

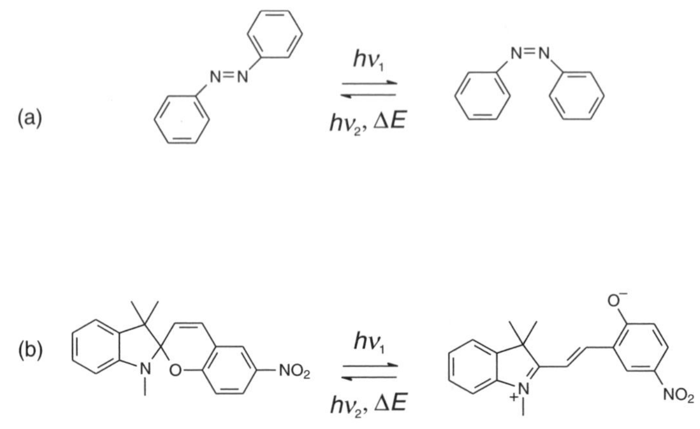

Figure 11.28 Two classes of photochromic compounds: (a) azobenzenes; (b) spiropyrans.

Azobenzene derivatives in the form of LB films have been widely studied. In these, UV illumination of the *trans* isomer at the wavelength of the $\pi \rightarrow \pi^*$ transition near 330 nm causes this band to diminish and a less intense $n\pi^*$ band of the *cis* isomer near 440 nm to grow. Heat, UV irradiation at 250 nm and irradiation at wavelengths greater than 400 nm (i.e. radiation of the $n\pi^*$ band) recovers the *trans* isomer. One problem is that photochromic *cis–trans* isomerization reactions which occur in solution are often suppressed in the semi-rigid environment of a thin solid film. Some novel solutions have been proposed to circumvent this difficulty. Most of these rely on methods to increase the area-per-molecule available for switching, for example by incorporating the azobenzene molecules into the cylindrical cavity provided by amphiphilic β-cyclodextrin molecules.

Figure 11.28(b) shows a particularly important class of photochromic compounds, the spiropyrans. The photochromism results from the presence of a conjugated ring, which is photochemically opened and closed – the process of cyclization. The majority of spiropyran compounds are positively photochromic; under UV irradiation they change from a pale yellow to a coloured form, as shown in Figure 11.29 for a spiropyran compound [56]. In LB films, spiropyrans are of interest because of their narrow H- and J-aggregate absorption bands (Chapter 4, Section 4.4.2), which cover a range of wavelengths and offer some potential for the realization of memory devices.

In certain photochromic compounds, a reversible photochemical transformation can lead to a change in their chirality (Chapter 2, Section 2.4.2); these can form the basis of *chiroptical* switches [54]. In a chiral photochromic system, the left-handed and right-handed forms of a chiral compound represent two distinct states in a binary logic element.

The term electrochromic is applied to compounds that can be interconverted by reversible redox (reduction–oxidation) processes between two different absorption spectra. Whereas photochromic systems usually involve two forms of a molecule, in electrochromic systems several successive switching processes can occur. A number of different electrochromic technologies have been developed. First, there are solution-based systems, which rely on organic electrochromic species dissolved in the electrolyte compartment of an electrochemical cell comprising at least one transparent electrode. The most commonly used

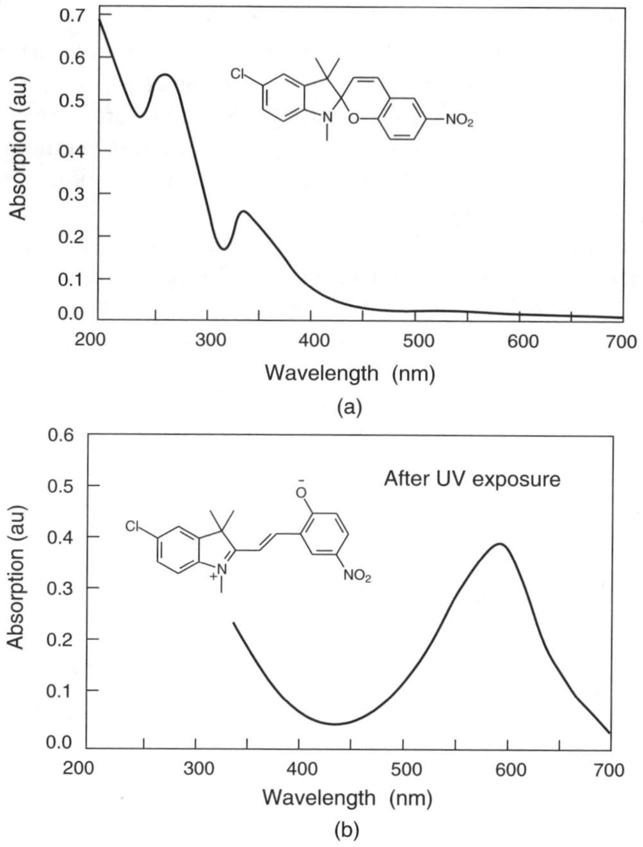

Figure 11.29 Absorption spectra and molecular structures of (a) the closed and (b) the open forms of a spiropyran compound. From *Functional Organic Polymeric Materials*, TH Richardson (Editor), pp. 273–293. Copyright (2000) John Wiley & Sons Limited. Reproduced with permission.

compounds are the viologens, salts of 4, 4′-bipyridines. These are synthetically tuneable, which allows for different colours, and have intrinsically high extinction coefficients, yielding excellent colouration intensities. The switching speed depends on the diffusion of these and other redox active species in the electrolyte to the electrodes and is typically of the order of seconds. Because the redox-active species are dissolved in an electrolyte, these mobile molecules will diffuse to both electrodes once an appropriate electrical potential has been applied to the circuit. When the potential has been removed, the charged species mix, transfer their charges and the colour dissipates from the system. Therefore, there is no open-circuit memory in these devices and power must be applied continuously to maintain colouration.

A different approach relies on the intercalation, or insertion and bonding, of ions into the crystal lattice of materials. This technology is based on electrochromic cells in which at least one of the electrodes is a thin but compact layer of a metal oxide, usually tungsten oxide (WO_3), which exhibits electrochromic properties. Colouration is achieved when ions, typically H^+ or Li^+, and electrons are injected electrochemically into a WO_3 layer by means of an applied voltage. An advantage of this intercalation electrochromic technology compared with solution-based technologies is that it offers a memory capability in devices. The condition is that the complementary redox species are bound to the counter electrode surface or a charge storage layer is available to support the charging of the device (without dissipation upon removal of the applied potential). The intercalation process is slow due to the diffusion of the ions in the electrolyte and into the thin film. In addition, the metal oxides used do not form very highly coloured complexes, so the contrast ratio is intrinsically lower than for solution-based organic systems.

An important application is the electrochromic window. This can block the glare of the sun or provide instant privacy with the flip of a switch. Electrochromic windows are part of a new generation of technologies called switchable glazing, or 'smart' windows, which change the light transmittance, transparency or shading of windows in response to an environmental signal.

The spectroscopic technique of *hole burning* has been studied as a method to store data on the molecular level. In this technique, either a photochemical or a photophysical process leads to a change in the electronic structure, and hence in the optical absorption spectrum of a molecule in a polymer matrix. Since the molecule is just one member of an ensemble which gives rise to an inhomogeneously broadened absorption spectrum, the change reduces the absorption at a specific wavelength within the band, i.e. 'burns' a hole in the absorption band, as shown in Figure 11.30.

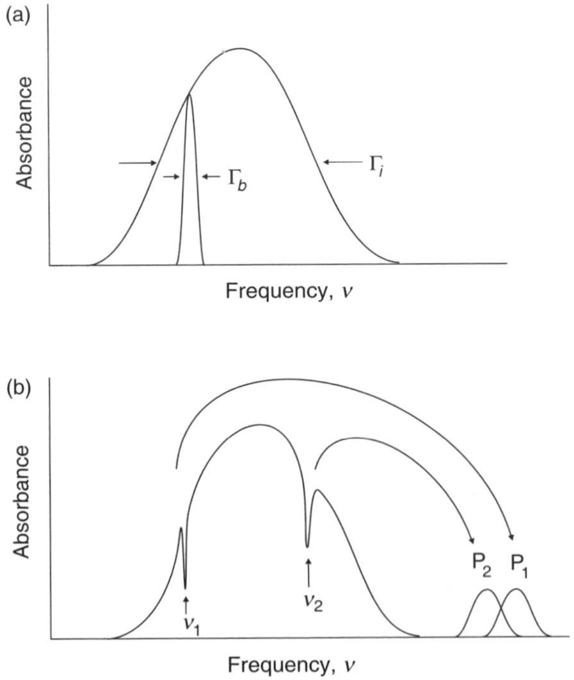

Figure 11.30 (a) Spectral profile of an inhomogeneously broadened absorption band of width Γ_i is composed of numerous homogeneous bands of width Γ_b but different centre frequencies. (b) Hole burning causes the absorption intensity to be removed from bands at frequencies ν_1 and ν_2 and transferred to product bands P_1 and P_2.

11.8.3 Chemical Control

Chemical stimuli can be exploited in a variety of ways to achieve switching. For example, electronic communication between metal centres across a bridging molecular wire can be controlled by protonation/deprotonation of an acid or basic site on the bridge. Supramolecular species comprising donor and acceptor units connected by means of noncovalent forces can be disassembled and reassembled by modulating the reactions that keep the two components together [54].

There have been a number of suggestions that utilize the properties of the hydrogen bond. This is nearly unique in its range of energies (approximately 0.02–1.3 eV per atom) and is ubiquitous in biological systems. Figure 11.31 shows a proposed switch based on hydrogen bonding in a hemiquinone molecule [57]. The two isomers of the molecule depicted differ only in the placement of the protons and are called *tautomers.* The hydrogen atoms that are involved in the tautomerism are suspended in a perfectly symmetrical double well potential and are not completely localized in one particular well, or in one part of the molecule, but move to the left and right, giving rise to an oscillation of the structure shown in Figure 11.31. In the proposed operation of the device, an applied electric field perturbs the double well potential, which becomes asymmetric and therefore allows the two states of the molecule to be distinguished. The protons move between the potential wells by tunnelling.

Many organic molecules, such as the fullerene C_{60}, exhibit multiple reduction/oxidation states, offering the possibility of 'multibit' information storage [58]. An example of this is

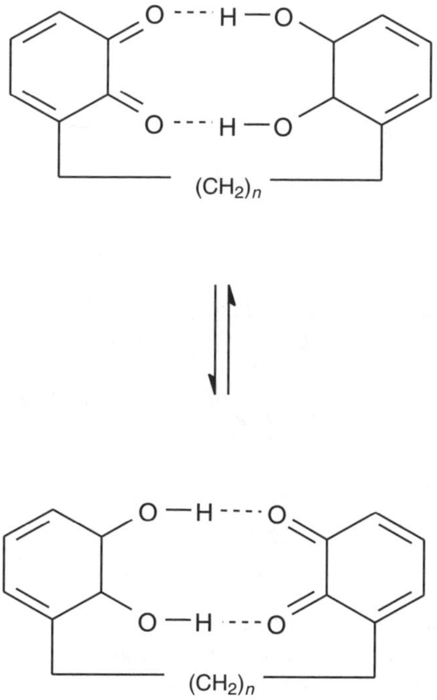

Figure 11.31 Tautomeric forms of a hemiquinone derivative proposed as a molecular switch [57].

shown Chapter 5, Section 5.5.1, Figure 5.11. One bit of data can be associated with each redox state.

11.9 NANOMAGNETIC SYSTEMS

Magnetic behaviour depends on the dimensionality of the system and low-dimensional materials can give rise to phenomena not observed in isotropic solids. In this respect, there have been a number of studies on the magnetic behaviour of ultra-thin film systems, particularly Langmuir–Blodgett films [59]. The initial experiments, in the late 1970s, focused on a study of manganese stearate; the motivation was to investigate theoretical predictions of magnetic order in a two-dimensional system. However, such a magnetic monolayer might be considered the ultimate magnetic memory storage device.

Other compounds that have been studied in LB and self-assembled films include metallic phosphonates, derivatives of phosphonic acid and charge-transfer compounds such as TTF and BEDT-TTF (Chapter 5, Section 5.4) and bimetallic compounds with the general formula $A_k[B(CN)_6].nH_2O$, where A and B can be divalent or trivalent transition metals. The CN ligands are good mediators of magnetic exchange, and examples of these mixed valence bimetallic compounds show ferrimagnetic ordering or ferromagnetic ordering (Chapter 5, Section 5.7.1) at critical temperatures, which range from liquid helium temperatures to above room temperature. Langmuir–Blodgett films can be formed by using a positively charged lipid spread on a subphase containing a dilute colloidal suspension of the inorganic compound. Figure 11.32 shows a hysteresis loop at 3 K for

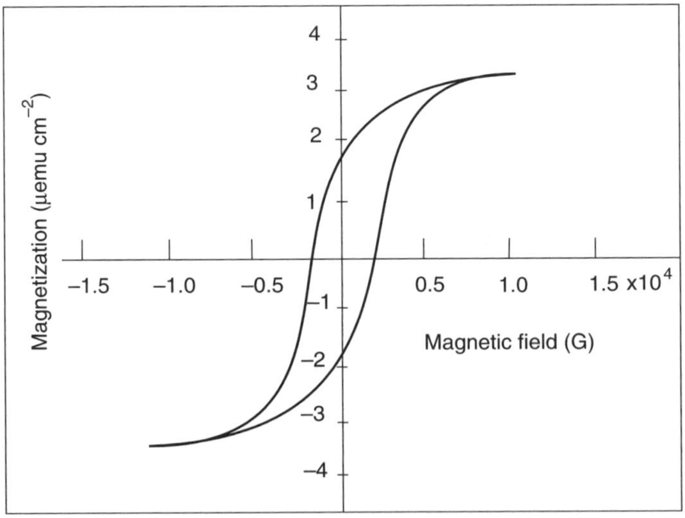

Figure 11.32 Hysteresis loop at 3 K in the plot of magnetization versus field for an LB film based on an amphiphilic lipid and $Cu_3[Fe(CN)_6]_2$. Reprinted from *Magnetism: Molecules to Materials. Molecule- Based Materials*, Vol. 2, JS Miller (Editor), pp. 457–484. Copyright (2001), with permission from Wiley-VCH.

such an LB film based on $Cu_3[Fe(CN)_6]_2$; the remanent magnetization is approximately 2.3 μemu cm^{-2} and the coercive field is 2100 G [59]

Interest is also focused on 'high-spin' molecules. Cluster of such molecules exhibit large magnetic hysteresis and may provide the means for achieving bistability and information storage at the molecular level (see Section 11.14). Furthermore, these nano-magnets provide interesting architectures for observing quantum magnetization and electron tunnelling through a potential barrier from one molecular state to another. The most thoroughly studied single-molecule magnets are the mixed valence manganese clusters based on the $Mn_{12}O_{12}$ core. These are formed by an internal tetrahedron of four Mn^{IV} (i.e. valency of four) ions ($S = \frac{3}{2}$) surrounded by eight Mn^{III} (valency of three) ions ($S = 2$). Exchange interactions within the cluster results in a ground state with a large spin ($S = 10$), which encounters a thermal barrier for reversal in the direction of magnetisation along the uniaxial magnetic axis. In the crystalline state, these neutral clusters, such as $Mn_{12}O_{12}$ $(acetate)_{18}$, exhibit a stable hysteresis loop with a coercive field as large as 1.5 T at 2 K [59]. Such neutral clusters can be incorporated into Langmuir–Blodgett films by mixing them with a fatty acid matrix. Plots of magnetization versus applied magnetic field for LB films exhibit a hysteresis loop at 2 K. Other approaches to the formation of self-organized magnetic materials have included layer-by-layer electrostatic deposition and methods based on classical colloidal chemistry.

11.10 NANOTUBE ELECTRONICS

Carbon nanotubes are unique materials (Chapter 5, Section 5.5.2) and a vast number of papers are continuing to appear on the properties and applications of these compounds [60–65].

Both experiments and theory have shown that single-walled carbon nanotubes (SWNTs) can be either metals or semiconductors. The remarkable electrical properties of SWNTs originate from the unusual electronic structure of the two-dimensional material graphene. When wrapped to form a SWNT, the momentum of the electrons moving around the circumference of the tube is quantized. The result is either a one-dimensional metal or a semiconductor, depending on how the allowed momentum sates (i.e. wavevector or k states) compare with the preferred directions for conduction.

The physics and technology of electron transport in nanodevices are fully explained by the seminal work of the late Rolf Landauer, an eminent IBM scientist, who derived the following equation for the conductance G of a one-dimensional ballistic conductor (Chapter 3, Section 3.2.1):

$$G = \frac{2e^2}{h} MT \tag{11.9}$$

where M is the number of modes or channels and T is the transmission coefficient, or the average probability that an electron injected at one end of the conductor will reach (be transmitted to) the other end. The factor of 2 in Equation (11.9) takes into account the spin degeneracy of the electrons. Note the appearance again of the term h/e^2, the resistance quantum. In the case of a SWNT, taking into account sublattice degeneracy of graphene and assuming that the contacts are perfect $(T = 1)$ gives [62]

$$G = \frac{4e^2}{h} = 155\,\mu\text{S} \tag{11.10}$$

This corresponds to a resistance of about 6.5 kΩ. Experiments reveal that the resistance of SWNTs varies considerably, from approximately 6 kΩ to many megohms, the differences being mostly due to contact resistances between the electrodes and the nanotubes. When contact resistances are eliminated, the measurements reveal a resistance per length of $4\,\text{k}\Omega\,\mu\text{m}^{-1}$, a mean free path of 2 μm and a room temperature resistivity of approximately $10^{-6}\,\Omega\,\text{cm}$. The conductivity of metallic nanotubes can therefore be equal to, or even exceed, the conductivity of metals such as copper at room temperature.

Semiconducting nanotubes can possess very high charge carrier mobilities, e.g. hole mobilities of 10^3–$2 \times 10^4\,\text{cm}^2\,\text{V}^{-1}\,\text{s}^{-1}$ for tubes grown by chemical vapour deposition [62]. These values should be contrasted with the figures for inorganic and organic semiconductors given in Chapter 3, Section 3.4.1, Table 3.1. The conductivity of SWNTs is mainly p-type, but n-type conduction is also observed. Chemical doping can be achieved in a number if ways. For example, doping with alkali metals, which donate electrons to the nanotubes, can be used to achieve n-type material.

Figure 11.33(a) shows a schematic diagram of a field effect transistor based on an SWNT; an AFM image of a nanotube in such a device is shown in Figure 11.33(b). The drain current versus drain voltage characteristics for an SWNT FET are shown in Figure 11.34. In this particular case, a thin layer of a high dielectric material, ZrO_2, was used as the gate dielectric [62, 63]. Standard FET behaviour is evident (Chapter 9, Section 9.4), with the drain current initially increasing in a linear fashion with the drain voltage, and then saturating. The nanotube exhibits excellent electrical properties, with a maximum transconductance $g_m = 12\,\mu\text{A}\,\text{V}^{-1}$ at $V_g = 0.4$ V. The performance of such devices is superior, in some respects, to that of state-of-the-art silicon MOSFETs. The remarkable behaviour can be attributed, in part, to the lack of surface states (Chapter 2, Section 2.5.7) in SWNT devices.

Under certain conditions (e.g. with a thin gate insulator), *ambipolar* FETs can be fabricated [64]. For these devices, the conduction is by electrons when a positive bias is applied to the gate, but by holes when a negative voltage is applied. It is therefore possible to inject electrons and holes simultaneously from opposite ends of the SWNT channel. These injected carriers are confined to the nanotube and, when they meet they recombine with the emission of radiation; the experiment is depicted in Figure 11.35 [64]. The diameter of the nanotube defines the wavelength of the emitted light, typically in the infrared region, and the position of the emitting spot along the length of the nanotube can be varied by changing the gate bias. This three-terminal device is essentially a single-molecule, electrically controlled light source.

It has already been noted that (Section 11.6.1) memory devices might be made from nanotubes. One scheme uses a crossbar arrangement of rows and columns of nanotubes separated by supporting blocks [38]. The application of an appropriate voltage between the desired column and row bends the top tube into contact with the bottom nanotube. Van der Waals forces maintain the contact, even after the voltage has been removed. Separation of the SWNTs is then achieved by the application of a voltage pulse of the same potential. Assuming a minimum cell size to be a square of 5 nm length, a packing density of 10^{12} elements cm^{-2} can be achieved. The inherent switching time is estimated to be in the 100 GHz range.

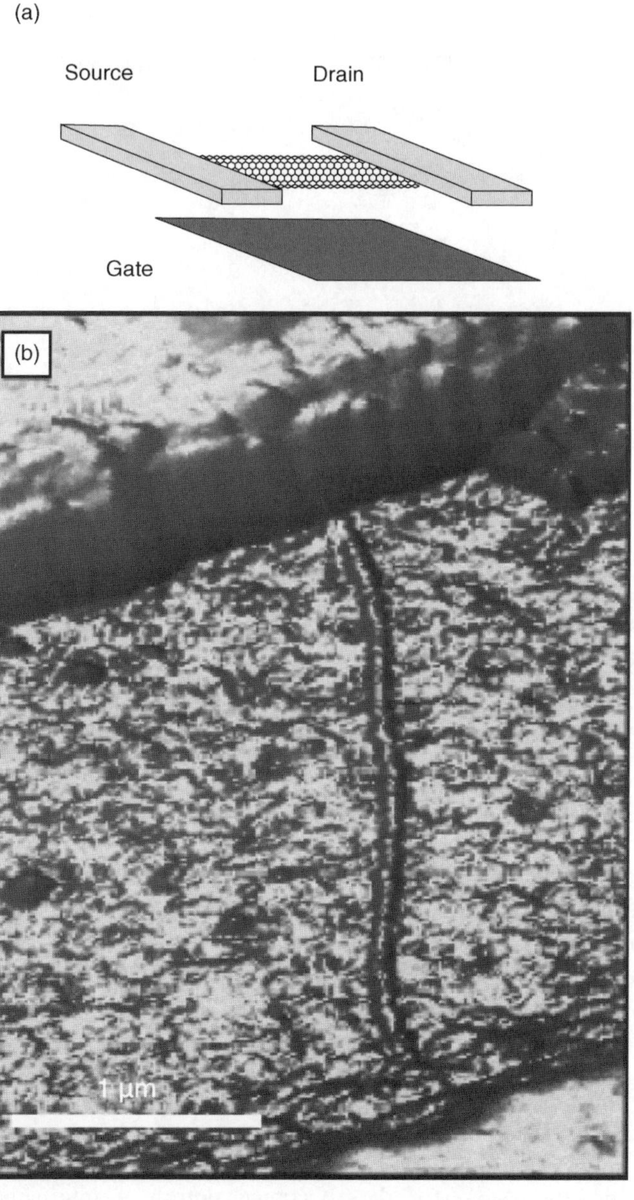

Figure 11.33 (a) Schematic diagram of a single-walled carbon nanotube transistor. (b) Atomic force microscope image of the device showing a nanotube connecting the source and drain electrodes. From PL McEuen, JY Park, 'Electron transport in single-walled carbon nanotubes', *MRS Bull.*, **29** (2004), pp. 272–275. Reproduced by permission of the MRS Bulletin.

The prospect therefore exists of fabricating microelectronic circuitry in which all the necessary elements – transistors, resistors, memories and interconnects – are fabricated from carbon nanotubes [65]. However, reliable methods are needed to control their electrical properties and means must be devised to position the nanotubes accurately on flat surfaces.

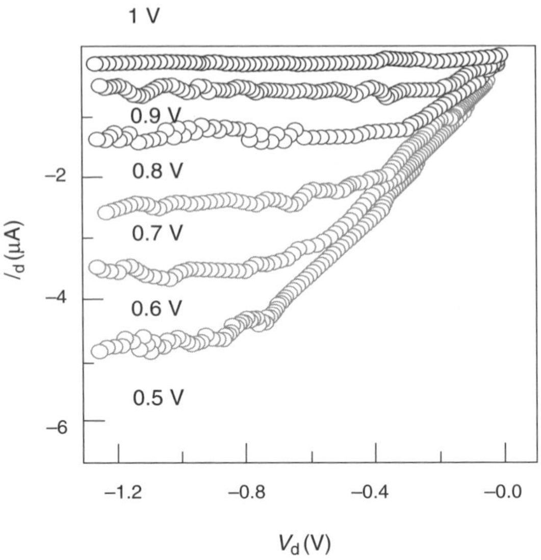

Figure 11.34 Drain current I_d versus drain voltage V_d for different values of gate bias for a SWNT field effect transistor with a ZrO_2 gate insulator. Reprinted by permission from Macmillan Publishers Ltd: *Nature Materials*, 2002, **1**, A Javey, H Kim, M Brink, Q Wang, A Ural, J Guo, P McIntyre, P McEuen, M Lundstrom, H Dai, 'High-κ dielectrics for advanced carbon-nanotube transistors and logic gates', pp. 241–246.

11.11 MOLECULAR ACTUATION

Mechanical movement is essential, both in Nature and in the technological world. Enzymes such as myosin and kinesin are natural linear motors that convert chemical energy into mechanical work. Rotary movements are also observed in Nature. These 'devices' are explored in more detail in Chapter 12, Section 12.11. However, systems exhibiting simple movements can be developed using the materials and fabrication techniques described in the earlier chapters of this book.

11.11.1 Dynamically Controllable Surfaces

An electrically driven system for the dynamic control of interfacial properties (e.g. wettability) is illustrated in Figure 11.36 [66]. This exploits a transition between a straight and bent conformation of a self-assembled surfactant molecule on the application of a voltage. Self-assembly generally leads to a dense packing arrangement of the aliphatic chains, leaving little room for conformation transitions (Chapter 7, Section 7.3.2). To create a less densely packed monolayer, a long-chain thiol molecule with a very bulky terminal group (at the opposite end from the sulfur termination) is used. Subsequent cleavage of the space-filling end groups leaves a relatively low density of the self-assembled monolayer. Upon application of a positive potential to the gold substrate, the negatively charged carboxylate groups experience an attractive force towards the substrate, causing the

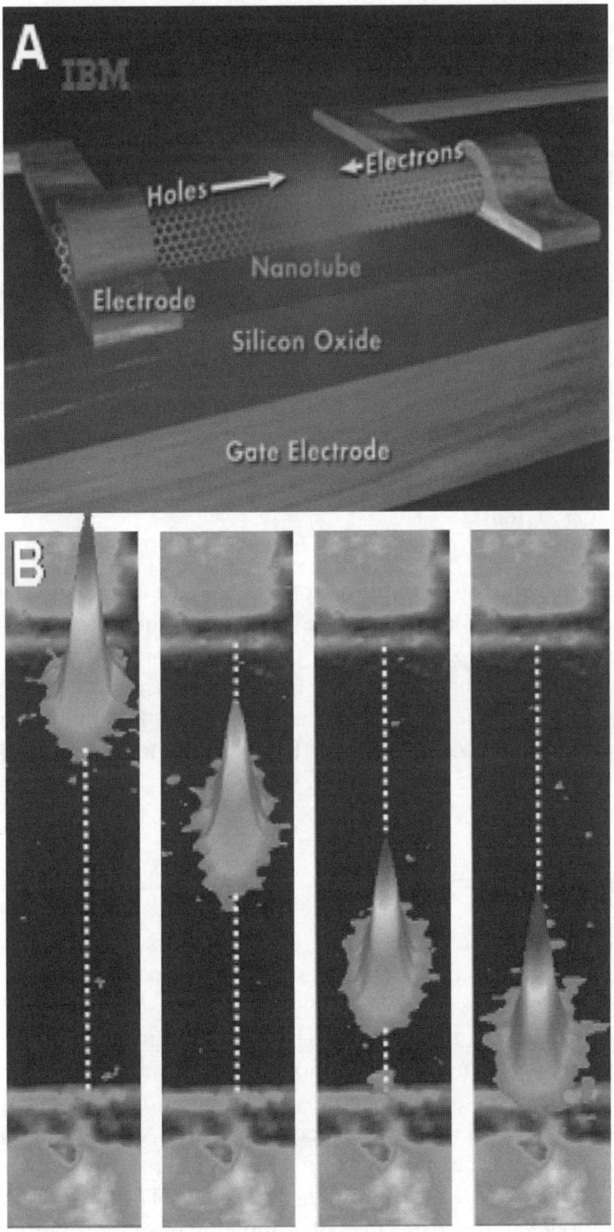

Figure 11.35 (A) Electrons and holes can be injected from opposite ends of a carbon nanotube to create a single molecule, electrically controlled light source. (B) The light emission can be moved between the two metal electrodes by varying the gate voltage. Reprinted from Avouris P, Apenzeller J, *The Industrial Physicist*, **10**, 'Electronics and optoelectronics with carbon nanotubes', pp. 18–21. Copyright (2004) American Institute of Physics.

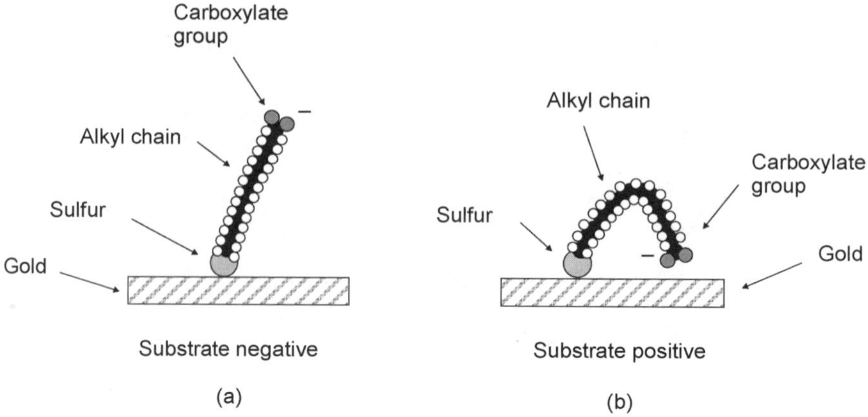

Figure 11.36 Idealized representation of the transition between (a) a straight (hydrophilic) and (b) bent (hydrophobic) molecular conformations of a long-chain thiol molecule self-assembled on to a gold surface. After Lahann *et al.* [66].

hydrophobic chains to undergo conformational changes. Reversible transitions can be confirmed at the macroscopic level using contact angle measurements. Other switchable surfaces have also been developed [67, 68]. For example, bipyridinium units tethered to an electrode surface by long-chain thiols can be electrochemically oxidized and reduced, providing changes in the conformation of the molecules. Potential applications for such smart surfaces may be as diverse as substrates to study important biochemical processes and functional units for microfluidic devices (e.g. valves).

11.11.2 Rotaxanes

Rotaxane molecules have already been met earlier in this chapter, in Section 11.6.1. A rotaxane is a large molecule that comprises of a macrocyclic and a dumbbell-shaped component [69]. The latter is encircled by the macrocycle, which is prevented from disengagement by two bulky end groups. The two components of the rotaxane cannot therefore disengage from one another, even though they are not linked by covalent chemical bonds. However, the two components may move relative to one another. Simple motions are depicted in Figure 11.37 [54]. When two recognition sites on the dumbbell differ in their chemical properties, a rotaxane can exist in two different conformations and under an appropriate stimulus may switch between theses states. This is shown schematically in Figure 11.37. The model is therefore one of a molecular shuttle, with the 'train' moving between 'stations'. Examples of two recognition sites are provided in Figure 11.15, where the two sites have polar and nonpolar properties. In this compound, the transition from one state to the other is achieved by the application of an electric field. However, in other rotaxanes, the linear movement can be controlled by chemical or photochemical stimuli.

Rotation of the 'wheel' component of the rotaxane [Figure 11.37(b)] is usually a spontaneous process. However, this can be artificially controlled. For example, chemical and electrochemical reactions can be used be make the ring component of a rotaxane oscillate between two positions [54].

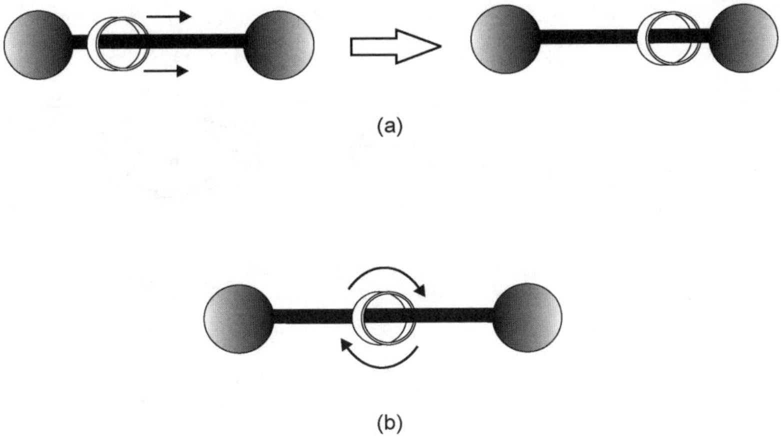

Figure 11.37 Schematic diagram showing movements that can occur in a rotaxane molecule. Reprinted from *Molecular Devices and Machines*, Balzani V, Gredi A, Venturi M, p. 391. Copyright (2003), with permission from Wiley-VCH.

11.11.3 Optical Tweezers

Molecular tweezers are a simple and elegant example of a molecular 'machine.' Figure 11.38 shows a classic example of tweezers that are actuated optically [70]. In the *trans* configuration, the azobenzene compound shown has weak coordinating capacity for large cations. Excitation with light causes *trans* to *cis* isomerization, leading to a molecular configuration suitable for enclosing large metal ions between the two crown ethers, with a strong increase in the coordination capacity. Somewhat related to tweezer-like movements are the conformational changes caused by electrostatic attractive forces generated by photoinduced electron transfer in dyads (bivalent molecules) with semi-flexible bridges. In these cases, photoinduced electron transfer can result in the formation of charges of opposite sign. As a consequence of the resulting electrostatic attraction, conformational distortion occurs and the two ends of the molecule approach each other – a molecular 'harpoon' [54].

11.12 LOGIC CIRCUITS

In most of the experiments on molecular-scale electronics, discrete devices are studied, i.e. the inorganic materials associated with silicon microelectronic devices are replaced by organic counterparts, albeit at the molecular scale. Such individual devices need to be connected together to form the logic circuitry that is the foundation of digital computers. In the latter, information is encoded in electrical signals. Threshold values and logic conventions are established for each signal. In a positive logic convention a '0' is used to represent a signal that is below the threshold value and a '1' indicates that a signal is above the threshold. The logic circuits of silicon microprocessor systems process such binary data through a sequence of logic gates. Although it is not necessarily true that the components of a molecular computer will have to operate in an analogous manner to a silicon-based

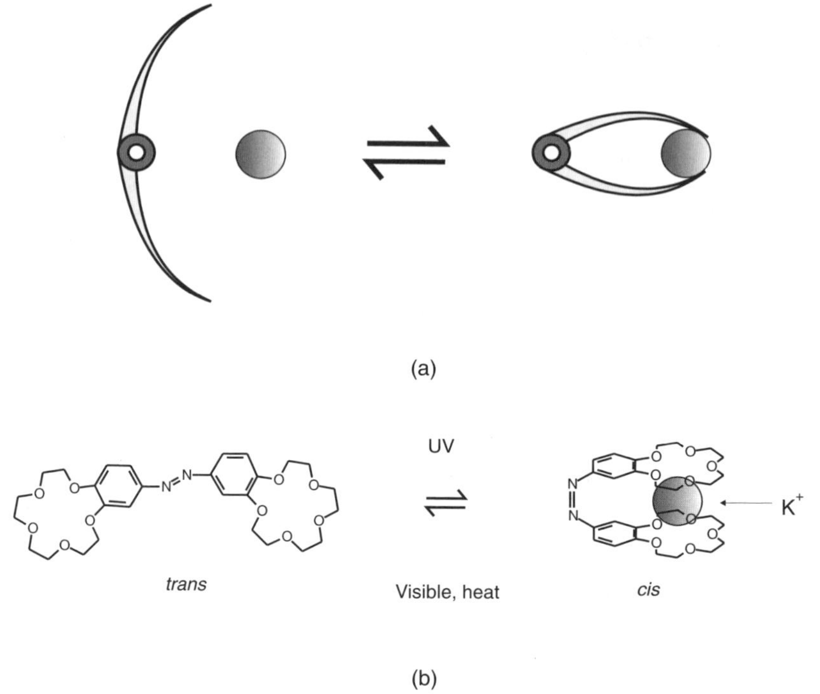

Figure 11.38 (a) Tweezers on the macroscopic scale contrasted with (b) molecular scale tweezers based on the *cis–trans* isomerization of an azobenzene derivative. Adapted from Shinkai *et al.* [70].

computer, much effort is being directed to the design, synthesis and characterization of molecular systems which mimic conventional logical operations [54].

The three basic types of logic gate are the NOT, AND and OR gates. Other gates, which are a combination of these basic three gates, also exist: examples are the NAND (AND gate plus a NOT gate) and the NOR (OR plus NOT). Each has been designed to perform according to a set of rules delineated in a so-called *truth table*, which is a list of outputs that the gate should give in response to the complete range of input combinations. The NOT gate simply inverts the signal at its input, i.e. a logic 0 input results in a logic 1 output, and vice versa. In chemical systems this function is fairly common. For example, a luminescence output can be quenched by a chemical input.

A simple example of a NOT chemical gate is shown in Figure 11.39, together with the circuit symbol used for the NOT function and the truth table [71]. In a mixture of methanol and water, the fluorescence of the compound shown is relatively high. However, this is quenched in the presence of acid (protons). The process is related to the occurrence of photoinduced electron transfer from the central pyrazoline unit to the benzoic acid.

A molecular AND gate is shown in Figure 11.40 [72]. This logic function has two inputs, A and B, and one output. Logic 1 is only obtained on the output when the input signals are both at 1, i.e. the output is 1 if, and only if, input A *and* input B are at 1. The molecular AND gate is based on an anthracene derivative. In methanol, and in the presence of H^+ and Na^+, the fluorescence quantum yield of this compound is high (output state 1 in the truth table). However, the three output states 0 have a low fluorescence output. The photoinduced

(a)
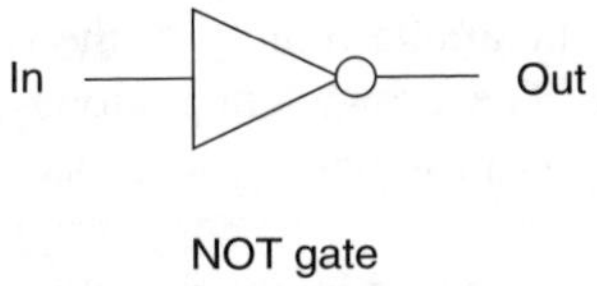

(b)
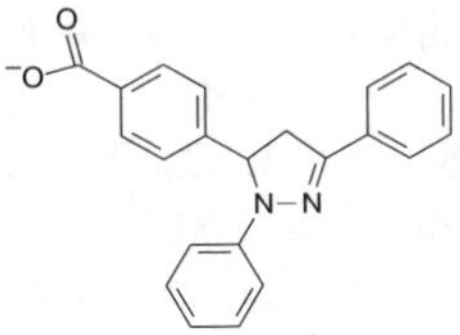

(c)

In (H^+)	Out (Fluorescence)
0	1
1	0

Figure 11.39 A chemically-based NOT logic gate. (a) Circuit symbol; (b) pyrazoline derivative; (c) truth table for NOT function. Reprinted from *Molecular Devices and Machines*, Balzani V, Gredi A, Venturi M, p. 241. Copyright (2003), with permission from Wiley-VCH.

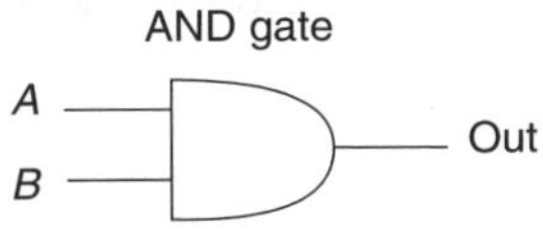

(a)

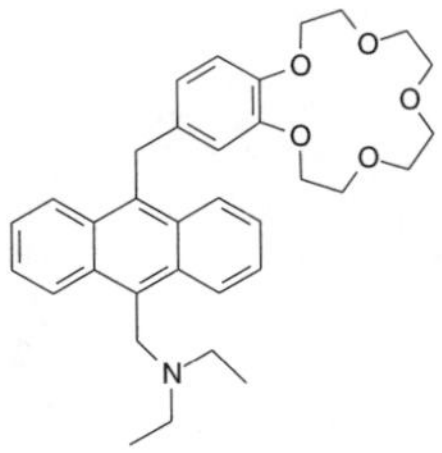

(b)

A (H^+)	*B* (Na^+)	Out (Fluorescence)
0	0	0
0	1	0
1	0	0
1	1	1

(c)

Figure 11.40 A chemically based AND logic gate. (a) Circuit symbol; (b) anthracene derivative; (c) truth table for AND function. Reprinted from *Molecular Devices and Machines*, Balzani V, Gredi A, Venturi M, p. 243. Copyright (2003), with permission from Wiley-VCH.

electron-transfer process involves the amine moiety in the first two states of the truth table and the crown ether in the third. The crown ether alone cannot quench the anthracene fluorescence, but when the amine is protonated the process becomes thermodynamically allowed and does occur.

Many other examples of chemically based logic operations exist [54]. In some of the early examples, it was proposed to exploit conformational changes in molecules. The *soliton switch*, introduced by Carter [73, 74], is perhaps the best known example of this. A soliton in a conjugated polymer such as polyacetylene is a defect that separates chain segments with different bond alternation (Chapter 3, Section 3.4.3). These segments may be considered as logical states and a passing soliton 'switches' the chain from one state to another. Figure 11.41 illustrates how solitons in polymer networks might be switched and steered – soliton valving. The figure shows the change of state of a three-state network following the passage of a soliton. The passage of the soliton from A to B (or from B to A) moves the double bond at the branch carbon from the A chain to the B chain. In the upper right of Figure 11.41, we note that a soliton moving from B to C moves the double bond to the C chain.

Other approaches to molecular logic exploit the electrical properties of organic molecules and propose to construct circuitry by chemically linking the various components (e.g. molecular rectifier molecules, Section 11.5). These and other ideas have attracted their fair share of criticism for being impractical to realize. The problems of interconnecting individual devices, be these molecular diodes, switches or simple logic circuits, are formidable. There is also the issue of how to read-in and read-out data at the molecular level. For example, in the case of the soliton switching, in order to 'read' the state of a chain segment, two adjacent carbon atoms have to be marked and the bond between them has to be inspected.

There have also been many speculations about self-assembling, self-repairing, fault-tolerant molecular systems; none have yet been well-specified or fabricated. The steps needed have been outlined [73]. First, a high-level logic function associated with a molecule

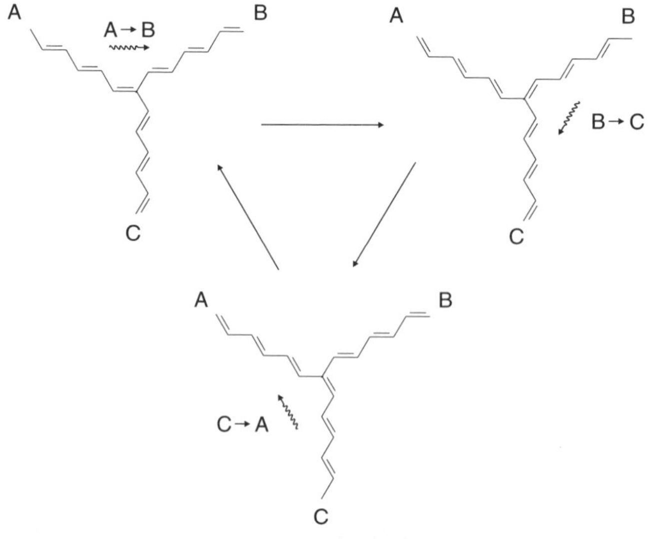

Figure 11.41 Soliton valving. The propagation of a soliton from chain A to chain B corresponds to a clockwise rotation by 120° of the upper left-hand configuration [57].

must be identified. A means is then required to place such molecules, with precision, to make an interface with the molecules and to verify both the position and communication. It is then essential to be able to interconnect and/or isolate the molecules without destroying their functionality. Finally, an efficient assembly scheme for very large numbers of such molecules must be developed.

11.13 COMPUTING ARCHITECTURES

Most computers use the stored-program concept designed by John von Neumann. The *von Neumann architecture* is a model for a computing machine that uses a single storage structure to hold both the set of instructions on how to perform the computation and the data required or generated by the computation. Such machines are also known as stored-program computers. The separation of storage from the processing unit is implicit in this model. In it, programs and data are stored in a slow-to-access storage medium (such as a hard disk) and work on these is undertaken in a fast-access, volatile storage medium (RAM).

The brain (Nature's molecular computer) utilizes parallel processing, instead of the serial approach in von Neumann systems. This means that the brain can send a signal to hundreds of thousands of other neurons (Chapter 12, Section 12.7) in less than 20 ms, even though it takes a million times longer to send a signal than a computer switch. The brain works mainly by nonlinear computation using the rate of pulse production by a neuron or nerve cell as the information signal being sent to another cell. There are about 10^{11} neurons in the human brain, and each is connected to 10^3–10^4 others. This gives a crude 'bit count' of 10^{11}–10^{15}. An equivalent artificial 'brain' might therefore be built from $10^5 \times 8$ Gbit chips, with a power dissipation of many megawatts!

Neural logic may be either analogue or digital. In the latter form, the neuron is designed to respond to the sum of its N inputs (which may be inhibitory or excitory). Figure 11.42 shows a schematic diagram of such a gate. Provided that the sum exceeds a given threshold T the neuron will output logic 1, otherwise it outputs logic 0. By combining neurons into totally connected networks – neural networks – it is possible to construct adaptive learning systems, control systems and pattern recognition systems. In Chapter 10, Section 10.6, we saw how artificial neural networks could be used in an electronic nose, emulating the human olfactory system. Some elementary properties of neurons can be mimicked by simple chemical systems, although no computational system exploiting these has been built. Neural networks are very similar to *cellular automata*. These are regular arrays of *finite-state machines* (an excellent discussion on this subject can be found in the book by Amos [76]) and provide a useful architectural target from the standpoint of fabrication and logical issues.

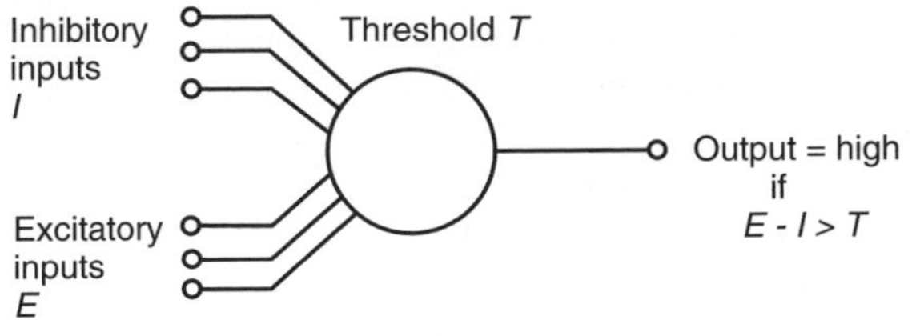

Figure 11.42 Representation of neural logic.

Figure 11.43 A cellular automaton. An individual cell, shown on the right, is a finite-state machine.

A cellular automaton is a regular array of identical finite-state machines, each connected to a finite number of its neighbours using the same interconnection net. Such systems were first investigated by von Neumann as models for completely discrete physical dynamic systems such as the brain. Figure 11.43 shows an example of a two-dimensional cellular automaton layout. It consists of a regular spatial array of identical cells, each characterized by a finite discrete-valued state function. At successive discrete intervals of time, each cell makes a transition to a new state which depends on the previous state values of the connected neighbours.

The best-known example of a cellular automaton is the *Game of Life*, devised by the mathematician John Conway. The game is played on a field of cells, as in Figure 11.43, each of which has eight neighbours (adjacent cells). In the standard Game of Life, a cell is 'born' if it has exactly three neighbours, stays alive (survival) if it has two or three alive neighbours, and dies otherwise. Ever since its publication, it has attracted much interest because of the surprising ways the patterns can evolve. The cellular automaton is an example of emergence and self-organization.

A major issue associated with the manufacture of any molecular electronic device or system will almost certainly be their relatively poor fabrication yield resulting from problems such as the influence of background charge, difficulties in making reliable contacts to these devices and lithographic inaccuracies when it comes at the nanometre level. Although it is expected that solutions to the above problems can be found through research, it seems intuitively difficult to find such solutions at the device level without an increase in the fabrication cost. A reasonable consequence of the above is that the search for an alternative to the expensive but very robust MOS technology in the form of a much less expensive but less robust technology should be accompanied by fault-tolerant schemes, permitting the functionality of the chips even if some of the devices are not even operational. A good example of such an approach is the Teramac computer that it was designed and fabricated by Hewlett-Packard in the USA [77]. This system was based on a high redundancy of interconnecting wires that connect 'unreliable' switching devices, so that even if some of the paths contain nonoperational devices there is always one functional path. Once built, the system is checked by software and the operational paths are identified. Consequently, the problem of expensive fabrication costs to make perfect devices is transferred to the time-consuming procedure of identifying the operational paths by software. The above scheme was mainly designed with wires and switching devices that permit a reconfiguration of the system when a nonidentified, nonfunctional path is found.

For molecular electronics to be competitive with silicon, the technology will clearly have to offer something other than speed and bit density. Three-dimensional memory/logic is often mooted as a potential advantage for molecular systems and that certainly could lead to higher bit densities than for two-dimensional architectures. However, it is much more likely that the *functional* properties of molecular materials will prove more significant. This is already obvious with liquid crystal and electroluminescent polymer displays. One might imagine a future molecular electronic technology that integrates electronic logic/functionality with sensing and/or display capability in a foldable or, in medical applications, implantable packaging.

11.14 QUANTUM COMPUTING

The unit of classical information is the bit, which takes one of the two possible values, 0 or 1. Any amount of classical information can be expressed as a sequence of bits. A classical computer executes a series of simple operations (gates), each of which acts on a single pair of bits. By executing many gates in succession, the computer can evaluate more complex mathematical operations on a set of input bits. Quantum information can also be reduced to elementary units, called *quantum bits* or *qubits*. A qubit is a two-level quantum system, such as the spin on an electron (exploited in the emerging technology of spintronics described in Section 11.6.3). The idea behind *quantum computing* is to use an array of such quantum 'particles' to perform mathematical operations [78, 79].

A simple quantum computer might use an array of quantum dots. Such a system is very similar to the cellular automata described in the previous section. To transfer the principles of cellular automata from the software level to real computing systems, two steps must be taken. First, the quantum dots, e.g. nanoparticles, must be arranged into regular two-dimensional arrays. Connections must then be established between the quantum dots by means of some physical interaction, for example, optical, magnetic or electrical. Finally, the matrix of quantum dots has to be connected to input and output devices.

A simple example of arrays of quantum dots interacting electrostatically might be used as the basis of a quantum computer [78, 80]. In the array of nanoparticles, which are firmly placed on the substrate and separated by small gaps comparable to the electron tunneling distance, ranging from 2 to 3 nm, the electron charge can propagate along the chain following the relay mechanism, as shown in Figure 11.44(a). The driving force of such electron transfer is the electrostatic interaction between two negatively charged neighbouring particles. As a result of this electrostatic repulsion, the electron will be transferred to the next particle. However, the electron relay transfer in such a simple chain of nanoparticles is not reliable. Much better stability can be achieved using a cell containing five particles, depicted in Figure 11.44(b). This cell has two stable configurations, which can be assigned the logical states '1' and '0.' As shown in Figure 11.44(c), the situation of two neighbouring cells having different configurations '1' and '0' (or '0' and '1') is not energetically favourable (and therefore not stable) and should be transformed to the more energetically situations of '1' and '1' (or '0' and '0'). Therefore, information can be transported along the row of

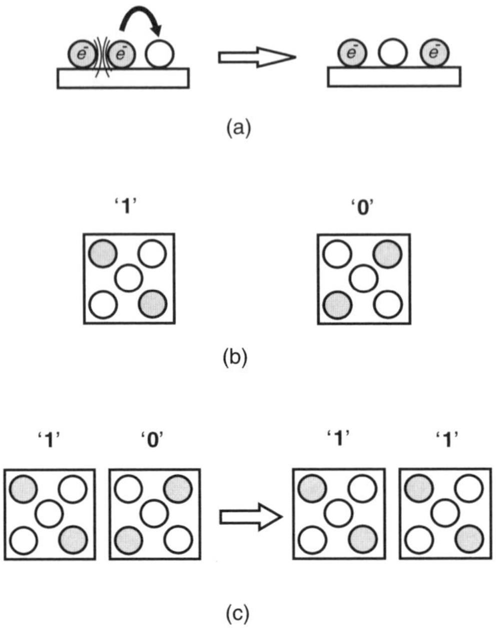

Figure 11.44 (a) Electrostatic interaction in a linear array of quantum dots. (b) Two stable logic states in a five-dot cell. (c) Switching process in the five-dot cell.

five-dot cells. This computing scheme is still binary and suitable to sequential computing, but it does not require the wiring of every cell.

BIBLIOGRAPHY

Amos M. *Genesis Machines*. London: Atlantic Books; 2006.

Anantram MP, Léonard F. Physics of carbon nanotube electronic devices. *Rep. Prog. Phys.* 2006; **69**: 507–561.

Balzani V, Credi A, Venturi M. *Molecular Devices and Machines*. Weinheim: Wiley-VCH; 2003.

Bar-Cohen Y (Editor). *Biomimetics: Biologically Inspired Technologies*. Boca Raton, FL: Taylor and Francis; 2006.

Birge RR (Editor). *Molecular and Biomolecular Electronics*. Washington, DC: American Chemical Society; 1994.

Carter FL (Editor). *Proceedings of the Molecular Electronic Devices Workshop*. Naval Research Laboratory Memorandum Report 4662. Washington, DC: Naval Research Laboratories; 1981.

Claverie A, Tsoukalas D, King T-J, Slaughter JM (Editors). *Materials and Processes for Nonvolatile Memories. Materials Research Society Symposium Proceedings*, Vol. 830. Warrendale, PA: Materials Research Society; 2005.

Cuniberti G, Fagas G, Richer K. (Editors). *Introducing Molecular Electronics*. Berlin: Springer; 2005.

Drexler EK. *Nanosystems*. New York: John Wiley & Sons, Inc.; 1992.

Edwards C. The wages of spin. *Eng. Technol.* 2006; **1**: 36–40.

Feringa BL (Editor). *Molecular Switches*. Weinheim: Wiley-VCH; 2001.

Göpel W, Ziegler Ch (Editors). *Nanostructures Based on Molecular Materials*. Weinheim: VCH; 1992.

Greer J, Korkin A, Labanowski J (Editors). *Nano and Giga Challenges in Microelectronics*. Amsterdam: Elsevier; 2003.
Jones RAL. *Soft Machines*. Oxford: Oxford University Press; 2004.
Mansoori GA. *Principles of Nanotechnology*. Singapore: World Scientific; 2005.
Nabook A. *Organic and Inorganic Nanostructures*. Boston: Artech House; 2005.
Nicolini C (Editor). *Molecular Manufacturing*. New York: Plenum Press; 1996.
Nicolini C. *Molecular Bioelectronics*. Singapore: World Scientific; 1996.
Nicolini C (Editor). *From Neural Networks and Biomolecular Engineering to Bioelectronics*. New York: Plenum Press; 1995.
Poole CP Jr, Owens FJ. *Introduction to Nanotechnology*. Hoboken, NJ: John Wiley & Sons, Inc.; 2003.
Reed MA, Takhee L (Editors). *Molecular Nanoelectronics*. California: American Scientific Publishers; 2003.
Reitman EA. *Molecular Engineering of Nanosystems*. New York: Springer; 2001.
Schwoerer M, Wolf HC. *Organic Molecular Solids*. Berlin: Wiley-VCH; 2006.
Timp G (Editor). *Nanotechnology*. New York: Springer; 1999.
Tour JM. *Molecular Electronics*. Singapore: World Scientific; 2003.
Waser R (Editor). *Nanoelectronics and Information Technology*. 2nd ed. Weinheim: Wiley-VCH; 2005.
Wolf EL. *Nanophysics and Nanotechnology*. Weinheim: Wiley-VCH; 2004.

REFERENCES

[1] Drexler EK. *Nanosystems*. New York: John Wiley & Sons, Inc.; 1992.
[2] Kasap SO. *Principles of Electrical Engineering Materials and Devices*. Boston: McGraw-Hill; 1997.
[3] Reitman EA. *Molecular Engineering of Nanosystems*. New York: Springer; 2001.
[4] Christie P, Roberts GG, Petty MC. Spontaneous polarization in organic superlattices. *Appl. Phys. Lett.* 1986; **48**: 1101–1103.
[5] Jones CA, Petty MC, Roberts GG. Langmuir–Blodgett films: a new class of pyroelectric materials. *IEEE Trans. Ultrasonics Ferroelectrics Frequency Control* 1988; **35**: 736–740.
[6] Petty MC. *Langmuir–Blodgett Films*. Cambridge: Cambridge University Press; 1996.
[7] Petty MC, Cresswell J. Deposition and characterization of multilayer films for nonlinear optics. In D Andrea, A Lapiccirella, G Marletta, A Viticoli, (Editors). *Materials for Photonic Devices*. Singapore: World Scientific; 1991. pp. 259–269.
[8] Ashwell GJ, Ranjan R, Whittam AJ, Gandolfo DS. Second-harmonic generation from alternate-layer and Z-type Langmuir–Blodgett films: optimization of the transparency/efficiency trade-off. *J. Mater. Chem.* 2000; **10**: 63–68.
[9] Pisignano D, Persano L, Gigli G, Visconti P, Stomeo T, De Vittorio M, Barbarella G, Favaretto L, Cingolani R. Planar organic photonic crystals fabricated by soft lithography. *Nanotechnology* 2004; **15**: 766–770.
[10] Galisteo JF, García-Santamaría F, Golmayo D, Juárez BH, López C, Palacios E. Self-assembly approach to optical metamaterials. *J. Opt. A: Pure Appl. Opt.* 2005; **7**: S244–S254.
[11] Bardosova M, Tredgold RH. Ordered layers of monodispersive colloids *J. Mater. Chem.* 2002; **12**: 2835–2842.
[12] Watkins NJ, Yan L, Gao Y. Electronic structure symmetry of interfaces between pentacene and metals. *Appl. Phys. Lett.* 2002; **80**: 4384–4386.
[13] Cui XD, Primak A, Zarate X, Tomfohr J, Sankey OF, Moore AL, Moore TA, Gust D, Harris G, Lindsay SM. Reproducible measurement of single-molecule conductivity. *Science* 2001; **294**: 571–574.

[14] Moore AJ, Goldenberg LM, Bryce MR, Petty MC, Monkman AP, Marenco C, Yarwood J, Joyce MJ, Port SN. Cation recognition by self-assembled layers of novel crown-annelated tetrathiafulvalenes. *Adv. Mater.* 1998; **10**: 395–398.

[15] Aviram A, Ratner MA. Molecular rectifiers. *Chem. Phys. Lett.* 1974; **29**: 277–283.

[16] Ashwell GJ, Stokes RJ. Do alkyl tunnelling barriers contribute to molecular rectification? *J. Mater. Chem.* 2004; **14**: 1228–1230.

[17] Metzger RM. Monolayer rectifiers. *J. Solid State Chem.* 2002; **168**: 696–711.

[18] Honciuc A, Jaiswal A, Gong A, Ashworth K, Spangler CW, Peterson IR, Dalton LR, Metzger RM. Current rectification in a Langmuir–Schaefer monolayer of fullerene–bis[4′-diphenylamino- 4″-(*N*-ethyl-*N*-2″-ethyl)amino-1,4-diphenyl-1,3-butadiene] malonate between Au electrodes. *J. Phys. Chem. B* 2005; **109**: 857–871.

[19] Ashwell GJ, Mohib A, Miller JR. Induced rectification from self-assembled monolayers of sterically hindered π-bridged chromophores. *J. Mater. Chem.* 2005; **15**: 1160–1166.

[20] Peterson IR. Langmuir–Blodgett films: a route to molecular electronics. In W Göpel, Ch Ziegler (Editors). *Nanostructures Based on Molecular Materials*. Weinheim: VCH; 1992, pp. 195–208.

[21] Yang Y, Ouyang J, Ma L, Tseng RJ-H, Chu C-W. Electrical switching and bistability in organic/polymeric thin films and memory devices. *Adv. Funct. Mater.* 2006; **16**: 1001–1014.

[22] Möller S, Perlov C, Jackson W, Taussig C, Forrest SR. A polymer/semiconductor write-once read-many-times memory. *Nature* 2003; **426**: 166–169.

[23] Vettiger P, Despont M, Drechsler U, Dürig U, Häberle W, Lutwyche MI, Rothuizen HE, Stutz R, Widmer R, Binnig GK. The ‘Millipede’ – more than one thousand tips for future AFM data storage. *IBM J. Res. Dev.* 2000; **44**: 323–340.

[24] Despont M, Drechsler U, Yu R, Pogge HB, Vettiger P. Wafer-scale microdevice transfer/interconnect: its application in an AFM-based data-storage system. *J. Microelectromech. Syst.* 2004; **13**: 895–901.

[25] Service RF. Next-generation technology hits an early midlife crisis. *Science* 2003; **302**: 556–559.

[26] Chen Y, Ohlberg DAA, Li X, Stewart DR, Williams RS, Jeppesen JO, Nielsen KA, Stoddart JF, Olynick DL, Anderson E. Nanoscale molecular-switch devices fabricated by imprint lithography. *Appl. Phys. Lett.* 2003; **82**: 1610–1612.

[27] Flood AH, Stoddart JF, Steuerman DW, Heath JR. Whence molecular electronics? *Science* 2004; **306**: 2055–2056.

[28] Diehl MR, Steuerman DW, Tseng H-R, Vignon SA, Star A, Celestre PC, Stoddart JF, Heath JR. Single-walled carbon nanotube-based molecular switch tunnel junctions. *Chem. Phys. Chem.* 2003; **4**: 1335–1339.

[29] Green JE, Choi JW, Boukai A, Bunimovich Y, Johnston-Halperin E, Delonno E, Luo Y, Sheriff BA, Xu K, Shin YS, Tseng H-R, Stoddart JF, Heath JR. A 160-kilobit molecular electronic memory patterned at 10^{11} bits per square centimetre. *Nature* 2007; **445**: 414–417.

[30] Bozano LD, Kean BW, Deine VR, Salem JR, Scott JC. Mechanism for bistability in organic memory elements. *Appl. Phys. Lett.* 2004; **84**: 607–609.

[31] Bozano LD, Kean BW, Beinhoff M, Carter KR, Rice PM, Scott JC. Organic materials and thin-film structures for cross-point memory cells based on trapping in metallic nanoparticles. *Adv. Funct. Mater.* 2005; **15**: 1933–1939.

[32] Simmons JG, Verderber RR. New conduction and reversible memory phenomena in thin insulating films. *Proc. R. Soc. London, Ser. A* 1967; **301**: 77–102.

[33] Ouyang J, Chu C-W, Szmanda CR, Ma L, Yang Y. Programmable polymer thin film and non-volatile memory device. *Nature Mater.* 2004; **31**: 918–922.

[34] Tseng RJ, Huang J, Ouyang J, Kaner RB, Yang Y. Polyaniline nanofibre/gold nanoparticle non-volatile memory. *Nano Lett.* 2005; **5**: 1077–1080.

[35] Tang W, Shi H, Xu G, Ong BS, Popovic ZD, Deng J, Ahao J, Rao G. Memory effect and negative differential resistance by electrode-induced two-dimensional single- electron tunnelling in molecular and organic electronic devices. *Adv. Mater.* 2005; **17**: 2307–2311.

[36] Lauters M, McCarthy B, Sarid D. Multilevel conductance switching in polymer films. *Appl. Phys. Lett.* 2006; **89**, 013507.
[37] Scott JC, Bozano LD. Nonvolatile memory elements based on organic materials. Adv. Mater. 2007; 19: 1452–1463.
[38] Rueckes T, Kim K, Joselevich E, Tseng GY, Cheung CL, Lieber CM. Carbon nanotube-based nonvolatile random access memory for molecular computing. *Science* 2000; **289**: 94–97.
[39] Terabe K, Hasegawa T, Nakayama T, Aono M. Quantised conductance atomic switch. *Nature* 2005; **433**: 47–50.
[40] Ruitenbeek J V. Silver nanoswitch. *Nature* 2005; **433**: 21–22.
[41] Liu Z, Yasseri AA, Lindsey JS, Bocian DF. Molecular memories that survive silicon device processing and real-world operation. *Science* 2003; **302**: 1543–1545.
[42] Fazio A. Flash memory scaling. *MRS Bull.* 2004; **29**: 814–817.
[43] Lee C, Meteer J, Narayan V, Kan EC. Self-assembly of metal nanocrystals on ultrathin oxide for non-volatile memory applications. *J. Electron. Mater.* 2005; **34**: 1–11.
[44] Paul S, Pearson C, Molloy A, Cousins MA, Green M, Kolliopoulou S, Dimitrakis P, Normand P, Tsoukalas D, Petty MC. Langmuir–Blodgett film deposition of metallic nanoparticles and their application to electronic memory structures. *Nano Lett.* 2003; **3**: 533–536.
[45] Kolliopoulou S, Dimitrakis P, Normand P, Zhang HL, Cant N, Evans SD, Paul S, Pearson C, Molloy A, Petty MC, Tsoukalas D. Integration of organic insulator and self-assembled gold nanoparticles on Si MOSFET for novel non-volatile memory cells. *Microelectron. Eng.* 2004; **73–74**: 725–729.
[46] Awschalom DD, Flatte, ME, Samarth N. Spintronics. *Sci. Am.* 2002; **286**: 66–73.
[47] Wolf SA, Chtchelkanova AY, Treger DM. Spintronics – a retrospective and perspective. *IBM J. Res. Dev.* 2006; **50**: 101–110.
[48] Rocha AR, García-Suárez VM, Bailey SW, Lambert CJ, Ferrer J, Sanvito S. Towards molecular spintronics. *Nature Mater.* 2005; **4**: 335–339.
[49] Wang FJ, Xiong ZH, Wu D, Shi J, Vardeny ZV. The case for Fe/Alq_3/Co spin-valve devices. *Synth. Met.* 2005; **155**: 172–175.
[50] Burrows PE, Donovan KJ, Wilson EG. Electron motion perpendicular to Langmuir–Blodgett multilayers of conjugated macrocyclic compounds: the organic quantum well. *Thin Solid Films* 1989; **179**: 129–136.
[51] Uchida K. Single-electron devices for logic applications. In R Waser (Editor). *Nanoelectronics and Information Technology*, 2nd ed. Weinheim: Wiley-VCH; 2005, pp. 423–441.
[52] Yano K, Ishii T, Sano T, Mine T, Murai F, Hashimoto T, Kobayashi T, Kure T, Seki K. Single-electron memory for giga-to-tera bit storage. *Proc. IEEE* 1999; **87**: 633–650.
[53] Oncel N, Hallbäck A-S, Zandvliet HJW, Speets EA, Ravoo BJ, Reinhoudt DN, Poelsema B. Coulomb blockade of small Pd clusters. *J. Chem. Phys.* 2005; **123**: 004703.
[54] Balzani V, Credi A, Venturi M. *Molecular Devices and Machines*. Weinheim: Wiley-VCH; 2003.
[55] Martin PJ. Photochromism. In MC Petty, MR Bryce, D Bloor (Editors). *An Introduction to Molecular Electronics*. London: Edward Arnold; 1995, pp. 112–141.
[56] Srinivasan MP. Organic photochromism: spiro compounds as functional molecules. In TH Richardson (Editor). *Functional Organic Polymeric Materials*. Chichester: John Wiley & Sons, Ltd 2000. pp. 273–293.
[57] Aviram A, Seiden PE, Ratner MA. Theoretical and experimental studies of hemiquinones and comments on their suitability for molecular storage elements. In FL Carter (Editor). *Proceedings of the Molecular Electronic Devices Workshop*. Naval Research Laboratory Memorandum Report 4662. Washington, DC: Naval Research Laboratoriers; 1981, pp. 3–16.
[58] Echegoyen L, Echegoyen LE. Electrochemistry of fullerenes and their derivatives. *Acc. Chem. Res.* 1998; **31**: 593–601.
[59] Mingotaud C, Delhaus P, Meisel MW, Talham DR. Magnetic Langmuir–Blodgett films. In JS Miller, M Drillon (Editors). *Magnetism: molecules to materials. Molecule- Based Materials* Vol. 2. Weinheim: Wiley-VCH; 2001, pp. 457–484.

[60] Avouris P. Molecular electronics with carbon nanotubes. *Acc. Chem. Res.* 2002; **35**: 1926–1034.

[61] Ouyang M, Huang J-L, Lieber CM. Fundamental electronic properties and applications of single-walled carbon nanotubes. *Acc. Chem. Res.* 2002; **35**: 1018–1025.

[62] McEuen PL, Park JY. Electron transport in single-walled carbon nanotubes. *MRS Bull.* 2004; **29**: 272–275.

[63] Javey A, Kim H, Brink M, Wang Q, Ural A, Guo J, McIntyre P, McEuen P, Lundstrom M, Dai H. High-κ dielectrics for advanced carbon-nanotube transistors and logic gates. *Nature Mater.* 2002; **1**: 241–246.

[64] Avouris P, Appenzeller J. Electronics and optoelectronics with carbon nanotubes. *Ind. Phys.* 2004; **10**: 18–21.

[65] Appenzeller J, Joselevich E, Hönlein W. Carbon nanotubes for data processing. In R Waser (Editor). *Nanoelectronics and Information Technology*, 2nd ed. Weinheim: Wiley- VCH; 2005, pp. 471–497.

[66] Lahann J, Mitragotri S, Tran TN, Kaido H, Sundaram J, Choi IS, Hoffer S, Somorjai GA, Langer R. A reversibly switching surface. *Science* 2003; **299**: 371–374.

[67] Lahann JR Langer R. Smart materials with dynamically controllable surfaces. *MRS Bull.* 2005; **30**: 185–188.

[68] Liu Y, Mu L, Liu B, Kong J. Controlled switchable surface. *Chem. Eur. J.* 2005; **11**: 2622–2631.

[69] Sauvage JP, Dietrich-Buchecker CO (Editors). *Molecular Catenanes, Rotaxanes and Knots*. Weinheim: Wiley-VCH; 1999.

[70] Shinkai S, Nakaji T, Ogawa T, Shigematsu K, Manabe O. Photoresponsive crown ethers. 2. Photocontrol of ion extraction and ion transport by a bis(crown ether) with a butterfly-like motion. *J. Am. Chem. Soc.* 1981; **103**: 111–115.

[71] De Silva AP, De Silva SA, Dissanayake AS, Sandanayake KRAS. Comparmental fluorescent pH indicators with nearly complete predictability of indicator parameters – molecular engineering of pH sensors. *J. Chem. Soc., Chem. Commun.* 1989: 1054–1056.

[72] De Silva AP, Gunaratne HQN, McCoy CP. Molecular photoionic AND logic gates with bright fluorescence and 'off–on' digital action. *J. Am. Chem. Soc.* 1997; **119**: 7891–7892.

[73] Carter FL (Editor). *Proceedings of the Molecular Electronic Devices Workshop*. Naval Research Laboratory Memorandum Report 4662. Washington, DC: Naval Research Laboratories; 1981.

[74] Roth S. *One-Dimensional Metals*. Weinheim: VCH; 1995.

[75] Barker JR. Molecular electronic logic and architectures. In MC Petty, MR Bryce, D Bloor (Editors). *An Introduction to Molecular Electronics*. London: Edward Arnold; 1995, pp. 345–376.

[76] Amos M. *Genesis Machines*. London: Atlantic Books; 2006.

[77] Heath JR, Kuekes PJ, Snider GS, Williams RS. A defect-tolerant computer architecture: opportunities for nanotechnology. *Science* 1998; **280**: 1716–1721.

[78] Nabook. A. *Organic and Inorganic Nanostructures*. Boston: Artech House; 2005.

[79] Preskill J. Battling decoherence: the fault-tolerant quantum computer. *Phys. Today* 1999; **52**: 24–30.

[80] Montemerlo MS, Love JC, Opiteck GJ, Goldhaber-Gordon DJ, Ellenbogen JC. *Technologies and Designs for Electronic Nanocomputers*. MITRE Technical Report No. 96W0000044. McLean, VA: MITRE; 1996.

12 Bioelectronics

Such stuff as dreams are made on

12.1 INTRODUCTION

Challenging and longer-term goals of molecular-scale electronics are to emulate some of the sophisticated processes occurring in nature and to produce operational biolectronic devices. In this chapter, we take a look into the biological world and its relationship, or possible relationship, with molecular electronics. A brief overview of the most important biomolecules is first provided. Examples of some biological processes that might be exploited in molecular electronic devices, particularly in the areas of sensors, optical memories and energy conversion, are described later in the chapter.

12.2 BIOLOGICAL BUILDING BLOCKS

The *cell* (Section 12.4) is the fundamental structural and functional unit of all living organisms. Many different molecules are found the cell. The detailed structure and conformation of each compound determines in which chemical reactions it can participate, and therefore its role in the life of the cell. Important classes of biomolecules include nucleic acids, proteins, carbohydrates and lipids. Other compounds perform functions such as transporting energy from one part of the cell to another, or utilizing the sun's energy to drive chemical reactions. All these molecules, and the cell itself, are in a state of constant change. A cell cannot remain healthy unless it is continually forming and breaking down proteins, carbohydrates and lipids, repairing damaged nucleic acids and using and storing energy. Such energy-linked reactions are collectively known as *metabolism*.

12.2.1 Amino Acids and Peptides

Amino acids are a class of organic compounds that contain both the amino (NH_2) and carboxyl (COOH) chemical groups. The primary building blocks of all proteins, regardless of their species of origin, are the group of 20 different amino acids listed in Table 12.1. The amino and carboxyl groups are both attached to a single carbon atom, called the α-carbon atom. The α-amino group is free or unsubstituted in all the amino acids except one, proline. A further variable group, R, is attached to the α-carbon; it is in their R groups that the molecules of the 20 amino acids differ from one another. The simplest of the acids, glycine, contains an R group that comprises a single hydrogen atom. Since the amino and carboxylic acid groups are basic and acidic, respectively (Chapter 5, Section 5.2.2), amino acids are

© 2007 John Wiley & Sons, Ltd

Table 12.1 The 20 amino acids common in proteins.

Amino acid	Three-letter symbol
Alanine	Ala
Arginine	Arg
Asparagine	Asn
Aspartic acid	Asp
Cysteine	Cys
Glutamine	Gln
Glutamic acid	Glu
Glycine	Gly
Histidine	His
Isoleucine	Ile
Leucine	Leu
Lysine	Lys
Methionine	Met
Phenylalanine	Phe
Proline	Pro
Serine	Ser
Threonine	Thr
Tryptophan	Trp
Tyrosine	Tyr
Valine	Val

zwitterions, with both negative and positive charges. The structure of the neutral (undissociated) and charged forms of glycine are shown in Figure 12.1. Zwitterions are highly polar substances for which intermolecular electrostatic attraction leads to strong crystal lattices. With the exception of glycine, all of the amino acids are chiral molecules (Chapter 2, Section 2.4.2).

Two amino acid molecules may be joined to yield a *dipeptide* through a *peptide bond*, formed by the removal of a water molecule from the carboxyl group of one amino acid and the α-amino group of the other by the action of strong condensing agents. Figure 12.2 shows the peptide formed from two molecules of glycine, glycylglycine. Like glycine, this exists as a zwitterion.

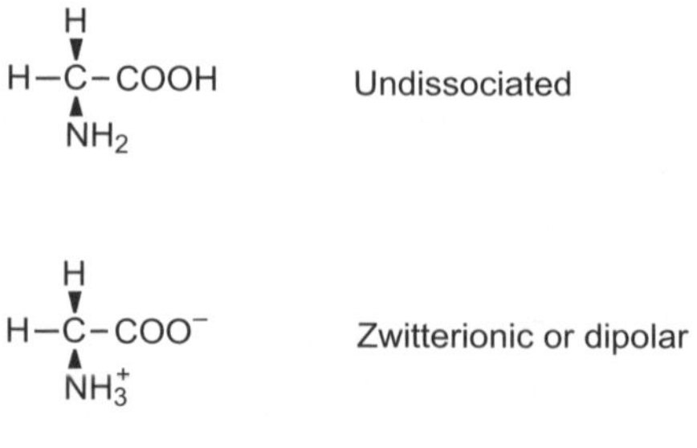

Figure 12.1 Undissociated and zwitterionic forms of the amino acid glycine.

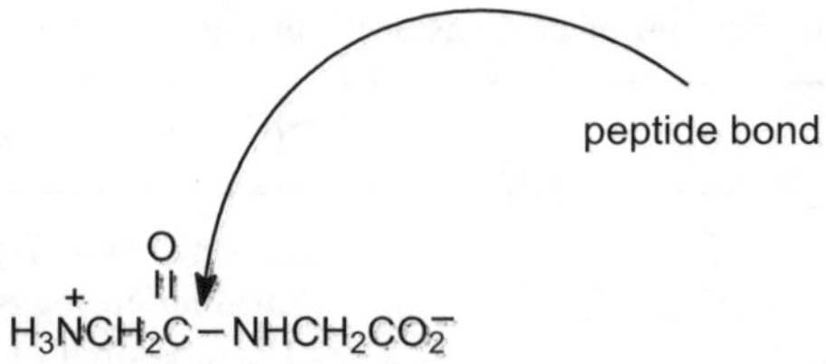

Figure 12.2 Glycylglycine – a dipeptide. The figure shows the peptide bond formed between the two molecules of glycine.

Higher peptides are also possible; a tripeptide contains three amino acids, a tetrapeptide four, and so on. Peptides are named from the sequence of their constituent amino acids, beginning from the amino-terminated end. When many amino acids are joined in a long chain, this is called a *polypeptide*. Such compounds contain only one free α-amino acid group and one free α-carboxyl group at their ends. An example is given in Figure 12.3, the pentapeptide serylglycltyrosinylalanylleucine.

In addition to the amino acids that form proteins, more than 150 other amino acids have been found in Nature, including some that have the carboxyl and amino groups attached to separate carbon atoms. These unusually structured amino acids are most often found in fungi and higher plants.

Amino-terminal end

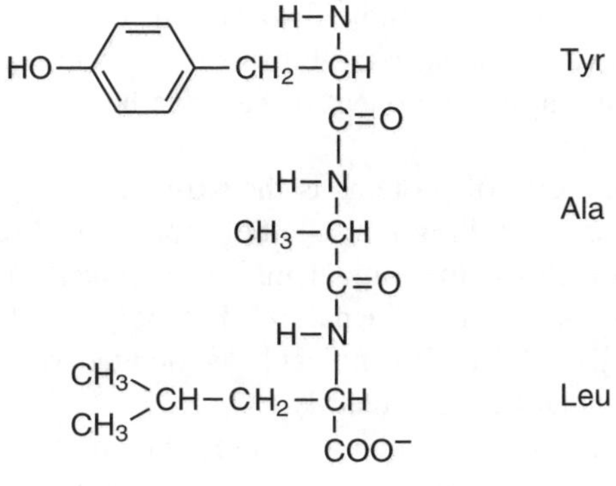

Carboxyl-terminal end

Figure 12.3 Structure of the polypeptide serylglycyltyrosinylalanylleucine. Peptides are named beginning with the NH_2-terminal amino acid.

Table 12.2 Classification of proteins according to their biological function.

Class	Examples
Enzymes	Ribonuclease Trypsin
Storage proteins	Ovalbumin (eggs) Casein (milk)
Transport proteins	Haemoglobin Serum albumin
Contractile proteins	Actin Myosin
Protective proteins of blood	Antibodies Fibrinogen
Toxins	Botulinus toxin Snake venoms
Hormones	Insulin Adreneocorticotrophin
Structure proteins	Keratins Collagen

12.2.2 Proteins

A *protein* is a complex, high molecular mass, organic compound consisting of amino acids joined by peptide bonds. These molecules are the most abundant species in most cells and generally constitute 50% of their dry weight. Proteins are versatile cell components, some are *enzymes*, some serve as structural components and some have *hormonal* activity. Table 12.2 provides examples of the different types of proteins, classified according to their biological function.

A protein may be a single polypeptide chain, or it may consist of several such chains held together by weak molecular bonds. The R groups of the amino acid subunits determine the final shape of the protein and its chemical properties. Using only 20 different amino acids, a cell constructs thousands of different proteins, each of which has a highly specialized function.

Proteins serve two important biological functions. First, they can act as a structural material. The structural proteins tend to be fibrous in nature, i.e. the polypeptide chains are lined up more or less parallel to each other and are joined to one another by hydrogen bonds. These are physically tough and are normally insoluble in water. Depending on the actual three-dimensional arrangement of the individual protein molecule and its interaction with other similar molecules, a variety of structural forms may result. Typical examples of fibrous proteins are α-*keratin*, the major component of hair, feathers, nails and skin, and *collagen*, the major component of tendons.

The other important property of proteins is their role as biological regulators. Here, the proteins are responsible for controlling the speed of biochemical reactions and the transport of various materials throughout the organisms. The catalytic proteins (enzymes) and transport proteins tend to be globular in nature. The polypeptide chain is folded around itself in such a way to give the entire molecule a rounded shape. Globular proteins are soluble in aqueous systems and diffuse readily.

Antibodies (also referred to as immunoglobulins) are produced by white blood cells. These special proteins respond to a specific *antigen* (bacteria, virus or toxin). Each antibody has a section in its chemical makeup that is sensitive to a particular antigen and binds to it in some way. An antibody consists of four polypeptides – two heavy chains and two light chains joined to form a Y-shaped molecule, as depicted in Figure 12.4. The amino acid sequences in the tips of the Y vary significantly among different antibodies. This variable region,

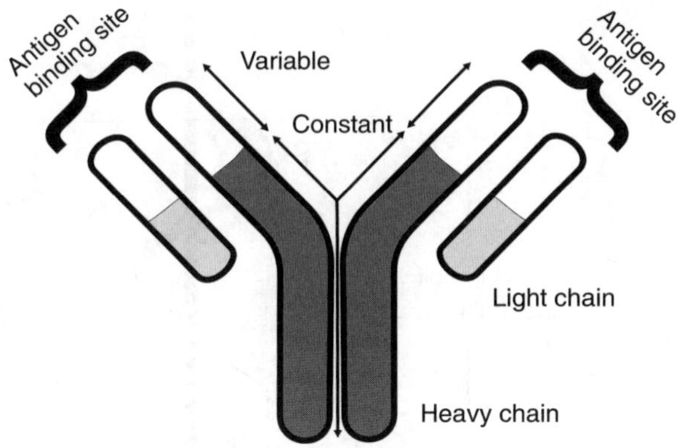

Figure 12.4 Schematic diagram showing the Y-shaped structure of an antibody.

composed of 110–130 amino acids, gives the antibody its specificity for binding an antigen. The variable region includes the ends of the light and heavy chains.

The sequence of amino acids in the covalent backbone of a protein is called its primary structure. The secondary structure refers to the specific geometric arrangement of the polypeptide chain along one axis. Two common arrangements are the α-*helix*, shown in Figure 12.5, and the β-*conformation*, Figure 12.6, in which the polypeptide chains are in an extended zigzag configuration called a *pleated sheet*. The α-helix is right-handed, with a pitch of 0.54 nm or 3.6 amino acid units. Both the helix and pleated sheet are very stable structures, held together by hydrogen bonding. The specific configurations of the polypeptide chains are stable because of particular amino acid sequences. For example, an α-helix tends to form spontaneously only in the case of polypeptide chains in which consecutive R groups are relatively small and uncharged, as in α-ketatin.

A tertiary structure for proteins also exists. This term is used to refer to the three-dimensional structure of globular proteins, in which the polypeptide chain is tightly folded and packed into a compact spherical form. The molecule tends to orient itself so that the nonpolar side chains lie inside the bulk of the structure, where they attract each other by van der Waals forces. The polar side chains are usually found on the surface of the molecule; consequently, these can hydrogen bond to the solvent molecules and confer the necessary water solubility. The process by which a linear protein chain forms its secondary and tertiary structure is called *protein folding*.

12.2.3 Enzymes

The enzymes make up the largest and most highly specialized class of proteins. They act as catalysts for thousands of chemical reactions. These reactions would otherwise occur at extremely low rates. Enzymes have traditionally been named according to the substance that they act on, called the *substrate* (NB: this word is used in a different context in other parts of this book, in particular Chapter 7) or according to the nature of the reaction catalyzed. Thus urease catalyzes the hydrolysis of urea and arginase catalyzes the hydrolysis of arginine. However, in many cases enzymes have been given names that are

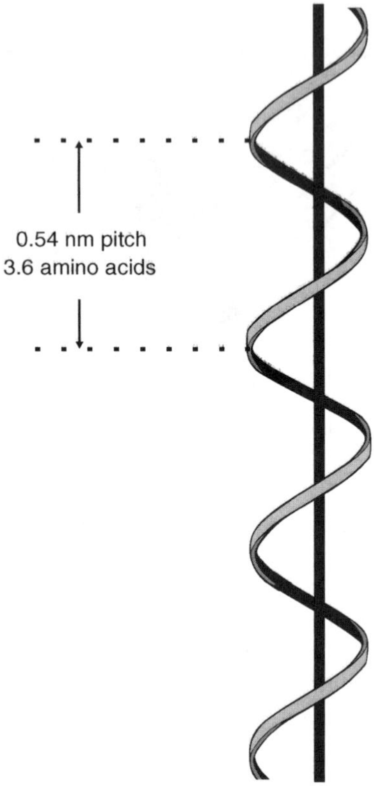

Figure 12.5 Right-handed helix.

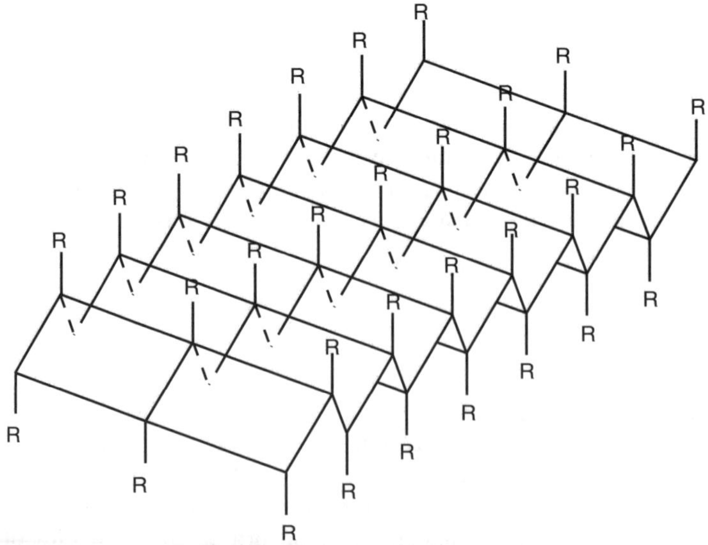

Figure 12.6 Schematic representation of three parallel chains in β structure, showing the pleated sheet arrangement. All the R groups project above or below the plane of the figure.

Table 12.3 Classification of enzymes.

Oxido-reductases (electron-transfer reactions)
Transferases (transfer of functional groups)
Hydrolases (hydrolysis reactions)
Lyases (addition to double bonds)
Isomerases (isomerization reactions)
Ligases (formation of bonds with ATP cleavage)

not so informative, such as pepsin or trypsin. Enzymes are grouped into six major classes, as indicated in Table 12.3.

Enzymes have molecular weights ranging from about 12 000 to over 10^6. Some enzymes consist only of one or more polypeptide chains, but others contain an additional component needed for activity, called a *cofactor.* This may be a metal such as Mg, Mn, Zn or Fe, or it may be a complex organic molecule, usually called a *coenzyme*. Enzymes use sophisticated molecular recognition and work via a 'lock–key' mechanism by which only a specific substrate will have the correct shape to fit within the enzyme's catalytic site. Therefore, most enzymes only serve a single well-defined catalytic function.

12.2.4 Carbohydrates

Carbohydrates are the basic fuel molecules of the cell. These molecules contain carbon, hydrogen and oxygen in approximately equal amounts. Green plants and some bacteria use *photosynthesis* to make simple carbohydrates (sugars) from carbon dioxide, water and sunlight (Section 12.10.2). Animals, however, obtain their carbohydrates from foods. Once a cell possesses carbohydrates, it may break these molecules down to yield chemical energy or use them as raw material to produce other biomolecules.

The term carbohydrate is used loosely to characterize an entire group of natural products that are related to simple sugars. There are three major classes of carbohydrates: *monosaccharides*, *oligosaccharides* and *polysaccharides*. Glucose is an example of a monosaccaride and is the most important fuel molecule for most organisms. This is a ring structure, as shown in Figure 12.7. The particular form of the molecule depicted is that used extensively by sugar chemists: the sugar ring is written as a planar hexagon with the oxygen

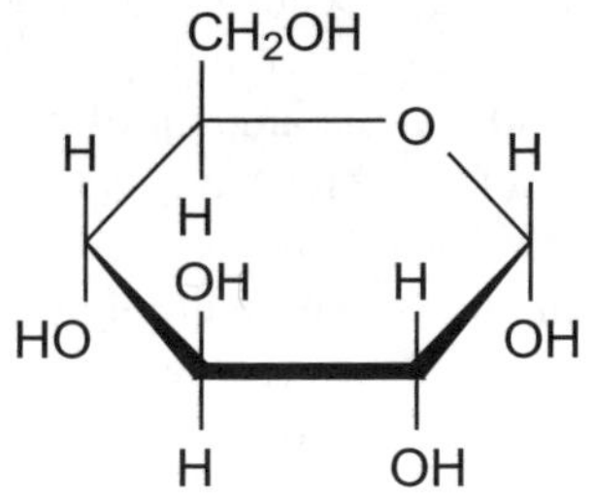

Figure 12.7 Structure of α-D-glucose.

in the upper right vertex and additional chemical groups are indicated by straight lines through each vertex, either above or below the plane. The carbohydrates occur in optically active form (generally true for natural products) and only one enantiomer (Chapter 2, Section 2.4.2) is found in Nature. The usual form of glucose found in Nature is dextrorotatory: D-glucose.

Sucrose, or cane sugar, is a disaccharide. The most abundant carbohydrates in nature are the plant polysaccharides cellulose and starch, which are polymers of glucose. Cellulose is the constituent of plant walls that provides stiffness and strength. It is important that plants do not lose this strength, thus they do not possess enzymes capable of digesting cellulose. In contrast, starch forms a staple food of many plants and animals; all plants and many animals possess enzymes that allow them to digest starch. Starch contains two types of polysaccharides: α-amylose and amylopectin. The former consists of long, unbranched chains of D-glucose units, whereas amylopectin is highly branched. Despite the widespread differences between cellulose and starch, these molecules have exactly the same chemical formula. The difference is in the way the rings are joined. The cellulose molecules are in such a geometric arrangement that they produce stiff molecules, packed tightly together, and are held in place by strong hydrogen bonds. In contrast, the orientation of the joints between the sugar rings in the starch molecules leads to a more open helical structure and fewer internal bonds. Although amylose and amylopectin molecules do pack together, they do so in a looser and weaker fashion and are easily separated, to be broken down by the specific enzymes that all plants and most animals possess for the purpose.

12.2.5 Lipids

Lipids are fatty substances that play a variety of roles in the cell. Some are held in storage for use as high-energy fuel; others serve as essential components of the cell membrane. Examples are shown in Figure 12.8. A simple type of lipid is a fatty acid [Figure 12.8(a)], which is used as the basis of Langmuir–Blodgett film deposition (Chapter 7, Section 7.3.1). This consists of a long hydrocarbon chain terminated in a carboxylic acid group. The chain may be saturated or it may contain one or more double bonds; a few fatty acids contain triple bonds. Nearly all fatty acids in nature have an even number of carbon atoms and have chains that are between 14 and 22 carbon atoms long.

The simplest and most abundant lipids, which contain fatty acids as building blocks, are the neutral lipids, also called *fats*, *triglycerides* or *triaclyglycerols* [Figure 12.8(b)]. These are esters of the alcohol glycerol, with three fatty acid molecules, and bear no net electrical charge. Triacylglycerols are the major component of fats in plant and animal cells. These compounds are excellent forms of energy because of the high number of reduced CH groups available for oxidation-dependent energy generation processes.

The melting point of a fat depends on the amount of unsaturation in the fatty acids. Fats with a preponderance of unsaturated fatty acids have melting points below room temperature and are referred to as oils (e.g. olive oil). However, the melting point of a natural fat may be increased by hydrogenation. The fat is exposed to hydrogen at a high temperature and in the presence of a catalyst. During this process some double bonds are converted into single bonds and other double bonds are converted from a *cis* to a *trans* configuration (Chapter 2, Section 2.4.3). In both cases, the effect is to straighten out the molecules so they can lie closer together and become solid rather than liquid.

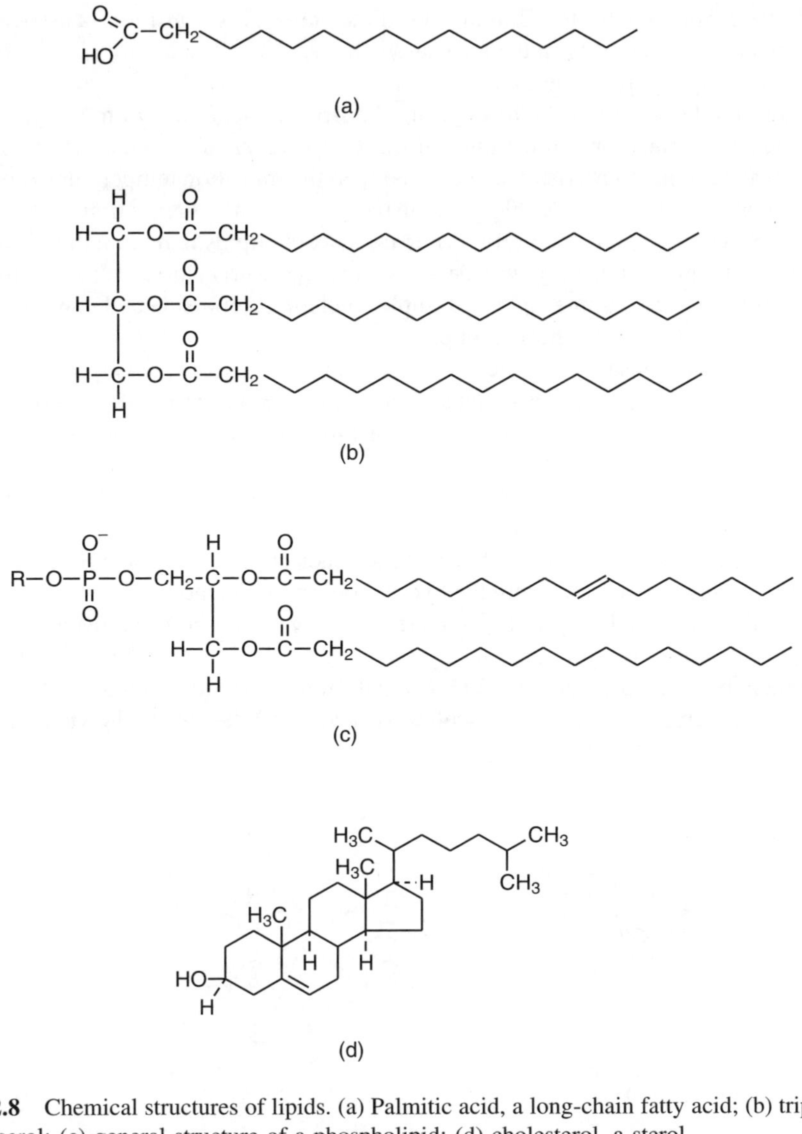

Figure 12.8 Chemical structures of lipids. (a) Palmitic acid, a long-chain fatty acid; (b) tripalmitin, a triacylglycerol; (c) general structure of a phospholipid; (d) cholesterol, a sterol.

Triacylglycerols undergo hydrolysis when boiled with acids or bases or when acted upon by lipase enzymes, such as present in pancreatic juices. Hydrolysis of triacylglycerols with an alkali, a process called *saponification*, yields a mixture of fatty acid soaps and glycerol. *Waxes* are related to the triacylglycerols. These compounds are naturally occurring esters of long-chain carboxylic acids (C_{16} or greater) with long-chain alcohols (C_{16} or greater).

Three major classes of lipids are found in biological membranes: *phospholipids*, *glycolipids* and *cholesterol*. Phospholipids [Figure 12.8(c)] can be derived from glycerol and include phosphorus in the form of phosphoric acid. The major phospholipids found in cells contain two fatty acid molecules, joined to the first and second hydroxyl groups of the glycerol. These molecules primarily act as structural elements. All phospholipids undergo a change of phase from a solid to a liquid crystalline state. This transition is associated with increased disorder and mobility in the fatty alkyl chains. As is the case for simple long-chain fatty acids, the temperature of the transformation depends on the chain length, the number of double bonds and the nature of the head group. Phosphatidylcholine, with two palmitic acid (C_{16}) chains, has a transition temperature of 42 °C. However, if the chain lengths are both reduced by two carbons (to myristic acid chains) then the transition temperature is reduced to 23 °C. The presence of one *cis* double bond also causes a marked reduction in the transition temperature. Since most of the naturally occurring phospholipids in mammalian cells have at least one such double bond, they will be above the phase transition at body temperature. Glycolipids are sugar-containing lipids; examples include cerebroside and ganglioside. They contain a polar, but uncharged head group.

Certain lipids are nonsaponificable, i.e. they cannot be hydrolyzed by heating with an alkali. There are two major groups of these lipids: *steroids* and *terpenes*. Although most of these occur in only trace amounts in cells, one type of steroid, the sterols, is extremely abundant. Cholesterol [Figure 12.8(d)] is the major sterol in animal tissues.

Polar lipids readily disperse to form micelles (Chapter 8, Section 8.2.2). These are spherical structures in which the hydrocarbon tails of the lipids are on the inside, hidden from the aqueous environment, and the charged hydrophilic heads are on the outside. If the head group of the lipid is not strong compared with the hydrophobic part, the molecules can form spherical *vesicles* in which the double layers form a shell, with water on both the inside and outside. Phospholipids also form bilayer structures, particularly at the interface between two aqueous surfaces. Such bilayers provide the fundamental framework for natural cell membranes (Section 12.6). Schematic diagrams of micelles, vesicles and bilayers are shown in Figure 12.9.

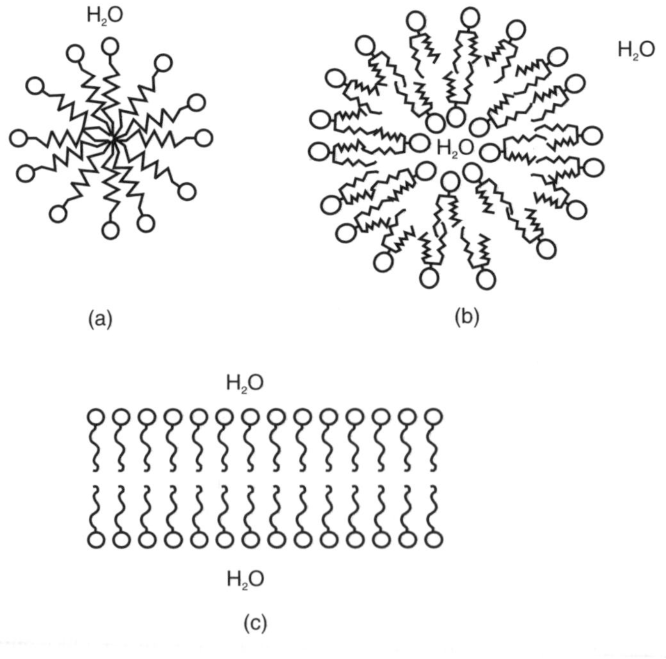

Figure 12.9 Cross-sections of (a) a micelle, (b) a vesicle and (c) a bilayer formed by amphiphilic molecules.

12.3 NUCLEOTIDES

Nucleotides are the recurring structural units of the nucleic acids. These molecules contain three characteristic components: a nitrogenous base, a five-carbon sugar and phosphoric acid. The nucleotides are the compounds responsible for storing and transferring genetic information and are enormous molecules made up of long strands of subunits, called bases, arranged in a precise sequence. The bases are 'read' by other components of the cell and used as a guide in making proteins. Two important types of nucleic acid are *ribonucleic acid* (RNA) and *deoxyribonucleic acid* (DNA).

12.3.1 Bases

The bases found in nucleotides are of two types. These are derivatives of two parent heterocyclic compounds pyrimidine and purine, which are themselves not found in Nature (Figure 12.10). Three pyrimidine bases are common in nucleic acids: uracil, thymine and cytosine, universally abbreviated as U, T and C, respectively. Uracil is generally found in

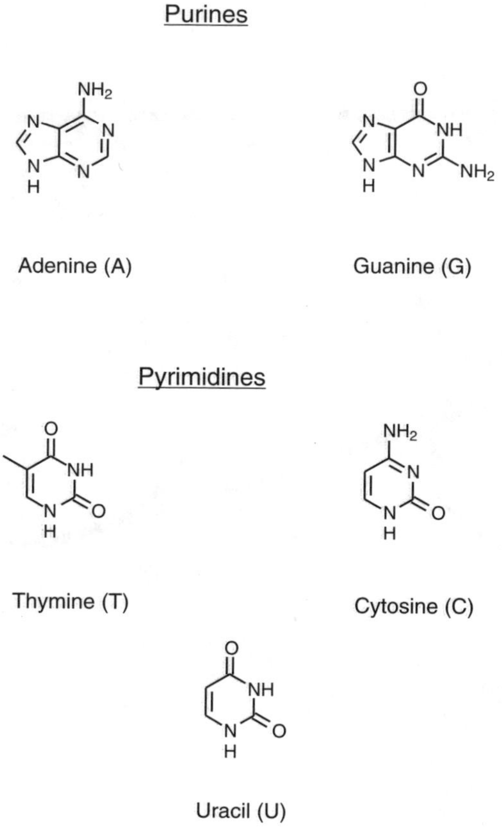

Figure 12.10 The bases in nucleic acids. Thymine occurs in DNA but not RNA and uracil in RNA but not DNA.

RNA and thymine in DNA; cytosine is found in both RNA and DNA. There are two common purine bases, found in both RNA and DNA: adenine (A) and guanine (G). The pyrimidine and purine bases are nearly flat molecules which are relatively insoluble in water.

12.3.2 DNA

In 1953, James Watson and Francis Crick established the structure of DNA, which in 1962 led to the award of the Nobel Prize for Medicine (together with Maurice Wilkins). The DNA molecule is composed of two long strands in the form of a double helix (Figure 12.11). The strands are made up of alternating phosphate and sugar molecules. The nitrogen bases provide links between these strands, holding them together. Each base is attached to a sugar molecule and is linked by a hydrogen bond to a complementary base on the opposite strand. Adenine always binds to thymine (A to T) and guanine always binds to cytosine (G to C). To

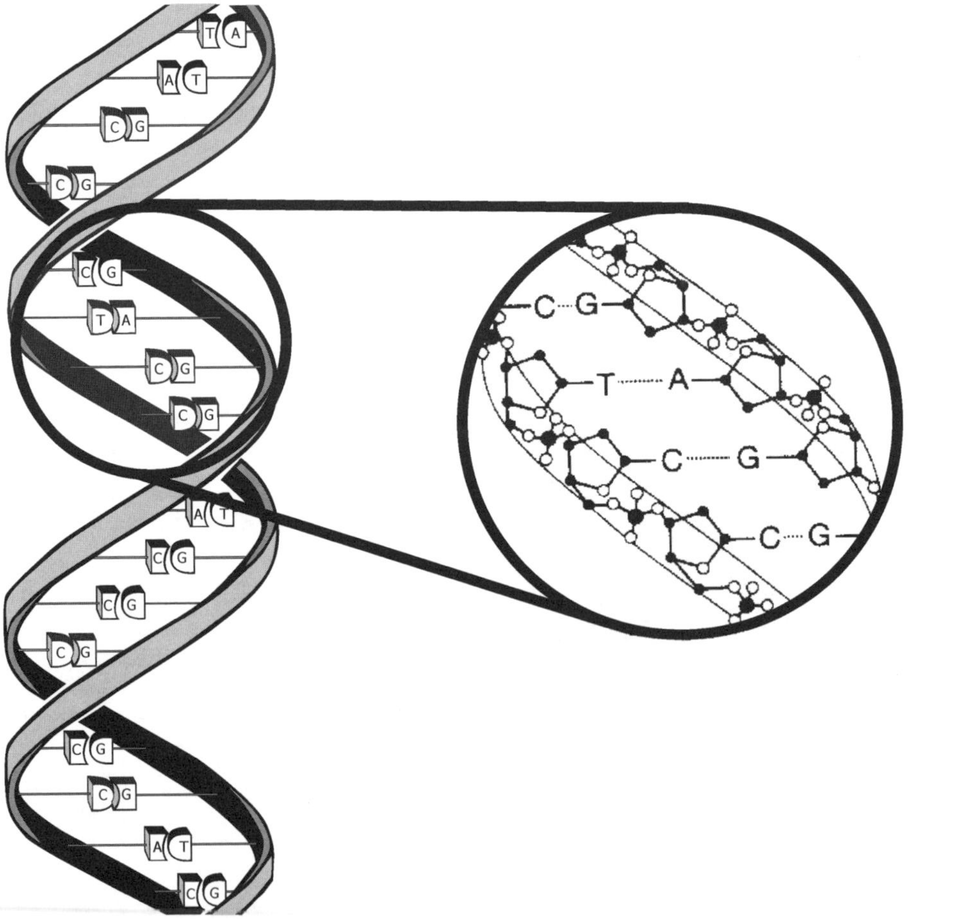

Figure 12.11 The DNA double helix. The two strands are held together by hydrogen bonding between complementary base pairs.

make a new, identical copy of the DNA molecule, the two strands need only unwind and separate at the bases (which are weakly bound); with more nucleotides available in the cell, new complementary bases can link with each separated strand and two double helixes result. If the sequence of bases were AGATC on one existing strand, the new strand would contain the complementary, or 'mirror image,' sequence TCTAG. In Nature, the DNA backbone is tightly coiled up. This packing is now known to be based on minute particles of protein known as *nucleosomes*. The DNA is wound around each nucleosome in succession to form a beaded structure. The structure is then further folded so that the beads associate in regular coils. Thus, the DNA has a coiled-coil configuration, like the filament of some electric light bulbs.

12.3.3 RNA

Translation of the genetic code from base sequences to amino acid sequences does not occur in a single step. This is the role of the other nucleic acid, RNA. RNA is a molecule that links the two worlds of DNA and proteins. DNA's letter Ts are made from RNA's letter Us. RNA makes up about 5–10% of the total weight of the cell. There are three major types of ribonucleic acids, *messenger RNA* (mRNA), *ribosomal RNA* (rRNA) and *transfer RNA* (tRNA). Messenger RNA contains only four bases, A, G, C and U. It is enzymatically synthesized in the cell nucleus in such a way that the base sequence of the mRNA molecule is complementary to the base sequence of one of the strands of the DNA molecule. Each mRNA molecule carries the code for one or more protein molecules. Transfer RNAs are relatively small molecules that act as carriers of specific amino acids during protein synthesis. Each of the 20 amino acids found in proteins has one or more corresponding tRNAs. The tRNA molecule may exist in its free form or attached to its specific amino acid. Finally, ribosomal RNA is the most abundant type of RNA. It plays an important role in the structure and biological function of *ribosomes* and constitutes up to 65% of their weight. Ribosomes are complexes of RNA and proteins and undertake the job of translating DNA 'recipes' into proteins.

12.3.4 ATP, ADP

Adenosine triphosphate (ATP) is a nucleotide that performs many essential roles in the cell. This molecule may be considered as the energy 'currency' of life. The ATP molecule is composed of three components, as shown in Figure 12.12. At the centre is a sugar molecule, ribose (the same sugar that forms the basis of DNA). Attached to one side of this is the base adenine. The other side of the sugar is attached to a string of phosphate groups. ATP is remarkable for its ability to enter into many coupled reactions and both to extract and provide energy. In animal systems, the ATP is synthesized in the tiny energy factories called *mitochondria*.

When the third phosphate group of ATP is removed by hydrolysis, a substantial amount of free energy is released; the exact amount depends on the conditions:

$$\text{ATP} + \text{H}_2\text{O} \rightarrow \text{ADP} + \text{phosphate} \tag{12.1}$$

where ADP is *adenosine diphosphate* (Figure 12.12).

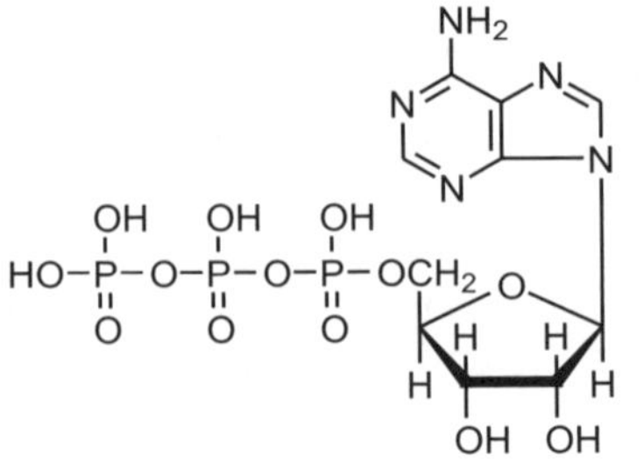

Adenosine triphosphate, ATP

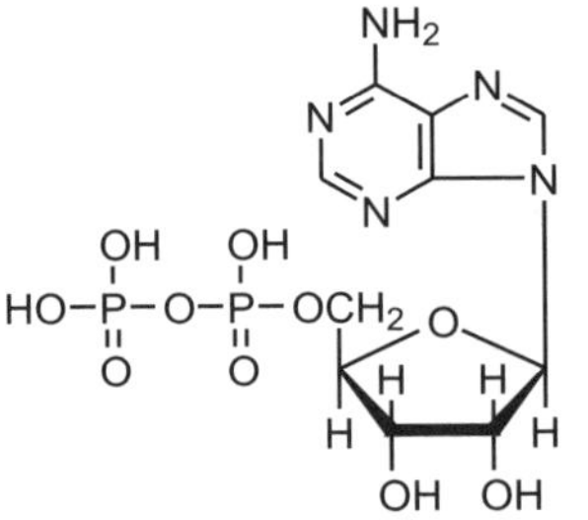

Adenosine diphosphate, ADP

Figure 12.12 Adenosine triphosphate (ATP) and adenosine diphosphate (ADP).

This conversion from ATP to ADP is a key reaction for the energy supply for life processes. Living things can extract energy from ATP like a battery. The ATP can power reactions by losing one of its phosphorous groups to form ADP, but then food energy in the mitochondria can be used to convert the ADP back to ATP, 'recharging' the battery. In plants, sunlight can be used to convert the less active compound back to the highly energetic form. For animals, the energy from high energy storage molecules is used to maintain life; these are then recharged to their high-energy state. The oxidation of glucose operates in a cycle called the *Krebs cycle* in animal cells to provide energy for the conversion of ADP to ATP.

12.4 CELLS

Some organisms, such as bacteria, are unicellular, consisting of a single cell. Other organisms, such as humans, consist of many cells (multicellular). Humans have an estimated 10^{14} cells; a typical cell has a size of 10 μm and a mass of 1 ng. Each cell is self-contained and self-maintaining: it can take in nutrients, convert these into energy, carry out specialized functions and reproduce as necessary. Each cell stores its own set of instructions for carrying out each of these activities. The contents of the cell are contained within membrane that contains proteins and a lipid bilayer (Section 12.6). This regulates what moves in and out, and maintains the electric potential of the cell.

There are two types of cells, *eukaryotic* and *prokaryotic*. These differ greatly in their size, internal structure and their genetic and metabolic organization. Prokaryotic cells, which include all the different species of bacteria, are comparatively primitive; these are very small and simple cells. On the other hand, eukaryotic cells, which include those of higher animals and plants, have volumes from 1000 to 10 000 times greater than those of prokaryotic cells.

Cells that have been attracting much attention are the *stem cells*. These have two important characteristics that distinguish them from other types of cells. First, they are unspecialized cells that renew themselves for long periods through cell division. The second is that under certain physiological or experimental conditions, they can be induced to become cells with special functions such as the beating cells of the heart muscle or the insulin-producing cells of the pancreas. Stem cells are primal cells common to all multi-cellular organisms that retain the ability to renew themselves through cell division and can differentiate into a wide range of specialized cell types.

Two broad categories of mammalian stem cells exist: *embryonic stem cells*, derived from *blastocysts* (a blastocyst is an early stage embryo – approximately 4–5 days old in humans and consisting of 50–150 cells) and *adult stem cells*, which are found in adult tissues. In a developing embryo, stem cells are able to differentiate into all of the specialized embryonic tissues. In adult organisms, stem cells and progenitor cells act as a repair system for the body, replenishing specialized cells. As stem cells can be readily grown and transformed into specialized tissues, such as muscles or nerves, through cell culture, their use in medical therapies has been proposed.

The framework of the cell is called the *cytoskeleton*. This helps to maintain the cell's shape during the uptake of external materials and the separation of daughter cells following cell division, and moves parts of the cell in processes of growth and mobility. The eukaryotic cytoskeleton is composed of microfilaments, intermediate filaments and microtubules. There are a great number of proteins associated with them, each controlling a cell's structure by directing, bundling, and aligning the filaments.

Inside the membrane, a salty *cytoplasm* takes up most of the cell volume. All cells possess DNA and RNA, containing the information necessary to build various proteins. The total length of the entire DNA in a single human cell is around 2 m equivalent to 5.5×10^9 base pairs. A human cell has genetic material in the *nucleus* (the nuclear genome). In humans, the nuclear genome is divided into 46 linear DNA molecules (23 pairs) called *chromosomes*. A *gene* is a segment of a chromosome that codes for a single polypeptide chain of a protein (Section 12.5). A gene may have anywhere from 300 to 6000 or more nucleotide pairs, giving molecular weights from about 10^5 to 2×10^6.

Within the cell, *organelles* (little organs) are discrete structures with specialized functions. There are many types of organelles, particularly in the eukaryotic cells of higher organisms. An organelle is to the cell what an organ is to the body. Organelles include mitochondria (Section 12.3.4) and *chloroplasts*, which are involved in photosynthesis. DNA is also present in certain cell organelles, particularly the mitochondria and chloroplasts.

12.5 GENETIC CODING

The cell's full complement of DNA, i.e. the organism's genetic blueprint, is called its *genome*. As noted above, this does not reside on one large, single strand of DNA but is divided up into several sections, each of which is contained in a chromosome. In the case of the human genome, the author Matt Ridley makes the analogy with a book [1]. If one supposes that the human genome is a book, then there are 23 chapters, called chromosomes. Each chapter contains several thousand stories, the genes. Each story is made up of paragraphs, called *exons*, which are interrupted by advertisements known as *introns*. Each paragraph is made up of words called *codons* and, finally, each word is written in letters called bases. The function of genes is to carry the information required to manufacture the enzyme proteins that orchestrate the body's chemistry.

The information within individual genes resides in a coded form. The chemical structure of a protein can be written down in terms of its constituent amino acids. DNA can be represented as a sequence of base pairs linked by the sugar and phosphate components of nucleotides, which are simply part of the scaffolding. The sequence of base pairs within a gene on a DNA molecule can then represent a protein molecule in coded form.

There are 20 different amino acids that occur in proteins, but only four DNA bases. Hence a complete DNA–protein code can be established by taking groups of bases to represent each amino acid. However, a code based on 'one base to one amino acid' will not work. A DNA code in which the characters are groups of just two bases will only provide $4 \times 4 = 16$ different elements – insufficient to encode all of the amino acids. However, with groups of three bases, there are 64 possible characters, which is more than enough. The DNA–protein code must therefore use at least three bases to represent each amino acid.

All living organisms employ the same code. Because there is some redundancy in the code – there are 64 different base triplets available to represent 20 amino acids – some amino acids are encoded by more than one triplet. Furthermore, some triplets do not represent amino acids at all, but are control codes, which signify the end of the protein-coding sequence in a gene. Table 12.4 shows this coding structure. Thus alanine (Ala) can be represented by the codons GCU, GCC, GCA or GCG, and lysine (Lys) can be represented by AAA or AAG.

12.5.1 Replication, Transcription and Translation

Under the right conditions, the genome can both photocopy and read itself. The photocopying is known as *replication* and the reading as *translation*. Between the replication and translation processes is a further step, *transcription*. Here, the message in the DNA is transcribed into the form of RNA, to be carried by the ribosomes. Replication works because of the fundamental property of the four bases: A likes to pair with T, and G with C. So a single strand of DNA can copy itself by assembling a complementary strand with Ts opposite all the As, As opposite all the Ts, Cs opposite all the Gs and Gs opposite all the Cs. The usual state of DNA is, of course, the double helix, consisting of the original strand and its complementary pair intertwined.

The first step in the progression from gene to protein is to produce an RNA version of the gene encoded on the DNA molecule, where the sequence of A, T, C and G nucleotides in the gene is reproduced as a complementary sequence of U, A, G and C nucleotides in the RNA. The relevant gene-bearing portion of the DNA double helix is unravelled and one of the single strands acts as a template for the construction of the mRNA molecule, as shown in Figure 12.13. The messenger RNA is then detached from the DNA strand, with the protein plan in the form of the sequence of bases.

Proteins are assembled on the mRNA template one amino acid at a time. Each three-base group on the mRNA, which corresponds to an amino acid, represents the codon. The tRNA

Table 12.4 Sequences of DNA bases encode the information required for the synthesis of proteins from amino acids. The amino acids are those listed in Table 12.1. 'Stop' is a codon that does not encode an amino acid but represents an instruction for protein synthesis to stop.

	Second position				
First position	U	C	A	G	Third position
U	Phe	Ser	Tyr	Cys	U
	Phe	Ser	Tyr	Cys	C
	Leu	Ser	Stop	Stop	A
	Leu	Ser	Stop	Trp	G
C	Leu	Pro	His	Arg	U
	Leu	Pro	His	Arg	C
	Leu	Pro	Gin	Arg	A
	Leu	Pro	Gin	Arg	G
A	Ile	Thr	Asn	Ser	U
	Ile	Thr	Asn	Ser	C
	Ile	Thr	Lys	Arg	A
	Met	Thr	Lys	Arg	G
G	Val	Ala	Asp	Gly	U
	Val	Ala	Asp	Gly	C
	Val	Ala	Glu	Gly	A
	Val	Ala	Glu	Gly	G

molecules have one end that becomes anchored to a specific codon; this anchoring region on tRNA consists of a triplet of the base pairs that are complementary to those on the mRNA codon, and it is therefore called an anticodon. For example, the base sequence CGA on the mRNA codon will become anchored to the tRNA anticodon GCU, via the complementary pairs C–G, G–C and A–U. The other end of the tRNA molecule binds specifically to the

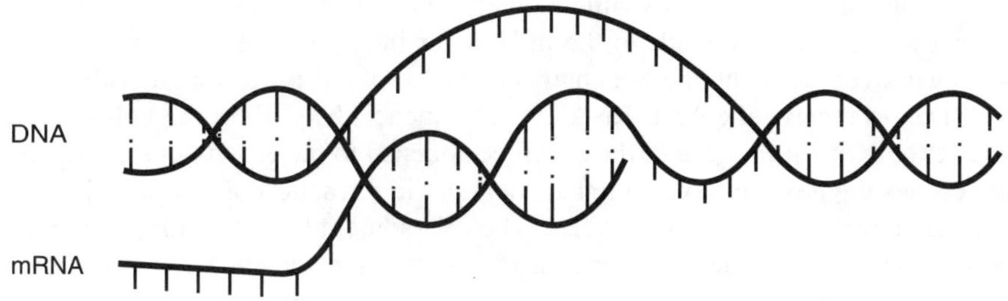

Figure 12.13 The genetic code is translated from DNA sequences to amino acid sequences via the mediation of RNA. RNA molecules containing the information in a single gene, called messenger RNA (mRNA) are constructed by unwinding sections of the DNA double helix and using one of its unwound strands as a template. From Ball P, *Designing the Molecular World.* Copyright (1994) Princeton University Press. Reprinted by permission of Princeton University Press.

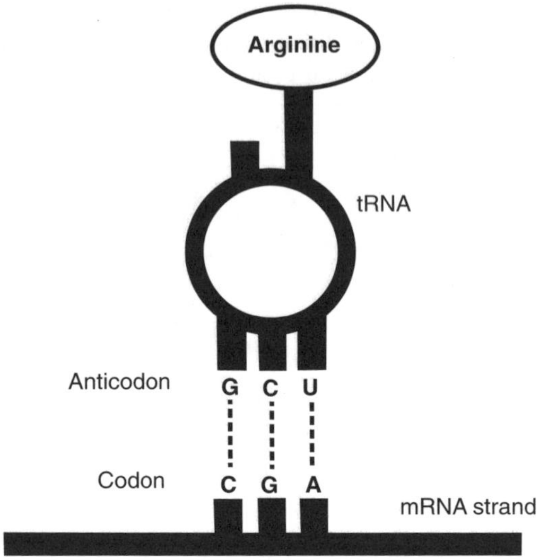

Figure 12.14 Function of transfer RNA or (tRNA). One end of the tRNA molecule binds the appropriate amino acid through a complex molecular recognition process. The other end contains a sequence of three bases called an anticodon, which docks into the complementary codon sequence on the mRNA molecule. Here, the codon is that for arginine (CGA); see Table 12.4. From Ball P, *Designing the Molecular World*. Copyright (1994) Princeton University Press. Reprinted by permission of Princeton University Press.

amino acid corresponding, in the genetic code, to the mRNA's codon. In the example above, the sequence CGA on mRNA produced from the sequence GCT in a gene on DNA, corresponds to the amino acid arginine (Table 12.4). Thus the tRNA molecule responsible for putting arginine into its place in the protein chain has the anticodon GCU at one end, and binds arginine at the other (Figure 12.14) [2].

The linking together of the tRNA-attached amino acids at the mRNA template requires the assistance of the rRNA, in conjunction with other enzymes and proteins. Several rRNA molecules are bound together with many more proteins in the ribosome, the function of which it is to control this linking process. The ribosome binds to the mRNA and first facilitates the docking of the tRNA anticodon on to the mRNA codon. As successive tRNAs bring their respective amino acids to the mRNA for incorporation into a protein molecule, the ribosome moves along the mRNA chain, one codon at a time, so that it is always ready in the right place to receive the next tRNA in the sequence. The ribosome holds in place two successive tRNAs at a time. One of these will be attached to the growing protein chain, while the other carries the next amino acid to be inserted, via formation of a peptide bond, into the chain, as depicted in Figure 12.15 [2]. The ribosome holds together the end of the polypeptide chain and the next amino acid in just the right position for a peptide bond to form. The formation of this bond transfers the chain to the new tRNA; the old one is then released by the ribosome, which then shunts along to the next codon and is ready to receive the next tRNA. There are base sequences at either end of the mRNA that do not correspond to codons, but instead act as signals to tell the ribosome where to begin and where to end protein synthesis. Once the protein chain is completed, it is detached from the ribosome–mRNA complex, and the mRNA, having accomplished its task, is destroyed by enzymes.

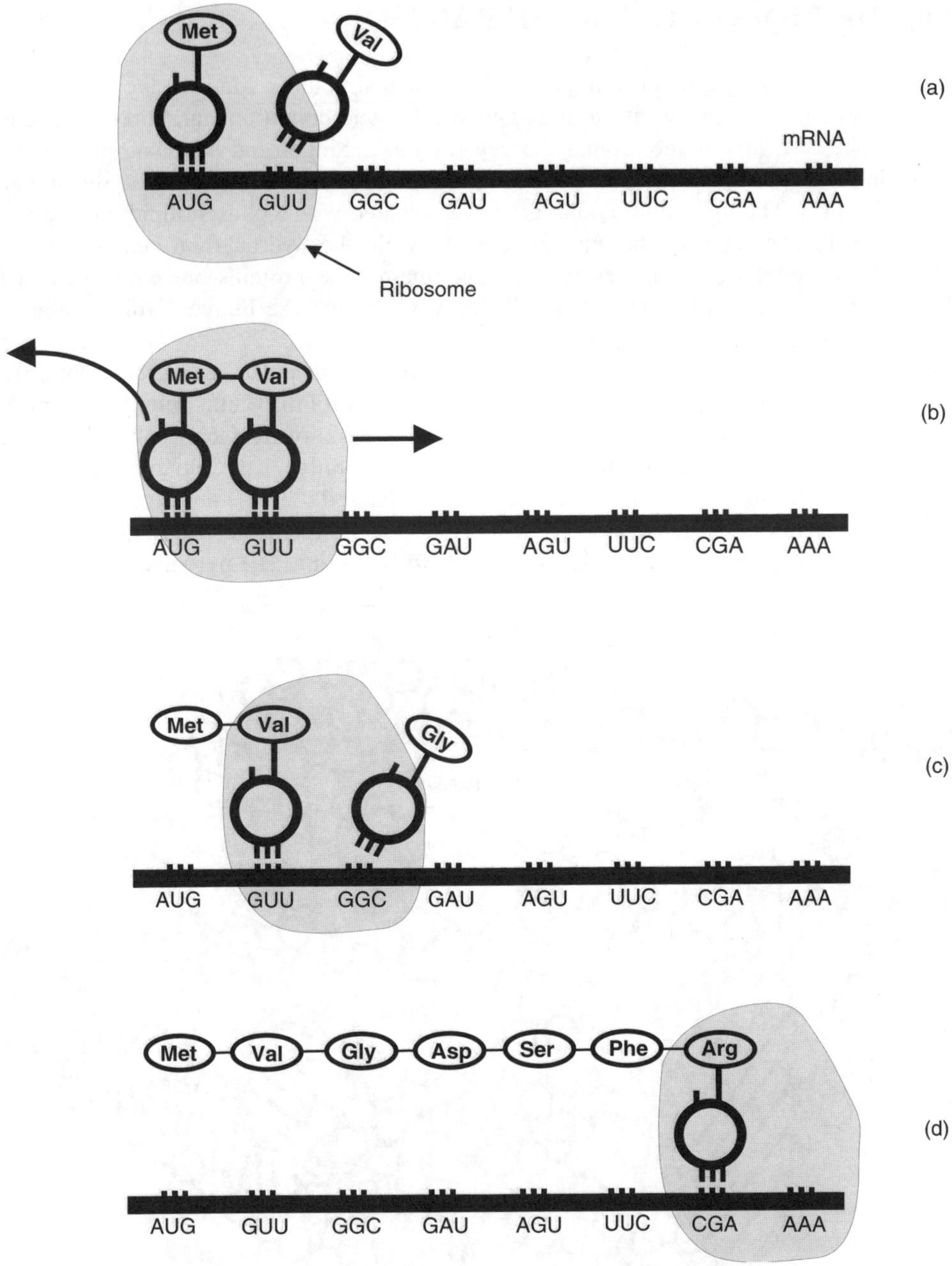

Figure 12.15 Protein synthesis on the mRNA template. A ribosome facilitates the docking of amino acid-charged tRNA on to the mRNA codon. The amino acid on the tRNA is then linked to the growing protein chain via a peptide bond. From Ball P, *Designing the Molecular World*. Copyright (1994) Princeton University Press. Reprinted by permission of Princeton University Press.

12.6 THE BIOLOGICAL MEMBRANE

Membranes are amongst the most important biological structures. Many of the key functions of living systems, e.g. the ability to maintain steady-state conditions, are directly linked to the existence of a membrane around the cell. In 1972, Singer and Nicolson proposed the now widely accepted *fluid mosaic model* of the structure of cell membranes, illustrated in Figure 12.16 [3]. The hydrophilic polar groups associated with a phospholipid bilayer are on the outside, in contact with the aqueous media, while the hydrocarbon chains are in the interior. The model also proposes that integral membrane proteins are embedded in the bilayer. Some of these proteins extend all the way through the bilayer, whilst some only partially across it.

The biological bilayer can be symmetric in terms of the polar head groups, but, more commonly, it exhibits asymmetry. Thus, for the red blood cell, sphingomyelin and phosphatidylcholine are disproportionately located in the outward-facing monolayer, while phosphatidylserine and phosphatidylethanolamine are mainly in the monolayer facing the inside of the cell. Most membranes are electrically polarized with the negative inside the cell (typically −60 mV). The asymmetry will give rise to piezoelectric and pyroelectric behaviour, as noted in Chapter 11, Section 11.3.1, for LB films. The pyroelectric coefficients

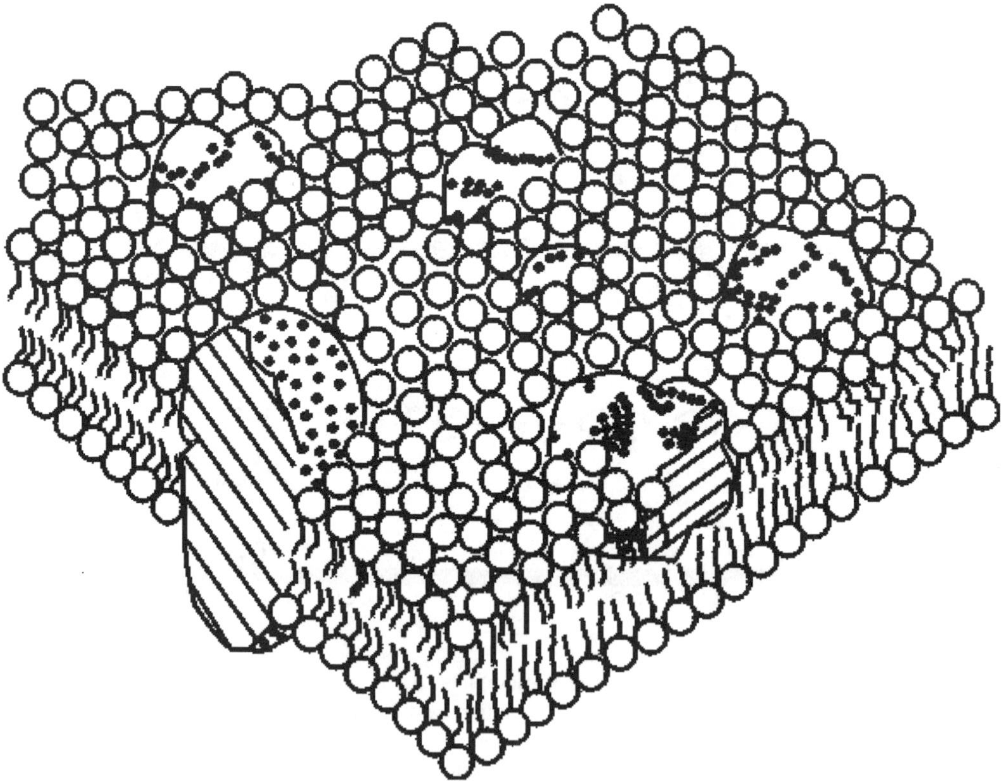

Figure 12.16 Model of a biological membrane. Protein molecules are shown embedded in and traversing the bilayer lipid structure.

found in alternating LB films of phospholipids are modest [4], and it is unclear if they are exploited in physiological processes.

The exact relationship between lipid composition, organization and function in biological systems has not yet been firmly established. Very few membrane proteins, for example, have an absolute requirement for a specific lipid, but their activity can vary considerably in different lipid mixtures. One important feature seems to be that the membranes should have a degree of fluidity. Biological membranes in their native state appear to permit rapid translational and rotational motion within the plane of the bilayer. In contrast, movement from one monolayer to the other is much more restricted. Lipids in cell membranes are therefore more likely to be in a liquid-crystalline rather than a condensed solid state. The most influential factor concerning fluidity is the nature of the esterified fatty acid. Shorter chain lengths and higher unsaturation lead to lower transition temperatures (solid to liquid) whereas longer and more saturated fatty acids have higher transition points (Section 12.2.5). The neutral lipid cholesterol [Figure 12.8(d)] reduces the lipid order in solid crystalline systems while decreasing disorder in the more fluid situations. This feature is thought to be significant in sustaining the integrity of the biological membrane whilst maintaining its dynamic behaviour.

12.6.1 Transport Across the Membrane

Molecular and Ionic Transport

Membranes are highly selective permeability barriers. The movement of molecules across the lipid bilayer consists of transfer from one aqueous environment to another and is restricted to solute molecules and water (Figure 12.17). Generally, the smaller and less polar the molecule, the more easily it passes through the bilayer. Water is an exception; the small size of H_2O molecules (which offsets their large polarity) leads to a very rapid exchange of water across the bilayer structure. Gases, such as oxygen and carbon dioxide, important in cell metabolism, pass in or out of the cell in a dissolved state and the rates of transfer are determined by the extent to which the gases are soluble in the aqueous environment. Carbon dioxide is very soluble in water and therefore passes freely through membranes. In contrast, oxygen is less soluble and this becomes a limiting factor in cellular metabolism.

The existence of a concentration gradient of solute molecules across a membrane tends to cause a net movement of solute molecules in the direction of this concentration gradient. Of course, transport occurs in both directions and that the net flux is the sum of these two movements.

In the simplest case, the rate of flow, the flux J ($\mathrm{mol\,m^{-2}\,s^{-1}}$) of uncharged molecules in the direction of the gradient can be described by Fick's law of diffusion (Chapter 2, Section 2.8).

Movements of ionized solutes are also influenced by electrical gradients. The flow of solute may still be described in simple terms by the *Nernst–Planck equation*:

$$J = -\mu c\left(\frac{k_B T}{ce}\frac{\mathrm{d}c}{\mathrm{d}x} + z\frac{\mathrm{d}V}{\mathrm{d}x}\right) \tag{12.2}$$

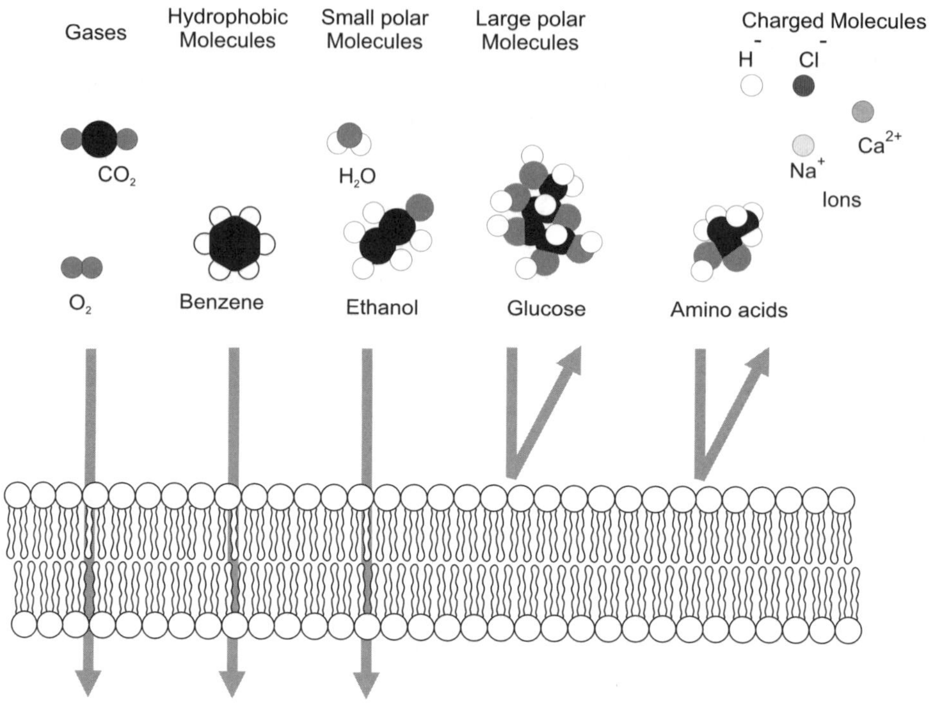

Figure 12.17 Schematic diagram showing the permeability of a lipid bilayer to different molecules. The smaller and less polar the molecule, the more easily it passes through the bilayer. Water is an exception.

where μ is the mobility of the ion ($m^2\,V^{-1}\,s^{-1}$), z is number of (electron) charges on the permeating molecule (its valence) and dV/dx is the electrical potential gradient across the molecule. The first term on the right-hand side of Equation (12.2) is due to diffusion and the second term originates from drift (electric field effect). The mobility of the ion can also be expressed in terms of the diffusion coefficient, D:

$$D = \frac{\mu k_B T}{e} \tag{12.3}$$

Equation (12.3) is known as the *Einstein relation*, after Einstein, who showed that diffusion and drift experience the same frictional resistance in solution.

In some instances, very rapid diffusion of material can take place across a membrane. Here, other constituents of the membrane, usually the proteins, play an important role. In a simple case, the 'transport' protein possesses a specific binding site which recognizes the substance to be translocated. Noncovalent association of the substance triggers structural changes in the protein which effectively allow it to move to the other side of the membrane, illustrated in Figure 12.18. The degree of movement of the transported entity on the protein surface may be fairly small – a few tenths of a nanometre – and it is probably not correct to envisage a permanent pore or hole through the protein. Where the flow is down a concentration gradient, no energy input is needed, but when the flow is up the gradient, some form of energy is needed to produce the desired conformational state in the protein.

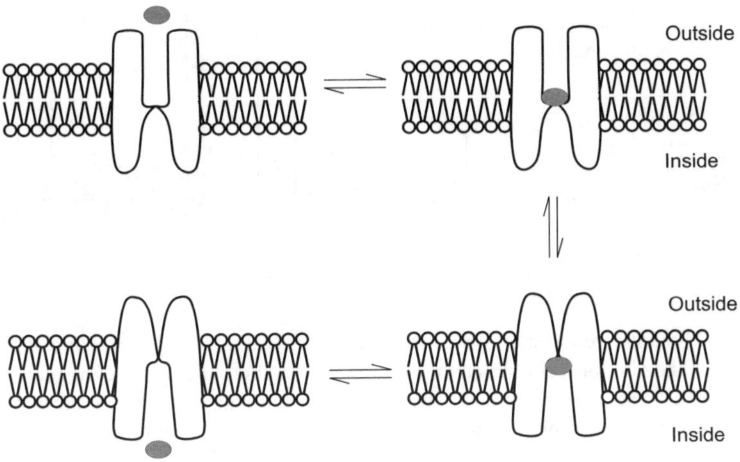

Figure 12.18 Mechanism of material transport across the bilayer. The noncovalent association of a substrate molecule with the transporter protein triggers structural changes that result in transport to the other surface of the membrane. Only a small structural change may be required for this.

A good example occurs with bacteriorhodopsin, a proton translocator from the membrane of *Halobacterium halobium* (Section 12.10.1). Like visual rhodopsin, the protein absorbs light energy via bound *cis*-retinal, but unlike the visual pigment, the energy is used to promote proton transfer across the lipid bilayer, through conformational changes in the protein molecule. A mechanism has been proposed whereby the proton is transferred from amino acid side chain to amino acid side chain through the protein. The three-dimensional nature of the protein is constructed such that this transfer occurs readily. A similar process may occur for other proton-transporting systems, but for larger substances such as glucose, this shuttle-type procedure is less likely.

Ion permeabilities of lipid bilayers and of natural membranes can be greatly increased by the incorporation of a range of small molecules called *ionophores* (simple crown ether ionophores have been described in Chapter 10, Section 10.4.6). Such materials have been used as *antibiotics* (a drug that kills or slows the growth of bacteria). Many ionophores form stable complexes with cations. The nonpolar groups of the ionophore molecule are directed outwards so that the ion becomes enclosed in a purse-like structure with a polar lining and a nonpolar exterior. One example is valinomycin, which is capable of selectively complexing with and transporting potassium ions across both biological and synthetic membranes. Complexation is associated with a change in conformation of the ionophore, facilitating this transport. A diagram of the valinomycin molecule and in its complexed state (with potassium), is shown in Figure 12.19. The molecule is a 12-membered macrocyclic ring of alternating D- and L-amino acids and α-hydroxy acids. There are three repeat units, each consisting of L-valine, L-lactic acid, D-valine and D-hydroxyisovaleric acid alternately joined by amide C=O and ester C=O linkages. On complexation with a potassium ion, the valinomycin molecule very subtly changes its conformation, and becomes more hydrophobic. During this process, the water molecules associated with the ions are replaced one by one as the ions become coordinated with the ionophore: this stepwise displacement of the water of hydration reduces the potential energy barrier to penetration into the membrane. The process of ionic transport across the membrane is illustrated in Figure 12.20.

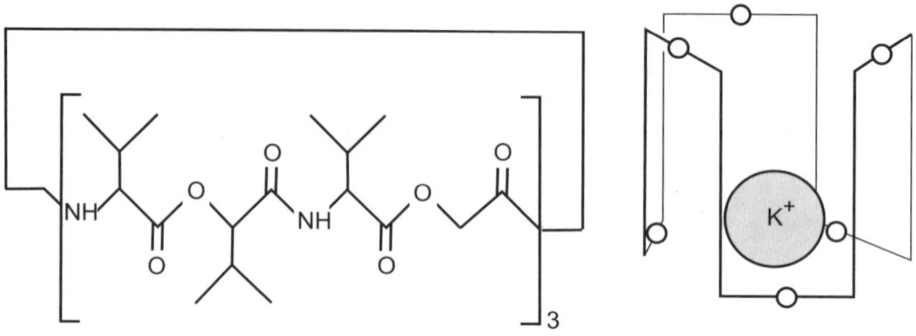

Figure 12.19 Schematic representation of the ionophore valinomycin (left) and its complex with potassium (right). The valinomycin molecule is composed to three identical segments linked to form a 36-membered ring. There are six amide groups and six ester groups arranged alternately around the ring. Nine isopropyl and three methyl side groups are also attached to the ring. The valinomycin–K^+ complex exhibits a so-called 'tennis-ball-seam' conformation.

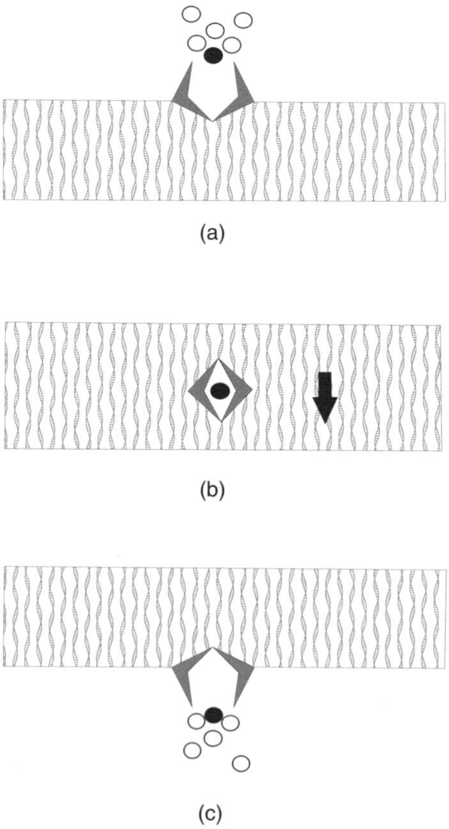

Figure 12.20 Transport of potassium ions (shown as filled circles) across a bilayer membrane. On complexation with a potassium ion, the valinomycin molecule subtly changes its conformation, and becomes more hydrophobic and moves across the membrane.

In other transport systems, ions moving by facilitated diffusion can traverse the cell membrane through channels created by proteins. These embedded transmembrane proteins allow the formation of a concentration gradient between the extracellular and intracellular contents. Ion channels are highly specific filters, even between ions of a similar character, e.g. Na over K, allowing only desired ions through the cell membrane. The specificity of an ion channel is a well-researched topic, although in some cases surprisingly little is known about the precise mechanism of ion channel filters. These ion channels are said to be 'gated' if they can be opened or closed. There are three types of gated ion channels: ligand gated, mechanically gated and voltage gated. Ligand gated channels open or close in response to the binding of a small signalling molecule or ligand. Some ion channels are gated by extra cellular ligands and some by intracellular ligands. In both cases, the ligand is not the substance that is transported when the channel opens. The binding of neurotransmitter acetylcholine opens sodium channels in certain synapses. Voltage gated channels are found in neurons (Section 12.7) and muscle cells. These open or close in response to voltage changes that occur across the membrane. For example, as an impulse passes down a neuron, the reduction in the voltage opens sodium channels in the adjacent portion of the membrane. This allows the influx of Na^+ into the neuron and thus the continuation of the nerve impulse.

Electron Transport Systems

Many biological processes require the efficient transport of electrons. Very often this occurs in conjunction with, but separate from, the transport of protons. Most of the electron transport systems in biology are involved in some way with bioenergetics, the best characterized structurally being the photosynthetic system (Section 12.10.2).

Electron-transferring reactions, such as those taking place in the respiratory chain, are more generally redox reactions (Chapter 3, Section 3.4.2). These proceed with a transfer of electrons from an electron donor (reducing agent) to an electron acceptor (oxidizing agent). In some oxidation–reduction reactions, the transfer of electrons is made via transfer of hydrogen atoms, each of which carries an electron. In this case, dehydrogenation is equivalent to oxidation and hydrogenation is equivalent to reduction. In other oxidation–reduction reactions, both an electron and a hydrogen atom may be transferred.

Important biological molecules involved in redox processes are the coenzymes nicotinamide adenine dinucleotide (NAD) and its phosphorylated derivative nicotinamide adenine dinucleotide phosphate (NADP). The reduced forms are usually abbreviated NADH and NADHP. These compounds catalyze the following general reactions

$$\begin{aligned}&\text{reduced substrate} + \text{NAD}^+ \rightleftharpoons \text{oxidized substrate} + \text{NADH} + \text{H}^+\\&\text{reduced substrate} + \text{NADP}^+ \rightleftharpoons \text{oxidized substrate} + \text{NADPH} + \text{H}^+\end{aligned} \tag{12.4}$$

The *cytochromes* are a group of iron-containing, electron-transferring proteins that act sequentially in the transport of electrons. These contain iron porphyrin groups and resemble haemoglobin and myoglobin; all are members of the class of haem proteins. Porphyrins (very similar to the phthalocyanine molecules described in Chapter 5, Section 5.4) are named and classified on the basis of their side-chain substituents. Protoporphyrin IX [Figure 12.21(a)] is the most abundant and contains four methyl groups, two vinyl groups and two propionic acid

(a)

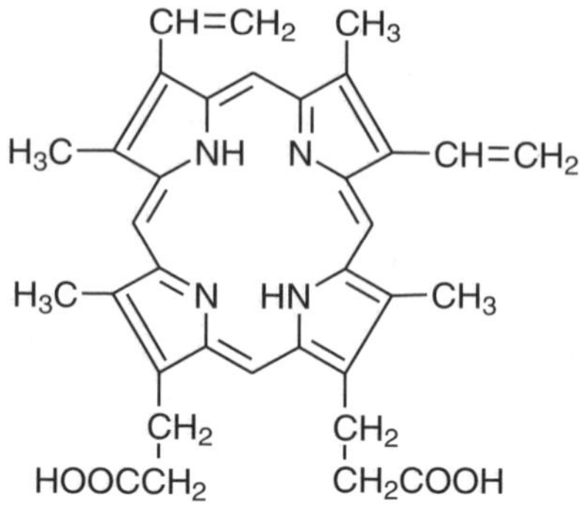

(b)

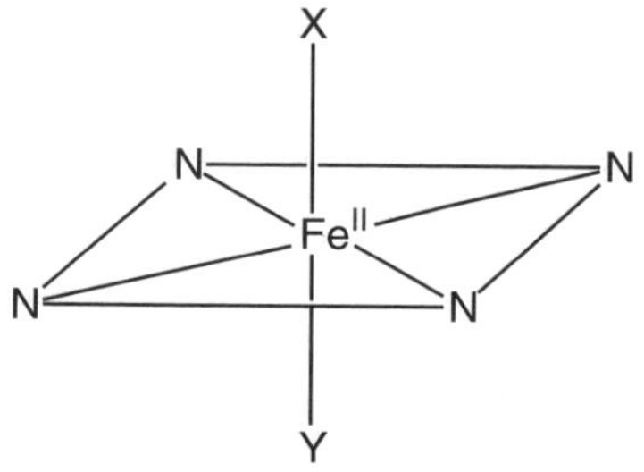

Figure 12.21 (a) Chemical structure of protoporphyrin IX. (b) Binding of an iron atom in cytochrome *c*. The four nitrogen atoms of the porphyrin ring bind to the iron in a planar arrangement. X and Y represent binding groups contributed by the protein.

groups. This is the porphyrin that is present in haemoglobin, myoglobin and most of the cytochromes. Protoporphyrin forms very stable complexes with di- and trivalent metal ions. Such a complex of photoporphyrin with Fe(II) is called haemin, or haematin. The cytochromes undergo Fe(II) $\rightarrow$ Fe(III) valence changes during their function as electron carriers. In cytochrome *c*, the single iron–protoporphyrin group is covalently linked to the protein, as illustrated in Figure 12.21(b).

12.7 NEURONS

Neurons (also spelled neurones or called nerve cells) are the electrically excitable primary cells of the nervous system. A human brain contains 10^{11} neurons. These register information from the environment, integrate and evaluate these data and then decide whether electrical signals are further transmitted. Many highly specialized types of neurons exist, and these differ widely in appearance. Neurons are highly asymmetric in shape, and a typical architecture is shown in Figure 12.22. The *soma*, or cell body, is the relatively large central part of the cell between the *dendrites* and the *axon*; this is the metabolic centre of the cell and the site of protein synthesis and production of energy (ATP). The cell body gives rise to two kinds of cellular extensions: several short dendrites and a single long axon. The dendrites branch out in a tree-like fashion and receive incoming signals from other neurons. Certain

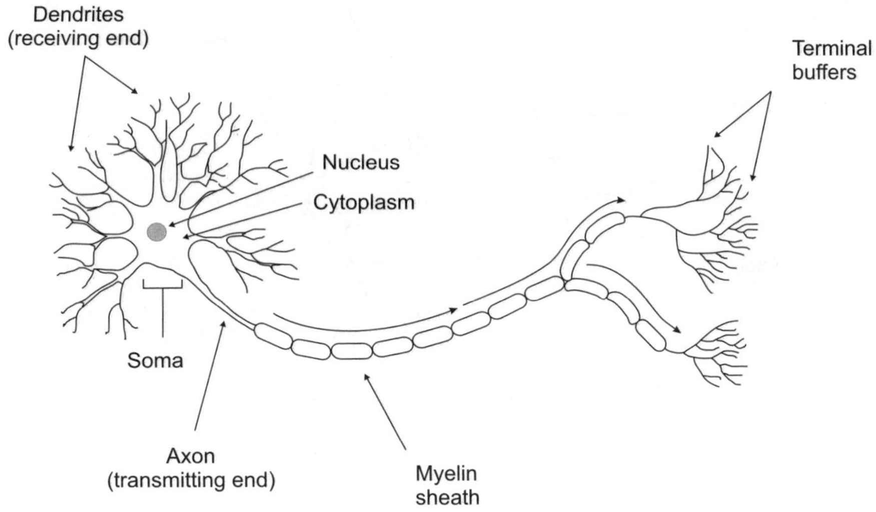

Figure 12.22 Schematic diagram showing the typical structure of a neuron.

neurons in mammals have over 1000 dendrites each, permitting connections with tens of thousands of other cells. The axon is a much finer, wire-like projection which may extend tens, hundreds or even tens of thousands of times the diameter of the soma in length. This is the structure which carries nerve signals away from the neuron. Each neuron has only one axon, but this axon may undergo extensive branching and thereby allow communication with many target cells. Axon and dendrites are typically only about 1 μm thick, whereas the soma is usually about 25 μm in diameter and not much larger than the cell nucleus that it contains. The axon of a human motoneuron can be over 1m long, reaching from the base of the spine to the toes.

Surrounding the axon is an electrically insulating layer called the *myelin sheath*. This is made up of protein and lipid. The purpose of the myelin sheath is to allow rapid and efficient transmission of impulses along the nerve cells If the myelin is damaged, the impulses are disrupted; this can cause diseases such as multiple sclerosis.

As is the case with every cell in the body, neurons are surrounded by a thin membrane, formed by a phospholipid bilayer. At rest, neurons maintain a difference in the electrical potential on either side of the membrane. The electrical signals are called *action potentials*, which constitute the signals by which the brain receives, analyses and conveys information. Action potentials are brief ($\sim$1 ms duration) and relatively large amplitude ($\sim$100 mV) electrical pulses. The action potential is either present or not. Whereas stimuli that do not reach a certain threshold value of the membrane potential produce no action potential, all stimuli above the threshold invariably generate the same signal.

The action potential is propagated without decay along the axon at a speed of up to $150\,\mathrm{m\,s^{-1}}$. Near to its end, the axon divides into fine branches that make contact to neighbouring neurons. The point of contact between two communicating cells is called the *synapse*, shown schematically in Figure 12.23. The nerve cell transmitting the signal is called the *presynaptic cell* and that receiving the signal is the *postsynaptic cell*. The arrival of an action potential at the tip of an axon triggers the release of neurotransmitter at a synaptic

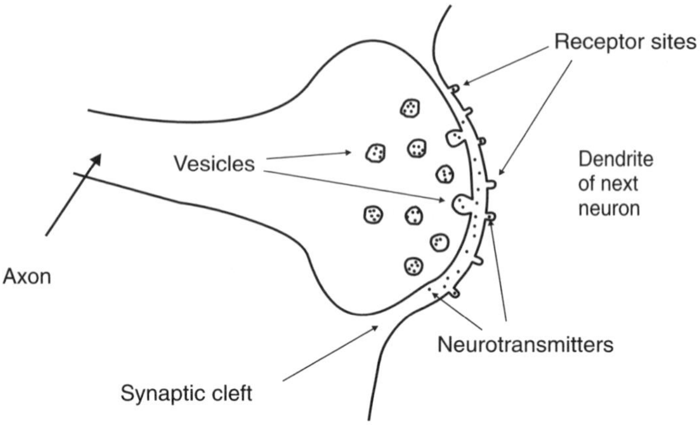

Figure 12.23 Schematic diagram of a synapse.

gap. Neurotransmitters can either stimulate or suppress (inhibit) the electrical excitability of a target cell. An action potential will only be triggered in the target cell if neurotransmitter molecules acting on their post-synaptic receptors cause the cell to reach its threshold potential.

The transmembrane voltage changes that take place during an action potential result from changes in the permeability of the membrane to specific ions, the internal and external concentrations of which cells maintain in an imbalance. In the axon fibres of nerves, depolarization results from the inward rush of sodium ions, whereas repolarization and hyperpolarization arise from an outward rush of potassium ions. Calcium ions make up most or all of the depolarizing currents at an axon's presynaptic terminus in muscle cells (including those in the heart) and in some dendrites. When a neuron has just generated an action potential, the cell is unable to generate another within a certain time span. This phase lasts a few millseconds and is caused by the time it takes for the ion channels to recover.

12.8 BIOSENSORS

Biosensors use biological or living materials to provide their sensing functions. In many ways they can be defined as a special type of chemical sensor (Chapter 10). Evolution of species by means of natural selection has led to extremely sensitive organs which can respond to the presence of just a few molecules. Artificial sensors exploit biologically active materials in combination with different physical sensing elements. The biorecognition element works like a bioreactor on the top of the conventional sensor. The response of the biosensor will be determined by the diffusion of the analyte, by the reaction products, by the co-reactants or interfering species and/or by the kinetics of the recognition process. Organisms, tissues, cells, membranes, enzymes, antibodies and nucleic acids can all be detected by means of a biosensor.

One of the key issues for biological sensing systems is the immobilization of the active element on the physical transducer. The biologically active material must be confined to the

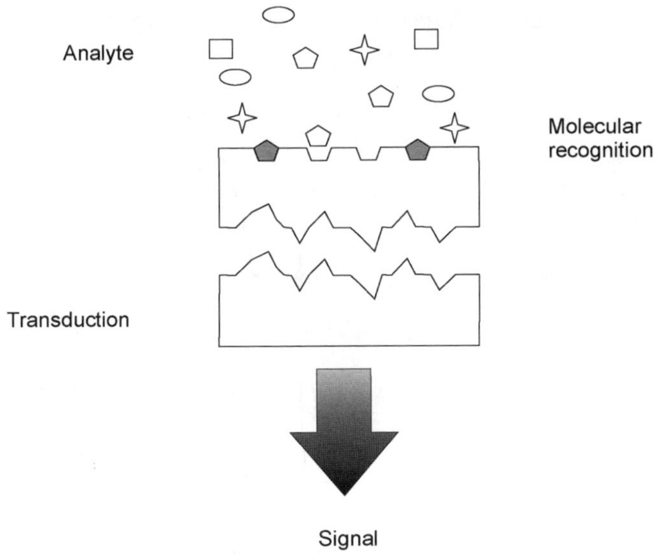

Figure 12.24 Molecular recognition and signal transduction parts of a biosensor.

sensing element and kept from 'leaking' while allowing contact with the analyte solution. Furthermore, the reaction products must readily diffuse out of the sensing layer so as not denature its biologically active characteristics. Many sensing materials are proteins or contain proteins in their chemical structure. Two techniques employed to immobilize the proteins are binding (adsorption or covalent binding) and retention. Physical retention involves separating the biologically active material from the analyte solution with a layer on the surface of the sensor which is permeable to the analyte and to any products of the recognition reaction, but not to the biologically active materials. A biosensor usually has two functional parts, for molecular recognition and signal transduction, illustrated in Figure 12.24. To achieve a high selectivity, either a biocatalyst or bioaffinity (immunological) material can be exploited as the molecular recognition element [5].

12.8.1 Biocatalytic Sensors

A biocatalyst recognizes the corresponding substrate (analyte) and immediately generates products by a specific reaction. The complex of the catalyst and the analyte remain stable in a transition state. A change in either the analyte or product is detected in the signal transducing device of the biosensor. Redox enzymes are recognized as the major material in constructing both biocatalytic and bioaffinity sensors. Biocatalytic sensors for glucose, lactate and alcohol utilize glucose oxidase, lactate dehydrogenase (lactate oxidase) and alcohol dehydrogenase (alcohol oxidase) as the molecular recognizable material. Since these redox enzymes are mostly associated with the generation of electrochemically active substances, many electrochemical enzyme sensors have been developed by linking redox enzymes for molecular recognition with electrochemical devices for signal transduction. Figure 12.25 shows a schematic representation of this process.

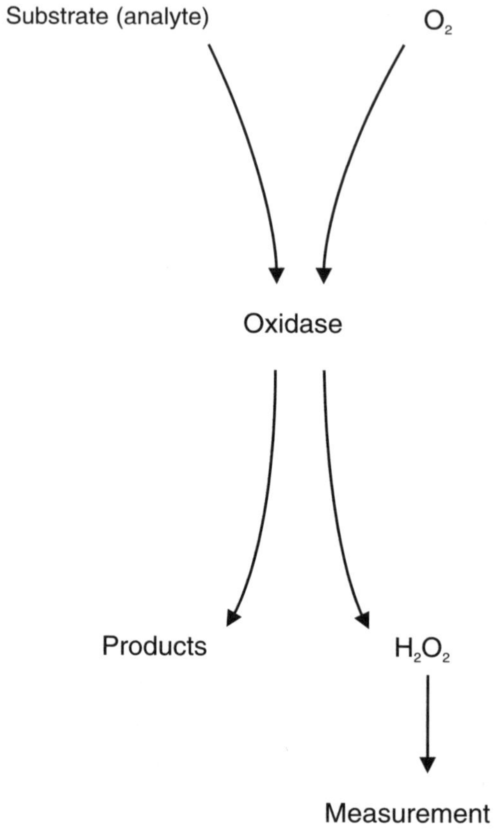

Figure 12.25 Principle of amperometric enzyme sensors.

12.8.2 Bioaffinity Sensors

A bioaffinity sensor involves an antibody, binding protein or receptor protein which forms a stable complex with a corresponding ligand. The bioaffinity protein–ligand complex formation is sufficiently stable to result in signal transduction. Immunosensors take advantage of the high selectivity provided by the molecular recognition of antibodies. Because of significant differences in affinity constants, antibodies may confer an extremely high sensitivity to immunosensors in comparison with enzyme sensors. Furthermore, antibodies may be obtained (in principle) for an unlimited number of determinants. Immunosensors are therefore characterized by high selectivity, sensitivity and versatility.

Immunosensors can be divided in principle into two categories: nonlabelled and labelled immunosensors. Nonlabelled immunosensors are designed so that the immunocomplex, i.e. the antigen–antibody complex, is directly determined by measuring the physical changes induced by the formation of the complex. An example is the use of the surface plasmon resonance technique, described in Chapter 4, Section 4.7.2, to monitor the change in thickness or optical constants on formation of the antigen–antibody complex.

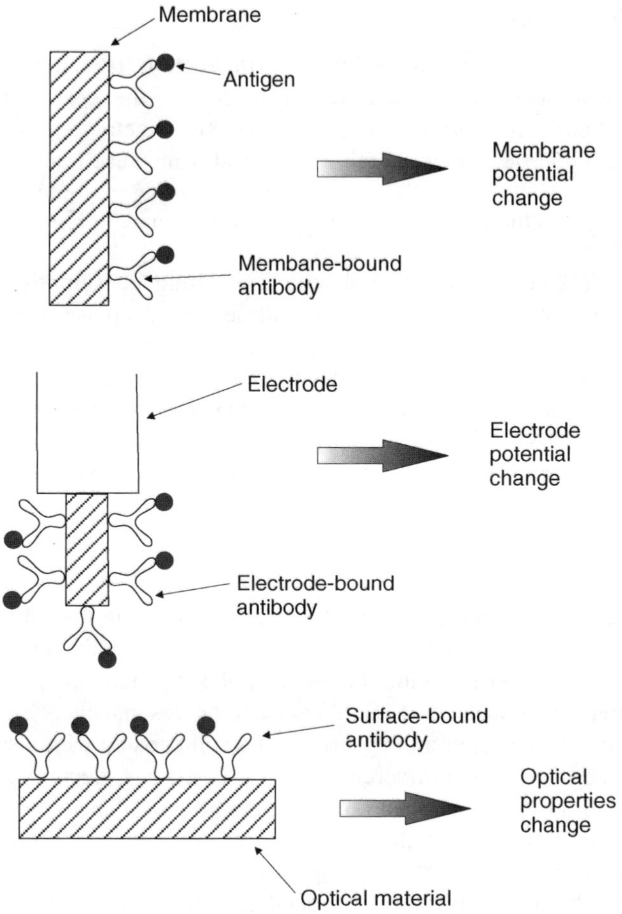

Figure 12.26 Principles of nonlabelled immunosensors.

Nonlabelled immunosensors are based on several principles, as illustrated in Figure 12.26. Either the antibody or the antigen can be immobilized on the solid matrix to form a sensing device. The solid matrix should be sensitive enough, in its surface characteristics, to detect the immunocomplex formation. Electrodes, membranes, piezoelectric materials or optically active surfaces may be used to construct nonlabelled immunosensors. The antigen or the antibody to be determined is dissolved in a solution and reacts with the complementary matrix-bound antibody or antigen to form an immunocomplex. This formation changes the physical properties of the surface, such as the electrode potential, the transmembrane potential, the piezoelectric resonant frequency (Chapter 10, Section 10.4.5) or the refractive index. A sufficiently high selectivity may be obtained with nonlabelled immunosensors, although such problems as non-specific adsorption on the matrix-bound antibody surface remain unresolved.

In a labelled immunosensor, a sensitively detectable label is incorporated; examples include enzymes, catalysts, fluorophores, electrochemically active molecules and liposomes.

Amperometric Glucose Sensors

An amperometric sensor, described in Chapter 10, Section 10.4.2, is an example of an electrochemical biosensor. The glucose sensor is one of the best examples. The basic structure consists of either an oxygen or hydrogen peroxide electrode, covered with a glucose oxidase membrane. Glucose oxidase catalyzes the following reaction:

$$\beta\text{-D-glucose} + O_2 + H_2O \rightarrow \beta\text{-D-gluconate} + H_2O_2 \qquad (12.5)$$

Glucose in a sample solution is oxidized with a resulting consumption of oxygen when contacted with the membrane-bound glucose oxidase. The decrease in dissolved oxygen is sensitively detected with the oxygen electrode. The output change of the sensor reflects the concentration of glucose in the solution. Hydrogen peroxide is generated in the glucose oxidase-catalyzed reaction and the detection of this can be performed using a platinum anode (polarized at about 0.7 V vs Ag/AgCl). The sensor responds linearly to the hydrogen peroxide, the output current being correlated with the analyte concentration.

Optical Biosensors

Optical sensors make use of the effect of chemical reactions on the optical properties of a material (Chapter 10, Section 10.4.6). If the solid surface is sufficiently sensitive to allow changes in its optical properties with immunocomplex formation, optical immunosensors without a label may be constructed. Surface plasmon resonance is so sensitive that an immunocomplex may be detected on the surface of a solid coated with a suitable metal. This principle is exploited in several commercial systems (Chapter 4, Section 4.7.2). Figure 12.27 shows an example of a biosensor based on an optical fibre [6]. The antibody is immobilized on the surface of the fibre core, while a fluorescence compound is used as the label. Both labelled antigen and free antigen react competitively with the bound antibody to form an immunocomplex on the core surface. The surface-bound label can be excited by an evanescent

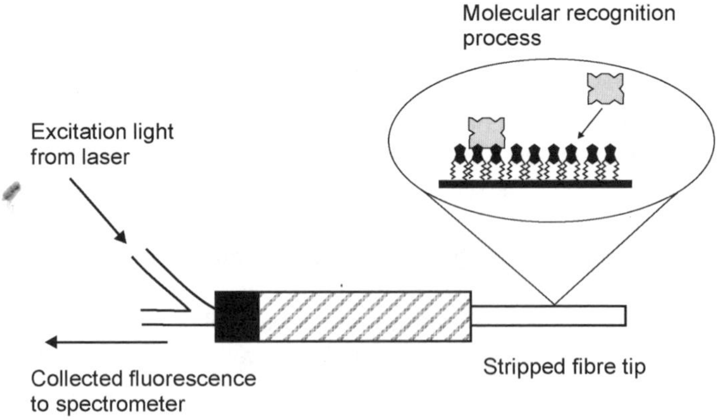

Figure 12.27 Schematic diagram of an evanescent fibre optic sensor used to excite and collect the fluorescence of surface bound avidin to a biotinylated phospholipid. Reprinted from *Langmuir*, **8**, Zhao S, Reichert WM. 'Influence of biotin lipid surface density and accessibility on avidin binding to the tip of an optical fibre sensor', pp. 2785–2791. Copyright (1992) American Chemical Society.

wave that passes through the optical fibre core. On the other hand, fluorescence labels in the bulk solution cannot be excited, even if the excitation beam comes through the optical fibre core. Labels attached to the surface-bound immunocomplex are thus discriminated from labels in solution.

12.9 DNA ELECTRONICS

The study of the electronic behaviour of organic compounds has led some scientists to work on the electrical properties of biological materials. DNA, described above in Section 12.3.2, is arguably the most significant molecule in nature. It may also be an important material for molecular electronics applications. Reports into the electronic properties of DNA have already generated controversy in the literature According to some, DNA is a molecular wire of very small resistance. Others, however, find that DNA behaves as an insulator. These seemingly contradictory findings can probably be explained by the different experimental conditions used to monitor the conductivity [7]. The DC resistivity of the DNA double helix over long length scales ($<10\ \mu m$) is very high ($\rho > 10^6 \Omega\, cm$). However, an appreciable AC conductivity can arise from the polarization of the surrounding water molecules [8].

'DNA chips' exploit the fact that short strands of DNA will bind to other segments of DNA that have complementary sequences, and can therefore be used to probe whether certain genetic codes are present in a given specimen of DNA. Micro-fabricated chips with many parallel DNA probes are becoming widespread in analytical and medical applications. Currently, the chips are read out optically, but further miniaturization might require new readout schemes, possibly involving the electron-transfer properties of DNA.

Computations by chemical or biological reactions overcome the problem of parallelism and interconnections in a classical system. If a string of DNA can be put together in the right sequence, it can be used to solve combinational problems. The 'calculations' are performed in test-tubes filled with strands of DNA. Gene sequencing is used to obtain the result. For example, Adleman and co-workers have solved the 'travelling salesman' and other problems to demonstrate the capabilities of DNA computing [9, 10]. Some excellent background reading on this can be found in the book by Amos [11]. DNA computing on parallel problems potentially provides 10^{14} MIPS (millions of instructions per second) and uses less energy and space than conventional supercomputers. Whereas CMOS supercomputers operate 10^9 operations per joule, a DNA computer could perform about 10^{19} operations per joule. The von Neumann limit [Chapter 1, Section 1.3.3, Equation (1.4)] predicts a limit of about 3×10^{20} operations per joule. Data could potentially be stored on DNA in a density of approximately 1 bit per nm^3 whereas existing storage media such as DRAMs require $10^{12}\, nm^3$ to store 1 bit.

12.10 PHOTOBIOLOGY

12.10.1 Bacteriorhodopsin

Bacteriorhodopsin (BR) is a compact molecular machine that pumps protons across a membrane powered by green sunlight. This molecule is the light-harvesting protein in the purple membrane of a microorganism, *Halobacterium salinarum* (formerly *Halobacterium*

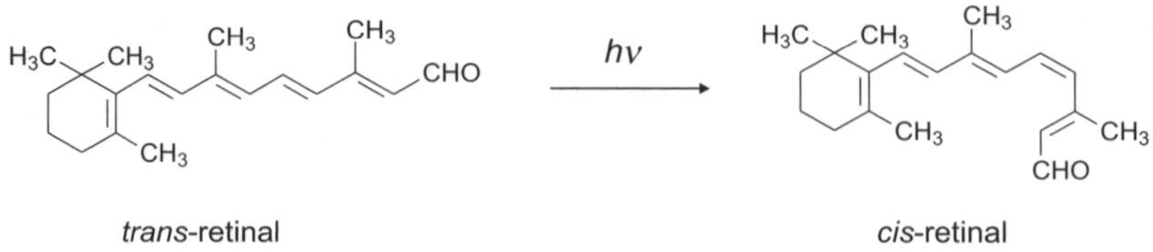

Figure 12.28 Light-induced transformation of retinal from a *trans* to *cis* configuration.

halobium), which lives in extreme conditions, in salt marches [12, 13]. The BR found in halobacteria is in the form of a two-dimensional crystal integrated into their cell membranes. The crystalline structure causes a substantial increase in BR's chemical and thermal stability (up to 140 °C for dry layers).

Sunlight interacts with this protein to pump protons outwards across the cell membranes, making the inside more alkaline than the outside. By this process, light energy is converted to chemical energy and a proton gradient across the membrane is established, which may be used for energy storage. A membrane-bound ATPase regenerates ATP from ADP, which then powers the cell.

Photocycle

Bacteriorhodopsin consists of 248 amino acids, arranged in seven α-helical bundles inside the lipid membrane and form a cage. At the heart of the BR is a molecule of all-*trans*-retinal (vitamin A aldehyde), which is bound deep inside the protein and connected through a lysine amino acid. Retinal contains a string of carbon atoms, and these strongly absorb light. When a photon is absorbed, it causes a change in the conformation of the molecule to its *cis* form (photoisomerization) as depicted in Figure 12.28. This change from a straight form to a bent form drives the pumping of protons.

After the photon absorption and proton transfer, the bacteriorhodopsin recovers to its initial conformation. The entire sequence of transitions is called the photocycle of the BR; a simplified version is depicted in Figure 12.29 [14]. There are a number of spectrally distinct intermediates. The initial resting state of the molecule is known as B. Green light transforms the initial B state to the intermediate state K. Next K relaxes, forms another intermediate state M then relaxes to O and finally back to B. The entire process takes about 10 ms. If the intermediate state O is irradiated with red light, it converts to a further state P (a branching reaction), which then relaxes to a very stable state Q. Blue light converts Q back to the initial B state.

The basic molecular functions of BR and their corresponding physical effects are depicted in Figure 12.30 [13]. The proton transport is initialized by proton absorption and a charge separation step on the picosecond timescale. After about 50 μs, the main colour change occurs and after about 10 ms the proton transport is completed.

Photochromism

The photoisomerisation from the B to K (Figure 12.29) occurs in less than 1ps. The *trans–cis* reaction is reversible if the protein is irradiated within the absorption band of the

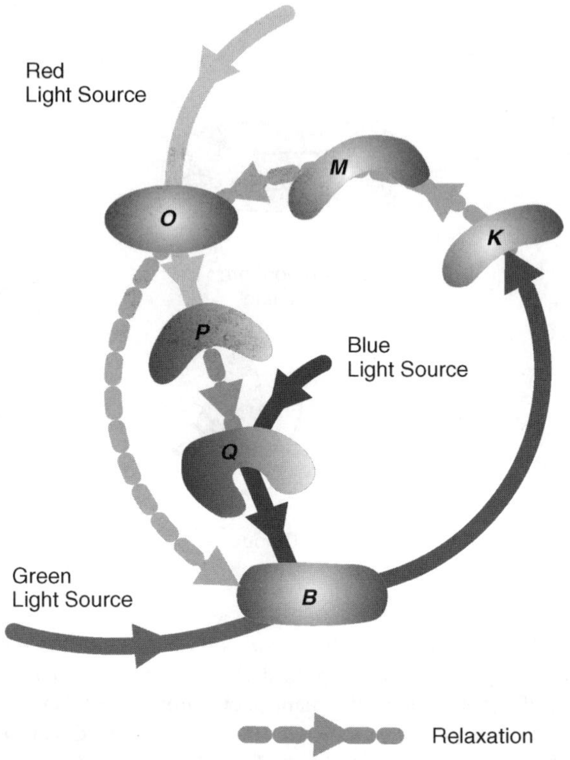

Figure 12.29 Photocycle of bacteriorhodopsin. Green light transforms the resting B state to the intermediate K. Next K relaxes, forming M and then O. If the O intermediate is exposed to red light, a so-called branching reaction occurs. Structure O converts to the P state, which quickly relaxes to the Q state – a form that remains stable almost indefinitely. Blue light, however, will convert Q back to B. Any two long-lasting states can be assigned the binary value 0 or 1, making it possible to store information Reprinted from Birge RR. 'Protein-based computers', *Sci. Am.* 1995, **272**, pp. 90–95. Copyright (1995) with permission from Tomo Narashima.

K state. Optical memory devices can exploit this effect, although the devices must be cooled below room temperature. The most significant photochemical intermediate is the blue light-absorbing M state. The formation of M occurs after a series of protein conformational changes occurring about 50 μs after the absorption of the initial light photon by B. Under normal conditions, M thermally reverts to the ground state with a time constant of about 10ms. Most importantly, B can also be photochemically regenerated from M by the absorption of blue light. This property of the material, where a ground-state photo-initiated reaction results in a relatively long-lived thermal intermediate, which can also be photochemically driven back to the ground state, is, of course, an example of photochromism (Chapter 11, Section 11.8.2). The $M \rightarrow B$ thermal transition is highly susceptible to temperature, chemical environment, genetic modification and chromophore substitution. This property is under study for exploitation in many optical devices based on bacteriorhodopsin. The cyclicity of the protein (i.e. the number of times that the protein can be photochemically cycled before denaturing) exceeds 10^6, a result of its evolution in a harsh environment of the salt marsh.

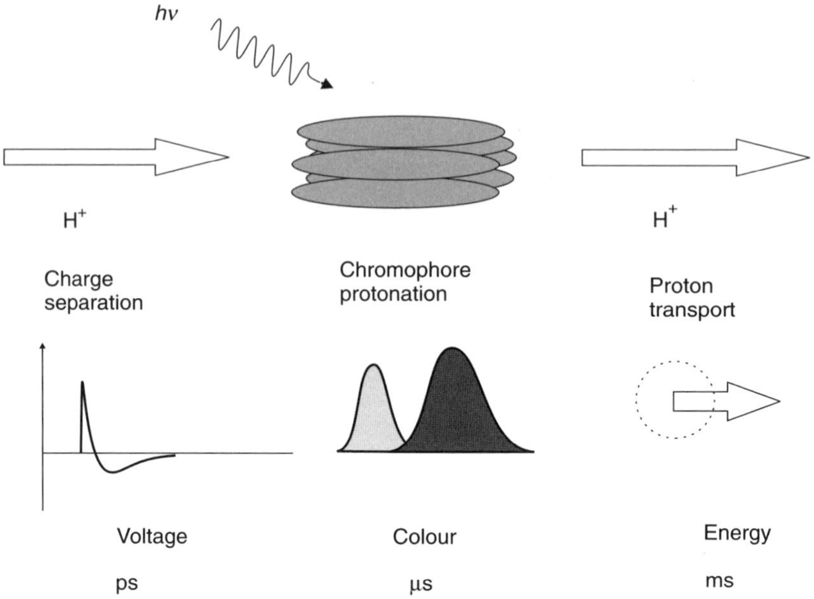

Figure 12.30 Basic molecular functions of bacteriorhodopsin. The proton transport is initialized by photon absorption and a charge separation step on the picosecond time scale. After about 50 μs, the deprotonation of the Schiff base leads to the main photochromic shift during the photocycle. After about 10 ms, the proton transport is completed. Reprinted from *Chem. Rev.*, **100**, Hampp N, 'Bacteriorhodopsin as a photochromic retinal protein for optical memories', pp. 1755–1776. Copyright (2000) American Chemical Society.

Data Storage

For optical storage applications, the principle is to assign any two long-lasting states of the protein to the binary values of '0' and '1', in order to store the required information. For example, the *B* state and *Q* state of the molecule can remain stable during many years.

The prospects for three-dimensional data storage seem more promising than for two-dimensional storage (Chapter 11, section 11.6). There are three different types of volume storage under investigation. The first is page-oriented holographic storage, the second is based on the branched photocycle scheme and the third uses two-photon excitation of the individual data points in the volume of the material.

The last type is accomplished by the intersection of two laser beams. Each beam carries photons with just half the energy to switch BR from *B* to *M* or vice versa. Two-photon absorption depends on the product of the contributing intensities. During readout, the same procedure is applied.

A different approach uses two one-photon transitions in BR. Data can be written into the bacteriorhodopsin protein contained in a cuvette using a combination of red and green diode lasers. These sources are mounted perpendicular to each other. Light from the green source hits the cuvette via a spatial light modulator, converting selected areas, or 'pages', of the protein into the *O* state. Then 2 ms later, light from a red laser–which also passes through a spatial light modulator–irradiates the protein to form the *P* state, which subsequently decays to Q. This two-stage illumination process produces a protein volume where the '1' bits are represented by proteins in the P and Q states and '0' bits correspond to the O and B states. Reading the data back is performed in exactly the same manner, except that the red source has a reduced intensity. The green paging light is only absorbed by proteins in the O state and the low-intensity red light images the P and Q states on to a charged-coupled device chip rather than convert any O proteins. Blue light erases the encoded data. The P and Q states absorb the blue light to return to their original B state. Individual data can be erased using a blue laser or a global erase can be performed with an incoherent blue source. Erasure is a highly efficient process and the memory cuvette must be shielded for protection. This procedure potentially allows a mass storage system.

The storage capacity of this type of memory is very high (10 Gbyte). The limitation of capacity is mainly connected with problems of lens system and quality of protein. It remains unclear whether this type of memory will complete with hard disks or semiconductor memory as a number of problems in producing commercial optical memories remain. For example, the Q state's red absorption is inefficient in the natural form of bacteriorhodopsin, so a high light intensity is required to carry out the read cycle. However, it is possible to exploit genetic engineering to improve the efficiency of the Q absorption. If such problems can be solved, the advantages of the protein memory are enormous: low cost, wide temperature operating range and nonvolatile memory (i.e. no loss of data if the power is switched off; Chapter 11, Section 11.6).

Holographic Memory

Holography is a technique of optical recording that allows high data density together with maximum data access speed. This is achieved because the holographic image, the *hologram*, is coded in one large data block. In holographic storage, complete data pages are stored during each recording. Approximately, 10^6 binary data points can be processed in parallel. After recording one page, either the recording material is shifted or rotated or, alternatively, the angle between the reference and data beams is altered. Thousands of such digital pages, which contain up to a million bits each, might be put into a device measuring a cubic centimetre. Theoretically, data densities could reach 1 Tbyte cm^{-3}, but, in practice, storage capacities will probably be around 10 Gbytes cm^{-3}. These figures should be contrasted with the storage current available in a 12 cm diameter digital versatile disc (DVD) of 4.7 GBytes.

Thin films of bacteriorhodopsin fabricated by incorporating the protein into optically transparent polymers and polymer blends have shown good holographic performance and the capability of real-time optical processing. The holographic memories take advantage of the large change in refractive index that occurs upon formation of the M state. The principle (Figure 12.31) is to record a plane wave hologram in a thin film of bacteriorhodopsin. Two laser beams, of intensities I_1 and I_2, derived from the same laser are absorbed by the protein. Due to coherence of the laser light (Chapter 4, Section 4.4.5), a three-dimensional interference pattern (the hologram) is imposed on the film.

The resulting light intensity distribution in one dimension, $I(x)$, can be described by

$$I(x) = (I_1 + I_2)\left[1 + V\cos\left(\frac{2\pi x}{P}\right)\right] \tag{12.6}$$

$$V = 2(I_1 I_2)^{\frac{1}{2}}/(I_1 + I_2) \tag{12.7}$$

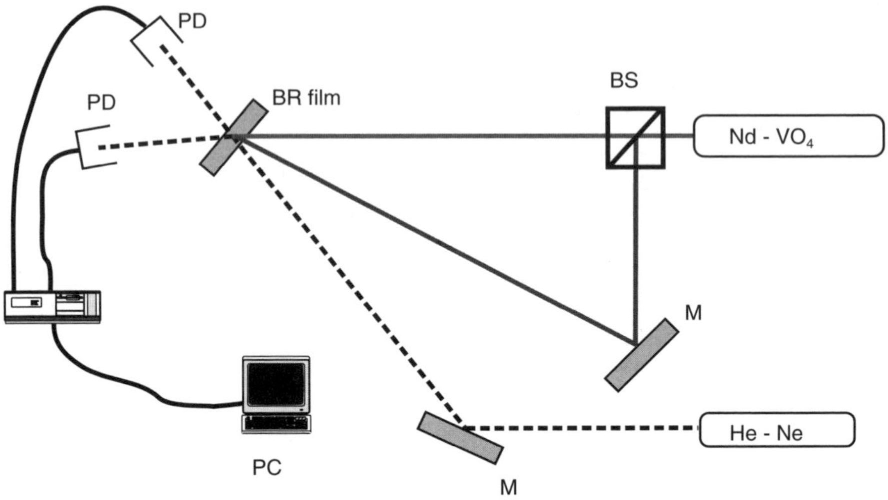

Figure 12.31 Schematic diagram of experimental set up for holography. M = mirror; BS = beam splitter; PD = photodetector.

where V is the contrast ratio of the interference pattern and P is the fringe spacing. In places of constructive interference, the B state is changed to M. The spatial concentration distribution of B and M can be conveniently viewed as a spatial modulation of the materials' absorption coefficient and refractive index. The hologram can be read by using a probe beam that is not strongly absorbed by the protein.

Thin films of bacteriorhodopsin can generate diffraction efficiencies of about 8%, which is more than enough for holographic data storage [12]. In addition, because mutants have enhanced its holographic properties, bacteriorhodopsin is competitive with photorefractive polymers but far less expensive [13]. Finally, both thin and thick films can be prepared using bacteriorhodopsin in whatever concentration is optimal for the optical architecture. This flexibility is the key to commercially competitive holographic data-storage systems.

Information Processing

Important technical applications of BR are in optical data processing. Here, the number of write–read–erase (WRE) cycles required is much higher. For example, in real-time image processing at video frame rate, about 10^6 WRE cycles for a single day of operation are required. This eliminates most of the synthetic organic materials. The reversibility of BR is extremely high, making it an important material in all optical processing situations where the number of WRE cycles is a critical issue. Such applications include nonlinear filtering, spatial light modulators, phase conjugation, neural networks, pattern recognition and interferometry.

Bacteriorhodopsin is certainly a fascinating natural material, combining the properties of photochromism with colour stability. However, despite significant progress, its full potential as a material for molecular electronics applications has yet to be realized in the commercial world.

12.10.2 Photosynthesis

All the food that we eat and all the fossil fuel that we use are products of photosynthesis, the process that converts the energy in sunlight into chemical forms of energy, which can then be used by biological systems. Photosynthesis takes place in many different organisms, from plants to bacteria. The best known form of photosynthesis is the one carried out by higher plants and algae, which are responsible for a major part of photosynthesis in oceans. All these organisms convert carbon dioxide to organic material by reducing the gas to carbohydrates in a rather complex set of reactions.

Sunlight is absorbed and converted initially to electronic excitation energy, which starts a chain of electron-transfer events leading to charge separation across a photosynthetic membrane. The resulting potential energy is then used to pump protons across the membrane, generating an osmotic and charge imbalance, which in turn powers the synthesis of ATP. For the sugar glucose (one of the most abundant products of photosynthesis), the relevant equation is

$$6CO_2 + 12H_2O \rightarrow C_6H_{12}O_6 + 6H_2O + 6O_2 \qquad (12.8)$$

Light provides the energy to transfer electrons from water to $NADPs^+$ forming NADPH, and to generate ATP. ATP and NADPH provide the energy and electrons to reduce CO_2 to organic molecules. Electrons for this reduction reaction ultimately come from water, which is then converted to oxygen and protons.

The first step in photosynthesis is the absorption of light by pigments such as primarily chlorophylls and carotenoids; the chemical structures of chlorophyll *a* and β-carotene are shown in Figure 12.32. Chlorophylls absorb blue and red light and carotenoids absorb

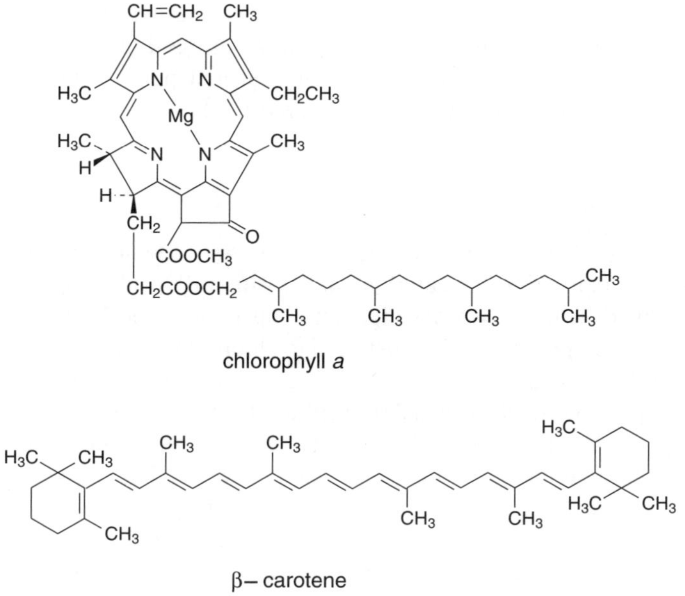

Figure 12.32 Chemical structures of chlorophyll a and β-carotene.

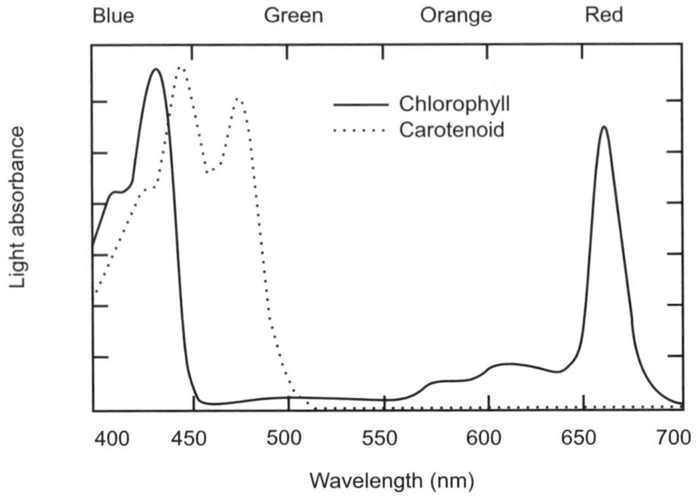

Figure 12.33 Optical absorption spectra of chlorophyll and carotenoids. From Arizona State University: Center for the Study of Early Events in Photosynthesis, http://photoscience.la.asu.edu/photosyn/.

blue–green light, as shown in Figure 12.33. However, green and yellow light are not effectively absorbed by photosynthetic pigments in plants; therefore, light of these colours is either reflected by leaves or passes through the leaves (which is why plants are green and carrots are orange).

Other photosynthetic organisms have additional pigments that absorb the colours of visible light which are not absorbed by chlorophyll and carotenoids. For example, some organisms contain bacteriochlorophyll, absorbing in the infrared in addition to the blue part of the spectrum. These bacteria do not evolve oxygen, but perform photosynthesis under *anaerobic* (oxygen-less) conditions using the infrared light.

Reaction Centres and Antennae

Photosynthetic pigments are normally bound to proteins, which provide the pigment molecules with the appropriate orientation and positioning with respect to each other. Light energy is absorbed by individual pigments, but is not used immediately for energy conversion. Instead, the light energy is transferred to chlorophylls that are in a special protein environment where the actual energy conversion event occurs. The pigments and proteins involved with this primary electron transfer event are together called the *reaction centre*. A large number of pigment molecules (100–5000), collectively referred to as antenna, 'harvest' light and transfer the light energy to the same reaction centre. The purpose is to maintain a high rate of electron transfer in the reaction centre, even at lower light intensities. Antennas permit an organism to increase significantly the absorption cross-section for light without having to build an entire reaction centre and associated electron transfer system for each pigment, which would be very costly in terms of cellular resources. The photosynthetic antenna system is organized to collect and deliver excited-state energy by means of

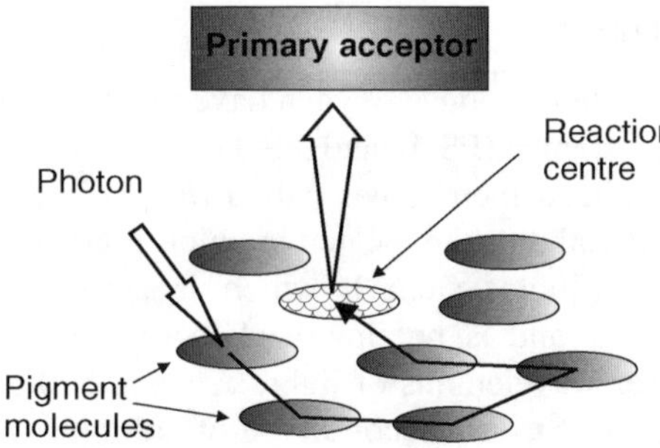

Figure 12.34 Schematic diagram of a photosynthetic reaction centre. A large number of pigment molecules 'harvest' the light and transfer the energy to the same reaction centre.

excitation transfer to the reaction centre complexes where photochemistry takes place. The process is illustrated in Figure 12.34.

Many antenna pigments transfer their light energy to a single reaction centre by transmitting the energy to another antenna pigment, and then to another, until the energy is finally trapped in the reaction centre. Each step of this energy transfer must be very efficient to avoid a large loss in the overall transfer process. The association of the various pigments with proteins ensures that transfer efficiencies are high by having the pigments close to each other, and by providing an appropriate molecular geometry of the pigments with respect to each other.

The overall process of photosynthesis is highly complex. Figure 12.35 shows the important reaction steps, which are described in the following sections.

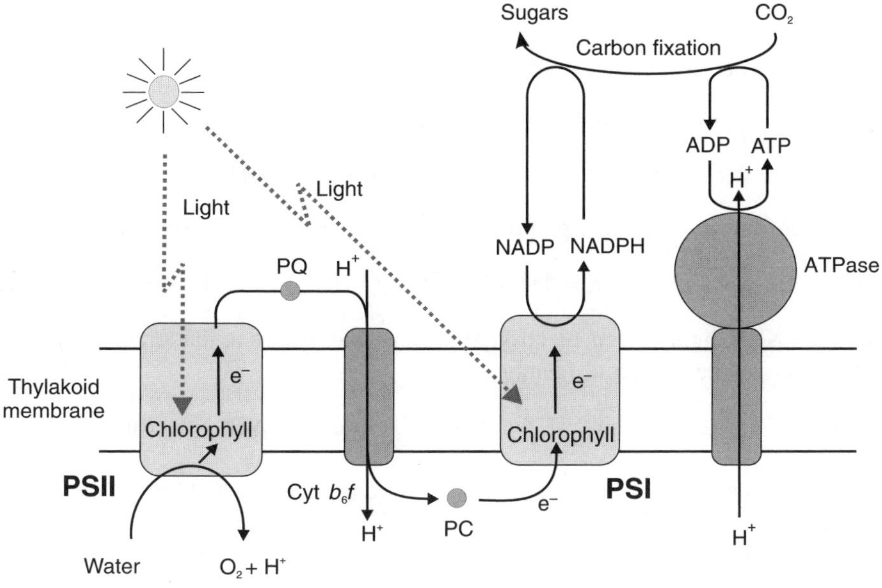

Figure 12.35 Process of photosynthesis.

Photosynthetic Electron Transfer

All photosynthetic organisms that produce oxygen have two types of reaction centres, named photosystem II and photosystem I (PS II and PS I), both of which are pigment/–protein complexes located in specialized membranes called *thylakoids*. In eukaryotes (plants and algae; Section 12.4), these thylakoids are located in chloroplasts (Section 12.4) and often are found in membrane stacks. Photosystem I, which is activated by light at 680 nm, is associated with chlorophyll *a* and is not involved in oxygen evolution. Photosystem II, which is activated by shorter wavelengths of light, between 500 and 600 nm, appears to be involved in oxygen evolution; it uses a second type of chlorophyll in addition to accessory pigments.

Upon oxidation of the reaction centre chlorophyll in PS II, an electron is transferred from a nearby amino acid (tyrosine), which is part of the surrounding protein, which in turn receives an electron from the water-splitting complex. From the PS II reaction centre, electrons flow to free electron-carrying molecules in the thylakoid membrane (plastoquinione; PQ in Figure 12.35) and from there to the cytochrome b_6f complex. Finally, the electrons are transported to the PS I centre by a small protein (plastocyanin; PC in Figure 12.35).

The other photosystem, PS I, also catalyzes light-induced charge separation in a fashion basically similar to PS II: light is harvested by an antenna, and light energy is transferred to a reaction centre chlorophyll, where light-induced charge separation is initiated. However, in PS I electrons are transferred eventually to NADP, the reduced form of which can be used for carbon fixation. The oxidized reaction centre chlorophyll eventually receives the electron from the cytochrome complex. Therefore, electron transfer through PS II and PS I results in water oxidation (producing oxygen) and NADP reduction, with the energy for this process provided by light (two quanta for each electron transported through the whole chain).

To summarize the rather complex process of photosynthesis:

(a) Light is absorbed by pigments, e.g. chlorophyll.

(b) Electron transport is accomplished by membrane-bound protein complexes.

(c) Photosystem II absorbs light, oxidizes water and reduces plastoquinone.

(d) The cytochrome b_6f complex pumps H^+ across the thylakoid during electron transport.

(e) Photosystem I absorbs light, reduces $NADP^+$ (to NADPH) and oxidizes plastocyanin.

(f) The process of light-driven electron transport and water oxidation generates an H^+ gradient across the thylakoid membrane.

(g) The flow of H^+ through an ATP synthase protein drives ATP synthesis.

The conversion of carbon dioxide into organic compounds during photosynthesis is called *carbon fixation*. Electron flow from water to NADP requires light and is coupled to the generation of a proton gradient across the thylakoid membrane. This proton gradient is used for the synthesis of ATP. ATP and reduced NADP that result from the light reactions are used for CO_2 fixation in a process that is independent of light.

Artificial Photosynthetic Systems

Increased understanding of photosynthetic energy conversion and advances in chemical synthesis have made it possible to create artificial nanoscale devices and semi-biological hybrids that carry out many of the functions of the natural processes. For example, reaction centres are simply assemblies of cofactors (Section 12.2.3) held in the appropriate position and orientation by the protein environment. This natural system has been used as a model to design organic molecules where the equivalents of the different cofactors are linked together by covalent bonds of various lengths. The result is the creation of a number of sophisticated molecules which serve as artificial reaction centres. The more advanced molecules consist of two chlorophyll-type molecules linked together (one serves as the electron donor, the other as the acceptor) with the electron-accepting molecule linked to two quinones, which serve as

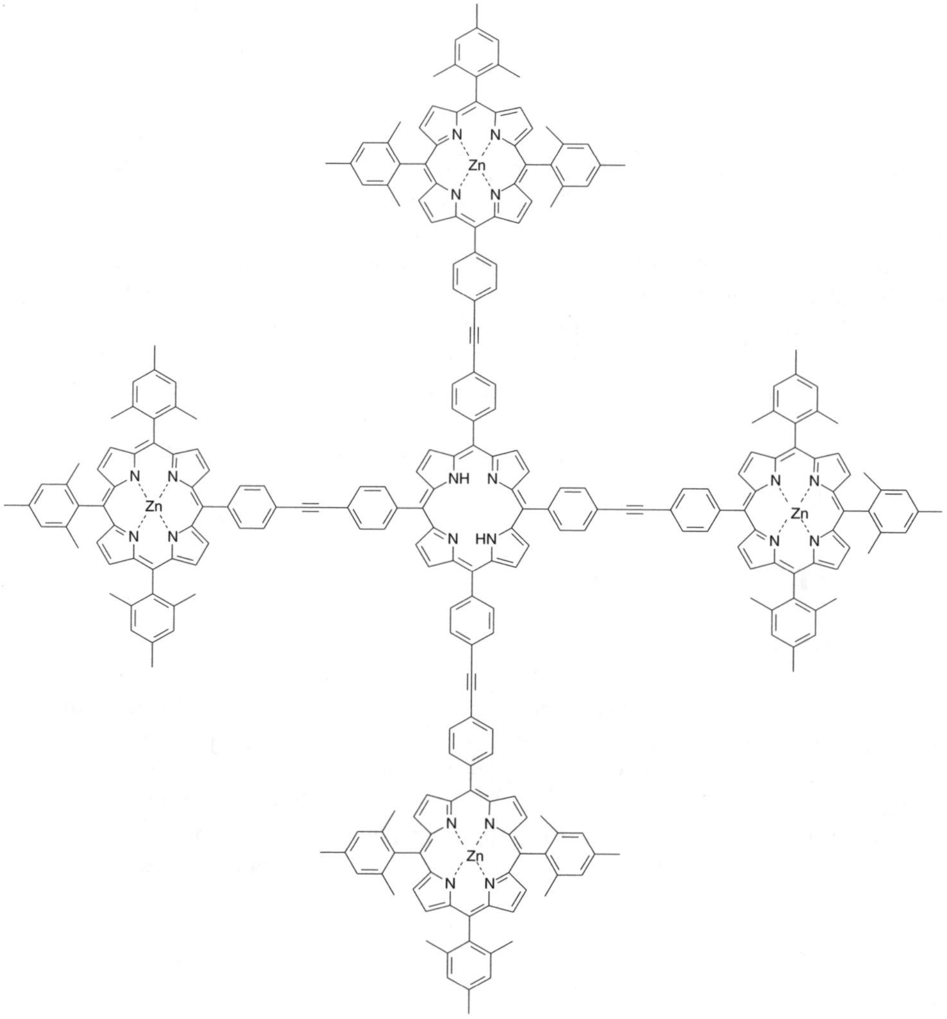

Figure 12.36 A star-like pentameric array of porphyrins for light harvesting [16].

electron acceptors in the natural system. The electron-donating chlorophyll analogue is linked covalently to a carotenoid, and this can donate an electron to the oxidized chlorophyll. Upon excitation of the chlorophyll, a charge separation occurs, resulting in an oxidized carotenoid and a reduced quinone. This charge-separated state is formed with high efficiency. In other artificial reaction centres, porphyrin moieties have been used in place of chlorophyll and fullerenes have been exploited used as the electron acceptor [15].

Current inorganic and organic photovoltaic devices for the harvesting solar energy require an energy-intensive production process (Chapter 9, Section 9.7) and, even though they have improved significantly over the years in terms of their efficiency, the development of photosynthesis-based technologies for energy collection may prove advantageous. However, photosynthesis and related processes can be applied to areas other than solar energy conversion. For example, there are many possible applications for artificial reaction centres and related molecules in nanotechnology. Synthetic pigments also have found biomedical uses in tumour detection, as they tend to accumulate preferentially in tumours and are highly fluorescent (and thus easily detectable in a patient whom is being operated on to surgically remove a tumour).

Porphyrins [Section 12.6.1, Figure 12.21(a)], the main chromophores of natural photosynthesis, are obvious candidates for the design of artificial antenna systems. Arrays containing porphyrin molecules are certainly the largest class of artificial antennae. In the pentameric array shown in Figure 12.36, efficient energy transfer from the peripheral Zn-containing units to the free-base core has been observed [16].

In biotechnology, photosynthetic organisms are likely to play an increasing role in the production of enzymes and pharmaceuticals, which until now are produced primarily by genetically modified heterotrophic microorganisms such as yeast and selected bacteria. A major advantage of photosynthetic organisms is that no fixed-carbon source needs to be added for growth and, therefore, production costs are lower and the chances of contamination with other microorganisms are less. Other potential applications of photosynthetic organisms are in *bioremediation*, i.e. the clean-up of environmental (soil or water) pollutants by biological means, and in the production of clean-burning fuels [17].

12.11 MOLECULAR MOTORS

12.11.1 Nature's Motors

There are a number of examples of both linear and rotating motors to be found in the natural world. Linear motors include the motor protein myosin (in fact, there is a family of proteins labelled myosin I, myosin II, etc.), which is involved in muscle contraction [18]. This is a sophisticated process that utilizes actin, a globular structural protein, which polymerizes in a helical fashion to form an actin filament. These form the cytoskeleton inside an eukaryotic cell. The motion is a result of actin filaments sliding on myosin heads. Actin filaments are formed of 375 amino acid long subunits that are associated with ATP. Myosin transport along actin is not limited to muscle contraction, and is involved in several cellular transport processes. While muscle myosin works in a non-processive manner, a single myosin V protein takes many steps along the actin filament without dissociating. Figure 12.37 shows a proposal of a model for the stepping kinetics of myosin V [18, 19]. The protein dwells with both heads attached to the actin filament, the leading

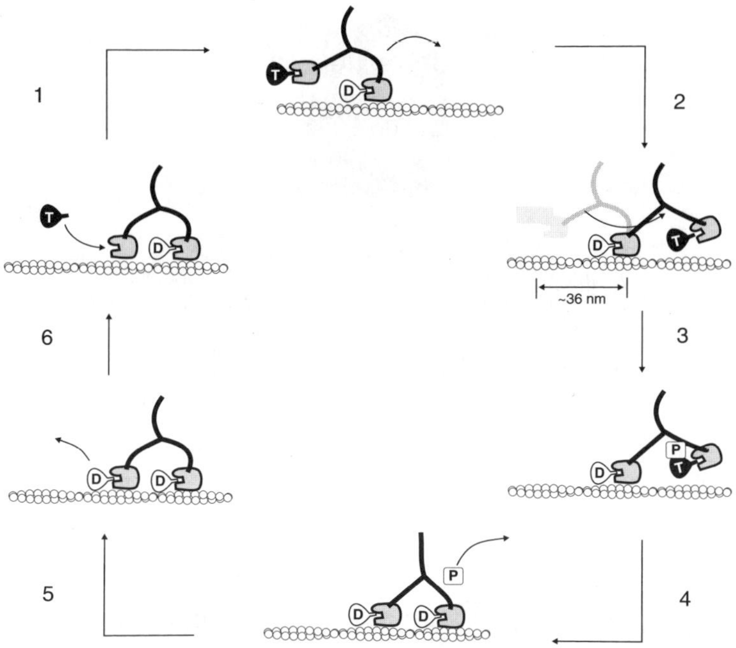

Figure 12.37 Model for myosin-V stepping [19]. D = ADP. T = ATP. P = phosphate. Reprinted from *Molecular Devices and Machines*, Balzani V, Credi A, Venturi M, p. 243, Copyright (2003), with permission from Wiley-VCH.

head with ADP and the trailing head stretched out (top left in Figure 12.37). ATP binding to the trailing head, step 1, leads to its dissociation from actin and forward movement of the released head, discharging intramolecular strain; the previous leading head then becomes the trailing head, step 2. During this process, the trailing head moves 72 nm to reach its new site of attachment, but this movement results in a 36 nm step only of the body of myosin V. The new detached leading head quickly hydrolyses ATP, step 3, and then binds to the actin, step 4. Force generation follows either actin binding or phosphate release, step 5. At this point (bottom left in Figure 12.37), both heads are bound to actin and the molecule is in its kinetically dominant state, because subsequent ADP release, step 6, is slow. In this state, ADP release, step 6, is slow. The leading head is stressed, which encourages subsequent ADP release to occur at the trailing head and not the leading head.

Much like myosin, kinesin also moves along a track, in this case microtubules. This is responsible for transporting cellular components such as organelles and signalling molecules. Kinesins typically consist of two large globular heads that allow attachment to microtubules, a central coiled region, and a region termed light-chain, which connects the kinesin to the intracellular component to be moved. The protein takes 8.3 nm steps along microtubules in a hand-over-hand walking manner.

Perhaps the most useful molecule for nanoscale applications is DNA. DNA replication is a required step before cells divide. For example, the processes involved as a ribosome shunts along an mRNA chain (Section 12.5.1 and Figure 12.15) represent a highly coordinated and efficient linear actuation mechanism.

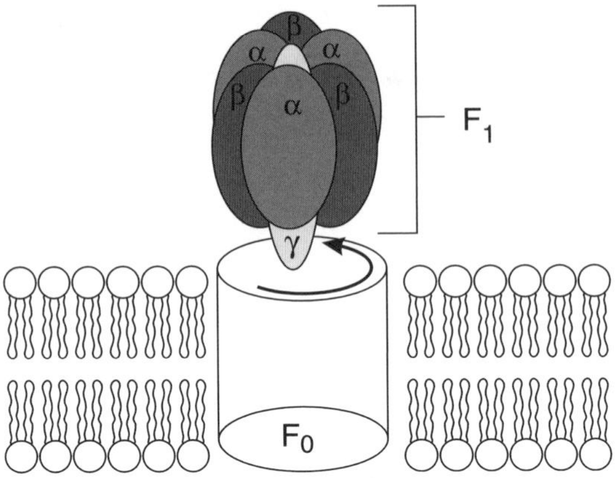

Figure 12.38 Structure of F_0F_1-ATP synthase. This enzyme consists of two principal parts. The asymmetric membrane-spanning F_0 part contains a proton channel whereas the soluble F_1 part contains three catalytic sites which cooperate in the synthetic reactions.

Two examples of biomolecular rotary motors are F_1-adenosine triphosphate synthase (F_1-ATPase) and ATP synthase (F_0F_1) [20]. These are similar to motors that rotate *flagella* (whip-like organelles that many unicellular organisms, and some multicellular organisms, use to move about) in bacteria and appear to have originated in cells about a billion years ago. They have evolved and serve many different purposes. These motors occur mounted in the wall of a cell and transmit rotational torque and power across the cell wall.

The ATPase molecular motors are found in the membranes of mitochondria, the microscopic bodies in the cells of nearly all living organisms, and in chloroplasts of plant cells, where the enzyme is responsible for converting food to usable energy (Section 12.4). The moving part of an ATPase is a central protein shaft (or rotor, in electric-motor terms) that rotates in response to electrochemical reactions with each of the molecule's three proton channels (comparable to the electromagnets in the stator coil of an electric motor), as depicted in Figure 12.38. During ATP hydrolysis, the tail rotates in an anticlockwise direction; it rotates clockwise during ATP synthesis from ADP. Both the F_1 and F_0 portions are rotary motors. The F_1 has been structurally characterized and is made up of α and β subunits arranged around the central γ unit. This contacts the F_0 part, which is membrane bound. The rotary activity of F_1-ATPase has been harnessed to develop motor prototypes in a number of laboratories.

Finally, it should be noted that biomotors are not restricted to animal cells. *Forisomes* are plant motors that permit long-distance transport in a natural microfluidics system. These elongated protein bodies of length up to 30 μm are found in the highly specialized cells in the phloem of vascular plants that form a continuous microtube system allowing pressure-driven long-distance transport of photo-assimilates. The forisomes act as reversible stopcocks by undergoing rapid conformational changes which involve more than three-fold increases in volume. The conformational switch is controlled by Ca^{2+} ions with a threshold concentration in the nanomolar range. These proteins have potential as reversible switches in integrated microfluidic systems.

12.11.2 Artificial Motors

Although the mechanism of some natural motors is fairly well understood, the greatest challenge lies in adapting these systems into usable devices. The functions carried out within a cell by these motors are not very different from what we would expect man-made motors to perform, such as load carrying or rotational movement. Thus mimicking these motors or adapting them in novel applications may have significant benefits. The broad principles of

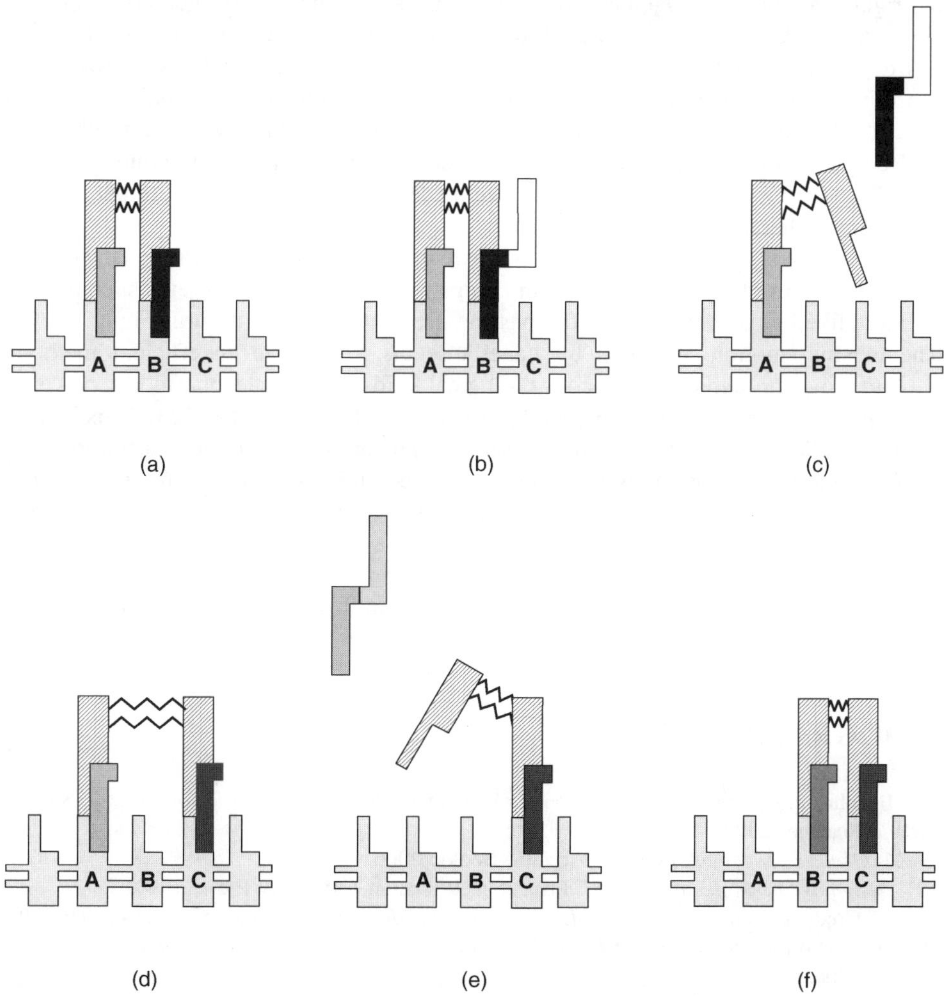

Figure 12.39 Schematic diagrams depicting the movement of a DNA robot. (a) Two DNA-based legs are held fast to DNA footholds A and B by anchors. (b) A free DNA strand attaches to the right anchor. (c) This strand strips the right anchor off, freeing the right foot. (d) An anchor strand for foothold C secures the floating foot to it. (e) Another free DNA strand strips the anchor from the left foot, detaching it. (f) An anchor strand secures the left foot to foothold B. Reprinted from *Nano Letters*, **4**, Sherman WB, Seeman NC. 'A precisely controlled DNA walking device', pp. 1203–1207. Copyright (2004) American Chemical Society.

molecular machines have been outlined by Eric Drexler in his book *Engines of Creation* [21]. Drexler introduced a nanorobot, which he called an *assembler*. Such a device could, in principle, build almost anything – including copies of itself – atom by atom.

Using techniques from biotechnology, it is possible to make DNA molecules with a sequence of base pairs chosen at will. For example, DNA has been used to build a two-legged walking motor or biped [22]. The motion is depicted in Figure 12.39. Each of the legs in the walker is 36 bases long and is made from two strands of DNA that pair up to form a double helix. At the top, a springy portion of each DNA strand runs across from the left leg to the right, linking them together. At the bottom, one of the two strands pokes out of the helix to serve as a sticky foot. The track that the walker travels along is also made of DNA, and is designed so that unpaired sections of DNA strands stick up likes spikes along its length. These act as footholds for the walker. The feet attach to the footholds via 'anchor' strands of DNA that match up with the foot sequence at one end and with the foothold at the other. Because the left and right foot/foothold sequences are unique, each requires a different anchor. To make the walker take a step, a free piece of DNA is introduced to peel away one of the anchors, thereby releasing the foot.

The movement is rather like an inchworm, with one foot edging forwards and the other then being dragged up to the same position. In contrast, natural walking proteins such as kinesin move more like humans, with each foot passing ahead of the other in turn (Figure 12.38).

The twisting motion produced by flagella seems to provide an efficient method for bacteria to move through highly viscous biological fluids. A biomolecular motor, powered by ATP synthase (see above), with tiny metal propellers has been described [23]. This motor can spin the propeller at eight revolutions per second. The ability of a molecular motor to draw its energy directly from the body has been considered an important step in the development of a new generation of ultrasmall medical devices. The molecular motors have propellers about 750 nm long and 150 nm in diameter (a human hair is about 1000 nm in diameter). One turn of the motor produces about $120\,\text{pW}\,\text{nm}^{-1}$. The energy released from the three ATP molecules needed to rotate the motor once is $240\,\text{pN}\,\text{nm}^{-1}$, giving the motor an efficiency of about 50%.

BIBLIOGRAPHY

Alberts B, Johnson A, Lewis J, Raff M, Roberts K, Walter P. *Molecular Biology of the Cell*. New York: Garland Science; 2002.

Amos M. *Genesis Machines*. London: Atlantic Books; 2006.

Ball P. *Designing the Molecular World*. Princeton: Princeton University Press; 1994.

Balzani V, Credi A, Venturi M. *Molecular Devices and Machines*. Weinheim: Wiley-VCH; 2003.

Bar-Cohen Y (Editor). *Biomimetics: Biologically Inspired Technologies*. Boca Raton, FL: Taylor and Francis; 2006.

Birge RR (Editor). *Molecular and Biomolecular Electronics*. Washington, DC: American Chemical Society; 1994.

Blankenship RE. *Molecular Mechanisms of Photosynthesis*. Oxford: Blackwell Science; 2001.

Drexler KE. *Engines of Creation*. London: Fourth Estate; 1990.

Jones RAL. *Soft Machines*. Oxford: Oxford University Press; 2004.

Kumar CSSR (Editor). *Nanomaterials for Biosensors*. Weinheim: Wiley-VCH; 2006

Nicolini C (Editor). *Molecular Manufacturing*. New York: Plenum Press; 1996.

Nicolini C. *Molecular Bioelectronics*. Singapore: World Scientific; 1996.

Nicolini C (Editor). *From Neural Networks and Biomolecular Engineering to Bioelectronics*. New York: Plenum Press; 1993.
Pessarakli M (Editor). *Handbook of Photosynthesis*. Boca Raton, FL: Taylor and Francis; 2005.
Ridley M. *Genome*. London: Fourth Estate; 1999.

REFERENCES

[1] Ridley M. *Genome*. London: Fourth Estate; 1999.
[2] Ball P. *Designing the Molecular World*. Princeton: Princeton University Press; 1994.
[3] Findlay JBC. The biological membrane. In MC Petty, MR Bryce, D Bloor (Editors). *An Introduction to Molecular Electronics*. London: Edward Arnold; 1995, pp. 279–294.
[4] Petty M, Tsibouklis J, Petty MC, Feast WJ. Pyroelectric behaviour of synthetic biomembrane structures. *Thin Solid Films* 1992; **210–211**: 320–323.
[5] Aizawa M. Biosensors. In MC Petty, MR Bryce, D Bloor (Editors). *An Introduction to Molecular Electronics*. London: Edward Arnold; 1995, pp. 295–314.
[6] Zhao S, Reichert WM. Influence of biotin lipid surface density and accessibility on avidin binding to the tip of an optical fibre sensor. *Langmuir* 1992; **8**: 2785–2791.
[7] Dekker C, Ratner MA. Electronic properties of DNA. *Phys. World* 2001; **14**: 29–33.
[8] Briman M, Armitage NP, Helgren E, Grüner G. Dipole relaxation losses in DNA. *Nano Lett.* 2004; **4**: 733–736.
[9] Adleman LM. Molecular computation of solutions to combinatorial problems. *Science* 1994; **266**: 1021–1024.
[10] Braich RS, Chelyapov N, Johnson C, Rothemund PWK, Adleman L. Solution of a 20-variable 3-SAT problem on a DNA computer. *Science* 2002; **296**: 499–502.
[11] Amos M. *Genesis Machines*. London: Atlantic Books; 2006.
[12] Birge RR, Gross RB. Biomolecular optoelectronics. In MC Petty, MR Bryce, D Bloor (Editors). *An Introduction to Molecular Electronics*. London: Edward Arnold; 1995, pp. 315–344.
[13] Hampp N. Bacteriorhodopsin as a photochromic retinal protein for optical memories. *Chem. Rev.* 2000; **100**: 1755–1776.
[14] Birge RR. Protein-based computers. *Sci. Am.* 1995; **272**: 90–95.
[15] Gust D, Moore TA, Moore AL. Mimicking photosynthetic solar energy transduction. *Acc. Chem. Res.* 2001; **34**: 40–48.
[16] Prathapan S, Johnson TE, Lindsey JS. Building-block synthesis of porphyrin light-harvesting arrays. *J. Am. Chem. Soc.* 1993; **115**: 7519–7520.
[17] Arizona State University: Center for the Study of Early Events in Photosynthesis. Available from URL http://photoscience.la.asu.edu/photosyn/.
[18] MRC Laboratory of Molecular Biology and the Cambridge Institute for Medical Research. *The Myosin Home Page*. Available from URL http://www.mrc-lmb.cam.ac.uk/myosin/myosin.html.
[19] Rief M, Rock RS, Mehta AD, Mooseker MS, Cheney RE, Spudich JA. Myosin-V stepping kinetics: a molecular model for processivity. *Proc. Natl. Acad. Sci. USA* 2000; **97**: 9482–9486.
[20] Balzani V, Credi A, Venturi M. *Molecular Devices and Machines*. Weinheim: Wiley-VCH; 2003.
[21] Drexler KE. *Engines of Creation*. London: Fourth Estate; 1990.
[22] Sherman WB, Seeman NC. A precisely controlled DNA walking device. *Nano Lett.* 2004; **4**: 1203–1207.
[23] Soong RK, Bachand GD, Neves HP, Olkhovets, AG, Craighead HG, Montemagno CD. Powering an inorganic nandevice with a biomolecular motor. *Science* 2000; **290**: 1555–1558.

Index

absorption coefficient 136
absorption of EM radiation 135–40
  effect of electric fields 143
acceptor atoms 90
acids 170
acoustic impedences 381
acoustic sensors 380–1
action potentials of neurons 481
active addressing 302–3
activity coefficient 367
activity of ionic species 367
actuators 360
  polymer actuators 392–5
addition polymerization 57
addressing, passive and active 302–3
adenosine triphosphate (ATP) 467–8
adensoine diphosphate (ADP) 467–8
admittance, electrical 123
admittance spectroscopy 125
adsorption 56
aerogels 257
alcohols 173–5
aldehydes 175
aliphatic compounds 179
alkanoic acids 176
alkoxides 175
alpha-helix 459, 460
ambipolar FETs 438
amides 176
amines 175–6
amino acids 455–8
amorphous solids 56
amperometric sensors 365
amphiphilic compounds 258
analyte 363
AND gate 444–5
angular frequency of waves 74
anions 170
anisotropic energy 206
anisotropic substances 46
anodization 253
anti conformation 39
antibiotics 477
antibodies 458
antibonding orbitals 31
antiferroelectric liquid crystal (AFLC) phase 306
antiferromagnetism 203, 204, 208
antigens 458
antistatic agents 351
anti-Stokes' radiation 231
aramid liquid crystal polymers 310
aromatic compounds 178–80
artificial neural networks (ANNs) 390
atomic force moscroscopy (AFM) 233
atomic magnetic dipole 201
atomic mass unit 19
atomic number 19
atomic structure 19–20
  electons 20–4
    orbitals 23–4
    periodic table 24–6
atomic weight 19
attenuated total reflection (ATR) 228, 385
Auger electron spectroscopy (AES) 232
Avogadro's number 20, 170
axial ratio 296
axons 480

back lighting 302
bacteriorhodopsin (BR) 487–8
  data storage 490–1
  holographic memory 491–2
  information processing 492
  photochromism 488–90
  photocycle 488
ballistic transport 67
bandwidth 362
based-centred unit cells 50

© 2007 John Wiley & Sons, Ltd

bases (alkalis) 170
bases (of nucleotides) 468–6
bathochromic shift 137
batteries 351–4
Beer–Lambert law 136–7
Beer's law 135–6
bimorphs 394
bioelectronics 455
  DNA electronics 487
  molecular motors
    artificial motors 501–2
    Nature's motors 498–500
biological cells 468–9
biological comparisons with molecular-scale electronics 12–13
biological membranes 474–5
  transport across membranes
    electrons 479–80
    molecular and ionic 475–9
biological molecules 455
  amino acids and peptides 455–8
  carbohydrates 461–2
  enzymes 459–61
  lipids 462–4
  nucleotides 465
    ATP and ADP 467–8
    bases 468–6
    DNA 466–7
    RNA 467
  proteins 458–9
biosensors 482–3
  see also sensors
  bioaffinity sensors 484–5
    amperometric glucose sensors 486
    optical biosensors 486–7
  biocatalytic sensors 483
bipolarons 110–11
birefringence 297–9
  birefringent media 133
blastocysts 469
Bloch walls 206
body-centred unit cells 50
Bohr atomic model 20
Boltzmann's constant 67
bonding orbitals 31
bottom-up nanotechnology 1
Bragg's law 218
Bravais lattices 50
Brewster angle 151–2
Brillouin zones 82–3
buckminsterfullerenes ('buckyballs') 188–9
calorimetric gas sensors 364
candelas 335
carbohydrates 461–2
carbon nanotubes 191–4
carbonyl group 175
carboxylate esters 177
carboxylates 176
carboxylic acids 175, 176
Casimir force 405
cations 170
cause and effect 202
cellular automata 447
centre of inversion 134
charge carrier mobility 69–70
charge-flow transistors 375
charge-transfer complexes 103–6, 185–8
chemical bonding 26, 35
  covalent bonds 29–32
  hydrogen bonds 34–5
  ionic bonds 27–8
  metallic bonds 32–3
  organic compounds
    double and triple bonds 42–5
    hybrid orbitals 35–7
    isomers 37–42
  principles 26–7
  van de Waals bonds 33
chemical force miscroscopy (CFM) 236
chemical self-assembly 268–9
chemical sensors *see* sensors
chemical vapour deposition (CVD) techniques 251–2
chemiresistors 369
chemisorption 56
chiral phase 290–2
chirality 40
chiroptical switches 432
chlorophyll 493–4, 496
chloroplasts 469
cholesterol 463
chromaticity diagrams 334
chromoionophores 383
chromosomes 469
circularly polarized radiation 129
cis-isomerism 40, 43
cladding 155
codons of DNA 470
coercive fields 199, 206
cofactors of enzymes 461
coherent radiation 147
cohesive energy 27

Cole–Cole diagram 125
collagen 458
compensated materials 91
complimentary MOS (CMOS)
transistors 328
computing architectures 447–9
condensation polymerization 57
conduction, intrinsic and extrinsic 88–9
disordered semiconductors 94
Fermi level position 92–3
low-dimensional solids 95–6
n-type doping 89–90
p-type doping 90–1
quantum wells 93
traps and recombination centres 92
conduction band 85, 87
conductive coatings 351
conductive polymers 180–5
conductivity, high-frequency 121
complex permittivity 121–5
impedance spectroscopy 125–7
conductivity, low-frequency 112–13
electronic versus ionic conductivity 113–14
quantum mechanical tunnelling 114–16
Schottky and Poole–Frenkel effects 120–1
space-charge injection 118–19
variable range hopping 116–17
conductors 85–6
configurational isomers 40–2
confocal microscopy 214
conformation of proteins 459
conformational isomers 38–9
conjugated chains 99
conservation of energy 76
contact potential 315
continuous random netword (CRN) 56
Cooper pairs 112
coordination number 27
copolymers 57
Coulomb blockade 427–8
Coulomb staircase 429
Coulombic repulsion 19
covalent bonds 29–32
critical angle of incidence 151
critical micelle concentration 294
crossed polarizers 298
crystal lattices 48–9
crystal planes 52
defects 52
line defects 53–4
plane defects 54–5
point defects 52–3
surface defects 55–6
crystal systems 49–50
Miller indices 50–1
crystallographic axes 48
cubic crystals 49
Curie law 205
Curie temperature 197, 205
Curie–Wiess law 207–8
current density 66
cyanide group 178
cytochromes 479
cytoplasm 469
cytoskeleton 469

dangling bonds 56
Davydov splitting 140
de Broglie relationship 75
Debye equations 125
degenerate energy states 109
degree of polymerization 59
Del vector operator 398
dendrites 480
density of states function 72
deoxyribonucleic acid (DNA) 465, 466–7
depletion 320
depletion region 315
deposition ratio 263
Dexter transfer 147–8
dextrarotatory isomers 41
diagonalizing a matrix 132
diamagnetism 203, 204, 205
dielectric constant 122
dielectric elastomers 395
dielectric sensors 376–80
dielectrophoresis 397–8
differential capacitance 320
diffraction limited optics 215
diffusion 63–4
dimers 137
diodes, organic 313
Schottky diode 313–18
dipeptides 456
dipolar polarization 122
dipole moment 131
dip-pen nanolithography (DPN) 280–2
director of liquid crystals 288
disclinations 307
discotic phase 292–3
dislocations 53–4
disordered semiconductors 94

dispersion curves 158
dispersion energy 140
displacement current 113
donor atoms 89
doping 107–8
  n-type 89–90
  p-type 90–1
double bonds 42–5
drain electrodes 321
drift velocity 67
dye-sensitized solar cells 348–50
dynamic random access (DRAM) 417
dynamic range 363

eclipsed conformation 38
effective mass 87
eigenenergies 77
eigenfunctions 77
Einstein relation 117, 476
electret 197
electric dipoles 33
electric displacement 132
electric field strength 66
electric potential 66
electrical conductivity 65
  classical theory 65–8
    charge carrier mobility 69–70
    Fermi level 70–2
    Ohm's law 68–9
electroabsorption 143
electroactive organic compounds 169
  see also organic compounds
  magnetic materials 208–10
  piezoelectric, pyroelectric and ferroelectric substances 197–200
electrochemical cells 365–9
electrochemical vapour deposition techniques 252–3
electrochromism 431
electroless deposition 253
electromagnetic radiation 129–30
  interactions with organic molecules
    absorption processes 135–40
    electric fields and absorption 143
    emission processes 144–7
    energy transfer 147–9
  linear and nonlinear optics 133–5
  permittivity tensor 132–3
  photonic crystals 162–4
    subwavelength optics 165–6
  reflection and refraction 149–50
    ellipsometry 152
    Fresnel equations 150–2
    thin films 152–4
  refractive index 130–2
  surface plasmons 156
    evanescent field 156–7
    resonance 157–62
  waveguiding 154–6
electromagnetic waves 73–4
electron affinity 98
electron-beam lithography 277
electron density profile 221–2
electron spin resonance (ESR) 109
electronegativity 25–6
electronic noses 390–1
Electronic Numerical Integrator and Computer (ENIAC) 6
electronic paper 345
electronic tongue 390
electrons 19, 86–8
  structure in atoms 20–4
    orbitals 23–4
    periodic table 24–6
  velocity in silicon 10
electro-orientation effects 398
electro-osmotic pumping 397
electrophoresis 397
electropositive elecments 25
electro-rotation effects 398
electrostatic layer-by-layer (LbL) deposition 270–4
electrostrictive Mawxell stress 395
ellipsometry 152, 237–8
elliptically polarized radiation 129
emission of EM radiation 144–7
emulsions 63
enantiomers 40
energy bands in solids 72–3
energy gaps 82
enthalpy 47
entropy 47
enzymes 459–61
epitaxial deposition 241
epithelium 389
equivalent weight 252
esters 177
ethers 177
eukaryotic cells 469
evanescent fields 156
  surface plasmons 156–7
excimers 145

excited vibrational states 139
excitons 140, 141–3
exons 470
extended X-ray absorption fine structure (EXAFS) 233
external quantum efficiency 335
extraordinary rays 133
extreme ultraviolet radiation 276
extrinsic conduction 88–9

face-centred unit cells 50
Faraday's constant 253
far-field optical microscopy 215
fats 462
fatty acid salts 264
Fermi level 70–2, 92–3
Fermi level pinning 318
Fermi wavelength 95
Fermi–Dirac quantum statistics 71
fermions 24
ferrimagnetism 203, 204, 208
ferroelectric liquid crystals 305–6
ferroelectricity 196–7
  organic compounds 197–200
ferromagnetism 203, 204, 205–8
Feynman, Richard 1–2
Fick's first law 64
field effect transistors 321–7
field emission 121
fingerprint region of IR spectra 227
finite-state machines 447
first-order transitions 47
flagellae 500
flash evaporation 245
flash memory 417, 421–4
flat band conditions 319
fluid mosaic model 474
fluorescence 144
foams 63
force modulation microscopy (FMM) 236
forisomes 500
Förster transfer 147
forward bias 315
Fourier summation 221
Fowler–Nordheim tunnelling 121
Franck–Condon principle 139
Freedericksz transition 299–300
Frenkel defects 53
Frenkel excitons 141–2
Fresnel equations 150–2
fuels cells 351–4
fullerenes 188–91
functional groups 173
  alcohols 173–5
  aldehydes 175
  amides 176
  amines 175–6
  carboxylic acids 176
  esters 177
  ethers 177
  ketones 177–8
  nitriles 178
  thiols 178

Game of Life 448
gas sensors
  calorimetric 364
  resistive 369–76
gate effect 321
gauche conformation 39
gelation 255
gels 63
genetic coding 469–70
  replication, transcription and translation 470–3
genome 469
giant magnetoresistance (GMR) 424
Gibbs function 47
glass transition temperature 62
glasses 56
glomeruli 389
glycolipids 463
grain boundaries 54
granular cells 389
graphene sheets 188
gravimetric sensors 380
gravure 282
gray scales 303
ground state for electrons 24
ground-state dipole 131
group velocity 75
gustation 390

H-band shift 140
Heisenberg model of ferromagnetism 206
Heisenberg's uncertainty principle 79–80
helielectric phase 305
Helmholtz function 47
heteroatoms 179, 181
heterogeneous crystal growth 264
hexagonal crystals 49
hexatic phases 290

highest occupied molecular orbital (HOMO) 97
historical perspective 1–2
hole accumulation 319
hole burning 434
holes 86–8
holographic memory 491–2
homeotropic texture 299
homogeneous crystal growth 264
homogeneous texture 299
hopping, variable range 116–17
hot carriers 70
Hückel rule 178
Hund's rule 24
hybrid orbitals 35–7
hydration shell 172
hydrogen bonds 34–5
hydroxyl group 173
hyperpolarizabilities 134
hypsochromic shift 137
hysteresis loop 199

ideality factor 317
image force effect 120
impedance, electrical 123
impedance spectroscopy 125–7
incoherent radiation 146
indium tin oxide (ITO) 392
information content 11–12
inkjet printing deposition techniques 253–5
insulators 85–6
integrated circuits (ICs) 6–7, 327
integrated organic circuits 327–8
integrated sensors 361
interatomic forces 404–5
interfacial polarization 122
intermolecular conductors 106
internal energy 47
internal quantum efficiency 335
intersystem crossing 144
intramolecular conductors 106
intrinsic conduction 88–9
introns 470
inverse micelles 295
inverse opal structure 411
inverter device 327
ion-beam projection lithography 277
ionic bonds 27–8
ionic polarization 122
ionization energy 28
ionophores 383, 477
ions 170, 171–2
ion-selective electrodes 366
ion-sensitive field effect transistors (ISFETs) 368
isomers 37–8
  configurational 40–2
  conformational 38–9
isosbestic point 383
isotopes 19
isotropic substances 46

Jablonski diagrams 144
J-aggregates 141

keratin 458
Kerr effect 134
ketones 177–8
Kevlar 310–11
kinetic theory of gases 65
Knudsen cells 249
Kraft temperature 294
Krebs cycle 468
Kretschmann configuration 159
Kronig–Penney model 80–5

lab-on-a-chip 395–9
Langmuir expression 243
Langmuir isotherm 373
Langmuir trough 263
Langmuir–Blodgett films 115, 148, 161
  production technique 258–68
Langmuir–Schaefer technique 268
lasers 147
lateral force microscopy (LFM) 236
lavorotatory isomers 41
layer-by-layer (LbL) deposition 270–4
Lewis acids 171
ligands 187
light-emitting displays, organic (OLEDs) 331–4
  device efficiency 335–6
  efficiency improvement methods 336
    blended and doped layer OLEDs 341–3
    electrodes 336–7
    microcavity structures 340–1
    multilayer structures 337–8
    shaped substrates 340
    triplet emission 338–40
limbic system 389
line defects 53–4
Linear Combination of Atomic Orbitals 30
linear electro-optic effect 134
linear optics 133

lipids 462–4
liquid crystals 2, 287
  display devices 297
    birefringence 297–9
    Freedericksz transition 299–300
    full-colour displays 303
    passive and active addressing 302–3
    super twisted nematic displays 304
    twisted nematic displays 300–2
  ferroelectric 305–6
  lenses 308–9
  other applications 309–11
  phases 287
    lyotropic liquid crystals 293–5
    thermotropic liquid crystals 287–93
  polymer-dispersed 306–8
  polymers 295–7
lithium ion batteries 351
lithium ion polymer batteries 351–3
logic circuits 443–7
lone pair electrons 37
Love wave 381
low-dimensional solids 95–6
lowest unoccupied molecular orbital (LUMO) 97
Lowry–Brønsted theory 170
lumens 335
luminescence 144
luminescent concentrator 350–1
luminosity 336
luminous efficiency 335
luminous flux 335
luminous intensity 335
luminous power efficiency 336
lyophillic 293
lyophobic 293
lyotropic liquid crystal phases 293–5

magetization 201
magnetic domains 206
magnetic force microscopy (MFM) 236
magnetic materials
  organic magnets 208–10
  principles 201–8
magnetic random access memory (MRAM) 424
magnetic susceptibility 202
magnetoelectronics 424
main-chain polymer liquid crystals 295
majority carriers 316
Maxwell–Boltzmann distribution 65, 66–7
mean free path 67
measurand 360
measurements
  energy change 9
  power dissipation 9
  switching energy 9
Meissner effect 112
mercaptans 178
mesophases 287
metal–insulator–semiconductor (MIS) structures 318–19
  idealized devices 319–20
  organic structures 320–1
metal–insulator–semiconductor field effect transistors (MISFETs) 318–19, 421–2
  drain current–voltage characteristics 322–3
metallic bonds 32–3
metallorganic CVD (MOCVD) 251–2
metal–oxide–semiconductor transistor (MOSFET) 5–6, 8–9
metastable states 47
micellar structures 293
micro-electro-mechanical systems (MEMS) 391
microprocessors 7
Miller indices 50–1
mitochondria 467
mitral cells 389
mixed valence states 105
mobility edges 94
mobility gaps 94
mode dispersion 156
mode of waves 155
molecular beam epitaxy (MBE) 248–9
molecular crystals 96–8
molecular electronics tools 213–14
  electron reflection 223–4
  film thickness measurement 236–8
  imaging, direct
    electron microscopy 216–18
    optical microscopy 214–16
  infrared spectroscopy 224–31
    Raman scattering 231–2
  neutron reflection 223
  scanning probe microscopies 233–6
  surface analytical techniques 232–3
  X-ray reflection 218–21
    electron density profile 221–2
    in-plane measurements 223
    Keissig fringes 222–3
molecular magnets 208–9
molecular materials for electronics 1, 2–5

molecular motors
artificial motors 501–2
Nature's motors 498–500
molecular weight 20
polymers 57–9
molecular-scale electronics 1, 5, 403
actuation 440
dynamically controllable surfaces 440–2
optical tweezers 443
rotaxanes 442–3
computing architectures 447–9
contemporary situation 8–12
device architectures 411–14
electronic switching and memory devices 417–18
flash memory 421–4
resistive bistable devices 418–21
spintronics 424–5
three-dimensional architectures 426
engineering materials 405
nonlinear optical materials 408–10
photonic crystals 410–11
polar materials 406–8
evolution of microelectronics 5–7
timeline 6
interatomic forces 404–5
logic circuits 443–7
Moore's Laws 7–8
nanomagnetic systems 436–7
nanosystems 403
scaling laws 403–4
nanotube electronics 437–40
optical and chemical switches 429–30
chemical control 435–6
fluorescence switching 430
photochromic systems 431–4
quantum computing 449–50
rectification 415–17
single-electron devices 427–9
molecules 169–70
moles 20, 169–70
monoclinic crystals 49
monomers 57
monosaccharides 461
Moore's Laws 7–8, 14–15
Morse curves 139
multiplicity 144
multivariate evaluation 390
myelin sheath 481

NAND gate 444–5
nanofabrication 275
dip-pen nanolithography 280–2
nanometre pattern definition 276–7
other methods 282
photolithography 275–6
scanning probe manipulation 278–80
soft lithography techniques 278
nanometre pattern definition 276–7
nanotubes 191–4, 437–40
near-edge X-ray absorption fine structure (NEXAFS) 233
Néel temperature 208
negative differential resistance (NDR) 420
nematic phase 288
Nernst equation 366
Nernst–Palnck equation 475–6
neurons 480–2
neurotransmitters 481–2
neutrons 19
Newman projections 39
nitriles 178
nits 335
nonbonding pairs of electrons 37
non-degenerate ground states 110
nonlinear optical materials 408–10
nonlinear optics (NLO) 134
nonpolar solvents 173
nonvolatile memory 417
NOR gate 444–5
NOT gate 444–5
n-type doping 89–90
nuclear magnetic resonance (NMR) spectroscopy 179
nucleosomes 467
nucleotides 465

ohmic contacts 318
Ohm's law 68–9
olfaction, biological 388–9
olfaction, electronic 390–1
olfactory cortex 389
oligosaccharides 461
on-wire lithography 278
opals, synthetic 410
optical microcavity 163
optical phenomena 129
see also electromagnetic radiation
optical tweezers 443
OR gate 444–5
orbital magnetic moment 201

orbitals 20–2, 23–4
  antibonding 31
  covalent bonds 29–32
  double and triple bonds 42–5
  hybridized 35–7
order of reflection 219
order parameter 289
orders of diffraction 161
ordinary rays 133
organelles 469
organic compounds
  see also electroactive organic compounds
  band structure 96
    charge-transfer complexes 103–6
    molecular crystals 96–8
    Peierls distortion 99–103
    polymers 98–9
  doping 107–8
  double and triple bonds 42–5
  hybrid orbitals 35–7
  interactions with EM radiation
    absorption processes 135–40
    aggregate formation 140–1
    electric fields and absorption 143
    emission processes 144–7
    energy transfer 147–9
    excitons 141–3
  isomers 37–42
organic diodes *see* diodes, organic
organic semiconductor solar cells 346–8
orientational polarization 122
orthorhombic crystals 49
Otto arrangement 160
oxygen acids 177

panchromatic photoconductive response 355
paraelectric phase 197
paramagnetism 203, 204, 205
parametric sensors 360
partition coefficient 376–7
passive addressing 302–3
Pauli Exclusion Principle 24
Pauli model of paramagnetism 205
Pauling scale of electronegativity 25
Peierls distortion 99–103
pellistors 364
peptides 455–8
periodic table 24–6
permeability, relative 132
permeability of free space 202
permittivity 121
permittivity of free space 123
permittivity tensor 132–3
phase changes 27, 47
phase contrast microscopy 214
phase problem 221
phase velocity 73, 219
phenol 174
phonons 112
phospholipids 463
phosphorescence 145
photobiology
  bacteriorhodopsin 487–8
    data storage 490–1
    holographic memory 491–2
    information processing 492
    photochromism 488–90
    photocycle 488
photolithography 275–6
photoluminescence 144
photometric LEDs 335
photonic crystals 162–4
photons 74–5
photoptic response 335
photosynthesis 461, 493–4
  artificial systems 497–8
  electron transfer 496
  reaction centres and antennae 494–5
physical vapour deposition techniques
  molecular beam epitaxy (MBE) 248–9
  sputtering 249–51
  thermal evaporation 243–8
pi bonds 42–4
piezoelectric coefficients 195
piezoelectricity 194–6
  organic compounds 197–200
pinch-off 323
planar texture 299
planar waveguide 155
Planck's constant 75
plane defects 54–5
plane polarized radiation 73, 129
plasma frequency 157
plasma-enhanced CVD (PECVD) 251
plasmons, surface 156
  evanescent field 156–7
  resonance 157–62
plastic crystal 287
plastic electronics 313
  field effect transistors 321–7
  integrated organic circuits 327–8

plastic electronics *(continued)*
RF identification (RFID) tags 328–31
metal–insulator–semiconductor (MIS) structures 318–19
idealized devices 319–20
organic structures 320–1
organic diodes 313
ohmic contacts 318
Schottky diode 313–18
organic light-emitting displays (OLEDs) 331–4
device efficiency 335–6
efficiency improvement methods 336–43
electronic paper 345
full-colour displays 343–5
other applications
batteries and fuel cells 351–4
conductive coatings 351
xerography 354–6
photovoltaic cells 345–6
dye-sensitized solar cells 348–50
luminescent concentrator 350–1
organic semiconductor solar cells 346–8
pleated sheets 459, 460
Pockels' effect 134, 408–9
point defects 52–3
point lattice 48
Poiseuille flow 396
polar aprotic solvents 173
polar protic solvents 173
polarizability 134
polarization 33, 133
polarons 110–11
poling 62
polyacetylene 2
polyelectrolytes (PEs) 270
polyethers 177
poly(3,4-ethylenedioxythiophene) (PEDOT) 183–5, 418
polymer actuators 392–5
polymer-dispersed liquid crystals 306–8
polymeric semiconductors 3–4
polymers 2, 57
band structure 98–9
conductive *see* conductive polymers
crystallinity 60–3
molecular weight 57–9
structure 59–60
polypeptides 457
polysaccharides 461
Poole–Frenkel effect 121
population inversion 147
porphyrins 497–8
postsynaptic cells 481
potentiometric sensors 365
potentiostatic feedback 365
p-polarized radiation 129
presynaptic cells 481
principal axes of a matrix 132
principal component analysis 390
prokaryotic cells 469
propagation constant 74
proteins 458–9
protein folding 459
protons 19, 170
p-type doping 90–1
pumped energy 147
pyroelectric coefficient 196
pyroelectricity 196–7
organic compounds 197–200

Q bands 382–3
quadratic electro-optic effect 134
quantum bits 449
quantum computing 449–50
quantum dots 95
quantum mechanical tunnelling 80, 114–16
quantum mechanics 20, 73
electromagnetic waves 73–4
electron wavefunction 75–6
Heisenberg's uncertainty principle 79–80
photons as particles 74–5
Schrödinger wave equation 76–9
quantum numbers for electrons 20–2
quantum wells 93, 93
qubits 449
quinoid structure 182–3

radicals 100
radiofrequency identification (RFID) tags 328–31
radiometric LEDs 335
Raleigh wave 381
Raman scattering 231
Raman spectroscopy 231–2
random walk 404
Rayleigh scattering 231
reaction centres 494–5
recombination centres 92
recovery time 361
redox reactions 107
reduced wavevector representation 83
reflectance 151

reflection 149–50
Fresnel equations 150–2
reflection absorption infrared spectroscopy (RAIRS) 228
reflection coefficient 150
refraction 149–50
refractive index 130–2
relative permeability 132
relative permittivity 121
reliability 361
remnant magnetization 206
remnant polarization 199
replication of genetic codes 470–3
residual polarization 199
resistance, electrical 68
resistive gas sensors 369–76
resolution 363
resonance hybrids 44–5
response time 361
reversed bias 315
reversibility 361
Reynolds number 396
rhombohedral crystals 49
ribonucleic acid (RNA) 465, 467
ribosomes 467
rotaxanes 442–3

saponification 463
Sauerbrey equation 380
scaling laws 403–4
scanning capacitance microscopy (SCM) 236
scanning electron miscroscopy (SEM) 216
scanning near-field optical microscopy (SNOM) 214–16
scanning probe manipulation 278–80
scanning tunnelling microscope (STM) 233
Schottky barrier 120, 313–14
Schottky barrier height 315
Schottky defects 53
Schottky diode 313–18
Schottky effect 120–1
Schottky emission 120
Schrödinger wave equation 76–9
secondary-ion mass spectrometry (SIMS) 232–3
second-harmonic generation 134
second-rank tensors 132
selectivity 361
self-assembly techniques 268–9
semiconductors 85–6
disordered 94
doping 107–8
polymeric 3–4
quantum wells 93
semi-metals 193
sensitivity 361
sensors 4, 359
see also biosensors
chemical sensors 363–4
acoustic devices 380–1
calorimetric gas sensors 364
dielectric sensors 376–80
electrochemical cells 365–9
optical sensors 382–8
resistive gas sensors 369–76
definition of terms 361–3
electronic noses 390–1
physical sensors and actuators 391
lab-on-a-chip 395–9
polymer actuators 392–5
touch sensors 391–2
sensing systems 360–1
smart textiles and clothing 399
seven-segment arrays 302
sheet resistance 69
shells 20
side-chain polymer liquid crystals 295
sigma bonds 31
signal-to-noise ratio 361
single-electron devices 427–9
single-electron transistors (SETs) 429
singlet states 139
single-wall nanotubes (SWNTs) 192–3, 437–8
skeletonization 267
smart sensors 361
smart textiles 399
smectic phase 290
Snell's law 132, 150, 151
soft lithography techniques 278
solar cells
dye-sensitized 348–50
organic semiconductor 346–8
sol–gel deposition techniques 255–7
solitons 109–10
soliton switch 446
solvatochromism 137
solvents 173, 174
soma 480
Soret band 382
source electrodes 321
space-charge injection 118–19
space lattice 48

spacers 295
spatial filtering 214
spin-coating 242–3
spin-forbidden transition 144
spin magnetic moment 201
spin-momentum transfer effect 424
spin valve 424
spintronics 424–5
spontaneous emission 147
sputtering 249–51
stability 361
stacking faults 55
staggered conformation 38
Stark effect 143
states of matter 45–6
  crystal lattices 48–9
  crystal systems 49–50
  Miller indices 50–1
  phase changes 47
static random access (SRAM) 417
stem cells 469
steradians 335
stereoisomerism 40–1
steric hindrance 296
steroids 464
stimulated emission 147
stoichiometric composition 91
Stokes radiation 231
Stokes shift 145
stop gap 165
strong force 19
structure factor 221
subshells 20
substrates of enzymes 459
subthreshold conduction 324
super twisted nematic displays 304
superconductivity 4, 112
supercritical drying 256–7
surface acoustic wave (SAW) 381
surface defects 55–6
surface plasmon polaritons (SPPs) 157
surface plasmon resonance (SPR) 157–62, 385–8
surface plasmons 156
  evanescent field 156–7
surface selection rule 229
surface states 318
surface-enhanced Raman spectroscopy 232
surfactants 293
susceptibility tensor 134
synapses 481
synthetic opals 410

tautomers 435
tensile strength 194
terpenes 464
tetragonal crystals 49
thermal velocity of electrons 66
thermistors 360
thermochromism 431
thermodynamic equilibrium 47
thermoelectric power 182
thermotropic liquid crystal phases 287
  chiral phase 290–2
  discotic phase 292–3
  nematic phase 288
  smectic phase 290
thin film processing 241
  deposition methods, established
    chemical vapour deposition (CVD) 251–2
    electrochemical methods 252–3
    inkjet printing 253–5
    other techniques 258
    physical vapour deposition 243–51
    sol–gel processing 255–7
    spin-coating 242–3
  molecular architectures 258
    chemical self-assembly 268–9
    electrostatic layer-by-layer (LbL) deposition 270–4
    Langmuir–Blodgett technique 258–68
  nanofabrication 275
    dip-pen nanolithography 280–2
    nanometre pattern definition 276–7
    other methods 282
    photolithography 275–6
    scanning probe manipulation 278–80
    soft lithography techniques 278
thin films 152–4
thioacetals 178
thioesters 178
thioethers 178
thiols 178
third rank tensors 195
third-harmonic generation 134
three-dimensional architectures 426
thylakoids 496
time constant 404
top-down nanotechnology 1
total internal reflection 151
touch sensors 391–2

trans-isomerism 40, 43
transconductance 323
transcription of genetic codes 470–3
transducers 360
transfer integral 101
transfer ratio 263
transition dipole moment 138
translation of genetic codes 470–3
transmission coefficient 150, 153
transmission electron microscopy (TEM) 216
transmittance 151
transverse-magnetic (TM) radiation 129
trap-filled limit 118
traps (in semiconductors) 92
travelling-wave dielectrophoresis 399
triacyglycerols 462
triboelectric effect 351
triclinic crystals 49
triglycerides 462
trimers 137
triple bonds 42–5
triplet states 144–5
truth tables 444–5
tunnelling, quantum mechanical 80, 114–16
twisted nematic displays 300–2
twisted nematic displays, super 304

unit cells of crystals 50
universal gas constant 170
universal law for dielectrics 124

vacuum level 314
valence band 87
valence electrons 24
valency 25
van de Waals bonds 27, 33
vesicles 464
virtual states 231
volatile memory 417
von Klitzing constant 95
von Neumann architecture 447

Wannier–Mott excitons 141–2
wave vector 74
wavefunction 22–3, 75–6
  bonding electrons 32
waveguiding 154–6
wave-mechanical model of
    atoms 20
wavenumber 226
Wilhemy plate 261
work functions 313–14

xerogel 255
xerography 354–6
X-ray absorption near-edge structure
    (XANES) 233
X-ray lithography 277
X-ray photoelectron spectroscopy
    (XPS) 232–3
X-type deposition 262

Young's modulus 194

Z-type deposition 262
zwitterions 456

Constants

velocity of light in free space	c	2.998×10^{8} m s^{-1}
permittivity of free space	ε_0	8.854×10^{-12} F m^{-1}
electronic charge	e	1.602×10^{-19} C
Planck's constant	h	6.626×10^{-34} J s
Boltzmann's constant	k_B	1.381×10^{-23} J K^{-1}
Avogadro's number	N_A	6.022×10^{26} kilomole^{-1} (= 6.022×10^{23} mole^{-1})
universal gas constant	R	8.314×10^{3} J kilomole^{-1} K^{-1}
Faradays constant	F	9.649×10^{7} C kilomole^{-1}
acceleration due to gravity	g	9.807 m s^{-2}

Useful Relationships

1 electronvolt (eV) = 1.602×10^{-19} J
For vacuum, energy in eV = 1.240/(wavelength in μm)
1 eV = 8066 cm^{-1}
1 eV per particle = 23,060 kcal kilomole^{-1} (= 23.06 kcal mole^{-1})
1 calorie = 4.186 J
At 300 K, kT $\approx$ 1/40 eV
1 atmosphere = 1.013×10^{5} N m^{-2}

The Electromagnetic Spectrum

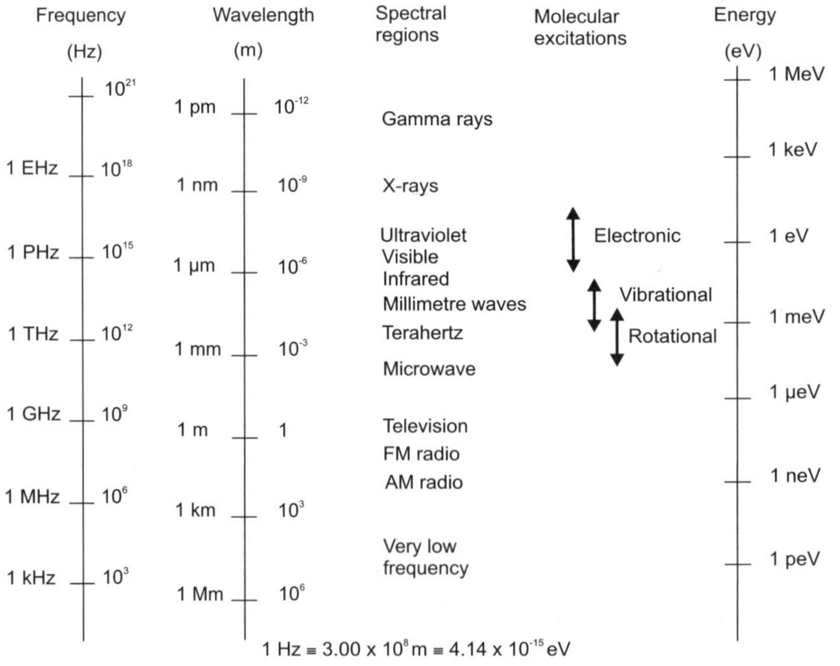

Properties of selected elements

Element	Symbol	Atomic number	Atomic weight [amu]	Outer electron configuration	Density of solid, ~20 °C [kg m^{-3}] $\times 10^3$	Crystal structure	Melting point [°C]
Hydrogen	H	1	1.008	1s	–	–	–259
Helium	He	2	4.003	$1s^2$	–	–	–272 (at 26 atm)
Lithium	Li	3	6.94	2s	0.534	bc cubic	181
Beryllium	Be	4	9.012	$2s^2$	1.85	hexagonal	1287
Boron	B	5	10.81	$2s^22p$	2.34	rhombohedral	2076
Carbon	C	6	12.011	$2s^22p^2$	2.27	hexagonal	sublimes at 3727
Nitrogen	N	7	14.007	$2s^22p^3$	–	–	–210
Oxygen	O	8	16.00	$2s^22p^4$	–	–	–219
Fluorine	F	9	19.00	$2s^22p^5$	–	–	–220
Neon	Ne	10	20.18	$2s^22p^6$	–	–	–249
Sodium	Na	11	22.99	3s	0.968	bc cubic	98
Magnesium	Mg	12	24.31	$3s^2$	1.74	hexagonal	650
Aluminium	Al	13	26.98	$3s^23p$	2.70	fc cubic	660
Silicon	Si	14	28.09	$3s^23p^2$	2.33	fc cubic	1414
Phosphorus	P	15	30.97	$3s^23p^3$	1.82	triclinic	44
Sulphur	S	16	32.06	$3s^23p^4$	2.07	orthorhombic	115
Chlorine	Cl	17	35.45	$3s^23p^5$	–	–	–102
Argon	Ar	18	39.95	$3s^23p^6$	–	–	–189
Potassium	K	19	39.10	4s	0.89	bc cubic	63
Calcium	Ca	20	40.08	$4s^2$	1.55	fc cubic	842
Titanium	Ti	22	47.87	$3d^24s^2$	4.51	hexagonal	1668
Vanadium	V	23	50.94	$3d^34s^2$	6.0	bc cubic	1910
Chromium	Cr	24	52.00	$3d^54s$	7.15	bc cubic	1907
Manganese	Mn	25	54.94	$3d^54s^2$	7.21	bc cubic	1246
Iron	Fe	26	55.85	$3d^64s^2$	7.86	bc cubic	1538
Cobalt	Co	27	58.93	$3d^74s^2$	8.9	hexagonal	1495
Nickel	Ni	28	58.69	$3d^84s^2$	8.91	fc cubic	1455
Copper	Cu	29	63.55	$3d^{10}4s$	8.96	fc cubic	1085
Zinc	Zn	30	65.41	$3d^{10}4s^2$	7.14	hexagonal	420
Gallium	Ga	31	69.72	$4s^24p$	5.91	orthorhombic	30
Germanium	Ge	32	72.64	$4s^24p^2$	5.32	fc cubic	938

Properties of selected elements (*Continued*)

Element	Symbol	Atomic number	Atomic weight [amu]	Outer electron configuration	Density of solid, ~20 °C [kg m^{-3}] × 10^3	Crystal structure	Melting point [°C]
Arsenic	As	33	74.92	$4s^24p^3$	5.73	rhombohedral	817
Selenium	Se	34	78.96	$4s^24p^4$	4.81	hexagonal	221
Bromine	Br	35	79.90	$4s^24p^5$	–	–	−7
Rubidium	Rb	37	85.47	5s	1.53	bc cubic	39
Niobium	Nb	41	92.91	$4d^45s$	8.57	bc cubic	2477
Molybdenum	Mo	42	95.94	$4d^55s$	10.2	bc cubic	2623
Palladium	Pd	46	106.42	$4d^{10}$	12.0	fc cubic	1555
Silver	Ag	47	107.87	$4d^{10}5s$	10.5	fc cubic	962
Cadmium	Cd	48	112.41	$4d^{10}5s^2$	8.65	hexagonal	321
Indium	In	49	114.82	$5s^25p$	7.31	tetragonal	157
Tin	Sn	50	118.71	$5s^25p^2$	7.27	tetragonal	232
Antimony	Sb	51	121.76	$5s^25p^3$	6.70	rhombohedral	631
Tellurium	Te	52	127.6	$5s^25p^4$	6.24	hexagonal	450
Iodine	I	53	126.90	$5s^25p^5$	4.93	orthorhombic	114
Caesium	Cs	55	132.91	6s	1.93	bc cubic	28
Barium	Ba	56	137.33	$6s^2$	3.51	bc cubic	727
Tantalum	Ta	73	180.95	$5d^36s^2$	16.7	bc cubic	3017
Tungsten	W	74	183.84	$5d^46s^2$	19.3	bc cubic	3422
Platinum	Pt	78	195.08	$5d^96s$	21.5	fc cubic	1768
Gold	Au	79	196.97	$5d^{10}6s$	19.3	fc cubic	1064
Mercury	Hg	80	200.59	$5d^{10}6s^2$	–	–	−39
Lead	Pb	82	207.2	$6s^26p^2$	11.3	fc cubic	327
Bismuth	Bi	83	208.98	$6s^26p^3$	9.78	rhombohedral	272